Studies in Logic

Volume 21

The Many Sides of Logic

Studies in Logic Series Editor
Dov Gabbay   dov.gabbay@kcl.ac.uk

The Many Sides of Logic

Edited by
Walter Carnielli
Marcelo E. Coniglio
Itala M. Loffredo D'Ottaviano

ISBN 978-1-904987-78-9

College Publications
Scientific Director: Dov Gabbay
Managing Director: Jane Spurr
Department of Computer Science
King's College London, Strand, London WC2R 2LS, UK

http://www.collegepublications.co.uk

Original cover design by orchid creative   www.orchidcreative.co.uk
Printed by Lightning Source, Milton Keynes, UK

Preface

From May 11th to May 17th, 2008, the Centre for Logic, Epistemology and the History of Science (CLE) of the State University of Campinas (UNICAMP) and the Brazilian Logic Society (SBL) hosted a conference in Paraty in the state of Rio de Janeiro, Brazil. This volume contains the proceedings of that conference, "CLE 30 Years/XV Brazilian Logic Conference/XIV Latin-American Symposium".

The Brazilian Logic Conferences (EBL) were started in 1977, organized by the Brazilian Logic Society and CLE. The Latin American Symposia on Mathematical Logic (SLALM) were started in 1970 with the first symposium held at the Catholic University of Chile in Santiago. Volumes of the proceedings of some of those have been published.

From May 7th to May 9th, as is traditional with the Latin American Symposia, a Logic School was held, aimed at advanced graduate students and other people interested in the study of logic and related areas. It was hosted by CLE at UNICAMP, and six tutorials were organized:

"Aristotle's Underlying Logic", by John Corcoran
"Translations between Logics", by Itala M. Loffredo D'Ottaviano
"Possible-Translations Semantics", by Walter Carnielli
"Combining Logics", by Marcelo E. Coniglio
"Algebraic Aspects of Substructural Logics" by Roberto Cignoli
"Logic and Information" by Jaakko Hintikka

Approximately 180 scholars, researchers, and students took part in conference, with participation from throughout Brazil, as well as Argentina, Belgium, Colombia, the United States of America, the Netherlands, Italy, Mexico, Poland, Uruguay, and Venezuela. This was the largest participation for any conference held by EBL or SLALM.

At the conference, Newton C. A. Da Costa was honored for his pioneering contribution to logic in Brazil and Latin America, and in particular for his development of paraconsistent logic. Raymond Smullyan offered a talk about Gödel's Theorems. Jaakko Hintikka, Raymond Smullyan, and John Corcoran gave interviews for the Historical Archives of CLE.

The papers presented at the conference covered most aspects of contemporary logic, including philosophical logic and applications. Extended abstracts of those papers are available electronically in the series *CLE e-*

Prints, http://www.cle.unicamp.br/e-prints/vol_8,n_6,2008.html, as edited by CLE. A report of the event with shorter abstracts appears in the Bulletin of Symbolic Logic, volume 15, issue 03, September 2009, pages 332–376.

The papers here have been selected by a careful refereeing process, which has resulted in a compilation of contributions in many areas of logic and their interconnections. This volume presents original research on several aspects of non-classical logics, proof theory, model theory, algebraic logic, philosophical logic, and connections between logic and mathematics as well as between logic and computer science.

This volume is divided into eight sections: Paraconsistent Logics and Adaptive Logics; Logic and Meaning; Logic and Mathematics; Logic, Proofs and Games; Logic and its Algebraic Side; Logic and the Question of Truth-Values; Logic in Abductive and Defeasible Reasoning; and Logic and Computing, though the classification is somewhat arbitrary, especially considering the connections between these topics.

From the 27 keynote speakers, 12 contributed papers to this volume: Jaakko Hintikka, Arnon Avron, Diderik Batens, John Corcoran, David Miller, Claudio Pizzi, Edgar G. K. Lopez-Escobar, Francisco Miraglia, Jairo José da Silva, Jean-Yves Béziau, Atocha Aliseda and José M. Turull-Torres (in the order their papers appear in this volume). Their contribution, together with the other 19 selected papers, witnesses the vigour and excellence of Brazilian and Latin-American tradition in Logic.

The success of CLE 30 Years/XV Brazilian Logic Conference/XIV Latin-American Symposium owes much to the work of the organizing committee and the scientific committee of the event. Its relevance for the development of logic in Brazil and Latin-America, reflected in the papers in this volume, owes much to the academic and financial sponsorship of CLE, SBL, Fundo do Apoio ao Ensino, à Pesquisa e à Extensão (FAEPEX/UNICAMP/Brazil), the State of São Paulo Research Foundation (FAPESP/Brazil), the National Council for Scientific and Technological Development (CNPq/Brazil), Coordenação de Aperfeiçoamento de Pessoal de Nível Superior (CAPES/Brazil), the Association for Symbolic Logic (ASL/USA), and the National Science Foundation (NSF/USA).

The three editors, who were chairs of the conference, are grateful for the assistance of Juan Carlos Agudelo and Anderson de Araújo on LATEX editing. We would also like to acknowledge the valuable support of the staff of CLE for all its contribution and work, and also to thank Mrs. Rosemarie Q. Donelson of the Department of Mathematics of the University of Texas at San Antonio.

We would like to thank especially José Iovino for representing the conference to the National Science Foundation and for his work on the meeting, and Charles Steinhorn, Secretary-Treasurer of the Association for Symbolic Logic for his assistance and support and personal participation in the conference.

Acknowledgement to referees

The editors would like to extend their gratitude and appreciation to the following colleagues whose (otherwise anonymous) effort helped to select the papers that appear in this book and to improve their contents:

João Fernando Alcântara
Atocha Aliseda
Horacio Arló-Costa
Jean-Yves Béziau
Xavier Caicedo
Martin Caminada
Oswaldo Chateaubriand
John Corcoran
Flavio S. Corrêa da Silva
Newton C.A. da Costa
Jairo José da Silva
Hendrik Decker
Francisco Antonio Dória
Phan Minh Dung
Eduardo Fermé
Marcelo Finger
Josep M. Font
Joanna Golinska-Pilarek
Ricardo Gonçalves
Alessio Guglielmi
Petr Hájek
Antonis C. Kakas
Michal Krynicki
Hugh Lacey
E.G.K. López-Escobar
Zoran Majkic
Janos A. Makowsky
Hugo Mariano
Ana Teresa Martins
Maria da Paz Medeiros
Norman Megill
David Miller
Daniele Mundici
Marcelino Pequeno
Gabriella Pigozzi
Jaime Ramos
João Rasga
Marco Ruffino
Ralph Schindler

Peter Schroeder-Heister
Barry Hartley Slater
Luca Spada
Charles Stewart
Frieder Stolzenburg
Miroslaw Truszczynski
Constantine Tsinakis
Jørgen Villadsen
Heinrich Wansing
Timothy Williamson

Walter A. Carnielli
Marcelo E. Coniglio
Itala M.L. D'Ottaviano
Editors

Centre For Logic, Epistemology and the History of Science
and
Department of Philosophy
State University of Campinas (UNICAMP)
Brazil

SECTION 1

PARACONSISTENT AND ADAPTIVE LOGIC

IF Logic Meets Paraconsistent Logic

JAAKKO HINTIKKA

1 The uniqueness of IF logic

My title might at first seem distinctly unpromising. Why should anyone think that one particular alternative logic could be relevant to another one? The most important part of a response to this question is to remind the reader of the fact that independence friendly (IF) logic is not an alternative or "nonclassical" logic. (See here especially Hintikka, "There is only one logic", forthcoming.) It is not calculated to capture some particular kind of reasoning that cannot be handled in the "classical" logic that should rather be called the received or conventional logic. No particular epithet should be applied to it. IF logic is not an alternative to our generally used basic logic, the received first-order logic, aka quantification theory or predicate calculus. It replaces this basic logic in that it is identical with this "classical" first-order logic except that certain important flaws of the received first-order logic have been corrected.

But what are those flaws and how can they be corrected? To answer these questions is to explain the basic ideas of IF logic. Since this logic is not as well known as it should be, such explanation is needed in any case. I will provide three different but not unrelated motivations for IF logic.

2 Quantifiers express dependence

One motivation comes from a closer look at the semantical function of quantifiers. This function is often taken to be exhausted by the ranging of quantifiers over a class of values. Frege even proposed construing the existential and the universal quantifiers as higher-order predicates of lower-order predicates, expressing their nonemptyness and exceptionlessness respectively. The same view of quantifiers underlies the entire theory of so-called generalized quantifiers which is therefore subject to the same limitations as the "ungeneralized" first-order logic.

This idea nevertheless captures only a part of the semantical job description of quantifiers. Their other important function is to express through the formal (syntactical) dependence or independence of a quantifier (Q_2y)

on another one (say (Q_1x)) the dependence or independence of the variable y bound to the former of the variable x bound to the latter. Now much of the scientific enterprise consists in establishing dependencies and independencies of variables on each other. Hence any satisfactory logical language ought to be able to express all logically possible patterns of dependence and independence between variables.

A language using the received first-order logic does not serve this purpose fully. The reason for this failure is the way in which the formal dependence of a quantifier (Q_2y) on another one, say (Q_2x) is expressed in it. As everybody who understands the notation of first-order logic knows, this dependence is expressed by the fact that the scope of (Q_2y) is included in the scope of (Q_1x), as in

(1) $(Q_1x)(-(Q_2y)(\cdots)-)$

But this inclusion relation is of a rather special kind. It is among other things antisymmetric and transitive. Hence only some possible patterns of dependence and independence can be expressed by means of the received first-order logic. In view of the importance of the dependence relations between variables this is a highly significant failure.

This shortcoming of the received "classical" first-order logic is eliminated in IF logic. This can be done in different ways. We could merely relax the conventions governing the scope relation, without any new notation. A more practicable way is to introduce a slash notation that exempts a quantifier (Q_2y) occurring within the formal scope of (Q_1x) from its dependence on (Q_1y) by writing it as (Q_2y/Q_1x). The result is IF first-order logic. It could also be formulated without any new notation, merely by relaxing the conventions governing the (formal) scope of quantifiers, that is, governing the use of parentheses. Even though such a formulation is too clumsy to be feasible in practice, it illustrates the fact that IF logic is nothing but the received "classical" first-order logic extended so as to make it more flexible.

3   Some characteristics of IF logic

Moreover, some important propositions turn out to be logical truths of IF logic even though they are not logical truths of the received first-order logic. Among them there are the axiom of choice and König's lemma. Likewise, a number of crucially important concepts can be defined by means of IF logic without resorting to higher-order quantification. They include equicardinality, infinity, topological continuity and truth in a sufficiently rich first-order language in terms of the same language. In the received logic, they can be expressed only by quantifying over higher-order entities.

Needless to say, the dependence relations expressible in IF logic (or in the received first-order logic) can also be expressed on the second-order level by asserting the existence of the functions that mediate the dependencies in question. In technical logic, they are known as Skolem functions. They play an important role in the theory (and applications) of first-order logic. For one thing they are the functions whose existence is affirmed by the so-called axiom of choice. Thus to assert a quantificational sentence is to assert the existence of a full set of its Skolem functions This does not take us from the scope of IF first-order logic, for this logic is equivalent with the Σ_1^1 (sigma one-one) fragment of second-order logic in which the existence of Skolem functions can be explicitly expressed.

One of the main facts about IF logic is that, although all its semantical rules are in a suitable formulation precisely the old "classical" ones, the negation $\sim$ defined by these rules is a strong dual negation not obeying the law of excluded middle. This is corrected by adding to IF logic a sentence-initial contradictory negation $\neg$. The result is called extended IF logic (EIF logic), and it can be considered as the true basic logic. By IF logic I will in the following refer as much to EIF logic as to the original "independence-friendly" logic.

4 IF logic and game-theoretical semantics

As was indicated, IF logic (and EIF logic) can be motivated in other ways, too. One intuitive and philosophically interesting way is to formulate first the semantics of the received first-order logic in terms of suitable games in the sense of the mathematical theory of games. Each interpreted first-order sentence S defines such game $g(S)$. They are called semantical games, and they can be thought of as games in which one player ("the verifier") defends the truth of S against an opponent ("the falsifier"). This is philosophically interesting because of the precise form of the definition of truth in them. *Pace* Dummett, truth does not mean a win in a semantical game. It means the existence of a winning strategy for the verifier. This is equivalent with the earlier characterization of truth for the Skolem functions of S are nothing but ingredients of winning strategies for the verifier.

From this game-theoretical semantics (GTS) for the traditional first-order logic we obtain a semantic for IF logic simply (and naturally) by allowing semantical games to be games with imperfect information. Since it is natural to define falsity as the existence of a winning strategy for the falsifier, the law of excluded middle is tantamount to the determinateness of semantical games. This determinateness fails in IF logic as a matter of (game-theoretical) course which shows that the failure of tertium non datur in IF logic should not be any surprise at all.

The upshot is that not all IF first-order sentences are either true or false: Some have an indefinite truth-value. It turns out to be possible to assign probabilities to them in an eminently natural way.

5   IF logic and the definability of truth

A third "transcendental deduction" of IF logic turns on an analysis of the notion of truth for quantificational sentences. When is such a sentence S true? One's first impulse - and a sound one - is to say that S is true when the "witness individuals" exist that show (in the sense of displaying) its truth.

A sentence of the form $(\exists x)F[x]$ is thus true if and only if a "witness individual" b exists that makes $F[b]$ true. A sentence of the form $(\forall x)(\exists y)F[x,y]$ is true if for any individual a there exists a "witness individual" b which makes $F[a,b]$ true and so on. The latter example shows that witness individuals may depend on other individuals. Hence the existence of all the truth-displaying witness individuals means the existence of all the functions that provide us with them as functions of other individuals. But a moment's reflection shows that these functions are nothing but the Skolem functions that we have already encountered repeatedly. Hence we arrive in this way at the same semantics as was reached by the first two ways of approaching the semantics of IF logic (as well as of "classical" logic).

Now the existence of Skolem functions for an IF sentence S turns out to be expressible by another IF sentence $T(S)$. I will not prove this result here, but I will illustrate the procedure of reaching $T(S)$ from S. This procedure is illustrated by the equivalence of a sentence of the form

(2) $(\exists f)(\forall x)F[A(f(x))]$

(where A(f(x)) is the only context in which f occurs in F) with

(3) $(\forall x_1)(\forall x_2)(\exists y_1/\forall x_2)(\exists y_2/\forall x_1)(((x_1 = x_2) \supset (y_1 = y_2))$ & $F[A(y_1)]$

(As defined in Section 2, the slash / expresses the independence of a quantifier of another one.) In order to see this, we can translate (3) into a second-order form, which obviously is

(4) $(\exists f_1)(\exists f_2)(\forall x_1)(\forall x_2)((x_1 = x_2) \supset f_1(x_1) = f_2(x_2))$ & $F[A(f_1(x_1))]$

Here the first conjunct says that f_1 and f_2 are the same function. By identifying them in (3) we get back to (2).

Other kinds of context in which f can occur in $(\exists f)F(f)$ require a different but equally feasible treatment.

The upshot is the definability of a truth-predicate for a suitable IF first-order language in the very same language. This puts the entire philosophical discussion of different so-called "theories of truth" in a new light. In logical theory, it opens the extremely important possibility of doing (at least some) of first-order logic by means of the same logic. (See here Hintikka, IF logic in a wider context, forthcoming.)

6   A link between IF logic and paraconsistent logic

What relevance does such an IF logic have for paraconsistent logics? I will not try to answer this question fully here. Instead, I will show how IF logic suggests a large number of questions concerning paraconsistent logic. Their answers will eventually show what the total impact IF logic on the theory of paraconsistent logics will be.

The two have a different initial motivation, and they may look incommensurable. This prima facie incommensurability can be largely overcome in a simple way suggested by the failure of tertium non datur in IF logic. It makes possible a comparison which in the first place involves only a change in terminology.

Let us assume that we decide to change the terms for the basic truth-values, as follows:

OLD LOCUTION	NEW LOCUTION
true or indefinite	true
false or indefinite	false
true	true but not false
false	false but not true
indefinite	true and false
tertium not datur	law of non-contradiction

When a semantically characterized IF logic is expressed in the new terminology, we obtain an apparently new logic that reads in many ways just like a paraconsistent logic. To honor the host town of the meeting that inspired this paper, perhaps we may call this logic paratyconsistent logic. Its prima facie status as a paraconsistent logic prompts a large number of questions many of which point to possible directions of future research or logical theory. Here the more technical questions are considered first.

So far, paratyconsistent logic has been characterized only semantically. What would it mean to axiomatize it? It would mean giving a recursive enumeration of all formulas that are true in the paratyconsistent sense which corresponds to "not false" in IF logic terms. This can be done, for IF logic has a complete disproof procedure. Hence we can ask: What does a complete axiomatization of paratyconsistent logic look like? How is it related to axiomatic systems of paraconsistent logic? This possibility of a complete axiomatization of para(ty)consistent logic might nevertheless easily create a seriously mistaken impression. Such an "axiomatization" means simply a recursive enumeration of logically true formulas. In IF logic, there exists a similar "axiomatization" of logically false (inconsistent) sentences, which is why in a paratyconsistent logic logical truths allow a complete "axiomatization" (i.e. recursive enumeration). But inconsistent formulas are not any longer recursively enumerable in paratyconsistent logic. In EIF logic the fragment corresponding to Σ_1^1 second-order logic allows for a complete axiomatization of logical inconsistencies but not of logical truths, whereas its Π_1^1 fragment allows for a recursive enumeration of logical truths but not of logical inconsistencies.

What all this shows methodologically is that the usual approach to different logics by means of construing them as "logical systems" of axioms and (recursive) rules of inference is not capable of doing justice to IF logic or to paratyconsistent logic. Hence there is no reason to expect that it can do justice to rightly understood paraconsistent logics, either. Formal axiomatization has to be supplemented by semantical methods. This is an important methodological suggestion of our comparison between IF logic and paraconsistent logics.

7   Challenges to and opportunities for paraconsistent logics

If IF logic and paraconsistent logic are at bottom identical or can be made identical by developing a suitable new paraconsistent system, then we should be able to reproduce all the achievements of IF logic in paraconsistent logic. We should be able to define the same mathematical concepts by means of paraconsistent logic as can be defined with the help of IF logic although they cannot be defined in conventional first-order logic, including equicardinaltity and infinity. Paraconsistent logic should turn the "axiom" of choice into a logical truth, as in IF logic. A merger of paratyconsistent logic and paraconsistent logic should also make it possible to formulate some kind of truth condition for a paraconsistent first-order language in the same language. What could it look like?

These question are intertwined with yet others. Especially basic are questions concerning negation. What is the natural treatment of negation in paratyconsistent logic? Should the negation that is in fact used in paraconsistent logics be interpreted as the contradictory negation or as some kind of stronger (dual) negation? Some of the problems listed above can only be solved by introducing a second negation into one's logic. How can this be done in the context of paraconsistent logic?

Other questions concerning the semantical basis of paraconsistent logics. In order to use it as its own metalogic, the concept of truth for a paraconsistent language must be expressible by means of paraconsistent logic. The analogous truth condition for an IF first-order language can be expressed in the same IF language. But there is a price to be paid for this achievement. The price is that the semantics of IF first-order logic is unavoidably noncompositional. This noncompositionality is due to the fact that the force of a quantifier is affected by its dependence and independence of the quantifiers in whose scope it occurs. (Compositionality is equivalent with semantical context-independence.) Now to the best of my knowledge noncompositional semantics for paraconsistent logic has not been seriously contemplated. Hence something has to be fundamentally changed if in semantical treatments of paraconsistent logic paraconsistent logicians want to use the bridge provided by paratyconsistent logic for the purpose of reaching a self-applicable truth condition and for the wider purpose of enabling a paraconsistent language to serve as its own metalanguage. In a general theoretical perspective, having to give up compositionality is likely to be the deepest methodological change needed for paraconsistent logicans to make use of the kinship of paraconsistent logic with IF logic.

These questions and suggestions mean challenges to theorists of paraconsistent logic. Since I cannot speak for them, I must leave actual attempts to follow up these leads to paraconsistent logicians.

8   On the interpretation of paraconsistent logics

This does not exhaust the questions that can be raised here. The most important aspect of the comparison between IF logic and paraconsistent logic concerns their interpretation. In one perspective, the impact of IF logic looks like the best thing that could have happened to paraconsistent logic. One of its main weak spots has been its interpretation. We seem to have a reasonable pretheoretical idea of what it means for a proposition to be true and of what it means for it to be false. But what on earth can it mean for a proposition to be true and false? If Cain tells in a cross-examination that it is both true and false that he killed Abel, he might be cited for contempt of the court. And it does not seem to be any less strange to say that a proposition and its negation can both be true. The existing discussions of paraconsistent logics do not yield a satisfactory unique account of the concrete down-to-earth meaning of self-contradictory but yet not disprovable propositions. And it would seem that it will be equally difficult to give a fully operationalized account of how a proposition and it negation can both be true.

Here the semantics of IF logic, transposed into a paratyconsistent logic, seems to be the answer to paraconsistent semanticist's prayers. The problem of interpreting propositions of paraconsistent logic that are both true and false corresponds to the problem of interpreting the propositions in an IF language that have the truth-value "indefinite". And as was pointed out above, the ascription of such a truth-value to a proposition amounts to an assertion of a certain objective fact about the world. It is a mere prejudice that the only way of conveying information about the world is to assert the truth of a sentence. Game-theoretical semantics shows convincingly that this is not the only way. I can equally well inform you about reality by confiding to you that a certain sentence is false or that it has an indefinite truth-value.

Thus here IF logic and its game theoretical semantics seem to provide paraconsistent logic with the concrete semantical interpretation it has been missing. If this interpretation (for brevity, I will call it the game-theoretical or GT interpretation) is accepted and put to work, we can expect all the nice results achieved by means of IF logic to be available also to paraconsistant logicians. What else could a logician hope for here?

9   Paraconsistent logic and epistemology

A GT interpretation of paraconsistent logic along these lines should thus satisfy all logicians. However, it does not by itself satisfy a critical philosopher of logic. Such an interpretation is objective in that it does not involve any epistemic element. It relies entirely on the meaning relations (rules for semantical games) that are independent of any particular use of language and logic. This objectivity is a great merit of the GT interpretation. But it does not seem to be possible to reconcile with the more or less official motivation of the entire idea of paraconsistent reasoning. The most interesting and most convincing kinds of motivation of paraconsistent logic have been epistemic, not purely logico-semantical. Paraconsistent logic has been supposed to provide us with means of coping with situations in which the different items of information we have received do not agree with each other. The importance of paraconsistent logic is supposed to derive from the importance and frequency of such informational conflicts in our epistemic life. This motivation now seems to be entirely lost, which would might seem to leave paraconsistent logic philosophically high and dry.

One possible response here is to suggest that the correlation between the semantics for IF logic and the semantics for paraconsistent logic explained above can apparently provide an epistemic pragmatics for paraconsistent logic and hence vindicate its rationale. One of the most striking recent applications of logical ideas to epistemology is the discovery of a near-identity of the optimal strategies of logical deduction and the optimal strategies of interrogative inquiry, at least in contexts of pure discovery. This strategic significance of logic for inquiry in general can presumably be preserved in the transition from IF logic to paraconsistent logic. In this correlation, the gradual elimination of de facto contradictions which paraconsistent logics are supposed to facilitate corresponds to a gradual elimination of truth-value gaps. This is indeed an eminently viable way of (albeit not the only one) of conceptualizing our epistemological enterprise. It does in fact provide a coherent and realistic account of the epistemological services that paraconsistent logic can provide. In this sense, one can agree with the claim of paraconsistent logicians that contradictions can be a rich source of information.

This strategic perspective hence provides us with a viable pragmatic and epistemological *raison d'être* of paraconsistent logic. However, this vindication does not agree fully with the "official" epistemological motivation of paraconsistent logics. What the gradual filling of truth value gaps means epistemologically is the reconciliation of apparently discrepant items of information or their integration with a coherent total account of reality. This happens through the acquisition of further information. The most important function of logic in epistemology is to provide strategies for this search for new information.

In so far as the "paratyconsistent" interpretation provides an account of how paraconsistent logics can help this reconciliation process by providing strategies for it, it vindicates the paraconsistency idea epistemologically. However, in order for them to serve this purpose, the strategic uses of paraconsistent logics have to be spelled out more fully than before.

Furthermore, and equally significantly, such strategic advice applies only if the reconciliation process does not involve the outright (or temporary) rejection of any items of information. About the strategies of such rejection, paraconsistent logics do not tell anything non-trivial. But this is not a flaw of paraconsistent logics. It is a fact of a logician's life. No logic in a genuine sense of the term can provide strategies for deciding which putative information to reject or not to reject in a context of actual material inquiry. The reason is that the advisability of rejection cannot be judged fully on the basis of information so far reached in an empirical inquiry. It depends on the total reality itself including its so far unknown parts and aspects. What they are can only be guessed at, not gathered from the strategies of any logic. Hence paraconsistent logic can serve legitimate epistemological purposes, but not as a logic of belief revision in a sense that would provide strategies for the outright rejection of any items of prima facie information. If this implies a change in the current philosophical motivation of paraconsistent logic, then that change is unavoidable.

Perhaps there is also something of a logical confusion threatening here.

Perhaps the illusion that paraconsistent logics could tell us something about
the rejection of information is due to an uncertainty about the meaning of
negation in paraconsistent logic. It seems that some logicians and philosophers think of it as being basically the contradictory negation. Yet what
seems to be the only really viable semantics for paraconsistent logic involves
inevitably also the dual (strong) negation, explicitly or implicitly.

10   Different Games

One suggestion that may clarify the motivation of paraconsistent logic is
to emphasize distinctions between different epistemologically and semantically relevant activities, alias "games". The semantical games explained in
Section 4 above in terms of which truth can be defined are not games of
truth-seeking. The truth of a sentence S means the existence of a winning
strategy, for the "verifier" in the semantical game $G(S)$ correlated with S.
It does not mean an actual win in any one play of the semantical game
$G(S)$. Hence to try to find the truth of S is to try to find such a winning
strategy. Such enterprises are essentially and deeply different from semantical games. I have urged that they should be conceptualized as questioning
games. Such interrogative games involve deduction as one component, but
they are neither semantical games nor "games" of formal proofs of logical
truth.

It seems to me eminently salutary in general to distinguish these three
types of games from each other. For instance, establishing the logical truth
(truth in all interpretations) of S through a formal proof is something cat-
egorically different from establishing the actual truth of S (i.e. truth in the
actual world). In particular, the problem of coping with contradictory information belongs squarely to interrogative games, not to semantical games
or to the formal games of theorem-proving. Hence paraconsistent logicians
might have to reconsider how their ostensive arm of dealing with contradictory information might best be realized. Whether or not a deductive
logic can be paraconsistent is one problem. A different question which is
not even controversial is that the logic of interrogative games should be
paraconsistent in the sense of admitting situations where one's prima facie
information involves contradictions. What is significant about the interrogative games that codify empirical reasoning is that contradictions are in
them dealt with by means of strategic rules and not by means of move-by-
move rules like rules of inference in deductive logic. Maybe the basic idea
of paraconsistency should be developed in the context of a strategy theory
for interrogative games. This theory is the true logic of "belief revision".

BIBLIOGRAPHY

[1] Walter A. Carnielli. Possible-translations semantics for paraconsistent logics. In D. Batens, et. al., editors, *Frontiers of Paraconsistent Logic: Proceedings of the I World Congress on Paraconsistency*, Logic and Computation Series, pages 149–163. Baldock: Research Studies Press, King's College Publications, 2000.

[2] Walter A. Carnielli, Marcelo E. Coniglio, and João Marcos. Logics of formal inconsistency. In D. Gabbay and F. Guenthner, editors, *Handbook of Philosophical Logic*, volume 14, pages 15–107. Springer, 2nd edition, 2007. Preprint available at *CLE e-Prints* vol 5, n. 1, 2005. `http://www.cle.unicamp.br/e-prints/vol\_5,n\_1,2005.html`.

[3] Jaakko Hintikka. There is only one logic (but it is not the one you know). Forthcoming.

[4] Jaakko Hintikka. *Principles of Mathematics Revisited*. Cambridge University Press, Cambridge, 1996.

[5] Jaakko Hintikka. *Socratic Epistemology: Explorations of Knowledge-Seeking by Questioning*. Cambridge University Press, New York, 2007.

[6] Jaakko Hintikka, I. Halonen, and A. Mutanen. Interrogative logic as a general theory of reasoning. In R. Johnson and J. Woods, editors, *Handbook of Applied Logic*. Kluwer Academic Publishers, Dordrecht, 1999.

[7] Jaakko Hintikka and Gabriel Sandu. Game-theoretical semantics. In Johan van Benthem and Alice ter Meulen, editors, *Handbook of Logic and Language*. Elsevier, Amsterdam, 1996.

Jaakko Hintikka
Department of Philosophy
Boston University
Boston 02215, U.S.A
E-mail: hintikka@bu.edu

Modular Semantics for Some Basic Logics of Formal Inconsistency

ARNON AVRON

ABSTRACT. We construct a modular semantic framework for LFIs (logics of formal (in)consistency) which extends the framework developed in previous papers, so it now includes all the basic axioms considered in the literature on LFIs, plus a few more. In addition, the paper provides another demonstration of the power of the idea of non-deterministic semantics, especially when it is combined with the idea of using truth-values to encode data concerning propositions.

1 Introduction

One of the oldest and best known approaches to the problem of designing useful paraconsistent logics (i.e. logics which allows nontrivial inconsistent theories — see [8, 14, 10, 9]) is da Costa's approach ([15, 16]), which seeks to allow the use of classical logic whenever it is safe to do so, but behaves completely differently when contradictions are involved. da Costa's approach has led to the family of LFIs (Logics of Formal (In)consistency — see [13]). In [3] and [1] we developed a semantic framework for this family, in which it is possible to provide simple semantics for almost all the propositional LFIs considered in the literature. This semantics is based on the use of non-deterministic matrices (Nmatrices). These are multi-valued structures (introduced in [6, 7]) where the value assigned by a valuation to a complex formula can be chosen non-deterministically out of a certain nonempty set of options.

The semantic framework for LFIs which is based on Nmatrices has two crucial properties that previous semantic frameworks used for this task (the bivaluations semantics and the possible translations semantics described in [12, 13, 17]) in general lack: [1]

---

[1] In [11] it was argued that the semantics of Nmatrices is a particular case of possible translations semantics. This observation is irrelevant to our claims concerning it, because it is precisely this generality which is the source of weakness of the possible translations semantics, and the reason why *in general* it lacks the two properties described below. Similarly, possible translations semantics (as well as practically any other type of semantics) can be viewed as a particular (but superior) case of the bivaluations semantics, and the two-valued semantics of classical logic is a particular (but superior) case of the more general semantics of Boolean algebras. In all these cases, as well as in many others, the isolation of a useful particular case of a more general framework is of crucial importance.

- It is *analytic*[2] in the sense that for determining whether $T \vdash_\mathcal{M} \varphi$ (where $\mathcal{M}$ is an Nmatrix) it always suffices to check only *partial* valuations, defined only on subformulas of $T \cup \{\varphi\}$. It follows that a logic which has a finite characteristic Nmatrix is necessarily decidable.

- It is *modular*: each axiom has its own semantics, and the semantics of a system is obtained by joining the semantics of its axioms. As demonstrated in [1, 2, 3, 4], this fact makes it possible to simultaneously prove soundness and completeness theorems for thousands of systems (this paper includes another striking example of this phenomenon).

Now [1, 3] have left one major gap: no semantics was provided there to Marcos' axiom (denoted below by (**m**)). [3] This axiom is crucial in one of the two LFIs which are considered as basic in [13]: Marcos' system **mCi**, to which the whole of section 4 of [13] is devoted. This gap was partially closed in [5], where a 5-valued Nmatrix which is characteristic for **mCi** has been given. [4] However, this Nmatrix does not provide an independent semantics for Marcos' axiom (in the form of a semantic condition that corresponds specifically to this axiom), but only to its *combination with another axiom* (axiom (**i**) below, to which, in contrast, an independent semantics *has* been provided in [1, 3]). As a result, the full modularity of the semantics was lost in [5]. The main goal of this paper is to restore it by extending the framework developed in [1, 3], so it includes systems with Marcos' axiom too (not only those that have already been investigated in [13], but also some new ones that naturally arise). The extended framework provides semantics in a modular way to practically *all* the axioms and systems considered in [13] — plus a few more. In addition, the paper provides another demonstration of the power of the idea of non-deterministic semantics, especially when it is combined with the idea of using truth-values to encode data concerning propositions (see section 3).

2 Preliminaries

2.1 A Taxonomy of LFIs

Let $\mathcal{L}_{cl}^+ = \{\wedge, \vee, \supset\}$, $\mathcal{L}_{cl} = \{\wedge, \vee, \supset, \neg\}$, and $\mathcal{L}_C = \{\wedge, \vee, \supset, \neg, \circ\}$. [5]

DEFINITION 1. Let **HCL**$^+$ (Hilbert-style positive Classical Logic) be some Hilbert-type system which has Modus Ponens as the only inference rule, and is sound and strongly complete for the $\mathcal{L}_{cl}^+$-fragment of CPL (classical propositional logic) [6]. The logic **B** [7] is the logic in $\mathcal{L}_C$ obtained from **HCL**$^+$

---

[2]In previous papers we use the term *"effective"* for this property, but now we believe that "analytic" is more appropriate.

[3]We have chosen the name "Marcos' axiom" for this axiom, because it was first introduced in Marcos' paper [17].

[4]A possible-translation semantics for **mCi** has been provided in [17].

[5]The intuitive meaning of $\circ\varphi$ is "φ is a consistent formula" or "φ behaves classically".

[6]I.e.: for every sentence φ and theory **T** in $\mathcal{L}_{cl}^+$, **T** $\vdash_{\mathbf{HCL}^+} \varphi$ iff **T** $\vdash_{CPL} \varphi$.

[7]The logic **B** is called mbC in [13]

by adding the following schemata (where φ and ψ vary through formulas):

(t) $\neg\varphi \vee \varphi$

(p) $\circ\varphi \supset ((\varphi \wedge \neg\varphi) \supset \psi)$

DEFINITION 2. For $n \geq 0$, let $\neg^0\varphi = \varphi$, $\neg^{n+1}\varphi = \neg(\neg^n\varphi)$.

DEFINITION 3. Let Ax be the set consisting of the following schemata:

(m) $\circ\neg^n\circ\varphi$ (for every $n \geq 0$)

(c) $\neg\neg\varphi \supset \varphi$

(e) $\varphi \supset \neg\neg\varphi$

(k1) $\circ\varphi \vee \varphi$

(k2) $\circ\varphi \vee \neg\varphi$

(k) $\circ\varphi \vee (\varphi \wedge \neg\varphi)$

(i1) $\neg\circ\varphi \supset \varphi$

(i2) $\neg\circ\varphi \supset \neg\varphi$

(i) $\neg\circ\varphi \supset (\varphi \wedge \neg\varphi)$

For $S \subseteq Ax$, $\mathbf{B}[S]$ is the extension of $\mathbf{B}$ by the axioms in S.

Notation: We'll usually denote $\mathbf{B}[S]$ by $\mathbf{B}s$, where s is a string consisting of the names of the axioms in S (thus we denote $\mathbf{B}[\{(\mathbf{i}), (\mathbf{e})\}]$ by $\mathbf{Bie}$). [8]

2.2   Non-deterministic Matrices

Our main semantic tool in what follows is the following generalization (from [6, 7, 1, 2]) of the concept of a matrix:

DEFINITION 4.

1. A *non-deterministic matrix* (*Nmatrix* for short) for a propositional language $\mathcal{L}$ is a tuple $\mathcal{M} = \langle \mathcal{V}, \mathcal{D}, \mathcal{O} \rangle$, where:

    (a) $\mathcal{V}$ is a non-empty set of *truth values*.

    (b) $\mathcal{D}$ is a non-empty proper subset of $\mathcal{V}$.

    (c) For every n-ary connective $\diamond$ of $\mathcal{L}$, $\mathcal{O}$ includes a corresponding n-ary function $\tilde{\diamond}$ from $\mathcal{V}^n$ to $2^{\mathcal{V}} - \{\emptyset\}$.

    We say that $\mathcal{M}$ is *(in)finite* if so is $\mathcal{V}$.

---

[8] In the literature on LFIs one usually writes $\mathbf{C}s$ instead of our $\mathbf{B}cs$ when S includes the axiom (c). Note also that what we call (m) is called (cc) in [13].

2. A *valuation* in an Nmatrix $\mathcal{M}$ is a function v from the set of formulas of $\mathcal{L}$ to $\mathcal{V}$ that satisfies the following condition for every n-ary connective $\diamond$ of $\mathcal{L}$ and $\psi_1, \ldots, \psi_n \in \mathcal{L}$:

$$v(\diamond(\psi_1, \ldots, \psi_n)) \in \widetilde{\diamond}(v(\psi_1), \ldots, v(\psi_n))$$

3. A valuation v in an Nmatrix $\mathcal{M}$ is a *model* of (or *satisfies*) a formula ψ in $\mathcal{M}$ (notation: $v \models^{\mathcal{M}} \psi$) if $v(\psi) \in \mathcal{D}$. v is a *model* in $\mathcal{M}$ of a set $\mathbf{T}$ of formulas (notation: $v \models^{\mathcal{M}} \mathbf{T}$) if it satisfies every formula in $\mathbf{T}$.

4. $\vdash_{\mathcal{M}}$, the consequence relation induced by the Nmatrix $\mathcal{M}$, is defined as follows: $\mathbf{T} \vdash_{\mathcal{M}} \varphi$ if for every v such that $v \models^{\mathcal{M}} \mathbf{T}$, also $v \models^{\mathcal{M}} \varphi$.

5. A logic $\mathbf{L} = \langle \mathcal{L}, \vdash_{\mathbf{L}} \rangle$ is *sound* for an Nmatrix $\mathcal{M}$ (where $\mathcal{L}$ is the language of $\mathcal{M}$) if $\vdash_{\mathbf{L}} \subseteq \vdash_{\mathcal{M}}$. $\mathbf{L}$ is *complete* for $\mathcal{M}$ if $\vdash_{\mathbf{L}} \supseteq \vdash_{\mathcal{M}}$. $\mathcal{M}$ is *characteristic* for $\mathbf{L}$ if $\mathbf{L}$ is both sound and complete for it (i.e.: if $\vdash_{\mathbf{L}} = \vdash_{\mathcal{M}}$). $\mathcal{M}$ is *weakly-characteristic* for $\mathbf{L}$ if for every formula φ of $\mathcal{L}$, $\vdash_{\mathbf{L}} \varphi$ iff $\vdash_{\mathcal{M}} \varphi$.

DEFINITION 5. Let $\mathcal{M}_1 = \langle \mathcal{V}_1, \mathcal{D}_1, \mathcal{O}_1 \rangle$ and $\mathcal{M}_2 = \langle \mathcal{V}_2, \mathcal{D}_2, \mathcal{O}_2 \rangle$ be Nmatrices for a language $\mathcal{L}$. $\mathcal{M}_2$ is called a *simple refinement* [9] of $\mathcal{M}_1$ if $\mathcal{V}_2 \subseteq \mathcal{V}_1$, $\mathcal{D}_2 = \mathcal{D}_1 \cap \mathcal{V}_2$, and $\widetilde{\diamond}_{\mathcal{M}_2}(\vec{x}) \subseteq \widetilde{\diamond}_{\mathcal{M}_1}(\vec{x})$ for every n-ary connective $\diamond$ of $\mathcal{L}$ and every $\vec{x} \in \mathcal{V}_2^n$.

The following proposition can easily be proved:

PROPOSITION 6. *If $\mathcal{M}_2$ is a simple refinement of $\mathcal{M}_1$ then $\vdash_{\mathcal{M}_1} \subseteq \vdash_{\mathcal{M}_2}$. Hence if $\mathbf{L}$ is sound for $\mathcal{M}_1$ then $\mathbf{L}$ is also sound for $\mathcal{M}_2$.*

2.3   General Non-deterministic Semantics for extensions of B

DEFINITION 7.

- A basic **B**-Nmatrix is an Nmatrix for the language $\mathcal{L}_C$ such that:

  1. $\mathcal{V} = \mathcal{T} \uplus \mathcal{I} \uplus \mathcal{F}$, where $\mathcal{T}, \mathcal{I},$ and $\mathcal{F}$ are disjoint nonempty sets.

  2. $\mathcal{D} = \mathcal{T} \cup \mathcal{I}$

  3. $\mathcal{O}$ is defined by:

$$a \widetilde{\vee} b = \begin{cases} \mathcal{D} & \text{if either } a \in \mathcal{D} \text{ or } b \in \mathcal{D}, \\ \mathcal{F} & \text{if } a, b \in \mathcal{F} \end{cases}$$

$$a \widetilde{\supset} b = \begin{cases} \mathcal{D} & \text{if either } a \in \mathcal{F} \text{ or } b \in \mathcal{D} \\ \mathcal{F} & \text{if } a \in \mathcal{D} \text{ and } b \in \mathcal{F} \end{cases}$$

$$a \widetilde{\wedge} b = \begin{cases} \mathcal{F} & \text{if either } a \in \mathcal{F} \text{ or } b \in \mathcal{F} \\ \mathcal{D} & \text{otherwise} \end{cases}$$

---

[9] A more general notion of a refinement was used in [2]. However, here we shall need only simple refinements.

$$\tilde{\neg} a = \begin{cases} \mathcal{F} & \text{if } a \in \mathcal{T} \\ \mathcal{D} & \text{otherwise} \end{cases}$$

$$\tilde{\circ} a = \begin{cases} \mathcal{V} & \text{if } a \in \mathcal{F} \cup \mathcal{T} \\ \mathcal{F} & \text{if } a \in \mathcal{I} \end{cases}$$

- A **B**-Nmatrix is an Nmatrix for $\mathcal{L}_C$ which is a simple refinement of some basic **B**-Nmatrix.

The following theorem from [3] can easily be proved:

THEOREM 8. **B** *is sound for any* **B***-Nmatrix.*

3 The Nmatrix $\mathcal{M}_{10}^B$ and Its Simple Refinements

The main semantic idea used in [1] is that truth-values can be used to encode the data concerning sentences which determine the consequence relation of a given logic. For all the LFIs considered in [1] and [3], the data needed about a sentence φ was whether φ is true or false, whether $\neg\varphi$ is true or false, and whether $\circ\varphi$ is true or false. Accordingly, for most of the systems considered there we used as truth-values triples in $\{0,1\}^3$ (or sometimes $\{0,1\}^2$, in case this sufficed). Now Marcos' axiom (**m**) is concerned with formulas of a certain particular *syntactic* form. Therefore the key idea in handling it is to add one more bit to the truth-values, indicating whether the sentence has this particular form, or at least "behaves" as if it has such a form (this modification is needed because atomic formulas may be assigned any truth value, even if they do not have the special syntactic form that this truth value is meant to signify). Accordingly, we shall use in what follows elements of $\{0,1\}^4$ as our truth-values. The intuitive meaning of the forth bit is: the sentence belong to a certain class A of sentences which is closed under negation and includes every formula of the form $\circ\psi$ (the identity of the class A is application-dependent, but here it is usually the class of sentences of the form $\neg^n\circ\psi$).

Notation For $1 \leq i \leq 4$ we let $P_i(\langle x_1, x_2, x_3, x_4 \rangle) = x_i$. We shall usually write $x_1 x_2 x_3 x_4$ instead of $\langle x_1, x_2, x_3, x_4 \rangle$ when the latter is in $\{0,1\}^4$.

DEFINITION 9. The Nmatrix $\mathcal{M}_{10}^B = \langle \mathcal{V}_{10}, \mathcal{D}_{10}, \mathcal{O}_{10} \rangle$ is defined as follows:

- $\mathcal{V}_{10} = \{1101, 1100, 1011, 1010, 1001, 1000, 0111, 0110, 0101, 0100\}$. In other words: $\mathcal{V}_{10}$ is the set of tuples in $\{0,1\}^4$ which satisfy the following two conditions:

  C(**t**): If $P_1(a) = 0$ then $P_2(a) = 1$

  C(**b**): If $P_1(a) = 1$ and $P_2(a) = 1$ then $P_3(a) = 0$

- $\mathcal{D}_{10} = \{a \in \mathcal{V}_{10} \mid P_1(a) = 1\}$

- Let $\mathcal{V} = \mathcal{V}_{10}$, $\mathcal{D} = \mathcal{D}_{10}$, $\mathcal{F} = \mathcal{V}_{10} - \mathcal{D}$. The operations in $\mathcal{O}_{10}$ are:

$$\tilde{\neg}a = \{b \in \mathcal{V} \mid P_1(b) = P_2(a),\ \text{and if}\ P_4(a) = 1\ \text{then}\ P_4(b) = 1\}$$

$$\tilde{\circ}a = \{b \in \mathcal{V} \mid P_1(b) = P_3(a)\ \text{and}\ P_4(b) = 1\}$$

$$a\tilde{\vee}b = \begin{cases} \mathcal{D} & \text{if either } a \in \mathcal{D} \text{ or } b \in \mathcal{D} \\ \mathcal{F} & \text{if } a, b \in \mathcal{F} \end{cases}$$

$$a\tilde{\supset}b = \begin{cases} \mathcal{D} & \text{if either } a \in \mathcal{F} \text{ or } b \in \mathcal{D} \\ \mathcal{F} & \text{if } a \in \mathcal{D} \text{ and } b \in \mathcal{F} \end{cases}$$

$$a\tilde{\wedge}b = \begin{cases} \mathcal{F} & \text{if either } a \in \mathcal{F} \text{ or } b \in \mathcal{F} \\ \mathcal{D} & \text{otherwise} \end{cases}$$

Note. It can easily be checked that

$$\tilde{\circ}a = \begin{cases} \{0101, 0111\} & \text{if } P_3(a) = 0 \\ \{1101, 1011, 1001\} & \text{if } P_3(a) = 1 \end{cases}$$

DEFINITION 10.

1. The general refining conditions induced by the conditions in Ax are:

   C(m): $\tilde{\neg}\, 0111 = \{1011\}$, $\tilde{\neg}1011 = \{0111\}$, and:

   $$\tilde{\circ}a = \begin{cases} \{0111\} & \text{if } P_3(a) = 0 \\ \{1011\} & \text{if } P_3(a) = 1 \end{cases}$$

   C(c): If $P_1(a) = 0$ then $\tilde{\neg}a \subseteq \{x \mid P_1(x) = 1 \text{ and } P_2(x) = 0\}$.

   C(e): If $P_1(a) = P_2(a) = 1$ then $\tilde{\neg}a \subseteq \{x \mid P_1(x) = 1 \text{ and } P_2(x) = 1\}$.

   C(k1): If $P_1(a) = 0$ then $P_3(a) = 1$ (equivalently: 0101 and 0100 should be deleted).

   C(k2): If $P_2(a) = 0$ then $P_3(a) = 1$ (equivalently: 1001 and 1000 should be deleted).

   C(k): Both C(k1) and C(k2) should be satisfied.

   C(i1): If $P_1(a) = 0$ then $\tilde{\circ}a \subseteq \{x \mid P_2(x) = 0\}$ (equivalently: C(k1) should be satisfied, and $\tilde{\circ}(a) \subseteq \{1011, 1001\}$ for $a \in \{0111, 0110\}$).

   C(i2): If $P_2(a) = 0$ then $\tilde{\circ}a \subseteq \{x \mid P_2(x) = 0\}$ (equivalently: C(k2) should be satisfied, and $\tilde{\circ}(1011) = \tilde{\circ}(1010) = \{1011\}$).

   C(i): Both C(i1) and C(i2) should be satisfied.

2. For $S \subseteq Ax$, let $C(S) = \{Cr \mid r \in S\}$, and let $\mathcal{M}_S$ be the weakest simple refinement of $\mathcal{M}_{10}^B$ in which the conditions in $C(S)$ are all satisfied (it is easy to check that this is well-defined for every $S \subseteq Ax$).

4   The Soundness and Completeness Theorem

The following is the main result of this paper:

THEOREM 11. *For $S \subseteq Ax$, $\mathcal{M}_S$ is characteristic for $\mathbf{B}[S]$.*

Proof. *Soundness:* Obviously, for each $S \subseteq Ax$, $\mathcal{M}_S$ is simple refinement of the basic **B**-Nmatrix in which $\mathcal{V} = \mathcal{V}_{10}$, $\mathcal{T} = \{a \in \mathcal{V} \mid P_2(a) = 0\}$, $\mathcal{F} = \{a \in \mathcal{V} \mid P_1(a) = 0\}$, and $\mathcal{I} = \{a \in \mathcal{V} \mid P_1(a) = P_2(a) = 1\}$. Therefore by Theorem 8 it follows that **B** is sound for $\mathcal{M}_S$. It remains to show that if $\mathbf{s} \in S$ then the axiom $\mathbf{s}$ is valid in $\mathcal{M}_S$. We do here the case of $(\mathbf{m})$ (handling the other cases is straightforward, and is very similar to the way they were handled in [1, 2] and [3]). So let φ be a sentence, and v an assignment in $\mathcal{M}_S$ (where $(\mathbf{m}) \in S$). Then by the third part of $C(\mathbf{m})$, $v(\circ\varphi) \in \{0111, 1011\}$. Accordingly, the first two parts of $C(\mathbf{m})$ entail that $v(\neg^n\circ\varphi) \in \{0111, 1011\}$ for every $n \geq 0$. Again by the third part of $C(\mathbf{m})$, $v(\circ\neg^n \circ \varphi) = 1011$. Since 1011 is designated, this means that $\circ\neg^n\circ \varphi$ is valid in $\mathcal{M}_S$.

Completeness: Assume that **T** is a theory and φ_0 a sentence such that $\mathbf{T} \not\vdash_{\mathbf{B}[S]} \varphi_0$. We construct a model of **T** in $\mathcal{M}_S$ which is not a model of φ_0. For this extend **T** to a maximal theory $\mathbf{T}^*$ such that $\mathbf{T}^* \not\vdash_{\mathbf{B}[S]} \varphi_0$. $\mathbf{T}^*$ has the following properties:

1. $\psi \notin \mathbf{T}^*$ iff $\psi \supset \varphi_0 \in \mathbf{T}^*$.

2. If $\psi \notin \mathbf{T}^*$ then $\psi \supset \varphi \in \mathbf{T}^*$ for every sentence φ of $\mathcal{L}_C$.

3. $\varphi \vee \psi \in \mathbf{T}^*$ iff either $\varphi \in \mathbf{T}^*$ or $\psi \in \mathbf{T}^*$.

4. $\varphi \wedge \psi \in \mathbf{T}^*$ iff both $\varphi \in \mathbf{T}^*$ and $\psi \in \mathbf{T}^*$.

5. $\varphi \supset \psi \in \mathbf{T}^*$ iff either $\varphi \notin \mathbf{T}^*$ or $\psi \in \mathbf{T}^*$.

6. For every sentence φ of $\mathcal{L}_C$, either $\varphi \in \mathbf{T}^*$ or $\neg\varphi \in \mathbf{T}^*$.

7. If both $\varphi \in \mathbf{T}^*$ and $\neg\varphi \in \mathbf{T}^*$ then $\circ\varphi \notin \mathbf{T}^*$.

The proofs of Properties 1–7 are *exactly* as in the proof of Theorem 1 of [3]: Property 1 follows from the deduction theorem (which is obviously valid for $\mathbf{B}[S]$) and the maximality of $\mathbf{T}^*$. Property 2 is proved first for $\psi = \varphi_0$ as follows: by 1, if $\varphi_0 \supset \varphi \notin \mathbf{T}^*$ then $(\varphi_0 \supset \varphi) \supset \varphi_0 \in \mathbf{T}^*$. Hence $\varphi_0 \in \mathbf{T}^*$ by the tautology $((\varphi_0 \supset \varphi) \supset \varphi_0) \supset \varphi_0$. A contradiction. Property 2 then follows for all $\psi \notin \mathbf{T}^*$ by 1 and the transitivity of implication. Properties 3–5 are easy corollaries of 1, 2, and the closure of $\mathbf{T}^*$ under positive classical inferences (for example: suppose $\varphi \vee \psi \in \mathbf{T}^*$, but neither $\varphi \in \mathbf{T}^*$, nor $\psi \in \mathbf{T}^*$. By property 1, $\varphi \supset \varphi_0 \in \mathbf{T}^*$ and $\psi \supset \varphi_0 \in \mathbf{T}^*$. Since φ_0 follows in positive classical logic from $\varphi \vee \psi$, $\varphi \supset \varphi_0$, and $\psi \supset \varphi_0$, we get $\varphi_0 \in \mathbf{T}^*$. A contradiction). Finally, Property 6 is immediate from Property 3 and Axiom $(\mathbf{t})$, and Property 7 follows from Axiom $(\mathbf{p})$.

define a valuation v in $\mathcal{M}_S$ by $v(\varphi) = \langle x_1(\varphi), x_2(\varphi), x_3(\varphi), x_4(\varphi) \rangle$, where:

$$x_1(\varphi) = \begin{cases} 1 & \varphi \in \mathcal{T}^* \\ 0 & \varphi \notin \mathcal{T}^* \end{cases}$$

$$x_2(\varphi) = \begin{cases} 1 & \neg\varphi \in \mathcal{T}^* \\ 0 & \neg\varphi \notin \mathcal{T}^* \end{cases}$$

$$x_3(\varphi) = \begin{cases} 1 & \circ\varphi \in \mathcal{T}^* \\ 0 & \circ\varphi \notin \mathcal{T}^* \end{cases}$$

$$x_4(\varphi) = \begin{cases} 1 & \varphi \text{ is of the form } \neg^n \circ \psi \\ 0 & \text{otherwise} \end{cases}$$

Now we show that v is a valuation in $\mathcal{M}_S$. Properties 6 and 7 of $\mathcal{T}^*$ together ensure that v takes values in $\mathcal{V}_{10}$. From the definition of v it immediately follows that v is a $\mathcal{M}_{\mathbf{B}}$-valuation (i.e. $P_1(v(\circ\psi)) = P_3(v(\psi))$, $P_4(v(\circ\psi)) = 1$, $P_1(v(\neg\psi)) = P_2(v(\psi))$, and if $P_4(v(\psi)) = 1$ then $P_4(v(\neg\psi)) = 1$). It remains to show that v respects the conditions corresponding to the axioms in S.

- Suppose $(\mathbf{m}) \in S$. Then $\circ\circ\varphi \in \mathcal{T}^*$ for every φ. Hence the definition of v entails that $P_3(v(\circ\varphi)) = P_4(v(\circ\varphi)) = 1$. This, the definition of v, and the fact that if $P_3(v(\varphi)) = 1$ then either $P_2(v(\varphi)) = 0$ or $P_2(v(\varphi)) = 0$ (but not both), together imply that v satisfies the part concerning $\tilde{\circ}$ in $C(\mathbf{m})$. Now assume that $v(\varphi) = 0111$. Then φ is of the form $\neg^n \circ \psi$. Hence so is $\neg\varphi$. This, axiom $(\mathbf{m})$, and the fact that $P_2(v(\varphi)) = 1$, imply that $P_1(v(\neg\varphi)) = P_3(v(\neg\varphi)) = P_4(v(\neg\varphi)) = 1$. Hence $v(\neg\varphi) = 1011$. That if $v(\varphi)) = 1011$ then $v(\neg\varphi) = 0111$ is proved similarly. It follows that v respects $C(\mathbf{m})$.

- Suppose $(\mathbf{c}) \in S$, and that $P_1(v(\varphi)) = 0$. Then $\varphi \notin \mathcal{T}^*$. By property 6 of $\mathcal{T}^*$ and axiom $(\mathbf{c})$, this implies that $\neg\varphi \in \mathcal{T}^*$, while $\neg\neg\varphi \notin \mathcal{T}^*$. Hence $P_1(v(\neg\varphi)) = 1$ and $P_2(v(\neg\varphi)) = 0$, and so v respects $C(\mathbf{c})$.

- Suppose $(\mathbf{k1}) \in S$, and that $P_1(v(\varphi)) = 0$. Then $\varphi \notin \mathcal{T}^*$. It follows that $\circ\varphi \in \mathcal{T}^*$, by $(\mathbf{k1})$ and property 3 of $\mathcal{T}^*$. Hence $P_3(v(\varphi)) = 1$, and so v respects $C(\mathbf{k1})$.

- Suppose $(\mathbf{i2}) \in S$, and that $P_2(v(\varphi)) = 0$. Then $\neg\varphi \notin \mathcal{T}^*$. It follows by $(\mathbf{i2})$ that $\neg\circ\varphi \notin \mathcal{T}^*$. Hence $P_2(v(\circ\varphi)) = 0$, and so v respects $C(\mathbf{i2})$.

We leave the other cases to the reader.

Obviously, $v(\psi) \in \mathcal{D}_S$ for every $\psi \in \mathcal{T}^*$, while $v(\varphi_0) \notin \mathcal{D}_S$. Hence v is a model of $\mathcal{T}$ in $\mathcal{M}_S$ which is not a model of φ_0. ∎

5   Examples and Applications

Without the axiom (**m**) all the systems considered so far have characteristic
Nmatrices with at most 5 truth-values (see [1, 3]) in which the truth-values
are just triples (or sometimes even pairs) of 0's and 1's. Accordingly, we
concentrate in our examples below on systems which contain (**m**), and in
which this axiom is not derivable from the other axioms of the system.[10]

5.1   The System Bm

Theorem 11 provides a ten-valued characteristic Nmatrix for **Bm**. However,
its proof actually uses only seven of them, since the valuation v constructed
there does not use the values 1101,1001, and 0101. Hence we get a 7-valued
refinement of $\mathcal{M}_{\mathbf{m}}$ which (by Proposition 6) is characteristic for **Bm**. In this
7-valued Nmatrix the interpretation of ∘ is fully deterministic (as dictated
by $C(\mathbf{m})$), but the other operations are not. It is easy to see that none
of the other axioms is valid in that Nmatrix (or in $\mathcal{M}_{\mathbf{m}}$). Hence none of
these axioms is provable in **Bm** (this can also be seen directly from the
corresponding conditions).

5.2   The System mCi

The fundamental system which is called **mCi** in [13] is what is called here
Bmi. Theorem 11 provides a characteristic 6-valued Nmatrix for this logic,
whose set of the truth-values is $\{1101, 1100, 1011, 1010, 0111, 0110\}$ (of which
the first four are designated). In this Nmatrix the interpretation of ∘ is
again the one dictated by $C(\mathbf{m})$, the interpretations of the classical positive
connectives are like in Definition 7, and the interpretation of ¬ is as follows:

$$\tilde{\neg} a = \begin{cases} \{1101, 1011\} & a = 1101 \\ \{1101, 1100, 1011, 1010\} & a = 1100 \\ \{0111\} & a = 1011 \\ \{0111, 0110\} & a = 1010 \\ \{1011\} & a = 0111 \\ \{1101, 1100, 1011, 1010\} & a = 0110 \end{cases}$$

Now it can easily be checked that Theorem 11 provides exactly the same
characteristic Nmatrix for **Bmk**. It followed that **Bmi** = **Bmk** = **mCi**.
Similarly, it can easily be seen from the corresponding conditions that in
the presence of axiom (**m**), axioms (**k1**) and (**i1**) are equivalent, and axioms
(**k2**) and (**i2**) are equivalent (none of these facts is true relative to **B**).

Again a close examination of the proof of Theorem 11 reveals that 1101
is not really used in the refutations it provides in $\mathcal{M}_{\mathbf{mi}}$ of formulas not
provable in **mCi**. Hence this proof actually provides a 5-valued Nmatrix
which is characteristic for this system (and is a simple refinement of the

---

[10]Using the semantics introduced here or the simpler ones introduced in [1, 3], it can
easily be seen that if $S \subseteq Ax$ and (**m**) $\notin S$, then (**m**) is provable in **B**[S] iff the latter is
an extension of **Bci**.

official $\mathcal{M}_{\mathbf{mi}}$). This 5-valued Nmatrix is isomorphic to the characteristic 5-valued Nmatrix which was provided for **mCi** in [5].

5.3   The System Bmce

The values 1101,1001, and 0101 are actually not used in the proof of Theorem 11 for *any* of the systems which includes axiom (**m**). [11] Hence this proof provides characteristic Nmatrices with at most seven values for *all* extensions of **Bm**. As an example we take **Bmce**. In this Nmatrix the interpretation of ∘ and the interpretations of the classical positive connectives are defined like in the case of **Bmi** (with different $\mathcal{D}$ and $\mathcal{F}$, of course). The interpretation of ¬ is this time as follows:

$$\tilde{\neg}a = \begin{cases} \{1100\} & a = 1100 \\ \{0111\} & a = 1011 \\ \{0111, 0110, 0100\} & a = 1010 \\ \{0111, 0110, 0100\} & a = 1000 \\ \{1011\} & a = 0111 \\ \{1011, 1010, 1000\} & a = 0110 \\ \{1011, 1010, 1000\} & a = 0100 \end{cases}$$

Note that the presence of 0100 and 1000 respectively mean that (**k1**) and (**k2**) are not provable in **Bmce**.

6   Other Axioms

We concentrated above on the set of axioms Ax for the sake of illustration and because of the particular importance of these axioms (especially with connection to Marcos' axiom). However, it is easy to apply the 10-valued framework developed here to handle in a modular way systems which are constructed from a much bigger set of axioms, practically using the same conditions that have been used in [1, 3]. Here is a list of 22 other axioms that we could easily have included in Ax:

($\mathbf{a}_\neg$) $\circ\varphi \supset \circ\neg\varphi$

($\mathbf{a}_\diamond$) $\circ\varphi \supset (\circ\psi \supset \circ(\varphi \diamond \psi))$ ($\diamond \in \{\wedge, \vee, \supset\}$)

($\mathbf{o}_\diamond^1$) $\circ\varphi \supset \circ(\varphi \diamond \psi)$ ($\diamond \in \{\wedge, \vee, \supset\}$)

($\mathbf{o}_\diamond^2$) $\circ\psi \supset \circ(\varphi \diamond \psi)$ ($\diamond \in \{\wedge, \vee, \supset\}$)

($\neg \supset$)$_1$ $\neg(\varphi \supset \psi) \supset \varphi$

($\neg \supset$)$_2$ $\neg(\varphi \supset \psi) \supset \psi$

($\neg \supset$)$_3$ $\varphi \supset (\neg\psi \supset \neg(\varphi \supset \psi))$

---

[11]However, they *are* needed for providing a comprehensive framework that can *modularly* handle both extensions of **Bm**, and extensions of **B** in which (**m**) is not derivable.

$(\neg\vee)_1$ $\neg(\varphi \vee \psi) \supset \neg\varphi$

$(\neg\vee)_2$ $\neg(\varphi \vee \psi) \supset \neg\psi$

$(\neg\vee)_3$ $(\neg\varphi \wedge \neg\psi) \supset \neg(\varphi \vee \psi)$

$(\neg\wedge)_3$ $\neg(\varphi \wedge \psi) \supset (\neg\varphi \vee \neg\psi)$

$(\neg\wedge)_1$ $\neg\varphi \supset \neg(\varphi \wedge \psi)$

$(\neg\wedge)_2$ $\neg\psi \supset \neg(\varphi \wedge \psi)$

$(\mathbf{K})$ $\circ(\varphi \supset \psi) \supset (\circ\varphi \supset \circ\psi)$

$(\mathbf{4})$ $\circ\varphi \supset \circ\circ\varphi$

$(\mathbf{T})$ $\circ\varphi \supset \varphi$

Note. There are 3 more axioms which have central role in LFIs (see [13]):

$(\mathbf{l})$ $\neg(\varphi \wedge \neg\varphi) \supset \circ\varphi$

$(\mathbf{d})$ $\neg(\neg\varphi \wedge \varphi) \supset \circ\varphi$

$(\mathbf{b})$ $(\neg(\varphi \wedge \neg\varphi) \vee \neg(\neg\varphi \wedge \varphi)) \supset \circ\varphi$

As shown in [3], these axioms almost never can be handled in a finite-valued framework. However, it is not difficult to combine the method used in [3] (for providing semantics for LFIs which include one or more of these three axioms, but not axiom $(\mathbf{m})$) with the method used in this paper (for providing semantics for LFIs which do include axiom $(\mathbf{m})$) to get an effective infinite-valued semantic framework in which characteristic Nmatrices can modularly be constructed for *any* LFI which is based on some subset of the set of axioms that have been mentioned in this paper.

Acknowledgement

This research was supported by THE ISRAEL SCIENCE FOUNDATION founded by The Israel Academy of Sciences and Humanities.

BIBLIOGRAPHY

[1] A. Avron, *Non-deterministic Matrices and Modular Semantics of Rules*, in **Logica Universalis** (J.-Y. Beziau, ed.), 149–167, Birkhäuser Verlag, 2005.

[2] A. Avron, *Logical Non-determinism as a Tool for Logical Modularity: An Introduction*, in **We Will Show Them: Essays in Honor of Dov Gabbay**, (S. Artemov, H. Barringer, A. S. d'Avila Garcez, L. C. Lamb, and J. Woods, eds.), vol. 1, 105–124, College Publications, 2005.

[3] A. Avron, *Non-deterministic Semantics for Logics with a Consistency Operator*, International Journal of Approximate Reasoning, Vol. 45 (2007), 271–287.

[4] A. Avron, *Non-deterministic Semantics for Families of Paraconsistent Logics*, in **Handbook of Paraconsistency** (J.-Y. Beziau, W. Carnielli, and D. M. Gabbay, eds.), 285–320, Studies in Logic 9, College Publications, 2007.

[5] A. Avron, *5-valued Non-deterministic Semantics for The Basic Paraconsistent Logic* **mCi**, Studies in Logic, Grammar and Rhetoric, Vol. 14 (2008), 127–136

[6] A. Avron and I. Lev, *Canonical Propositional Gentzen-Type Systems*, in **Proceedings of the 1st International Joint Conference on Automated Reasoning (IJCAR 2001)** (R. Goré, A Leitsch, and T. Nipkow, eds), LNAI 2083, 529–544, Springer Verlag, 2001.

[7] A. Avron and I. Lev, *Non-deterministic Multiple-valued Structures*, Journal of Logic and Computation, Vol. 15 (2005), 241–261.

[8] D. Batens, C. Mortensen, G. Priest, and J. P. Van Bendegem (eds.), **Frontiers of Paraconsistent Logic**, King's College Publications, Research Studies Press, Baldock, UK, 2000.

[9] J. Béziau, W. Carnielli, and D. Gabbay (eds.), **Handbook of Paraconsistency**, Studies in Logic and Cognitive Systems, Vol. 9, College Publications, 2007.

[10] M. Bremer **An Introduction to Paraconsistent Logics**, Peter Lang GmbH, 2005.

[11] W. A. Carnielli and M. E. Coniglio, *Splitting Logics*, in **We Will Show Them: Essays in Honor of Dov Gabbay**, (S. Artemov, H. Barringer, A. S. d'Avila Garcez, L. C. Lamb, and J. Woods, eds.), vol. 1, 389–414, College Publications, 2005.

[12] W. A. Carnielli and J. Marcos, *A Taxonomy of C-systems*, in [14], 1–94.

[13] W. A. Carnielli, M. E. Coniglio, and J. Marcos, *Logics of Formal Inconsistency*, in **Handbook of Philosophical Logic, 2nd edition** (D. Gabbay and F. Guenthner, eds), Vol. 14, 1–93, Kluwer Academic Publishers, 2007.

[14] W. A. Carnielli, M. E. Coniglio, and I. L. M. D'Ottaviano (eds.), **Paraconsistency — the logical way to the inconsistent**, Lecture Notes in Pure and Applied Mathematics, Marcel Dekker, 2002.

[15] N. C. A. da Costa, *On the theory of inconsistent formal systems*, Notre Dame Journal of Formal Logic, Vol. 15 (1974), 497–510.

[16] N. C. A. da Costa, D. Krause and O. Bueno, *Paraconsistent Logics and Paraconsistency: Technical and Philosophical Developments*, in **Philosophy of Logic** (D. Jacquette, ed.), 791–911, North-Holland, 2007.

[17] J. Marcos, *Possible-translations Semantics for Some Weak Classically-based Paraconsistent Logics*, Journal of Applied Non-classical Logics, Vol. 18 (2008), 7–28.

Arnon Avron
School of Computer Science
Tel-Aviv University
Ramat Aviv, Tel-Aviv, Israel
E-mail: aa@math.tau.ac.il

Adaptive $\mathbf{C}_n$ Logics

DIDERIK BATENS

ABSTRACT. Newton da Costa's $\mathbf{C}_n$ logics suggest a certain stratagem, which is related to the motivation behind inconsistency-adaptive
logics. An adaptive approach turns out to allow for a more controlled
solution than the stratagem. Two kinds of logics will be presented.
Those of the first kind offer a maximally consistent interpretation of
the premise set in as far as this is possible in view of logical considerations. At the same time, they indicate at which points further choices
may be made on extra-logical grounds. The logics of the second kind
allow one to introduce those choices in a defeasible way and handle
them.

1 Aim of This Paper

Both the structure of the $\mathbf{C}_n$ logics and certain statements of da Costa's
seem to suggest a specific application for those logics, viz. to apply a certain
stratagem—see Section 3—to theories that turned out inconsistent. Even
if da Costa did not have this application in mind, the stratagem is clearly
interesting and suggested by the $\mathbf{C}_n$ logics. This makes it worthwhile to
develop inconsistency-adaptive logics that have the $\mathbf{C}_n$ systems as their
lower limit. Indeed, the adaptive logics by themselves accomplish most
of the task that is served by the stratagem. To be more precise, they
accomplish that part of the task which can be accomplished in view of
logical considerations.

There is a further reason to devise adaptive logics that have $\mathbf{C}_n$ logics
as their lower limit—this term is explained in Section 4. It is in principle
possible to do so for any paraconsistent logic. The $\mathbf{C}_n$ logics are the oldest
paraconsistent logics that were presented in a direct form, that is by an
axiomatic system and not by a translation. So, as one may expect, to use
them as lower limit logics has been on the agenda of adaptive logicians for
a long time now. The delay is caused by a technical complication.

$\mathbf{C}_n$ logics introduce dependencies between inconsistencies. Where this is
the case, the flip-flop danger lurks. As we shall see in Section 5, flip-flop
logics are rather uninteresting adaptive logics. Until recently, no general
method was available for avoiding flip-flops. Today this problem is solved by

semantic means. The method will be applied to the $\mathbf{C}_n$ systems in Section 6. This will enable me to formulate the inconsistency-adaptive logics in Section 7.

If the suggestion from the first paragraph is right, da Costa had the idea to allow the person applying the $\mathbf{C}_n$ logics to add consistency statements as non-logical axioms to a theory. Technically, this is made possible by the definability of classical negation within the logics. The justification for such moves is a guess, viz. that a certain contradiction is not derivable from the theory. If the guess is right, the addition seriously enriches the inconsistent theory, bringing it closer to its original intention. If the guess is wrong, however, the addition causes disaster, viz. triviality. Where this happens another technical device may be invoked: the hierarchy of $\mathbf{C}_n$ logics.

An approach in terms of adaptive logics has several advantages. First, inconsistency-adaptive logics by themselves add to a (inconsistent or consistent) theory all consistency statements that can be added on the basis of logical considerations. On top of this, combined adaptive logics enable one to add further consistency statements in a *defeasible* way. This will be discussed in Section 8. In Section 9, I shall present some further clarifying comments and a generalization of the result to logics of formal inconsistency.

2   The $\mathbf{C}_n$ Logics

The axiomatization, devised by da Costa, consists of the following elements. $\mathbf{C}_\omega$ is (predicative) positive intuitionistic logic extended with the axiom schemas $A \lor \neg A$ and $\neg\neg A \supset A$ and with the rule "if $A \equiv^c B$, then $\vdash A \equiv B$", in which $A \equiv^c B$ denotes that A and B are congruent in the sense of Kleene *or* that one formula results from the other by deleting vacuous quantifiers— Kleene [19, p. 153] summarizes his definition as follows: "two formulas are congruent, if they differ only in their bound variables, and corresponding bound variables are bound by corresponding quantifiers."

In order to obtain the $\mathbf{C}_n$ logics ($1 \leq n < \omega$), we need some abbreviations. Let A° abbreviate[1] $\neg(A \land \neg A)$. Next, let A^1 abbreviate A°, let A^2 abbreviate $A^{\circ\circ}$, etc. Finally, let $A^{(n)}$ abbreviate $A^1 \land A^2 \land \ldots \land A^n$. The logic $\mathbf{C}_n$ is obtained by extending $\mathbf{C}_\omega$ with the following axiom schemas:

$$B^{(n)} \supset ((A \supset B) \supset ((A \supset \neg B) \supset \neg A))$$

$$(A^{(n)} \land B^{(n)}) \supset (A \dagger B)^{(n)} \quad \text{where } \dagger \in \{\lor, \land, \supset\}$$

$$\forall x(A(x))^{(n)} \supset (\forall x\, A(x))^{(n)}$$

---

[1] While $\neg A \land A$ is $\mathbf{C}_1$-equivalent to $A \land \neg A$, $\neg(\neg A \land A)$ and $\neg(A \land \neg A)$ are not $\mathbf{C}_1$-equivalent. Which of the latter two is taken to express the consistency of A in $\mathbf{C}_1$ is a conventional matter.

$$\exists x (A(x))^{(n)} \supset (\exists x\, A(x))^{(n)}$$

A formula of the form $A^{(n)}$ will be called a consistency statement in $\mathbf{C}_n$. It expresses that A behaves consistently—see for example [17]—in the sense that $A^{(n)}, A, \neg A \vdash_{\mathbf{C}_n} B$ is derivable from the first displayed axiom schema. The other displayed axiom schemas spread consistency statements. It is natural to see $A^{(0)}$ as an empty string and $\mathbf{C}_0$ as $\mathbf{CL}$. Incidentally, $A^{(n)} \supset (\neg A)^{(n)}$ is a theorem of each $\mathbf{C}_n$. It is also provable that $\neg^{(n)} A =_{df} \neg A \wedge A^{(n)}$ defines classical negation in $\mathbf{C}_n$.

It is easily seen that the $\mathbf{C}_n$ logics form a hierarchy of logics: if $\Gamma \vdash_{\mathbf{C}_n} A$, then $\Gamma \vdash_{\mathbf{C}_{n+1}} A$; and also $\vdash_{\mathbf{C}_n} A^{n+1}$ (but $\nvdash_{\mathbf{C}_n} A^{(n+1)}$). The logic $\mathbf{C}_\omega$ forms a limit of this hierarchy, although not a very natural one. In view of the subsequent sections, it is useful to note that an equally sensible limit is the logic $\mathbf{C}_{\overline{\omega}}$, which contains predicative positive $\mathbf{CL}$ (Classical Logic) together with the axiom schemas $A \vee \neg A$ and $\neg\neg A \supset A$ and the rule "if $A \equiv^c B$, then $\vdash A \equiv B$".[2] Moreover, it will be useful to have classical negation available even in $\mathbf{C}_{\overline{\omega}}$. So let us extend the language with the symbol $\dot{\neg}$ and give it the meaning of classical negation (by introducing the usual axioms). Note that the standard negation, $\neg$, is still paraconsistent. Note also the difference between $\neg^{(n)}$ and $\dot{\neg}$. The first is definable within the standard language and behaves like classical negation in $\mathbf{C}_n$ or, more precisely, in the logics $\mathbf{C}_1, \ldots, \mathbf{C}_n$. The second symbol does not belong to the standard language, and hence does not occur in the premises, but is added to the language for technical reasons.

Two features of the $\mathbf{C}_n$ logics may cause some wonder. First, what is the use of having classical negation, viz. the symbol $\neg^{(n)}$, definable within paraconsistent logics? Next, what is the use of the hierarchy of $\mathbf{C}_n$ logics? The subsequent section offers a possible answer.

3 A Possible Stratagem in Terms of $\mathbf{C}_n$ Logics

Suppose that a theory T has $\mathbf{C}_1$ as its underlying logic and contains the consistency statement $A^{(1)}$ (for some specific formula A). As $A, \neg A, A^{(1)} \vdash_{\mathbf{C}_1} B$, it is excluded that both A and $\neg A$ are theorems of T on penalty of triviality—this is the outlook taken in [15], where the function of $A^{(n)}$ is served by $\circ A$, in which $\circ$ is the consistency operator, implicitly defined by, for example, the axiom schema $\circ A \supset ((A \wedge \neg A) \supset A)$.

The definability of classical negation in the logics $\mathbf{C}_n$ $(1 \le n < \omega)$ is a striking feature, which distinguishes these logics from most other paraconsistent logics. Moreover, one may wonder which precise purpose it is

---

[2] An interesting study of limits of the hierarchy is presented in [16]. The logic $\mathbf{C}_{\overline{\omega}}$ is there called $\mathbf{C}_{min}$.

supposed to serve. Let Γ be the non-logical axioms of a theory that was intended as consistent but turns out to be inconsistent—Frege's set theory is an obvious example. As Γ was intended and believed to be consistent, it will not comprise any explicit consistency statements—these are **CL**-theorems. So what is the use of consistency statements?

Seen from the perspective of inconsistency-adaptive logics, the answer to this question seems obvious. Suppose that $A \vee B$ and $\neg A$ are $\mathbf{C}_1$-derivable from Γ. As Γ was intended to be consistent, one would expect B to be derivable as well. But $A \vee B, \neg A \nvdash_{\mathbf{C}_1} B$. So, if A is not $\mathbf{C}_1$-derivable from Γ, one might add the consistency statement $A^{(1)}$ as a new axiom of the theory. This delivers the desired result because $A \vee B, \neg A, A^{(1)} \vdash_{\mathbf{C}_1} B$. Exactly the same situation arises if $B \supset A$ and $\neg A$ are $\mathbf{C}_1$-derivable from Γ.

The possibility to extend an inconsistent theory with consistency statements has dramatic effects: within the paraconsistent context, it leads to a theory that is drastically richer than the original theory. As a result, the extended theory approaches the theory as it was originally intended, viz. as fully consistent.

Adding consistency statements involves a danger. If one reformulates an inconsistent theory T_0 in terms of $\mathbf{C}_1$ and adds, for one or more specific A, $A^{(1)}$ to T_0, it may turn out that triviality results. When this happens, one may retract the added consistency statements. There is, however, another possibility.

At this point, the use of the other $\mathbf{C}_n$ becomes apparent. We have seen that one may replace **CL** by $\mathbf{C}_1$ in an attempt to save the original ideas behind T_0. By adding consistency statements, the theory is brought closer to T_0 as originally intended. If the resulting theory T_1 turns out trivial, one may replace $\mathbf{C}_1$ by $\mathbf{C}_2$. By this move, triviality is avoided because the statements of the form $A^{(1)}$, are not consistency statements in the context of $\mathbf{C}_2$. Moreover, relying on the insights from the failed previous attempt, one may again enrich T_1 by adding consistency statements of the form $A^{(2)}$, which have the desired effect in the context of $\mathbf{C}_2$. This process may be repeated. If T_n has $\mathbf{C}_n$ as its underlying logic, comprises no statements $A^{(m)}$ for which $m > n$,[3] and is trivial, replacing $\mathbf{C}_n$ by $\mathbf{C}_{n+1}$ restores non-triviality because no $A^{(m)}$ occurring in T_n is a consistency statement with respect to $\mathbf{C}_{n+1}$.

I do now know whether da Costa ever had such stratagem in mind. I can only note that the structure of the $\mathbf{C}_n$ systems and the definability of classical negation makes the stratagem possible and even suggests it. Some phrases used by da Costa go in the same direction. Thus he sometimes

---

[3]It shouldn't because $\vdash_{\mathbf{C}_n} A^m$ whenever $m > n$.

states that $\mathbf{C}_n$ logics isolate inconsistencies. In $\mathbf{C}_n$-theories he distinguishes between 'good' and 'bad' theorems, the latter being those whose negation is also a theorem. Note that, in order to take advantage of the 'good' quality of one of the former, one needs to add a consistency statement to the theory. Also da Costa notes, [17, p. 501], that $\mathbf{NF}_1$, a specific inconsistent variant of Quine's $\mathbf{NF}$, contains elementary arithmetic and is apparently arithmetically consistent. Yet, in order that this contained arithmetic be as strong as elementary arithmetic, consistency statements will have to be added in some or other way.

It actually is worthwhile to comment on the notions of 'good' and 'bad' theorems of a theory. Actually, several such notions were introduced over the years. So let me distinguish between them by adding subscripts. In [17], a theorem A of a theory $\langle \Gamma, \mathbf{C}_1 \rangle$ is called $good_1$ if $A \in Cons_{\mathbf{C}_1}(\Gamma)$ and $\neg A \notin Cons_{\mathbf{C}_1}(\Gamma)$ and is called bad_1 if $A, \neg A \in Cons_{\mathbf{C}_1}(\Gamma)$.[4] In [18], A is called $good_2$ if $A^{(1)} \in Cons_{\mathbf{C}_1}(\Gamma)$ and bad_2 if $A^{(1)} \notin Cons_{\mathbf{C}_1}(\Gamma)$. Consider a premise set $\Gamma = \{\neg p \vee q, p, r \vee s, \neg r, r\}$. Obviously, p is $good_1$, r is bad_1, and both are bad_2. There obviously are still different notions: A is $good_3$ if $\Gamma \cup \{A^{(1)}\}$ is non-trivial and A is bad_3 if $\Gamma \cup \{A^{(1)}\}$ is trivial. With respect to the aforementioned premise set, p is $good_3$, and r is bad_3. That neither of all these notions is as significant as they might seem to be is obvious from the following premise set: $\Gamma = \{q, \neg p \vee \neg q, p, r \vee s, r\}$. With respect to this Γ, p, q, and r are all $good_1$, bad_2, and $good_3$. Yet, $\Gamma \cup \{p^{(1)}, r^{(1)}\}$ and $\Gamma \cup \{q^{(1)}, r^{(1)}\}$ are non-trivial, whereas $\Gamma \cup \{p^{(1)}, q^{(1)}\}$ is trivial. So it seems advisable to define good and bad with respect to sets, and this is exactly the outlook taken by the adaptive approach, as we shall see below.

4   Adaptive Logics

Several introductions to adaptive logics are available, for example [5, 6, 9, 13]. So I shall be very brief here. Adaptive logics 'interpret' a premise set 'as normally as possible' with respect to some standard of normality. In particular, inconsistency-adaptive logics 'interpret' a premise set 'as consistently as possible'. It is worth mentioning that, while inconsistencies may be naturally seen as abnormalities with respect to a classical framework, standard of normality is a general notion which is technical in nature, and can only be given a philosophical interpretation with respect to a specific application context.

If adaptive logics are tailored from the $\mathbf{C}_n$ logics, these logics—let us call them $\mathbf{C}_n{}^m$—are inconsistency-adaptive. They should have a manifold of properties, among which the following. First, the adaptive logics

---

[4]As usual, $Cons_{\mathbf{L}}(\Gamma) = \{A \mid \Gamma \vdash_{\mathbf{L}} A\}$. I write $Cons$ instead of the more common Cn or C for the sake of readability in the present context.

should extend the $\mathbf{C}_n$ logics: $Cons_{\mathbf{C}_n}(\Gamma) \subseteq Cons_{\mathbf{C}_n\mathbf{m}}(\Gamma)$. If Γ is non-trivial, $Cons_{\mathbf{C}_n\mathbf{m}}(\Gamma)$ should be non-trivial and, border cases aside, the extension should be proper. Next, $Cons_{\mathbf{C}_n\mathbf{m}}(\Gamma)$ should be a fixed point, $Cons_{\mathbf{C}_n\mathbf{m}}(\Gamma) = Cons_{\mathbf{C}_n\mathbf{m}}(Cons_{\mathbf{C}_n\mathbf{m}}(\Gamma))$, and should be closed under $\mathbf{C}_n$, $Cons_{\mathbf{C}_n\mathbf{m}}(\Gamma) = Cons_{\mathbf{C}_n}(Cons_{\mathbf{C}_n\mathbf{m}}(\Gamma))$.

Some logicians believe in the existence of a *true* (deductive) logic, which then serves as their standard of deduction. If John takes **CL** (or another explosive logic) to be his standard of deduction and the theory $T = \langle \Gamma, \mathbf{CL} \rangle$ was meant as consistent, but turns out to be inconsistent and hence trivial, he will want to move to a theory that interprets Γ *as consistently as possible*, viz. as close as possible to the original intention. This theory will be $T' = \langle \Gamma, \mathbf{AP} \rangle$, in which **AP** is an inconsistency-adaptive logic that has the paraconsistent logic **P** as its lower limit. For John, **AP** is a *corrective* logic: the standard of deduction leads to disaster and, for the time being and possibly in preparation of an improved consistent theory, John moves on to an approximation of the original theory that is non-trivial, locates the inconsistencies, but approximates the original theory in as far as possible. The approximation requires that inconsistencies are taken to be false unless and until proven otherwise. Next, consider Mary, who considers a paraconsistent logic **P** as her standard of deduction but agrees that most inconsistencies are false and hence can be considered as false unless and until proven true. So her original theory is $T = \langle \Gamma, \mathbf{P} \rangle$, but, relying on the extra-logical consideration that most inconsistencies are false, she upgrades to $T' = \langle \Gamma, \mathbf{AP} \rangle$. So for her the inconsistency-adaptive logic **AP** is an *ampliative* logic: it delivers a stronger consequence set than the standard of deduction.

Whatever the position taken, adaptive logics are not competitors for the standard of deduction. They are formal systems characterizing defeasible reasoning forms; they are instruments, formally characterized methods, and the like. Note that there are also people (like me) who do not believe in the existence of a unique standard of deduction, for example because they take the standard to vary with the context or problem-solving situation.

Let us now briefly look at the technicalities involved in adaptive logics. An adaptive logic **AL** (in standard format) is a triple:

- a *lower limit logic* **LLL**: a compact Tarski logic for which there is a positive test and that has a characteristic semantics

- a *set of abnormalities* Ω: a set of formulas characterized by a (possibly restricted) logical form

- a *strategy*: Reliability, Minimal Abnormality, ...

Every adaptive logic defines an upper limit logic **ULL**, which is a Tarski logic obtained by extending the lower limit logic with an axiom or rule that trivializes abnormalities. Semantically the **ULL** models are the **LLL** models that verify no abnormality. Note that **ULL** extends **LLL** with some further rules, requiring that *all* abnormalities are false, whereas **AL** extends **LLL** with certain *applications* of **ULL**-rules, requiring that as many abnormalities are false as the premises permit.

Let $\mathcal{F}$ be the set of open and closed formulas of the standard predicative language and $\exists(A \wedge \neg A)$ the existential closure of $A \wedge \neg A$. An example of a specific adaptive logic in standard format is $\mathbf{C}_1{}^m$, viz.

- *lower limit logic*: $\mathbf{C}_1$

- *set of abnormalities* $\Omega = \{\exists(A \wedge \neg A) \mid A \in \mathcal{F}\}$

- *strategy*: Minimal Abnormality[5]

The need for a strategy is best illustrated by an example. Let the premise set be $\{\neg p, \neg q, p \vee r, q \vee s, p \vee q\}$. From this $(p \wedge \neg p) \vee (q \wedge \neg q)$ is $\mathbf{C}_1$-derivable. So we need to decide in which way this disjunction of abnormalities will affect our maximally normal 'interpretation' of the premise set.

The standard format provides the adaptive logic with (i) a dynamic proof theory—the systematic study of this is available in [10], (ii) a preferential semantics, and (iii) most of the metatheory. The dynamic proofs consist of lines that have a *condition* attached to them, are marked or unmarked (in function of their condition and of formulas derived at other lines), and are governed by rules (to add lines) and a *marking definition* (to settle which lines are marked at a stage of the proof). The preferential semantics selects lower limit models of the premise set. The metatheory includes soundness, completeness, and many properties (cautious monotonicity, cautious transitivity, . . .).

Incidentally, the aim of the adaptive logic program is to characterize *all* forms of defeasible reasoning by (combinations of) adaptive logics in standard format, possibly under a translation.

5   Adaptive Logics and Flip-Flops

The stratagem described in Section 3 requires human interference, viz. adding consistency statements. Precisely this is avoided by adaptive logics. In the previous section, I introduced $\mathbf{C}_1{}^m$. What does this logic come to? The $\mathbf{C}_1{}^m$-consequence set of a premise set Γ offers a maximally consistent

---

[5]A strategy is defined (semantically) by the selection of models—see Section 5—and (proof-theoretically) by the marking definition associated with it—see Section 7.

interpretation of Γ. Applying the standard adaptive semantics for the minimal abnormality strategy gives the following result. The $\mathbf{C_1}^m$-consequence set of Γ comprises the formulas that are verified by all minimally abnormal $\mathbf{C_1}$-models of Γ. Where M is a $\mathbf{C_1}$-model, let $Ab(M)$ be the set of the abnormalities, that is members of Ω, that are verified by M. A $\mathbf{C_1}$-model M of Γ is a *minimal abnormal* model of Γ iff there is no $\mathbf{C_1}$-model M' of Γ for which $Ab(M') \subset Ab(M)$. If a $\mathbf{C_1}$-model M verifies $A \vee B$ as well as $\neg A$, and does not verify $A \wedge \neg A$, then M verifies B. So, if no minimal abnormal $\mathbf{C_1}$-model of Γ verifies $A \wedge \neg A$, then the $\mathbf{C_1}^m$-consequence set of Γ is identical to the $\mathbf{C_1}^m$-consequence set of $\Gamma \cup \{A^{(1)}\}$. Seen from the stratagem from Section 3, one might see the $\mathbf{C_1}^m$-consequence set of Γ as the $\mathbf{C_1}$-consequence set of $\Gamma \cup \Gamma'$, in which Γ' is the set of all consistency statements that can be added to Γ without resulting in triviality. This description needs to be a refined, as we shall see in Section 8, but for the time being it will do.

Although an inconsistency-adaptive logic with a $\mathbf{C_n}$ logic as its lower limit seems an attractive alternative for the 'handwork' required by the stratagem, there is a problem. The $\mathbf{C_n}$ logics introduce dependencies between contradictions. For example, $\neg A \wedge \neg\neg A$ entails $A \wedge \neg A$ in view of the axiom schema $\neg\neg A \supset A$. If there are such dependencies, there is a particular difficulty for defining the set of abnormalities. If this set is not defined in a sufficiently restrictive way, a flip-flop logic results. This demands some explanation.

A flip-flop logic is an adaptive logic, but a rather uninteresting one.[6] A flip-flop $\mathbf{L}$ displays the following behaviour. If Γ has models M for which $Ab(M) = \emptyset$ (so Γ is 'normal'), $\mathbf{L}$ behaves like its upper limit logic—in the present context this is $\mathbf{CL}$. If, however, $Ab(M) \neq \emptyset$ for all models M of Γ, then $\mathbf{L}$ behaves like its lower limit logic—in the present case $\mathbf{C_1}$. Obviously, this is rather uninteresting. Suppose we apply $\mathbf{C_1}$ to the premise set $\{\neg p, p \vee r, \neg q \wedge p, q \vee s\}$. The reason for going adaptive is that we want to derive s because $q \wedge \neg q$ is not $\mathbf{C_1}$-derivable from the premises, while avoiding the consequence r because $p \wedge \neg p$ is $\mathbf{C_1}$-derivable from the premises. A flip-flop logic, however, will assign to that premise set exactly the same consequences as its lower limit, $\mathbf{C_1}$, and hence will not deliver s.

Let us consider some well known examples to illustrate the problem. The logic $\mathbf{CLuN}$ comprises full positive logic together with excluded middle.[7]

---

[6]Astounding as it may seem, some flip-flops have interesting application contexts, for example in the context of inductive generalization, in case one wants to completely reject certain background theories that are contradicted by the data.

[7]The propositional fragment of $\mathbf{CLuN}$ was first studied in [3] (with the name $\mathbf{PI}$). It is a basic paraconsistent logic, and is taken as the basis for constructing the basic logic of formal inconsistency, $\mathbf{mbC}$, in [15]. The predicative version was first presented in [4].

The suitable set of abnormalities for an adaptive logic that has **CLuN** as its lower limit is $\{\exists(A \wedge \neg A) \mid A \in \mathcal{F}\}$. The logic **CLuNs** (see [11]), which actually is the most popular paraconsistent logic (under sundry names), is obtained from **CLuN** by adding both directions of double negation, de Morgan theorems, and similar axioms to push negation inwards. If an adaptive logic has **CLuNs** as its lower limit and the set of abnormalities is defined as $\{\exists(A \wedge \neg A) \mid A \in \mathcal{F}\}$, a flip-flop results. A decent adaptive logic is obtained by restricting this set to $\{\exists(A \wedge \neg A) \mid A \in \mathcal{F}^a\}$ in which $\mathcal{F}^a$ is the set of (open and closed) atomic (or primitive) formulas—those in which occurs no logical symbol except possibly for identity.

As $\mathbf{C}_1$ spreads inconsistencies, one might fear that the logic $\mathbf{C}_1{}^m$ from Section 4 is a flip-flop. So the question is whether, in order to avoid this, the set of abnormalities should be restricted if $\mathbf{C}_1$ is the lower limit. I failed to obtain an answer for many years, but today the problem is solved because a general criterion has been devised. I apply this criterion to $\mathbf{C}_1$ in the next section.

6   The Semantic Criterion

The criterion ties up the abnormalities to the occurrence of gluts (and possibly gaps) in an indeterministic semantics—see [1, 2] on indeterministic semantics. There are certain restrictions on the indeterministic semantics, but the easiest approach is to present a semantics for $\mathbf{C}_1$ and to explain it. So here we go.

The semantics is provably characteristic for $\mathbf{C}_1$. In order to avoid assigning values to open formulas (containing free occurrences of members of the set $\mathcal{V}$ of variables), I extend the standard predicative language $\mathcal{L}$ to a pseudo-language $\mathcal{L}_{\mathcal{O}}$ by adding, next to the set of individual constants $\mathcal{C}$, a set $\mathcal{O}$ of pseudo-constants. $\mathcal{O}$ should have the cardinality of the largest set, which is the largest domain of the models considered. Let $\mathcal{S}$ be the set of sentential letters, $\mathcal{P}^r$ the set of predicative letters of rank r, $\mathcal{W}$ the set of closed formulas of $\mathcal{L}$, $\mathcal{W}_{\mathcal{O}}$ the set of closed formulas of $\mathcal{L}_{\mathcal{O}}$, and $\mathcal{W}_{\mathcal{O}}^a$ the set of atomic formulas in $\mathcal{W}_{\mathcal{O}}$.

In a model $M = \langle D, v \rangle$, D is the domain and v the assignment function defined by:

C1 $v \colon \mathcal{W}_{\mathcal{O}} \rightarrow \{0,1\}$
C2 $v \colon \mathcal{C} \cup \mathcal{O} \rightarrow D$ (where $D = \{v(\alpha) \mid \alpha \in \mathcal{C} \cup \mathcal{O}\}$)
C3 $v \colon \mathcal{P}^r \rightarrow \wp(D^r)$

Later the Ghent group standardly extended **CLuN** with classical negation. Replacement of Identicals is invalid in **CLuN**: it does not apply within the scope of a negation. Obviously, **CLuN** can be extended with Replacement of Identicals. Replacement of Equivalents is also invalid in **CLuN**, as in many other paraconsistent logics.

Clause C1 assigns a truth value to every closed formula. Only the values assigned to sentential letters matter for the present semantics. The other values are important to turn the indeterministic semantics into a deterministic one, an exercise that will be skipped in the present paper. The restriction in clause C2 ensures that every member of D is named by a constant or pseudo-constant.[8] In C3, $\mathcal{P}^r$ is the set of predicative letters of rank r and $\wp(D^r)$ is the power set of the r-th Cartesian product of D.

The pre-valuation $v_M : \mathcal{W_O} \to \{t, u, f\}$, with t (true) and u (glut) designated, is characterized by the following tables:

Where $A \in \mathcal{S}$:

$v(A)$	A
1	t
0	f

Where $\alpha_1, \ldots, \alpha_r \in \mathcal{C} \cup \mathcal{O}$ and $\pi \in \mathcal{P}^r$:

$\langle v(\alpha_1), \ldots, v(\alpha_r) \rangle, v(\pi)$	$\pi\alpha_1 \ldots \alpha_r$
$\in$	t
$\notin$	f

Where $\alpha, \beta \in \mathcal{C} \cup \mathcal{O}$:

$v(\alpha), v(\beta)$	$\alpha = \beta$
$=$	t
$\neq$	f

Where $A \in \mathcal{W}^a_{\mathcal{O}}$:

A	$\neg A$
t	$[f, u]$
f	t

Where $A * B$ is not of the form $C \wedge \neg C$:

$A * B$	$A^{(1)}$	$B^{(1)}$	$\neg(A * B)$
t	t	t	f
t	(other)		$[f, u]$
f	(any)		t

Where $Q \in \{\forall, \exists\}$ and $\alpha \in \mathcal{V}$:

$Q\alpha A(\alpha)$	$\{v_M(A(\beta)^{(1)}) \mid \beta \in \mathcal{C} \cup \mathcal{O}\}$	$\neg Q\alpha A(\alpha)$
t	$\{t\}$	f
t	(other)	$[f, u]$
f	(any)	t

---

[8]So some models are ω-incomplete with respect to the standard language, but the extended language allows for a transparent handling of the quantifiers.

The other tables apply to all members of $\mathcal{W}_\mathcal{O}$:

$\neg A$	$\neg\neg A$
t	f
u	$[f,u]$
f	t

$\neg A$	$A^{(1)}$
t	t
u	f
f	t

$\wedge$	t	u	f
t	t	t	f
u	t	t	f
f	f	f	f

$\vee$	t	u	f
t	t	t	t
u	t	t	t
f	t	t	f

$\supset$	t	u	f
t	t	t	f
u	t	t	f
f	t	t	t

$\equiv$	t	u	f
t	t	t	f
u	t	t	f
f	f	f	t

$\{v_M(A(\alpha)) \mid \alpha \in \mathcal{C} \cup \mathcal{O}\}$	$\forall\alpha A(\alpha)$	$\exists\alpha A(\alpha)$
$\in \wp\{t,u\}$	t	t
$\{f\}$	f	f
(other)	f	t

A pre-valuation v_M is a valuation iff $v_M(A) = v_M(B)$ whenever $A \equiv^c B$ (see Section 2). This means that, at points where the semantics is inde-terministic, the same choice was made for A and B. If some pre-valuation assigns the value t to all members of a set of formulas, then so does some valuation.

The expression $[f,u]$ indicates that the value may be f or u—this is an indeterministic semantics. Note that $\emptyset$ is not a possible value on the first line of the table for the quantifiers. The "(other)" on the third line of that table abbreviates sets that contain a designated as well as an non-designated value. In the table for $\neg(A * B)$, "(other)" means that either $A^{(1)}$ or $B^{(1)}$ does not have the value t (they cannot have the value u); "(any)" indicates that the values of $A^{(1)}$ or $B^{(1)}$ do not matter—this line summarizes four lines. In the table for $\neg Q\alpha A(\alpha)$, "(other)" means that some $A(\beta)^{(1)}$ does not have the value t; "(any)" indicates that the values of the $A(\beta)^{(1)}$ do not matter.

The pre-valuation and the valuation assign a value to all closed formulas of $\mathcal{L}$, which is what we are interested in. Validity and semantic consequence are defined as usual.

This particular three-valued indeterministic semantics is constructed from the two-valued one. Typical for this semantics is that the value u is only assigned where a glut *originates* (in comparison to $\mathbf{CL}$). Thus if both A and B have a designated value, the truth of $A \wedge B$ agrees with $\mathbf{CL}$ *at this point* and hence is not a glut. For example, if all of A, $\neg A$ and $\neg(A \wedge \neg A)$ have a designated value, then $v_M(\neg A) = u$, $v_M(A \wedge \neg A) = t$, $v_M(\neg(A \wedge \neg A)) = u$, and $v_M((A \wedge \neg A) \wedge \neg(A \wedge \neg A)) = t$.

Transforming the above semantics to any logic $\mathbf{C}_n$ $(n < \omega)$ is an easy exercise left to the reader—the formulation of the tables for $\mathbf{C}_1$ and the plot described in the previous paragraph indicate the road. For $\mathbf{C}_{\overline{\omega}}$ (see Section 2), one replaces the tables for negation by the left and middle table below, and adds the table to the right below for the (added) classical negation:

A	$\neg A$		$\neg A$	$\neg\neg A$		A	$\bar{\neg} A$
t	$[f, u]$		t	f		t	f
f	t		u	$[f, u]$		u	f
			f	t		f	t

We have seen that the abnormalities of inconsistency-adaptive logics have the form $\exists(A \wedge \neg A)$. It can be shown, in terms of this particular type of indeterministic three-valued semantics, that an inconsistency-adaptive logic is not a flip-flop if the abnormalities are restricted to the case where (i) $A \wedge \neg A$ is an abnormality if its truth requires $\neg A$ to have the value u and (ii) $\exists(A \wedge \neg A)$ is an abnormality if it requires that there is an instance $B \wedge \neg B$ (obtained by systematically replacing every free variable in $A \wedge \neg A$ by a constant or pseudo-constant) for which $\neg B$ has the value u. This solves the flip-flop problem for the $\mathbf{C}_n$ logics: defining the set of abnormalities as $\Omega = \{\exists(A \wedge \neg A) \mid A \in \mathcal{F}\}$ does not cause a flip-flop. Some of the abnormalities are logically impossible in $\mathbf{C}_1$, but that does not cause any trouble.

7 The Inconsistency-Adaptive Logics

The result is even simpler than one might expect. The inconsistency-adaptive logic $\mathbf{C}_n{}^m$ from Section 7 is not a flip-flop. The result may be generalized. The following adaptive logics are not flip-flops: $\mathbf{C}_n{}^m$ $(1 \leq n \leq \omega)$ defined as the triple[9] consisting of (i) $\mathbf{C}_n$, (ii) $\Omega = \{\exists(A \wedge \neg A) \mid A \in \mathcal{F}\}$, and (iii) Minimal Abnormality. So there is no need to vary Ω for any $\mathbf{C}_n{}^m$ logic. Given that these logics are in standard format, their proof theory and semantics are at once defined, the soundness and completeness of the proof theory with respect to the semantics is warranted (in view of the soundness and completeness of the lower limit logics with respect to their semantics), and most metatheoretic properties of the logics are known (because they follow from the standard format).

A striking specific feature reveals itself. If a theory was meant to be consistent and its underlying logic is explosive, then, as we have seen before, it will not contain any consistency statement $A^{(n)}$. In this case all $\mathbf{C}_n{}^m$

---

[9]I only consider the Minimal Abnormality strategy for lack of space; the result generalizes, for example, to Reliability.

($n \in \{1, 2, \ldots, \overline{\omega}\}$) assign the same consequence set to Γ. In general, if Γ is not $\mathbf{C}_k$-trivial, then all $\mathbf{C}_n{}^m$ ($n \in \{k, k+1, \ldots, \overline{\omega}\}$) assign the same consequence set to Γ. The astonishing result is that $\mathbf{C}_{\overline{\omega}}{}^m$ can be used for all premise sets. So while adaptive logics involve some complexity problems, they avoid the complication of the hierarchy of $\mathbf{C}_n$ logics that is needed for implementing the stratagem.

For the proof theory, semantics, etc., I refer to [9]. Here I can at best present a simple propositional $\mathbf{C}_{\overline{\omega}}{}^m$-proof. The strategy determines the marking definition while the inference rules depend on the lower limit logic and the set of abnormalities. Let the premise set be $\{\neg\neg p, \neg q, \neg p, p \vee r, q \vee s\}$.

1	$\neg\neg p$	Prem	$\emptyset$
2	$\neg q$	Prem	$\emptyset$
3	$\neg p$	Prem	$\emptyset$
4	$p \vee r$	Prem	$\emptyset$
5	$q \vee s$	Prem	$\emptyset$
6	r	3, 4; RC	$\{p \wedge \neg p\}$ ✓[8]
7	s	2, 5; RC	$\{q \wedge \neg q\}$
8	$p \wedge \neg p$	1, 3; RU	$\emptyset$

The generic rules are common to all adaptive logics. The unconditional rule RU corresponds to **LLL**-derivability and carries over to the derived formula the conditions of the formulas from which it is derived—this is not illustrated by the proof. The conditional rule RC functions similarly, except that it introduces a new condition. Thus line 6 is added because $r \vee (p \wedge \neg p)$ is $\mathbf{C}_{\overline{\omega}}$-derivable from 3 and 4; line 7 because $s \vee (q \wedge \neg q)$ is $\mathbf{C}_{\overline{\omega}}$-derivable from 2 and 5. The superscripted number 8 on line 6 indicates that this line is marked at stage 8 of the proof, that is immediately after line 8 was added. In whichever way the proof (from these premises) is extended, the marks of lines 1-8 are stable from this point on (the marked line remains marked and the unmarked ones remain unmarked). So s is a final consequence of the premise set and r is not (because line 6 is marked). Precise formulations of the rules and of the marking definition occur in dozens of published papers, for example [9].

The adaptive logics solve certain problems that may arise if the stratagem from Section 3 is applied. Thus even the infinite set of consistency statements $\{\neg(A \wedge \neg A) \mid A \in \mathcal{F} - \{p\}\}$ is insufficient to obtain by $\mathbf{C}_n$ all $\mathbf{C}_{\overline{\omega}}{}^m$-final consequences of the original premise set (an example is $\forall x P x \supset (\exists x \neg P x \supset r)$). If a non-recursive set of minimal disjunctions of contradictions[10] is $\mathbf{C}_n$-derivable from the premise set, no recursive set of consistency

---

[10]The role played by minimal disjunctions of abnormalities becomes clear in the next section.

statements can be added to the original premises to obtain by a $\mathbf{C}_n$ logic the consequences that $\mathbf{C}_{\overline{\omega}}{}^m$ delivers from the original premise set.

8   Defeasible Guesses

While the inconsistency-adaptive logics $\mathbf{C}_n{}^m$ do in most respects better than the stratagem outlined in Section 3, there is one respect in which they do worse. The $\mathbf{C}_n{}^m$ offer a maximally consistent interpretation of the premises, but only in as far as *logical considerations* allow for a justified choice. This may be illustrated by the premise set $\{\neg p, \neg q, p \vee r, q \vee s, p \vee q\}$. Note that $(p \wedge \neg p) \vee (q \wedge \neg q)$ is $\mathbf{C}_{\overline{\omega}}$-derivable from these premises. The upshot is that $r \vee s$ is a $\mathbf{C}_{\overline{\omega}}{}^m$-consequence of the premises. However, neither $p^{(1)}$ nor $q^{(1)}$ is derivable, which is related to the fact that no logical considerations enable one to prefer one over the other.

Suppose that $(p \wedge \neg p) \vee (q \wedge \neg q)$, $(p \wedge \neg p) \vee (r \wedge \neg r)$, and $s \wedge \neg s$ are $\mathbf{C}_{\overline{\omega}}$-derivable from a premise set, but that neither $p \wedge \neg p$ nor $q \wedge \neg q$ nor $r \wedge \neg r$ have been so derived. The premises then apparently inform us that either both $p \wedge \neg p$ and $s \wedge \neg s$ are true or that all of $q \wedge \neg q$, $r \wedge \neg r$, and $s \wedge \neg s$ are true. The inconsistency-adaptive logic cannot possibly 'chose' between both possibilities. The person that applies the logic might, however, have a reason to make a choice. Thus both q and r may concern well-entrenched properties that may be taken to behave consistently. In this case, the person applying the stratagem would add the new premises $\neg(q \wedge \neg q)$ and $\neg(r \wedge \neg r)$ in the context of $\mathbf{C}_1$. So she would obtain a theory that is more consistent than the one provided by $\mathbf{C}_{\overline{\omega}}{}^m$. The logic cannot make this choice, because the reasons for making it are extra-logical.

The choice involves a danger. If later $q \wedge \neg q$ would turn out to be $\mathbf{C}_1$-derivable from the premise set, the resulting theory would be $\mathbf{C}_1$-trivial and one would have to move to $\mathbf{C}_2$ in order to make another try.

However, there is a way to eat your cake and still have it, viz. by replacing the adaptive $\mathbf{C}_{\overline{\omega}}{}^m$ by a specific combined adaptive logic. Indeed, this logic retains all the $\mathbf{C}_{\overline{\omega}}{}^m$-consequences, allows one to add consistency statements on extra-logical grounds, but circumvents the danger because the consistency statements are introduced in a defeasible way. In other words, a combined adaptive logic allows one to make choices that will have no effect if they would run one into triviality in the context of $\mathbf{C}_{\overline{\omega}}$.

Two important remarks are in place at this point. First when does it make sense to defeasibly introduce a consistency statement? Given the result on $\mathbf{C}_{\overline{\omega}}$, the answer is obvious: where one has reasons to believe that a choice can be made. If one has derived a disjunction of abnormalities that is, by present insights, minimal, it makes sense to eliminate some (obviously not all) of the disjuncts as not abnormal. In other words, if one has (unconditionally) derived a disjunction of (existentially quantified) contradictions, and no stronger disjunction has been derived, then one may posit, preferably on good grounds, that some of the disjuncts are false. This agrees nicely with Wiśniewski's erotetic logic—see, for example, [20, 21, 22]. A set of declarative statements *generates* a question if the disjunction of the direct answers to the question is derivable from the statements whereas no direct answer is derivable.

The second remark is that if one wants to consider defeasible consistency statements, one better introduces them in a prioritized way. So if one has found that $(p \wedge \neg p) \vee (q \wedge \neg q) \vee (r \wedge \neg r)$ is $\mathbf{C}_{\overline{\omega}}$-derivable from the premise set, one may want to stipulate that both $p \wedge \neg p$ and $r \wedge \neg r$ are false. Yet, it is possible that one is more certain about the falsehood of $p \wedge \neg p$ than about that of $r \wedge \neg r$. So it is intuitively appealing to defeasibly reject $p \wedge \neg p$ in a stronger way than $r \wedge \neg r$. Note that, if it would turn out at a later point that $(p \wedge \neg p) \vee (r \wedge \neg r)$ is $\mathbf{C}_{\overline{\omega}}$-derivable from the premise set, then $p \wedge \neg p$ would still be taken to be false whereas $r \wedge \neg r$ would be taken to be true.

Let $!A$ abbreviate $\exists(A \wedge \neg A)$, whence $!!A$ abbreviates $\exists(A \wedge \neg A) \wedge \neg \exists(A \wedge \neg A)$, etc. Next let $!^n A$ abbreviate whatever is abbreviated by n exclamation marks followed by A. Finally, let $\mathbf{i}^n A$ abbreviate $\neg!^1 A \wedge \neg!^2 A \wedge \ldots \wedge \neg!^n A$.

The guesses that are introduced as new premises have the form $\mathbf{i}^n A$. The priority assigned to a guess is directly proportional to n. To handle the guesses, we consider a combined adaptive logic, viz. a specific combination of a set of adaptive logics. For each of the latter, the lower limit is $\mathbf{C}_{\overline{\omega}}$ and the strategy is Minimal Abnormality. The sets of abnormalities are defined by $\Omega^i = \{!^i A \mid A \in \mathcal{F}\}$. Let the resulting adaptive logics be called $\mathbf{C}_{\overline{\omega}}{}^{mi}$, in which i determines Ω^i. Note that $\mathbf{C}_{\overline{\omega}}{}^{m1}$ is identical to $\mathbf{C}_{\overline{\omega}}{}^m$ from Section 7.

The consequence set of the combined adaptive logic of level n is identical to $Cn_{\mathbf{C}_{\overline{\omega}}{}^{m1}}(Cn_{\mathbf{C}_{\overline{\omega}}{}^{m2}}(\ldots(Cn_{\mathbf{C}_{\overline{\omega}}{}^{mn}}(\Gamma))\ldots))$. In semantic terms, the combined logic is easiest described as follows. From the $\mathbf{C}_{\overline{\omega}}$-models of a premise set Γ, it first selects the minimal abnormal models with respect to Ω^n,[11] from these the minimal abnormal models with respect to Ω^{n-1}, and so on up to Ω^1. The proof theory of the combined logic has an interesting property: the rules of all combining logics may be applied together. At every stage, the marking definition of the combined logic proceeds first in terms of minimal disjunctions of Ω^n-abnormalities that have been derived in the proof on the condition $\emptyset$, next in terms of the unmarked minimal disjunctions of Ω^{n-1}-abnormalities that have been derived in the proof on a condition that comprises at most members of Ω^n, next in terms of the unmarked minimal disjunctions of Ω^{n-2}-abnormalities that have been derived in the proof on a

---

[11]The value of n is determined by the premises.

condition that comprises at most members of $\Omega^n \cup \Omega^{n-1}$, etc.—the marking definition is identical to Definition 13 of [8].

Note that a guess $¡^n A$ may be strengthened by introducing a new premise $¡^m A$ with $m > n$. This means that guesses may be corrected, where it is desirable, in view of insights obtained from an ongoing proof.

A small digression is in place at this point. If no free variable occurs in A, $¡^n A$ is identical to $A^{(n)}$. So at the propositional level, the story may be told in terms of the original consistency statements of the $\mathbf{C}_n$ logics. This is a most astonishing fact. The original construction forged by da Costa contained the means required, at the propositional level, to handle defeasible guesses and to overrule them if there are reasons for doing so. According to the stratagem, the logic has to be replaced, for example $\mathbf{C}_n$ has to be replaced by $\mathbf{C}_{n+1}$ in order to overrule the earlier guess. The combined adaptive logic, to the contrary, is the same throughout the whole process. Apart from this and apart from the absence of the existential closure in da Costa's $A^{(n)}$, $¡^n A$ and $A^{(n)}$ serve the same function and are identical where no free variables occur in A.

9 Some Concluding Remarks

All by itself, $\mathbf{C}_{\overline{\omega}}^m$ *restores consistency* where logical reasons permit and indicates the road to obtain further consistency. The combined adaptive logic *handles* attempts to make the result even more consistent, reveals points at which choices may be made, and prevents the enterprise from running into triviality.

That the adaptive logics have $\mathbf{C}_{\overline{\omega}}$ as their lower limit seems to make classical negation unavailable. As was described in Section 2, this may be repaired by introducing classical negation as a symbol, $\check{\neg}$, that does not belong to the standard language and hence does not occur in the premises—actually, this has been the standard adaptive approach for many years now. The upshot is that, if only one of A and $\neg A$ is a member of the adaptive consequence set, then so will be one of $\check{\neg}\neg A$ and $\check{\neg} A$.

The adaptive logic $\mathbf{C}_{\overline{\omega}}^m$ defines a maximal consistent interpretation of a premise set Γ. The price to pay is obviously that the $\mathbf{C}_{\overline{\omega}}^m$-consequence set of Γ is not in general decidable; there even is no positive test for it (the set is not recursively enumerable). This is unavoidable at the predicate level[12] because there is no positive test for consistency.

There are some consolations. The first is that there are proof procedures that form criteria for final derivability.[13] If the procedure is applied to some

---

[12] At the propositional level adaptive logics are decidable in the same sense as $\mathbf{CL}$ is. The same holds for certain fragments of the predicative logic.

[13] The procedure for Reliability was presented in [7]; that for Minimal Abnormality was

Γ and A and it stops, it answers the question whether A is finally derivable from Γ. Moreover, if a finite proof establishes that A is finally derivable from Γ—see [12] for a more precise formulation—then the procedure will stop with that answer.

A very different consolation is that the introduction of defeasible guesses may circumvent the problem even where final derivability cannot be established. If the user feels to have a sufficient insight in the studied theory she will often try to phrase a consistent replacement. In order to do so, the logic should be able to isolate inconsistencies and should *in principle* be able to locate all inconsistencies. But even an incomplete analysis may permit one to attain a consistent replacement. An illustration is that many (apparently) consistent replacements for Frege's set theory were formulated before the Curry paradox was discovered. This paradox apparently does not affect any of those set theories.

The results presented in this paper may obviously be generalized to all logics of formal inconsistency in the sense of [15]. Given such a logic $\mathbf{L}$, an adaptive logic $\mathbf{AL}$ is articulated as follows. Take $\mathbf{L}$ as the lower limit logic. Formulate a two-valued indeterministic semantics for $\mathbf{L}$ and turn it into a 3-valued or 4-valued indeterministic semantics with the values t, f, u, and a as described in Section 6. On the basis of the insights gained from this, define $\Omega = \{\exists(A \wedge \neg A) \mid A \in \mathcal{F}^x\}$, in which x is the restriction required for avoiding a flip-flop. Finally, chose an adaptive strategy. The resulting logic restores consistency where this is possible on the basis of logical considerations; it indicates how further consistency may be obtained. From this logic, one defines the combined adaptive logic along the lines followed in Section 8. The combined logic handles attempts to obtain an even more consistent result.

10   Appendix: Comparison of Approaches

One of the referees introduced a host of demands concerning the comparison between the adaptive approach and its 'competitors'. Some of these are strange because they concern papers hardly referred to by anyone except for their authors. I briefly comment on the demands in the present section.

Two points should be clear from the outset. The aim of this paper was to turn the stratagem into a viable logical instrument. I do not know of any comparable contributions. Next, this is certainly not the place to compare the adaptive approach to alternatives.

The adaptive logic program aims at characterizing all defeasible reasoning forms into a single framework, which at once provides a semantics, a

studied by Peter Verdée (paper soon forthcoming).

proof theory, and most of the metatheory. Note that not all defeasible reasoning is non-monotonic—Rescher's Weak consequence relation is an often quoted example. No specific *logic* is an alternative for the *program*. The comparative work consists in the adaptive characterization; many papers dealing with this are available, listed up in [14], and more are forthcoming.

Some distinctive features of adaptive logics are the following. (i) They are *formal* logics in the strict sense of the term (unlike default logics, circumscription, etc.). (ii) They have a proof theory (unlike default logics, circumscription, etc.). (iii) Propositional adaptive logics extend naturally to the predicative level. This is not so for many defeasible logics characterized otherwise. Obviously the degree of complexity of the adaptive consequence relation is drastically lowered by restricting to the propositional level or to finite premise sets.

Acknowledgments

Research for this paper was supported by subventions from Ghent University and from the Fund for Scientific Research – Flanders. I am indebted to Christian Straßer, to Peter Verdée, and to the two referees for comments on a former draft of this paper.

BIBLIOGRAPHY

[1] Arnon Avron. Non-deterministic matrices and modular semantics of rules. In Jean-Yves Béziau, editor, *Logica Universalis*, pages 149–167. Birkhäuser Verlag, Basel – Boston – Berlin, 2005.

[2] Arnon Avron and Beata Konikowska. Multi-valued calculi for logics based on non-determinism. *Journal of the Interest Group of Pure and Applied Logic*, 10:365–387, 2005.

[3] Diderik Batens. Paraconsistent extensional propositional logics. *Logique et Analyse*, 90–91:195–234, 1980.

[4] Diderik Batens. Inconsistency-adaptive logics. In Ewa Orłowska, editor, *Logic at Work. Essays Dedicated to the Memory of Helena Rasiowa*, pages 445–472. Physica Verlag (Springer), Heidelberg, New York, 1999.

[5] Diderik Batens. A general characterization of adaptive logics. *Logique et Analyse*, 173–175:45–68, 2001. Appeared 2003.

[6] Diderik Batens. The need for adaptive logics in epistemology. In Dov Gabbay, S. Rahman, J. Symons, and J. P. Van Bendegem, editors, *Logic, Epistemology and the Unity of Science*, pages 459–485. Kluwer Academic Publishers, Dordrecht, 2004.

[7] Diderik Batens. A procedural criterion for final derivability in inconsistency-adaptive logics. *Journal of Applied Logic*, 3:221–250, 2005.

[8] Diderik Batens. Narrowing down suspicion in inconsistent premise sets. In Jacek Malinowski and Andrzej Pietruszczak, editors, *Essays in Logic and Ontology*, volume 91 of *Poznań Studies in the Philosophy of the Sciences and the Humanities*, pages 185–209. Rodopi, Amsterdam/New York, 2006.

[9] Diderik Batens. A universal logic approach to adaptive logics. *Logica Universalis*, 1:221–242, 2007.

[10] Diderik Batens. Towards a dialogic interpretation of dynamic proofs. In Cédric Dégremont, Laurent Keiff, and Helhe Rückert, editors, *Dialogues, Logics and Other*

Strange Things. Essays in Honour of Shahid Rahman. College Publications, London, 2009 (in print).

[11] Diderik Batens and Kristof De Clercq. A rich paraconsistent extension of full positive logic. *Logique et Analyse*, 185–188:227–257, 2004. Appeared 2005.

[12] Diderik Batens, Kristof De Clercq, Peter Verdée, and Joke Meheus. Yes fellows, most human reasoning is complex. *Synthese*, 166:113–131, 2009.

[13] Diderik Batens and Joke Meheus. Recent results by the inconsistency-adaptive labourers. In Jean-Yves Béziau, Walter Carnielli, and Dov Gabbay, editors, *Handbook of Paraconcistency*, volume 9 of *Studies in Logic: Logic and Cognitive Systems*, pages 81–99. College Publications, London, 2007.

[14] Diderik Batens, Christian Straßer, and Peter Verdée. On the transparency of defeasible logics: Equivalent premise sets, equivalence of their extensions, and maximality of the lower limit. To appear.

[15] Walter A. Carnielli, Marcelo E. Coniglio, and João Marcos. Logics of formal inconsistency. In D. Gabbay and F. Guenthner, editors, *Handbook of Philosophical Logic*, volume 14, pages 1–93. Springer, Springer, 2007.

[16] Walter A. Carnielli and João Marcos. Limits for paraconsistent calculi. *Notre Dame Journal of Formal Logic*, 40:375–390, 1999.

[17] Newton C.A. da Costa. On the theory of inconsistent formal systems. *Notre Dame Journal of Formal Logic*, 15:497–510, 1974.

[18] Newton C.A. da Costa and Walter Carnielli. On paraconsistent deontic logic. *Philosophia*, 16:293–305, 1986.

[19] Stephen Cole Kleene. *Introduction to Metamathematics.* North-Holland, Amsterdam, 1952.

[20] Andrzej Wiśniewski. *The Posing of Questions. Logical Foundations of Erotetic Inferences.* Kluwer, Dordrecht, 1995.

[21] Andrzej Wiśniewski. The logic of questions as a theory of erotetic arguments. *Synthese*, 109:1–25, 1996.

[22] Andrzej Wiśniewski. Erotetic search scenarios, problem-solving, and deduction. *Logique et Analyse*, 185–188:139–166, 2004. Appeared 2005.

Diderik Batens
Centre for Logic and Philosophy of Science
Universiteit Gent
Belgium
E-mail: Diderik.Batens@UGent.be

An Adaptive Logic for Rational Closure

CHRISTIAN STRAßER

ABSTRACT. In [12] Lehmann and Magidor study a strong non-monotonic, so-called *rational* consequence relation, which extends the preferential consequence relation of [10] by also validating the rule of rational monotonicity. Every rational consequence relation can be semantically represented by a ranked model, and vice versa. To answer for a conditional assertion $a \mathrel{|\!\sim} b$ the question whether it is entailed by a set of conditional assertions K, it is not sufficient to check if it is derivable by the rules for rational consequence relations, or semantically, to check if it is valid in all ranked models of K, as it can be shown that the intersection of all ranked models does not in general satisfy rational monotonicity. The authors therefore define a semantic selection in order to obtain the so-called *rational closure*. However, a proof theory for rational closure is missing.

This paper will fill the syntactical gap for a finite language by defining an adaptive logic **ARC$^{\mathbf{s}}$** such that an assertion $a \mathrel{|\!\sim} b$ is derivable from a knowledge base K containing conditional assertions and negated conditional assertions iff it is in the rational closure of K.

1 Introduction

Monotonicity is an essential property of classical logic. For instance for a mathematician it would be rather inefficient (and demotivating) if a proof of a statement φ from a set of statements Γ would be invalidated by the addition of other statements to Γ. Of course, the situation in everyday life is quite different. The derivative power of human (and other intelligent) beings depends to a high degree on methods for drawing consequences which can be invalidated with the arrival of new information. Modelling ways of nonmonotonic reasoning in a formal framework is therefore very interesting for philosophers, as well as for the artificial intelligence community. Many systems were proposed in the literature to get a better grip on some of them, such as negation as failure [5], circumscription [13], default logic [16], or autoepistemic logic [14]. A more general view enabling comparative studies of nonmonotonic logics was introduced on the one hand by Gabbay [7], who focused on consequence relations, and on the other hand by Shoham (e.g. [18], [17]), who proposed a general model theory for nonmonotonic inference based on preference relations on worlds. The adaptive logic

program (e.g. [2]) offers another promising generic framework for defeasible and nonmonotonic reasoning forms (cp. [3]).[1] These logics are able to
adapt themselves to given premise sets, interpreting them "as normally as
possible" with respect to some criterion for normality. The proofs of these
logics are dynamic: some rules are conditionally applied, leaving open the
possibility that they are invalidated in view of results derived, or in view of
new information coming in at a later stage of the proof.

In [10] Kraus, Lehmann and Magidor give an account of nonmonotonic
consequence relations by isolating a list of properties and putting them in
the form of conditions on consequence relations. These properties have
become well-known as the KLM-properties and the basis of active research.
They have been put in the context of a more generic framework in [6] and
[1]. The authors of [10] define, among other things, *preferential* consequence
relations, represented by so-called *preferential models*, which prove to be a
powerful tool for studying what could be considered as reasonable inference
procedures. In [12] Lehmann and Magidor specify an even more restricted
notion, that of *rational relations* satisfying rational monotonicity, which can
semantically be represented by specific preferential models, the so-called
ranked models. The main question that paper discusses is: What are the
reasonable consequences of a conditional knowledge base? A conditional
knowledge base is a set of statements $a \mathrel{\vtop{\hbox{$\sim$}}} b$, which can be read as "if a is
the case, then normally b". It represents the set of the in general defeasible
information an agent may have. The considered question may be answered
with the notion of *rational closure*. This is achieved by a specific selection
based on a preference relation $\prec$ on rational consequence relations. A major
obstacle for giving an axiomatic account of rational closure is the fact that
the intersection of all ranked models of a given premise set (i.e. a conditional
knowledge base) is identical to the intersection of all preferential models,
which does not in general satisfy rational monotonicity. Therefore it is of
no use to simply employ a logic **R** composed of the axioms for rational
consequence relations. Nevertheless, especially in the tradition of Gabbay
and [10], it would be highly desirable to obtain a non-semantic account of
rational closure.

In this paper we are able to represent rational closure for finite languages
from a syntactical point of view by means of an adaptive logic **ARC^s** in
the standard format which has a simple extension of **R** as the lower limit
logic. As abnormalities we will consider a specific class of statements of the
kind "a is less normal than b". This opens the door to a proof theory for
the rational closure of a knowledge base. The proof theory is of a dynamic
nature, which is suitable for modelling actual reasoning processes. As it has
been shown in [9], rational closure is equivalent to 1-entailment as defined
in the context of Pearl's **Z**-system [15]. Therefore we additionally gain a
proof theory for 1-entailment as well.

---

[1] Adaptive logics are more thoroughly introduced in Section 4.

This paper can also be considered as another step in demonstrating that adaptive logics provide a very powerful and general framework to represent logics with nonmonotonic consequence relations and defeasible reasoning (as has recently been argued for in [3]).

2 Rational closure

Let a propositional language $\mathcal{L}_p$ be characterized as follows. We denote the classical connectives by $\neg, \vee, \wedge, \rightarrow$ and $\equiv$. Furthermore, we use the primitive symbols $\top$ and $\bot$. Lower case Greek letters α, β, γ and δ denote formulas in $\mathcal{V}_p$, the set of propositional formulas over the set of propositional letters $\mathcal{P} = \{p_1, \ldots, p_n\}$.

The authors in [10] introduce the symbol $\vdash\!\!\!\sim$ for conditional consequence relations. Statements of the form "$\alpha \vdash\!\!\!\sim \beta$", where $\alpha, \beta \in \mathcal{V}_p$, are called *conditional assertions* and can be interpreted as "from α sensibly conclude β" or "α usually implies β".

DEFINITION 1 (cp. [10], Def. 15). A relation $\vdash\!\!\!\sim \,\subseteq \mathcal{V}_p \times \mathcal{V}_p$ is called a *preferential consequence relation* iff it satisfies the following properties:[2]

$$a \vdash\!\!\!\sim a \qquad \text{(Reflexivity)}$$

$$\frac{\vdash \alpha \leftrightarrow \beta \quad a \vdash\!\!\!\sim \gamma}{\beta \vdash\!\!\!\sim \gamma} \qquad \text{(Left Logical Equivalence)}$$

$$\frac{\vdash \alpha \rightarrow \beta \quad \gamma \vdash\!\!\!\sim \alpha}{\gamma \vdash\!\!\!\sim \beta} \qquad \text{(Right Weakening)}$$

$$\frac{\alpha \vdash\!\!\!\sim \beta \quad \alpha \vdash\!\!\!\sim \gamma}{\alpha \vdash\!\!\!\sim \beta \wedge \gamma} \qquad \text{(And)}$$

$$\frac{\alpha \vdash\!\!\!\sim \beta \quad \alpha \vdash\!\!\!\sim \gamma}{\alpha \wedge \beta \vdash\!\!\!\sim \gamma} \qquad \text{(Cautious Monotonicity)}$$

$$\frac{\alpha \vdash\!\!\!\sim \gamma \quad \beta \vdash\!\!\!\sim \gamma}{\alpha \vee \beta \vdash\!\!\!\sim \gamma} \qquad \text{(Or)}$$

Another way to look at the properties is to interpret them as inferential rules where $\vdash\!\!\!\sim$ is a logical connective and does therefore not represent a consequence relation. We call the system constituted by the above rules $\mathbf{P}_{\mathsf{KLM}}$.

REMARK 2. The following properties have shown to be valid for preferential consequence relations in [10]:

$$\frac{\alpha \vdash\!\!\!\sim \beta \quad \beta \vdash\!\!\!\sim \alpha \quad \alpha \vdash\!\!\!\sim \gamma}{\beta \vdash\!\!\!\sim \gamma} \qquad \text{(Equivalence)}$$

$$\frac{\alpha \wedge \beta \vdash\!\!\!\sim \gamma \quad \alpha \vdash\!\!\!\sim \beta}{a \vdash\!\!\!\sim \gamma} \qquad \text{(Cut)}$$

---

[2]This is the characterization presented in [12]. In [10] the authors give an alternative (equivalent) characterization replacing the (And) condition with (Cut).

$$\frac{\alpha \mathrel{|\!\sim} \beta \to \gamma \quad \alpha \mathrel{|\!\sim} \beta}{\alpha \mathrel{|\!\sim} \gamma} \tag{MPC}$$

$$\frac{\alpha \wedge \beta \mathrel{|\!\sim} \gamma}{\alpha \mathrel{|\!\sim} \beta \to \gamma} \tag{S}$$

$$\frac{\alpha \wedge \neg\beta \mathrel{|\!\sim} \gamma \quad \alpha \wedge \beta \mathrel{|\!\sim} \gamma}{\alpha \mathrel{|\!\sim} \gamma} \tag{D}$$

The following properties are in general not valid for preferential consequence relations:

$$\frac{\alpha \wedge \gamma \mathrel{|\!\not\sim} \beta \quad \alpha \wedge \neg\gamma \mathrel{|\!\not\sim} \beta}{\alpha \mathrel{|\!\not\sim} \beta} \qquad \text{(Negation Rationality)}$$

$$\frac{\alpha \mathrel{|\!\not\sim} \gamma \quad \beta \mathrel{|\!\not\sim} \gamma}{\alpha \vee \beta \mathrel{|\!\not\sim} \gamma} \qquad \text{(Disjunctive Rationality)}$$

$$\frac{\alpha \wedge \beta \mathrel{|\!\not\sim} \gamma \quad \alpha \mathrel{|\!\not\sim} \neg\beta}{\alpha \mathrel{|\!\not\sim} \gamma} \qquad \text{(Rational Monotonicity)}$$

DEFINITION 3 ([12], Def. 13). A preferential consequence relation is a *rational consequence relation* if it satisfies rational monotonicity. The corresponding inferential system we call $\mathbf{R}_{\mathsf{KLM}}$.

A *knowledge base* $\mathbf{K}$ is a set of conditional assertions. We call $\mathrel{|\!\sim}$ a *preferential (resp. rational) extension of* $\mathbf{K}$ iff $\mathrel{|\!\sim}$ is a preferential (resp. rational) consequence relation and it satisfies all the conditional assertions in $\mathbf{K}$.

REMARK 4 (cp. [12], Lem. 10, Lem. 12). "Rational Monotonicity" implies "Disjunctive Rationality" and this implies "Negation Rationality".

Let worlds be assignments of truth values to the propositional variables. With $\mathcal{U}$ we denote the set of all worlds. We use the standard definitions for the satisfaction of (sets of) formulas by worlds and for the validity of formulas. We write $w \models \alpha$ iff α is valid in world w.

DEFINITION 5. Let $\prec$ be a partial order on a set U and $V \subseteq U$. We say that $x \in V$ is *minimal* in V iff there is no $y \in V$, such that $y \prec x$.

We shall say that V is *smooth* iff for all $x \in V$, either there is a y minimal in V, such that $y \prec x$ or x is itself minimal.

DEFINITION 6 ([12], Def. 4). A *preferential model* M is a triple $\langle S, l, \prec \rangle$ where S is a set, the elements of which will be called states, $l : S \to \mathcal{U}$ assigns a world to each state and $\prec$ is a strict partial order on S satisfying the following *smoothness condition*: for all $\alpha \in \mathcal{V}_p$, the set of states $\hat{\alpha} =_{\mathrm{df}} \{s : s \in S, s \models \alpha\}$ is smooth, where $\models$ is defined as $s \models \alpha$ (read s satisfies α) iff $l(s) \models \alpha$. The model M will be said to be finite iff S is finite. It will be said to be well-founded iff $\langle S, \prec \rangle$ is well-founded, i.e., iff there is no infinite descending chain of states.

DEFINITION 7 ([12], Def. 5). Suppose a model $M = \langle S, l, \prec \rangle$ and $\alpha, \beta \in \mathcal{V}_p$ are given. The consequence relation defined by M will be denoted by $\mathord{\succ}_M$ and is defined by: $\alpha \mathrel{\succ}_M \beta$ iff for any s minimal in $\hat{\alpha}$, $s \models \beta$.

In case $\alpha \mathrel{\succ}_M \beta$ we say that M is a model of $\alpha \mathrel{\succ} \beta$, that M satisfies $\alpha \mathrel{\succ} \beta$ or that $\alpha \mathrel{\succ} \beta$ is valid in M. Furthermore, if $\mathord{\succ}_M$ extends some knowledge base **K**, we say that M is a model of **K**.

If $\hat{\alpha}$ is empty we have $\alpha \mathrel{\succ}_M \beta$ for any $\beta \in \mathcal{V}_p$, i.e. α represents something so abnormal, that we say that literally anything would happen if α were the case.

The next result shows that the characterization of preferential consequence relations in terms of preferential models is adequate. Furthermore, the intersection of all preferential consequence relations extending a given knowledge base is itself a preferential consequence relation. This enables us to effortlessly introduce a notion of preferential entailment.

THEOREM 8 ([12], Thm. 1). *A binary relation $\mathord{\succ} \subseteq \mathcal{V}_p \times \mathcal{V}_p$ is a preferential consequence relation iff it is the consequence relation defined by some preferential model.*

DEFINITION 9 ([12], Def. 6). A conditional assertion $\alpha \mathrel{\succ} \beta$ is *preferentially entailed* by a conditional knowledge base **K** iff it is satisfied by all preferential models of **K**. The set of all conditional assertions that are preferentially entailed by **K** is denoted by $\mathbf{K}^p$. The relation $\mathbf{K}^p$ is called the *preferential closure* of **K**.

THEOREM 10 (cp. [12], Thm. 2). *Let **K** be a conditional knowledge base. The following conditions are equivalent:*

1. $\alpha \mathrel{\succ} \beta \in \mathbf{K}^p$

*2. for all preferential models M of **K**, $\alpha \mathrel{\succ}_M \beta$*

*3. $\alpha \mathrel{\succ} \beta$ has a proof from **K** in $\mathbf{P}_{\mathsf{KLM}}$.*

COROLLARY 11 (cp. [12], Cor. 2). *For a conditional knowledge base **K** the preferential closure $\mathbf{K}^p$ defines a preferential consequence relation.*

Let us proceed by giving a semantics for rational consequence relations. This is not difficult: specific preferential models are suitable for our purpose, namely the ones for which the ordering on the states $\prec$ is modular.

DEFINITION 12. A partial order $\prec$ over V is called *modular* iff it satisfies the following equivalent conditions:

- there is a totally ordered set Υ (the strict order on Υ is denoted by $\prec'$) and a function $\mathsf{r} : V \to \Upsilon$ (the *ranking function*) such that $x \prec y$ iff $\mathsf{r}(x) \prec' \mathsf{r}(y)$.

- for any $x, y, z \in V$ if $x \not\prec y, y \not\prec x$ and $z \prec x$, then $z \prec y$

- for any $x, y, z \in V$ if $x \prec y$, then either $z \prec y$ or $x \prec z$

- for any $x, y, z \in V$ if $x \not\prec y, y \not\prec z$, then $x \not\prec z$

DEFINITION 13 ([12], Def. 14). A *ranked model* is a preferential model $\langle S, l, \prec \rangle$ for which the strict partial order $\prec$ is modular. Therefore we can associate a ranking function r with each ranked model.

This is a good place to introduce some insights which help to greatly simplify the technical requirements for our further reasoning. Using states and mapping states into worlds in the definition of a preferential model $M = \langle S, l, \prec \rangle$ allowed for one and the same world w to be associated with different states, i.e. there are $s, t \in S$ such that $l(s) = l(t)$. However, if $\prec$ is modular, either both states have the same rank or one state is higher than the other, e.g. $s \prec t$. In the first case it is due to the modularity of $\prec$ obvious that $N = \langle S', l, \prec \rangle$, where $S' = S \setminus \{t\}$, is such that $\vdash_N = \vdash_M$. In the second case t is such that for any $\alpha \in V_p$ for which $t \models \alpha$, t is not minimal in $\hat{\alpha}$ (since $s \prec t$ and $s \models \alpha$). What we have just seen is that, whenever for a ranked model two states are mapped into the same world, then we can do without one of them. But then, in case $\prec$ is modular, the talk about states is superfluous. Similar to [8] and to [4] we use therefore the following simplification:

DEFINITION 14. A *rational model* M is a pair $\langle W, \prec \rangle$ where $W \subseteq \mathcal{U}$ and $\prec$ is a strict modular order on W. We define $\alpha \vdash_M \beta$ iff for all $w \in \min_\prec(\sigma_M(\alpha)) : w \models \beta$, where $\sigma_M(\alpha) =_{\mathrm{df}} \{w \in W : w \models \alpha\}$ and for all $W' \subseteq W$

$$\min_\prec(W') =_{\mathrm{df}} \{w \in W' : \text{there is no } w' \in W' \text{ such that } w' \prec w\}.$$

Note that due to the modularity of $\prec$ we have $w \prec w'$ for all $w \in \min_\prec(W')$ and for all $w' \in W' \setminus \min_\prec(W')$. Note further that the modularity of $\prec$ gives rise to equivalence classes of worlds which are mutually incomparable, namely the ones which have the same ranking. The equivalence classes themselves are totally ordered.

For a rational model $M = \langle W, \prec \rangle$ we define the *canonical ranking function* $\mathsf{rank}_\prec : W \to \mathbb{N}$ ($\mathbb{N}$ being the natural numbers incl. 0) as the unique epimorphism from $\langle W, \prec \rangle$ onto $\langle \{0, \ldots, n\}, < \rangle$ where $n \in \mathbb{N}$. Obviously for every equivalence class $[w]$ there is a unique natural number i assigned to it by $\mathsf{rank}_\prec$, i.e. for all $w' \in [w], \mathsf{rank}_\prec(w') = \mathsf{rank}_\prec(w) = i$ and for each $w' \notin [w], \mathsf{rank}_\prec(w') \neq i$. It is obvious that all minimal worlds in $\sigma_M(\alpha)$ belong to the same equivalence class. We define another useful related mapping $\mathsf{Rank}_\prec : V_p \to \{0, \ldots, n, \omega\}$ on basis of $\mathsf{rank}_\prec$.[3] For a $\alpha \in V_p$, $\alpha \mapsto h$ iff $\sigma_M(\alpha) \neq \emptyset$ and $\mathsf{rank}_\prec(w) = h$ for any $w \in \min_\prec(\sigma_M(\alpha))$. If $\sigma_M(\alpha) = \emptyset$ then $\alpha \mapsto \omega$.

---

[3]The reader should be warned at this place not to mistake our $\mathsf{Rank}_\prec$ function with the notion of the rank of a formula defined in [12].

The following abbreviation will be very useful in the remainder of the paper:

DEFINITION 15. $\alpha \lessdot \beta =_{\mathrm{df}} \alpha \vee \beta \hspace{0.2em}\mid\hspace{-0.55em}\sim \neg\beta$.[4] For a rational model M we write $\alpha \lessdot_M \beta$ for $\alpha \vee \beta \hspace{0.2em}\mid\hspace{-0.55em}\sim_M \neg\beta$.

$\alpha \lessdot \beta$ can be read as "α is more normal than β". Some properties:

LEMMA 16. *Let* $M = \langle W, \prec \rangle$ *be a rational model. We have*

(i) $\alpha \lessdot_M \beta$ *iff (*$\mathsf{Rank}_\prec(\alpha) < \mathsf{Rank}_\prec(\beta)$ *or* $\mathsf{Rank}_\prec(\alpha) = \mathsf{Rank}_\prec(\beta) = \omega$*);*

(ii) $\alpha \hspace{0.2em}\mid\hspace{-0.55em}\sim_M \beta$ *iff (*$\mathsf{Rank}_\prec(\alpha) < \mathsf{Rank}_\prec(\alpha \wedge \neg\beta)$ *or* $\mathsf{Rank}_\prec(\alpha) = \omega$*);*

(iii) If $\alpha \lessdot_M \beta$ *and* $\beta \lessdot_M \gamma$*, then* $\alpha \lessdot_M \gamma$.

Proof. Ad (i): Let $\alpha \lessdot_M \beta$. Then $\alpha \vee \beta \hspace{0.2em}\mid\hspace{-0.55em}\sim_M \neg\beta$. In case $\sigma_M(\alpha \vee \beta) \neq \emptyset$, for all $w \in \min_\prec(\sigma_M(\alpha \vee \beta))$, $w \models \neg\beta$. Hence $\min_\prec(\sigma_M(\alpha \vee \beta)) = \min_\prec(\sigma_M(\alpha \wedge \neg\beta))$. But then obviously for any $w \in \min_\prec(\sigma_M(\alpha))$ and any $w' \in \sigma_M(\beta)$, $\mathsf{rank}_\prec(w) < \mathsf{rank}_\prec(w')$ or $\sigma_M(\beta) = \emptyset$. Therefore $\mathsf{Rank}_\prec(\alpha) < \mathsf{Rank}_\prec(\beta)$. The case $\sigma_M(\alpha \vee \beta) = \emptyset$ is trivial, since then $\sigma_M(\alpha) = \sigma_M(\beta) = \emptyset$.

Let $\mathsf{Rank}_\prec(\alpha) < \mathsf{Rank}_\prec(\beta)$. Then $\sigma_M(\alpha) \neq \emptyset$. Furthermore, either $\sigma_M(\beta) = \emptyset$ or for all $w \in \sigma_M(\beta)$, $\mathsf{rank}_\prec(w) > \mathsf{rank}_\prec(w')$ where w' is any minimal world in $\sigma_M(\alpha)$. But then obviously $w' \models \neg\beta$. Hence $\min_\prec(\sigma_M(\alpha)) = \min_\prec(\sigma_M(\alpha \wedge \neg\beta)) = \min_\prec(\sigma_M(\alpha \vee \beta))$ and therefore $\alpha \vee \beta \hspace{0.2em}\mid\hspace{-0.55em}\sim_M \neg\beta$. The case $\mathsf{Rank}_\prec(\alpha) = \mathsf{Rank}_\prec(\beta) = \omega$ is trivial, since then $\sigma_M(\alpha \vee \beta) = \emptyset$.

(ii) is an immediate consequence of (i) since $\alpha \hspace{0.2em}\mid\hspace{-0.55em}\sim \beta$ is equivalent to $\alpha \lessdot \alpha \wedge \neg\beta$, as can easily be shown.

(iii) Applying (i) we have $\mathsf{Rank}_\prec(\alpha) < \mathsf{Rank}_\prec(\beta) \leq \mathsf{Rank}_\prec(\gamma)$ or $\mathsf{Rank}_\prec(\alpha) = \omega = \mathsf{Rank}_\prec(\beta) = \mathsf{Rank}_\prec(\gamma)$. The rest follows by (i). ∎

THEOREM 17 (cp. [12], Thm. 5). *A binary relation* $\hspace{0.2em}\mid\hspace{-0.55em}\sim \subseteq \mathcal{V}_p \times \mathcal{V}_p$ *is a rational consequence relation iff it is the consequence relation defined by some rational model.*

Unfortunately, for rational models we don't have an analogous result to Theorem 11 which makes it a much more difficult task to define an adequate notion of rational entailment. This is shown in the following theorem.

THEOREM 18 ([12], Thm. 6). *If an assertion* $\alpha \hspace{0.2em}\mid\hspace{-0.55em}\sim \beta$ *is satisfied by all rational models of a conditional knowledge base* **K***, then it is satisfied by all preferential such models. Furthermore,* $\alpha \hspace{0.2em}\mid\hspace{-0.55em}\sim \beta$ *can be derived from* **K** *by the rules of* $\mathbf{R}_{\mathsf{KLM}}$ *iff it can be derived from* **K** *by* $\mathbf{P}_{\mathsf{KLM}}$.

The theorem shows that the simple idea to define rational entailment analogously to the way preferential entailment was defined, i.e. in terms of the intersection of all rational models, is ill-founded. Lehmann and Magidor

---

[4]The authors in [12] use '<' do denote both, this connective and the canonical ordering relation on the ordinals. In order to avoid confusion we denote by < exclusively the latter, while we use $\lessdot$ for the former.

conclude that "the intersection of all rational extensions of $\mathbf{K}$ is exactly $\mathbf{K}^p$ and therefore not in general rational and highly unsuitable" ([12], p. 27). So the authors introduce another, more sophisticated concept: the rational closure of a conditional knowledge base.

DEFINITION 19 ([12], Def. 20). Let $\vdash_0$ and $\vdash_1$ be two rational consequence relations. We shall say that $\vdash_0$ is *preferable* to $\vdash_1$ ($\vdash_0 \sqsubset \vdash_1$) iff:[5]

1. there exists an assertion $\alpha \vdash_1 \beta$ such that $\alpha \not\vdash_0 \beta$ and for all γ for which $\gamma \prec \alpha$ for $\vdash_0$, and for all δ such that $\gamma \vdash_0 \delta$, we also have $\gamma \vdash_1 \delta$, and

2. for any assertion $\gamma \vdash_0 \delta$ such that $\gamma \not\vdash_1 \delta$, there are γ' and δ' for which $\gamma' \vdash_1 \delta'$, $\gamma' \not\vdash_0 \delta'$ and $\gamma' \prec \gamma$ for $\vdash_1$.

Since we cannot, as in case of preferential relations, simply intersect all rational relations, we define a preference order on them. Let us illustrate this rather abstract notion with a small example: Sue and Peter are both from California and they agree that usually the sun shines, $\top \vdash s$. Furthermore they agree that whenever they wear sunglasses, usually the sun is shining as well,—$g \vdash s$. So their shared knowledge base $\mathbf{K}$ is $\{\top \vdash s, g \vdash s\}$. We present the rational extensions of Sue, $\vdash_S$, and Peter, $\vdash_P$, by two rational models $M_S = \langle \mathcal{U}, \prec_S \rangle$ and $M_P = \langle \mathcal{U}, \prec_P \rangle$. The following truth-table characterizes four classes of worlds W_1, W_2, W_3 and W_4.

	W_1	W_2	W_3	W_4
s	0	0	1	1
g	0	1	0	1

Read the table e.g. as follows: W_1 is the class of worlds w for which $w \models \neg s, \neg g$. Analogous for W_2, W_3 and W_4. As remarked above, rational models are characterized by a strict ordering on the equivalence classes of worlds with the same ranking. Models M_P and M_S are represented in the following illustration (M_P to the left and M_S to the right):

$$
\begin{array}{cc}
W_1 \cup W_2 \cup W_3 & W_1 \cup W_2 \\
| & | \\
W_4 & W_3 \cup W_4
\end{array}
$$

For instance we have for all $w \in W_4$ and for all $w' \in W_1 \cup W_2 \cup W_3$, $w \prec_P w'$. The reader can easily verify that for Peter we have for instance $g \vdash_P s$ and $\top \vdash_P g$, whereas $\neg g \not\vdash_P s$. For Sue we have for instance $\neg g \vdash_S s$ but $s \not\vdash_S g$, $\top \not\vdash_S g$. The consequence relation defined by Sue's extension of

---

[5]In [12] the authors use the "$\prec$" symbol for this relation. In order to disambiguate the usage of the various symbols for ordering relations in this article and therefore to serve readability, in this paper "$\sqsubset$" is being used exclusively for the preferability relation of Definition 19.

K is preferable to the one defined by Peter's extension of **K**. Indeed, Peter derives too much. There are various statements Sue is not willing to accept, for instance $s \mathrel{\vdash\mkern-9mu\sim} g$. It is in no way normal to wear sunglasses whenever the sun is shining. Consider cases in which you are sitting in an office or any other closed spaces. With the information at hand, **K**, $s \mathrel{\vdash\mkern-9mu\sim} g$ should not be in the rational closure. Indeed, $s \mathrel{\vdash\mkern-9mu\sim}_P g$ satisfies criterion 1 of Definition 19 since there is no proposition more normal in Sue's knowledge base than s. Still, there are statements in Sue's extension that Peter does not accept, for instance $\neg g \mathrel{\vdash\mkern-9mu\sim} s$. However, Sue is able to point out that Peter states something for more normal circumstances (according to his reasoning) that is unacceptable for her, for instance $s \mathrel{\vdash\mkern-9mu\sim} g$. Note here that $g <_P \neg g$. We have shown that both criteria are met and hence $\mathrel{\vdash\mkern-9mu\sim}_S \sqsubset \mathrel{\vdash\mkern-9mu\sim}_P$.

It should by now have become obvious that the preferability relation $\sqsubset$ can be explained in terms of an argumentation game. In order to win, Sue had to be able to attack Peter in such a way that Peter cannot respond (criterion 1) and for every attack of Peter, Sue had to be able to respond (criterion 2). To attack somebody with an assertion $\alpha \mathrel{\vdash\mkern-9mu\sim} \beta$ is to demonstrate that $\alpha \mathrel{\vdash\mkern-9mu\sim} \beta$ is in the knowledge base of the one attacked but not in ones own knowledge base. To defend oneself against an attack with an assertion $\alpha \mathrel{\vdash\mkern-9mu\sim} \beta$ by person H is to counterattack with an assertion $\gamma \mathrel{\vdash\mkern-9mu\sim} \delta$ such that γ is more normal than α for H.

DEFINITION 20 ([12], Def. 21). Let **K** be an arbitrary knowledge base. If there is a rational extension $\mathrel{\vdash\mkern-9mu\sim}$ of **K** that is preferable to all other rational extensions of **K**, then $\mathrel{\vdash\mkern-9mu\sim}$ is called the *rational closure* of **K**. For finite languages the rational closure does exist.

As our ultimate goal is to develop a logic, the consequences of which, for a given conditional knowledge base, correspond to the rational closure, the following logic **R** which generalizes the inferential system **R**$_{\mathsf{KLM}}$ will be very useful for purposes. We are operating in a language $\mathcal{L}$ consisting of Boolean combinations of conditional assertions and propositional formulas (while the language of **R**$_{\mathsf{KLM}}$ only consisted of conditional assertions). Let $\mathcal{V}_p$ be again the set of propositional formulas, $\mathcal{P} = \{p_1, \ldots, p_n\}$ the set of propositional letters, and $\mathcal{V}_{\mathrel{\vdash\mkern-9mu\sim}}$ is the set of conditional assertions. $\mathcal{V}$ is the set of all Boolean combinations of the formulas in $\mathcal{V}_{\mathrel{\vdash\mkern-9mu\sim}} \cup \mathcal{V}_p$. Some notational conventions: as done so far we use α, β, γ and δ to refer to propositional formulas, φ and ψ are reserved for formulas in $\mathcal{V}$. Logic **R** consists of all axioms and rules of the propositional calculus together with the following axioms and rules:

$$\alpha \mathrel{\vdash\mkern-9mu\sim} \alpha \qquad \text{(REF)}$$

$$\text{If } \vdash \alpha \leftrightarrow \beta, \text{ then } \vdash (\alpha \mathrel{\vdash\mkern-9mu\sim} \gamma) \rightarrow (\beta \mathrel{\vdash\mkern-9mu\sim} \gamma) \qquad \text{(LLE)}$$

$$\text{If } \vdash \alpha \rightarrow \beta, \text{ then } \vdash (\gamma \mathrel{\vdash\mkern-9mu\sim} \alpha) \rightarrow (\gamma \mathrel{\vdash\mkern-9mu\sim} \beta) \qquad \text{(RW)}$$

$$((\alpha \mathrel{\vdash\mkern-9mu\sim} \beta) \wedge (\beta \mathrel{\vdash\mkern-9mu\sim} \gamma)) \rightarrow (\alpha \mathrel{\vdash\mkern-9mu\sim} \beta \wedge \gamma) \qquad \text{(AND)}$$

$$((\alpha \mathrel{\vdash\mkern-9mu\sim} \gamma) \wedge (\alpha \mathrel{\vdash\mkern-9mu\sim} \gamma)) \rightarrow (\alpha \vee \beta \mathrel{\vdash\mkern-9mu\sim} \gamma) \qquad \text{(OR)}$$

$$((\alpha \mathrel{\vdash\mkern-9mu\sim} \beta) \wedge (\alpha \mathrel{\vdash\mkern-9mu\sim} \gamma)) \rightarrow (\alpha \wedge \beta \mathrel{\vdash\mkern-9mu\sim} \gamma) \qquad \text{(CM)}$$

$$((\alpha \mathrel{\vdash\mkern-9mu\sim} \beta) \wedge \neg(\alpha \mathrel{\vdash\mkern-9mu\sim} \neg\gamma)) \rightarrow (\alpha \wedge \gamma \mathrel{\vdash\mkern-9mu\sim} \beta) \qquad \text{(RM)}$$

The logic was more thoroughly studied in [6], for instance completeness and soundness was demonstrated for various semantics. In [8] a tableau calculus was developed for **R**. Like the authors in [8], we use a semantics based on rational models. For a rational model $M = \langle W, \prec \rangle$ we define $\models$ as follows:

- for $\alpha \in \mathcal{V}_p$: $M \models \alpha$ iff $w \models \alpha$ for all $w \in W$;

- for $\alpha \mathrel{\vdash\mkern-9mu\sim} \beta \in \mathcal{V}_{\mathrel{\vdash\mkern-9mu\sim}}$: $M \models \alpha \mathrel{\vdash\mkern-9mu\sim} \beta$ iff $\alpha \mathrel{\vdash\mkern-9mu\sim}_M \beta$;

Furthermore, for Boolean combinations $\models$ is defined in the usual way (e.g. $M \models \varphi \vee \psi$ iff $M \models \varphi$ or $M \models \psi$, etc.).

3   Constructing a model for rational closure

As shown in [12], the rational closure of a given knowledge base does exist for a finite language $\mathcal{L}$. Furthermore, there are various model-theoretic approaches to construct a model for it (e.g. in [12], Section 5.7). The following proposal presented in [4] is interesting for our purposes. Let **K** be a conditional knowledge base. Let $\mathcal{M}$ be the set of all rational models of **K**. Define

$$U_0 =_{\text{df}} \bigcup_{\langle W, \prec \rangle \in \mathcal{M}} \{w \in W \mid \mathsf{rank}_{\prec}(w) = 0\}$$

$$U_i =_{\text{df}} \bigcup_{\langle W, \prec \rangle \in \mathcal{M}} \{w \in W \mid \mathsf{rank}_{\prec}(w) = i\} \setminus \bigcup_{j<i} U_j, \text{ where } i > 0$$

We are interested in $M = \langle U, \prec \rangle$ where $U = \bigcup_{i \in \mathbb{N}} U_i$ and $\prec$ is defined by: for all natural numbers i, j and for all w, w' for which $w \in U_i$ and $w' \in U_j$, $w \prec w'$ iff $i < j$. As proven in [4], M is a rational model defining the rational closure of **K**.

4   What are Adaptive Logics?

The logic **ARC**$^{\mathbf{s}}$ for rational closure we are about to present in Section 5 belongs to the class of adaptive logics. The mechanism of adaptive logics has been presented in various papers. Space limitations require that we refer the reader interested in a detailed description of them to [2]. Here we will only mention some key features.

An adaptive logic in standard format is a triple consisting of (i) a lower limit logic (henceforth **LLL**), which is a reflexive, transitive, monotonic, and compact logic that has a characteristic semantics and contains **CL** (classical logic), (ii) a set of abnormalities Ω, characterized by a (possibly restricted) logical form, and (iii) an adaptive strategy. Formulating an adaptive logic in standard format provides the logic with all of the important metatheoretic features, such as soundness and completeness (as is shown in [2]). As the name itself suggests, the idea underlying adaptive logics is that they adapt themselves to specific premise sets, interpreting them "as normally as possible" with respect to some criterion for normality. Their dynamic proofs make them very useful for modelling defeasible reasoning, since a formula derivable at one stage of the proof may turn out to be underivable at a later stage.

In this section we use φ and ψ as metavariables for well-formed formulas of a given language. The proof dynamics is governed by marking conditions for proof lines. A line of a proof consists of a line number, a formula, a justification, and a condition. The latter is a new element which is not part of standard proofs. Conditions are finite subsets of the set of abnormalities. We abbreviate $\bigvee_{\varphi \in \Delta} \varphi$ with $\mathrm{Dab}(\Delta)$ for some finite set Δ of abnormalities. The proof dynamics require the following generic rules, where "$\varphi \quad \Delta$" abbreviates that φ occurs in the proof on the condition Δ:

PREM   If $\varphi \in \Gamma$:

$$\frac{\cdots \quad \cdots}{\varphi \quad \emptyset}$$

RU   If $\varphi_1, \ldots, \varphi_n \vdash_{\mathbf{LLL}} \psi$:

$$\frac{\begin{array}{cc} \varphi_1 & \Delta_1 \\ \vdots & \vdots \\ \varphi_n & \Delta_n \end{array}}{\psi \quad \Delta_1 \cup \cdots \cup \Delta_n}$$

RC   If $\varphi_1, \ldots, \varphi_n \vdash_{\mathbf{LLL}} \psi \vee \mathrm{Dab}(\Theta)$:

$$\frac{\begin{array}{cc} \varphi_1 & \Delta_1 \\ \vdots & \vdots \\ \varphi_n & \Delta_n \end{array}}{\psi \quad \Delta_1 \cup \cdots \cup \Delta_n \cup \Theta}$$

Premises can be introduced on the empty condition by PREM. Everything which can be derived from previous steps by the **LLL** can be derived in an adaptive logic without adding new conditions. Conditions of lines used in a derivational step are carried forward. The most interesting generic rule is RC: if we can derive ψ in the **LLL** in disjunction with a disjunction of abnormalities, $\mathrm{Dab}(\Theta)$, then we are able to derive ψ in the adaptive logic on condition Θ. Like for RU, we carry forward the conditions of the lines used for the derivational step.

It is the job of the marking definition to determine if lines are "in" or "out" at a certain stage, i.e. to govern the dynamics of the proof procedure. In this paper a highly unsophisticated strategy is used: the Simple Strategy. It is applied in cases in which the two standard strategies for adaptive logics coincide.[6] In these cases the three strategies characterize the same consequence relation.

DEFINITION 21 (Marking for the Simple Strategy). A line with condition Δ is marked at stage s iff a $\varphi \in \Delta$ has been derived on the empty condition.

DEFINITION 22. φ is *finally derived* from a premise set Γ on line i iff there is no extension of the proof in which line i is marked. We write $\Gamma \vdash_{\mathbf{AL}} \varphi$.

We use the abbreviation $\mathrm{Ab}(M) =_{\mathrm{df}} \{\varphi \in \Omega \mid M \models \varphi\}$ for the set of abnormalities verified by a **LLL**-model M. M is a minimal abnormal **LLL**-model M of Γ iff M is a **LLL**-model of Γ and for all **LLL**-models M' of Γ, $\mathrm{Ab}(M') \not\subset \mathrm{Ab}(M)$. We also use the notation: $Cn_{\mathbf{L}}(\Gamma) =_{\mathrm{df}} \{\varphi \mid \Gamma \vdash_{\mathbf{L}} \varphi\}$.

LEMMA 23. *Let Γ be a premise set. The following conditions are equivalent:*

- *M is a minimal abnormal **LLL**-model of Γ iff M is a **LLL**-model of Γ and $\mathrm{Ab}(M) = Cn_{\mathbf{LLL}}(\Gamma) \cap \Omega$,*

- *for all minimal abnormal **LLL**-models M, M' of Γ, $\mathrm{Ab}(M) = \mathrm{Ab}(M')$.*

DEFINITION 24. We call a **LLL**-model M of Γ *simple* iff $\mathrm{Ab}(M) = Cn_{\mathbf{LLL}}(\Gamma) \cap \Omega$. We write $\Gamma \models_{\mathbf{AL}} \varphi$ iff φ is valid in all simple **LLL**-models of Γ.

THEOREM 25. *For a class of premise sets $\mathbf{G}$ such that for all $\Gamma \in \mathbf{G}$ the conditions in Lemma 23 are valid, the following completeness result is valid for all $\Gamma \in \mathbf{G}$:*

$$\Gamma \vdash_{\mathbf{AL}} \varphi \; iff \; \Gamma \models_{\mathbf{AL}} \varphi$$

Lehmann and Magidor restrict themselves to premise sets consisting of conditional assertions. We are going to allow also for negated conditional assertions. Since for these classes of premise sets the conditions of Lemma 23 are fulfilled we will be able to make use of Theorem 25.

5   An adaptive logic for Rational Closure

The main idea underlying the adaptive logic **ARC^s** for rational closure which is going to be introduced in this section is that propositional formulas are interpreted as normal as the given knowledge base allows for.

---

[6]These strategies are the Minimal Abnormality and the Reliability Strategy. For details compare [2].

Semantically this means that worlds of rational models are ranked as low as possible. In order to make this description more precise and in order to define this logic we need to introduce an extended language $\mathcal{L}^+$ on basis of our language $\mathcal{L}$.

Therefore we extend the given set of propositional letters $\mathcal{P} = \{p_1, \ldots, p_n\}$ with the set of propositional letters $\{l_i \mid i \in \mathcal{N}\}$ where $\mathcal{N} = \{0, \ldots, 2^n\}$. We define $\mathcal{P}^+ =_{\mathrm{df}} \mathcal{P} \cup \{l_i \mid i \in \mathcal{N}\}$, $\mathcal{V}_p^+$ is the set of propositional formulas with propositional letters $\mathcal{P}^+$ and $\mathcal{V}_{\mid\sim}^+$ is the set of conditional assertions $\alpha \mid\sim \beta$ where $\alpha, \beta \in \mathcal{V}_p^+$. Finally let $\mathcal{V}^+$ be the set of all Boolean combinations of formulas in $\mathcal{V}_{\mid\sim}^+ \cup \mathcal{V}_p^+$.

The idea is that the l_i's represent different degrees of abnormality. $l_i \prec \alpha$ can be read as "α has a higher degree of abnormality than i", $\alpha \mid\sim l_i$ can be read as "α's degree of abnormality is at least i". The higher the index i, the higher is the degree of abnormality represented by l_i. For instance, if $\neg(l_0 \prec \alpha)$, then α is considered to be normally the case. We call a rational consequence relation realizing this idea a *rational$^+$ consequence relation*. It is supposed to satisfy the following additional conditions, where $\alpha \in \mathcal{V}_p^+$:

$$\alpha \mid\sim l_0 \qquad\qquad (\mathrm{C}^+1)$$

$$\frac{l_{i-1} \prec \alpha}{\alpha \mid\sim l_j}, \text{ for all } j \leq i, \text{ where } i \in \mathcal{N} \setminus \{0\}, \qquad (\mathrm{C}^+2)$$

$$l_{i-1} \prec l_i, \text{ for all } i \in \mathcal{N} \setminus \{0\} \qquad (\mathrm{C}^+3)$$

$$\frac{l_{2^n} \prec \alpha}{\alpha \mid\sim \bot} \qquad\qquad (\mathrm{C}^+4)$$

Note that due to the transitivity of $\prec$ (Lemma 16 (iii)) and (C^+3) we immediately have $l_i \prec l_j$ for all $i, j \in \mathcal{N}$ for which $i < j$. (C^+1) ensures that each α has at least abnormality degree 0. By (C^+2), if α is more abnormal than $i-1$, then it has at least abnormality degree j for $j \leq i$. Should α be more abnormal then 2^n, then we consider it as maximally abnormal by (C^+4), i.e. $\alpha \mid\sim \bot$.

REMARK 26. Note that the equivalence of l_0 and $\top$ (with respect to $\mid\sim$), namely $l_0 \mid\sim \top$ and $\top \mid\sim l_0$, can easily be shown. As a consequence we have by (Equivalence) for all $\alpha \in \mathcal{V}_p^+$: $\top \mid\sim \alpha$ iff $l_0 \mid\sim \alpha$. Also the following two facts can easily be proven: $\top \prec \alpha$ iff $l_0 \prec \alpha$, and $l_0 \mid\sim l_0 \vee \alpha$.

In order to avoid unnecessary confusions, we need to be very precise from now on concerning the language that is used in the context of models and consequence relations. Where $\mathbf{L} \in \{\mathcal{L}, \mathcal{L}^+\}$, rational models are called rational $\mathbf{L}$-models if the corresponding worlds are $\mathbf{L}$-worlds, namely assignments of truth values to the propositional letters in $\mathbf{L}$. Similarly we call a consequence relation $\mid\sim$ a $\mathbf{L}$-consequence relation if it is defined over the propositional formulas in $\mathbf{L}$. Finally we call a knowledge base consisting of conditional assertions over propositional formulas in $\mathbf{L}$ a $\mathbf{L}$-conditional knowledge base.

DEFINITION 27 (Rational$^+$ models). A *rational$^+$ $\mathcal{L}^+$-model* is a rational $\mathcal{L}^+$-model $M = \langle W, \prec \rangle$ that meets the following requirement (R):

> For each world $w \in W$, $\mathsf{rank}_\prec(w) \leq 2^n$ and where $\mathsf{rank}_\prec(w) = i$,
> (i) $w \models l_j$ for all $j \in \mathcal{N}$ such that $j \leq i$ and (ii) $w \models \neg l_j$ for all $j \in \mathcal{N}$ such that $j > i$.

REMARK 28. Where $M = \langle W, \prec \rangle$ is a rational$^+$ model, note that by the definition of rational$^+$ models we immediately have: $\mathsf{Rank}_\prec(l_i) = i$ if $\sigma_M(l_i) \neq \emptyset$, else $\mathsf{Rank}_\prec(l_i) = \omega$. Therefore, by Lemma 16, for all $\alpha \in \mathcal{V}_p^+$, $l_i <_M \alpha$ iff $\mathsf{Rank}_\prec(\alpha) > i$, and $\alpha \mathrel{|\!\!\sim}_M l_i$ iff $\mathsf{Rank}_\prec(\alpha) \geq i$. In this way the degree of abnormality of a world is mirrored on the one hand by its canonical ranking, and can, on the other hand, now directly be expressed in the language of the logic.

Lemma 30 is going to show that rational $\mathcal{L}$-models and rational$^+$ $\mathcal{L}^+$-models define the same rational $\mathcal{L}$-consequence relations. It is an immediate consequence that for each rational $\mathcal{L}$-consequence relation there is a rational$^+$ $\mathcal{L}^+$-model defining it and vice versa. Therefore, if a rational closure exists for a given $\mathcal{L}$-knowledge base, then there are rational$^+$ $\mathcal{L}^+$-models that define it. It will be the task of the adaptive logic **ARC^s** to pick out these models. In order to establish these results the following mappings λ and μ will be helpful.

DEFINITION 29. We define a mapping λ from the rational $\mathcal{L}$-models into the rational$^+$ $\mathcal{L}^+$-models in the following way. For $M = \langle W, \prec \rangle$, $M \mapsto \langle \pi(W), \prec' \rangle$, where π maps $\mathcal{L}$-worlds into $\mathcal{L}^+$-worlds such that [i] for all $\alpha \in \mathcal{P}$, $w \models \alpha$ iff $\pi(w) \models \alpha$; and [ii] if $\mathsf{rank}_\prec(w) = i$ then $\pi(w) \models l_j$ for all $j \in \mathcal{N}$ for which $j \leq i$, and $\pi(w) \models \neg l_j$ for all $j \in \mathcal{N}$ for which $j > i$. Furthermore, define $\prec'$ in the following way, $w \prec w'$ iff $\pi(w) \prec' \pi(w')$. It is an easy exercise to show that $\lambda(M)$ satisfies requirement (R) and is therefore a rational$^+$ model.

Note that by the construction we immediately have for $\alpha, \beta \in \mathcal{V}_p$, $\alpha \mathrel{|\!\!\sim}_M \beta$ iff $\alpha \mathrel{|\!\!\sim}_{\lambda(M)} \beta$. Furthermore, $\mathsf{rank}_\prec(w) = \mathsf{rank}_{\prec'}(\pi(w))$ for all $w \in W$.

We define a mapping μ from the rational$^+$ $\mathcal{L}^+$-models into the rational $\mathcal{L}$-models in the following way. For $M = \langle W, \prec \rangle$, $M \mapsto \langle \eta(W), \prec' \rangle$, where η maps $\mathcal{L}^+$-worlds into $\mathcal{L}$-worlds such that for all $\alpha \in \mathcal{P}$, $\eta(w) \models \alpha$ iff $w \models \alpha$. Furthermore, $\eta(w) \prec' \eta(w')$ iff $\min_\prec(\eta^{-1}(\{w\})) \prec \min_\prec(\eta^{-1}(\{w'\}))$.

Note that by the construction we immediately have for all $\alpha, \beta \in \mathcal{V}_p$, $\alpha \mathrel{|\!\!\sim}_M \beta$ iff $\alpha \mathrel{|\!\!\sim}_{\mu(M)} \beta$. Furthermore, $\mathsf{rank}_\prec(w) \geq \mathsf{rank}_{\prec'}(\eta(w))$ for all $w \in W$.

LEMMA 30. *(i) For each rational$^+$ $\mathcal{L}^+$-model M there is a rational $\mathcal{L}$-model N for which $\mathrel{|\!\!\sim}_M \cap (\mathcal{V}_p \times \mathcal{V}_p) = \mathrel{|\!\!\sim}_N$. And vice versa, (ii) for each rational $\mathcal{L}$-model N there is a rational$^+$ $\mathcal{L}^+$-model M for which $\mathrel{|\!\!\sim}_M \cap (\mathcal{V}_p \times \mathcal{V}_p) = \mathrel{|\!\!\sim}_N$.*

Proof. For (i) take $N = \mu(M)$, for (ii) take $M = \lambda(N)$. ∎

COROLLARY 31. *(i) If $\hspace{0.5pt}\vdash\hspace{-9pt}\sim$ is a rational$^+$ $\mathcal{L}^+$-consequence relation, then $\hspace{0.5pt}\vdash\hspace{-9pt}\sim \cap \, (\mathcal{V}_p \times \mathcal{V}_p)$ is a rational $\mathcal{L}$-consequence relation. (ii) If $\hspace{0.5pt}\vdash\hspace{-9pt}\sim$ is a rational $\mathcal{L}$-consequence relation, then there is a rational$^+$ $\mathcal{L}^+$-model M s.t. $\hspace{0.5pt}\vdash\hspace{-9pt}\sim_M \cap \, (\mathcal{V}_p \times \mathcal{V}_p) = \hspace{0.5pt}\vdash\hspace{-9pt}\sim$*

We have an analogous representation theorem for rational$^+$ consequence relations and rational$^+$ models as we had in the case of rational consequence relations and rational models. First we show that every rational model defining a rational$^+$ consequence relation is a rational$^+$ model.

LEMMA 32. *For all rational $\mathcal{L}^+$-models $M = \langle W, \prec \rangle$ defining a rational$^+$ consequence relation $\hspace{0.5pt}\vdash\hspace{-9pt}\sim$, M satisfies requirement* (R).

Proof. Let $w \in W$. We prove the lemma by induction over $\mathcal{N}$. "$n = 0$": Let $\mathsf{rank}_\prec(w) = 0$. Then there is a $\alpha_0 \in \mathcal{V}_p^+$ such that $w \models \alpha_0$. Since $\mathsf{rank}_\prec(w) = 0$, $w \in \min_\prec(\sigma_M(\alpha_0))$. Since by (C$^+$1) $\alpha \hspace{0.5pt}\vdash\hspace{-9pt}\sim l_0$ for any α, we have $w \models l_0$. Since $\mathsf{rank}_\prec(w) = 0$, $w \in \min_\prec(\sigma_M(l_0))$. Furthermore, since $l_0 \prec l_i$ for $\hspace{0.5pt}\vdash\hspace{-9pt}\sim$, for all $i \in \mathcal{N} \setminus \{0\}$, $w \models \neg l_i$.

"$n \to n + 1$": Let $\mathsf{rank}_\prec(w) = n + 1$. There is a $\alpha_{n+1} \in \mathcal{V}_p^+$ such that $w \in \min_\prec(\sigma_M(\alpha_{n+1}))$. By induction hypothesis we have $l_i \prec_M \alpha_{n+1}$ for all $i < n + 1$. But then, due to (C$^+$2), $\alpha_{n+1} \hspace{0.5pt}\vdash\hspace{-9pt}\sim l_i$ for all $i \leq n + 1$. Therefore $\alpha_{n+1} \hspace{0.5pt}\vdash\hspace{-9pt}\sim_M l_{n+1}$ and hence $w \models l_i$ for all $i \leq n+1$. It follows also immediately by the induction hypothesis that $w \in \min_\prec(\sigma_M(l_{n+1}))$. Since $l_{n+1} \prec l_j$ for all $j \in \mathcal{N}$ for which $j > n + 1$, $w \models \neg l_j$.

That $\mathsf{rank}_\prec(w) \leq 2^n$ follows now immediately by (C$^+$4). ∎

THEOREM 33. *A binary relation $\hspace{0.5pt}\vdash\hspace{-9pt}\sim$ is a rational$^+$ $\mathcal{L}^+$-consequence relation iff it is the consequence relation defined by some rational$^+$ $\mathcal{L}^+$-model.*

Proof. "$\Rightarrow$" Note that there is a rational $\mathcal{L}^+$-model for $\hspace{0.5pt}\vdash\hspace{-9pt}\sim$. This model is with Lemma 32 rational$^+$. "$\Leftarrow$" Let on the other hand $M = \langle W, \prec \rangle$ be a rational$^+$ $\mathcal{L}^+$-model. Trivially, the corresponding $\hspace{0.5pt}\vdash\hspace{-9pt}\sim_M$ is rational. An easy inspection of requirement (R) shows that $\hspace{0.5pt}\vdash\hspace{-9pt}\sim_M$ satisfies the additional rules for rational$^+$ consequence relations. In case $\sigma_M(\alpha) = \emptyset$ (C$^+$1) is trivial. Else we get (C$^+$1) by (R)(i) since for each $w \in \min_\prec(\sigma_M(\alpha))$, $\mathsf{rank}_\prec(w) \geq 0$. For (C$^+$2) let $l_i \prec_M \alpha$. Again, the case $\sigma_M(\alpha) = \emptyset$ is trivial. Else we have by Remark 28 $\mathsf{rank}_\prec(w) \geq i + 1$ for all $w \in \min_\prec(\sigma_M(\alpha))$ and hence by (R)(i), $\alpha \hspace{0.5pt}\vdash\hspace{-9pt}\sim_M l_j$ where $j \leq i + 1$. (C$^+$3) is immediately clear by Remark 28. (C$^+$4) is follows immediately by (R). ∎

As done in the rational case we can introduce a logic $\mathbf{R}^+$ on bases of the conditions for rational$^+$ consequence relations.

DEFINITION 34 (Logic $\mathbf{R}^+$). We define logic $\mathbf{R}^+$ by the axioms for $\mathbf{R}$ and the following:

$$\alpha \mathrel{\vert\!\!\sim} l_0 \qquad\qquad (\text{R}^+1)$$

$$(l_{i-1} \prec \alpha) \to (\alpha \mathrel{\vert\!\!\sim} l_j) \text{ for all } j \le i, \text{ where } i \in \mathcal{N} \setminus \{0\} \qquad (\text{R}^+2)$$

$$l_{i-1} \prec l_i \text{ for all } i \in \mathcal{N} \setminus \{0\} \qquad\qquad (\text{R}^+3)$$

$$(l_{2^n} \prec \alpha) \to (\alpha \mathrel{\vert\!\!\sim} \bot) \qquad\qquad (\text{R}^+4)$$

A semantics for $\mathbf{R}^+$ is given by rational$^+$ models. We define $\models$ for rational$^+$ models analogously to the way we defined $\models$ for rational models.

THEOREM 35. $\mathbf{R}^+$ *is complete and sound with respect to rational$^+$ models.*

Proof. This can be easily shown on basis of the soundness and completeness of $\mathbf{R}$. First we need some definitions: Let $\mathsf{Atoms} =_{\text{df}} \{\bigwedge_{i=1}^n p_i' \mid p_i' \in \{p_i, \neg p_i\}\}$, $\mathsf{D} =_{\text{df}} \{\bigvee_{\alpha \in A} \alpha \mid A \in \wp(\mathsf{Atoms})\}$. We transform a given premise set Γ to Γ^+ in the following way: $\Gamma^+ = \Gamma \cup \{\alpha \mathrel{\vert\!\!\sim} l_0 \mid \alpha \in \mathsf{D}\} \cup \{(l_{i-1} \prec \alpha) \to (\alpha \mathrel{\vert\!\!\sim} l_j) \mid \alpha \in \mathsf{D}, i \in \mathcal{N} \setminus \{0\}, j \le i\} \cup \{l_{i-1} \prec l_i \mid i \in \mathcal{N} \setminus \{0\}\} \cup \{(l_{2^n} \prec \alpha) \to (\alpha \mathrel{\vert\!\!\sim} \bot) \mid \alpha \in \mathsf{D}\}$.

We have: $\Gamma \vdash_{\mathbf{R}^+} \varphi$ iff $\Gamma^+ \vdash_{\mathbf{R}} \varphi$ iff all rational models of Γ^+ validate φ. Using (LLE) and (RW) and considering the way Γ^+ is defined, it is very easy to show that the first equivalence holds. Furthermore, each rational model validating Γ^+ is a rational$^+$ model validating Γ and vice versa. Let M be a rational model of Γ^+. Again, the way Γ^+ was defined together with (LLE) and (RW) ensures that $\mathrel{\vert\!\!\sim}_M$ fulfills the properties for rational$^+$ consequence relations and therefore by Lemma 32, M is a rational$^+$ model. Let now M be a rational$^+$ model of Γ. Obviously M is rational and by the definition of Γ^+ M satisfies Γ^+.   ■

We now define the adaptive logic $\mathbf{ARC^s}$ for rational closure. The lower limit logic is $\mathbf{R}^+$. Abnormalities are conditional assertions of the kind $l_i \prec \alpha$. The intention served by minimizing abnormalities of this form is to interpret a propositional formula α as normal as possible. Note here that, if for a model M, $M \models \neg(l_i \prec \alpha)$, then for all minimal worlds w in $\sigma_M(\alpha)$, $\mathrm{rank}_{\prec}(w) \le i$. Minimizing abnormalities can therefore in semantical terms be seen as ranking worlds as low as possible, hence as interpreting formulas as normal as possible.

DEFINITION 36. $\mathbf{ARC^s}$ is the adaptive logic in standard format defined by the following triple:

- **LLL:** $\mathbf{R}^+$

- Abnormalities: $\Omega =_{\text{df}} \{l_i \prec \alpha \mid \alpha \in \mathcal{V}_p, i \in \mathcal{N}\}$

- Strategy: simple strategy

We write $\mathrm{Ab}(M)$ for the set $\{\varphi \in \Omega \mid M \models \varphi\}$ where M is a **LLL**-model for a given premise set.

THEOREM 37. *Let $M = \langle W_M, \prec_M \rangle$ be the rational $\mathcal{L}$-model of the rational closure of a $\mathcal{L}$-knowledge base **K** from Section 3.*

(i) *$\lambda(M) = \langle \pi(W_M), \prec_{\lambda(M)} \rangle$ is a minimal abnormal rational$^+$ $\mathcal{L}^+$-model of **K**;*

(ii) *for all minimal abnormal rational$^+$ $\mathcal{L}^+$-models N of **K**, $\mathrm{Ab}(N) = \mathrm{Ab}(\lambda(M))$;*

(iii) *for all minimal abnormal rational$^+$ $\mathcal{L}^+$ models $N = \langle W_N, \prec_N \rangle$ of **K** and for all $\alpha \in \mathcal{V}_p$, $\mathsf{Rank}_{\prec_N}(\alpha) = \mathsf{Rank}_{\prec_M}(\alpha) = \mathsf{Rank}_{\prec_{\lambda(M)}}(\alpha)$;*

(iv) *for all minimal abnormal rational$^+$ $\mathcal{L}^+$-models N of **K** and for all conditional assertions $\alpha \mathrel{|\!\sim} \beta$ (where $\alpha, \beta \in \mathcal{V}_p$), $\alpha \mathrel{|\!\sim}_N \beta$ iff $\alpha \mathrel{|\!\sim}_{\lambda(M)} \beta$.*

Proof. Ad (i) and (ii): Suppose there is a rational$^+$ model $N = \langle W_N, \prec_N \rangle$ of **K** such that there is a $l_i \triangleleft \alpha$ valid in $\lambda(M)$ but not in N (for $i \in \mathcal{N}, \alpha \in \mathcal{V}_p$). Hence by Remark 28

$$\mathsf{rank}_{\prec_N}(w) \leq i \tag{1}$$

where w is any world in $\min_{\prec_N}(\sigma_N(\alpha))$. We know that $\mu(N) = \langle \eta(W_N), \prec_{\mu(N)} \rangle$ is a rational $\mathcal{L}$-model of **K**. Furthermore, by construction of M, $\eta(w) \in W_M$ and therefore $\pi(\eta(w)) \in \sigma_{\lambda(M)}(\alpha)$. Furthermore, since $l_i \triangleleft_{\lambda(M)} \alpha$ and by Remark 28, $\mathsf{rank}_{\prec_{\lambda(M)}}(w') > i$ where w' is any world in $\min_{\prec_{\lambda(M)}}(\sigma_{\lambda(M)}(\alpha))$. By construction of M we have

$$\mathsf{rank}_{\prec_M}(\eta(w)) \leq \mathsf{rank}_{\prec_{\mu(N)}}(\eta(w)) \tag{2}$$

We also know that

$$\mathsf{rank}_{\prec_{\lambda(M)}}(\pi(\eta(w)) = \mathsf{rank}_{\prec_M}(\eta(w)) \text{ and} \tag{3}$$

$$\mathsf{rank}_{\prec_{\mu(N)}}(\eta(w)) \leq \mathsf{rank}_{\prec_N}(w) \tag{4}$$

But (1), (2), (3) and (4) imply that $\mathsf{rank}_{\prec_{\lambda(M)}}(\pi(\eta(w)) \leq i$,—a contradiction.

Ad (iii): Suppose for a $\alpha \in \mathcal{V}_p$, $i = \mathsf{Rank}_{\prec_{\lambda(M)}}(\alpha) < \mathsf{Rank}_{\prec_N}(\alpha)$. But then by Remark 28 $l_i \triangleleft \alpha \in \mathrm{Ab}(N) \setminus \mathrm{Ab}(\lambda(M))$,—this contradicts (ii). Suppose for a $\alpha \in \mathcal{V}_p$, $i = \mathsf{Rank}_{\prec_N}(\alpha) < \mathsf{Rank}_{\prec_{\lambda(M)}}(\alpha)$. But then $l_i \triangleleft \alpha \in \mathrm{Ab}(\lambda(M)) \setminus \mathrm{Ab}(N)$,—this contradicts (ii).

Ad (iv): This is an immediate consequence of (iii) and Lemma 16. ∎

COROLLARY 38. *Let **K** be a $\mathcal{L}$-conditional knowledge base. We have for all $\alpha, \beta \in \mathcal{V}_p$: $\alpha \mathrel{|\!\sim} \beta$ is in the rational closure of **K** iff $\mathbf{K} \vdash_{\mathbf{ARC^s}} \alpha \mathrel{|\!\sim} \beta$.*

Proof. The following is an immediate consequence of Theorem 37 and Lemma 23: $\alpha \mathrel{|\!\sim} \beta$ is in the rational closure of $\mathbf{K}$ iff $\alpha \mathrel{|\!\sim} \beta$ is valid in all simple models of $\mathbf{K}$. The rest follows by Theorem 25 since by Theorem 37 the conditions of Lemma 23 are fulfilled for the class of conditional knowledge bases. ∎

Some examples are needed in order to demonstrate what we have achieved.

EXAMPLE 39. We return to our example from page 54: $\mathbf{K} = \{\top \mathrel{|\!\sim} s, g \mathrel{|\!\sim} s\}$.

1	$g \mathrel{|\!\sim} s$	PREM	$\emptyset$
2	$\top \mathrel{|\!\sim} s$	PREM	$\emptyset$
3	$l_0 \mathrel{|\!\sim} s$	2; RU	$\emptyset$
4	$\neg(l_0 < g)$	RC	$\{l_0 < g\}$
5	$l_0 \vee g \mathrel{|\!\not\sim} \neg g$	4; RU	$\{l_0 < g\}$
6	$l_0 \mathrel{|\!\not\sim} \neg g$	5; RU	$\{l_0 < g\}$
7	$\top \mathrel{|\!\not\sim} \neg g$	6; RU	$\{l_0 < g\}$
8	$\neg(l_0 < \neg g)$	RC	$\{l_0 < \neg g\}$
9	$l_0 \mathrel{|\!\not\sim} g$	8; RU	$\{l_0 < \neg g\}$
10	$\top \mathrel{|\!\not\sim} g$	9; RU	$\{l_0 < \neg g\}$
11	$\big((l_0 \mathrel{|\!\sim} s) \wedge (l_0 \mathrel{|\!\not\sim} \neg g)\big) \to (l_0 \wedge g \mathrel{|\!\sim} s)$	RU	$\emptyset$
12	$\big((l_0 \mathrel{|\!\sim} s) \wedge (l_0 \mathrel{|\!\not\sim} \neg g)\big) \to g \mathrel{|\!\sim} s$	11; RU	$\emptyset$
13	$\big((l_0 \mathrel{|\!\sim} s) \wedge (l_0 \mathrel{|\!\not\sim} g)\big) \to \neg g \mathrel{|\!\sim} s$	RU	$\emptyset$
14	$g \mathrel{|\!\sim} s$	3, 6, 11; RU	$\{l_0 < g\}$
15	$\neg g \mathrel{|\!\sim} s$	3, 9, 13; RU	$\{l_0 < \neg g\}$

As there is no way to derive $l_0 < \neg g$ or $l_0 < g$ on the empty condition, $\top \mathrel{|\!\not\sim} \neg g$, $\top \mathrel{|\!\not\sim} g$, $\neg g \mathrel{|\!\sim} s$ and $g \mathrel{|\!\sim} s$ are finally derivable. The reader might not immediately see what specific rules were used. Some remarks: Note that line 11 is an application of rational monotonicity. At line 12 we arrive by (Equivalence) (cp. Remark 2)—note that $g \mathrel{|\!\sim} l_0 \wedge g$ and $l_0 \wedge g \mathrel{|\!\sim} g$ can easily be shown. Line 13 is derived in an analogous way. Furthermore, at lines 3, 6, 7, 9 and 10 we use facts presented in Remark 26. In the following part of the proof we try to derive $\neg(\top < \neg s)$, namely that it is not abnormal that the sun is not shining.

[19]15	$\neg(l_0 < \neg s)$	RC	$\{l_0 < \neg s\}$
[19]16	$\neg(\top < \neg s)$	15; RU	$\{l_0 < \neg s\}$
17	$\neg(\top < \neg s) \vee (l_0 < \neg s)$	16; RU	$\emptyset$
18	$\top \vee \neg s \mathrel{|\!\sim} s$	2; RU	$\emptyset$
19	$l_0 < \neg s$	17, 18; RU	$\emptyset$
20	$\top < \neg s$	19; RU	$\emptyset$

Line 18 is achieved by applying (LLE) to line 2. Obviously we have derived the condition of lines 15 and 16 at line 19. Hence these lines are marked.

EXAMPLE 40. We consider the so-called *Nixon diamond* example where

$\Gamma = \{r \mathrel{|\!\sim} \neg p, q \mathrel{|\!\sim} p\}.$[7] The formulas can be interpreted by

- Being a republican usually implies not being a pacifist. $(r \mathrel{|\!\sim} \neg p)$

- Being a Quaker usually implies being a pacifist. $(q \mathrel{|\!\sim} p)$

It would for example be desirable to derive $w \wedge q \mathrel{|\!\sim} p$ (where w can be read as "being a worker").

1	$r \mathrel{|\!\sim} \neg p$	PREM	$\emptyset$
2	$q \mathrel{|\!\sim} p$	PREM	$\emptyset$
3	$\neg(l_0 \prec w \wedge q)$	RC	$\{l_0 \prec w \wedge q\}$
4	$\neg(\top \prec w \wedge q)$	3; RU	$\{l_0 \prec w \wedge q\}$
5	$\top \mathrel{|\!\!\not\sim} \neg(w \wedge q)$	4; RU	$\{l_0 \prec w \wedge q\}$
6	$\top \mathrel{|\!\!\not\sim} \neg w \vee \neg q$	5; RU	$\{l_0 \prec w \wedge q\}$
7	$(\top \mathrel{|\!\!\not\sim} \neg w \vee \neg q) \rightarrow (q \mathrel{|\!\!\not\sim} \neg w)$	RU	$\emptyset$
8	$q \mathrel{|\!\!\not\sim} \neg w$	6, 7; RU	$\{l_0 \prec w \wedge q\}$
9	$((q \mathrel{|\!\sim} p) \wedge (q \mathrel{|\!\!\not\sim} \neg w)) \rightarrow (q \wedge w \mathrel{|\!\sim} p)$	RU	$\emptyset$
10	$q \wedge w \mathrel{|\!\sim} p$	2, 8, 9; RU	$\{l_0 \prec w \wedge q\}$

At line 5 (LLE) is applied to line 4, at line 7 we arrive by (S), line 9 is an instance of rational monotonicity. It is easy to see that there is no way to derive $l_0 \prec w \wedge q$ on the empty condition,—$q \wedge w \mathrel{|\!\sim} p$ is therefore finally derivable.

6   Including negative knowledge

Knowledge bases were so far restricted to contain only the positive knowledge an agent may have. However, an agent might also have statements of the following kind in his knowledge base: if α then not normally β, i.e. $\alpha \mathrel{|\!\!\not\sim} \beta$. Adding statements of this kind to a knowledge base requires special attention since such knowledge bases might not be consistent anymore. Note here that $\mathrel{|\!\sim} = \mathcal{V}_p \times \mathcal{V}_p$ defines a rational consequence relation which extends any given (positive) knowledge base $\mathbf{K}$. However, take for instance $\mathbf{K} = \{a \mathrel{|\!\sim} b, a \mathrel{|\!\!\not\sim} \neg c, a \wedge c \mathrel{|\!\!\not\sim} b\}$. Applying rational monotonicity to $a \mathrel{|\!\sim} b$ and $a \mathrel{|\!\!\not\sim} \neg c$ leads to a contradiction.

Let a *general knowledge* base be a conditional knowledge base which may contain negated conditional assertions. As shown in [4], for a consistent general conditional knowledge base $\mathbf{K}$, $\mathrel{|\!\sim}_M$ is the rational closure of $\mathbf{K}$, where M is the model constructed in Section 3.

Since for a general conditional knowledge base $\mathbf{K}$, $\lambda(M)$ clearly is a rational$^+$ model of $\mathbf{K}$, we get by Theorem 37 the following result:

COROLLARY 41. *Let $\mathbf{K}$ be a consistent general $\mathcal{L}$-conditional knowledge base. We have for all $\alpha, \beta \in \mathcal{V}_p$: $\alpha \mathrel{|\!\sim} \beta$ is in the rational closure of $\mathbf{K}$ iff $\mathbf{K} \vdash_{\mathbf{ARC^s}} \alpha \mathrel{|\!\sim} \beta$.*

---

[7]This has been presented e.g. in [12].

7 Conclusion

The preferential closure of a conditional knowledge base **K** can be obtained by intersecting all supersets of **K** satisfying the conditions for preferential consequence relations. The proof theory of preferential closure is therefore simply given by a logic defined by these conditions, interpreted as rules. Adding Rational Monotonicity as a further condition, on the other hand, defines a family of consequence relations which is not closed under intersection. Although Lehmann and Magidor are able to characterize it semantically by a selection on ranked models, a proof theory for it was missing. This syntactical gap has been filled in this paper by the adaptive logic **ARC^s** for finite propositional languages for which the rational closure is guaranteed to exist.

It has been shown in [12] that the rational closure exists for a larger class of knowledge bases and infinite languages, namely the so-called admissible knowledge bases (e.g. knowledge bases for which a well-founded preferential model exists). If we want to proceed in a similar manner as in this paper we are in need of a set-theoretic construction for the rational closure of a knowledge base in this more general case. This is no trivial exercise: The one we used in Section 3 (cp. [4]) is restricted to finite languages. The authors in [12] presented another procedure. However, this construction requires another severe restriction: It only works on the basis of so-called well-founded preferential consequence relations.[8] Unfortunately, as has been shown in [12], even finite knowledge bases can give rise to a non well-founded preferential closure, and the existence of a well-founded preferential model M does not ensure that the consequence relation defined by M is well-founded.

It is a future challenge to develop a proof theory allowing for admissible knowledge bases in general and furthermore to take into account the predicative case (cp. [11]).

Acknowledgements

Research for this paper was supported by subventions from Ghent University and from the Fond for Scientific Research - Flanders. I am indebted to Diderik Batens, Joke Meheus and Dunja Šešelja for helpful comments on a former version of this paper.

BIBLIOGRAPHY

[1] O. Arieli and A. Avron. General patterns for nonmonotonic reasoning: from basic entailments to plausible relations. *Logic Journal of the IGPL*, 8(2):119–148, March 2000.

[2] Diderik Batens. A universal logic approach to adaptive logics. *Logica Universalis*, 1:221–242, 2007.

---

[8]The well-foundedness of a preferential consequence relation $\mid\sim$ should not be mistaken for well-foundedness of preferential models, e.g. a model defining $\mid\sim$.

[3] Diderik Batens, Christian Straßer, and Peter Verdée. On the transparency of defea-
     sible logics: Equivalent premise sets, equivalence of their extensions, and maximality
     of the lower limit. Forthcoming.
[4] R. Booth and J. B. Paris. A note on the rational closure of knowledge bases with
     both positive and negative knowledge. *Journal of Logic, Language and Information*,
     7(2):165–190, 1998.
[5] Keith L. Clark. Negation as failure. In H. Gallaire, J. Minker, and J. Nicolas, editors,
     Logic and Data Bases, pages 293–322. Plenum Press, 1977.
[6] Nir Friedman and Joseph Y. Halpern. Plausibility measures and default reasoning.
     Journal of the ACM, 48:1297–1304, 1996.
[7] D. Gabbay. Theoretical foundations for non-monotonic reasoning in expert systems.
     In *Logics and models of concurrent systems*, pages 439–457. Springer-Verlag New
     York, Inc., New York, NY, USA, 1985.
[8] Laura Giordano, Valentina Gliozzi, Nicola Olivetti, and Gian Pozzato. Analytic
     tableau calculi for KLM rational logic R. pages 190–202. 2006.
[9] M. Goldszmidt and J. Pearl. On the relation between rational closure and system
     Z. In *Third Int. Workshop Nonmonotonic Reasoning (South Lake Tahoe)*, pages
     130–140, 1990.
[10] Sarit Kraus, Daniel J. Lehmann, and Menachem Magidor. Nonmonotonic reasoning,
     preferential models and cumulative logics. *Artificial Intelligence*, 44:167–207, 1990.
[11] Daniel Lehmann and Menachem Magidor. Preferential logics: the predicate cal-
     culus case. In *TARK '90: Proceedings of the 3rd conference on Theoretical aspects
     of reasoning about knowledge*, pages 57–72, San Francisco, CA, USA, 1990. Morgan
     Kaufmann Publishers Inc.
[12] Daniel Lehmann and Menachem Magidor. What does a conditional knowledge base
     entail? *Artificial Intelligence*, 55(1):1–60, 1992.
[13] J. McCarthy. Circumscription — a form of non-monotonic reasoning. *Artificial
     Intelligence*, 13:27–29, 1980.
[14] Robert C. Moore. Possible-world semantics for autoepistemic logic. In *NMR*, pages
     344–354, 1984.
[15] Judea Pearl. System Z: a natural ordering of defaults with tractable applications
     to nonmonotonic reasoning. In *TARK '90: Proceedings of the 3rd conference on
     Theoretical aspects of reasoning about knowledge*, pages 121–135, San Francisco, CA,
     USA, 1990. Morgan Kaufmann Publishers Inc.
[16] Raymond Reiter. A logic for default reasoning. *Artificial Intelligence*, 1–2(13),
     1980.
[17] Y. Shoham. A semantical approach to nonmonotonic logics. In M. L. Ginsberg,
     editor, *Readings in Non-Monotonic Reasoning*, pages 227–249. Morgan Kaufmann,
     Los Altos, CA, 1987.
[18] Y. Shoham. *Reasoning about Change: Time and Causation from the Standpoint of
     Artificial Intelligence*. M.I.T. Press, Cambridge,Mass., 1988.

Christian Straßer
Centre for Logic and Philosophy of Science
Ghent University, Belgium
E-mail: Christian.Strasser@UGent.be

SECTION 2
LOGIC AND MEANING

Sentence, Proposition, Judgment, Statement, and Fact: Speaking about the Written English Used in Logic

JOHN CORCORAN

Dedication: *for Professor Newton da Costa, friend, collaborator, co-discoverer of the Truth-set Principle, and co-creator of the Classical Logic of Variable-binding Term Operators-on his eightieth birthday.*

ABSTRACT. The five ambiguous words—sentence, proposition, judgment, statement, and fact—each have meanings that are vague in the sense of admitting borderline cases. This paper discusses several senses of these and related words used in logic. It focuses on a constellation of recommended primary senses. A *judgment* is a *private* epistemic act that results in a new belief; a *statement* is a *public* pragmatic event involving an utterance. Each is executed by a unique person at a unique time and place. Propositions and sentences are timeless and placeless abstractions. A *proposition* is an intensional entity; it is a meaning composed of concepts. A *sentence* is a linguistic entity. A written sentence is a string of characters. A sentence can be used by a person to express meanings, but no sentence is intrinsically meaningful. Only propositions are properly said to be true or to be false—in virtue of *facts*, which are subsystems of the universe. The fact that two is even is timeless; the fact that Socrates was murdered is semi-eternal; the most general facts of physics—in virtue of which propositions of physics are true or false—are eternal. As suggested by the title, this paper is meant to be read aloud.

1 Introduction

The words—sentence, proposition, judgment, statement, and fact—are *ambiguous* in that logicians use each of them with multiple normal meanings. Several of their meanings are *vague* in the sense of admitting borderline cases. This paper juxtaposes, distinguishes, and analyzes several senses of these and related words, focusing on a constellation of recommended senses. According to the recommendation, a *judgment* is a private epistemic event that results in a new belief and a *statement* is a public pragmatic event, an act of writing or speaking. Both are made by a unique person at a unique time and place. Judgments are often not voluntary, although they cannot be coerced. Statements are usually voluntary, although they can be coerced. In contrast, propositions and sentences are timeless and placeless abstractions. A *proposition* is an intensional entity; in some cases it is a meaning of a sentence: it is a meaning composed of concepts, a complex sense composed of simpler senses. A [declarative] *sentence* is a linguistic entity. A written sentence is a string of characters; it is composed of character-strings—usually words or "symbols"[1] that can be used to express meanings, but which are not in themselves meaningful. A spoken sentence is composed of [articulate] sounds. Only a proposition is properly said to be true or to be false—although in certain contexts, or with suitable qualifications, judgments, statements, or even sentences, may be said to be true or false in appropriate derivative senses. Propositions are true or false in virtue of *facts*, which are either timeless or temporal subsystems of the universe.[2]

A fact is *timeless* if it is only about timeless entities such as numbers. The fact that two is even is timeless. The proposition that two is even can only be expressed using the *timeless-present* tense.[3] It is incoherent to say that two is presently even or that two is still even. A fact is *temporal* if it is about temporal entities such as material objects. Temporal facts are semi-eternal or eternal. A semi-eternal fact comes into being in an interval of time and it persists eternally thereafter. The fact that Socrates was murdered can never be erased. An eternal fact exists without having come into being. The most general facts of physics—in virtue of which propositions of physics are true or false-are eternal. In the sense intended here, a fact never changes. Facts are prior to propositions in the same sense that the past is prior to

---

[1]'Character', 'sign', and 'symbol' are often exact synonyms in logic usage. Many logicians show no awareness of the awkwardness of the fact that, in the senses they prefer, characters do not characterize, signs do not signify, and symbols do not symbolize—except under an interpretation, i.e., extrinsically, never intrinsically. See Corcoran, Frank, and Maloney *1974*.

[2]The proposition that Plato taught Aristotle, which is composed of concepts or senses, is true in virtue of a fact composed of historical entities that are not concepts. The fact in question is the fact that Plato taught Aristotle, which has Plato and Aristotle as constituents. The word 'fact' has been used in many other senses. For example, Frege (*1918/1956*, 307) takes facts to be true propositions, in a sense of 'proposition' very close to the one recommended in the current article. Later in the same article on page 311, he uses the word 'fact' in yet another sense: he says that thinking, judging, understanding, and the like are "facts of human life". Austin (*1961*, 91) disapproves of taking 'fact' as synonymous with 'true statement'.

[3]Frege made a similar point (*1918/1956*, 309–310).

historical truths and nature is prior to laws of nature.

People use sentences to express propositions. People express propositions they state when making statements; they also express propositions they do not state. For example, the proposition that zero exists is expressed both in stating that zero exists and in asking whether zero exists.[4] A proposition can be entertained in many ways. What we investigate, doubt, assume, postulate, believe, and know are often propositions. One and the same sentence is routinely used on different occasions to express different propositions. Likewise, one and the same proposition is commonly expressed by different sentences in different languages and, quite often, even in the same language.[5]

Differences between sentences and propositions are involved in describing differences between direct and indirect quotation. In some cases, when we quote directly we intend to give the exact sentence quoted without conveying the proposition expressed. Abe said: "She loves him". In some cases, when we quote indirectly we intend to give the proposition expressed without conveying the sentence used. Abe said that Betty loves Carl. We can quote a person directly and without awkwardness add "But I did not understand what was meant". Quoting indirectly presumes understanding the proposition indirectly quoted; quoting directly does not.[6]

Although both propositions and sentences are abstractions, some sentences can be seen: when we read an *inscription* or *token* of a sentence we see the sentence—not the proposition, if any, the sentence is used to express. Sometimes we can look at an inscription for some time before seeing the sentence inscribed. Most sentences have not been inscribed and are thus not seen: such sentences will not be seen until inscribed and they will not be seen when all inscriptions are obliterated.

Ontology is important in this paper as shown in the following considerations. Socrates taught Plato. My judgment that Socrates taught Plato was made many years ago. My statement of the proposition that Socrates taught Plato was made seconds ago using the sentence 'Socrates taught Plato', which involves three English words but which contains no meanings and no persons. The proposition that Socrates taught Plato involves three meanings but it contains no words and no persons. The fact that Socrates taught Plato involves two persons but it contains no meanings and no words. In the senses recommended, facts, propositions, judgments, sentences, and

---

[4]To the best of my knowledge, this type of observation is due to Frege (*1918/1956*, 294).

[5]The last few points have been standard for years. The result of adding 'in fact' and a comma to the front of a sentence expressing a given proposition is a different sentence expressing the same proposition. See Cohen-Nagel *1934/1962/1993*, xxii–xxv, 27–35.

[6]Indirect quotation is a subspecies of indirect discourse (Audi *1999*, 424).

statements comprise mutually exclusive ontological categories. However, questions of "ontological status" play no role in this paper: it is irrelevant whether sentences, propositions, or facts—or for that matter, characters, numbers, or truth-values, are "real entities"—are "idealizations", or are "theoretical constructs".[7] As is suggested by the title, this paper is written to be read aloud.

2   Sentences, Judgments, Statements

> A sentence is made up of words; a statement is made in words.—Austin *1961*, 88.
> Statements are made; ... sentences are used.—Austin *1961*, 88.

Since context is important in speech, it might help to begin with examples. Along the 3000-mile northern border of the United States with Canada, the weather is changeable, to say the least. The weather is a frequent topic of conversation. There is keen interest not only in the weather itself, but also in the nature of communication about weather, in weather discourse.

Perhaps interest in weather discourse is increased by the fact that Americans report weather in one system of units while Canadians report it in another. Americans report temperature in Fahrenheit degrees, where thirty-two is freezing. Canadians report temperature in Celsius degrees, where zero is freezing but thirty-two is very hot. People on one side of the border converse with friends and relatives on the other. It is not unusual for someone to get a weather report in one system and convey its content to others in another system. This situation gives rise to sentences that may seem strange in other contexts.

Here are some examples.[8]

---

[7] I have discussed such questions elsewhere (e.g. *2006s*).

[8] As indicated in the abstract, this paper is primarily meant to be read aloud and then later to be discussed as a written text. As usual, the reader must "impersonate" the author. In order to make the text look "normal", instead of displaying a sentence as such, e.g. 'ten is fifty' (no uppercase first letter, no period), I display the so-called *assertoric* form of it which begins with an uppercase tee and ends with a period, 'Ten is fifty.'(sic) in this case. Where logic matters, this becomes important. The sentence 'ten is fifty' is literally a part of its own negation 'it is not the case that ten is fifty', but the assertoric form of the sentence is never a part of the negation of the sentence, or of the assertoric form of the negation of the sentence. Moreover, the sentence is part of its quotes-name and the assertoric form of the sentence is part of the quotes-name of the assertoric form of the sentence. But neither is part of the quotes-name of the other. Tarski (e.g. *1969/1993*, 103) is one of the few writers who usually keep the sentence apart from its assertoric form. And he never explains what he is doing or why. See Corcoran *2006s* for more details.

Zero is thirty-two.

Ten is fifty.

Twenty is more than sixty-five.

Twenty is between sixty-five and seventy.

Zero is to one hundred as thirty-two is to two-twelve.

Both hundreds are hot but only one is boiling.

Even without conversions between Celsius and Fahrenheit, weather discourse sometimes involves sentences that are otherwise unusual.

Zero is freezing.

Ten is cold.

Twenty is comfortable.

Thirty is hot.

Forty is sweltering.

Freezing is zero.

Freezing is thirty-two.

Thirty-two is freezing.

Thirty-two is hot.

One-hundred is boiling.

We can swim in eighteen.

In doing conversions from Celsius to Fahrenheit, it helps to know that ten Celsius degrees is eighteen Fahrenheit degrees. Zero, ten, twenty, and thirty degrees Celsius are respectively thirty-two, fifty, sixty-eight, and eighty-six degrees Fahrenheit. This is calculated by adding eighteen Fahrenheit degrees whenever ten Celsius degrees have been added.

Notice that ten degrees Celsius is a definite single point on the Celsius scale whereas ten Celsius degrees is not a point but an interval-size that has no location on the scale. It would be *incoherent*, a so-called *category mistake*, to say that today's temperature is ten Celsius degrees or that the difference between yesterday's high and low was ten degrees Celsius.[9] A person saying something incoherent is sometimes making a category mistake.

I have been using sentences made up of English words to make statements based on judgments. In the words of Austin quoted above: "A sentence is made up of words; a statement is made in words". In several places Frege

---

[9]Of course, if someone writes the expression 'the current temperature is ten Celsius degrees' we should not jump to the conclusion that there was a category mistake and not just an inadvertent writing error, a transposition of words. We could ask whether they intended to write 'The current temperature is ten degrees Celsius'. An affirmative answer would suggest a writing mistake and not a category mistake. Notice that the quotes name of the assertoric form of a sentence does not have a period at the end unless it occurs at the end of the sentence using it.

said that the answer to a question is a statement based on a judgment (*1918,
1919, 1952*, 117; *1997*, 299).

The statements, or assertions[10], that I have just made aloud in this spoken delivery were based on judgments that I made earlier, silently of course.
Some judgments were made as I was planning the paper; some were made
long before that. Some judgments involved little beyond confirmation of
memory; some involved perception; some calculation; some deduction; some
induction. Every objective judgment is made in reference to the fact it concerns (Corcoran *2006e*). The statements were all public and were all made
in public today. The judgments were all private and were all made in private
before today. No two persons can make the same judgment.

In judging I form a fresh belief, often a completely new belief in the truth
of a proposition not previously believed by me—perhaps even previously
disbelieved by me. In stating per se new beliefs are not formed: some
statements are not based on belief at all; some are contrary to their speaker's
judgments—a fact repeatedly overlooked by Frege.

In the senses recommended, 'judgment' and 'statement' are common
nouns whose extensions are classes of events: epistemic events in the case
of 'judgment'; pragmatic events in the case of 'statement' (cf. Austin *1961*,
87). But the ending 'ment' brings other senses. The two words can be used
as proper names in various senses. For example, 'judgment' can be used to
name the human faculty by which judgments are made or with qualification
it can be used to name a person's faculty as 'Abe's judgment improved in
time'. Both can be used as common nouns for the objects or results of
the eponymous acts in various senses: Abe's judgment (statement) [sc. the
proposition Abe judged (stated)] was the same as Ben's.

3   People Use Words to Mention Things.

> When saying something of an object, one
> always uses a name of this object and not the
> object itself, even when dealing with linguistic
> objects.—Alfred Tarski *1969/1993*, 104.

> Premises rhymes with nemesis, not with
> nemeses.—Frango Nabrasa (per. comm.)

---

[10]I use the word 'assertion' reluctantly as a synonym for 'statement' in the recommended sense. However, the fact that it has drastically different senses even in logic calls
for caution. Some logicians use it to referred theorems, statements having the highest
level of warranted assertibility. But it is widely used outside of logic for statements with
the lowest level of warrant. At the end of a famous address before the United Nations,
Colin Powell asserted that his previous statements were not assertions. Other uses occur,
e. g. Quine *1970/1986* and Goldfarb *2003*. See Quine's insightful quip (ibid. p. 2).

After this section of the paper, I may seem to be at a slight disadvantage. Up to this point I have been using some spoken sentences to mention other spoken sentences. The sentence I am now reading is a spoken sentence.

After this paragraph I will of course continue to *use* only spoken sentences, but I will be *mentioning* mainly written sentences. From here on, most of the sentences that will be mentioned are written sentences.[11] The sentence I am now reading is a written sentence.

I will not be able to use what I will be mentioning. Isn't that the usual case? No one uses the freezing point to mention the freezing point.

However, we routinely use expressions that we are mentioning. As I noted above, I used spoken sentences to mention spoken sentences. Before the invention of writing there was no other way to mention spoken sentences.

As we saw above, it is common to encounter an ambiguous sentence, one that could normally be used to express two or more propositions. When I read an ambiguous sentence to an audience, one listener might have a *thought* of one proposition at the precise time that another listener has a thought of another proposition. When I said 'The sentence I am now reading is a spoken sentence', some readers thought that the proposition thereby stated was false and some thought that the proposition stated was true. The expression 'the sentence I am now reading' is ambiguous. Some readers might have thought that I was talking about a sentence written on the paper I am reading from; others might have thought that I was talking about the sentence I was enunciating.

Using one ambiguous expression twice in consecutive paragraphs intending different meaning each time is a practice to be avoided, other things being equal.

Of course, although no two persons can have the same *thought*, thankfully nothing prevents their separate thoughts from being thoughts of the same proposition. Otherwise, all agreements and disagreements would be merely "verbal" and communication would be of a very restricted nature, if indeed it could go on at all.

Occasionally, two of the propositions expressed by one ambiguous sentence are of such a nature that in expressing one we use a word that we mention when expressing the other. For example, consider the three-word sentence 'Santiago is Spanish'. In one sense, this expresses a proposition to the effect that the city of Santiago de Compostela is in Spain. The eight-

---

[11]Thus, every sentence is a string of characters, counting the space or blank as a character. But, of course, not every string of characters is a sentence. The terminology varies, even within the writings of one author. Tarski says that a sentence is a series of printed signs (*1956/1983*, 156). Church writes of a sentence as being a succession of letters (*1956*, 27). Boolos writes of sequences of symbols and of strings of symbols on the same page (*2002*, 103).

letter word 'Santiago' is used (to mention the city), but the word itself is
not mentioned. In another sense, the same three-word sentence expresses a
proposition to the effect that the one-word expression 'Santiago' is part of
the Spanish language; it translates as 'Saint James'. Here the word 'Santiago' is mentioned (by using the same word, 'Santiago' itself). In a third
sense, it expresses the proposition that Saint James is Spanish, which is
false, of course. Here again the word 'Santiago' is used (but to mention
the saint). Before leaving this important topic let us see what Church says
(*1956i*, 61).

> Following the convenient and natural phraseology of Quine, we may distinguish
> between *use* and *mention* of a word or symbol. In 'Man is a rational animal', the
> word 'man' is used but not mentioned. In 'The English translation of the French
> word *homme* has three letters', the word 'man' is mentioned but not used. In
> 'Man is a monosyllable', the word 'man' is both mentioned and used, though in an
> anomalous manner, namely autonymously.[12]

Notice that the last spoken word quoted is ambiguous; the sense intended
by Church was coined by Rudolf Carnap in 1934 (*1934/1937*, 17). The
third syllable of its written counterpart is spelled en-wye-em as is the case
in the rhyming word 'synonymously'. It should not be confused with the
phonetically similar and older 'autonomously' whose third syllable is spelled
en-oh-em (Chateaubriand *2005*, 150). The two contextually homophonic
third syllables *nym* and *nom* are not etymologically related - any more
than the number-word 'two' is related to the preposition 'to' with which it
is a homophone.[13]

---

[12]Church (*1956*, 61). The italicization is Church's, but Church uses double quotation
marks where single quotation marks are used above in keeping with the style of this
article. In the same book, actually on the next page (*1956*, 62), Church uses single
quotation marks in mentioning words and he even explicitly notices the advantage of
using single quotations in this way.

[13]As far as I know, no meaning has yet been assigned to 'synonomy', 'synonomous', or
'synonomously'.

4   Written English is not a phonetic transcription of Spoken English.

> Now spoken sounds are symbols of mental ideas, and written marks symbols of spoken sounds. Just as written marks are not the same for all humans, neither are spoken sounds. But what the latter are symbols of - mental ideas - are the same for all; and what these ideas are likenesses of - actual things - are also the same—Aristotle, *On Interpretation*.[14]

The adjectives 'spoken' and 'written' are ambiguous: each has a broad and a narrow sense—like the word animal. In the broad sense, every human is an animal; in the narrow sense, no human is an animal. There are many spoken sentences (broad sense) that may never have been spoken or uttered by anyone and which therefore are not spoken sentences (narrow sense). Likewise, many written sentences (broad sense) have never been written or inscribed by anyone and which therefore are not written sentences (narrow sense).

For example, the two-word sentence 'zero sleeps' made up of the four-letter word 'zero', or zee-ee-ar-oh, and the six-letter word 'sleeps', or es-el-ee-ee-pee-es, may never have been written or inscribed by anyone before, but even before it was written it was a written sentence.[15] Of course, any proper name can be written or inscribed one space in front of an inscription of an intransitive verb in order to make an inscription of a written sentence. You can picture the following examples of written sentences.

Arithmetic boils.

Geometry freezes.

---

[14]This translation of *On Interpretation*, I, 16a3–8 is essentially the one used by N. Kretzmann (*1974*, 3–4) in his informative and accessible discussion of this passage, which he calls "the most influential text in the history of semantics". Kretzmann credits the translation largely to Ackrill *1963*.

[15]This is the first spoken sentence today illustrating two of the ways of making spoken proper names of words: the appositional method is to say the two-word expression 'the word' just before pronouncing the word itself, the phonetic-orthographic method is to "spell-out" the word, i.e. to pronounce the names of the word's letters in the order in which they occur in the word. Thus, the word 'zero' is one and the same thing as zee-ee-ar-oh. These two spoken alternatives correspond somewhat to the two written ones in Tarski's truth-definition paper: the quotation-mark name and the structural-descriptive name (*1956*, 156).

Have inscriptions of these sentences ever existed before? Who could know except someone who saw such an inscription?

A written sentence is not necessarily a sentence that has been written but rather a sentence of the kind that *could* be written. In other words, a written sentence is a string of characters.[16] In our case, the characters are all either letters of the Latin alphabet or punctuation marks. Is the space that occurs between consecutive words or between consecutive sentences a character? Or is it a non-character? Or is it a borderline case?

It is certainly significant; it is used meaningfully even though it is not used to express a constituent of a proposition, a concept. The word 'character' is in many ways typical of the words we will be focusing on: it is ambiguous and some of its meanings are vague. In the sense recommended in this article, the space is a character. Ambiguity (or polysemy) and vagueness (or indefiniteness) are two of the most pervasive of linguistic phenomena. My advice is to recognize them and to become accustomed to them.[17] A writer might use an ambiguous expression without being equivocal. A writer might use a vague concept without being guilty of imprecision. Aristotle tells us not to strive to be more precise than the subject-matter permits (*Nicomachean Ethics*, I, 3). Describing a sunrise requires vague concepts.

Written sentences are things I will be mentioning but not using in this spoken presentation. The sentences that I will be using are strings of sounds, in one sense of this ambiguous word. Using the perspicacious terminology coined by Charles Sanders Peirce, a written sentence is one that can only be *embodied* visually and a spoken sentence is one that can only be *embodied* audibly. The linguists who know about such matters conclude that spoken English and written English are such disparate systems of communication

---

[16]The problem of finding an exact definition of "Written English sentence" is far from settled; it is one of the famous unsolved problems in contemporary linguistics. Noam Chomsky brought this and related problems to the forefront of linguistic research. Likewise, even the problem of finding an exact definition of "Written English word" is far from settled (Lyons *1977*, Vol. I, Ch.1). The word 'word' is of course ambiguous but, in the "typographical" sense used in this article, a word is nothing more than a string of characters. Thus, there are no such things as homonyms, or more precisely, homo*graphs*. No *two* words have the same spelling, or the same succession of characters. A word is its spelling, so to speak; "two" words spelled the same are one. And no *two* sentences have the same wording, or the same succession of words and spaces.

[17]The "logically perfect" or "formalized" languages studied in mathematical logic exhibit neither ambiguity nor vagueness—this is an advantage for some purposes, and a disadvantage for other purposes. (Cf. Church *1956*, 2, 3, 32, 50.) There are of course needless ambiguities, sometimes called misnomers, which are exasperating because they entered the language by way of a mistake involving what had been a misuse of a "less ambiguous" word. The stock example is the word 'Indian' whose misuse became so widespread that it had to be recognized as a use. There are many other examples, several in philosophy and logic. Church cites 'sentence' as a possible example (*1956p*, 6).

that the two should not be regarded as different forms of the same language but rather as two different languages. My language, [American] English, is actually two languages.[18]

In keeping with the usual convention of capitalizing proper names of languages, I will be using Spoken English (written as a proper name) to discuss Written English (also with capitals); henceforth, unless explicitly indicated otherwise, the single word 'sentence' is elliptical for the two-word expression 'Written sentence'—in the broad sense of a writable sentence, not one that has necessarily been written.[19]

There is no such thing as an English sentence that has two "forms"—a "Spoken form" and a "Written form". Every English sentence is a sentence of Spoken English or of a Written English, and not both, of course.

Using the terminology of modern logic, we can say that in the discussion to follow, Spoken English is the *metalanguage* and Written English is the *object-language*. Actually I should say '*an* object-language' or 'one of the object-languages', not '*the* object-language', because I will be mentioning Spoken English sentences also, as well as sentences of a few foreign written languages. In our case, Spoken sentences of *the* metalanguage will be used by a speaker, namely me, to mention mainly Written sentences of one object-language, Spoken sentences of another object-language, and written sentences of other object-languages. More importantly, I will be using sentences to make statements about object-language sentences to the members of the audience, namely you, who are responsible for critical evaluation. The members of the audience routinely make judgments[20] about the speaker's

---

[18]It is obvious that Written English is not a phonetic representation of Spoken English. There are no Spoken sight rhymes and no Written sound rhymes. Moreover, Written English can represent logical groupings with parentheses and other typographical devices where Spoken English is without equally adequate devices. On the other hand, there are propositions expressible in a small number of words of Spoken English that can not be expressed in Written English at all. In these cases, every Written sentence expressing a proposition implying one intended by such a Spoken sentence also implies propositions not implied by the intended one. The gulf between the two languages becomes even more evident if we try to find a pair of sentences one Written and one Spoken which are uniformly translations of each other in the sense that they express exactly the same range of messages. To the question of whether there is a peculiarly English system of messages that can only be exactly conveyed in one or both of the two media Spoken English and Written English, I would answer yes, tentatively and with qualifications.

[19]Accordingly, the two two-word expressions 'Spoken sentence' and 'Written sentence' are used as ellipses respectively for the two three-word expressions 'Spoken English sentence' and 'Written English sentence'.

[20]Gottlob Frege, the meticulous logician regarded as the founder of Analytic Philosophy, uses the word 'judgment' in a sense close to that used here. He says (*1980*, 20): "According to my way of speaking, we think by grasping a thought, we judge by recognizing a thought as true, and we assert by making a judgment known [to others]". Also see Frege *1918/1956*, 294. However, not everyone is this careful. For example, David

statements and about the subject matter of the speaker's statements. My next statement is a good example to reflect on.

> It is obvious that the Written English sentence 'zero is cold' can be used to express a true proposition if and only if zero—in some sense—is cold—in some sense.

By using this Spoken sentence, I mention but do not use the Written sentence 'zero is cold'; and I use but do not mention the Spoken sentence *zero is cold*.

There is no way to confuse a statement with a sentence or with a proposition, since it is only the statement that inherently has a speaker, who is responsible for its accuracy, and it is only the statement that per se has an audience, which is responsible for its critical evaluation. Every statement is an event. Like every other event, every statement has effects, ramifications which are also events, not propositions; and like any other action, it changes the world in which it is made. No proposition is an event. Every proposition has consequences, implications which are propositions, not events; the proposition per se is causally inert. No statement is a proposition. No proposition is a statement. Here I am using the standard, traditional terminology that Church so adequately describes in his classic 1956 article "Propositions and Sentences", perhaps the clearest thing ever written on the topic, certainly the best thing I know of (*1956p*, 3–11). To be explicit, the word 'statement' is used in the *transactional* sense, the word 'proposition' in the *intensional* sense, and the word 'sentence' in the *syntactic* sense.

5 Sentences Express Propositions and People Make Judgments.

> When a sentence expressing a proposition is asserted we shall say that the proposition itself is thereby asserted.—Church, *1956i*, 27.

> It seems to me that [by imagination vs. judgment] you have in view the difference between grasping a thought [sc. proposition] and recognizing a thought as true. The latter is what I call judging.—Frege to Russell in 1904. *1980*, 163.

Any one sentence—made of words and ultimately of letters and characters—can be used to express two or more different propositions or, more generally,

---

Hilbert uses the word 'judgment' to mean "proposition", in a sense close to that of this paper (Luce *1950*, 165–6, Hilbert-Ackermann *1928/38/50*, 171).

different *messages*—made up of meanings or concepts. The number of words in a sentence is not usually the same as the number of concepts in a message it expresses; it could be more but it is often less. Aristotle had already said that although sentences are not the same for all humans, what they express is the same for all (*On Interpretation*, Ch. 1, 16a3-18).

The three-word sentence 'thirty-two is freezing' is typical. It can be used to express the true proposition that thirty-two degrees Fahrenheit is the freezing point of water. Thirty-two *is* freezing. I just made a statement that thirty-two degrees Fahrenheit is the freezing point of water by using the three-word sentence. The very same sentence can also be used to express the false proposition that thirty-two degrees Celsius is the freezing point of water. I never made a statement to that effect. It can be used to express a proposition about one of the athletes known as thirty-two.

And it can be used to express an incoherent *message* to the effect that the number thirty-two is in the process of becoming frozen, or perhaps that the number thirty-two feels cold. I am using the word 'message' for the genus of what is expressed by a complete sentence. Thus a message is either a proposition, whether true or false, or an incoherency, which is neither. An incoherent message cannot be coherently said to be a mistake any more than a false proposition can coherently be said to be a lie. Some incoherencies can be poetic.[21]

Uncomfortable Numbers: A Poem.

Thirty-two is freezing.
Thirty-two is hot.
Thirty-two is sweating.
Thirty-two is not.

The above paragraph including the poem illustrates several facts. First, sentences are ambiguous, or polysemous: rarely if ever is a sentence unambiguous, or monosemous. Second, propositions are about something and it

---

[21]There are several fallacies that arise in discussions of incoherencies, i.e. incoherent messages. Perhaps the most common is to think that there are sentences that somehow by nature can only be used to express incoherencies. The fact is that people can make conventions instituting new coherent uses for any given sentence. Power corrupts. Glory addicts. Fame fades. Logic confuses. Logic clarifies. Another common fallacy is to think that a sentence expressing an incoherency is meaningless. The expression 'meaningless sentence expressing' itself understood in the intended way is an oxymoron: every sentence expressing something means that which it expresses. Another common fallacy is to confuse contradictory messages with incoherent messages. The message that some odd number is not odd is coherent but contradictory and thus false. The message that some odd number trisects a right angle is incoherent, thus not false, and thus not contradictory.

is in reference to their subject-matter that they are true, or false, as the case may be.[22] Third, not every meaningful use of a sentence involves using it to express a proposition; a sentence can be used to express an incoherency, a message that is neither true nor false.[23] Fourth, not every pair of sentences that could be used to express a contradiction are so used in a given discourse—there are "verbal contradictions" that are not contradictory. We have already seen that it is not necessarily self-contradictory or even false to say that not all written sentences are written sentences, nor is it necessarily tautological or even true to say that all written sentences are written sentences. Logicians and grammarians might legislate that no one expression may be used in two or more senses in one text. Even though their advice is generally sound, their jurisdiction hardly ever extends beyond the classroom.

The established notational convention already applied above is to use single quotes for making names of sentences and other expressions, for example, names of characters, words, phrases, and sentences. These are the so-called single-quotes-names.[24] Thus 'one plus two is three' is a five-word English sentence and 'square' is a six-letter English word, both recognized by me, neither of which would have been recognized by Aristotle. Following Bertrand Russell, Rudolf Carnap, Morris Cohen, Ernest Nagel, John Lyons and others, double quotes are used in naming propositions, incoherent messages, and other meanings.[25] These are the so-called double-quotes-names. Thus, "one plus two is three" is a true proposition known both to me and to Aristotle and "square" is a concept also well known to both.

---

[22]According to this traditional conception of proposition it does not seem coherent to say that a proposition is true *in*, *at*, *on* or *of* a possible world, a time, a circumstance, a situation or anything else; a proposition is either true or false. This is of course not to say that the word could not be used in some other sense according to which it is coherent to say, e. g. that some propositions true in this world are false in some other world, although in this case it would be hard to imagine how such an apparently inaccessible fact could be known. Frege, Moore, Church, and Austin are silent on these senses of 'proposition'.

[23]The question arises whether incoherent messages have the same objective existence that belongs to propositions or whether they are entirely subjective, merely mental constructs. This is a difficult issue which may turn out to be as much a matter of convention as of fact.

[24]Church (*1956p*, 61) says that Frege used single quotes to indicate that the quoted material was being used autonymously; this is *not* using single-quotes-names. In diametrical opposition, both Tarski and Quine regard a sentence using an occurrence of a quotes name as not using the quoted expression at all. Consider the next sentence in the body of the paper; it mentions 'square'. Frege regarded it as using the 6-letter word 'square'; Tarski and Quine regarded it as using the 8-character quotes-name of the 6-letter word but not using the 6-letter word at all.

[25]*Bertrand Russell 1903*, 53ff, *1905/1967*, 99; *Carnap 1934/1937*, 14; *Cohen-Nagel 1934/1962/1993*, xxii; *Lyons 1977*, Vol. I, x.

In simple cases, expressions [are used to] express meanings or senses and they [are used to] name, or mention, entities or things. Thus the number-word 'three' expresses the individual concept "three" and it names the number three, the number named 'three'. In some cases the situation is more complicated: the sentence 'one plus two is three' expresses the proposition "one plus two is three"; but it is incoherent, even ungrammatical, to say that it names one plus two is three. The expression 'three' expresses the concept "three". This is a pattern often safe to follow. The expression 'three' names the entity three. This pattern is less safe; it must be used with care.[26]

The expression 'no number' denotes no number, but 'some number' does not denote some number. In fact, no number is named 'some number', or for that matter 'no number'. Of course, in a heteronymous sense, the expression 'no number' does not denote at all, a fortiori it does not denote something named 'no number'. There is nothing named 'no number' except the expression itself, which is no number. In the autonymous sense of 'no number', no number denotes no number, which is an expression not a number. Likewise in the autonymous sense of 'no number', no number denotes nothing, thus not an expression, a number, or anything else.[27]

There are several other terminological systems that are in use; some may be more appropriate in certain contexts. All must be used with care. Some writers say that an expression expresses, connotes, or signifies an intension, a connotation, or a meaning, or a sense. And, in contrast, they say that an expression names, refers to, or denotes an extension, a referent, or a denotation. Some writers take "the connotation" of an expression to be its emotional overtones and explain that it may vary from person to person even when there is agreement on what is expressed and on what is named. The expression 'seven' taken by two people to express "seven" and name seven may connote good luck for one person and bad luck for another.

There is little chance to confuse a judgment with a sentence or with a proposition. Each judgment is an event; it is an act done at a particular time by a particular person. The person judging that a certain proposition is true usually grasps the proposition before judging that it is true. I remem-

---

[26]Church (*1956p*, 25f.) follows Frege in holding that, in a logically perfect or formalized language totally devoid of ambiguity, all sentences expressing true propositions name an entity *truth* and all expressing false propositions name an entity *falsehood*. This "discovery" by Frege was unprecedented in the entire history of logic and linguistics and, as far as I can tell, it has yet to be widely accepted. My guess is that in the fullness of time it will be relegated to a footnote in the history of ideas.

[27]Contrary to the impression given by Russell in his 1905 paper "On Denoting", most of the expressions he calls denoting phrases do not denote in any sense of 'denote' I know of—certainly not in the sense of Church *1956p* used in this paper—except when used autonymously.

ber clearly that before concluding that thirty degrees Celsius is eighty-six Fahrenheit, I considered the proposition and wondered whether it was true. Before doing the mental process of verifying by calculation the premise that thirty-two plus three times eighteen is eighty-six, I considered the conclusion as a hypothesis, consciously suspending judgment. You might have done the same. As mentioned above, neither the proposition nor the sentence used to express it has a time of occurrence.[28] These facts are independent of the fact that the word 'judgment' exhibits what is known as process/product ambiguity—besides being used for the action or process of concluding that a proposition is true it is also used for the result or product, for the proposition concluded, the conclusion of a judging. The same ambiguity belongs to the word 'conclusion'.

There is little chance to confuse a judgment or conclusion with a statement or assertion although they are both actions. Both require a proposition but the judgment does not require a sentence. Perhaps you made a judgment or arrived at a conclusion and then later decided which sentence or language to use to express it. Both require an agent but no judgment can have an audience.

When the words are not used in the senses recommended there is much room for confusion. The same process/product ambiguity belongs to the word 'statement'—besides being used for the action or process of stating that a proposition is true it is also used for the result or product, for the proposition stated, or even the sentence used. When the words are both used in the product sense, we have the sorry spectacle of saying correctly that the judgment is a proposition, the statement is a proposition, and the judgment is a statement. Even though there is nothing objectively wrong about using ambiguous words, or even using them in multiple senses in one and the same paragraph, nevertheless many able writers have confused their readers by doing so.

In the above product senses of 'judgment' and 'statement' there is nothing incoherent about saying that a judgment or statement is true or is false. There are other such senses as well: the sentence 'The statement you made is true' can be taken as elliptical for 'The [proposition asserted in the] statement you made is true'.

---

[28] Absurdly, the word 'judgment' (French 'jugement') has been used in logic since at least the early 1600s for the fictitious act of simultaneously constructing a proposition, determining that it is true, and asserting it (*Arnauld and Nicole 1662*, Part II, Ch. 2 and esp. Ch. 3,: *Whately 1826*, 57, 75–77). Whately can not conceive of a non-judged proposition; he thinks that the copula 'is' or 'is not' "indicates the act of judgment" (*1826*, 57). These writers overlook the fact that every day we consider propositions not known to be true and not known to be false. The non-judged hypothesis that the defendant is the murderer is often grasped by the jury even before any evidence can be presented.

6   Sentences, Not Propositions, Are Ambiguous or Unambiguous, Extrinsically; Propositions, Not Sentences, Are True or Untrue, Intrinsically.

> A proposition is composed not of words, nor yet of thoughts, but of concepts.—G. E. Moore *1899*, 179.

The words 'sentence' and 'proposition' are both ambiguous. Besides the syntactic, typographical, or morphological meaning explained above, the word 'sentence' can also be used as elliptical for 'meaningful sentence' in the hybrid, or composite, sense of a sentence expressing a certain one of the messages it normally expresses. This would be a sentence meant or taken in a certain way; in some cases, one might say a *propositional sentence*. Likewise, besides the intensional sense explained above, the word 'proposition' can also be used in a hybrid sense to indicate a message or a proposition expressed in words in a certain way. The latter would be a proposition expressed a certain way, a *sentential proposition*. There is little more than emphasis to separate the hybrid sense of 'sentence' from the hybrid sense of 'proposition'. This is brought out by the fact that 'sentence' in its hybrid sense was used by Mahoney to translate the German word 'Satz' as used by Frege (*1884/1964*), while 'proposition' in its hybrid sense was used by Austin to translate the same passages (*1884/1964*). Frege himself recognized an ambiguity as did Hans Kaal, the translator of the Frege correspondence, who agrees with Austin. Frege's letter to Russell of 20 October 1902 (*1980*, 149) contains the following passage:

> Your example ... prompts me to ask the question: What is a proposition? German logicians understand by it the expression of a thought [sc. proposition], a group of audible or visible signs expressing a thought [sc. proposition].[29] But you evidently mean the thought [sc. proposition] itself. This is how mathematicians tend to use the word. I prefer to follow the logicians in their usage.

In fact, Tarski often used 'sentence' in the sense "meaningful sentence", i.e., for propositional sentence. Church discussed the use of 'proposition' in the sense "sentential proposition". In this sense, a proposition is "a composite entity, sentence plus proposition" (Church *1956p*, 6). But he

---

[29]I am reluctantly following the established practice of reading 'proposition' where a strict and literal translation of Frege would require the grossly inappropriate word 'thought'. However, it is clear to me that Frege's "thoughts" are not my propositions. Two sentences expressing the same thought need not express the same proposition. Consider the following scenario. Abe called yesterday. Also yesterday, I used the sentence 'Abe called today' to report the fact that Abe called yesterday. The proposition expressed using 'today' is not the same as the one expressed using 'yesterday' but the thoughts are the same (*1918*, 296).

used the word only in the abstract intensional sense of this paper, i.e. for a combination of meanings that is either true or false. Boolos (*2002*, 112, 114) used the word 'sentence' as a technical term only in the purely syntactic, typographical sense stipulated here in which a sentence per se is a "syntactic entity", a string of characters without regard to whether it has been used to express a message or to which meaning has been or will be assigned to it by writers.[30] He also used it casually in the hybrid sense (*2002*, 103).

In order to eliminate one source of confusion, slashes (virgules) can be used to refer to composite entities. The composite word /one/ is a two-part system composed of the string 'one' and the sense "one". The number one is denoted by the word /one/ and determined by the sense "one". The composite /one plus two is three/ is "a composite entity, sentence plus (abstract) proposition" (Church *1956p*, 6), that expresses or *contains* the proposition that one plus two is three. As Chateaubriand (*2001*, 379) points out, composites such as /one/ give rise to what has been called a *semiotic triangle*: the three vertices are respectively a string, a sense, and a referent and the three sides are respectively the relations of string-to-sense, string-to-referent, and sense-to-referent. He says that a composite expression "contains" its sense.

I used the word /Santiago/ autonymously above in this article. As already implied, the word /autonymous/ coined by Rudolf Carnap in 1934 should not be confused with the phonetically similar and older /autonomous/ which is not etymologically related (Chateaubriand *2005*, 150).The slash notation cannot be used unless it is clear in the context which of the various interpretations is intended as in:

The number one is denoted by the composite word /one/.

A string per se is meaningless: nothing is denoted by the string 'one' except under an interpretation by a person. In many ways it would be graceful to use something more familiar such as italics to mention composites.[31]

The composite word *one* composed of 'one' and "one" denotes the number one.

In both of the two senses mentioned the word 'proposition' is a technical term of logic and has been for centuries. Church wrote (*1956p*, 3):

---

[30]The logician Warren Goldfarb used the word 'statement' for something very close to an idealized sentence-proposition (Goldfarb *2003*, 5ff.). Quine (*1970/1986*, 13–14) used the expression 'eternal sentence' for this. See the comments below about Quine's most mature terminology.

[31]The disadvantages include the ambiguity resulting from the fact that italic font has other uses and the fact that italic can not be italicized: //one// is /*one*/ or */one/* and is composed of '/one/' and "/one/".

In Latin, *propositio* was originally a translation of the Greek *protasis*, and seems to have been used at first in the sense of premiss.

In either technical sense it has no meaningful connection with a proposing or a proposal—just as the sense of the word 'work' as a technical term of physics has no meaningful connection with employment or labor. Some logicians avoid using the word 'proposition' as a technical term but nevertheless say many things that I would say using that word. Some simply substitute a different word, and not always the same different word. Tarski sometimes used 'thought' (*1956/1983*, 160), sometimes 'judgment' (*1969/1993*, 106), and sometimes 'fact' (ibid. 104). Sometimes Tarski simply avoided the concept "proposition" by studiously using the word 'sentence' in his favored sense of "propositional sentence" (*1986*, 143–154, *1996*, 119–130).

According to Tarski: the sentence 'zero is even' is true if and only if zero is even. But that would be incoherent in the terminology of this paper. According to this paper: the proposition "zero is even" is true if and only if zero is even. Likewise, the sentence 'zero is even' is used to express a true proposition if and only if it is used in a sense in which zero is even.

In the purely intensional sense of 'proposition', it is incoherent to speak of the meaning or of the wording of a proposition—but in the hybrid sense both make perfect sense. Likewise for 'sentence', in the pure syntactic sense, it is incoherent to speak of the meaning or of the wording of a sentence—whereas in the hybrid sense both make perfect sense. Accordingly, in the hybrid senses it is incoherent to speak of ambiguous propositions or of ambiguous sentences.

As indicated above, in the purely syntactical sense advocated in this work, a sentence is ambiguous if there are two or more meanings that it normally used to express. And there is no such thing as a *pair* of sentences composed of exactly the same characters in exactly the same order: a sentence *is* a series of characters. However, in the hybrid sense, often but not always used by Tarski, a sentence is ambiguous if there is *another* sentence composed of exactly the same characters in exactly the same order but having a different meaning. And there is no such thing as one sentence having two meanings: a sentence *is* a meaningful unity. A hybrid sentence is a compound unity uniting two parts: it is a pure sentence plus one of its meanings. Many modern logicians seem to use the word 'sentence' usually in the pure sense as is suggested in the common locution that a sentence is only true or false under an interpretation. Many philosophers follow Tarski in usually using the word in the composite sense as suggested in the common locution that not every true sentence is known to be true. But, there are other important senses to be recognized.[32]

---

[32] Frege (*1918*, 292–4, passim) uses it in a sense very close to "spoken statement". He

The notions of sentence and proposition are both useful for discussing various phenomena including translation and ellipsis. In order for a sentence used to express a certain proposition to be elliptical it is sufficient for the sentence to omit a word corresponding to a constituent of the proposition. The following is elliptical under two of its normal interpretations: Russell read Frege more than Peano. Under one interpretation, it expresses the proposition that Russell read Frege more than Peano *did* [sc. read Frege]. Under another interpretation, it expresses the proposition that Russell read Frege more than *he did* [sc. Russell read] Peano.

7 Tokens, Not Occurrences, Embody Types. Occurrences, Not Tokens, Are in Types.

> A string-type has instances which are string-tokens or string-inscriptions composed of instances (not occurrences) of characters; the string-tokens are ultimately composed of character-tokens.—John Corcoran et al. *1974*, "String Theory".

Let us try a little thought experiment. Imagine that we have one copy of a Greek edition of Euclid's *Elements* on the left end of the table open to Proposition 47 of Book I, now known as the Pythagorean Theorem. In a line spreading to the right, imagine one copy each of ten perfect translations into ten different languages, say Arabic, Chinese, English, French, German, Portuguese, Spanish and three other languages that you may choose. Of course we imagine them all open to Proposition 47 of Book I. The English translation by Thomas Heath has the following: *In right-angled triangles the square on the side subtending the right angle is equal to the squares on the sides containing the right angle.*

We have one proposition which is expressed by the Greek sentence written by Euclid. And we have eleven sentences each in a different language, all of which express one and the same proposition, namely the Pythagorean Theorem. The Pythagorean Theorem is a proposition in the preferred intensional sense used above. It is made up of concepts not words, not letters, not characters. This one proposition, called 'the Pythagorean Theorem', is that in virtue of which the ten translations of Euclid's Greek sentence are all perfect translations not only of Euclid's Greek sentence but also of each other. All eleven sentences express one and the same thing—and that thing is the proposition. In the hybrid sense, there are eleven propositions. A

---

says that an indicative sentence is a series of sounds (ibid. 292) that contains not only a proposition but also the assertion of the proposition (ibid. 294).

person who does not recognize propositions in the intensional sense should not say '*the* Pythagorean Theorem' but rather '*a* Pythagorean Theorem' or '*one of the* Pythagorean Theorems (plural)'even when it is clear from the context that English is the language being used. After all there are many different ways of expressing a given proposition in a given language.

If persons who knew none of the eleven languages were to carefully examine all eleven sentences expressing the Pythagorean Theorem, they would find nothing to indicate a common meaning. Not even one character is found in all eleven. However, were someone to look at two copies of one and the same translation, say the English, turned to the same page, it would be easy to find an exact duplicate of the sentence token located in my copy of the Heath translation embodying a sentence expressing the Pythagorean Theorem.

In one sense of the word 'sentence' we have two English sentences, one in each of the two copies of the book—two concrete, visible things made of scattered bits of dried ink on paper. In another sense there is only one sentence, but it has been printed twice—one abstract ideal thing capable of being printed but not itself visible. The one thing printed twice (and capable of being printed indefinitely often) is what Peirce called a *type*, specifically a *sentence-type*. The two printings he called *tokens*, specifically *sentence-tokens*. The two sentence-tokens, in Peirce's terminology, are *embodiments* of the sentence-type. The two sentence-tokens came into existence at a certain definite time—maybe we could find out from the publisher the exact day. Sometimes the year of the printing is in the front of the book. But, when did the sentence-type, the abstract thing that there is only one of, come into existence?

Again, the two sentence-tokens can be destroyed by a fire in one library, or maybe by two fires in two libraries. But how could a sentence-type be destroyed? If someone destroyed all of the books, would that destroy the sentence-type? These questions are raised only to clarify the differences between types and tokens. Before exploring the type-token ambiguity[33] further, let us read Peirce's exact words. The following is from his 1906 *Monist* article (*1906*, 504–5) quoted in Ogden-Richards *1923*, 280–281.

---

[33]Many nouns and noun phrases are subject to type-token ambiguity in that they can be used to refer to types or to tokens. See the previous paragraph. The type-token distinction is used to explain type-token ambiguity. Incidentally, a language that makes the type-token distinction need not have expressions that have the type-token ambiguity, and conversely. A sufficiently rich logically perfect metalanguage could have symbols for types and ways of mentioning tokens of those types but, of course, have no ambiguous expressions. A sufficiently primitive language could suffer from type-token ambiguity while lacking resources to discuss the distinction.

> A common mode of estimating the amount of matter in a ... printed book is to count the number of words. There will ordinarily be about twenty 'thes' on a page, and, of course, they count as twenty words. In another sense of the word 'word', however, there is but one word 'the' in the English language; and it is impossible that this word should lie visibly on a page, or be heard in any voice ... . Such a ... Form, I propose to term a *Type*. A Single ... Object ... such as this or that word on a single line of a single page of a single copy of a book, I will venture to call a *Token*. ... In order that a Type may be used, it has to be embodied in a Token which shall be a sign of the Type, and thereby of the object the Type signifies.

In my copy of Euclid's *Elements*, the concrete visible token of the sentence-type expressing the Pythagorean Theorem contains six concrete tokens of the word-type 'the', tee-aitch-ee. The sentence-token is spread out over three lines of print with two *thes* on the first line, three on the second, and one on the third. This means that the abstract sentence-type contains six *thes*, to use Peirce's clever word. Does this mean that a sentence-type is made up of sentence-tokens? Does this mean that something abstract, a sentence-type, is made up of concrete embodiments, sentence-tokens? We need another distinction here to deal with those questions and to make the point that, although the type 'the' has only one *occurrence* of the type 'e', the type 'e' *occurs* twice in the type 'these' and the type 'e' is instantiated, betokened, or embodied (to use Peirce's term) twice in every token of the type 'these', tee-aitch-ee-ess-ee. One and the same word-type 'the' occurs six times in the one sentence type expressing the Pythagorean Theorem. Every sentence-type is made up of character-types. Typically a sentence-type has multiple *occurrences* of at least one of it character types. Every token of a sentence type has the same number of tokens of a given word-type as the sentence-type itself has occurrences of the given word-type.

Some authors explicitly make the three-part type-token-occurrence distinction without introducing special terminology for the occurrence relation. Lyons (*1977*, Vol. 1, 13–18) has a section called *Type and token* which discusses the three-way distinction while using the same expressions for token and occurrence. However, the present terminology is familiar to logicians, as pointed out in my "Meanings of word: type-occurrence-token", *Bulletin of Symbolic Logic* 11(2005) 117.

At some point in the history of logic the type-token dichotomy gave way to the type-token-occurrence trichotomy. Given Peirce's penchant for trichotomies and his logical creativity, one is led to speculate, even hope, that it was Peirce who made this discovery.[34]

Another difference between sentence-tokens and sentence-types is that every sentence-token has a length in inches or centimeters, whereas the sentence-type, being abstract, has no length—not zero length, *no* length—

---

[34]Substantially the same points have been made recently (Corcoran *2005*).

in inches or centimeters. Of course every expression-type has a "length" in character-occurrences. Even though the Nile is longer than the Niagara, in inches, 'the Nile' is "shorter" than 'the Niagara', in character-occurrences.

Moreover the sentence-token may be squeezed onto one line of a book or spread out over several lines, but these formattings do not apply to sentence-types. Probably the most important difference is that, as Peirce implies in the above passage, sentence-tokens are visible whereas sentence-types are invisible—a point rarely made. Church (*1956p*, 8) makes a closely related point without explicitly noting that they are invisible. From an ontological and epistemological point of view the difference between scientific treatment of string tokens and scientific treatment of string-types is dramatic. String tokens are studied in physics; string types are studied in mathematics—in the field known as string theory (Corcoran et al. *1974*, "String Theory", *J. Sym. Logic.* 39: 625–37).

Let us consider an application. Consider, e.g., the word 'letter'. In one sense there are exactly twenty-six *letters* (letter-types or ideal letters) in the English alphabet and there are exactly four letters in the word-type 'letter'. In another sense, there are exactly six *letters* (letter-repetitions or letter-occurrences) in the word-type 'letter'. In yet another sense, every new inscription (act of writing or printing) of 'letter' brings into existence six new *letters* (letter-tokens or ink-letters) and one new word that had not previously existed. The number of letter-occurrences (occurrences of a letter-type) in a given word-type is the same as the number of letter-tokens (tokens of a letter-type) in a single token of the given word. Many logicians fail to distinguish "token" from "occurrence" and a few actually confuse the two concepts. It is almost a rule that any article or book that explicitly mentions two of the three concepts without mentioning the third confuses the third with one of the other two.

There is a kind of cold war of words in the literature between two camps of logician philosophers, neither of which "officially" recognizes the type-token ambiguity of the word 'sentence'. Tarski may be taken as representing one camp, Quine the other. Whenever it is relevant for Tarski to clarify his use of the word 'sentence', he makes it a point to say that sentences are "inscriptions". He never mentions Peirce's work in this area. He almost never uses the word 'type' in Peirce's sense and he never uses the word 'token'. In the single place (*1930/1983*, 31) where he mentions "the type of a sentence", he says that it is "the set of all sentences [inscriptions]" in the same shape as the sentence [token][35]. This is not Peirce's view at all.

---

[35] Although Tarski repeatedly says that shape is the determining factor, it is clear that he has not explored the details very far. Don't the left and right parentheses have the same shape? How about the plus and times signs, the less-than and greater than signs,

Tarski seems to imply repeatedly and flatly that it is "not strictly correct", an "error" or even a "widespread error", to use to the word 'sentence' in the sense of "sentence-type" (*1969/1993*, 114, *1930/1983*, 30, 31, *1933/1983*, 156).

Quine, however, in his last comprehensive statement of his philosophy of logic explicitly mentioned Peirce's work; and he used both 'type' and 'token' in Peirce's senses. But he never mentioned the type-token ambiguity of the word 'sentence': for Quine a sentence is a sentence-type-exactly the opposite of Tarski's usage. Let us look at Quine's own words[36] (*1970/1986*, 13–14): "In Peirce's terminology, utterances and inscriptions are *tokens* of the sentence or other linguistic expression concerned; and this [sentence or other] linguistic expression is the *type* of those utterances and inscriptions. In Frege's terminology, truth and falsity are the two *truth-values*. Succinctly, then, an eternal sentence is a sentence whose tokens all have the same truth-value [i.e. are all true or all false]."

In the recommended senses, both sentences and propositions are timeless. In contrast, each inscription or embodiment of a sentence comes into existence at a unique time and, in the fullness of time, will become illegible. Likewise, each thought of a proposition comes into existence as it is being thought of and goes out of existence as soon as the person thinking it turns to the next thought. Euclid and each of his translators thought of the Pythagorean Theorem, one and the same proposition which is timeless. But Euclid's thought of the Pythagorean Theorem was long defunct before his translators' thoughts of it came into being.

As I wrote this paper I reread an English translation of the Pythagorean Theorem and, despite its awkwardness and distracting ambiguity, I managed to have a thought of the Theorem. My thought was private; your belief that it ever existed is based on "mere hearsay". Some people are inclined to suppose that a particular thought of the Pythagorean Theorem is to the abstract proposition as a concrete inscription of a sentence expressing it is to the abstract sentence. There might be a sense in which this is true. But even so, it would be a stretch to conclude that a thought of a proposition is an embodiment of the proposition. A thought of a proposition seems to be an action, a kind of performance, more akin to an act of inscribing than to a static inscription.

---

the vee and the caret, the zee and the en? Besides, there are arbitrarily "long" sentence-types but there is evidently a limit to the total number of character-tokens ever made in the history of writing. Of course, every argument against any given attempt to reduce types to tokens can be answered by an ever more elaborate "epicyclical construction".

[36]The italicizations are Quine's and the bracketed interpolations are mine.

8   Sentential Functions: Roles Sentences Play

A proposition grasped by a person for the first time can be put in a form of words which will be understood by someone to whom the proposition is completely new. This would be impossible were we not able to distinguish parts in the proposition corresponding to the parts in the sentence, so that the structure of the sentence serves as a model for the structure of the proposition.—Frege *1979*, 1923.

The topic of sentential functions—the functions, roles, or uses of sentences—is one of the most interesting subjects in logic. A sentence can be mentioned. A sentence can be used to make a statement. A sentence can be used to make a promise, to make a request, to give an order, to refuse an order, to make an apology, to make a prediction, to ask a question, to insult, to inform, to deceive, to retract a previous statement. A sentence can be used ironically to make a statement that contradicts what would be "expected". Plato tells us that Cephalus, then quite old, "welcomed" the decline in his appetite. Continuing, Plato adds that Socrates "admired" Cephalus for saying so (*Republic* I, 329). A sentence can be used poetically or fictionally where the speaker is not responsible for the accuracy of what is expressed, but only for its poetic or literary qualities. The "pedagogical license" enjoyed by the teacher is warranted by the fact that the teacher's role demands combining the role of witness responsible for accuracy with the role of poet or storyteller responsible for something else (Corcoran *1999*).

In fact, of all the topics in logic suitable for discussion with an audience of logicians and non-logicians, this topic, despite its attractiveness, even its fascination, is hardly ever discussed. Consequently, the literature of logic and of non-logic abounds in confusions and fallacies that would be corrected by a small amount of concentrated attention to this topic.

The three-word sentence 'Ten is fifty', as peculiar as it may be, is in many ways typical. Of course, ten is not fifty. The number ten is only one-fifth of fifty. But, ten degrees Celsius equals fifty degrees Fahrenheit. The sentence 'Ten is fifty' can be used to express a true proposition concerning temperature conversions and to express a false proposition about numbers. It is exactly the same with corresponding sentences in other languages. We could just as easily consider the Portuguese sentence 'Dez é cinqüenta', the Spanish 'Diez es cincuenta', the German 'Zehn ist fünfzig', or the French 'Dix est cinquante'.

The English sentence mentioned above can also be used to express many other propositions, some true and some false; and it can be used to express incoherent *messages*. Thus when someone has used the sentence to make an utterance in writing, we cannot, without further investigation, conclude that a true proposition or a false proposition was stated.

Examples of other propositions that can be expressed with this sentence can be discovered by imagining different contexts in which it might be written. The sentence 'ten is fifty' might be used to answer a question about the price of handkerchiefs; it might mean that the price of ten handkerchiefs is fifty cents, or fifty dollars, or fifty pounds, or fifty euros. In such cases the grammarian would explain that the three-word sentence is elliptical for an eight-word sentence by writing, e.g.: '[the price of] ten [handkerchiefs] is fifty [cents]'—using the brackets to indicate "restored ellipsis". It is only in a logically perfect language that grammatical structure of sentences models logical structure of the respective propositions expressed.

In order to determine the message, if any, that a given inscription of a sentence was intended to convey, it is often necessary to consider the context, including the discourse preceding the inscription in time and sometimes the past history of communication among the participants. Some interpret Frege as having puzzled his readers by saying that a word did not have meaning in isolation but only as part of a sentence. He asserted that the word gets its meaning from the sentence, and he denied that the sentence gets its meaning from the meanings of the words in it.[37] For years this was regarded as an extreme viewpoint.[38] Now we see to the contrary that Frege did not go far enough. Using the same exaggerated and elliptical style that Frege used, we can say that no sentence has meaning in isolation but only in the context of a communication—among several people, between two people, or by a single person, perhaps making a memorandum. We can affirm that determining what was meant by a sentence in a given inscription requires grasping the context and we can deny that it is sufficient to consider

---

[37] In his *Grundlagen der Arithmetik* Frege wrote: "... we must always consider a complete sentence. Only in [the context of] the latter do words really have a meaning. ... It is enough if the sentence as a whole has a sense; by means of this its parts also receive their content." (Frege *1884/1964*, §60, tr. M. Mahoney). Later writers have followed him on this issue, not always giving due acknowledgement. Over a quarter century later Wittgenstein wrote: "Only sentences make sense; only in the context of a sentence does a name signify anything." (Wittgenstein *1921/1998*, §3.3 tr. D. Kolak).

[38] In view of the strangeness and implausibility of this view, not to mention the fact that neither Frege nor Wittgenstein gives a scintilla of evidence for it, we should not be surprised to find it being ignored by practicing linguists. For example, John Lyons does not mention it in his comprehensive, two-volume treatise *Semantics*. In fact, like many other semanticists, he takes a kind of diametrical opposite view to be obvious. He writes: "[It] ... is obvious enough ... that the meaning of a ... sentence is a product of the meanings of the words of which it is composed." (Lyons *1977*, Vol. I, 4). Also see the above quote from Frege *1923*.

only the sentence *per se*.[39]

Frege tells us "never to ask for the meaning of a word in isolation" (*1884/1959*, Intro., X). But he thinks that it is coherent to inquire as to the meaning of a sentence-type. If this paper has been successful, we have learned that we should never ask for the meaning or meanings of any expression-type, we should never ask "what does it mean to say such-and-such?" Rather, we should ask "what did so-and-so mean when writing such-and-such in thus-and-so text?"

9   Propositions Imply; Statements Implicate: Propositions Have Implications; Statements Have Implicatures

> The fact that a person defines an expression in a certain sense, in and of itself, is no evidence that the person uses the expression in that sense.—Frango Nabrasa, *2001*.

The proposition "Socrates taught Plato, who taught Aristotle" *implies* the proposition "Plato, who was taught by Socrates, taught Aristotle", which in turn implies the proposition "Plato taught Aristotle". But it does not imply the proposition "Socrates taught Aristotle". The relation in question is not transitive–even if it were, there would still be no implication.[40] Nor does it imply "Corcoran believes that Plato taught Aristotle".

Let us compare the above *proposition* with the *statement* I will now make.

Socrates taught Plato, who taught Aristotle.

---

[39]Similar points have been made before by Quine, who says that the sentence per se is not what carries the truth or falsity but rather each token, more accurately each "event of utterance" (*1970/1986*, 13). Quine does not reveal it, but his view was anticipated by his teacher C. I. Lewis over thirty years earlier. Lewis wrote: "it is ... the word as read, the visual impression received, which arouses in the mind of the reader a corresponding meaning, and it is only when a mark ... gives rise to meaning that it is operating as a symbol" (*1932*, 311). Michael Scanlan observed that Lewis never follows up on or develops this interesting idea (personal communication). Other philosophers have made points similar to the one just attributed to Frege, Lewis, and Quine. But in many cases their works are so labored or so cryptic that it is hard to be confident that this point has even been grasped, let alone articulated. It is important to note that this thesis does not exclude the proposition that different tokens of the same sentence have shared significance.

[40]I am using 'implies' in the information-containment sense as elaborated in Corcoran *1998*; it is the second of twelve senses listed in Corcoran *1973*. In many cases, the information in a proposition involving a relational concept that determines a transitive relation does not contain the information that the relation is transitive. "One precedes two" does not imply "Every number preceding a given number precedes every number the given number precedes", which is logically equivalent to transitivity of number-theoretic precedence. Thinking otherwise is fallacious, perhaps the fallacy of premise-smuggling.

The last statement I made *implicates* "Corcoran believes that Socrates taught Plato, who taught Aristotle". Moreover, it implicates "Corcoran believes that Plato taught Aristotle". Every statement implicates every proposition implied by a proposition it implicates. It also implicates the proposition that I, Corcoran, understand the sentence / Socrates taught Plato, who taught Aristotle/. It also implicates that I believe that you understand the sentence.

As a rule, *propositions* about other people do not imply informative propositions about me. As a rule, statements implicate that the speaker believes the proposition stated, that the speaker understands the composite sentence uttered, and that the speaker believes that the audience understands it also.

The *implications* of a given proposition are the propositions whose information content is included in that of the given proposition.[41] Clearly, the information that Corcoran believes something is not contained in the information that Socrates taught Plato, who taught Aristotle. The *implicatures* of a given statement are the propositions it implicates; they include many of the propositions to which the speaker is morally committed by making the statement. Every proposition implies its *implications* and every statement implicates its *implicatures*. Propositions do not have implicatures.

According to the above conventions "imply" and "implicate" require non-human subjects: the former a proposition; the latter a statement. However, it is natural to used both elliptically so that human subjects are available. A person can be said to *imply* the propositions implied by a proposition the person stated; a person can be said to *implicate* a propositions if it is implicated by a statement the person made.[42]

It took years to figure out satisfactory ways to say what a proposition's implications are. We should not forget that both 'implication' and 'implicature' are ambiguous and that some of their meanings are vague. Maybe someone knows a satisfactory explanation of what a statement's implicatures are (Borchert *1996*, 225); I do not. Roughly, perhaps, we can say that a statement also implicates what the audience can take the statement as evidence for—assuming that the speaker is sincere, sane, responsible, etc.

---

[41]This traditional view is elaborated in Corcoran *1998*.

[42]For these important conventions, which correspond to observed usage, I am indebted to the late Kenneth Barber (Barber and Corcoran *2009*). However, this terminology is not universally accepted. What I call a person's *implicatures* (singular *implicature*) others call a person's *implicata* (singular *implicatum*)—which is far too stuffy for my taste. Moreover, some restrict implicatures to persons only without allowing coherent mention of the implicatures of a statement, contrary to the usage found here and elsewhere (e. g., Borchert *1996*, 225). I thank my colleague David Braun for many helpful points including the need for this footnote.

A statement *of* a proposition is made *in* a language, *by* a speaker, *to* an audience, *at* a time and place. My statements today were all made in English *using* English sentences that can be used tomorrow to make other statements with other implicatures. A statement (*affirms / denies*) the propositions (implied / contradicted) by the proposition it states. If we define the *implications of a statement* to be the implications of the proposition it states, we can say that a statement affirms exactly what it implies and it denies exactly the propositions whose negations it implies. I do not recommend this definition which would burden the word 'implication' with yet another meaning having another range of applicability.

Normally, statements do not implicate all they affirm. My statement "Socrates taught Plato, who taught Aristotle" affirms but does not implicate the proposition "Plato taught Aristotle". Moreover, even when we can be confident that the speaker is being sincere, sane, responsible, etc., the statement need not be evidence for its affirmations. A statement of the axioms and definitions of arithmetic affirms the Goldbach Hypothesis or it affirms the Goldbach Hypothesis's negation.

A statement is *self-denying (SD)* if it *denies* one of its own *implicatures* (or equivalently, *implicates* a proposition it *denies*). Examples of SD propositions are: a statement to the effect that the speaker is not making a statement or is (using / not using) a word that in fact (is not / is) being used. A statement is *self-affirming (SA)* if it *affirms* one of its own *implicatures* (or, equivalently, it *implicates* a proposition it *affirms*). Examples of SA statements are: a statement to the effect that the speaker is making a statement or is (addressing / not addressing) an audience that in fact (is / is not) being addressed. A statement is (*correct / incorrect*) if (every / some) proposition it affirms or implicates is (true / false). Every SD statement is incorrect. Every statement of a contradiction is SD. Nevertheless, it is not the case that every SD statement affirms a contradiction: the proposition stated in a statement to effect that the speaker disbelieves a certain one of the statement's own affirmations need not contradict itself. As has been noted in discussions of "Moore's Paradox", such statements can be made using sentences such as 'Abe died but I *disbelieve* it'. Not every SA statement is correct. In fact, every SA lie is incorrect. A half-truth is a statement implicating a falsehood but affirming only truths.

10   Conclusion

> The proposition must be distinguished from
> the sentence, the combination of words or
> signs though which it is expressed; from the
> fact, the actual complex situation whose ex-
> istence renders it true or false; and from the
> judgment, which affirms or denies the propo-
> sition the proposition.—Eaton *1931*, 12.[43]

The five ambiguous words—sentence, proposition, judgment, statement,
and fact—have received distinct recommended meanings that are needed
for any sufficiently full discussion of logical phenomena such as self-denying
and self-affirming statements. This paper has sketched a conceptual framework necessary for discussing issues that have concerned logicians over the
years. I hope that most objective and informed logicians will agree that this
framework is indispensable for comprehensive and comparative treatment
of the history of logic. A historian inadvertently or willfully ignorant of
any given one of these five concepts will be unable to give a full and fair
treatment of a historical figure that uses it. I believe that appreciation of
the place of this framework in logical thought will reveal how crude, evasive,
and incomplete many important logic papers are. I hope that this awareness
does not lessen the credit attributed to the authors of those works. If this
paper succeeds, every future logic book should be affected.

Acknowledgements

This paper clarifies, qualifies, and—in only a few cases I am glad to say—
retracts views I previously expressed. I started writing it in 2003 partly as
an independent philosophical preface to my 1973 *Dialogos* article "Meanings
of Implication". As I became more and more clear about the content of this
paper, my dissatisfaction with 1973 paper grew. There is still much to like
about that paper despite the deficiencies in its conceptual framework made
evident by this paper. The first published fruits of my rethinking are in
"Sentential Functions: the Functions of Sentences" in Corcoran, Griffin, et
al. *2004*.

For bringing errors and omissions to my attention, for useful suggestions,
and for other help, I gladly acknowledge the following scholars : G. Fulugo-
nio (Argentina), O. Chateaubriand, F. Nabrasa, J. da Silva, and L. Weber

---

[43]Eaton seems to be combining the statement with the judgment. This is in keeping
with other logicians who call Frege's "judgment-stroke" the "assertion sign" (Eaton *1931*,
369–374). But, the situation is actually much worse, this sign, which resembles a counter-
clockwise rotated tee, is often taken to indicate a logically cogent judgment—a judgment
based entirely on "laws of logic", a logical inference.

(Brazil); D. Hitchcock and J. Van Evra (Canada); R. Torretti (Chile); K. Miettinen (Finland and USA); B. Smith (Germany and USA); H. Masoud (Iran), C. Penco (Italy), A. Visser (Netherlands); R. Santos (Portugal); J. Sagüillo (Spain); J. Gasser (Switzerland); I. Grattan-Guinness (UK); W. Abler, R. Barnes, D. Braun, D. Brewer, B. Decker, J. Foran, J. Gould, C. Guignon, I. Hamid, F. Hansen, P. Hare, L. Jacuzzo, C. Jongsma, J. Kearns, J. Miller, M. Moore, M. Mulhern, M. Murphey, S. Nambiar, S. Newberry, P. Penner, A. Preus, M. Scanlan, K. Shockley, T. Tracy, J. Yu, and J. Zeccardi (USA); and others. Earlier versions were presented at the Buffalo Logic Colloquium, the University of Buffalo Philosophy Colloquium, Canisius College, the University of Santiago de Compostela, the University of South Florida, and the 14th Latin-American Symposium on Mathematical Logic in Paraty, Brazil.

My greatest debt is to Professor José Miguel Sagüillo, *Catedrático de Lógica* of the University of Santiago de Compostela, Spain. He discussed these ideas with me over many years and he inspired me to write this article. He has done more than any other person toward clarifying for me the synergistic relations between the teaching of logic and the epistemology of logic.

BIBLIOGRAPHY

[1] Arnauld, A. and P. Nicole. 1662. *Port Royal Logic*. Tr. and ed. J. Buroker. 1996. Cambridge: Cambridge UP.

[2] Aristotle. *Aristotle's Categories and De Interpretatione*. Tr. J. Ackrill. 1963. Oxford: Oxford UP.

[3] Aristotle. *Nicomachean Ethics*. Tr. T. Irwin. 1999. Indianapolis: Hackett.

[4] Audi, R., Ed. 1995/1999. *The Cambridge Dictionary of Philosophy*. Second edition. Cambridge: Cambridge University Press.

[5] Austin, J.L. 1950. Truth. *Proceedings of the Aristotelian Society*. Supp. Vol. 24. Reprinted Austin *1961*.

[6] Austin, J.L. 1961. *Philosophical Papers*. Ed. J. Urmson and G. Warnock. Oxford: Oxford UP.

[7] Barber, K. and J. Corcoran. 2009. Agent and Premise Implication. *Bulletin of Symbolic Logic 15*: 235

[8] Beaney, M., ed. 1997. *The Frege Reader*. Oxford: Blackwell.

[9] Bochenski, I., A. Church, N. Goodman. 1956. *The Problem of Universals*. Notre Dame, IN: Notre Dame UP.

[10] Boolos, G., J. Burgess, and R. Jeffrey. 2002. *Computability and logic*. Cambridge: Cambridge UP.

[11] Borchert, D. , ed. 1996. *Encyclopedia of Philosophy*. New York: Simon & Schuster.

[12] Carnap, R. 1934/1937. *The Logical Syntax of Language*. Tr. A. Smeaton. London: Routledge & Kegan Paul.

[13] Chateaubriand, O. 2001. *Logical Forms. Part I*. Campinas: CLE Unicamp.

[14] Chateaubriand, O. 2005. *Logical Forms. Part II*. Campinas: CLE Unicamp.

[15] Church, A. 1956i. *Introduction to Mathematical Logic*. Princeton: Princeton UP.

[16] Church, A. 1956p. Propositions and Sentences. Bochenski et al. *1956*. Notre Dame, IN: Notre Dame UP.

[17] Cohen, M., and E. Nagel. 1934/1962/1993. *Introduction to Logic*. Indianapolis: Hackett.

[18] Corcoran, J. 1973. Meanings of Implication. *Dialogos* 9:59-76. Spanish translation by José M. Sagüillo, "Significados de la Implicación", *Agora* 5 (1985) 279–294, updated reprint in Hughes *1993*.

[19] Corcoran, J., ed. 1974. *Ancient Logic and its Modern Interpretations*. Reidel: Dordrecht.

[20] Corcoran, J. 1998. Information-theoretic logic, in *Truth in Perspective* edited by C. Martínez, U. Rivas, L. Villegas-Forero. Aldershot, England: Ashgate Publishing Limited, 113–135.

[21] Corcoran, J. 1999. Critical thinking and pedagogical license. *Manuscrito*. XXII: 109–116.

[22] Corcoran, J. 2004. Sentential Functions: the Functions of Sentences. In Corcoran, Griffin, et al. *2004*.

[23] Corcoran, J. 2005. Meanings of word: type-occurrence-token. *Bulletin of Symbolic Logic 11*: 117.

[24] Corcoran, J. 2006e. An Essay on Knowledge and Belief. *The International Journal of Decision Ethics*. II.2, 125–144.

[25] Corcoran, J. 2006s. Schemata: the Concept of Schema in the History of Logic. *Bulletin of Symbolic Logic 12*: 219–240.

[26] Corcoran, J., W. Frank, and M. Maloney. 1974. String Theory. *Journal of Symbolic Logic 39*: 625–637.

[27] Corcoran, J., J. Griffin, et al. 2004. *Discursos da Investidura de D. John Corcoran y D. James Griffin como Doutores Honoris Causa*. Santiago: Universidade de Santiago de Compostela.

[28] Eaton, R. 1931. *General Logic*. New York: Charles Scribner's Sons.

[29] Frege, G. 1884/1959. *The Foundations of Arithmetic*. Breslau: Koebner. Tr. J. L. Austin. Oxford: Basil Blackwell.

[30] Frege, G. 1884/1964. *The Foundations of Arithmetic*. Breslau: Koebner. Excerpted tr. M. Mahoney in Eds. P. Benacerraf and H. Putnam. *Philosophy of Mathematics*. Cambridge: Cambridge UP.

[31] Frege, G. 1918/1956. The Thought: a Logical Inquiry. Trans. A. and M. Quinton. *Mind 65* (1956) 289–311.

[32] Frege, G. 1919/1952. Negation. Trans. P. Geach. Reprinted in Geach-Black *1952/1966*: 117–135.

[33] Frege, G. 1979. *Posthumous Writings*. Ed. H. Hermes et al. Trs. P. Long and R. White. Chicago: University of Chicago Press.

[34] Frege, G. 1980. *Philosophical and Mathematical Correspondence*. Chicago: University of Chicago Press.

[35] Frege, G. 1997. *The Frege Reader*. Ed. M. Beaney. Oxford: Blackwell.

[36] Geach, P. and M. Black, trs. 1952/1966. *Translations from the Philosophical Writings of Gottlob Frege*. Oxford: Basil Blackwell.

[37] Goldfarb, W. 2003. *Deductive Logic*. Indianapolis: Hackett.

[38] Grice, P. 1989. *Studies in the Way of Words*. Cambridge, MA: Harvard UP.

[39] Hilbert, D. and W. Ackermann. 1928/1938/1950. *Principles of Mathematical Logic*. Providence, RI: American Mathematical Society.

[40] Hughes, R., ed. 1993. *Philosophical Companion to First-order Logic*. Indianapolis: Hackett.

[41] Kretzmann, N. 1974. Aristotle on Spoken Sound Significant by Convention. In Corcoran *1974*.

[42] Lewis, C. I. and C. H. Langford. 1932/1959. *Symbolic Logic*. New York: Century. Reprinted New York: Dover.

[43] Luce, R. 1950. Notes. In Hilbert and Ackermann *1928/1938/1950*.

[44] Lyons, J. 1977. *Semantics*. 2 Vols. Cambridge: Cambridge UP.

[45] Moore, G. E. 1899. The Nature of Judgment. *Mind*. n. s. 8: 176–193.

[46] Ogden, C. K. and Richards, I.A. 1923. *The Meaning of Meaning*. Reprint of eighth edition. Harcourt: New York.

[47] Peirce, C. 1906. Prolegomena to an Apology for Pragmaticism. *Monist 16*: 492–546.

[48] Peirce, C.S. 1933. *Collected Papers of Charles Sanders Peirce*. Eds. C. Hartshorne and P. Weiss. Cambridge: Harvard UP.

[49] Plato. *Complete Works*. Ed. J.Cooper. Indianapolis: Hackett.

[50] Quine, W. 1945. On the Logic of Quantification. *Journal of Symbolic Logic 10*: 1–12.

[51] Quine, W. 1951. *Mathematical Logic*. Cambridge MA: Harvard UP.

[52] Quine, W. 1970/1986. *Philosophy of logic*. Cambridge MA: Harvard UP.

[53] Russell, B. 1903. *The Principles of Mathematics*. Cambridge: Cambridge UP.

[54] Russell, B. 1905. On denoting. *Mind*, n. s. 14: 479-493. Reprinted in Copi and Gould *1967*, 93–104.

[55] Shapiro, S., ed. 1996. *The Limits of Logic*. Aldershot, UK: Dartmouth.

[56] Tarski, A. 1969/1993. Truth and proof. *Scientific American*. June 1969. Reprinted Hughes 1993.

[57] Tarski, A. 1956/1983. *Logic, Semantics, Metamathematics*. Trans. J. H. Woodger. Indianapolis: Hackett.

[58] Tarski, A. 1986. What are Logical Notions? *History and Philosophy of Logic* 7:143–154. Reprinted in Shapiro 1996.

[59] Whately, R. 1826. *Elements of Logic*. Ed. P. Dessi. 1988. Bologna: Editrice CLLEB.

[60] Wittgenstein, L. 1921/1998. *Tractatus Logico-Philosophicus*. Tr. D. Kolak. Mountain View, CA: Mayfield.

John Corcoran
Department of Philosophy
University of Buffalo
Buffalo, NY 14260-4150 USA
email: corcoran@buffalo.edu

Truth Defined

DAVID MILLER

ABSTRACT. Tarski's theorem advises us that no completely satisfactory definition of the term *true sentence* can be expected. It is here shown that, nevertheless, it is possible to formulate within a fragment $\mathbb{T}$ of Zermelo set theory $\mathbb{Z}$ a definition of the truth of sentences that is *materially adequate, formally correct, explicit, universal, versatile,* and *modestly paraconsistent.* The definition is extended from sentences to deductive theories, and an investigation is begun into why the antinomy of the liar fails to arise.

1 Introduction

It is the purpose of this paper to offer an explicit and materially adequate definition of truth for any denumerable language that contains, in addition to the standard elementary theory of identity, enough of its own syntax to name all its expressions and to distinguish among them effectively. The languages for which the definition is appropriate include the language of Zermelo–Fraenkel set theory $\mathbb{ZF}$, and any applied language $\mathcal{L}^+$ constructed from it by adding items of extralogical vocabulary. The *background theory* $\mathbb{T}$ of the investigation, which is fixed, is a fragment of Zermelo set theory $\mathbb{Z}$. A detailed discussion and defence of the definition are offered in Miller (2009).

A *structural-descriptive name* $\ulcorner y \urcorner$ of the formula y is a name, such as a quotation name or a numeral for the Gödel number of y, from which the syntactic structure of y may be deduced. The most important demand imposed on any structural-descriptive names $\ulcorner x \urcorner$ and $\ulcorner z \urcorner$ given to the formulas x, z, is that the sentence $\ulcorner x \urcorner = \ulcorner z \urcorner$ (that is, the sentence consisting of $\ulcorner x \urcorner$, the identity sign, and $\ulcorner z \urcorner$, in that order) be demonstrable or refutable in the background theory $\mathbb{T}$.

Tarski called a definition of the term 'true sentence' *materially adequate* for the language $\mathcal{L}$ if within the background theory $\mathbb{T}$ it logically implies every instance of the **T**-scheme

$$P \text{ is a true sentence if \& only if } p, \tag{\textbf{T}}$$

where 'p' is replaced by a sentence y of $\mathcal{L}$ and 'P' is replaced by a structural-descriptive name $\ulcorner y \urcorner$ of y. Tarski observed that for a language $\mathcal{L}$ with only finitely many distinct sentences $x_0, \ldots, x_{j-1}$, the problem of the definition of truth, as he conceived it, that is, the problem of providing a materially adequate definition of the term 'true sentence', can be fully solved, provided that each sentence x_i of $\mathcal{L}$ is furnished with a structural-descriptive name $\ulcorner x_i \urcorner$ (Tarski 1933a, p. 188). We may define 'y is a true sentence', or $\mathbf{Tr}(y)$, by either of the equivalences

$$\mathbf{Tr}(y) \Leftrightarrow_{\mathrm{Df}} (y = P_0 \wedge p_0) \vee \ldots \vee (y = P_{j-1} \wedge p_{j-1}), \quad (1.0)$$

$$\mathbf{Tr}(y) \Leftrightarrow_{\mathrm{Df}} (y = P_0 \to p_0) \wedge \ldots \wedge (y = P_{j-1} \to p_{j-1}), \quad (1.1)$$

where, for each $i < j$, the sentence x_i replaces 'p_i', and the name $\ulcorner x_i \urcorner$ replaces 'P_i'. It is evident that, provided that the connectives $\vee$, $\wedge$, and $\to$ are governed by the usual (classical) rules, both (1.0) and (1.1) are materially adequate definitions of truth. It is evident too that each of these definitions is longer than any sentence of the language $\mathcal{L}$, and has to be formulated outside $\mathcal{L}$.

It is the burden of *Tarski's theorem* that no language $\mathcal{L}$ whose deductive structure is of a complexity that allows forms of self-ascription can consistently include within itself a definition of the term 'true sentence' that is materially adequate for $\mathcal{L}$. Languages to which Tarski's theorem applies include in particular the languages of elementary arithmetic and set theory.

Tarski's principal achievement was to show how, for each elementary language $\mathcal{L}_0$, it is possible to prepare, within a richer *metalanguage* $\mathcal{L}_1$, an explicit definition, materially adequate for $\mathcal{L}_0$, of the term 'true sentence'. The metalanguage $\mathcal{L}_1$ must be richer than $\mathcal{L}_0$ in the sense that it must incorporate stronger set-theoretical postulates than $\mathcal{L}_0$ does, not in the sense that it must boast a wider vocabulary. To generate a definition of the term 'true sentence' that is materially adequate for $\mathcal{L}_1$, a yet richer metametalanguage $\mathcal{L}_2$ must be presumed on. An unending hierarchy of languages is in this way initiated, and the concept of 'true sentence' is never completely defined.

Tarski's theorem will not be contested here. But the conclusion customarily drawn from it, that there can be no fully satisfactory definition of the concept of truth, for arithmetic or for set theory, will be constructively contested (if not refuted). Truth, as it is to be defined in this paper, is not a property $\mathbf{Tr}$ of sentences but a function $\boldsymbol{\theta}$, from terms (names, descriptions, variables) that stand for sentences, to deductive theories. Described informally, $\boldsymbol{\theta}(\mathfrak{y})$, *the truth of* $\mathfrak{y}$, is the theory that states the conditions under which the sentence named by the term $\mathfrak{y}$ is true. In particular, if x is any sentence of the extended language $\mathcal{L}^+$, and $\ulcorner x \urcorner$ is a structural-descriptive name of x, then $\boldsymbol{\theta}(\ulcorner x \urcorner)$ turns out to be identical with the theory $\mathbf{Cn}(\{x\})$ that consists of all the logical consequences of x; that is to say, $\boldsymbol{\theta}(\ulcorner x \urcorner)$ and x are logically interderivable. The definition of truth to be presented is materially adequate for the entire language $\mathcal{L}^+$ of applied set theory.

For most of the paper, a somewhat different function, $\boldsymbol{\vartheta}$, will be investigated alongside the function $\boldsymbol{\theta}$. Only in § 7 will a genuine preference be expressed between them as definitions of truth. Everything said in the previous paragraph concerning $\boldsymbol{\theta}$ applies equally to the alternative definiens for truth, $\boldsymbol{\vartheta}$.

2 Deductive Theories

This section summarizes the principal results of Tarski's *general metamathematics* (or *calculus of deductive systems*) that are initially required for an understanding of the definition of truth to be presented in §3 below. These results, together with the less ordinary results presented in §4 below, can be found, in most cases without proofs, in Tarski (1935–1936). In addition to the transposition (now common) of Tarski's original connotations of the terms 'deductive theory' and 'deductive system', two variations deserve remark. The first is that we shall allow sets of well-formed formulas, and not only sets of sentences (closed formulas), to count as deductive theories. This change makes little difference until we come in §7 to define the truth of theories themselves. The second variation is that we shall invert the ordering on theories (by set-theoretical inclusion) that Tarski uses throughout his works, and use instead the ordering by logical derivability. Where Tarski writes $\mathbf{X} \subseteq \mathbf{Z}$, we shall often write $\mathbf{Z} \vdash \mathbf{X}$; the operations of product (intersection) and sum in Tarski's treatment, both finite and infinite, become disjunction and conjunction in the treatment here, with the awkward repercussion that a set-theoretically expanding sequence of theories will here be called *decreasing* (see Theorem 4 below). A more significant discrepancy is that whereas the complement $\overline{\mathbf{Y}}$ that Tarski defines for the theory $\mathbf{Y}$ is a pseudocomplement, which obeys the law of non-contradiction but not the law of excluded middle, the complement $\mathbf{Y}'$ that we shall define (in 4.3) is an authocomplement (that is, a dual pseudocomplement), which obeys the law of excluded middle but not the law of non-contradiction. This all needs to be said explicitly lest anyone should compare the present report with Tarski (1935–1936) and conclude with a sigh that everything here is upside down. It is intended that everything be upside down.

Let S be the set of formulas of a denumerable language $\mathcal{L} \subseteq \mathcal{L}^+$. An operation $\mathbf{Cn} : \wp(S) \mapsto \wp(S)$ is a *consequence operation* if it fulfils the conditions of *idempotence* (2.0), and of *finitariness* (2.1):

$$X \subseteq S \implies X \subseteq \mathbf{Cn}(X) = \mathbf{Cn}(\mathbf{Cn}(X)) \subseteq S \tag{2.0}$$

$$X \subseteq S \implies \mathbf{Cn}(X) = \bigcup \{\mathbf{Cn}(Y) \mid Y \subseteq X \text{ and } |Y| < \aleph_0\}. \tag{2.1}$$

By (2.1), the operation $\mathbf{Cn}$ must also be monotone:

$$X \subseteq Z \subseteq S \implies \mathbf{Cn}(X) \subseteq \mathbf{Cn}(Z). \tag{2.2}$$

The pair $\langle S, \mathbf{Cn} \rangle$ may sometimes be called a *deductive system* or (to use an expression of Tarski 1930b, Introduction) a *deductive discipline*, or simply a *logic*. Reference to the set S will usually be omitted. With each consequence

operation **Cn** is associated a *derivability* relation $\vdash$ such that $y \in \mathbf{Cn}(Y)$ if
& only if $Y \vdash y$. The derivability notation will often, for the sake only
of flexibility, be read 'Y implies y'. Note that the logic **Cn** need not be
identical with the logic that regulates the background theory $\mathbb{T}$. It may be
stronger, it may be weaker. Explicit assumptions about **Cn** that will made
in the course of the paper include: (i) the conjunction $x \wedge z$ of two formulas
is always defined; (ii) their disjunction $x \vee z$ is also defined; (iii) both are
defined, and the distributive law (2.7) and its dual hold for all formulas;
(iv) the conditional $x \to z$ is defined for some or all pairs of formulas (if it is
defined for all pairs, then the distributive law (2.7) holds); (v) **Cn** includes
at least the whole of classical sentential logic. When we come, in § 3, to
define truth, it will be mandatory that (vi) **Cn** includes, in addition to
the standard logic of identity, structural-descriptive names for all formulas
in S (and the capacity to prove $\ulcorner x \urcorner \neq \ulcorner z \urcorner$ whenever x and z are distinct
formulas), not forgetting a rule of substitution for free variables, so that,
for example, if y and w are free then $\mathbf{Cn}(y \neq w)$ contains each identity of
the form $\ulcorner x \urcorner \neq \ulcorner z \urcorner$. In §§ 5f. we shall consider what happens when (vii)
the logic **Cn** is, like elementary Peano arithmetic and $\mathbb{ZF}$, both ω-consistent
and incompletable. It is easily proved in $\mathbb{T}$ that if **Cn** is a logic, and $X \subseteq S$,
then $\mathbf{Cn}_X(Y) = \mathbf{Cn}(X \cup Y)$ is also a logic.

If $\mathbf{Y} = \mathbf{Cn}(\mathbf{Y})$, the set $\mathbf{Y} \subseteq S$ is a (deductive) *theory*. It is a (finitely)
axiomatizable theory if $\mathbf{Y} = \mathbf{Cn}(Y)$ for some finite $Y \subseteq S$, in which case
we may write **y** in place of **Y**. In classical logic, where the operation $\wedge$
of conjunction exists and, moreover, $\mathbf{Cn}(\varnothing) = \mathbf{Cn}(\top)$ for any theorem $\top$,
finite axiomatizability is the same as axiomatizability by a single formula:
$\mathbf{Cn}(\mathbf{Y}) = \mathbf{Cn}(\{y\})$. It is customary (and does no harm) to identify formulas
x, z for which $\mathbf{Cn}(\{x\}) = \mathbf{Cn}(\{z\})$, and to identify the formula y with
the axiomatizable theory $\mathbf{Cn}(\{y\})$, which is more commonly written $\mathbf{Cn}(y)$.
Such an identification is innocuous in regard to (finite) axiomatizability:
$\mathbf{Cn}(\mathbf{Y}) = \mathbf{Cn}(Y)$ for a finite set Y of formulas if & only if $\mathbf{Cn}(\mathbf{Y}) = \mathbf{Cn}(\bigcup \mathcal{Y})$,
where $\mathcal{Y}$ is a finite set, for example $\{\mathbf{Cn}(y) \mid y \in Y\}$, of finitely axiomatizable
theories. Tarski's calculus of deductive theories is of independent interest
only when the deductive discipline $\langle S, \mathbf{Cn} \rangle$ is *non-trivial* in the sense that,
for each formula $x \in S$, there are infinitely many formulas z for which
$\mathbf{Cn}(x) \neq \mathbf{Cn}(z)$.

As earlier noted, $\mathbf{Z} \vdash \mathbf{X}$ is defined to mean $\mathbf{X} \subseteq \mathbf{Z}$. It must be borne in
mind that this extended derivability relation $\vdash$ is not finitary; for evidently,
$\mathbf{X} \vdash \mathbf{X}$ does not imply that there is a finite subset $\mathbf{Y} \subseteq \mathbf{X}$ for which $\mathbf{Y} \vdash \mathbf{X}$. A
theory $\mathbf{Y}$ is *consistent* provided that $\mathbf{Y} \neq S$, and is *maximal* (or *complete*)
if it is consistent and has no consistent proper extension. Lindenbaum's
theorem (Tarski 1935–1936, p. 366), whose proof requires of the logic **Cn**
only that $\mathbf{Cn}(S)$ be axiomatizable, states that if $\mathbf{Y}$ is a consistent theory
then it has at least one maximal extension Ω.

The *disjunction* $\mathbf{X} \vee \mathbf{Z}$ and *conjunction* $\mathbf{X} \wedge \mathbf{Z}$ of the theories $\mathbf{X}$ and $\mathbf{Z}$

are defined like this:

$$\mathbf{X} \vee \mathbf{Z} \;=\; \mathbf{X} \cap \mathbf{Z}, \qquad\qquad (2.3)$$

$$\mathbf{X} \wedge \mathbf{Z} \;=\; \mathbf{Cn}(\mathbf{X} \cup \mathbf{Z}). \qquad\qquad (2.4)$$

Proving that $\mathbf{X} \vee \mathbf{Z}$ is a theory requires little work; proving that $\mathbf{X} \wedge \mathbf{Z}$ is a theory requires none. The *join* $\bigvee \mathcal{K}$ and the *meet* $\bigwedge \mathcal{K}$ of a family $\mathcal{K}$ of theories are defined in a similar way:

$$\bigvee \mathcal{K} \;=\; \bigcap \mathcal{K}, \qquad\qquad (2.5)$$

$$\bigwedge \mathcal{K} \;=\; \mathbf{Cn}(\bigcup \mathcal{K}). \qquad\qquad (2.6)$$

If $\mathcal{T}$ is the class of all deductive theories in $\mathcal{L}$, then the meet $\bigwedge \mathcal{T}$ is identical with the set S of all formulas and is itself a theory. To emphasize this, we sometimes write $\mathbf{S}$ instead of S. Under the derivability relation $\vdash$ associated with $\mathbf{Cn}$, the *absurd* theory $\mathbf{S}$ is the logically strongest of all, while the weakest of all is (by (2.2)) the *trivial* theory $\mathbf{L}$, which is identical with $\bigvee \mathcal{T}$ or $\mathbf{Cn}(\varnothing)$. $\mathbf{L}$ is always axiomatizable. Note that the disjunction $\mathbf{X} \vee \mathbf{Z}$ and conjunction $\mathbf{X} \wedge \mathbf{Z}$ of two theories are well defined even if there exist no operations of disjunction $x \vee z$ and conjunction $x \wedge z$ on formulas, and the theories $\mathbf{S}$ and $\mathbf{L}$ are well defined even if the consequence operation $\mathbf{Cn}$ recognizes no minimal and maximal formulas $\perp$ and $\top$. Even when the consequence operation $\mathbf{Cn}$ does not support $\vee$ and $\wedge$, we shall freely mix in a single expression terms for formulas and terms for theories; $x \vee \mathbf{X}$, for example, is to be regarded as shorthand for $\mathbf{Cn}(x) \vee \mathbf{X}$.

It is a routine task to verify that if the logic $\mathbf{Cn}$ includes the standard elimination and introduction rules for disjunction $\vee$ and conjunction $\wedge$, and in addition the distributive law

$$y \vee (x \wedge z) \;=\; (y \vee x) \wedge (y \vee z) \qquad\qquad (2.7)$$

(or its dual, which is equivalent) holds for all formulas x, y, z, then the class $\mathcal{T}$ of all theories forms a complete distributive lattice with join $\vee$ and meet $\wedge$ defined as in (2.3) and (2.4). We have in particular the absorption laws (2.8) and the distributive law (2.9) for theories:

$$\mathbf{X} \vee (\mathbf{X} \wedge \mathbf{Z}) \;=\; \mathbf{X} \;=\; (\mathbf{X} \vee \mathbf{Z}) \wedge \mathbf{X}, \qquad\qquad (2.8)$$

$$\mathbf{Y} \vee (\mathbf{X} \wedge \mathbf{Z}) \;=\; (\mathbf{Y} \vee \mathbf{X}) \wedge (\mathbf{Y} \vee \mathbf{Z}). \qquad\qquad (2.9)$$

Only a handful of the many possible distributive laws are valid for infinite joins and meets, even if $\mathbf{Cn}$ is a classical consequence operation. If the distributive law (2.7) holds for $\mathbf{Cn}$, then

$$\bigwedge \{\mathbf{X} \vee \mathbf{Z} \mid \mathbf{Z} \in \mathcal{K}\} \;=\; \mathbf{X} \vee \bigwedge \mathcal{K}, \qquad\qquad (2.10)$$

$$\bigvee \{\mathbf{X} \wedge \mathbf{Z} \mid \mathbf{Z} \in \mathcal{K}\} \;\vdash\; \mathbf{X} \wedge \bigvee \mathcal{K}, \qquad\qquad (2.11)$$

with equality in (2.11) if the family $\mathcal{K}$ is finite. If in addition, the consequence operation **Cn** admits a conditional $x \to z$ for each two formulas x, z (in which case (2.9) holds too) then:

$$\bigvee \{x \wedge z \mid z \in \mathcal{K}\} \;=\; x \wedge \bigvee \mathcal{K}. \tag{2.12}$$

It deserves to be noted that although the meet $\bigwedge \mathcal{K}$ behaves very much as we expect an infinite conjunction to behave, the join $\bigvee \mathcal{K}$ is somewhat anomalous in that it can hold ('be true' intuitively) even if no element of $\mathcal{K}$ holds. This oddity may be illustrated by the language of classical elementary logic with identity and no other predicates or individual names. In this language we may, for every $j > 0$, formulate a sentence ω_j that states that the universe contains exactly j elements. Each theory $\omega_j = \mathbf{Cn}(\omega_j)$ is consistent and categorical, and hence maximal. The theory $\Omega = \bigwedge \{\mathbf{Cn}(\neg \omega_j) \mid j > 0\}$, which states that the universe is infinite, is consistent (by (2.1)) and maximal (by Vaught's test). If $\omega_j \vdash y$ for every $j > 0$, then $\neg y \vdash \neg \omega_j$ for every j, and so $\neg y \vdash \Omega$. Since Ω is maximal, and not axiomatizable (as shown in Corollary 6 below), $\neg y$ is inconsistent. In other words, y is a theorem, and the join $\bigvee \{\omega_j \mid j > 0\}$ is the trivially true theory **L**, despite its not exhausting all the possibilities.

More advanced aspects of Tarski's calculus of deductive theories are expounded in § 4 below.

3   The Definition of Truth

We henceforth identify expressions of the language $\mathcal{L}^+$ with their Gödel numbers, or other set-theoretical codes, with the result that all expressions are sets, and all variables are variables for sets; that is, they may be replaced by names of sets. For ease of exposition we shall often use the locution 'name of' rather loosely in place of the perhaps more correct words 'expression, constant or variable, standing for', so that a variable (such as $\mathfrak{y}$ below) that may be replaced by a name (such as $\ulcorner 0 = 1 \urcorner$) of a formula may also be called a name of a formula. We may reasonably suppose that the set $\mathfrak{N}$ of all names of sets is definable in the background theory $\mathbb{T}$.

As announced earlier, it will be assumed that the logic embodied in the operation **Cn** includes the standard elementary logic of identity. We shall use $=$ and $\neq$ as names of the equality and the inequality signs (and later $\in$ to name the membership sign). Structural-descriptive names for all elements of S must also be available, and $\ulcorner x \urcorner \neq \ulcorner z \urcorner$ must be demonstrable (that is, $\ulcorner x \urcorner \neq \ulcorner z \urcorner \in \mathbf{L}$) whenever x and z are distinct formulas in S, and refutable whenever $x = z$. The extent to which the logic **Cn** needs to include also some sentential or quantificational logic is left open. The background theory $\mathbb{T}$, which is a fragment of the theory $\mathbb{Z}$, is assumed to contain as much elementary logic as it needs.

In the remainder of the paper we shall, for the sake of clarity, adopt '$\mathfrak{x}$', '$\mathfrak{y}$', '$\mathfrak{z}$', ... as variables for names of formulas, open to substitution, and revert to 'u', 'v', 'w', 'x', 'y', 'z', ... as variables for formulas. As is usual in formal work, we shall rarely, outside this paragraph, display any of the symbols of which we speak, but only their names. If α and γ are strings of symbols, then we shall use '$\alpha\gamma$' as a name for the concatenation $\alpha^\frown\gamma$ (in that order) of α and γ.

We now proceed rigorously to generalize the right-hand sides of the two equivalent finite definitions (1.0) and (1.1), by replacing the finite disjunction in (1.0), for example, by a generalized join. We may define a function $\vartheta : \mathfrak{N} \mapsto \mathcal{T}$ from expressions for formulas to deductive theories

$$\vartheta(\mathfrak{n}) \quad =_{\mathrm{Df}} \quad \bigvee \{ \mathsf{Cn}(\mathfrak{n} = \ulcorner y \urcorner \wedge y) \mid y \in S \}. \tag{3.0}$$

In the same way we may generalize the alternative definition (1.1) of '$\mathsf{Tr}(y)$' in the finite case:

$$\boldsymbol{\theta}(\mathfrak{n}) \quad =_{\mathrm{Df}} \quad \bigwedge \{ \mathsf{Cn}(\mathfrak{n} = \ulcorner y \urcorner \to y) \mid y \in S \}. \tag{3.1}$$

It will be helpful later, especially in § 5, to have the definition of the function:

$$\boldsymbol{\delta}(\mathfrak{n}) \quad =_{\mathrm{Df}} \quad \bigvee \{ \mathsf{Cn}(\mathfrak{n} = \ulcorner y \urcorner) \mid y \in S \}. \tag{3.2}$$

The aim of this paper is to give a definition of truth that requires as little deviation from classical logic as possible. We have implicitly eschewed recourse to infinitary logic, and now we eschew recourse to intensional logic. It is evident that if the definiendum of the functions ϑ and $\boldsymbol{\theta}$ (more exactly, of the functors 'ϑ' and '$\boldsymbol{\theta}$') are to be names of deductive systems, then the substituends for $\mathfrak{n}$ in (3.0) and (3.1) must be not formulas but expressions standing for formulas.

Provided that due caution is exercised, however, both '$\vartheta(\mathfrak{n})$' and '$\boldsymbol{\theta}(\mathfrak{n})$' may be read intensionally as 'the *semantical value* of the formula (named by) $\mathfrak{n}$'; and when $\mathfrak{n}$ is the name of a sentence, both '$\vartheta(\mathfrak{n})$' and '$\boldsymbol{\theta}(\mathfrak{n})$' may be read also as 'the truth of the sentence (named by) $\mathfrak{n}$'. In at least two places it will be rewarding to have the functions ϑ and $\boldsymbol{\theta}$ defined for all formulas: in § 7, where we shall discuss the truth of theories, and in Miller (2009), where the compositionality of the semantic values will be investigated. Speaking very roughly, we should like it to be the case that the semantic value of a quantified sentence is in some way related to the semantic value of the open formula within the scope of the quantifier. These matters will be made properly precise in the appropriate places.

THEOREM 1.   Both (3.0) and (3.1) are materially adequate definitions of truth; that is,

$$\vartheta(\ulcorner x \urcorner) = \quad \mathsf{Cn}(x) \quad = \boldsymbol{\theta}(\ulcorner x \urcorner) \tag{3.3}$$

whenever $\ulcorner x \urcorner$ is the structural-descriptive name of a formula $x \in S$.

Proof. If $\mathfrak{y}$ is replaced by a structural-descriptive name $\ulcorner x \urcorner$, then $\mathfrak{y} = \ulcorner y \urcorner$ is either demonstrable or refutable in the logic **Cn**, and hence the term $\mathbf{Cn}(\mathfrak{y} = \ulcorner y \urcorner \wedge y)$ is either $\mathbf{Cn}(x)$ or **S**, and $\mathbf{Cn}(\mathfrak{y} = \ulcorner y \urcorner \to y)$ is either $\mathbf{Cn}(x)$ or **L**. We conclude that $\vartheta(\ulcorner x \urcorner) = \mathbf{Cn}(x) = \theta(\ulcorner x \urcorner)$.   ∎

For a defence of the assumption here that material adequacy is rendered as well in terms of interderivability (or equivalence) as in terms of the usual biconditionals, see Miller (2009).

THEOREM 2.   If the formula named by $\mathfrak{x}$ is implied in **Cn** by the formulas with names $\{\mathfrak{z}_i \mid i < k\}$, then $\vartheta(\mathfrak{x}) \subseteq \bigwedge\{\vartheta(\mathfrak{z}_i) \mid i < k\}$.

Proof. Immediate using (2.6).   ∎

This theorem may be taken to say (in a slightly odd way) that in a valid inference the truth of the premises is transferred to the truth of the conclusion. A similar result holds for θ.

We now show that ϑ and θ may not be the same function, even if S is finite. This difference becomes evident if for $\mathfrak{y}$ is substituted a term $\mathfrak{v}$ that demonstrably names no formula $y \in S$, for then $\vartheta(\mathfrak{v})$ is, by (3.0), the join of theories all identical with **S**, while $\theta(\mathfrak{v})$ is, by (3.1), the meet of theories all identical with **L**. In short, $\vartheta(\mathfrak{v}) = \mathbf{S} \neq \mathbf{L} = \theta(\mathfrak{v})$. The theory $\delta(\mathfrak{v})$, which expresses something like the idea that $\mathfrak{v}$ demonstrably names some formula of S, is also identical with **S**. The divergence between ϑ and θ is, however, sometimes more dramatic than this. We begin with a lemma that is a kind of generalization of the rules for the conditional.

THEOREM 3.   If the logic **Cn** constrains enough conditional sentences into obeying both the rule of modus ponens and the rule of conditional proof, then

$$\vartheta(\mathfrak{y}) \quad = \quad \theta(\mathfrak{y}) \wedge \delta(\mathfrak{y}); \tag{3.4}$$

that is to say,

$$\bigvee\{\mathbf{Cn}(\mathfrak{y} = \ulcorner y \urcorner \wedge y) \mid y \in S\} \tag{3.5}$$

$$= \quad \bigwedge\{\mathbf{Cn}(\mathfrak{y} = \ulcorner y \urcorner \to y) \mid y \in S\} \wedge \bigvee\{\mathbf{Cn}(\mathfrak{y} = \ulcorner y \urcorner) \mid y \in S\}.$$

Proof. Let z be any formula in S. Then each formula of the form $\mathfrak{y} = \ulcorner y \urcorner \wedge y$ implies the formula $\mathfrak{y} \neq \ulcorner z \urcorner$ if $z \neq y$ and the formula z otherwise; in short, each formula of the form $\mathfrak{y} = \ulcorner y \urcorner \wedge y$ implies each formula of the form $\mathfrak{y} = \ulcorner z \urcorner \to z$. This means that $\bigvee\{\mathbf{Cn}(\mathfrak{y} = \ulcorner y \urcorner \wedge y) \mid y \in S\}$ implies $\bigwedge\{\mathbf{Cn}(\mathfrak{y} = \ulcorner y \urcorner \to y) \mid y \in S\}$. That it implies also $\bigvee\{\mathbf{Cn}(\mathfrak{y} = \ulcorner y \urcorner) \mid y \in S\}$ is trivial, and hence the theory named on the left of (3.5) implies the theory named on the right. For the converse, suppose that for each formula y, the formula $\mathfrak{y} = \ulcorner y \urcorner \wedge y$ implies the formula w. Then by modus ponens, $(\mathfrak{y} = \ulcorner y \urcorner \to y) \land (\mathfrak{y} = \ulcorner y \urcorner)$ implies w for each y. By conditional proof, $\mathfrak{y} = \ulcorner y \urcorner$ implies $(\mathfrak{y} = \ulcorner y \urcorner \to y) \to w$ for each y. It follows that $\bigvee \{ \mathbf{Cn}(\mathfrak{y} = \ulcorner y \urcorner) \mid y \in S \}$ implies $(\mathfrak{y} = \ulcorner y \urcorner \to y) \to w$ for each y, and hence that $\bigwedge \{ \mathbf{Cn}(\mathfrak{y} = \ulcorner y \urcorner \to y) \mid y \in S \} \land \bigvee \{ \mathbf{Cn}(\mathfrak{y} = \ulcorner y \urcorner) \mid y \in S \}$ implies w. The theory named on the right of (3.5) is proved to imply the theory named on the left.   ∎

If $\mathfrak{y}$ is a structural-descriptive name of a formula y, then one of the elements in the final join $\boldsymbol{\delta}(\mathfrak{y}) = \bigvee \{ \mathbf{Cn}(\mathfrak{y} = \ulcorner y \urcorner) \mid y \in S \}$ in (3.4) and (3.5) is $\mathbf{L}$, and so the entire join disappears: $\boldsymbol{\delta}(\mathfrak{y}) = \mathbf{L}$ and $\boldsymbol{\vartheta}(\mathfrak{y}) = \boldsymbol{\theta}(\mathfrak{y})$. But when $\mathfrak{y}$ demonstrably names no formula, the final term $\bigvee \{ \mathbf{Cn}(\mathfrak{y} = \ulcorner y \urcorner) \} = \mathbf{S}$.

More interesting than these limiting cases are names of formulas that may loosely be called contingent or factual (though they exist also in incomplete mathematical theories, such as Zermelo set theory $\mathbb{Z}$). An example that has been given in § 1 is $\mathfrak{p} = $ 'the first sentence of *Pride & Prejudice*'. Although $\mathfrak{p} \in \mathfrak{N}$ may, for the purposes of this discussion, be assumed to be demonstrable in $\mathbb{T}$, it does not follow that there is any structural-descriptive name $\ulcorner y \urcorner$ for which the formula $\mathfrak{p} = \ulcorner y \urcorner$ belongs to $\mathbf{L}$. As already observed, it is because Tarski's scheme ($\mathbf{T}$) does not permit the elimination of the word 'true' from such sentences as 'the first sentence of *Pride & Prejudice* is true' that it is not a complete definition of the term 'true sentence'. Such blind ascriptions of truth, however, present no difficulty for the genuine definitions (3.0) and (3.1).

Let us abbreviate the sentence 'It is a truth universally acknowledged, that a single man in possession of a good fortune, must be in want of a wife' by 'p'. It is evident that $\boldsymbol{\vartheta}(\mathfrak{p}) \neq \boldsymbol{\vartheta}(\ulcorner p \urcorner) = \mathbf{Cn}(p) = \boldsymbol{\theta}(\ulcorner p \urcorner) \neq \boldsymbol{\theta}(\mathfrak{p})$, since only a few of the disjuncts in the join $\bigvee \{ \mathbf{Cn}(\mathfrak{p} = \ulcorner y \urcorner \land y) \mid y \in S \}$ imply the sentence p (one of the successful disjuncts is $\mathbf{Cn}(\mathfrak{p} = \ulcorner p \urcorner \land p)$), and equally, only a few of the conjuncts in the meet $\bigwedge \{ \mathbf{Cn}(\mathfrak{p} = \ulcorner y \urcorner \to y) \mid y \in S \}$ are implied by p. It will be shown in § 5 that in most cases in which $\mathfrak{y}$ is, in the sense explained, a contingent name of a sentence, each of $\boldsymbol{\vartheta}(\mathfrak{y})$, $\boldsymbol{\theta}(\mathfrak{y})$, and $\boldsymbol{\delta}(\mathfrak{y})$ is an unaxiomatizable theory. In order to prepare for this work, we shall return in the next section to the study of Tarski's general metamathematics.

At this point it is an open question which of the definitions (3.0) and (3.1) is a better definition of truth. As far as material adequacy is concerned, they are on a par (Theorem 1). In favour of the function $\boldsymbol{\theta}$ is the advisability of not making a definition any stronger than necessary. It must be admitted also that infinite meets are easier to work with, and more intuitive, than are infinite joins (as illustrated in the penultimate paragraph of § 2). On the other side, since = (the name of identity sign), for example, is demonstrably the name of no formula, it is satisfactory that $\boldsymbol{\vartheta}(=)$ is the logically false theory $\mathbf{S}$, less satisfactory that $\boldsymbol{\theta}(=)$ is the logically true theory $\mathbf{L}$.

From the point of view of general metamathematics or abstract logic, the definition (3.0) has, however, the decided advantage of (3.1) that it requires, in addition to the elementary logic of identity, the existence among formulas only of the operation $\wedge$ of conjunction, and not the conditional operation $\to$. It is easily seen, indeed, that even this requirement can be suspended. For the conjunction $x \wedge z$ of two formulas, where it exists, has the same logical force as the meet $\mathbf{x} \wedge \mathbf{z}$ of the axiomatizable systems $\mathbf{x} = \mathbf{Cn}(x)$ and $\mathbf{z} = \mathbf{Cn}(z)$ that x and z axiomatize. The definition (3.0) may therefore be generalized to

$$\vartheta(\mathfrak{y}) \quad = \quad \bigvee \{\mathbf{Cn}(\mathfrak{y} = \ulcorner y \urcorner) \wedge \mathbf{Cn}(y) \mid y \in S\}. \tag{3.6}$$

This definition is widely applicable, for example to many higher-order logics, intensional logics, and paraconsistent logics.

Definition (3.1) is evidently well formed if enough conditionals exist. There is an old result of Skolem's that if the logic $\mathbf{Cn}$ includes the standard rules for the conditional, and disjunction $\vee$ and conjunction $\wedge$ are also present, then the distributive law (2.7) holds. But in the absence of the conditional operation $\to$ among elements of S, it is not easy, without some concessions to distributivity, to simulate it, or even negation, at the level of deductive theories; it is not easy, that is, to define the conditional of two theories, or the complement of a theory. When $\mathbf{Cn}(x)$ is identical with either $\mathbf{S}$ or $\mathbf{L}$, the theory $\mathbf{Cn}(x \to z)$ can be readily represented as $\mathbf{Cn}(x)' \vee \mathbf{Cn}(z)$, where $\mathbf{Y}'$ is the authocomplement of the theory $\mathbf{Y}$ (which is defined in (4.3) below), and the identities (4.5) and (4.6) are provable without any assumption of distributivity. The definition

$$\tau(\mathfrak{y}) \quad =_{\mathrm{Df}} \quad \bigwedge \{((\mathbf{Cn}(\mathfrak{y} = \ulcorner y \urcorner))' \vee \mathbf{Cn}(y) \mid y \in S\}, \tag{3.7}$$

is applicable where (3.1) is not, and is still materially adequate: $\tau(\ulcorner x \urcorner) = \mathbf{Cn}(x)$ for all formulas x. But in non-distributive logics its import for contingent names such as $\mathfrak{p}$ is less transparent.

Two features of the definitions (3.0) and (3.1) merit special remark. The first is the scantiness of the resources of Zermelo set theory $\mathbb{Z}$ that are needed in the background theory $\mathbb{T}$ for the formulation of (3.0) and (3.1), whatever may be the logical strength of the system for which truth is being defined. The denumerably many formulas of S, and their structural-descriptive names, can be represented by natural numbers or hereditarily finite sets. S therefore belongs to $\mathcal{V}_{\omega+1}$. Deductive theories are (in general) denumerable sets of formulas, and they too belong to $\mathcal{V}_{\omega+1}$. An ordered pair of elements of $\mathcal{V}_\nu$ belongs to $\mathcal{V}_{\nu+2}$, and therefore the operation $\mathbf{Cn} : \wp(S) \mapsto \wp(S)$, which is a set of ordered pairs of subsets of S, belongs to $\mathcal{V}_{\omega+4}$. A deductive system $\langle S, \mathbf{Cn} \rangle$ is an ordered pair of an element of $\mathcal{V}_{\omega+1}$ and an element of $\mathcal{V}_{\omega+4}$, and this belongs to $\mathcal{V}_{\omega+6}$. The entire construction fits comfortably into a short extension of the hereditarily finite sets $\mathcal{V}_\omega$.

The other most significant feature of the definitions (3.0) and (3.1) is that they are universal. According to many philosophers, one of the chief shortcomings of Tarski's approach is that, although his method of definition can be universalized, there is no universal definition of truth. Each deductive system requires its own work. This disadvantage is plainly overcome here, since both (3.0) and (3.1) contain explicit references to the two components of the variable deductive system $\langle S, \mathbf{Cn} \rangle$. Note also that although the system $\langle S, \mathbf{Cn} \rangle$ for which truth is defined may be distinguished from the system (the background theory $\mathbb{T}$) in which the definition is conducted, there is no need for any hierarchy of distinct metalanguages, syntactical or semantical.

4   Unaxiomatizable Theories

This section is devoted to several characterizations of unaxiomatizable deductive theories. The first characterization (Theorem 4 and its corollaries) is well known, and is effective for every consequence operation $\mathbf{Cn}$. On the way to the final characterization (Theorem 18), which requires that $\mathbf{Cn}$ include the whole of classical logic, is a result (Theorem 10) for distributive consequence operations that relates the axiomatizability of some theories with that of others.

4.1   Sequences of Theories

An infinite sequence $\{\mathbf{y}_j \mid j \in \mathcal{N}\}$ of axiomatizable theories is (for reasons outlined in §2 above) called *decreasing* if $\mathbf{y}_i \subseteq \mathbf{y}_k$ whenever $i < k$, and *strictly decreasing* if in addition $\mathbf{y}_k \not\subseteq \mathbf{y}_i$ whenever $i < k$. An infinite sequence $\{y_j \mid j \in \mathcal{N}\}$ of formulas is likewise called *decreasing* if $y_k \vdash y_i$ whenever $i < k$, and *strictly decreasing* if also $y_i \not\vdash y_k$ whenever $i < k$.

THEOREM 4.   The theory $\mathbf{Y}$ is unaxiomatizable if & only if $\mathbf{Y} = \mathbf{Cn}(\bigcup\{\mathbf{y}_j \mid j \in \mathcal{N}\})$ for some strictly decreasing sequence $\{\mathbf{y}_j \mid j \in \mathcal{N}\}$ of axiomatizable theories.

Proof. Suppose that $\mathbf{Y} = \mathbf{Cn}(\bigcup\{\mathbf{y}_j \mid j \in \mathcal{N}\})$ is axiomatized by the finite set $\{y_0, \ldots, y_{k-1}\}$. Then by (2.1) there is for each $l < k$ some finite i_l for which $\mathbf{Cn}(\mathbf{y}_{i_l}) \vdash y_l$. If i is the maximum of these i_l then, since the sequence $\{\mathbf{y}_j \mid j \in \mathcal{N}\}$ is strictly decreasing, $\mathbf{y}_i \vdash y_l$ for each $l < k$, and so $\mathbf{Y} \subseteq \mathbf{y}_i$. But $\mathbf{y}_{i+1} \subseteq \mathbf{Y}$, and hence $\mathbf{y}_{i+1} \subseteq \mathbf{y}_i$, contrary to assumption. To represent as $\mathbf{Cn}(\bigcup\{\mathbf{y}_j \mid j \in \mathcal{N}\})$ any given unaxiomatizable theory $\mathbf{Y}$, enumerate its consequences $\{y_i \mid i \in \mathcal{N}\}$, and then define $\mathbf{y}_0$ as $\mathbf{Cn}(\{y_0\})$ and $\mathbf{y}_{j+1}$ as $\mathbf{Cn}(\{y_0, \ldots, y_k\})$ where k is the least number for which $\mathbf{Cn}(\{y_0, \ldots, y_k\}) \not\subseteq \mathbf{y}_j$. It is clear that the theories $\mathbf{y}_j$ form a strictly decreasing sequence.   ∎

COROLLARY 5 (A. Robinson).   If the logic $\mathbf{Cn}$ includes conjunction, the theory $\mathbf{Y}$ is unaxiomatizable if & only if $\mathbf{Y} = \mathbf{Cn}(\{y_j \mid j \in \mathcal{N}\})$ for some strictly decreasing sequence $\{y_j \mid j \in \mathcal{N}\}$.

Proof. The proof is essentially the same as the proof of the theorem.   ∎

COROLLARY 6.   In elementary logic with identity, let ω_j say that the universe has cardinality j. The maximal theory $\bigwedge\{\mathbf{Cn}(\neg\omega_j) \mid j > 0\}$ identified at the end of §2 is unaxiomatizable.

Proof. Define the sentence y_j as the conjunction $\neg\omega_1 \wedge \cdots \wedge \neg\omega_j$, so that $y_0 = \top$. The import of y_j is that the universe has cardinality greater than j. The sequence $\{y_j \mid j \in \mathcal{N}\}$ is obviously strictly decreasing. ∎

COROLLARY 7. Let $\{y_j \mid j \in \mathcal{N}\}$ be any enumeration of a set Y of formulas, $X_i = \{y_j \mid j < i\}$, and $\mathbf{x}_i = \mathbf{Cn}(X_i)$ for each $i \in \mathcal{N}$. Then $\mathbf{Cn}(Y)$ is axiomatizable if & only if the decreasing sequence $\{\mathbf{x}_i \mid i \in \mathcal{N}\}$ of axiomatizable theories is eventually constant.

Proof. $\mathbf{Cn}(Y)$ is axiomatizable if & only if $\mathbf{Cn}(Y) = \mathbf{x}_i$ for some $i \in \mathcal{N}$. Since the sequence $\{\mathbf{x}_i \mid i \in \mathcal{N}\}$ is decreasing, and each $\mathbf{x}_k \subseteq \mathbf{Cn}(Y)$, this holds if & only if $\mathbf{x}_i = \mathbf{x}_k$ for all $k \geq i$. ∎

4.2   A Consequence of Modularity

The least intuitive of the results that concern disjunctions and conjunctions only, holding whenever the logic $\mathbf{Cn}$ includes the distributive (in fact, the modular) law, is Theorem 18 of Tarski (1935–1936). It leads to an alternative criterion of unaxiomatizability in classical logic, in terms of failure of the law of non-contradiction. Since there appear not to exist published proofs of most of these results, we shall prove them here. Three preparatory lemmas are required. Corollary 9 and Lemma 10 are included in Lemmas 1 and 2 on pp. 113f. of Birkhoff (1973).

LEMMA 8.   Suppose that $x \wedge z$ is always defined. If $\mathbf{X} \vee \mathbf{Z}$ is axiomatizable, there are axiomatizable theories $\mathbf{x} \subseteq \mathbf{X}$ and $\mathbf{z} \subseteq \mathbf{Z}$ such that $\mathbf{X} \vee \mathbf{Z} = \mathbf{x} \vee \mathbf{z}$.

Proof. If $\mathbf{X} \vee \mathbf{Z} = \varnothing$ then we may identify both $\mathbf{x}$ and $\mathbf{z}$ with $\varnothing$. If $\mathbf{X} \vee \mathbf{Z} \neq \varnothing$, the existence of conjunctions implies that $\mathbf{X} \vee \mathbf{Z} = \mathbf{Cn}(y)$ for some formula $y \in \mathbf{X} \vee \mathbf{Z}$, from which it follows that $\mathbf{X} \vee \mathbf{Z} = \mathbf{y} \vee \mathbf{y}$, where $\mathbf{y} = \mathbf{Cn}(y)$ is an axiomatizable subtheory of both $\mathbf{X}$ and $\mathbf{Z}$. ∎

COROLLARY 9 (Birkhoff).   Suppose that $x \vee z$ is always defined. Then

$$\mathbf{X} \vee \mathbf{Z} \;=\; \{x \vee z \mid \mathbf{X} \vdash x \text{ and } \mathbf{Z} \vdash z\}, \tag{4.0}$$

Proof. If $\mathbf{X} \vdash x$ and $\mathbf{Z} \vdash z$ then both $\mathbf{X}$ and $\mathbf{Z}$ imply $x \vee z$, and so $\mathbf{X} \vee \mathbf{Z} \vdash x \vee z$. Since y is equivalent to $y \vee y$, every formula in $\mathbf{X} \vee \mathbf{Z}$ is a disjunction of formulas from $\mathbf{X}$ and from $\mathbf{Z}$. ∎

LEMMA 10 (Tarski).   If $\mathbf{X} \wedge \mathbf{Z}$ is axiomatizable, then there are axiomatizable theories $\mathbf{x} \subseteq \mathbf{X}$ and $\mathbf{z} \subseteq \mathbf{Z}$ such that $\mathbf{X} \wedge \mathbf{Z} = \mathbf{x} \wedge \mathbf{z}$.

Proof. If $\mathbf{X} \wedge \mathbf{Z} = \mathbf{Cn}(\{u_0, \ldots u_{j-1}\})$, then by (2.1) there are axiomatizable theories $\mathbf{y}_0, \ldots \mathbf{y}_{j-1}$, subsets of $\mathbf{X}$, and axiomatizable theories $\mathbf{w}_0, \ldots \mathbf{w}_{j-1}$, subsets of $\mathbf{Z}$, such that $u_l \in \mathbf{Cn}(\mathbf{y}_l \cup \mathbf{w}_l) = \mathbf{y}_l \wedge \mathbf{w}_l$ for each $l < j$. Let $\mathbf{y} = \mathbf{y}_0 \wedge \ldots \wedge \mathbf{y}_{j-1}$ and $\mathbf{w} = \mathbf{w}_0 \wedge \ldots \wedge \mathbf{w}_{j-1}$. It is immediate that $\mathbf{y} \subseteq \mathbf{X}$ and $\mathbf{w} \subseteq \mathbf{Z}$ are axiomatizable, and that $\mathbf{Cn}(\{u_0, \ldots u_{j-1}\}) \subseteq \mathbf{y} \wedge \mathbf{w} \subseteq \mathbf{X} \wedge \mathbf{Z} = \mathbf{Cn}(\{u_0, \ldots u_{j-1}\})$. We may conclude that $\mathbf{X} \wedge \mathbf{Z} = \mathbf{y} \wedge \mathbf{w}$. ∎

LEMMA 11. Suppose that $x \wedge z$ is always defined. If $\mathbf{X} \vee \mathbf{Z}$ and $\mathbf{X} \wedge \mathbf{Z}$ are axiomatizable, then there are axiomatizable theories $\mathbf{x} \subseteq \mathbf{X}$ and $\mathbf{z} \subseteq \mathbf{Z}$ such that $\mathbf{X} \vee \mathbf{Z} = \mathbf{x} \vee \mathbf{z}$ and $\mathbf{X} \wedge \mathbf{Z} = \mathbf{x} \wedge \mathbf{z}$.

Proof. By Lemmas 8 and 10, there are axiomatizable theories $\mathbf{u}, \mathbf{y} \subseteq \mathbf{X}$ and axiomatizable theories $\mathbf{v}, \mathbf{w} \subseteq \mathbf{Z}$ for which $\mathbf{X} \vee \mathbf{Z} = \mathbf{u} \vee \mathbf{v}$ and $\mathbf{X} \wedge \mathbf{Z} = \mathbf{y} \wedge \mathbf{w}$. Set $\mathbf{x} = \mathbf{u} \wedge \mathbf{y}$ and $\mathbf{z} = \mathbf{v} \wedge \mathbf{w}$. Since $\mathbf{u}$ and $\mathbf{y}$ are subsets of $\mathbf{X}$, so is $\mathbf{x}$, and likewise $\mathbf{z} \subseteq \mathbf{Z}$. This implies that $\mathbf{x} \vee \mathbf{z} \subseteq \mathbf{X} \vee \mathbf{Z}$. It is clear too that $\mathbf{u} \subseteq \mathbf{x}$ and $\mathbf{v} \subseteq \mathbf{z}$, so that $\mathbf{X} \vee \mathbf{Z} = \mathbf{u} \vee \mathbf{v} \subseteq \mathbf{x} \vee \mathbf{z}$. We may conclude that $\mathbf{X} \vee \mathbf{Z} = \mathbf{x} \vee \mathbf{z}$. In the same way, $\mathbf{x} \wedge \mathbf{z} \subseteq \mathbf{X} \wedge \mathbf{Z} = \mathbf{y} \wedge \mathbf{w} \subseteq (\mathbf{u} \wedge \mathbf{y}) \wedge (\mathbf{v} \wedge \mathbf{w}) = \mathbf{x} \wedge \mathbf{z}$. We may conclude that $\mathbf{X} \wedge \mathbf{Z} = \mathbf{x} \wedge \mathbf{z}$. ∎

THEOREM 12 (Tarski). Suppose that the distributive law (2.7) holds in the logic $\mathbf{Cn}$. Then if $\mathbf{X} \vee \mathbf{Z}$ and $\mathbf{X} \wedge \mathbf{Z}$ are both axiomatizable, $\mathbf{X}$ and $\mathbf{Z}$ are both axiomatizable.

Proof. By Lemma 11 there are axiomatizable theories $\mathbf{x} \subseteq \mathbf{X}$ and $\mathbf{z} \subseteq \mathbf{Z}$ for which $\mathbf{X} \vee \mathbf{Z} = \mathbf{x} \vee \mathbf{z}$ and $\mathbf{X} \wedge \mathbf{Z} = \mathbf{x} \wedge \mathbf{z}$. It is plain that $\mathbf{x} \vee \mathbf{Z} \subseteq \mathbf{X} \vee \mathbf{Z}$; and moreover, if $\mathbf{X} \vee \mathbf{Z} \vdash y$, then $\mathbf{x} \vee \mathbf{z} \vdash y$, and so $\mathbf{x} \vee \mathbf{Z} \vdash y$. It follows that $\mathbf{X} \vee \mathbf{Z} = \mathbf{x} \vee \mathbf{Z}$. In the same way, $\mathbf{X} \wedge \mathbf{Z} = \mathbf{x} \wedge \mathbf{Z}$. Using (2.9), which is a consequence of (2.7), and the absorption laws (2.8), we continue

$$
\begin{aligned}
\mathbf{X} = \mathbf{X} \vee (\mathbf{X} \wedge \mathbf{Z}) \;&=\; \mathbf{X} \vee (\mathbf{x} \wedge \mathbf{Z}) & (4.1)\\
&=\; [\mathbf{X} \vee \mathbf{x}] \wedge [\mathbf{X} \vee \mathbf{Z}] \\
&=\; \mathbf{x} \wedge (\mathbf{x} \vee \mathbf{Z}) \\
&=\; \mathbf{x}.
\end{aligned}
$$

$\mathbf{Z} = \mathbf{z}$ is proved likewise. It should be noted that, since $\mathbf{x} \subseteq \mathbf{X}$, at line (4.1) the modular law

$$\mathbf{Y} \vdash \mathbf{X} \;\;\Leftrightarrow\;\; \mathbf{Y} \vee (\mathbf{X} \wedge \mathbf{Z}) = (\mathbf{Y} \vee \mathbf{X}) \wedge (\mathbf{Y} \vee \mathbf{Z}) \qquad (4.2)$$

(for deductive theories) suffices in place of the full distributive law (2.9). ∎

COROLLARY 13. If the modular law (4.2) holds for the logic $\mathbf{Cn}$, then $\mathbf{X} \vee \mathbf{Z}$ and $\mathbf{X} \wedge \mathbf{Z}$ are both axiomatizable if & only if $\mathbf{X}$ and $\mathbf{Z}$ are both axiomatizable.

Proof. The modular law is not needed for the converse of Theorem 12. Suppose that $\mathbf{X}$ and $\mathbf{Z}$ are axiomatizable. It is evident that no conditions on $\mathbf{Cn}$ are needed to guarantee that $\mathbf{X} \wedge \mathbf{Z}$ is axiomatizable. The existence of both disjunction and conjunction is plainly sufficient (but not necessary) for $\mathbf{X} \vee \mathbf{Z}$ to be axiomatizable. In the absence of conjunction, however, there may be no axiomatization of $\mathbf{X} \vee \mathbf{Z}$ when $\mathbf{X} = \mathbf{Cn}\{x_0, x_1\}$ and $\mathbf{Z} = \mathbf{Cn}(z)$; and in the absence of disjunction, there may be no axiomatization of $\mathbf{X} \vee \mathbf{Z}$ even if $\mathbf{X} = \mathbf{Cn}(x)$ and $\mathbf{Z} = \mathbf{Cn}(z)$. ∎

4.3   Authocomplementation

When we come to define the complement $\mathbf{Y}'$ of an unaxiomatizable theory $\mathbf{Y}$ we inevitably enter non-classical territory. For by Theorem 12, if $\mathbf{Y}$ is unaxiomatizable then at least one of $\mathbf{Y} \vee \mathbf{Y}'$ and $\mathbf{Y} \wedge \mathbf{Y}'$ is unaxiomatizable, no matter which theory we decide to identify $\mathbf{Y}'$ with. Since both $\mathbf{L}$ and $\mathbf{S}$ are axiomatizable in classical logic, it follows that either the law of excluded middle $\mathbf{Y} \vee \mathbf{Y}' = \mathbf{L}$ or the law of non-contradiction $\mathbf{Y} \wedge \mathbf{Y}' = \mathbf{S}$ must fail for each unaxiomatizable theory $\mathbf{Y}$. In fact, it is always the latter, classically and elsewhere. Following Tarski (1935–1936), Theorem 12(a), we therefore define what we shall call the *authocomplement*

$$\mathbf{Y}' \;=\; \bigwedge \{\mathbf{Z} \mid \vdash \mathbf{Y} \vee \mathbf{Z}\} \qquad (4.3)$$

of the theory $\mathbf{Y}$. If we assume the infinite distributive law (2.10), or the more basic law (2.7), we may derive from (4.3) some simple consequences (including the law of excluded middle (4.4), one law of contraposition (4.8), one De Morgan law (4.9), and the law of triple negation (4.11)).

THEOREM 14.   If the consequence operation $\mathbf{Cn}$ obeys the distributive law (2.7), then the following identities, implications, and equivalences hold for all theories $\mathbf{X}$, $\mathbf{Y}$, $\mathbf{Z}$, and families $\mathcal{K}$.

$$\mathbf{Y} \vee \mathbf{Y}' \;=\; \mathbf{L} \qquad (4.4)$$

$$\mathbf{S}' \;=\; \mathbf{L} \qquad (4.5)$$

$$\mathbf{L}' \;=\; \mathbf{S} \qquad (4.6)$$

$$\mathbf{X}' \vdash \mathbf{Z} \;\Leftrightarrow\; \vdash \mathbf{X} \vee \mathbf{Z} \qquad (4.7)$$

$$\mathbf{X}' \vdash \mathbf{Z} \;\Leftrightarrow\; \mathbf{Z}' \vdash \mathbf{X} \qquad (4.8)$$

$$(\bigwedge \mathcal{K})' \;=\; \bigvee \{\mathbf{Y}' \mid \mathbf{Y} \in \mathcal{K}\} \qquad (4.9)$$

$$\mathbf{Y}'' \;\vdash\; \mathbf{Y} \qquad (4.10)$$

$$\mathbf{Y}''' \;=\; \mathbf{Y}'. \qquad (4.11)$$

Proof. These results are all straightforward consequences of (4.3) and (2.7). The proof of (4.4) is much facilitated by resort to (2.10).   ∎

COROLLARY 15.   Suppose that the logic $\mathbf{Cn}$ includes the distributive law (2.7), and that $\mathbf{S}$ is axiomatizable. If $\mathbf{Y} \wedge \mathbf{Y}' = \mathbf{S}$, then $\mathbf{Y}$ is axiomatizable.

Proof. Since $\mathbf{L} = \mathbf{Cn}(\varnothing)$ is always axiomatizable, it follows from (4.4) and Theorem 12 that both $\mathbf{Y}$ and $\mathbf{Y}'$ are axiomatizable. Since it is an immediate consequence of (4.5) that $\mathbf{S} \wedge \mathbf{S}' = \mathbf{S}$, the condition that $\mathbf{S}$ be axiomatizable is essential if the corollary is to hold.   ∎

COROLLARY 16.   Suppose that the logic $\mathbf{Cn}$ obeys the distributive law (2.7), and that $\mathbf{S}$ is axiomatizable. If Ω is a theory that is both maximal and unaxiomatizable, then $\Omega' = \mathbf{L}$.

Proof. By the previous corollary, $\Omega \wedge \Omega' \neq \mathsf{S}$. By the definition of maximality, $\Omega \vdash \Omega'$. This means that $\Omega' = \Omega \vee \Omega'$, which is L by (4.4). ∎

As in the case of disjunction noted in the proof of Corollary 13, the absence of the corresponding connective in the system $\langle S, \mathsf{Cn} \rangle$ may render compounds of axiomatizable theories unaxiomatizable. If S is infinite then it may not be axiomatizable (for example, if $\mathsf{Cn}(Y) = Y$ for every $Y \subseteq S$), even though by (4.5) it is the complement of an axiomatizable theory L; and if the logic Cn does not support a properly behaved negation $\neg y$ of each formula y, then Y' may be unaxiomatizable even if $\mathsf{Y} = \mathsf{Cn}(y)$. As might have been expected, if $\neg$ is classical, all is well.

LEMMA 17. If the logic Cn includes the classical negation operation $\neg$, then $\mathsf{Cn}(y)' = \mathsf{Cn}(\neg y)$.

Proof. Classical logic guarantees that $y \wedge \neg y \vdash \mathsf{Z}$ for any y and any Z. Now if $\vdash \mathsf{Cn}(y) \vee \mathsf{Z}$ then $\vdash y \vee \mathsf{Z}$, and hence also $\neg y \vdash \mathsf{Z}$, and hence $\mathsf{Cn}(\neg y) \vdash \mathsf{Z}$. It follows that $\mathsf{Cn}(\neg y)$ implies the meet $\bigwedge \{\mathsf{Z} \mid \vdash \mathsf{Cn}(y) \vee \mathsf{Z}\} = \mathsf{Cn}(y)'$. Classical logic guarantees too that $\vdash y \vee \neg y$ for any y. That is, $\vdash \mathsf{Cn}(y) \vee \mathsf{Cn}(\neg y)$, from which we may conclude by (4.7) that $\mathsf{Cn}(y)' \vdash \mathsf{Cn}(\neg y)$. ∎

THEOREM 18. If Cn is classical, then Y is an axiomatizable theory if & only if $\mathsf{Y} \wedge \mathsf{Y}' = \mathsf{S}$.

Proof. To establish the converse to Corollary 15, it is necessary to note only that if $\mathsf{Y} = \mathsf{Cn}(y)$ then $\mathsf{Y}' = \mathsf{Cn}(\neg y)$, and hence $\mathsf{Y} \wedge \mathsf{Y}' = \mathsf{Cn}(y) \wedge \mathsf{Cn}(\neg y) = \mathsf{Cn}(y \wedge \neg y) = \mathsf{S}$. ∎

To sum up: whereas in intuitionistic logic the negation $\neg y$ of a formula y contradicts y, but does not always complement it (the law of excluded middle sometimes fails), in the calculus of theories based on distributive consequence operations the theory Y' is an authentic complement of Y, but does not always contradict it (the equation $\mathsf{Y} \wedge \mathsf{Y}' = \mathsf{S}$ sometimes fails). Since the algebraic counterpart of intuitionistic negation is usually called *pseudocomplementation*, it is appropriate to call the operation $'$ on theories based on a distributive consequence operation Cn an operation of *authocomplementation*. The conditional (relative pseudocomplement) $x \rightarrow z$ of intuitionistic logic likewise does not survive unscathed between theories, though it is always possible to define a conditional $\mathsf{X} \twoheadrightarrow \mathsf{Z}$ if the antecedent theory X is finitely axiomatizable.

LEMMA 19. If Cn is classical and $\mathsf{X} = \mathsf{Cn}(x)$, then the theory $\mathsf{Y} = \mathsf{Cn}(\neg x) \vee \mathsf{Z}$ complies with both the laws of *modus ponens*: $\mathsf{X} \wedge \mathsf{Y} \vdash \mathsf{Z}$, and *conditional proof*: if $\mathsf{X} \wedge \mathsf{W} \vdash \mathsf{Z}$ then $\mathsf{W} \vdash \mathsf{Y}$.

Proof. For modus ponens use Lemma 17 and Theorem 18. For conditional proof, note that if $x \wedge \mathsf{W}$ implies each $z \in \mathsf{Z}$, then W implies each $x \rightarrow z$, that is, each $\neg x \vee z$. But the formulas of the form $\neg x \vee z$ (where $z \in \mathbf{Z}$) are exactly the consequences of $\mathbf{Cn}(\neg x) \vee \mathbf{Z}$, which is $\mathbf{Y}$. ∎

Other conditionals sometimes exist. The theory $\delta(\mathfrak{y})$ defined in (3.2), for example, is the conditional $\theta(\mathfrak{y}) \twoheadrightarrow \vartheta(\mathfrak{y})$ (and symmetrically, $\theta(\mathfrak{y}) = \delta(\mathfrak{y}) \twoheadrightarrow \vartheta(\mathfrak{y}))$. But if Ω is an unaxiomatizable maximal theory that does not imply z, then there is no conditional $\Omega \twoheadrightarrow \mathbf{z}$. For $\Omega \vdash x \to z$ if & only if $\Omega \vdash \neg x$, and (since Ω is unaxiomatizable) there is no logically weakest such x.

The *remainder* or relative authocomplement $\mathbf{Z} - \mathbf{X}$ may be defined for any two theories by a generalization of (4.3): $\mathbf{X} - \mathbf{Y} =_{\mathrm{Df}} \bigwedge \{\mathbf{Z} \mid \mathbf{X} \vdash \mathbf{Y} \vee \mathbf{Z}\}$. But it will not be needed in this paper.

5   The Unaxiomatizability of Truth

The present section establishes that if $\mathfrak{y}$ is replaced by a variable name of a formula, the theories $\vartheta(\mathfrak{y})$ and $\theta(\mathfrak{y})$ are unaxiomatizable when $\mathbf{Cn}$ is classical logic; and that, provided that $\mathbf{Cn}$ is recursively presented and rich enough to include Peano arithmetic, $\delta(\mathfrak{y})$ is unaxiomatizable. It must be remembered that we are silently assuming that the logic $\mathbf{Cn}$ includes the whole of the standard elementary logic of identity, and also enough apparatus to allow construction of, and effective manipulations with, structural-descriptive names. The three proofs to be given differ, and only the results concerning $\delta(\mathfrak{y})$ require that, in addition, $\mathbf{Cn}$ be stronger than elementary classical logic.

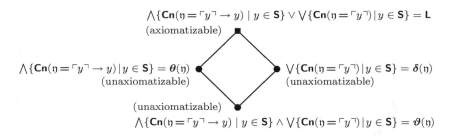

$$\bigwedge \{\mathbf{Cn}(\mathfrak{y} = \ulcorner y \urcorner \to y) \mid y \in \mathbf{S}\} \vee \bigvee \{\mathbf{Cn}(\mathfrak{y} = \ulcorner y \urcorner) \mid y \in \mathbf{S}\} = \mathbf{L}$$
(axiomatizable)

$\bigwedge \{\mathbf{Cn}(\mathfrak{y} = \ulcorner y \urcorner \to y) \mid y \in \mathbf{S}\} = \theta(\mathfrak{y})$ $\bigvee \{\mathbf{Cn}(\mathfrak{y} = \ulcorner y \urcorner) \mid y \in \mathbf{S}\} = \delta(\mathfrak{y})$
(unaxiomatizable)   (unaxiomatizable)

(unaxiomatizable)
$$\bigwedge \{\mathbf{Cn}(\mathfrak{y} = \ulcorner y \urcorner \to y) \mid y \in \mathbf{S}\} \wedge \bigvee \{\mathbf{Cn}(\mathfrak{y} = \ulcorner y \urcorner) \mid y \in \mathbf{S}\} = \vartheta(\mathfrak{y})$$

Figure 1. The theories $\vartheta(\mathfrak{y})$, $\theta(\mathfrak{y})$, $\delta(\mathfrak{y})$, and $\mathbf{L}$

We show first that if the logic $\mathbf{Cn}$ contains enough of classical logic, then $\theta(\mathfrak{y}) \vee \delta(\mathfrak{y}) = \mathbf{L}$ (Theorem 20). In these circumstances $\delta(\mathfrak{y}) = \theta(\mathfrak{y}) \twoheadrightarrow \vartheta(\mathfrak{y})$, and the four theories $\vartheta(\mathfrak{y})$, $\theta(\mathfrak{y})$, $\delta(\mathfrak{y})$, and $\mathbf{L}$ form a lattice quadrilateral, as shown in Figure 1. We then use Theorem 4 to prove that if $\mathbf{Cn}$ is nontrivial and classical and contains the rule of existential introduction, the theory $\theta(\mathfrak{y})$ is unaxiomatizable (Theorem 24). An application of Theorem 3 and Theorem 12 then shows that, provided $\mathbf{Cn}$ is distributive, the theory $\vartheta(\mathfrak{y})$ is also unaxiomatizable (Theorem 25). These are the main results. More work, and more assumptions, are needed for the results, admittedly less central, concerning the unaxiomatizability of the theory $\delta(\mathfrak{n})$ (Lemma 28, Theorem 30).

THEOREM 20. In every distributive logic **Cn** in which $(x \to z) \vee x$ is a theorem, $\boldsymbol{\theta}(\mathfrak{n}) \vee \boldsymbol{\delta}(\mathfrak{n}) = \mathbf{L}$.

Proof. The join $\boldsymbol{\theta}(\mathfrak{n}) \vee \boldsymbol{\delta}(\mathfrak{n}) \vdash w$ if & only if both $\boldsymbol{\theta}(\mathfrak{n}) \vdash w$ and $\boldsymbol{\delta}(\mathfrak{n}) \vdash w$. Thanks to finitariness (2.1), the former implies that there exists some finite set of formulas $\{y_i \mid i < k\} \subseteq S$ such that $(\mathfrak{n} = \ulcorner y_0 \urcorner \to y_0) \wedge \cdots \wedge (\mathfrak{n} = \ulcorner y_{k-1} \urcorner \to y_{k-1}) \vdash w$, and so $\bigwedge \{\mathfrak{n} = \ulcorner y_i \urcorner \to y_i \mid i < k\} \vee \boldsymbol{\delta}(\mathfrak{n}) \vdash w$. It follows by (2.10) that $\bigwedge \{(\mathfrak{n} = \ulcorner y_i \urcorner \to y_i) \vee \boldsymbol{\delta}(\mathfrak{n}) \mid i < k\} \vdash w$, and hence that the set of formulas $\{(\mathfrak{n} = \ulcorner y_i \urcorner \to y_i) \vee (\mathfrak{n} = \ulcorner y_i \urcorner) \mid i < k\} \vdash w$. The assumption that $(x \to z) \vee x$ is a theorem of the logic **Cn** now allows us to conclude that $w \in \mathbf{L}$, and in consequence that $\boldsymbol{\theta}(\mathfrak{n}) \vee \boldsymbol{\delta}(\mathfrak{n}) = \mathbf{L}$. ■

COROLLARY 21. In every distributive logic **Cn**, $\boldsymbol{\tau}(\mathfrak{n}) \vee \boldsymbol{\delta}(\mathfrak{n}) = \mathbf{L}$.

Proof. According to its definition (3.7), $\boldsymbol{\tau}(\mathfrak{n}) = \bigwedge \{(\mathbf{Cn}(\mathfrak{n} = \ulcorner y \urcorner))' \vee \mathbf{Cn}(y) \mid y \in S\}$, where $'$ is the authocomplementation operation on theories defined in (4.3); and so by (2.10) and (3.2),

$$\boldsymbol{\tau}(\mathfrak{n}) \vee \boldsymbol{\delta}(\mathfrak{n}) = \bigwedge \{(\mathbf{Cn}(\mathfrak{n} = \ulcorner y \urcorner))' \vee \mathbf{Cn}(y) \vee \boldsymbol{\delta}(\mathfrak{n}) \mid y \in S\}.$$

This implies that if $\boldsymbol{\tau}(\mathfrak{n}) \vee \boldsymbol{\delta}(\mathfrak{n}) \vdash w$ then

$$\bigwedge \{(\mathbf{Cn}(\mathfrak{n} = \ulcorner y \urcorner))' \vee \mathbf{Cn}(y) \vee \mathbf{Cn}(\mathfrak{n} = \ulcorner y \urcorner) \mid y \in S\} \vdash w.$$

But by (4.4), $(\mathbf{Cn}(\mathfrak{n} = \ulcorner y \urcorner))' \vee \mathbf{Cn}(\mathfrak{n} = \ulcorner y \urcorner) = \mathbf{L}$, and so $w \in \mathbf{L}$, and in consequence $\boldsymbol{\tau}(\mathfrak{n}) \vee \boldsymbol{\delta}(\mathfrak{n}) = \mathbf{L}$. ■

LEMMA 22. Suppose that **Cn** contains the classical rules for the conditional $\to$ and for existential introduction. If y does not contain x free then $x = \ulcorner y \urcorner \to y$ is a theorem (that is, belongs to $\mathbf{L}$) if & only if y is a theorem.

Proof. If the formula $x = \ulcorner y \urcorner \to y \in \mathbf{L}$, then $x = \ulcorner y \urcorner \vdash y$, and accordingly $\exists x(x = \ulcorner y \urcorner) \vdash y$. That is to say, if $x = \ulcorner y \urcorner \to y$ is a theorem then so is y. The converse is immediate. ■

LEMMA 23. If **Cn** contains the classical rules for the conditional $\to$, and the rule of *ex falso quodlibet* (from z and $\neg z$ every formula may be derived), and Y is a subset of $\{\mathfrak{n} = \ulcorner y \urcorner \to y) \mid y \in S\}$ that contains no theorems, then no proper subset of Y implies any other element of Y.

Proof. Thanks to (2.1) we need consider only a finite subset $X = \{\mathfrak{n} = \ulcorner y_i \urcorner \to y_i \mid i < k\} \subseteq Y$, and an element $\mathfrak{n} = \ulcorner y_k \urcorner \to y_k \in Y \backslash X$. If $\bigwedge \{\mathfrak{n} = \ulcorner y_i \urcorner \to y_i \mid i < k\} \vdash \mathfrak{n} = \ulcorner y_k \urcorner \to y_k$, then $\bigwedge \{\mathfrak{n} \neq \ulcorner y_i \urcorner \mid i < k\} \vdash \mathfrak{n} = \ulcorner y_k \urcorner \to y_k$. But $\mathfrak{n} = \ulcorner y_k \urcorner \vdash \mathfrak{n} \neq \ulcorner y_i \urcorner$ for each $i < k$ (since the structural-descriptive names name distinct formulas), and hence $\mathfrak{n} = \ulcorner y_k \urcorner \vdash \mathfrak{n} = \ulcorner y_k \urcorner \to y_k$, and hence $\vdash \mathfrak{n} = \ulcorner y_k \urcorner \to y_k$. This is an impossible conclusion if Y contains no theorems. ■

THEOREM 24. If **Cn** is non-trivial and distributive, and contains the rules assumed in Lemma 22 and Lemma 23, then the theory $\theta(\mathfrak{n}) = \bigwedge\{\mathbf{Cn}(\mathfrak{n} = \ulcorner y \urcorner \to y) \mid y \in S\}$ is unaxiomatizable.

Proof. Since **Cn** is non-trivial, there are infinitely many formulas y that are not theorems. By Lemma 22 there are infinitely many y such that the formula $x = \ulcorner y \urcorner \to y$ is not a theorem, and hence the subset Y of $\{\mathfrak{n} = \ulcorner y \urcorner \to y) \mid y \in S\}$ that contains no theorems is infinite; and since these two sets differ only in the theorems they contain, $\mathbf{Cn}(Y) = \theta(\mathfrak{n})$. Since S is countable, the elements of Y can be listed as $\{y_i \mid i < k\}$, which is equivalent to the sequence $\{z_i \mid i < k\}$ defined by $z_0 = y_0$ and $z_{k+1} = z_k \wedge y_{k+1}$ for every $k \in \mathcal{N}$. By Lemma 23, this sequence is strictly decreasing. By Theorem 4, $\theta(\mathfrak{n}) = \mathbf{Cn}(Y)$ is unaxiomatizable. ∎

THEOREM 25. If **Cn** is non-trivial and distributive, and contains the rules assumed in Lemma 22 and Lemma 23, then the theory $\vartheta(\mathfrak{n})$ is unaxiomatizable.

Proof. Since $\mathbf{L} = \theta(\mathfrak{n}) \vee \delta(\mathfrak{n})$ is axiomatizable in all logics and, by Theorem 24, $\theta(\mathfrak{n})$ is unaxiomatizable in all sufficiently strong non-trivial logics, we may apply Theorem 12 to conclude that $\vartheta(\mathfrak{n}) = \theta(\mathfrak{n}) \wedge \delta(\mathfrak{n})$ is unaxiomatizable in all these subsystems of classical logic. ∎

At this stage nothing has been proved about the theory $\delta(\mathfrak{n})$. For the remainder of this section it will be assumed that **Cn** contains the whole of classical logic. We shall show in Corollary 27 that in an ω-complete logic **Cn** (that is to say, one in which the set **L** is ω-complete), $\delta(\mathfrak{n}) = \mathbf{L}$, and accordingly $\vartheta(\mathfrak{n}) = \theta(\mathfrak{n})$. But in some richer logics, including those that are ω-consistent as well as essentially incomplete, $\delta(\mathfrak{n})$ is unaxiomatizable. The properties of ω-completeness and ω-consistency are discussed by Tarski (1933b), and also in many modern textbooks.

A formula $z(\mathfrak{n})$ in S will be called $\mathfrak{n}$-*universal* if each instance $z(\ulcorner y \urcorner)$ obtained by substituting for $\mathfrak{n}$ a structural-descriptive name $\ulcorner y \urcorner$ of a formula y in S is a theorem in **Cn**. If the variable $\mathfrak{n}$ is not free in z, then z is $\mathfrak{n}$-*universal* if & only if it is a theorem in **Cn**, an element of **L**.

LEMMA 26. The theory $\delta(\mathfrak{n})$ is identical with the meet of all the $\mathfrak{n}$-universal formulas of S:

$$\bigvee\{\mathbf{Cn}(\mathfrak{n} = \ulcorner y \urcorner) \mid y \in S\} = \bigwedge\{z(\mathfrak{n}) \mid z(\mathfrak{n}) \in S \ \& \ z \text{ is } \mathfrak{n}\text{-universal}\}.$$

Proof. Let $z(\mathfrak{n})$ be a $\mathfrak{n}$-universal formula. It is plain that $\mathfrak{n} = \ulcorner y \urcorner \vdash z(\mathfrak{n})$ for any $y \in S$, and hence that $\delta(\mathfrak{n}) = \bigvee\{\mathbf{Cn}(\mathfrak{n} = \ulcorner y \urcorner) \mid y \in S\} \vdash z(\mathfrak{n})$. For the converse, suppose that $z(\mathfrak{n})$ is derivable from each formula of the form $\mathfrak{n} = \ulcorner y \urcorner$. It is immediate that each $z(\ulcorner y \urcorner)$ is derivable from $\ulcorner y \urcorner = \ulcorner y \urcorner$, and accordingly is a theorem. That is to say, z is a $\mathfrak{n}$-universal formula. ∎

COROLLARY 27. If the logic **Cn** is ω-complete, then $\delta(\mathfrak{y}) = \bigvee\{\textbf{Cn}(\mathfrak{y} = \ulcorner y \urcorner)$ $\mid y \in \textbf{S}\} = \textbf{L}$.

Proof. This is immediate, using Lemma 26. If $z(\mathfrak{y})$ is $\mathfrak{y}$-universal in an ω-complete logic, then $\forall \mathfrak{y} z(\mathfrak{y})$ is a theorem. It follows that

$$\bigvee\{\textbf{Cn}(\mathfrak{y} = \ulcorner y \urcorner) \mid y \in \textbf{S}\} = \textbf{L}.$$

∎

THEOREM 28. If the logic **Cn** is ω-incomplete, then $\delta(\mathfrak{y}) = \bigvee\{\textbf{Cn}(\mathfrak{y} = \ulcorner y \urcorner)$ $\mid y \in \textbf{S}\} \neq \textbf{L}$.

Proof. This is also practically immediate, using Lemma 26. If $\bigwedge\{z(\mathfrak{y}) \mid z(\mathfrak{y}) \in S \ \& \ z$ is $\mathfrak{y}$-universal$\} = \textbf{L}$, then every $\mathfrak{y}$-universal formula $z(\mathfrak{y})$ is a theorem, which implies that **Cn** is ω-complete.   ∎

For an example, let g be an effective $1-1$ association in $\mathbb{ZF}$ of sentences with proofs, and let u be an undecidable sentence in $\mathbb{ZF}$ that is interderivable with the sentence $\ulcorner u \urcorner$ *is not a theorem of* $\mathbb{ZF}$. Then for each $y \in S$, the formula $g(\ulcorner y \urcorner)$ *is not a proof of* $\ulcorner u \urcorner$ is a theorem of $\mathbb{ZF}$. This implies that $g(\mathfrak{y})$ *is not a proof of* $\ulcorner u \urcorner$ follows from $\mathfrak{y} = \ulcorner y \urcorner$ for each $y \in S$; and hence from $\bigvee\{\textbf{Cn}(\mathfrak{y} = \ulcorner y \urcorner) \mid y \in S\}$. But if $g(\mathfrak{y})$ *is not a proof of* $\ulcorner u \urcorner$ were a theorem, so would be its universal generalization $\ulcorner u \urcorner$ *is not a theorem of* $\mathbb{ZF}$, and so also would be the sentence u.

Theorem 28 can be considerably improved if we choose for **Cn** a logic based on an ω-consistent and incompletable theory such as elementary Peano arithmetic. In such logics the set **L** of theorems is (we hope) ω-consistent, but incompletable in the sense that there exists no finitely (or even recursively) axiomatizable theory that is maximal. We begin with a simplifying lemma.

LEMMA 29. Suppose that **Cn** is ω-consistent. If $\delta(\mathfrak{y}) \vdash w(\mathfrak{y})$, then $\forall \mathfrak{y} w(\mathfrak{y})$ is consistent.

Proof. By (2.1), $w(\mathfrak{y})$ is implied by some finite conjunction $z_0(\mathfrak{y}) \wedge \cdots \wedge z_{k-1}(\mathfrak{y})$ of $\mathfrak{y}$-universal formulas, and hence $\forall \mathfrak{y} z_0(\mathfrak{y}) \wedge \cdots \wedge \forall \mathfrak{y} z_{k-1}(\mathfrak{y}) \vdash \forall \mathfrak{y} w(\mathfrak{y})$. In an ω-consistent theory, the universal quantification of a $\mathfrak{y}$-universal formula is consistent; ergo, $\forall \mathfrak{y} w(\mathfrak{y})$ is consistent (in **Cn**).   ∎

THEOREM 30. If **Cn** is ω-consistent and incompletable then the theory $\delta(\mathfrak{y})$ is unaxiomatizable.

Proof. Suppose that there is some formula $w(\mathfrak{y})$ for which $\delta(\mathfrak{y}) = \textbf{Cn}(w(\mathfrak{y}))$. By Lemma 29, the sentence $u = \forall \mathfrak{y} w(\mathfrak{y})$ is consistent in **Cn**. It will be shown that $\textbf{Cn}(u)$ is a maximal theory.

The quantifier-free sentences of the language of Peano arithmetic are all decidable (either such a sentence x is a theorem, or its negation $\neg x$ is a theorem). If the extended language $\mathcal{L}^+$ contains quantifier-free sentences that are not decidable, then one of each pair $\{x, \neg x\}$ may be added in a constructive manner to the stock of theorems. In other words, we may safely assume that if $x(\mathfrak{y})$ is a formula of one free variable, then u decides each quantifier-free sentence $x(\ulcorner y \urcorner)$.

Now let $x(\mathfrak{y})$ be a formula that contains free at least the variable $\mathfrak{y}$. If $u \vdash x(\ulcorner y \urcorner)$ for some y, then $u \vdash \exists \mathfrak{y} x(\mathfrak{y})$. The alternative is that $u \vdash \neg x(\ulcorner y \urcorner)$ for every y, and so $u \to \neg x(\ulcorner y \urcorner) \in \mathbf{L}$ for every y, and hence $u \to \neg x(\mathfrak{y})$ is $\mathfrak{y}$-universal. It follows that $w(\mathfrak{y}) \vdash u \to \neg x(\mathfrak{y})$. But $u \vdash w(\mathfrak{y})$, and hence $u \vdash \neg x(\mathfrak{y})$. Since u is a sentence, $u \vdash \forall \mathfrak{y} \neg x(\mathfrak{y})$, so $u \vdash \neg \exists \mathfrak{y} x(\mathfrak{y})$.

For all formulas x containing $\mathfrak{y}$ free, that is to say, u decides $\exists \mathfrak{y} x(\mathfrak{y})$ if it decides each instance $x(\ulcorner y \urcorner)$, and so u decides also $\exists \mathfrak{y} \neg x(\mathfrak{y})$, and hence also $\forall \mathfrak{y} x(\mathfrak{y})$. This means that a universal formula $\forall \mathfrak{y} x(\mathfrak{y})$ is decided by u if all its instances $x(\ulcorner y \urcorner)$ are. The base of the induction was settled two paragraphs ago, and we may therefore conclude that every universal formula is decided by u. In short, $\mathbf{Cn}(u)$ is a maximal theory. This is impossible if $\mathbf{Cn}$ is incompletable.   ∎

Theorems 24, 25, and 30 establish the unaxiomatizability, under appropriate conditions, of the theories $\vartheta(\mathfrak{y})$, $\theta(\mathfrak{y})$, $\delta(\mathfrak{y})$, where $\mathfrak{y}$ is a variable for names of formulas. With care they can be extended to other terms, including most names for formulas that are not structural-descriptive names. By Lemma 23, Theorem 24, and Theorem 25, for example, the theories $\theta(\mathfrak{p})$ and $\vartheta(\mathfrak{p})$, where $\mathfrak{p} =$ 'the first sentence of *Pride & Prejudice*', are unaxiomatizable in classical logic, provided that there are infinitely many non-theorems in the set $\{\mathfrak{p} = \ulcorner y \urcorner \to y) \mid y \in S\}$.

6   Falsehood and Untruth

In this section we assume that the consequence operation $\mathbf{Cn}$ is non-trivial and includes the whole of classical logic. Our task is to establish that, undeterred by their classical origins, the function ϑ exhibits truth-value gaps and the functions ϑ and θ both exhibit truth-value gluts. In parallel with the use of the two lower-case forms of the Greek letter *theta*, ϑ and θ, for the two functions defining truth, we shall henceforth use the two lower-case forms of *phi*, φ and ϕ, for the corresponding functions defining falsity, and the two lower-case forms of *pi*, ϖ and π, for the corresponding functions defining untruth.

The falsity of a formula, anyway in classical logic, should be little more than the truth of its negation. One way to express this identity is to employ, (or, if necessary, introduce), for any named formula, a name for the formula that would be obtained by some canonical insertion of a negation sign. But rather than take such a route, which will taken in Miller (2009), when we deal systematically with questions about compositionality, we shall proceed as follows. We define

$$\varphi(\mathfrak{y}) \quad = \quad \bigvee \{\mathbf{Cn}(\mathfrak{y} = \ulcorner y \urcorner \wedge \neg y) \mid y \in S\}, \tag{6.0}$$

$$\phi(\mathfrak{y}) \quad = \quad \bigwedge\{\mathbf{Cn}(\mathfrak{y} = \ulcorner y \urcorner \to \neg y) \mid y \in S\}, \tag{6.1}$$

as the *falsity* operations that correspond to ϑ and θ, and alongside them define also two *untruth* (or *perjury*) operations $\varpi(\mathfrak{y})$ and $\pi(\mathfrak{y})$ by means of theory complementation (4.3):

$$\varpi(\mathfrak{y}) \quad =_{\mathrm{Df}} \quad \vartheta(\mathfrak{y})', \tag{6.2}$$

$$\pi(\mathfrak{y}) \quad =_{\mathrm{Df}} \quad \theta(\mathfrak{y})'. \tag{6.3}$$

THEOREM 31. For classical $\mathbf{Cn}$, (6.0) and (6.2), and (6.1) and (6.3), are materially adequate:

$$\varphi(\ulcorner x \urcorner) = \quad \mathbf{Cn}(\neg x) \quad = \varpi(\ulcorner x \urcorner), \tag{6.4}$$

$$\phi(\ulcorner x \urcorner) = \quad \mathbf{Cn}(\neg x) \quad = \pi(\ulcorner x \urcorner), \tag{6.5}$$

whenever $\ulcorner x \urcorner$ is the structural-descriptive name of a formula $x \in S$.

Proof. The proof is immediate, using the assumed properties of structural-descriptive names. ■

LEMMA 32. If $\mathbf{Cn}$ includes classical logic, and $\mathbf{Y}'$ stands for the complement of the theory $\mathbf{Y}$,

$$\varpi(\mathfrak{y}) \quad\quad = \quad\quad \varphi(\mathfrak{y}) \vee \delta(\mathfrak{y})', \tag{6.6}$$

$$\pi(\mathfrak{y}) \quad = \quad \varphi(\mathfrak{y}) \quad = \quad \phi(\mathfrak{y}) \wedge \delta(\mathfrak{y}). \tag{6.7}$$

Proof. By (3.4), $\vartheta(\mathfrak{y}) = \theta(\mathfrak{y}) \wedge \delta(\mathfrak{y})$, and so by (3.0), (4.9), Lemma 17, (6.2), and (6.0),

$$
\begin{aligned}
\varpi(\mathfrak{y}) \quad = \quad \vartheta(\mathfrak{y})' \quad &= \quad (\bigwedge\{\mathbf{Cn}(\mathfrak{y} = \ulcorner y \urcorner \to y) \mid y \in S\} \wedge \delta(\mathfrak{y}))' \\
&= \quad \bigvee\{(\mathbf{Cn}(\mathfrak{y} = \ulcorner y \urcorner \to y))' \mid y \in S\} \vee \delta(\mathfrak{y})' \\
&= \quad \bigvee\{\mathbf{Cn}(\neg(\mathfrak{y} = \ulcorner y \urcorner \to y)) \mid y \in S\} \vee \delta(\mathfrak{y})' \\
&= \quad \bigvee\{\mathbf{Cn}(\mathfrak{y} = \ulcorner y \urcorner \wedge \neg y) \mid y \in S\} \vee \delta(\mathfrak{y})' \\
&= \quad \varphi(\mathfrak{y}) \vee \delta(\mathfrak{y})'.
\end{aligned}
$$

The identity of $\pi(\mathfrak{y})$ and $\varphi(\mathfrak{y})$ is a simple consequence of (6.2), (3.1), (4.9), and (6.0). To prove that $\varphi(\mathfrak{y}) = \phi(\mathfrak{y}) \wedge \delta(\mathfrak{y})$, it should suffice to repeat the proof of Theorem 3.4. ■

According to (6.6) and (6.7), $\varphi(\mathfrak{y})$ implies $\varpi(\mathfrak{y})$, and $\pi(\mathfrak{y})$ implies $\phi(\mathfrak{y})$. For structural-descriptive names, falsity and untruth coincide, but in general the implications are not reversible.

THEOREM 33. If the logic $\mathbf{Cn}$ is ω-incomplete (that is, according to Theorem 28, $\delta(\mathfrak{y}) \neq \mathbf{L}$), then $\varpi(\mathfrak{y})$ does not imply $\varphi(\mathfrak{y})$, and $\phi(\mathfrak{y})$ does not imply $\pi(\mathfrak{y})$.

Proof. If $\varpi(\mathfrak{n})$ implies $\varphi(\mathfrak{n})$, then by (6.6), $\delta(\mathfrak{n})'$ implies $\varphi(\mathfrak{n})$. It is evident from their definitions (3.2) and (6.0) that $\delta(\mathfrak{n})$, which is identical with $\bigvee\{\mathbf{Cn}(\mathfrak{n}=\ulcorner y\urcorner) \mid y \in S\}$, is implied by $\varphi(\mathfrak{n})$, which is identical with $\bigvee\{\mathbf{Cn}(\mathfrak{n}=\ulcorner y\urcorner \wedge \neg y) \mid y \in S\}$, and hence that $\delta(\mathfrak{n}) \vee \varphi(\mathfrak{n}) = \delta(\mathfrak{n})$. It follows that $\delta(\mathfrak{n}) \vee \delta(\mathfrak{n})'$ implies $\delta(\mathfrak{n})$, and so, by (4.4), $\delta(\mathfrak{n}) = \mathbf{L}$.

If $\phi(\mathfrak{n})$ implies $\pi(\mathfrak{n})$, then by (6.7) and Lemma 26, $\phi(\mathfrak{n})$ implies every $\mathfrak{n}$-universal formula $z(\mathfrak{n})$. By (2.1), there is, for each $z(\mathfrak{n})$, some finite set $\{y_i \mid i < k\}$ of formulas such that the conjunction $(\mathfrak{n}=\ulcorner y_0\urcorner \to \neg y_0) \wedge \cdots \wedge (\mathfrak{n}=\ulcorner y_{k-1}\urcorner \to \neg y_{k-1})$ implies $z(\mathfrak{n})$. It follows that the conjunction $\mathfrak{n} \neq \ulcorner y_0\urcorner \wedge \cdots \wedge \mathfrak{n} \neq \ulcorner y_{k-1}\urcorner$ implies $z(\mathfrak{n})$. But since $z(\mathfrak{n})$ is $\mathfrak{n}$-universal, the finite disjunction $\mathfrak{n}=\ulcorner y_0\urcorner \vee \cdots \vee \mathfrak{n}=\ulcorner y_{k-1}\urcorner$ also implies $z(\mathfrak{n})$. In short, every $\mathfrak{n}$-universal formula $z(\mathfrak{n})$ is demonstrable. This cannot be the case if $\delta(\mathfrak{n}) \neq \mathbf{L}$, as has been assumed.   ∎

LEMMA 34. If the logic **Cn** is ω-incomplete (more generally, if $\delta(\mathfrak{n}) \neq \mathbf{L}$), then $\vartheta(\mathfrak{n}) \vee \varphi(\mathfrak{n}) \neq \mathbf{L}$.

Proof. If both $\vartheta(\mathfrak{n})$ and $\varphi(\mathfrak{n})$ imply the formula x, then each formula of the form $\mathfrak{n}=\ulcorner y\urcorner \wedge y$ implies x and each formula of the form $\mathfrak{n}=\ulcorner y\urcorner \wedge \neg y$ implies x. It follows that each formula $\mathfrak{n}=\ulcorner y\urcorner$ implies x, and hence that $\delta(\mathfrak{n})$ implies x. If $\delta(\mathfrak{n}) \neq \mathbf{L}$ then $\vartheta(\mathfrak{n}) \vee \varphi(\mathfrak{n}) \neq \mathbf{L}$.   ∎

LEMMA 35. If the logic **Cn** is ω-consistent and incompletable (more generally, if $\delta(\mathfrak{n})$ is unaxiomatizable), then $\vartheta(\mathfrak{n}) \wedge \varphi(\mathfrak{n}) \neq \mathbf{S}$.

Proof. By (3.0), (6.0), and Theorem 3 we have for any $x \in S$

$$\vartheta(\mathfrak{n}) \wedge \varphi(\mathfrak{n}) = \bigvee\{\mathbf{Cn}(\mathfrak{n}=\ulcorner y\urcorner \wedge y) \mid y \in S\} \wedge \qquad (6.8)$$
$$\wedge \bigvee\{\mathbf{Cn}(\mathfrak{n}=\ulcorner y\urcorner \wedge \neg y) \mid y \in S\}$$
$$= \bigwedge\{\mathbf{Cn}(\mathfrak{n}=\ulcorner y\urcorner \to y) \mid y \in S\} \wedge \delta(\mathfrak{n}) \wedge$$
$$\wedge \bigwedge\{\mathbf{Cn}(\mathfrak{n}=\ulcorner y\urcorner \to \neg y) \mid y \in S\} \wedge \delta(\mathfrak{n}).$$

Now $(\mathfrak{n}=\ulcorner y\urcorner \to y) \wedge (\mathfrak{n}=\ulcorner y\urcorner \to \neg y)$ is logically equivalent to $\mathfrak{n} \neq \ulcorner y\urcorner$. It follows that $\vartheta(\mathfrak{n}) \wedge \varphi(\mathfrak{n}) = \bigwedge\{\mathfrak{n} \neq \ulcorner y\urcorner \mid y \in S\} \wedge \delta(\mathfrak{n}) = \bigwedge\{\mathfrak{n} \neq \ulcorner y\urcorner \mid y \in S\} \wedge \bigvee\{\mathfrak{n}=\ulcorner y\urcorner \mid y \in S\}$, which by (4.9), is identical with $\bigwedge\{\mathfrak{n} \neq \ulcorner y\urcorner \mid y \in S\} \wedge (\bigwedge\{\mathfrak{n} \neq \ulcorner y\urcorner \mid y \in S\})'$. By Theorem 18, $\vartheta(\mathfrak{n}) \wedge \varphi(\mathfrak{n}) = \mathbf{S}$ if & only if $\bigwedge\{\mathfrak{n} \neq \ulcorner y\urcorner \mid y \in S\}$ is axiomatizable. In classical logic, the complement $\mathbf{Y}'$ of an axiomatizable theory $\mathbf{Y}$ is also axiomatizable. It follows that if $\vartheta(\mathfrak{n}) \wedge \varphi(\mathfrak{n}) = \mathbf{S}$ then $\delta(\mathfrak{n}) = (\bigwedge\{\mathfrak{n} \neq \ulcorner y\urcorner \mid y \in S\})'$ is axiomatizable. The result follows by contraposition.   ∎

THEOREM 36. In general, $\vartheta(\mathfrak{n}) \vee \varpi(\mathfrak{n}) = \theta(\mathfrak{n}) \vee \pi(\mathfrak{n}) = \theta(\mathfrak{n}) \vee \phi(\mathfrak{n}) = \mathbf{L}$, but $\vartheta(\mathfrak{n}) \vee \varphi(\mathfrak{n}) \neq \mathbf{L}$.

Proof. The law of excluded middle (4.4) holds for all deductive theories, and hence $\vartheta(\eta) \vee \varpi(\eta)$ and $\theta(\eta) \vee \pi(\eta)$ are both **L**. By (6.7), $\theta(\eta) \vee \phi(\eta) =$ **L**. By Lemma 34, $\vartheta(\eta) \vee \varphi(\eta) \neq$ **L**.   ∎

THEOREM 37. In general, none of the four theories $\vartheta(\eta) \wedge \varpi(\eta)$, $\theta(\eta) \wedge \pi(\eta)$, $\theta(\eta) \wedge \phi(\eta)$, $\vartheta(\eta) \wedge \varphi(\eta)$ is identical with **S**.

Proof. By Theorems 18, 24, and 25, $\vartheta(\eta) \wedge \varpi(\eta) \neq$ **S** and $\theta(\eta) \wedge \pi(\eta) \neq$ **S**. It then follows from (6.7) that $\theta(\eta) \wedge \phi(\eta) \neq$ **S**, and Lemma 35 says that $\vartheta(\eta) \wedge \varphi(\eta)$ may differ from **S**.   ∎

Loosely stated: only ϑ admits *truth-value gaps*, but both ϑ and θ admit *truth-value gluts*.

7 The Truth of Theories

A deductive theory $\mathbf{Y} = \mathbf{Cn}(\mathbf{Y})$ is true if & only if every sentence y in $\mathbf{Y}$ is true. That is what we are tempted to say. But on the present account, truth is not a predicate, and we cannot say what we are tempted to say. Yet the truth of a sentence is a theory, and the truth of that theory is presumably no more than the truth of the sentence. If $\mathfrak{Y}$ is an expression (variable or name) that stands for a theory, or more generally any set of formulas of $\mathcal{L}^+$, our aim is to extend the definitions of $\vartheta(\eta)$ and $\theta(\eta)$ in such a way that if $\mathfrak{Y} = \mathbf{Cn}(\eta)$ is demonstrable in the background theory $\mathbb{T}$ then $\vartheta(\mathfrak{Y}) = \vartheta(\eta)$ and $\theta(\mathfrak{Y}) = \theta(\eta)$ are also demonstrable. Like $\vartheta(\eta)$ and $\theta(\eta)$, the sets $\vartheta(\mathfrak{Y})$ and $\theta(\mathfrak{Y})$ are to be deductive theories.

The least that we must hope for in addition is that the extended definition of truth be materially adequate. Theorem 38 of Tarski (1935–1936) states that in a non-trivial classical deductive system there are continuum many deductive theories, which implies that only a handful of them, in addition to the axiomatizable theories, can be endowed with structural-descriptive names. It is perhaps reasonable to hope that a structural-descriptive name $\ulcorner\mathbf{Y}\urcorner$ can be given to each recursively axiomatizable theory $\mathbf{Y}$. But if we insist that the primary demands on structural-descriptive names, that $\ulcorner\mathbf{X}\urcorner = \ulcorner\mathbf{Z}\urcorner$ be demonstrable if $\mathbf{X} = \mathbf{Z}$ and refutable if $\mathbf{X} \neq \mathbf{Z}$, and that $\ulcorner x\urcorner \in \ulcorner\mathbf{Z}\urcorner$ be demonstrable if $x \in \mathbf{Z}$ and refutable if $x \notin \mathbf{Z}$, be achieved within the logic **Cn**, then the investigation will have to be restricted to logics sufficiently powerful to allow this.

The line of approach to be adopted here has recourse to a simple extension of Tarski's calculus of deductive theories. It is disappointing that, although it works well for θ, it fails badly for ϑ.

We have seen in (2.3)–(2.6) how Tarski proposed to define finite and infinite analogues of Boolean operations on theories. To my knowledge he never explicitly considered defining analogues of quantification, though he explored in detail elsewhere the closely related operation of cylindrification (for a brief survey of his work on cylindric algebras see Monk 1986, pp. 902–905).

Let **Y** be a theory (or any other set of formulas drawn from S) in which
η is a variable that is sometimes free. We shall define $\exists\eta\mathbf{Y}$ and $\forall\eta\mathbf{Y}$ as
theories, and then extend the definition of truth θ from formulas to theories
so as to deliver the wanted analogue of material adequacy. But it does not
appear to be possible to obtain a corresponding result for the function ϑ.

Every theory **Y** contains formulas (for example, $\eta = \eta$) in which the variable η occurs free. We are interested at present in those theories in which η
has no essential occurrence. **Y** will accordingly be called η-*less* (rather than
η-free, which might be too confusing) if $\mathbf{Y} = \mathbf{Cn}(Y)$ for some set Y none of
whose elements contains η free. We proceed to define, rather obviously,

$$\exists\eta\mathbf{Y} \quad =_{\text{Df}} \quad \bigwedge\{\mathbf{X} \mid \mathbf{X} \text{ is } \eta\text{-less and } \mathbf{Y} \vdash \mathbf{X}\}, \qquad (7.0)$$

$$\forall\eta\mathbf{Y} \quad =_{\text{Df}} \quad \bigvee\{\mathbf{X} \mid \mathbf{X} \text{ is } \eta\text{-less and } \mathbf{X} \vdash \mathbf{Y}\}. \qquad (7.1)$$

Taking for granted that these definitions license the usual quantifier inferences, we may define:

$$\vartheta(\mathfrak{Y}) \quad =_{\text{Df}} \quad \forall\eta[\mathbf{Cn}(\eta \in \mathfrak{Y}) \twoheadrightarrow \vartheta(\eta)], \qquad (7.2)$$

$$\theta(\mathfrak{Y}) \quad =_{\text{Df}} \quad \forall\eta[\mathbf{Cn}(\eta \in \mathfrak{Y}) \twoheadrightarrow \theta(\eta)]. \qquad (7.3)$$

It will be recalled from Lemma 19 that in classical logic the conditional
$\mathbf{X} \twoheadrightarrow \mathbf{Z}$ exists whenever $\mathbf{X}$ is axiomatizable, and is identical with $\mathbf{X}' \vee \mathbf{Z}$; if
$\mathbf{X} = \mathbf{Cn}(x)$ then $\mathbf{X} \twoheadrightarrow \mathbf{Z} = \neg x \vee \mathbf{Z}$. Similar extensions to deductive theories
may be given for the functions φ, ϕ, ϖ, and π.

There exist direct generalizations of Theorem 2 for θ and ϑ, and of Theorem 3.3 for θ.

THEOREM 38.   If the theory named by $\mathfrak{X}$ is implied in **Cn** by the theory
named by $\mathfrak{Z}$, then $\theta(\mathfrak{X})$ is implied by $\theta(\mathfrak{Z})$.

Proof. Immediate using (2.6).   ■

THEOREM 39.   If $\ulcorner\mathbf{Y}\urcorner$ is a structural-descriptive name for the η-less theory
Y, then

$$\theta(\ulcorner\mathbf{Y}\urcorner) \quad = \quad \mathbf{Y}. \qquad (7.4)$$

Proof. If $y \in \mathbf{Y}$, then the formula $\ulcorner y\urcorner \in \ulcorner\mathbf{Y}\urcorner$ is demonstrable in **Cn**, and
hence the theory $\mathbf{Cn}(\ulcorner y\urcorner \in \ulcorner\mathbf{Y}\urcorner) \twoheadrightarrow \theta(\ulcorner y\urcorner)$, which is identical with the
theory $\mathbf{Cn}(\ulcorner y\urcorner \notin \ulcorner\mathbf{Y}\urcorner) \vee \theta(\ulcorner y\urcorner)$, is demonstrably identical with $\theta(\ulcorner y\urcorner)$.
Since, by (7.3), $\theta(\ulcorner\mathbf{Y}\urcorner)$ implies the theory $\mathbf{Cn}(\ulcorner y\urcorner \in \ulcorner\mathbf{Y}\urcorner) \twoheadrightarrow \theta(\ulcorner y\urcorner)$ for
every $y \in S$, it implies $\theta(\ulcorner y\urcorner)$ for every $y \in \mathbf{Y}$. By Theorem 1, $\theta(\ulcorner y\urcorner) =$
$\mathbf{Cn}(y)$. In short, $\theta(\ulcorner\mathbf{Y}\urcorner)$ implies **Y**. (If $y \notin \mathbf{Y}$ then $\ulcorner y\urcorner \in \ulcorner\mathbf{Y}\urcorner$ is refutable,
and $\mathbf{Cn}(\ulcorner y\urcorner \in \ulcorner\mathbf{Y}\urcorner) \twoheadrightarrow \theta(\ulcorner y\urcorner) = \mathbf{L}$.)

For the converse implication, note that if $y \in \mathbf{Y}$ then **Y** implies y,
while if $y \notin \mathbf{Y}$ then $\mathbf{Cn}(\ulcorner y\urcorner \in \ulcorner\mathbf{Y}\urcorner) = \mathbf{S}$. In other words, for each $y \in$

S, the theory $\mathbf{Y} \wedge \mathbf{Cn}(\ulcorner y \urcorner \in \ulcorner \mathbf{Y} \urcorner)$ implies y, and therefore, for each $y \in S$, the theory $\mathbf{Y} \wedge \mathbf{Cn}(\mathfrak{y} \in \ulcorner \mathbf{Y} \urcorner) \wedge \mathbf{Cn}(\mathfrak{y} = \ulcorner y \urcorner)$ implies y. By Lemma 19, $\mathbf{Y} \wedge \mathbf{Cn}(\mathfrak{y} \in \ulcorner \mathbf{Y} \urcorner)$ implies every (axiomatizable) theory $\mathbf{Cn}(\mathfrak{y} = \ulcorner y \urcorner) \twoheadrightarrow y$, where $y \in S$, and therefore implies their meet $\bigwedge \{ \mathbf{Cn}(\mathfrak{y} = \ulcorner y \urcorner \to y) \mid y \in S \}$, which is $\theta(\mathfrak{y})$. By Lemma 19 again, $\mathbf{Y}$ implies the theory $\mathbf{Cn}(\mathfrak{y} \in \ulcorner \mathbf{Y} \urcorner) \twoheadrightarrow \theta(\mathfrak{y})$. Since $\mathbf{Y}$ is $\mathfrak{y}$-less, we may conclude that $\mathbf{Y}$ implies $\theta(\ulcorner \mathbf{Y} \urcorner)$. ∎

THEOREM 40.   If $\ulcorner \mathbf{Y} \urcorner$ is a structural-descriptive name for the $\mathfrak{y}$-less theory $\mathbf{Y}$, then

$$\vartheta(\ulcorner \mathbf{Y} \urcorner) \;\vdash\; \mathbf{Y}. \tag{7.5}$$

Proof. By the definition (7.2), $\vartheta(\ulcorner \mathbf{Y} \urcorner)$ implies $\mathbf{Cn}(\ulcorner y \urcorner \in \ulcorner \mathbf{Y} \urcorner) \twoheadrightarrow \vartheta(\ulcorner y \urcorner)$ for each y; and, on the assumption (made use of throughout the previous theorem) that formulas such as $\ulcorner y \urcorner \in \ulcorner \mathbf{Y} \urcorner$ are correctly decidable in $\mathbf{Cn}$, it follows that $\vartheta(\ulcorner \mathbf{Y} \urcorner)$ implies $\vartheta(\ulcorner y \urcorner)$ for each $y \in \mathbf{Y}$. By (3.3), $\vartheta(\ulcorner \mathbf{Y} \urcorner)$ implies $\mathbf{Cn}(y)$ for each $y \in \mathbf{Y}$; that is, $\vartheta(\ulcorner \mathbf{Y} \urcorner) \vdash \mathbf{Y}$. ∎

Let us inquire a little further why the converse of Theorem 40 is unattainable. If $\forall \mathfrak{y} [\mathbf{Cn}(\mathfrak{y} \in \ulcorner \mathbf{Y} \urcorner) \twoheadrightarrow \vartheta(\mathfrak{y})]$ were implied by $\mathbf{Y}$, then $\vartheta(\mathfrak{y})$ would be implied by $\mathbf{Y} \wedge \mathbf{Cn}(\mathfrak{y} \in \ulcorner \mathbf{Y} \urcorner)$. But then by (3.4), $\theta(\mathfrak{y}) \wedge \delta(\mathfrak{y})$ would be so implied. The first conjunct is unproblematic. But write $\mathfrak{p}$ for $\mathfrak{y}$, and let $\mathbf{Y}$ be $\mathbf{Cn}(p)$. Then $\mathbf{Cn}(\mathfrak{p} \in \ulcorner \mathbf{Y} \urcorner) = \mathbf{L}$, and hence $\mathbf{Cn}(p)$ implies $\delta(\mathfrak{p})$; which means, by (26), that the sentence 'It is a truth universally acknowledged, that a single man in possession of a good fortune, must be in want of a wife', which we called 'p', implies $z(\mathfrak{p})$ where z is any $\mathfrak{y}$-universal formula of S. It is one thing for $\vartheta(\mathfrak{p})$ to imply $z(\mathfrak{p})$, but quite another thing for p to do so. If the implication were to hold, indeed, the definition (7.2) would be a creative one.

These results (Theorem 39, and the absence of a corresponding identity for ϑ) seem to settle decisively the question of which theory, $\vartheta(\mathfrak{y})$ or $\theta(\mathfrak{y})$, is to be preferred for the task of defining the truth of the formula named by $\mathfrak{y}$. As was made clear in § 3, the function ϑ is much more generally applicable amongst weak deductive disciplines. But in disciplines such as arithmetic and set theory, where unaxiomatizable theories abound, the function θ, unlike ϑ, provides a definition of truth that meets the minimum standard of material adequacy. To avoid minor irregularities, noted just before (3.0) above, the domain of the function θ must be restricted to names of formulas (and theories). Within that domain it succeeds in doing a satisfactory job.

8   The Antinomy of the Liar

Having introduced no truth predicate, the present discussion is immune from the liar antinomy in its customary forms. Although we have presented two materially adequate definitions of truth, ϑ and θ, that assign to each name of a formula a theory that encapsulates its truth, it is plain that there can exist at best an axiomatizable theory, but never a sentence or formula, that asserts its own untruth. The liar antinomy beloved of philosophers is unceremoniously blocked.

The antinomy seems to be reborn, however, if a theory $\mathbf{Y}$ can be constructed for which $\mathbf{Y} = \pi(\ulcorner\mathbf{Y}\urcorner) = \theta(\ulcorner\mathbf{Y}\urcorner)'$. For according to Theorem 39, $\mathbf{Y} = \theta(\ulcorner\mathbf{Y}\urcorner)$, and according to (4.4), $\theta(\ulcorner\mathbf{Y}\urcorner) \vee \theta(\ulcorner\mathbf{Y}\urcorner)' = \mathbf{L}$. It follows that both $\theta(\ulcorner\mathbf{Y}\urcorner)$ and its complement $\theta(\ulcorner\mathbf{Y}\urcorner)'$ are identical with $\mathbf{L}$, and therefore that $\mathbf{L} = \mathbf{S}$. The antinomy of the liar is not expunged merely by the adoption of a paraconsistent logic in which truth and falsity (and truth and untruth) are not always incompatible (Theorem 37). The version of the liar just sketched presumes to exhibit a theory that is not just both true and false but both logically true and logically false.

This cannot, of course, be right, since the definition (7.3) of $\theta(\mathfrak{Y})$, and its ancestors, are all explicit definitions within the background theory $\mathbb{T}$ (which is surely consistent). It is simple enough to pinpoint the invalid step: it must lie in the supposed construction of the theory $\mathbf{Y} = \pi(\ulcorner\mathbf{Y}\urcorner)$. For although it is perfectly possible, by parodying its definition, to provide a theory such as $\pi(\mathfrak{Y})$ with a name, $\langle\pi(\mathfrak{Y})\rangle$ say, that in some sense reveals its structure, and then apply the diagonal construction, such names are not of a kind that permits proof of an analogue of (7.4), that $\mathbf{Y} = \theta(\langle Y\rangle)$. To take a simple example, consider $\pi(\mathfrak{y})$, which is defined in (6.3) as $\theta(\mathfrak{y})'$, or equivalently (by (4.9) and (3.1)) as $\bigvee\{\mathfrak{y} = \ulcorner y\urcorner \wedge \neg y \mid y \in S\}$. The structure of this theory — let us call it $\mathbf{Z}$ — is not well enough known for a name $\langle\mathbf{Z}\rangle$ to provide information about which formulas $\mathbf{Z}$ does and does not imply. We cannot expect, that is, to prove $\ulcorner y\urcorner \in \langle\mathbf{Z}\rangle$ when $\mathbf{Z}$ implies y, or (especially) to prove $\ulcorner y\urcorner \notin \langle\mathbf{Z}\rangle$ when $\mathbf{Z}$ does not imply y. But it is exactly these features of names of formulas that are needed for material adequacy (7.4) to be established.

The above three paragraphs give only a hint of why and how the antinomy of the liar is bypassed in this theory of truth. A more elaborate treatment is planned for Miller (2009).

Acknowledgements

This paper is based on a lecture given at the triple congress CLE 30 – XV EBL – XIV SLALM [30TH ANNIVERSARY OF THE CENTRE FOR LOGIC, EPISTEMOLOGY AND THE HISTORY OF SCIENCE, UNICAMP – 15TH BRAZILIAN LOGIC CONFERENCE – 14TH LATIN-AMERICAN SYMPOSIUM ON MATHEMATICAL LOGIC] held in Paraty RJ in May 2008. Together with its philosophical companion 'Truth Defended' (Miller 2009), the paper originated in a talk in the fringe symposium POPPER ON TRUTH organized by Alain Boyer at the centenary congress KARL POPPER 2002 in Vienna in July 2002. Some of the material, now much revised, was presented under the title 'Infinite Truth' at meetings of the BRITISH LOGIC COLLOQUIUM in Birmingham in September 2002, and of the CALCUTTA LOGIC CIRCLE in January 2003, and at the THIRD WORLD CONGRESS ON PARACONSISTENCY in Toulouse in July 2003; and some closely related material was presented at the 12TH INTERNATIONAL CONGRESS OF LOGIC, METHODOLOGY AND PHILOSOPHY OF SCIENCE in Oviedo in August 2003. I am indebted to John Collins, Newton da Costa, Jeffrey Ketland, †Peter Madden, and Diego Rosende, for their encouragement as much as for their lively criticism; and also to Amita Chatterjee and her colleagues, Ivor Grattan-Guinness, Jeff Paris, Peter Roeper, Patrick Suppes, and Jan Woleński, for the interest that they have shown, and the criticism that they have provided. Responsibility for errors and misjudgements is strictly reserved.

The two papers are dedicated to the memory of Karl Popper, whose devotion to truth never wavered.

BIBLIOGRAPHY

[1] Birkhoff, G. (1973). *Lattice Theory*. 3rd edition. *AMS Colloquium Publications*, Volume 25. Providence: American Mathematical Society.

[2] Miller, D. W. (2009). Truth Defended. In preparation.

[3] Monk, J. D. (1986). The Contributions of Alfred Tarski to Algebraic Logic. *The Journal of Symbolic Logic*, 51:899–906.

[4] Tarski, A. (1930a). Über einige fundamentale Begriffe der Metamathematik. *Comptes Rendus des séances de la Societé des Sciences et des Lettres de Varsovie*, cl. iii, 23:22–29. References are to the English translation. 'On Some Fundamental Concepts of Metamathematics', Chapter III, pp. 30–37, of Tarski (1956).

[5] Tarski, A. (1930b). Fundamentale Begriffe der Methodologie der deduktiven Wissenschaften. I'. *Monatshefte für Mathematik und Physik*, 37:361–404. References are to the English translation. 'Fundamental Concepts of the Methodology of the Deductive Sciences', Chapter V, pp. 60–109, of Tarski (1956).

[6] Tarski, A. (1933a). Pojęcie prawdy w językach nauk dedukcyjnych. Warsaw: Prace Towarzystwa Naukowego Warszawskiego, Wydiał III Nauk Matematyczno–Fizycznyzch, 34. References are to the English translation. 'The Concept of Truth in Formalized Languages', Chapter VIII, pp. 152–278, of Tarski (1956).

[7] Tarski, A. (1933b). Einige Betrachtungen über die Begriffe ω-Widerspruchsfreiheit und der ω-Vollständigkeit. *Monatshefte für Mathematik und Physik*, 40:97–112. References are to the English translation. 'Some Observations on the Concepts of ω-Consistency and ω-Completeness', Chapter IX, pp. 279–295, of Tarski (1956).

[8] Tarski, A. (1935–1936). Grundzüge des Systemenkalkül. *Fundamenta Mathematicae*, 25(4):503–526, and 26(2):283–301. References are to the English translation. 'Foundations of the Calculus of Systems', Chapter XII, pp. 342–383, of Tarski (1956).

[9] Tarski, A. (1956). *Logic, Semantics, Metamathematics*. Oxford: Clarendon Press. Translated by J. H. Woodger. 2nd edition 1983. Edited by J. Corcoran. Indianapolis: Hackett Publishing Company.

David Miller
Department of Philosophy
University of Warwick
COVENTRY CV4 7AL UK
E-mail: dwmiller57@yahoo.com

The Problem of Existential Import in First-Order Consequential Logics

CLAUDIO PIZZI

ABSTRACT. The paper aims at discussing the problem of existential import in the framework of logics of so-called consequential implication, which is a modal reinterpretation of connexive implication. In developing this line of inquiry it is argued that two ways are open: (a) to resort to the language of the first order extensions of consequential logic; (b) to translate modal operators into first order quantifiers, so to define a special operator of consequential implication in terms of quantificational language itself. The two approaches are explored in some of their logical and philosophical aspects. It is stressed that more than one operator satisfies the properties pertaining to consequential implication, so that more than one quantificational translation of consequential operators may be introduced, pointing out the legitimacy of different intuitions concerning the implicative relations among quantified propositions.

§1. The discussion about existential import has been characterized since the beginning by a confusion over the exact nature of its object. The problem of existential import consists, in fact, in a couple of questions which are frequently conflated, namely

(a) whether universal propositions such as, for instance, "All unicorns are Italian" implies what traditional logicians call the corresponding "particular" proposition "Some unicorns are Italian".

(b) whether a particular proposition such as "Some unicorns are Italian" has the same meaning as the proposition "Italian unicorns *exist*", which implies the existence of unicorns.

In concern with (a) a well- known fact is that, if we use standard formalization, the inference from $\forall x(Px \supset Qx)$ to $\exists x(Px \land Qx)$ is not valid in first-order logic. So it appears that the answer to (a) should be in the negative. However, the weaker $(\forall x(Px \supset Qx) \land \exists xPx) \supset \exists x(Px \land Qx)$ is a first order theorem. This theorem, however, appears also to be a source of problems. In fact $\exists xPx$ may be, say, the logical truth that something is identical with Pegasus (derived from the obvious identity

Pegasus = Pegasus). So from the truth "Every object which is identical with Pe-gasus is a winged horse" in conjunction with the logical truth that something is identical with Pegasus, we derive the truth "Something is identical with Pegasus and is a winged horse ". From such statement it seems that we could legitimately draw the conclusion that something is a winged horse and, if we give a positive reply to (b), also the conclusion that a winged horse exists. So-called "free" log-ics have been worked out with the aim of blocking this argument by putting some restriction on the inference which leads from α to $\exists x\alpha$.

Since, in contemporary logic, the logical form of "Some A are B" is provided by $\exists x(Ax \wedge Bx)$, the answer which is normally given to question (b) is in the posi-tive. So it is often supposed that the controversy about existential import concerns simply question (a). There are strong reasons, however, to underline the distinc-tion of the two questions and to raise doubts both about the alleged impossibility of inferring "Some" from "All" and about the existential import of particular state-ments.

§2. In 1967 Storrs McCall wrote an important paper on the subject of the exis-tential import of universal statements.[1]

The key idea of McCall's paper is that the problem of existential import is not originated by the properties of standard quantifiers but by the properties of stan-dard (truth-functional) implication[2]. In his words: "The whole question of what inferences are logically valid ought to be entirely independent of what things may or may not exist - it would be ridiculous to think that the discovery of unicorns in the mountains of the moon would affect the validity of an inference."([5], p.347).

McCall rightly quotes the original contributions of Hugh MacCall, who has been one of the ancestors of contemporary modal logic and the first proponent of a system of strict implication. MacCall was the first logician to define class inclusion in terms of implication. In his system of logic a universal statement like "All unicorns are Italians" is translated into an implication $U \rightarrow I$, where U stands for "an arbitrary u belongs to the class of unicorns". The problem of existential import was solved by MacCall by postulating that no class is empty (in propositional terms, this amounts to saying that no proposition is impossible) while admitting, on the other hand, the existence of classes consisting of imaginary or inconsistent beings[3].

S. McCall and R. B. Angell, inspired by some ideas anticipated by E. Nelson in the Thirties (see [12] and [13]), in the Sixties proposed a logic of an implication

---

[1]See [5]. In his [1] R. B. Angell formulated an analogous line of thougth, apparently with no knowledge of McCall's paper published earlier in the same journal.

[2]See [4]. See also the issue dedicated to Hugh Mac Call by *Nordic Journal of Philosophical Logic*, vol.3/1-2, 1998.

[3]It is of some interest to recall that some suggestions in the same direction were anticipated by Lewis Carroll, who was also a defender of the existential import of universal statements.

named "*connexive*", whose basic principle was the so-called Aristotle's Thesis[4]

(AT) $\neg(\alpha \to \neg\alpha)$

The key idea of this logic is that nothing implies its own negation: in particular, contradictions should imply only contradictions and tautologies should imply only tautologies. In connexive systems AT is interdependent with another law, which here will be called "Strong Boethius' Thesis":

(SBT) $(\alpha \to \beta) \to \neg(\alpha \to \neg\beta)$

If *cotenability* is defined as

(Def /°) $\alpha/°\beta \overset{\text{def}}{=} \neg(\alpha \to \neg\beta),$

SBT may be rewritten as

(SBT') $(\alpha \to \beta) \to \alpha/°\beta$

Little attention has been devoted to the quantificational extensions of such systems. According to McCall, if $\to$ is the connexive arrow, "All men are mortal" should be formalized as

(1) $\forall x(Ax \to Bx)$

while "Some man is mortal" should be represented not by[5]

(2) $\exists x(Ax \wedge Bx)$ but by
(3) $\exists x\neg(Ax \to \neg Bx)$

Since $\forall x\alpha$ implies $\exists x\alpha$ and $Ax \to Bx$ implies $\neg(Ax \to \neg Bx)$, the Aristotelian square now takes the following form:

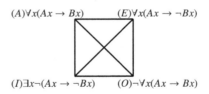

$(A)\forall x(Ax \to Bx)$ $(E)\forall x(Ax \to \neg Bx)$

$(I)\exists x\neg(Ax \to \neg Bx)$ $(O)\neg\forall x(Ax \to Bx)$

Figure 1.

---

[4]For a recent survey on connexive logics see [23].
[5]This symbolization is attributed by McCall to Hugh MacCall and to the Polish logician Czezowski.

A generalized version of the square would be provided by putting $\forall x(\alpha(x) \rightarrow \beta(x))$ in place of $\forall x(Ax \rightarrow Bx)$ (where $\alpha(x)$ of course means that x is a free variable in an arbitrary proposition α), but this kind of generalization will be neglected for sake of simplicity also in the remaining part of the paper.

Thus in this new framework "All unicorns are Italian" entails "Some unicorns are Italian", but this does not imply that something is a unicorn and something is Italian. In this logic, in fact, $\exists x \neg(Ax \rightarrow \neg Bx)$ does not logically imply neither $\exists x(Ax \wedge Bx)$ nor $\exists x Ax$. A countermodel can be devised simply by interpreting A and B as denoting a set of individuals which are, say, round squares (RS). So in this interpretation $\forall x(RS\,x \rightarrow RS\,x)$ is true, the consequent $\exists x \neg(RS\,x \rightarrow \neg RS\,x)$ is true by first order logic and Aristotle's Thesis, but $\exists x\ RS\,x$ is false since the set covered by a contradictory property is an empty set. The solution offered in this way to problem (a), then, is such that problem (b) becomes irrelevant. In any case, in fact, we do not reach the conclusion that something is a round square or, alternatively, that there exists a round square, even if we may be willing to accept as true the proposition that all round squares are round squares.

Some problems are suggested by McCall's analysis.

The first may be stated as follows. If we read $\exists x \neg(Ax \rightarrow \neg Bx)$ as "Some A is B", how have we to read the non-equivalent formula $\exists x(Ax \wedge Bx)$[6]? What is the meaning of the existential quantifiers in the latter statement? It seems that we cannot read $\exists x(Ax \wedge Bx)$ as "*there exists* an x which is A and also B" since the same quantifier $\exists x$ appears in $\exists x \neg(Ax \rightarrow \neg Bx)$ and cannot be submitted to a different reading. The problem concerns also the universal quantifiers. In fact, if we read $\forall x(Ax \rightarrow Bx)$ as "Every A is B", as McCall proposes, how should we read the truth-functional formula $\forall x(Ax \supset Bx)$?

A possible reply is that even if we choose to read $\exists x$ as "There exists an x", and $\exists x(Ax \wedge Bx)$ as "There exists an x that is A and is B", we have not to change this reading when passing to the wff $\exists x \neg(Ax \rightarrow \neg Bx)$ since a plausible reading of the latter formula in fact could be as follows: "There exists an x such that x's being A does not exclude x's being B", or indifferently "Some x is such that its being A does not exclude its being B". There is no problem in saying that there exists an x such that x's being a round square does not exclude x's being a round square, since this statement is an instance of a connexive theorem and does not imply that there is an x which is a round square.

Passing to the universal quantifier, we may retain the reading "All A is B" for $\forall x(Ax \supset Bx)$, but $\forall x(Ax \rightarrow Bx)$ should be read as "For every x, x's being A *implies* x's being B". In other words, we simply have to stress the well known distinction between (truth functional) conditional and implication. If "Every cat is such that

---

[6]Notice that in every connexive system in which $\alpha \rightarrow \beta$ implies $\neg(\alpha \wedge \neg\beta)$, $(\alpha \wedge \beta)$ implies $\neg(\alpha \rightarrow \neg\beta)$, so the formula $\exists x(Ax \wedge Bx)$ is stronger than $\exists x \neg(Ax \rightarrow \neg Bx)$. In what follows the definition $(\alpha \supset \beta) \overset{\text{def}}{=} \neg(\alpha \wedge \neg\beta)$ will be taken for granted.

being a cat implies its mortality " appears to be not very conversational, we may resort to the simpler reading "Being a cat implies being mortal" or also to "For every cat, being a cat implies being mortal". Consequently, the corresponding existential statement could be read "For some cat, being a cat does not exclude being mortal".

The problem of the semantic interpretation to be associated to quantifiers (objectual *versus* substitutional) will not be treated here. It is also unessential in the present context to mark a distinction between merely quantificational existence and "real" existence. In order to avoid taking care of this difficult question, the examples we will give of non-existing objects will be examples of entities defined by contradictory descriptions, which are non-existing in every conceivable sense of existence.

A second difficulty has been discussed by McCall himself. If we say " No miners have tunnelled to the center of the earth", this is true and its logical form is represented by

(4) $\forall x(Mx \rightarrow \neg Tx)$

But this formula by contraposition becomes

(5) $\forall x(Tx \rightarrow \neg Mx)$

so by quantification theory and Boethius' Thesis

(6) $\exists x \neg (Tx \rightarrow Mx)$

and again by contraposition

(7) $\exists x \neg (\neg Mx \rightarrow \neg Tx)$

According to McCall's reading, (7) means "some non-miners have tunnelled to the center of the earth". This conclusion is *prima facie* false. McCall's reply is, however, that "some non-miners have tunnelled to the center of the earth" is true inasmuch there are some non-miners with the mentioned property, i.e. the non-miners who have tunnelled to the center of the earth (an empty subclass of non-miners). From our viewpoint there could be however a different solution of the problem. According to our proposed interpretation, the correct reading of $\forall x(Mx \rightarrow \neg Tx)$ is "For every miner, being a miner implies not-having tunnelled to the center of earth", while the proper reading of $\exists x \neg (\neg Mx \rightarrow \neg Tx)$ is "some non-miners are such that their being non-miners does not exclude having tunnelled to the center of earth". This appears to be true since among the objects which are non-miners one can identify something - for instance imaginary superpowerful engines- which actually do not exist but *might have* (i.e. do not exclude having) the mentioned extraordinary property.

A further serious problem which should be taken into account is that connexive logic in the formulation which received in the Sixties is actually untenable. Many flaws of such systems have been devised by Montgomery and Routley in a quite desctructive paper published in JSL (see [22]). In this connection it is enough here to quote a paper by Angell published in 1978 in italian translation (see [3]). In this work Angell examined the flaws of system **PA1** and of its extension **CC1** due to McCall. McCall's system may be proved to be Post-complete with respect of Angell's 4-valued matrix, but turns out to be not functionally complete. In **CC1** it is possible to prove such desirable laws lacking in **PA1** as $p \rightarrow (p \wedge p \rightarrow p)$, but unfortunately also such laws as $(p \rightarrow q \wedge r \rightarrow q) \rightarrow (p \rightarrow r)$ which seem to be counterintuitive ("if a dog is a mammal and a cat is a mammal then a dog can be a cat").

Both **PA1** and **CC1** exclude the wffs $p \wedge p \rightarrow p$ and the converse $p \rightarrow p \wedge p$. **CC1** however includes $p \wedge p \wedge p \rightarrow p$, so creating an unjustified distinction between antecedents with even and odd numbers of conjuncts of the same kind. The treatment of conjunction in such systems cannot be easily corrected, since Routley and Montgomery showed that the extending even subsystems of **CC1** with the law $p \wedge p \rightarrow p$ leads to an inconsistency. Attempts to modify the 4-valued matrix for conjunction have not been successful since they would lead to reject some intuitively acceptable laws .

In his 1978 paper Angell was inclined to think that every connexive system not granting the equivalence between $p \wedge p$ and p should be considered logically implausible. As a matter of fact, the equivalence holds in Angell's system **PA2** outlined in the same paper and also in more recent contributions due to McCall (see for instance [7]).

One might be willing to add that the Factor law $p \rightarrow q \supset (p \wedge r \rightarrow q \wedge r)$, which is a connexive thesis, is questionable just from a connexive viewpoint. Since r may be exactly $\neg p$, we would have as an instance $p \rightarrow q \supset (p \wedge \neg p \rightarrow q \wedge \neg p)$. But $p \rightarrow q$ might be analytically true (e.g."a is a mammal $\rightarrow a$ is an animal"), so to be a theorem of a proper extension of first order logic. However, contrary to the connexive viewpoint, the consequent would assert that the absurdity $\perp$ implies $q \wedge \neg p$ (a is animal and non-mammal), which is not an absurdity.

§3. In more recent papers, McCall himself proposed a revision of original connexive logic by weakening BT, i.e. $(p \rightarrow q) \rightarrow \neg(p \rightarrow \neg q)$ into the formula sometimes known as "Strawson's Thesis" $(p \rightarrow q) \supset \neg(p \rightarrow \neg q)$. In what follows this formula will be called "Weak Boethius' Thesis"(WBT). The system **CFL** proposed by McCall in [7], however, cannot be considered fully satisfactory for at least two reasons: (i) nesting of arrows is forbidden; (ii) as proved by R.K. Meyer, $\alpha \rightarrow \beta$ boils down to the conjunction $(\alpha \equiv \beta) \wedge \alpha \dashv3 \beta$ (where $\alpha \dashv3 \beta \overset{\text{def}}{=} \square(\alpha \supset \beta)$) (see [7], p.450). More exactly, the set of provable first degree formulas containing

$\rightarrow$ in the systems **S1-S5** extended with the definition

(Def $\rightarrow$ 0) $\alpha \rightarrow \beta \overset{\text{def}}{=} (\alpha \equiv \beta) \wedge \alpha \dashv 3 \beta$

coincides with the set of theorems of **CFL**. This result is perplexing, since it implies that $\rightarrow$ - statements are a subclass of $\equiv$ - statements. Furthermore, given that, by virtue of (Def $\rightarrow$ 0), $\top \rightarrow \beta$ and $\Box\beta$ turn out to be coincident, we cannot avoid noticing that theorems of **CFL** such as $(\alpha \wedge \Box\beta) \supset (\alpha \rightarrow \beta)$ are far from the spirit of connexive logics.

In the Nineties a revised version of connexive logic was developed under the name of *consequential implication* (see [14], [15], [17]). According to this theory to assert $\alpha \rightarrow \beta$ is the same as saying that (i) $\alpha \dashv 3 \beta$ is true and (ii) that α and β have the same modal status, i.e. have the same position in the Aristotelian square:

(Def $\rightarrow$ 1)

$$\alpha \rightarrow \beta \overset{\text{def}}{=} \alpha \dashv 3 \beta \wedge (\Diamond\beta \equiv \Diamond\alpha) \wedge (\Box\beta \equiv \Box\alpha) \wedge (\neg\Diamond\beta \equiv \neg\Diamond\alpha) \wedge (\neg\Box\beta \equiv \neg\Box\alpha)$$

This boils down, in the minimal normal system **K**, to

(Def $\rightarrow$ 2) $\alpha \rightarrow \beta \overset{\text{def}}{=} \alpha \dashv 3 \beta \wedge (\Diamond\beta \supset \Diamond\alpha) \wedge (\Box\beta \supset \Box\alpha)$

An Italian historian of logic, Mauro Nasti de Vincentis, in his studies about Chrysippean logic (see [8], [9], [10], [11]) proposed the following definition of the arrow (where $\nabla\alpha$ stands for $\Diamond\alpha \wedge \Diamond\neg$ and $\alpha = \beta$ for $\alpha \dashv 3 \beta \wedge \beta \dashv 3 \alpha$):

(Def $\rightarrow$ 3) $\alpha \rightarrow n \beta \overset{\text{def}}{=} (\nabla\alpha \wedge \nabla\beta \wedge \alpha \dashv 3 \beta) \vee (\neg(\nabla\alpha \wedge \nabla\beta) \wedge \alpha = \beta)$

However, it is possible to show that the two definitions (Def $\rightarrow$ 1) and (Def $\rightarrow$ 1) are equivalent[7]. A recent proposal by McCall is the definition:[8]

(Def $\rightarrow$ 4) $\alpha \rightarrow m \beta \overset{\text{def}}{=} \alpha = \beta \vee (\alpha \dashv 3 \beta \wedge \Diamond\alpha \wedge \Diamond\neg\beta)$

It is possibile to show, however, that $\rightarrow m$ and $\rightarrow n$ are equivalent in every normal modal logic. This convergence of definitions encourages the idea of reconsidering the whole question of existential import in the framework of consequential logics.

On the one hand, $\rightarrow$ satisfies the most interesting properties of connexive logic: in fact the wff **WBT**, i.e. $(p \rightarrow q) \supset \neg(p \rightarrow \neg q)$ is a thesis and turns out to be equivalent to $\neg(p \rightarrow \neg p)$, $\bot(\top)$ consequentially implies and is implied only by

---

[7]For two different proofs of this fact see [19] and [9].

[8]Private correspondence. I thank Prof. McCall for granting me permission to publish this definition. The weaker definition $\alpha \equiv \beta \vee (\alpha \dashv 3 \beta \wedge \Diamond\alpha \wedge \Diamond\neg\beta)$ yields the unwelcome result that $\alpha \wedge \beta$ implies $\alpha \rightarrow m \beta$.

contradictions (tautologies). On the other hand, the interest of the new connective → is that one may prove the equivalence between systems of consequential
implication and systems of normal modal logic. In fact, $\Box \alpha$ may be defined in
consequential systems in terms of → as $\Box \alpha \stackrel{\text{def}}{=} \top \to \alpha$. The three basic systems of
consequential logic **CIw**, **CI**, **CI.0** may be proved to be definitionally equivalent
to **K**, **KD** and **KT** respectively (see Appendix). All the extensions of system **CI**
may be called Aristotelian since **CI** (but not **CIw**) contains the Aristotle's Thesis
$\neg(p \to \neg p)$.

Thanks to this translation into modal systems, it turns out that none of the preceding systems is Post-complete and each one of them is *tableaux*-decidable whenever the equivalent modal system has such a property.

Simplification, i.e. the wff $p \wedge q \to p$, is not a consequential thesis, but a weak-
ened version of it, i.e. $(\Diamond(p \wedge q) \wedge \neg \Box p) \supset (p \wedge q \to p)$ is derivable. Monotonicity
and Factor Law do not hold, even if weakened versions of them are provable:
$\Diamond(p \wedge r) \supset (p \to q \supset (p \wedge r \to p))$ and $\Diamond(p \wedge r) \supset (p \to q \supset (p \wedge r \to p \wedge q))$.

A remarkable property of such class of systems is that weak Boethius' Thesis
holds not only in standard form but also in a secondary version which we will call
"secondary Boethius' Thesis":

(SBT) $(p \to q) \supset \neg(\neg p \to q)$

Of course, we have also at our disposal a secondary Aristotle's Thesis:

(SAT) $\neg(\neg p \to p)$

A further feature of consequential systems which deserves attention is that there
is more than one implicative connective which has most of the properties which
characterize → and is also modally definable. The most interesting example is
offered by the following operator (truncated consequential implication):

(Def ⇒ 1) $\alpha \Rightarrow \beta \stackrel{\text{def}}{=} \Box(\alpha \supset \beta) \wedge (\Diamond \beta \equiv \Diamond \alpha)$

The same operator may also be defined in terms of → as

(Def ⇒ 2) $\alpha \Rightarrow \beta \stackrel{\text{def}}{=} \alpha \to \beta \vee (\Diamond \alpha \wedge \Box \beta)$

As a matter of fact, it may be easily proved that $\Box \beta \wedge \Diamond \alpha$ implies $\alpha \Rightarrow \beta$, and
this introduces an exception to the "equimodality principle" which is a feature of
→. For the truth of $\alpha \Rightarrow \beta$ we are simply asking that $\alpha \dashv 3 \beta$ is true and that
the modal status of the clauses α and β either are identical or, if they are not, the
modal status of the antecedent is logically weaker than the one of the consequent
($\Diamond \alpha$ being weaker than $\Box \alpha$). If such relation between modal operators is called
"*connection*", this condition amounts to asking that the modal operators of the

clauses are *connected*. Of course $\alpha \rightarrow \beta$ implies $\alpha \Rightarrow \beta$ but not *vice versa*. We may also define $\Rightarrow$ in terms of $\rightarrow$ as follows

(Def $\rightarrow$ 5) $\alpha \rightarrow \beta \overset{\text{def}}{=} \alpha \Rightarrow \beta \wedge \neg\beta \Rightarrow \neg\alpha$

This definition may be understood considering that $\Rightarrow$ is not contrapositive, even if we have the weakened contraposition law $\Diamond\neg q \supset (p \Rightarrow q \supset \neg q \Rightarrow \neg p)$.

If in **CI** we introduce Def $\Rightarrow$, **CI$\Rightarrow$** will be the name of the subsystem of **CI** axiomatizing $\Rightarrow$ (see [16] and [19]). If $/^{\circ}$' is the relation of cotenability defined as

(Def $/^{\circ}$') $\alpha/^{\circ}\beta \overset{\text{def}}{=} \neg(\alpha \Rightarrow \neg\beta)$

the following figure is one of the many possible Aristotle's cubes, where Aristotle's cubes are 3-dimensional figures in which two faces are Aristotelian squares, while the other squares contain propositions whose interrelations belong to the set of relations occurring in standard Aristotelian square (see [21]).

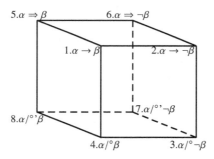

Figure 2.

§4. Let us now see how the problem of existential import can be treated in this new framework. It seems that two different approaches to the problems are viable.

The first one is to work with a first order extension of the Aristotelian system **CI**. Our language will have now as primitives all the truth functional operators, $\forall$ and $\rightarrow$. The axioms governing $\forall$ are the axioms of the first order logic **QL**. The weakest system in which we may analyze the question is then **QL+CI**, and it is obviously equivalent to **QL+KD**.

In the basic Aristotelian square there is a relation of subalternance between $\alpha \rightarrow \beta$ and $\alpha/^{\circ}\beta$. But since $\forall x\alpha$ also implies $\exists x\alpha$ in the Aristotelian square of quantifiers now $\forall x(Ax \rightarrow Bx)$ implies by subalternance three distinct statements. This introduces a complication in the Aristotelian square. In the next figure the vertical lines represent logical implication (in the direction from top to bottom). The lines from 1 to 4 and from 5 to 8 in the next figure are sides of the traditional square.

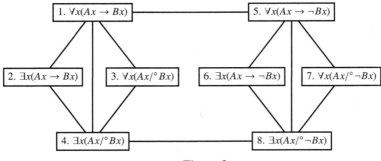

Figure 3.

Two questions should be discussed. The first concerns so-called "secondary Boethius". If we define secondary cotenability as

(Def \°) $\alpha\backslash°\beta \overset{\text{def}}{=} \neg(\neg\alpha \to \beta)$

we should take then into account in fact another variant of the left diamond in 3 which has 3bis.$\forall x(Ax\backslash°Bx)$ in place of 3.$\forall x(Ax/°Bx)$ and 4bis.$\exists x(Ax\backslash°Bx)$ in place of 4.$\exists x(Ax/°Bx)$

How have to read the items 3bis and 4bis? 3bis means "For every x, not-Ax does not imply Bx": e.g., for every x, x's not being a man does not imply x's mortality. Notice that both /° and \° are symmetric relations: so an equivalent statement would be "For every x, x's not being mortal does not imply x's being a man)" or, more simply, "immortality does not imply humanity".

A second problem concerns the modal translation of the quantified consequential formulas. Translating the formula $\forall x(Ax \to Bx)$ in terms of its modal definition we have $\forall x(\square(Ax \supset Bx) \land (XAx \equiv XBx))$, where X is one of the four modal operators belonging to the Aristotelian square. Let us now consider the statement "Necessarily, for every cat, being a cat implies being mortal". As always, we have the alternative between the *de dicto* and *de re* interpretation. In the first case the formalization is provided by

(i) $\square\forall x(Ax \to Bx)$, so its modal translation is $\square(\forall x(\square(Ax \supset Bx) \land (XAx \equiv XBx))$, where X is as above defined. Now, by applying the distributivity of $\square$ and $\forall$ int-elim rules, the above formula implies

(ia) $\forall x(\square\square Ax \equiv \square\square Bx \land \square\lozenge Ax \equiv \square\lozenge Bx)$.

The *de re* interpretation is provided by the wff (ii) $\forall x\square(Ax \to Bx)$, so by

(iia) $\forall x\square(\square(Ax \supset Bx) \land XAx \equiv XBx)$ and by the equivalent wff (iib) $\forall x(\Box\Box(Ax \supset Bx) \wedge \Box(XAx \equiv XBx))$

The meaning of (ia) and (iib) is not clear, however, unless we introduce some reduction axioms for iterated modalities, and the most simple choice is to resort to the **S5** axioms. As is well-known, **QL+S5** contains the Barcan Formula and its converse, so the equivalence

(BF) $\Box\forall xAx \equiv \forall x\Box Ax$

Then it is straightforward to prove in this system both

(iii) $\forall x\Box(Ax \to Bx) \equiv \forall x(Ax \to Bx)$ and

(iv) $\Box\forall x(Ax \to Bx) \equiv \forall x(Ax \to Bx)$

The simplification reached in this way is technically welcome but questionable. "Necessarily, being a cat implies being mortal" turns out to be equivalent to the simple "Being a cat implies being mortal". This equivalence is not philosophically obvious, since one could maintain that some implicative relations are to be considered contingently, not necessarily subsisting, such as the one expressed in such sentences as "being Hawaian implies being American"[9]. On the other hand, giving up **S5** reduction principles yields a certain amount of a difficulty in performing an intuitive reading of the relevant formulas.

§5. The inconveniences of the preceding joint treatment of consequential implication and quantifiers are not actually due to technical problems but to a difficulty in performing an intuitive reading of the formulas containing $\to$, $\Box$, $\forall$. In the light of the preceding considerations, we try to suggest an alternative solution. The idea is to interpret directly the modal operators which translate the consequential arrow in term of quantifiers.

Let us start from the propositional system **CI** , which is equivalent to **KD**.

For our purposes we may use a simplified version of the well-known translation function from modal language to first-order language as defined in Correspondence Theory. Let us suppose that there exists an enumeration of atomic variables p, p_1, p_2.... q, q_1, q_2 ... in the language of the propositional logic **PC** and a parallel enumeration of monadic predicates P, P_1, P_2... Q, Q_1, Q_2...in the language of **QL**. Let us introduce a function s from the former to the latter language which is defined as follows, where x is a fixed individual variable:

(i) $s(p_n) = P_n x$

(ii) $s(\neg\alpha) = \neg s(\alpha)$

(iii) $s(\alpha \supset \beta) = s(\alpha) \supset s(\beta)$

(iv) $s(\Box\alpha) = \forall y(s(\alpha)[x/y])$, where y is the first variable (in some fixed enumeration) of the individual variables not occurring in α.

The extension of the definition to wffs containing $\wedge$, $\vee$, $\equiv$ is obvious.

---

[9]We omit here treating the combination of $\Rightarrow$ or $\to$ with the circumstantial operator *, which allows to treat formally counterfactual conditionals (see [14]). If $A > B$ is defined, say, as $^*A \to B$, it is implausible to maintain that $A > B$ is equivalent to $\Box(A > B)$.

From (iv), it follows that $s(\Diamond\alpha) = \exists y(s(\alpha)[x/y])$, where y is as in (iv). For instance, $\Diamond\Box(p \wedge q)$ is translated into $\exists y \forall x(Px \wedge Qx)$, which is equivalent in **QL** to $\forall x(Px \wedge Qx)$. In this way we translate modal systems into subsystems of the monadic fragment of the first order logic **QL**.

Considering the simple modal sistem **KD**, it is easy to prove by induction on the length of proofs what follows:

(MT) For every α, $\vdash_{KD} \alpha$ implies $\vdash_{QL} s(\alpha)$

In fact, all s-images of **KD** axioms are **QL**-theorems, and the rules of **KD** preserve this property. (An analogous metatheorem may be proved also for systems which are extenxions of **KD**, but this subject will not be treated here).

Let us now perform a double translation: the first from consequential logic **CI** to the modal logic **KD** and the second from **KD** to the quantificational logic **QL**. The first mapping is the function φ defined in the Appendix, whose key step is contained in Def $\rightarrow$ 2 of p. 139. For p and q atomic we have for instance, given that $\varphi(p) = p$,

(t1) $\varphi(p \rightarrow q) = \Box(p \supset q) \wedge (\Box q \supset \Box p) \wedge (\Diamond q \supset \Diamond p)$

The second step will be

(t2) $s(\Box(p \supset q) \wedge (\Box q \supset \Box p) \wedge (\Diamond q \supset \Diamond p)) =$
$$\forall x(Px \supset Qx) \wedge (\forall x Qx \supset \forall x Px) \wedge (\exists x Qx \supset \exists x Px).$$

Thus, for p and q atomic,

(t3) $s\varphi(p \rightarrow q) = \forall x(Px \supset Qx) \wedge (\forall x Qx \supset \forall x Px) \wedge (\exists x Qx \supset \exists x Px)$.

The extension of the double mapping to formulas of arbitrary length is obvious and will be omitted.

It is useful now to introduce by definition a new kind of implication which is the quantificational counterpart of $\rightarrow$ and which we define as follows for arbitrary statements containing x:

(Def $\forall^{\rightarrow}$)
$$\alpha(x)\forall^{\rightarrow}\beta(x) \stackrel{\text{def}}{=} \forall x(\alpha(x) \supset \beta(x)) \wedge (\forall x\alpha(x) \equiv \forall x\beta(x)) \wedge (\exists x\alpha(x) \equiv \exists x\beta(x)).$$

Obviously, the dual auxiliary symbol is defined by

(Def $\exists/°$) $\alpha(x)\exists/°\beta(x) \stackrel{\text{def}}{=} \neg(\alpha(x)\forall^{\rightarrow}\neg\beta(x))$

As a consequence of Def $\forall^{\rightarrow}$, (t3) may be generalized to the following identity:

(t4) $s\varphi(\alpha \rightarrow \beta) = s\varphi\alpha(x)\forall^{\rightarrow} s\varphi\beta(x)$

Notice that in **KD** $\alpha/°\beta$ may be proved to be equivalent to $((\diamond\beta \vee \Box\alpha) \wedge (\Box\beta \vee \diamond\alpha)) \supset \diamond(\alpha \wedge \beta)$. So in a parallel way we have that in **QL** $Cx\exists/°Mx$ is equivalent to $((\exists xCx \vee \forall xMx) \wedge (\forall xCx \vee \exists xMx)) \supset \exists x(Cx \wedge Mx)$

Let us now call **CI$\forall^{\rightarrow}$** the fragment of **QL** which contains all the $s\varphi$-translations of formulas which are **CI**-theorems. In other words, if α stands for an arbitrary first-order wff and β for an arbitrary **CI**-wff, **CI$\forall^{\rightarrow}$** $= \{\alpha :$ there is a β such that $\vdash_{CI} \beta$ and $\alpha = s\varphi\beta\}$. A plausible conjecture is that **CI$\forall^{\rightarrow}$** is an axiomatizable and decidable theory.

Let us have a look at the meaning which can be associated to the quantificational formulas which are the output of this translation. For instance "x is a cat $\forall^{\rightarrow} x$ is mortal" means "(No cats are immortal)and (everything is mortal iff everything is a cat) and (something is mortal iff something is a cat)". Notice that such conjunction is true since "everything is a cat" is false and "everything is mortal" is also false, so their equivalence is true, while "something is a cat" and "something is mortal" are both true, so their equivalence is also true.

Given that $\alpha(x)\forall^{\rightarrow}\beta(x)$ is after all a quantificational formula, it is useful to find some suitable rendering of this new operator in ordinary language. To distinguish it from the standard reading of quantifiers we introduce the notion of "*non banal truth*" (*nb-truth*). So $Cx\forall^{\rightarrow}Mx$ will be read, say, "it is non-banally true (*nb-true*) that all cats are mortal", while the simple $\forall x(Cx \supset Mx)$ is to be read as "it is true that all cats are mortal", which is assumed here to be equivalent to the standard "all cats are mortal" in view of a redundancy conception of truth. By an obvious extension, $\neg(Cx\forall^{\rightarrow}\neg Mx)$, i.e. $C(x)\exists/°M(x)$, will be read "it is non-banally true that some cat is mortal". Since this statement is translated into $((\exists xCx \vee \forall xMx) \wedge (\forall xCx \vee \exists xMx)) \supset \exists x(Cx \wedge Mx)$, as we already remarked, it obviously that it does not imply $\exists x(Cx \wedge Mx)$, $\exists xCx$ or $\exists xMx$, so it has no existential bearing at all.

The new square of oppositions then will have this shape:

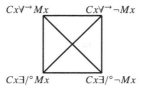

Figure 4.

Needless to say, we have among the theorems of **CI$\forall^{\rightarrow}$** the wff $(Cx\forall^{\rightarrow}Mx) \supset \neg(Cx\forall^{\rightarrow}\neg Mx)$, which is the translation of Boethius' Thesis and grants subalternance among the two left corners of the square[10].

---

[10]The wff $(Cx\forall^{\rightarrow}Mx) \supset \neg(Cx\forall^{\rightarrow}\neg Mx)$ may also be proved as a theorem of **QL** by the following

An open line of investigation is to extend **QL** with the definition Def $\forall^{\rightarrow}$ and to study the the the interrelations between $\forall^{\rightarrow}$-formulas and standard quantificational formulas. We have, for instance, the thesis $(Cx\forall^{\rightarrow}Mx) \supset \forall x(Cx \supset Mx)$, so also the BARBARA syllogism in the form $(Cx\forall^{\rightarrow}Mx \wedge Ca) \supset Ma$. It may be suggested that an extension of the consideration to wffs containing n-placed predicates and more than one variable is also of interest in the analysis of syllogistic reasoning.

§6. Some remarks on the new operator $\forall^{\rightarrow}$ are in order.

To begin with, the notion of non-banal truth does not exclude combining modalities and quantifiers. In first order quantification logic we may give a sense to the formula $\square(Cx\forall^{\rightarrow}Mx)$ reading it as "necessarily it is a nb-truth that all cats are mortals" or to $\lozenge(Cx\exists/^{\circ}Mx)$ as "It is possible that it is a nb-truth that some cat is mortal". Without having to resort to reduction postulates or even to the Barcan Formula we have to notice that $\square(Cx\forall^{\rightarrow}Mx)$ is anyway distinct from $\forall x(Cx \rightarrow Mx)$ and from its modal translation.

A second remark is more problematic. If RS stands for "being a round square" it is well-known that $\neg\exists x RS\,x$, i.e. $\forall x \neg RS\,x$, vacuously entails $\forall x(RS\,x \supset \neg Cx)$, which is equivalent to $\forall x(Cx \supset \neg RS\,x)$. So, for instance, "no cat is a round square" is a (vacuously) true statement. What have we to say about "it is nb-true that no cat is a round square? $\forall x Cx \equiv \forall x \neg RS\,x$ is a false equivalence given that $\forall x Cx$ ("Everything is a cat") is false while $\forall x \neg RS\,x$ is true. So we are forced to say that "No cat is a round square" is non-banally false, while it is true in standard sense.

A way to defend the plausibility of this conclusion could be in observing that "No dog is a round square", "No fish is a round square" and all statements belonging to this infinite pattern are all true, and this simply proves the irrelevance (disconnection) of the antecedents with respect to the consequents. By the same argument "no miners tunnelled to the center of the earth"is (vacuously) true, but it is nb-false simply because it implies the equivalence $\forall x Mx \equiv \forall x \neg Tx$, in which the first clause is false and the second is true.

We have to notice that not only "No cat is a round square" is nb-false, but also that the converse statement "No round square is a cat" is nb-false for the reason that the equivalence $\exists x RS\,x \equiv \exists x \neg Cx$ is false. It is to be remarked that other apparently analogous examples are nb-true, not nb-false. For instance "All married bachelors are round squares" turns out to be nb-true given that "Everything is a married bachelor", "Everything is a Round Square","Something is a married bachelor", "Something is a Round Square" are all false.

Such a variety of results may be a challenge to our intuitions, which are perhaps

---

argument. Let us suppose that in **QL** $Cx\forall^{\rightarrow}Mx$ and, by Reductio, $\neg(Cx\exists/^{\circ}Mx)$, which means $(\exists x Cx \vee \forall x Mx) \wedge (\forall x Cx \vee \exists x Mx) \wedge \forall x(Cx \supset \neg Mx)$. From the last conjunct we have $\forall x Cx \supset \forall x \neg Mx$. But $Cx\forall^{\rightarrow}Mx$ implies the equivalence $\forall x Cx \equiv \forall x Mx$, so by the property of $\equiv$, the wff $(\exists x Cx \vee \forall x Mx) \wedge (\forall x Cx \vee \exists x Mx) \wedge \forall x(Cx \supset \neg Mx)$ implies, via $\forall x Cx \equiv \forall x Mx$, $(\exists x Cx \vee \forall x Cx) \wedge (\forall x Mx \vee \exists x Mx) \wedge \forall x(\neg Cx \vee \neg Cx)$ i.e. $\exists x Cx \wedge \exists x Mx \wedge \forall x \neg Cx$ which is a contradiction. End of the proof.

not so clear as we might wish. One could be willing to admit, for instance, that according to some intuitions "No cat is a round square" should be accepted as true in some non- trivial sense while the converse "No round square is a cat",i.e. "Every round square is a non-cat", should be rejected being an instance of *ex falso quodlibet*.

It is of some interest to notice that consequential implication may provide a more articulated treatment of this class of statements. Let us recall, in fact, that beyond $\rightarrow$ we have other variants of the implicational arrow which may be classified as consequential inasmuch they satisfy Boethius and Aristotle. What happens if we select the weaker $\Rightarrow$ in place of $\rightarrow$ for a quantificational interpretation? We may define a connective weaker than $\forall^{\rightarrow}$ which is

(Def $\forall^{\Rightarrow}$) $\alpha(x)\forall^{\Rightarrow}\beta(x) \stackrel{\text{def}}{=} \forall x(\alpha(x) \supset \beta(x)) \wedge (\exists x\alpha(x) \equiv \exists x\beta(x))$.

To mark the distinction from what we called non-banal truth, we will speak now of *not-banal°-truth* (nb°-truth). An example of the difference between the two kinds of non- banal truth is provided by "No miners tunnelled to the center of the earth", which becomes $M(x)\forall^{\Rightarrow}\neg T(x)$, i.e. $\forall x(Mx \supset \neg Tx) \wedge (\exists xMx \equiv \exists x\neg Tx)$. Since the two conjuncts are true, the statement is nb°-true (while before we argued that it is nb-false) and the converse statement $\neg T(x)\forall^{\Rightarrow}M(x)$ is nb°-false. By a parallel argument, of course, we may prove that "No cat is a round square " is nb°-true (while we before argued that it is nb-false) while its converse "No round square is a cat" is nb°-false.

The connective $\Rightarrow$ is not a trivial duplicate of $\rightarrow$ but has some distinctive properties. The first is that Boethius in the secondary form $(\alpha \Rightarrow \beta) \supset \neg(\neg\alpha \Rightarrow \beta)$ does not hold, so we have not to take care of a double kind of subalternation. Another property of $\Rightarrow$ is that contraposition holds only in the weakened form

(WC$\Rightarrow$) $\Diamond\neg\beta \supset ((\alpha \Rightarrow \beta) \supset (\neg\beta \Rightarrow \neg\alpha))$

So, considering the first order translation, contraposition holds only for instances having this form:

(WC$\forall^{\Rightarrow}$) $\exists xUx \supset (Cx\forall^{\Rightarrow}\neg Ux \supset Ux\forall^{\Rightarrow}\neg Cx)$

For example, one can pass from "It is nb° -true that no Italian is a married bachelor " to "It is nb° -true that no married bachelor is Italian", provided that some married bachelor exists, which of course is false. We have in fact to record this difference: "It is nb° -true that no married bachelor is Italian" has the form $\forall x(MBx \supset \neg Ix) \wedge (\exists x\neg Ix \equiv \exists xMBx)$ and turns out to be false since $\exists xMBx$ is false while $\exists x\neg Ix$ is true. "It is nb-true that no Italian is a married bachelor ", however, is true. In fact $\forall x(Ix \supset \neg MBx) \wedge (\exists x\neg MBx \equiv \exists xIx)$ is true given that $\forall x(Ix \supset \neg MBx)$ is true and both clauses of the equivalence are true. The operator $\forall^{\rightarrow}$, however, is unrestrictedly contrapositive.

The qualities of $\forall^{\rightarrow}$ are remarkable also with reference to another logical property. In order to understand what is at stake here, we have to remark that the universal statement "All married bachelors are married", which is an instance of the law of Simplification, is nb-false and also nb° -false. If fact, we lack Simplification both for $\rightarrow$ and for $\Rightarrow$. However, as already remarked, for $\rightarrow$ we have a weakened Simplification in the form

(WS$\rightarrow$) $(\Diamond(p \wedge q) \wedge \neg\Box p) \supset (p \wedge q \rightarrow p)$

while a theorem we have for $\Rightarrow$ is the simpler

(WS$\Rightarrow$) $\Diamond(p \wedge q) \supset (p \wedge q \Rightarrow p)$.

The quantificational law corresponding to (WS$\rightarrow$) is then equivalent to $\exists x(Mx \wedge Bx) \supset (\forall x(Mx \wedge Bx) \supset Mx) \wedge (\exists xMx \supset \exists x(Mx \wedge Bx)))$, so to $\exists x(Mx \wedge Bx) \supset (Mx \wedge Bx\forall^{\rightarrow}Mx)$.

To go back to our example, what is a logical law is then the assertion that, provided there is something which is is a married bachelor (so in no case!), it is nb°-true that all married bachelors are married. As expected, "All married Italians are married" and "All married Italians are Italians" turn out to be both nb° -true since there is something which is a married Italian. This of course works also for the operator $\forall^{\rightarrow}$ but in correspondence with the restriction $\neg\Box p$ occuring in WS$\rightarrow$ we must consider the additional clause that the consequent Mx must not be universally true. This restriction may be judged to be troublesome. So, for instance, "All married people who are identical to themselves are identical to themselves" is nb false given that the consequent is universally true. However, it turns out to be nb° -true, what surely we are inclined to find more intuitive.

APPENDIX

Axioms of **CIw**

(PC) All truth-functional tautologies

(a) $(p \rightarrow q \wedge q \rightarrow r) \supset (p \rightarrow r)$

(b) $(\top \rightarrow (p \supset q) \wedge \neg(\top \rightarrow \neg p) \wedge \neg(\top \rightarrow q)) \supset (p \rightarrow q)$

(c) $\neg(\top \rightarrow \neg(p \wedge r)) \supset (p \rightarrow q \supset (p \wedge r \rightarrow q \wedge r))$

(d) $(\neg p \rightarrow \neg q) \supset (q \rightarrow p)$

(e) $(p \to \bot) \supset (\bot \to p)$

(f) $(\bot \to p) \supset (p \to \bot)$

(g) $p \to p$

Rules of inference:

(US) Uniform Substitution

(MP) Modus Ponens for $\supset$

(Eq) Replacement of proved material equivalents

   CIw may be proved to be definitionally equivalent to **K** via the following two mappings φ and ψ

(1a) $\varphi p = p$ (if p is a propositional variable)

(2a) $\varphi \bot = \bot$

(3a) $\varphi(\alpha \supset \beta) = \varphi\alpha \supset \varphi\beta$

(4a) $\varphi(\alpha \to \beta) = \Box(\varphi\alpha \supset \varphi\beta) \wedge (\Box\varphi\beta \supset \Box\varphi\alpha) \wedge (\Diamond\varphi\beta \supset \Diamond\varphi\alpha)$

(1b) $\psi p = p$ (if p is a propositional variable)

(2b) $\psi \bot = \bot$

(3b) $\psi(\alpha \supset \beta) = \psi\alpha \supset \psi\beta$

(4b) $\psi\Box\alpha = \top \to \psi\alpha$

For every α, in fact, it is possible to prove

$\vdash_{CIw} \alpha$ only if $\vdash_K \varphi(\alpha)$

$\vdash_K \alpha$ only if $\vdash_{CIw} \psi(\alpha)$

$\vdash_K \alpha \equiv \varphi\psi(\alpha)$ and $\vdash_{CIw} \alpha \equiv \psi\varphi(\alpha)$.

System **CI**, i.e. **CIw** extended with $(p \to q) \supset \neg(p \to \neg q)$ (BT) is equivalent to **KD**, i.e. **K** + $\Box p \supset \Diamond p$.

   System **CI.0**, i.e. **CI** extended with $(p \to q) \supset (p \supset q)$ is equivalent to **KT**. As a matter of fact, every normal modal system may be converted into an equivalent consequential modal system (see [20]).

BIBLIOGRAPHY

[1] Angell, R.B. Truth-Functional Conditionals and Modern vs. Traditional Syllogistic. *Mind*, 95:210–223, 1986.

[2] Angell, R.B. A Propositional Logic with Subjunctive Conditionals. *Journal of Symbolic Logic*, 17:327–343, 1962.

[3] Angell, R.B. Tre logiche dei condizionali congiuntivi. In C. Pizzi, editor, *Leggi di Natura, Modalità, Ipotesi*, pages 156–180. Feltrinelli, Milano, 1978.

[4] MacCall, H. The Existential Import of propositions. *Mind*, 14:401–402 and 578-579, 1905.

[5] McCall, S. Connexive Implication and the Syllogism. *Mind*, 76:346–356, 1967.

[6] McCall, S. Connexive Implication. *Journal of Symbolic Logic*, 31:415–433, 1967.

[7] McCall, S. Connexive implication. In A. R. Anderson and N. Belnap, editors, *Entailment. The Logic of Relevance and Necessity*, vol. I, pages 434–452. Princeton U.P, 1975.

[8] Nasti de Vincentis, M. Connexive implication in a Chrysippean Setting. In M. C. Di Maio, C. Cellucci and G. Roncaglia, editors, *Logica e Filosofia della Scienza: problemi e prospettive*, pages 595–603. ETS, Pisa, 1994.

[9] Nasti de Vincentis, M. Stoic implication and Stoic Modalities. In C. Mangione, G. Corsi and M. Mugnai, editors, *Le teorie delle modalità*, pages 259–263. CLUEB, Bologna, 1989.

[10] Nasti de Vincentis, M. *Logiche della connessività*. Haupt, Bern, 2002.

[11] Nasti de Vincentis, M. From Aristotle's Syllogistic to Stoic Conditionals: Holzwege or Detectable Paths? *Topoi*, 23:113–138, 2004.

[12] Nelson, E. Intensional Relations. *Mind*, 39:444–453, 1930.

[13] Nelson, E. The Square of Opposition. *The Monist*, 42:269–278, 1932.

[14] Pizzi, C. Decision procedures for logics of Consequential Implication. *Notre Dame Journal of Formal Logic*, 32:618–636, 1991.

[15] Pizzi, C. and Williamson, T. Strong Boethius' Thesis and Consequential Implication. *Journal of Philosophical Logic*, 26:569–588, 1997.

[16] Pizzi, C. Implicazione crisippea e dipendenza contestuale. *Dianoia*, 3:25–44, 1998.

[17] Pizzi, C. A modal framework for consequential implication and the Factor Law. *Contemporary Mathematics*, 235:313–332, 1999.

[18] Pizzi, C. Aristotle's Thesis between paraconsistency and modalization. *Journal of Applied Logic*, 3:119–131, 2004.

[19] Pizzi, C. Cotenability and the logic of consequential implication. *Logic Journal IGPL*, 12:561–579, 2004.

[20] Pizzi C. and Williamson, T. Conditional Excluded Middle in systems of Consequential Implication. *Journal of Philosophical Logic*, 34:333–362, 2005.

[21] Pizzi, C. Aristotle's Cubes and Consequential Implication. *Logica universalis*, 2:143–153, 2008.

[22] Routley, R. and Montgomery, H. On Systems Containing Aristotle's Thesis. *Journal of Symbolic Logic*, 33:82–96, 1968.

[23] Wansing, H. Connexive Logic. *Stanford Enciclopedia*, Internet resource, 2006.

Claudio Pizzi
Dipartimento di Filosofia e Scienze Sociali
Università di Siena
Via Roma 47 - 53100 Siena, Italy
E-mail: pizzic@msn.com

SECTION 3
LOGIC AND MATHEMATICS

Logic and Mathematics: Propositional Calculus with only three primitive terms

E. G. K. LÓPEZ-ESCOBAR

ABSTRACT. A Second Order Propositional Calculus, PM, with just three *primitive* terms, namely (i) *universal quantification*, (ii) *equivalence* and (iii) a binary operation named *subordination*, is presented. It has the property that it includes the full Second Order Intuitionistic Propositional Calculus and that by the addition of *finitely* many axioms (using the same three primitive terms), it also *includes* Impredicative Set Theory.

1 Introduction

In the first part of [7] I gave a plausible interpretation of A. P. Morse's unorthodox Set Theory ([11]) by allowing sets to have an additional attribute beyond the one of *cardinality* while in the second part I developed an alternate interpretation using a Second Order Propositional Calculus with additional propositional operators. The main emphasis in [7] was on proof-theoretical questions engendered by its various propositional subsystems. In the article [8] I further developed the idea of Morse's Set Theory as the full Second Order Propositional Calculus with just *two* additional binary propositional functions; one of them taking the place of Morse's *definor* which was essential in the definition of new symbols, the other was shown to have many of the *formal* properties of "∈". I also introduced two semantics for the calculus. The first relied on Boolean algebras with additional operators and the other was modeled after the Scott-Solovay Boolean valued models for ZF.

Having co-authored the monograph [10] in which we extended ideas of Tarski about Leśniewski's Protothetic (which could be loosely described as a *Third* Order Propositional Calculus, that is, with quantification over *propositions* and *propositional functions*), it was natural to attempt to do the same with the Second Order Propositional Calculus of [8].

I was successful in showing that in a Propositional Calculus with the universal quantifier over *propositions*, $\bigwedge$, the equivalence connective, $\equiv$, and a single additional binary operator $\lhd$, one can define the propositional connective of conjunction and thence the full system of [8]. A note was submitted to *Fundamenta Mathematicae*, [9], in which it is shown how three very simple axioms concerning $\lhd$ allow us to mimic enough of universal quantification over *propositional functions* to transfer Tarski's proof of the definition of conjunction in the Protothetic to a propositional system with quantification only over *propositions*.

This article is an outline of enhancements, simplifications and further developments of the the articles [8] and [9]; for example, I give a complete Beth type semantics for the system which could be also adapted to a more constructive interpretation.

The first thing that had to be decided is what to call the system under consideration, that is, a Second Order Propositional Calculus, whose *primitive terms* are: $\bigwedge$, $\equiv$, $\lhd$. Since it was S. Leśniewski who first required that definitions were to be theses of the Protothetic, I will use: **Protomathematics**, or **PM** for short[1].

1.1   The primitive term of *subordination*

Thanks to A. Church, [3], and G. Frege[2] it is now well known that a declarative English sentence has (or can be postulated to have) both a *truth-value* and a *sense* (or: *meaning*). *Propositions* can be considered as abstractions of the declarative sentences[3] and thus one would expect that in at least in some of the many *formalizations* of the various propositional calculi one would find some formal reference to the *meanings* of the propositions and not only to their *truth-values*. However, except for some minimal involvement of the meaning of propositions in some modal logics, there appear to be none.

PM goes a step beyond the modal logics in making use of the *meaning* of propositions[4] and this is achieved through the use of the binary propositional operation of *subordination* represented by the symbol: $\lhd$. Note, however, that our intent is not so much a philosophical explanation of *meaning* but rather to see whether a mathematical use could be made of it.

PM symbolizes not so much the *meaning* of a proposition but rather the condition for two propositions to have equivalent meanings—which we shall

---

[1]This is still at the development stage so it is conceivable that a better name might come to mind.

[2]For bibliographical references concerning Frege see [3], page 4, footnote 5.

[3]Much as the *natural numbers* can be considered to be abstractions of heaps of stones.

[4]Or more accurately, the abstraction of the concept of meaning in declarative sentences.

loosely refer as "having the same meaning"—and it is for the latter that ◁ enters the picture[5].

An informal interpretation of *subordination* is that a proposition $\mathcal{A}$ is subordinate to the proposition $\mathcal{B}$ just in case that the proposition $\mathcal{A}$ is involved in the meaning of proposition $\mathcal{B}$. We measure the degree of involvement by the truth-value of the proposition $(\mathcal{A} \triangleleft \mathcal{B})$. Now $(\mathcal{A} \triangleleft \mathcal{B})$ being itself also a proposition has both a *truth-value* and a *meaning*. Since PM is but a beginning into a mathematical analysis of *meaning*, we shall avoid any discussions on the *meaning of meaning* so that we include an axiom which allows us to prove:

$$p \triangleleft q \;\longrightarrow\; (\, (p \triangleleft (\mathcal{A} \triangleleft \mathcal{B})) \;\equiv\; (\mathcal{A} \triangleleft \mathcal{B}) \,) \,,$$

which gives us the intuitive interpretation that the *meaning* of the proposition $(\mathcal{A} \triangleleft \mathcal{B})$ may be taken to be its *truth-value*.

PM then uses the formula:

$$\bigwedge x (\, (x \triangleleft \mathcal{A}) \;\equiv\; (x \triangleleft \mathcal{B}) \,) \,,$$

to express that $\mathcal{A}$ and $\mathcal{B}$ have the same subordinate propositions. We take that condition as expressing that $\mathcal{A}$ and $\mathcal{B}$ have the *same meaning*[6]. Another of the axioms (Morse's axiom) of PM allows us to prove that if two propositions have the same *meaning* then they have the same *truth-value*.

2   The language of PM

2.1   Symbols of the language of PM

It has become the tradition that formal languages of Mathematical Logic are completely determined by their symbols to the extent that in some cases the *language* is simply defined as the set of predicate symbols, etc..

---

[5]This is analogous to Cantor's treatment of the cardinalities of sets for although Cantor did write something on the nature of the cardinality of the set M:

> [482] *We denote the result of this double act of abstraction, the cardinal number or power of M by... $\overline{\overline{M}}$. Since every single element m, if we abstract from its nature, becomes a "unit," the cardinal number $\overline{M}$ is a definite aggregate composed of units, and this number has existence in our mind as an intellectual image or projection of the given aggregate M.*

what was very useful in the development of Set Theory was the mathematical condition for two sets to be equivalent [with respect to their cardinality] as the existence of a bijection between them. See [1].

[6]As already mentioned it would be more accurate to say "equivalent meaning". Note also that the formula '$(\mathcal{A} \equiv \mathcal{B})$' may be considered as expressing that $\mathcal{A}$ and $\mathcal{B}$ have the same *truth-value*.

Consequently the addition of a new symbol changes the language and hence the system.

One of the advantages of PM is that new symbols can be introduced as a theory is being developed. The symbols are introduced both in an abbreviatory role as well as in a creative role[7]. Thus the language of PM is much more comparable to a modern computer language, for example to C^{++}, in that there is a (small) set of **reserved symbols** and a **protocol** for introducing additional symbols.

The reserved symbols of PM

The *reserved symbols of* PM are (i) the *propositional variables* $(x, y, z, \ldots)$, (ii) the *propositional parameters*[8] $(p, q, r, \ldots)$, (iii) the parsing symbols (parentheses) and (iv) the three *primitive terms*: $\bigwedge$, $\equiv$ and $\triangleleft$.

The Protocol for introducing additional symbols

PM has, by choice, a rather restrictive protocol for the type of symbols that may be introduced which, for a lack of a better name, we shall call *defined operators*. Only n-ary $(n \in \omega)$ *propositional operators* and *variable binding operators (vbo)* (a typical example of the latter being the existential quantifier) can be added to the language of PM.

Definitions are the means by which PM permits the introduction of additional operators. With each definition there is associated a *definitional postulate* which is adopted as a thesis of PM.

Definitional Labels and Postulates

In the Leśniewskian Protothetic the definition itself was used as the definitional axiom. For example, consider a familiar definition of *negation*:

$$\bigwedge x \left(\sim x \equiv (x \equiv \bigwedge xx) \right).$$

The above *sentence* could be used both as the *definition* and the *definitional axiom*. Considered just as a definition it tells us that $\sim$ is a unary propositional operator. Considered as a definitional axiom we obtain that $\sim$ is accepted as a (defined) operator of the Protothetic and then using all the other axioms and rules of inference of the Protothetic one can derive the required properties of negation.

In PM we consider a *definition* as having two parts: (i) the *definitional label* and (ii) the *definitional postulate*. The definitional label is, from a

---

[7]Not in the sense that the proof theoretic strength is changed but rather in the sense that Gauss' definition of the congruence relation "$a \equiv b \pmod{m}$" may be considered as a creative one.

[8]Since we are using a Gentzenian Natural Deduction style of formalization, it is advantageous to separate those variables that may occur free from those that may occur bound.

proof theoretical viewpoint, completely unnecessary; it is introduced as a mnemonic for the human reader and it is not even a formula of the language of PM. On the other hand the definitional postulate[9] is a thesis of PM which then can be used to obtain the required properties for the defined operator.

Using the above example, the *definitional label of* $\sim$ in PM would be:

$$\langle\; \sim p \;\overset{df}{=}\; (p \equiv \textstyle\bigwedge xx)\; \rangle,$$

and then the *definitional postulate* would be the *sentence*:

$$\textstyle\bigwedge x \bigwedge y \left(\; (y \triangleleft \sim x) \;\equiv\; (y \triangleleft (x \equiv \textstyle\bigwedge xx))\; \right).$$

Note that the definitional postulate may be interpreted as stating that $\sim p$ and $(p \equiv \bigwedge xx)$ have the same *meaning* and not just the same *truth-value*.

Let us next consider how the existential quantifier, $\bigvee$, is added to the symbols of PM. It should be clear that the property of *binding variables within its scope* cannot be achieved by using a single sentence, so that what is required is a definitional postulate *schema*. Since the definitional label is not a formula of PM we introduced Morse's *schemator*, $\underline{u}$, and thus still are able to have a single expression as a definitional label for vbo's.

The *definitional label for*[10] $\bigvee$ is:

$$\langle\; \textstyle\bigvee x\underline{u}x \;\overset{df}{=}\; \textstyle\bigwedge y \left(\; \textstyle\bigwedge x(\underline{u}x \rightarrow y) \;\rightarrow\; y \right)\; \rangle,$$

while for any formula $\mathcal{F}_p$, the universal closure of the following formula is an *instance of the definitional postulate schema for* $\bigvee$:

$$\textstyle\bigwedge z \left(\; (z \triangleleft \textstyle\bigvee w \mathcal{F}_w) \;\equiv\; (z \triangleleft \textstyle\bigwedge y(\textstyle\bigwedge w(\mathcal{F}_w \rightarrow y) \;\rightarrow\; y\;))\; \right),$$

subject of course to the usual requirements on the variables.

From the above paradigms and the instances of definitions given in this article, one can specify the precise requirements on the acceptable definitions, however for the sake of expediency I shall leave it to the reader to do it.

---

[9]The reason I prefer to use "definitional postulate" instead of "definitional axiom" is that I prefer to reserve the word "axiom" for those sentences which are absolutely essential in the description of PM.

[10]The existential quantifier is defined after we have defined the conditional ' $\rightarrow$ '.

2.2 Quasi-formulae, formulae and sentences of PM

The **Basic Language of** PM consists of the formulae of the Second Order Propositional Calculus[11] generated by the operators: $\bigwedge$ (the universal quantifier over propositions), $\equiv$ (the binary connective of equivalence) and $\lhd$ (the binary operator of subordination). The *quasi-formulae* are those well-formed expressions in which there may occur free occurrences of variables, the *formulae* are those quasi-formulae in which there are no free occurrences of variables and the *sentences* are those formulae in which there are no occurrences of parameters.

The moment one adds a definitional postulate to PM then the class of formulae of PM is enlarged. For example, suppose that following the protocol on the acceptable definitions, the ternary prefix operator, $\mathbb{O}$, had been added; then to the recursive characterization of the quasi-formulae the following clause would have been inserted:

> If $\mathcal{A}_1, \mathcal{A}_2, \mathcal{A}_3$ are quasi-formulae then so is: $\mathbb{O}\mathcal{A}_1\mathcal{A}_2\mathcal{A}_3$.

On the other hand suppose that the variable binding operator, $\mathbb{Q}$, had been introduced then we would add the following clause:

> If $\mathcal{A}$ is a quasi-formula and α a propositional variable, then $\mathbb{Q}\alpha\mathcal{A}$ is a quasi-formula in which all the occurrences of the variable α are bound occurrences.

3 Derivations in PM

PM is organized as a Gentzenian Natural Deduction System in which the *derivations* are trees of formulae. We use three sets of rules of inference. The traditional (intuitionistic) Introduction/Elimination rules for $\bigwedge$ and $\equiv$ and a third one which could be considered as an Introduction/Elimination pair for the combination of primitive symbols $(-\lhd\bigwedge\cdot-)$.

3.1 Rules of inference

Rules of inference for the equivalence connective

$$\frac{\begin{array}{cc}[\mathcal{A}] & [\mathcal{B}]\\ \Pi_1 & \Pi_2\\ \mathcal{B} & \mathcal{A}\end{array}}{(\mathcal{A} \equiv \mathcal{B})}$$

$$\frac{\begin{array}{cc}\Pi_1 & \Pi_2\\ \mathcal{A} & (\mathcal{A} \equiv \mathcal{B})\end{array}}{\mathcal{B}} \qquad \frac{\begin{array}{cc}\Pi_1 & \Pi_2\\ \mathcal{B} & (\mathcal{A} \equiv \mathcal{B})\end{array}}{\mathcal{A}}$$

---

[11]See [3], page 151, where it is called the *Extended propositional calculus*.

Rules of inference for the universal quantifier

$$\frac{\Pi \qquad \mathcal{A}_p^q}{\bigwedge x \mathcal{A}_p^x} \qquad\qquad \frac{\Pi \qquad \bigwedge x \mathcal{A}_p^x}{\mathcal{A}_p^{\mathcal{B}}}$$

where the propositional parameter q (the *eigenparameter*) does not occur free in any undischarged assumption formula and where $\mathcal{A}_p^{\mathcal{Q}}$ is the quasi-formula obtained by *proper substitution* of all the occurrences of the propositional parameter p in the formula $\mathcal{A}$ by the quasi-formula $\mathcal{Q}$.

Rules of inference for distributive quantification

$$\frac{\Pi \qquad \left(\mathcal{A} \vartriangleleft \mathcal{B}_p^q \right)}{\left(\mathcal{A} \vartriangleleft \bigwedge x \mathcal{B}_p^x \right)} \qquad\qquad \frac{\Pi \qquad \left(\mathcal{A} \vartriangleleft \bigwedge x \mathcal{B}_p^x \right)}{\left(\mathcal{A} \vartriangleleft \mathcal{B}_p^{\mathcal{C}} \right)}$$

where in the Introduction rule, the parameter q does not occur in any undischarged assumption formula nor in the conclusion formula.

3.2   The Axioms of PM

The fact that $\bigwedge$, $\equiv$, $\vartriangleleft$ are the only primitive terms of PM means that, in theory, the basic language of PM would suffice for the complete development of the system. Unfortunately simple logically valid formulae such as:

$$(p \rightarrow (q \rightarrow r)) \ \rightarrow \ ((p \wedge q) \rightarrow r),$$

when written as a formula of the basic language of PM would be so long that it is doubtful that anyone would recognize it as logically valid. Consequently I shall first list the axioms, in the basic language, which suffice to prove that the defined operator " $\rightarrow$ " has the (proof theoretic) properties of the intuitionistic conditional. Then I use the defined operator $\rightarrow$ in the statements of the remaining axioms.

The Rudimentary Axioms of PM

Morse's Axiom $\bigwedge x \big((\bigwedge x x \vartriangleleft x) \equiv x \big).$

Minimality Axiom $\bigwedge x \big((x \vartriangleleft \bigwedge x x) \equiv \bigwedge x x \big).$

Maximality Axiom $\bigwedge x \big((\bigwedge x (x \equiv x) \vartriangleleft x) \equiv \bigwedge x x \big).$

The definition of the conditional

Definitional Labels 1.

$$\langle \{p\}(q) \overset{df}{=} \bigwedge x\big(q \equiv ((x \triangleleft p) \equiv \bigwedge xx)\big) \rangle.$$

$$\langle (p \rightarrow q) \overset{df}{=} \big(p \equiv \bigwedge z(p \equiv (\{z\}(p) \equiv \{z\}(q))) \big) \rangle.$$

The Definitional Labels 1 tell us that $\rightarrow$ is now an acceptable binary operator of PM and thus it may be used in the following axioms.

The Distributive Axioms of PM

Following Leśniewski's example, the axioms (and postulates) are required to be sentences. However it is more convenient to omit the outermost universal quantifiers when listing them. Thus the actual axioms are the universal closures of the listed quasi-formulae.

$(\triangleleft , \triangleleft)$-*Axiom* $\big((x \triangleleft w) \rightarrow ((x \triangleleft (y \triangleleft z)) \equiv (y \triangleleft z)) \big)$.

$(\triangleleft , \equiv)$-*Axiom* $\big((x \triangleleft w) \rightarrow ((x \triangleleft (y \equiv z)) \equiv ((x \triangleleft y) \equiv (x \triangleleft z))) \big)$.

Leibniz' Axiom for PM
$$\Big((x \triangleleft w) \rightarrow \big(\bigwedge z((x \triangleleft z) \rightarrow (y \triangleleft z)) \equiv \bigwedge z((z \triangleleft x) \equiv (z \triangleleft y)) \big) \Big).$$

Notation for derivations in PM

" $\Gamma \vdash \mathcal{A}$ " informs us that there is a PM-derivation whose end-formula is $\mathcal{A}$ and whose undischarged top-formulae are either *axioms*, *definitional postulates* or *belong to* Γ.

We write " $\Gamma \vdash^R \mathcal{A}$ " when the only *axioms* used in the derivation are the Rudimentary axioms.

4   Rudimentary derivations

The first thing to observe is that: *if* $\langle \mathcal{D}_1 \overset{df}{=} \mathcal{D}_2 \rangle$ *is an acceptable definitional label then:*

$$\vdash^R (\mathcal{D}_1 \equiv \mathcal{D}_2).$$

The argument is as follows: From the definitional postulate one obtains, by $\bigwedge$-Elimination, that $((\bigwedge xx \triangleleft \mathcal{D}_1) \equiv (\bigwedge xx \triangleleft \mathcal{D}_2))$. Applying Morse's Axiom and making use of the rules of inference for $\equiv$ the result follows.

LEMMA 1.

1. $\quad \vdash^R \left(\{p\}(q) \;\; \equiv \;\; \bigwedge x \big(q \equiv ((x \triangleleft p) \equiv \bigwedge xx) \big) \right).$

2. $\quad \vdash^R \left((p \rightarrow q) \;\; \equiv \;\; \big(p \equiv \bigwedge z (p \equiv (\{z\}(p) \equiv \{z\}(q))) \big) \right).$

4.1 Tarski's Protothetical proof for the definability of conjunction

The definition of conjunction given by Tarski in the Protothetic, in which the propositional functions range over the *truth-functions*[12], is:

$$(p \wedge q) \;\; \equiv \;\; \bigwedge f \left(p \equiv (f(p) \equiv f(q)) \right),$$

where "f" is a variable for truth-functions, see [13].

An interesting point about Tarski's proof is that it only makes explicit use of *two* truth-functions, namely the *identity* function and the constantly *false* function. Now Tarski's definition for conjunction gives us the following Protothetical definition for the conditional:

$$(p \rightarrow q) \;\; \equiv \;\; \left(p \equiv \bigwedge f \left(p \equiv (f(p) \equiv f(q)) \right) \right)$$

and its proof of correctness will also explicitly use only the truth-functions of *identity* and *falsehood*.

4.2 The PM definability of the conditional

The way in which Tarski's proof is imported into PM is to replace the quantification over truth- functions:

$$\dots \bigwedge f(\dots f(p) \dots f(q) \dots),$$

by quantification over propositions:

$$\dots \bigwedge z(\dots \{z\}(p) \dots \{z\}(q) \dots).$$

Of course it has to be shown that (i) $\{z\}(\ )$ behaves as a *truth-function* and (ii) that $\{\bigwedge xx\}(\ )$ behaves as the *identity* and $\{\bigwedge x(x \equiv x)\}(\ )$ as the *falsehood* function.

LEMMA 2.

1. $\quad (p \equiv q) \; \vdash^R (\{s\}(p) \equiv \{s\}(q)).$

2. $\quad \vdash^R (\{\bigwedge xx\}(q) \equiv q).$

---

[12]That is those propositional functions invariant under $\equiv$.

3. $\vdash^R (\{\bigwedge x(x \equiv x)\}(q) \equiv \bigwedge xx)$.

Justifications: The proof of *(1.)* is a simple exercise in the $\bigwedge$, $\equiv$ calculus.
Re. (2.): Lemma 1.1 gives us that it suffices to prove that:

$$\left(\bigwedge x (q \equiv ((x \triangleleft \bigwedge xx) \equiv \bigwedge xx)) \equiv q \right) .$$

The *Minimality Axiom* completes the proof.
Re. (3.) It suffices to prove that:

$$\bigwedge x \left(q \equiv ((x \triangleleft \bigwedge x(x \equiv x)) \equiv \bigwedge xx) \right) \vdash^R \bigwedge xx.$$

Thus assume $\bigwedge x \left(q \equiv ((x \triangleleft \bigwedge x(x \equiv x)) \equiv \bigwedge xx) \right)$. Instantiating the outermost quantifier by '$\bigwedge x(x \equiv x)$' one obtains:

$$\left(q \equiv ((\bigwedge x(x \equiv x) \triangleleft \bigwedge x(x \equiv x)) \equiv \bigwedge xx) \right),$$

and then the *Maximality Axiom* leads to q . Instantiating with '$\bigwedge xx$' one obtains:

$$\left(q \equiv ((\bigwedge xx \triangleleft \bigwedge x(x \equiv x)) \equiv \bigwedge xx) \right)$$

and this time *Morse's Axiom* leads to $(q \equiv \bigwedge xx)$. And thus to $\bigwedge xx$.

Properties of $\rightarrow$ in PM

Theorems 1 and 2 below give the essential (intuitionistic) properties of the conditional.

THEOREM 3. p , $(p \rightarrow q)$ $\vdash^R q$.

Justification: Assume that $(p \rightarrow q)$ and p.

Then the following sequence of formulae can be derived from the Rudimentary axioms:

$$p \equiv \bigwedge z(p \equiv (\{z\}(p) \equiv \{z\}(q)))$$
$$\bigwedge z(p \equiv (\{z\}(p) \equiv \{z\}(q)))$$
$$p \equiv (\{\bigwedge xx\}(p) \equiv \{\bigwedge xx\}(q))$$
$$\{\bigwedge xx\}(p) \equiv \{\bigwedge xx\}(q)$$
$$p \equiv q$$
$$q$$

THEOREM 4.   *If* $p \vdash q$, *then* $\vdash (p \rightarrow q)$.
Justification: Assume that: $p \vdash q$.

(I) Now assume p. Then:

$$p \equiv q$$
$$\{s\}(p) \equiv \{s\}(q)$$
$$p \equiv (\{s\}(p) \equiv \{s\}(q))$$
$$\bigwedge z\,(p \equiv (\{z\}(p) \equiv \{z\}(q)))$$

Thus: $p \vdash \bigwedge z\,(p \equiv (\{z\}(p) \equiv \{z\}(q)))$.

(II) Now assume $\bigwedge z\,(p \equiv (\{z\}(p) \equiv \{z\}(q)))$. This time we obtain:

$$p \equiv (\{\bigwedge x(x \equiv x)\}(p) \equiv \{\bigwedge x(x \equiv x)\}(q))$$
$$p \equiv (\bigwedge xx \equiv \bigwedge xx)$$
$$p.$$

Thus: $\bigwedge z\,(p \equiv (\{z\}(p) \equiv \{z\}(q))) \vdash p$.

Combining (I) and (II) we obtain:

$$\vdash \quad (p \equiv \bigwedge z\,(p \equiv (\{z\}(p) \equiv \{z\}(q))))$$
$$\vdash \quad (p \rightarrow q).$$

Theorems 3 and 4 can be used to obtain, as derived rules, the Elimination/Introduction rules of inference for the conditional. Since PM has universal quantification over propositions we may use the traditional intuitionistic definitions for the remaining logical atoms[13].

Definitional Labels 2.

$$\langle \bot \overset{df}{=} \bigwedge xx \,\rangle.$$
$$\langle \top \overset{df}{=} \bigwedge x(x \equiv x) \,\rangle.$$
$$\langle \sim p \overset{df}{=} (p \rightarrow \bot) \,\rangle.$$
$$\langle (p \wedge q) \overset{df}{=} \bigwedge x(\,(p \rightarrow (q \rightarrow x)) \rightarrow x\,) \,\rangle.$$
$$\langle (p \vee q) \overset{df}{=} \bigwedge x(\,((p \rightarrow x) \wedge (q \rightarrow x)) \rightarrow x\,) \,\rangle.$$
$$\langle \bigvee x\,\underline{u}x \overset{df}{=} \bigwedge y(\,\bigwedge x\,((\underline{u}x \rightarrow y) \rightarrow y)\,) \,\rangle.$$

THEOREM 5.   PM *contains all of the Second Order Intuitionistic Propositional Calculus. That is, if a formula $\mathcal{F}$ is a theorem of the Second Order Intuitionistic Propositional Calculus then $\mathcal{F}$ is a theorem of* PM.

If we consider an extension of PM in which not only the definable operators may be added, but also derived rules of inference as rules of inference, then PM also contains all the *derivations* of Second Order Intuitionistic Propositional Calculus.

---

[13]See, for example, [12], page 67.

5　Derivations in Protomathematics

From this point on we shall place no restrictions on the *axioms* used in the derivations.

One of the advantages of having an open system is that one can use more than one symbol for the same operator, specially when one wants to emphasize different aspects of the operator.

Definitional Labels 3.

$$\langle\ \emptyset\ \overset{df}{=}\ \perp\ \rangle.$$
$$\langle\ V\ \overset{df}{=}\ \top\ \rangle.$$
$$\langle\ (p\ \varepsilon\ q)\ \overset{df}{=}\ (p\triangleleft q)\ \rangle.$$
$$\langle\ (p\subseteq q)\ \overset{df}{=}\ \bigwedge x((x\,\varepsilon\,p)\ \rightarrow\ (x\,\varepsilon\,q))\ \rangle.$$
$$\langle\ (p=q)\ \overset{df}{=}\ \bigwedge x((x\,\varepsilon\,p)\ \equiv\ (x\,\varepsilon\,q))\ \rangle.$$
$$\langle\ (p\cong q)\ \overset{df}{=}\ \bigwedge x((p\,\varepsilon\,x)\ \rightarrow\ (q\,\varepsilon\,x))\ \rangle.$$

The axioms and the rule of distributed quantification yield the following theorems in the alternate notation:

THEOREM 6.

1. $\vdash (\emptyset\,\varepsilon\,p)\ \equiv\ p.$

2. $\vdash\ \sim(p\,\varepsilon\,\emptyset).$

3. $\vdash\ \sim(V\,\varepsilon\,p).$

4. $\vdash (p\,\varepsilon\,V)\ \rightarrow\ (\ (p\,\varepsilon\,(q\,\varepsilon\,r))\ \equiv\ (q\,\varepsilon\,r)\ ).$

5. $\vdash (p\,\varepsilon\,V)\ \rightarrow\ (\ (p\,\varepsilon\,(q\equiv r))\ \equiv\ ((p\,\varepsilon\,q)\equiv(p\,\varepsilon\,r))\ ).$

6. $\vdash (p\,\varepsilon\,V)\ \rightarrow\ (\ (p\cong q)\ \equiv\ (p=q)\ ).$

7. $\vdash (p\,\varepsilon\,\bigwedge x\mathcal{F})\ \equiv\ \bigwedge x(p\,\varepsilon\,\mathcal{F}).$

The following are some of the theorems that are particular to PM:

THEOREM 7.

- $\vdash (p\subseteq q)\ \rightarrow\ (p\rightarrow q).$

- $\vdash (p=q)\ \rightarrow\ (p\equiv q).$

- $\vdash (p\,\varepsilon\,V)\ \rightarrow\ (\ (p\,\varepsilon\,(q\rightarrow r))\equiv((p\,\varepsilon\,q)\ \rightarrow\ (p\,\varepsilon\,r))\ ).$

- $\vdash (p\,\varepsilon\,(q\wedge r))\ \equiv\ ((p\,\varepsilon\,q)\wedge(p\,\varepsilon\,r)).$

- $\vdash (p\,\varepsilon\,(q \lor r)) \equiv ((p\,\varepsilon\,q) \lor (p\,\varepsilon\,r))$.

- $\vdash (p\,\varepsilon\,\bigvee x\,\mathcal{F}) \equiv \bigvee x(p\,\varepsilon\,\mathcal{F})$.

- $\vdash (p\,\varepsilon\,\mathrm{V}) \longrightarrow (\,(p\,\varepsilon\,(q = r)) \equiv (q = r)\,)$.

- $\vdash (p\,\varepsilon\,\mathrm{V}) \longrightarrow (\,(p\,\varepsilon\,(q \subseteq r)) \equiv (q \subseteq r)\,)$.

5.1   Equality and substitution in PM

From the definition of $=$ one immediately obtains that it is (or has the properties of) an equivalence relation. Just as easy is to verify that:
$$(p = q) \longrightarrow (\,(r\,\varepsilon\,p) \equiv (r\,\varepsilon\,q)\,).$$
A little more thought is required to obtain:
$$(p = q) \longrightarrow (\,(r\,\varepsilon\,p) = (r\,\varepsilon\,q)\,).$$
Leibniz' axiom is essential in order to obtain:
$$(p = q) \longrightarrow (\,(p\,\varepsilon\,r) \equiv (q\,\varepsilon\,r)\,)$$
and then that:
$$(p = q) \longrightarrow (\,(p\,\varepsilon\,r) = (q\,\varepsilon\,r)\,).$$

The above remarks are the principal cases in the Basis step of a proof by induction on the complexity of the formula that:

THEOREM 8.   *For all formulae $\mathcal{F}_r$:*

1. $\vdash (\,(p = q) \longrightarrow (\mathcal{F}_p = \mathcal{F}_q)\,)$.

2. $\vdash (\,(p = q) \longrightarrow (\mathcal{F}_p \equiv \mathcal{F}_q)\,)$.[14]

5.2   Equality and the Definitions in PM

THEOREM 9.   *If $\langle\,\mathcal{DM} \overset{df}{=} \mathcal{DS}\,\rangle$ is an acceptable definitional label of* PM *then the formula $(\,\mathcal{DM} = \mathcal{DS}\,)$ is a theorem of* PM.

Justification: Assume that $\langle\,\mathcal{DM} \overset{df}{=} \mathcal{DS}\,\rangle$ is an acceptable definitional label of PM. Then its definitional postulate yields:
$$\bigwedge x(\,(x\,\varepsilon\,\mathcal{DM}) \equiv (x\,\varepsilon\,\mathcal{DS})\,).$$
Instantiating the definitional postulate for $=$ and applying Morse's axiom one obtains:
$$(\mathcal{DM} = \mathcal{DS}) \equiv \bigwedge x(\,(x\,\varepsilon\,\mathcal{DM}) \equiv (x\,\varepsilon\,\mathcal{DS})\,).$$
Hence $(\mathcal{DM} = \mathcal{DS})$ is a theorem—but not an axiom nor postulate—of PM.

---

[14]The following is from Ishiguro's [4], page 17:

> *The third, which I shall discuss in this chapter, is Leibniz's claim that "those terms of which one can be substituted for the other without affecting truth are identical":* "Eadem sunt, quorum unum alteri substitui potest salva veritate."

6   A Definable Class constructor of PM

Definitional Labels 4.

$$\langle \operatorname{\mathbf{sng}}(p) \overset{df}{=} \bigwedge x(x \to (p \,\varepsilon\, x)) \rangle.$$
$$\langle \operatorname{\mathsf{E}} x \underline{u} x \overset{df}{=} \bigvee x((\emptyset \,\varepsilon\, \underline{u} x) \wedge \operatorname{\mathbf{sng}}(x)) \rangle.$$

The following formulae are theorems of PM (and of [7], [11]):

- $(p \,\varepsilon\, \mathrm{V}) \to ((q \,\varepsilon\, \operatorname{\mathbf{sng}}(p)) \equiv (q = p))$

- $\sim(p \,\varepsilon\, \mathrm{V}) \to (\operatorname{\mathbf{sng}}(p) = \emptyset)$

They are used to prove the following theorem schema characterizing the *Class Constructor.*

THEOREM 10. $\vdash ((p \,\varepsilon\, \operatorname{\mathsf{E}} x \mathcal{F}_x) \equiv ((p \,\varepsilon\, \mathrm{V}) \wedge \mathcal{F}_p)).$

Some familiar theorems of PM:

$(\operatorname{\mathsf{E}} x(x \,\varepsilon\, p) = p), (\operatorname{\mathsf{E}} x(x = p) = \operatorname{\mathbf{sng}}(p)), \sim (\operatorname{\mathsf{E}} x \sim (x \,\varepsilon\, x) \,\varepsilon\, \mathrm{V}).$

7   Set Theory in PM

7.1   General Class Theory in PM

General Class Theory is a theory of classes in which the Natural Numbers form a class; it is powerful enough to develop the theory of well-orderings and Generalized Recursive Definitions. A detailed account can be found in in Part 2 of Chuaqui's book on *"Axiomatic Set Theory. Impredicative Theories of Classes"*, [2].

Now although in PM one can define the class $\mathbb{N}$ of natural numbers, it is not strong enough to prove that $(\emptyset \,\varepsilon\, \mathbb{N})$. Hence the following:

Definitional Labels 5.

$$\langle \operatorname{\mathsf{LK}} \overset{df}{=} \bigwedge x \bigwedge y \bigwedge z ((x \equiv (y \equiv z)) \equiv ((x \equiv y) \equiv z)) \rangle.$$
$$\langle \operatorname{\mathbf{SC}} \overset{df}{=} \bigwedge x \bigwedge y ((x \,\varepsilon\, \mathrm{V}) \to ((x \vee \operatorname{\mathbf{sng}}(y)) \,\varepsilon\, \mathrm{V})) \rangle.$$
$$\langle \operatorname{\mathbf{GCT}} \overset{df}{=} (\operatorname{\mathsf{LK}} \wedge \operatorname{\mathbf{SC}}) \rangle.$$

$\operatorname{\mathsf{LK}}$, which is Łukasiewicz's axiom, renders PM into a Classical system while $\operatorname{\mathbf{SC}}$ allows us to prove that the *succesor* of an *ordinal* is an *ordinal.*

The relation between PM and *General Class Theory* is given by the following theorem whose proof is nothing more than a rewrite of the proofs in Chuaqui's book:

THEOREM 11. *If $\mathcal{F}$ is a theorem of General Class Theory, then:*

$$\operatorname{\mathbf{GCT}} \vdash \mathcal{F},$$

that is, there is a PM *derivation of $\mathcal{F}$ from the sentence* $\operatorname{\mathbf{GCT}}$.

7.2   Impredicative Set Theory in PM

In an analogous fashion one can show that there is a single sentence, **KMT**, of PM such that:

THEOREM 12. *If $\mathcal{F}$ is a theorem of Kelley-Morse-Tarski Impredicative Set Theory[15], then:*

$$\mathbf{KMT} \vdash \mathcal{F}.$$

8   A Beth-like Semantics for PM

8.1   The Binary Tree

The *Binary Tree*, $\mathfrak{T}$, is the set of all *infinite binary sequences, ibs*. If $\alpha \in \mathfrak{T}$ and k is a natural number, then $\bar{\alpha}k = (\alpha(0), \dots, \alpha(k-1))$ is a *node* of $\mathfrak{T}$. If $\vec{n}$ and $\vec{m}$ are nodes then $\vec{n} \ll \vec{m}$ if and only if $\vec{n}$ is an initial segment of $\vec{m}$.

From now on it will be assumed that: α, β and γ are ibs. If $\vec{n}$ is a node, then we write: "$\vec{n} \ll \alpha$," just in case that there is a k such that $\bar{\alpha}k = \vec{n}$.

8.2   Prime Formulae

The *atomic formulae* are the propositional parameters, $\perp$ and $\top$. The *nucleic formulae* are the formulae of the form $(\mathcal{F} \lhd A)$, where A is an atomic formula. A formula is a *prime formula* if and only if it is either an atomic formula or a nucleic formula.

8.3   Beth structures

By a *Beth structure*, $\mathfrak{B}$, we understand a monotonic function[16] on the *nodes* of $\mathfrak{T}$ to the collection of finite sets of prime formulae such that for all nodes $\vec{n}$:

- $\{(\mathcal{F} \lhd \perp), (\top \lhd A)\} \cap \mathfrak{B}(\vec{n}) = \emptyset$.

- If $(\mathcal{F} \lhd A) \in \mathfrak{B}(\vec{n})$ then $(\mathcal{F} \lhd \top) \in \mathfrak{B}(\vec{n})$.

- For any atomic formula $A: \quad A \in \mathfrak{B}(\vec{n}) \quad$ iff $\quad (\perp \lhd A) \in \mathfrak{B}(\vec{n})$.

8.4   Validation

The formula S *is validated at the node $\vec{n}$ of the Beth structure $\mathfrak{B}$, $\mathfrak{B} \Vdash_{\vec{n}} S$:*

- $\forall \alpha_{\vec{n} \ll \alpha} \exists t \left(A \in \mathfrak{B}(\bar{\alpha}t) \right);$ for atomic formulae A.

- $\mathfrak{B} \Vdash_{\vec{n}} (\mathcal{F} \equiv \mathcal{G})$ iff $\forall \vec{m}_{\vec{n} \ll \vec{m}} \left(\mathfrak{B} \Vdash_{\vec{m}} \mathcal{F} \text{ iff } \mathfrak{B} \Vdash_{\vec{m}} \mathcal{G} \right)$.

---

[15]That is, the Set Theory presented in the Appendix to J. L. Kelley's book [5].
[16]That is; if $\vec{n} \ll \vec{m}$ then $\mathfrak{B}(\vec{n}) \subseteq \mathfrak{B}(\vec{m})$.

- $\mathfrak{B} \Vdash_{\vec{n}} \bigwedge_x \mathcal{F}_x$ iff   for all atomic formulae A: $\mathfrak{B} \Vdash_{\vec{n}} \mathcal{F}_A$.

- $\mathfrak{B} \Vdash_{\vec{n}} \mathbb{O}\mathcal{A}_1 \ldots \mathcal{A}_k$ iff $\mathfrak{B} \Vdash_{\vec{n}} \mathcal{D}_{\mathcal{A}_1 \ldots \mathcal{A}_k}$; for a defined $\mathbb{O}$ with definiens $\mathcal{D}_{p_0 \ldots p_{k-1}}$[17].

- $\forall \alpha_{\vec{n} \ll \alpha} \exists t (\mathcal{N} \in \mathfrak{B}(\bar{\alpha}t))$;   for nucleic formulae $\mathcal{N}$.

- $\mathfrak{B} \Vdash_{\vec{n}} (\mathcal{F} \lhd (\mathcal{G} \lhd \mathcal{H}))$ iff $[\mathfrak{B} \Vdash_{\vec{n}} (\mathcal{F} \lhd \top)$ and $\mathfrak{B} \Vdash_{\vec{n}} (\mathcal{G} \lhd \mathcal{H})]$.

- $\mathfrak{B} \Vdash_{\vec{n}} (\mathcal{F} \lhd (\mathcal{G} \equiv \mathcal{H}))$ iff $[\mathfrak{B} \Vdash_{\vec{n}} (\mathcal{F} \lhd \top)$ and $\mathfrak{B} \Vdash_{\vec{n}} ((\mathcal{F} \lhd \mathcal{G}) \equiv (\mathcal{F} \lhd \mathcal{H}))]$.

- $\mathfrak{B} \Vdash_{\vec{n}} (\mathcal{F} \lhd \bigwedge x \mathcal{G}_x)$ iff   for all atomic formulae A: $(\mathfrak{B} \Vdash_{\vec{n}} (\mathcal{F} \lhd \mathcal{G}_A))$.

- $\mathfrak{B} \Vdash_{\vec{n}} (\mathcal{F} \lhd \mathbb{O}\mathcal{A}_1 \ldots \mathcal{A}_k)$ iff $(\mathfrak{B} \Vdash_{\vec{n}} (\mathcal{F} \lhd \mathcal{D}_{\mathcal{A}_1 \ldots \mathcal{A}_k}))$, for defined operator $\mathbb{O}$.

LEMMA 13. *For all formulae $\mathcal{F}$ and Beth structures $\mathfrak{B}$: if $\vec{n} \ll \vec{m}$ and $\mathfrak{B} \Vdash_{\vec{n}} \mathcal{F}$, then $\mathfrak{B} \Vdash_{\vec{m}} \mathcal{F}$.*

Making use of the fact that $\mathfrak{T}$ is a finitely branching tree one obtains:

THEOREM 14. *For all formulae $\mathcal{F}$ and Beth structures $\mathfrak{B}$:*

$$\mathfrak{B} \Vdash_{\vec{n}} \mathcal{F} \qquad iff \qquad \forall \alpha_{\vec{n} \ll \alpha} \exists t (\mathfrak{B} \Vdash_{\bar{\alpha}t} \mathcal{F}).$$

Because all the axioms and definitional postulates are *sentences* it can be verified that:

THEOREM 15. *If $\mathfrak{B}$ is a Beth structure, then all the definitional postulates and all the Rudimentary and Distributive axioms are valid in $\mathfrak{B}$.*

8.5   Beth models

An *Impredicative Beth Structure*, or simply a *Beth model*, is understood a Beth structure $\mathfrak{M}$ such that for all nodes $\vec{n}$ of $\mathfrak{T}$:

- $\mathfrak{M}$ *validates at $\vec{n}$ the Leibnizian axiom,*

- *and for all universal formulae: if $\mathfrak{M} \Vdash_{\vec{n}} \bigwedge x \mathcal{F}_x$, then for all formulae $\mathcal{G}$: $\mathfrak{M} \Vdash_{\vec{n}} \mathcal{F}_{\mathcal{G}}$.*

The formula $\mathcal{F}$ is an *$\mathfrak{M}$-consequence* at $\vec{n}$ of the set Σ of formulae iff for all nodes $\vec{m}$, such that $\vec{n} \ll \vec{m}$:

$$if \ for \ all \ \mathcal{A} \in \Sigma \ (\mathfrak{M} \Vdash_{\vec{m}} \mathcal{A}), \quad then \quad (\mathfrak{M} \Vdash_{\vec{m}} \mathcal{F}).$$

---

[17]And correspondingly for defined vbo's.

8.6   Soundness and Completeness of Beth models

The standard proof on the length of the derivations give us:

THEOREM 16.   *If Γ is a finite consistent set of formulae, $\Gamma \vdash \mathcal{F}$ and* $\mathfrak{M}$ *is a Beth model, then* $\mathcal{F}$ *is a* $\mathfrak{M}$-consequence of Γ.

The completeness takes the following strong form:

THEOREM 17.   *To every finite and consistent set* Γ *of formulae there corresponds a Beth model* $\mathfrak{M}_\Gamma$ *such that for all formulae* $\mathcal{F}$:

$$\Gamma \vdash \mathcal{F} \qquad \textit{if and only if} \qquad \mathfrak{M}_\Gamma \Vdash_{()} \mathcal{F}.$$

For the details we refer the reader to an analogous result in [10]. We shall content ourselves in just giving the instructions on how to develop $\mathfrak{M}_\Gamma$.

The first step is to take an enumeration, without repetition, of the propositional parameters: $p_0, p_1, p_2, \ldots$.

Then assume that Γ is a consistent finite set of formulae and let Par_{-1} be the finite set of propositional parameters that have an occurrence in the formulae of Γ. Then for each natural number k, set $\mathrm{Par}_k = \mathrm{Par}_{-1} \cup \{p_0, \ldots, p_k\}$.

Next obtain an enumeration $\mathcal{F}_0, \mathcal{F}_1, \mathcal{F}_2 \ldots$ of all the formulae such that:

- *each formula occurs infinitely often in the enumeration,*

- *The propositional parameters that occur free in the formula* $\mathcal{F}_k$ *are contained in the finite set* Par_k.

Using the above enumeration we define finite sets of formulae, $\Sigma_{\vec{n}}$, for each node $\vec{n}$ of $\mathfrak{T}$.

Basis step: $\Sigma_{()} = \Gamma$.

Inductive step: Assume that Σ_- has been defined for all nodes of length k (and are consistent). Let $(n_0, \ldots, n_{k-1}) = \vec{n}$ be a node of length k. Consider $\mathcal{F}_k$, the k-th formula in the enumeration of the formulae.

Case 1. $\Sigma_{\vec{n}} \vdash \mathcal{F}_k$.   Then : $\Sigma_{\vec{n}0} = \Sigma_{\vec{n}1} = \Sigma_{\vec{n}} \cup \{\mathcal{F}_k\}$.
Case 2. $\Sigma_{\vec{n}} \vdash (\mathcal{F}_k \equiv \bot)$.   Then : $\Sigma_{\vec{n}0} = \Sigma_{\vec{n}1} = \Sigma_{\vec{n}}$.
Case 3. Neither Case 1 or 2. Then : $\Sigma_{\vec{n}0} = \Sigma_{\vec{n}}, \quad \Sigma_{\vec{n}1} = \Sigma_{\vec{n}} \cup \{\mathcal{F}_k\}$.

Then define: $\mathfrak{M}_\Gamma^*(\vec{n}) = \{\mathcal{P} \mid \mathcal{P} \in \Sigma_{\vec{n}} \quad and \quad \mathcal{P} \quad is\ a\ prime\ formula\}$.

Finally let:

$$\mathfrak{M}_\Gamma(\vec{n}) = \mathfrak{M}_\Gamma^*(\vec{n}) \cup \{A \mid (\bot \vartriangleleft A) \in \mathfrak{M}_\Gamma^*(\vec{n})\} \cup \{(\mathcal{F} \vartriangleleft \top) \mid (\mathcal{F} \vartriangleleft A) \in \mathfrak{M}_\Gamma^*(\vec{n})\}.$$

BIBLIOGRAPHY

[1] Georg Cantor. *Contributions to the Founding of the Theory of transfinite Numbers, 1897.* Translation by P. E. B. Jourdain. Dover Publications, New York, 1955.

[2] Rolando B. Chuaqui. *Axiomatic Set Theory: Impredicative Theories of Classes,* volume 51 of *Notas de Matemática.* North-Holland, Amsterdam, 1981.

[3] Alonzo Church. *Introduction to Mathematical Logic.* Princeton University Press, New Jersey, 1956.

[4] Hidé Ishiguro. *Leibniz's Philosophy of Logic and Language,* Second Edition. Cambridge University Press, Cambridge, 1990.

[5] J. L. Kelley. *General Topology.* Van Nostrand, New York, 1955.

[6] St. Leśniewski. *Grundzüge eines neuen Systems der Grundlagen der Mathematik.* Fundamenta Mathematicae, 14:1–81, 1929.

[7] E. G. K. López-Escobar. *Set Theories as Extensions of Propositional Logics.* Manuscrito-Rev. Int. Fil., 28:417–448, 2005.

[8] E. G. K. López-Escobar. *The Logic of Classes.* Logic Journal of the IGPL, 15:689–706. Oxford University Press, 2007.

[9] E. G. K. López-Escobar. *Sur les termes primitifs de la logique mathématique.* Submitted to Fundamenta Mathematicae, 2007.

[10] E. G. K. López-Escobar and F. Miraglia. *Definitions: The primitice concept of Logics,* volume 401 of *Dissertationes Mathematics.* PWN, Warszawa, 2002.

[11] A. P. Morse. *A Theory of Sets,* volume 18 of *Pure and Applied Mathematics.* Academic Press Inc., New York, 1965.

[12] Dag Prawitz. *Natural Deduction. A Proof-Theoretical Study.* Dover Publications, Inc., New York, 2006.

[13] Alfred Tarski. *Sur le terme primitif de la logistique.* Fundamenta Mathematicae, volume 4, pp.196-204, 1923.

E. G. K. López-Escobar
Departament of Mathematics
University of Maryland, USA
CLE, Brazil

The Geometric Analogy and the Idea of Pure Logic

LUIS ESTRADA-GONZÁLEZ

ABSTRACT. The geometric analogy is often mentioned to make logical pluralism plausible, nonetheless, how far the analogy should be
taken is an issue hardly ever discussed. Rescher in [24] gave one of the
most detailed analyses of the analogy, and at the same time one of
the most severe critiques to the idea of taking the analogy seriously.
More recently Priest in [20] has argued for the legitimacy of the geometric analogy, trying to reject some of Rescher's criticisms, but I
will show that he failed. Here I will argue that Rescher's arguments
against the idea that the subject matter of logic is not necessarily
its canonical application, but a pure mathematical subject akin to
that of geometry, are not conclusive. In particular, I will show that
Rescher's denial of the analogy and his rejection of the idea of a pure
logic are grounded on some historical fallacies regarding the development of geometry, as well as on an uncritically assumption about the
nature of logic as an essentially applied theory.

1 Introduction

The philosophical import of the analogy of logic with geometry is at least
twofold. On one hand there is an *epistemological* analogy, namely on that
pluralism is available for both logic and geometry. Usually the influence
has gone from geometry to logic: Since in geometry one can drop certain
axioms in order to obtain different geometries, so in logic, as in the case
of maybe the earliest example of influence of non-Euclidean geometries in
logic, namely Vasiliev's development of an "imaginary logic", akin to the
"imaginary geometry" of Lobachevsky.[1] Intimately related to the former
is that both geometry and logic have had a similar *historical development*.
First there was just one geometry and just one logic; their theorems were
regarded as necessary truths about certain province of reality, (the structure

---

[1]'Pluralism' means here just the mere existence of many theories (be they pure or
applied), not necessarily their simultaneous correctness.

of) physical space in the case of geometry and (the structure of) norms that rules valid reasoning in the case of logic.

Nonetheless, how far the analogy should be taken is an issue hardly ever discussed. One of the most detailed assessments of the analogy is in Rescher's [24], which is at the same time one of the most severe critiques to the idea of regarding the analogy as a deep one. According to Rescher, epistemological and developmental similarities are not enough for claiming an analogy in more important issues, namely their respective treatment of rivalry, the question about their very nature or the characterization of their subject matter. Priest [20] has argued for the legitimacy of the analogy, trying to reject some of Rescher's criticisms, but as I will try to show, he does it in a wrong way.[2] I am not going to discuss the whole of connections between logic and geometry, but just the issue whether and how a distinction between the pure and the applied like in geometry can be drawn in logic, discussing mainly Rescher's arguments against this possibility. My strategy will be to show that Rescher's denial of the analogy and his rejection of the idea of a pure logic are grounded on some historical fallacies regarding the development of geometry and on an uncritically assumption about the nature of logic as an essentially applied theory.

The plan of the paper is as follows. In section 2 I examine Rescher's criticisms to the analogy between geometry and logic. In section 3 I discuss one of Rescher's premises, to wit the correctness of only one applied geometry but the correctness of just one applied logic. In section 4 I will try to show why Rescher's arguments against the idea of pure logic are misguided and that his master argument against the geometric analogy is based on a wrong view of the development of geometry during nineteenth century. In section V I present a plea for a distinction between pure and applied logic, trying to bring out a new different view of logic that could serve both as the fundamentals of a reliable account of what logic is as well as of a framework in which to understand contemporary logic. Some important philosophical questions quite related to the present ones are left for further work, like the assessments of criticisms to the idea that logic could be a branch of mathematic, like Barceló's [1], neither I can discuss here all the pertinent questions concerning the status of application and applicability of logics. See [24, pp. 462ff] on this last issue. Even if logic is going to be considered a branch of mathematics the problem of whether logics can be reduced to other mathematical branches, e.g. algebra remains; see [5] for a thorough discussion, but as I have said all this should be discussed in a separate work.

---

[2]Priest's arguments for the geometric analogy can be found also in others of his works. I mention just [20] because there is an explicit reference in it to Rescher's ideas on the subject.

2   Rescher on the geometric analogy

Despite the analogy between non-standard logics and non-Euclidean geometries is an important and interesting one, several people have denied it.
One objection to the geometric analogy in logic is due to the Kneales in
[15, p. 575]. They say that the geometric analogy fails because classical
logic contains all the other logics as fragments of itself and as Euclidean
geometry does not contain the other geometries, there are no alternatives
to classical logic in the sense in which, for example, Riemann's geometry
is alternative to Euclid's. There are at least two responses to the Kneales'
objection. First, this objection fails because not all non-standard logics are
contained in classical logic. For example, connexivist logics have as logical
truths formulae that are not classical logical truths.[3] Second, as Priest [20,
p. 443] has rightly pointed out, from the fact that a logic is maximal it
does not follow that another cannot be an alternative because the maximal
logic might entail too much. Moreover, the Kneales' argument exemplifies
the extended opinion that the radical turn of nineteenth-century geometry
is due to the apparition of non-Euclidean geometry. However, it should be
taken into account that this is only partially the case, since both projective
geometry and n-dimensional geometry played a chief role and represented
even deeper and far-reaching revolutions in geometry than the denial of the
parallel postulate (cf. [26, p. 149] and [28]).

Rescher accepts a general analogy between geometry and logic, namely
that in the same way as there are many geometries there are also many
logics. But this would be all the analogy. Rescher's arguments depart from
the following premises:

1. In geometry it is taken for granted that there is a distinction between
   "pure" and "applied" (or "physical") geometry.

2. It is also considered that formally all geometries are "right", or none
   can be formally "wrong" (because they are consistent, etc.), but only
   would exist one right physical geometry.

Rescher says that there is nothing like "pure logic" [24, p. 217], unlike
geometry in 1, and that there are good reasons to think that there can
be many right (applied) logics [24, p. 218], contrary to 2 in the case of
geometry. For him there is no pure logic because every system deserving
the name "logic" must satisfy, among others, the requirement of having
an interpretation which not only be a mathematical model for an abstract
calculus nor anything satisfying some formal axioms, but an interpretation

---

[3]Rescher in [24, p. 228] gives an interesting example based in a Post three-valued
logic which, like connexivist logics, validates Aristotle's thesis ($\neg(A \rightarrow A)$).

involving concepts such as those of 'sentences' (or another linguistic term), 'inferences', 'arguments', as well as the concepts of "meaning" and "truth" of sentences. That is, Rescher discards the distinction pure logic-applied logic using the traditional view of logic, because according to it logic is always applied, logic is identified with its canonical application. This is *Rescher's first argument* against the geometric analogy.

Rescher's criticizes an analogy with geometry based on the second point by saying that even when supposing that there are pure logics, all of them formally correct, it would be false that there is just one right applied logic. Rescher thinks that there are many correct logics. Even the mere possibility of the correctness of more than one logic is enough to regard logic as essentially different of geometry. This is *Rescher's second argument* against the geometric analogy.

But Rescher's ultimate argument for rejecting the geometric analogy is that in order to develop a logic one necessarily employs a "presystematic logic machinery". Rescher draws a distinction between "systematized logic", which would be a special branch of knowledge, and "presystematic logic", which would be a general instrument for the realization of knowledge throughout all its branches. The presystematic logic machinery would be the "preexisting idea of what logic is" [24, p. 231] which would determine the following "regulative principles" or, better, criteria of logicality: Precision, exactness, economy, simplicity, coherence and consistency, but "[a]bove all, one must (...) stress the regulative ideal of by-and-large *conformity* to the key features of the presystematic practice, of 'saving the phenomena' that are involved in the presystematic practice." [24, pp. 228 and also 234] Such regulative principles would serve as criteria of logicality because they "(...) will play a key role throughout the range of diverse 'logics' and their employment will effectively condition our view of such systematizations." [24, p. 224] Thus, systems of logic would be systematizations of the presystematic practices of reasoning, normatively regarded. In other words, Rescher claims that in order to develop a system of logic one needs the preexisting idea of what logic is, while for developing a geometry no "presystematic geometry" is needed (and according to him such presystematic geometry does not even exist). Then he says that this difference implies "(...) that the nature of the choice between alternative systems in the two cases will in fact have to be quite different." [24, p. 219]

3   On Rescher's second argument

Rescher's second argument is inconclusive since its main premise, that there can be many applied logics but just one applied geometry, can be seriously doubted. Applied logical monism and applied geometric pluralism are far more defensible than Rescher believes. Somebody who is a monist and rejects the traditional view of logic, i.e. an applied logical monist can adopt the distinction pure logic-applied logic without great difficulties. Some Australian inconsistency-tolerant logicians and philosophers are good examples of this kind of applied logical monists (see for example [20] and [25]). They take as basis for their proposal a distinction between *the* logic and *a* logic, or better between logic and logical systems. Logic would be the structure of norms that governs valid reasoning, reasoning valid in all situations. Such structure would be the subject of study of logical systems; said otherwise different logical systems would be theories about reasoning (cf. [21, pp. 257f]). The distinction logic-logical systems corresponds to the distinction object-theory, like in moving bodies-dynamics. According to these logical monists, there can be many pure logics without rivalry between them. The question of rivalry and correctness only makes sense when such logics are applied to something in the same way pure geometries are not rivals and no one can claim to be the right one until they are applied to, say, the study of physical space.[4]

Roughly, Priest's argument for logical monism is that "[e]ven if modes of legitimate inference do vary from domain to domain, there must be a common core determined by the intersection of all these." [20, p. 464] That intersection would be the correct logic because its laws would be valid in all, and independently of, every domain. It is possible that this intersection is empty, "(. . .) but I never heard a plausible argument to this effect." [20, pp. 464f].[5]

Concerning geometric pluralism it is often said that different geometries are appropriate for different contexts. For example, when we build a tower it is appropriate to use Euclidean geometry. But when we do surveying, it is appropriate to use spherical geometry. And when we do cosmology, it is appropriate to use Riemannian geometry. It is possible that the usage of different geometries in different regions of the universe is not only a matter of simplicity, but also a requirement of the structure of the universe; maybe there is no global geometry. Rescher's second argument does not make its point; it is at least incomplete.

---

[4]But it is only its canonical application; "one must always be able to say 'tables, chairs, beer mugs' instead of 'points, lines, planes'".

[5]Beall and Restall seem to make fun of Priest's argument when they say that the intersection of all logics might consist just of $A \rightarrow A$ (they cannot believe that, though; they have to include at least a schema for transitivity) and that that cannot be called a logic and much less *the correct logic* (see [3]). Arguments more serious against Priest's monism can be found in [10, pp. 545f] and in [3, p. 93] as well as in Mortensen's works on possibilism and trivialism (cf. [17] and [18]).

4   A short note on the development of geometry

Priest tries to nullify Rescher's third argument in the following terms:

> Rescher's observation [that the articulation of a logical system requires a presys-
> tematic logic whereas the articulation of a geometric system does not require a
> presystematic geometry] seems correct. But (...) it is difficult to see it as having
> significant import for the question. The formulation of a grammar for a language re-
> quires the employment of a metalanguage, and so a metagrammar. But this hardly
> entails that there cannot be rival grammars for a language, or that the question of
> which is correct is not *a posteriori*. The same could be true of logic. [20, p. 443]

However, it seems to me that Priest missed Rescher's remark. Rescher is
contending neither logical pluralism nor the existence of rivalry in his third
argument. Remember the second argument: He thinks that in the case of
geometry just one will be the correct among rival geometries; in the case of
logic there are rivals, genuine alternatives, but many of them may be right.[6]
Rescher tries to show that there is a fundamental disanalogy between logic
and geometry, that they are radically different disciplines, not that there
are no rival logics or that the question of which logic is correct should be
answered a priori. In the next section I will try to dispute Rescher's third
argument by suggesting his ideas on the distinction pure-applied and on the
inexistence of a presystematic geometry are based on a misinterpretation of
the historical development of geometry.

A large part of Rescher's conclusions, its orthodoxy, is mostly based in
what can be considered a historical fallacy. The granted distinction pure
geometry-physical geometry is very recent, and it also had detractors of the
"pure side". Until the mid nineteenth century, mathematics was mostly
ruled by empirical or more or less intuitive considerations. It was viewed
as a collection of exact observations about the physical universe. Most
mathematical problems arose from physics; in fact, there was no separation
between mathematics and physics. Proof was a helpful method for organizing facts and reducing the chance of errors, but each physical fact remained
true by itself regardless of any proof. This pre-nineteenth-century viewpoint
persists in many textbooks until now, because textbooks do not change
rapidly, and because a more sophisticated viewpoint may require higher
learning. Prior to the nineteenth century Euclidean geometry was seen as
the best known description of physical space. Some non-Euclidean axioms
for geometry were also studied, but not taken seriously; they were viewed
as works of fiction. Indeed, most early investigations of non-Euclidean axioms were carried out with the intention of proving those axioms wrong:
Mathematicians hoped to prove that Euclid's fifth postulate was a consequence of other axioms, by showing that a denial of the parallel postulate

---

[6]Rescher explicitly says that he believes in the existence of genuine rival, alternative
logics. Cf. [24, p. 234]

would lead to a contradiction. All such attempts were unsuccessful —the denial of the parallel postulate merely led to odd conclusions, not to outright contradictions— though sometimes errors temporarily led mathematicians to believe that they had succeeded in producing a contradiction. It is not wrong at all to say that Beltrami's and Klein's attempts to find models for non-Euclidian geometries were motivated to understand them from the intuitive, Euclidean geometry, though their research favored the mathematical autonomy of geometry. Around the turn of the century, Poincaré and Hilbert each made available an explanation of geometry that took the discipline to be an implicit definition of its concepts: Its terms could be applied to any system of objects that satisfies the required axioms. Each mathematician found vigorous opposition from a different logicist-Russell against Poincaré and Frege against Hilbert- who maintained the vanishing viewpoint that geometry is essentially concerned with physical space or spatial intuitions.

Alberto Coffa provides an enjoyable summary of the state of affairs on the ground at the end of nineteenth century and the beginning of twentieth century:

> During the second half of the nineteenth century, through a process still awaiting explanation, the community of geometers reached the conclusion that all geometries were here to stay (...). [A] community of scientists had agreed to accept in a not-merely-provisory way all the members of a set of mutually inconsistent theories about a certain domain (...). It was now up to philosophers (...) to make epistemological sense of the mathematicians' attitude toward geometry (...). The challenge was a difficult test for philosophers, a test which (sad to say) they all failed (...). [11, p. 8]

Some pages later he continues:

> For decades professional philosophers had remained largely unmoved by the new developments, watching them from afar or not at all (...). As the trend toward formalism became stronger and more definite, however, some philosophers concluded that the noble science of geometry was taking too harsh a beating from its practitioners. Perhaps it was time to take a stand on their behalf. In 1899, philosophy and geometry finally stood in eyeball-to-eyeball confrontation. The issue was to determine what, exactly, was going on in the new geometry. What was going on, I believe, was that geometry was becoming less the science of space or space-time, and more the formal study of certain structures. Issues concerning the proper application of geometry to physics were being separated from the status of pure geometry, the branch of mathematics. [11, p. 17][7]

---

[7]These quotations need some remarks. As Shapiro and Martin-Löf accurately point out (cf. [27, p. 63]), maybe Coffa is right on saying that philosophers failed in making epistemological sense of mathematicians' attitude toward geometry, but not all philosophers ignored and were opposed to the new trends in geometry. For example, Husserl had a great interest in geometry, as is revealed principally by his writings during 1886-1901, and his 1900 *Logische Untersuchungen* (Chapter 11, especially §§ 70–71) used the new perspective in the development of his metaphysics and philosophy of science. But the rest of Coffa's description is quite accurate.

These passages are useful to see two points. First, the idea of presystematic geometry is not so odd. Even contemporary philosophers speak as if there were such presystematic geometry: "A finite projective plane is not going to be used to model physical space, but it may be used to model something analogous to physical space." [2, p. 489][8] What happened, I believe, was that mathematicians and philosophers revised their presystematic idea of 'geometry' (and of 'algebra', etc.) removing any empirical content from it and replacing most constant elements by variables.[9] Second, I think that logic has suffered a similar transformation. Nevertheless, most logicians, philosophers and even mathematicians are still reticent to regard logic less as the science of evaluation of arguments and more as the formal study of certain structures. This leads me to the analysis of Rescher's first argument and the consideration of the distinction pure-applied in the case of logic.

5 The distinction pure logic-applied logic: An overview

There has been a long debate on the mathematical aspects of logic and its mathematical generalizations. Quine [22], Haack [14] and Rescher [24] among other eminent philosophers of logic, for different reasons, are some of whom are against an essential relation between logic and pure mathematics. This has been the orthodoxy in philosophy of logic, since despite the analogy of the development of logics different to classical logic with the development of non-Euclidean geometries virtually nobody has neglected the posited necessary relation between logic and knowledge, human reasoning, or certain eternal and unquestionable principles. Quine, for example, has suggested that many-valued semantics are an abstract meaningless generalization developed "après coup", "theory without an interpretation, abstract algebra" (Quine 1970: 84).[10]

Quine's and Haack's arguments to refuse the geometric analogy, especially the distinction pure logic-applied logic, are question-begging. They start from the claim that the subject-matter of logic is such and such and then conclude that its subject is precisely the assumed one, but what is being questioned is precisely that assumption, the traditional view of logic. Rescher's first argument falls short with that same assumption. Nonethe-

---

[8]However, I think that the similarity between different geometries or branches of geometry is not that they can be used to model analogous things, but rather a structural similarity and that this structural similarity allows to model similar objects.

[9]For a detailed exposition of the idea of the replacement of the constant by the variable as an essential process in mathematics see [4, p. 236]

[10]Rescher refers that Bocheński heard Leśniewski to say "(...) that all many-valued systems are merely games: there is but one single authentic system of logic: the orthodox, standard one." [24, p. 215]

less, far from being regarded as question-begging, it is taken as an argument departing from the very characterization of logic, a characterization which needs no further explanation since there is common agreement on it, it agrees tradition, etc.

It can be argued that logic is different from mathematics in that logic has as essential components some extra-structural features. Take the category-theoretic treatment of natural numbers: A natural numbers object can be anything satisfying the conditions for being such an object, conditions which do not include criteria of identity for each number (cf. [16]). In contradis-tinction, someone could say, a logic, a connective, is not anything satisfying some structural conditions. Logic has uniquely individuating properties that are not irrelevant; henceforth a genuine logic is different from electric circuits although they share a common structure. In the same way a con-nective is a very different thing from an operation on gates. Such differences cannot be skipped; it would be like claiming that (some parts of) physics is the same as functional analysis because they share a common structure. As in the case of physics, it is an essential component of logic to have a certain subject-matter, namely the study of right reasoning.

Even some more broad-minded proposals toward pure logic have remi-niscences of the traditional view of logic, and some logics are left out from the realm of logics by arguments invoking some ideas from the traditional view of logic. Beall and Restall take away non-reflexive logics since, they say, any consequence relation should be reflexive:

> The given kinds of non-transitive or irreflexive systems of 'logical consequence'
> are logics by courtesy and by family resemblance, where the courtesy is granted
> via analogy with logics *properly* so called. Non-transitive or non-reflexive systems
> of entailment may well model interesting phenomena, but they are not accounts
> of *logical consequence*. One must draw the line somewhere and, pending further
> argument, we (defeasibly) draw it where we have. We require transitivity and
> reflexivity in logical consequence. We are pluralist. It does not follow that *anything*
> goes. [3, p. 96, italics in the original])

Beall and Restall's position is worth noting. Remember that Quine made quite the same point against many-valued logics saying that they are abstract meaningless generalizations of two-valued logic developed "après coup", "theory without an interpretation, abstract algebra" [22, p. 84]. "Theory without interpretation, abstract algebra" seems quite the same as "they are modeling interesting model phenomena, but they are not accounts of logical consequence". Thus, Quine once used an argument to reject as logics some logics now accepted by Beall and Restall, but they use a very similar argument to reject other logics quite weird for today philosophers. On other hand, Quine used the notion of 'family resemblance' to explain why do we call "logics" such "abstract algebras", and Haack once used it in order to avoid a precipitated decision about whether some things called "logics" are indeed logics, and some analogy was enough for accepting some systems as logics (cf. [14, pp. 3–10]). Again, Beall and Restall opt for the Quinean way.

Thus, some axioms and principles seem sacred, and some readings of the operator (or relation) are not allowed. But, why monotonicity can be dropped while transitivity not? Why $\neg(A \wedge \neg A)$ can fail to be a theorem but $(A \rightarrow A)$ not? Why nonmonotonic logics are logics *tout court* and not merely systems analogous to, and resemblances of, true logics? Beall and Restall's claim suggests that an expression like '$A \vdash B$' has accepted readings, like "B is deduced from A", "every proof of A can be converted into a proof of B", "every input of type A results in an output of type B". A reading of '$A \vdash B$' like "A explains scientifically B" might lead to drop $A \vdash A$, since a notion of scientific explanation that allows explanations of phenomena by themselves seems not very useful. But if scientific explanation does not admit an inference like $A \vdash A$ surely, it is said, it does not count as a kind of logical consequence.[11]

Of course Beall and Restall make very strong arguments for their position, but on the long term all this is problematic: Once the principle of non-contradiction was taken as a defining feature of logic and if thought carefully, the boundary lines between acceptable and unacceptable readings of the turnstile are not clear. Of course, we are not allowed to made a pessimistic conclusion and think that any other principle would have the same fate that its predecessors, but I think that the traditional view of logic as a science of right reasoning such that both object and theory must accomplish some necessary constraints is not very illuminating on the nature of logic, and makes no well sense of contemporary research in logic.

The question involved here is, I guess, how one is going to conceive the absolute generality of logic. Many authors have defended the idea that logic is absolutely general and nonetheless is not pure. Logic has certain subject-matter, the study of right reasoning, which is different from, but common to, any other specific reasoning, say of biology, chemistry, law, etc. On this let me quote a very illuminating passage from Béziau [7, p. 4]:

> At the end of the twenties, he [Tarski] launched the theory of consequence operator (...). This theory is about an "operator", a function Cn defined on the power set of a given set S. Following the philosophical ideas of his master, Leśniewski, Tarski

---

[11] Read [23] resumes the problems of Beall and Restall's pluralism based on reflexivity and transitivity as necessary conditions of logicality saying that it falls down on the count that it does not respect the core motivation of some non-standard logics, which first prompted them as rivals to the classical orthodoxy.

calls these objects [the elements of S] "meaningful sentences". But in fact, the name does not matter, the important thing is that here Tarski is considering a very general theory, because the *nature* of these objects is not specified.

Tarski himself, as many others before and after him, thought that the unspecified nature of the elements of S means that they can be sentences of any kind of scientific language (since Tarski was studying the methodology of deductive sciences, not only metamathematics).[12] Nonetheless, I think that the unspecified nature of the elements of S involves more than the fact that sentences considered are of an extreme generality; the unspecified nature of the elements of S says that they may be arbitrary objects.

Someone may argue that what compels us to think of the elements of S as sentences or propositions comes from our presystematic requirements: We are trying to give an answer to the question about the validity of arguments, and sentences are part of arguments and thus some of these concepts are going to be central in an account of logic, no matter how abstract or general it could be. Notwithstanding how entrenched is the traditional view of logic, it appears to me as an unattractive view because it has the following shortcomings:

— It has been subject to continuous refutation: Some principles were regarded as essential to reasoning and hence to logic, but then we found that it was not so;

— It cannot explain why some similar structural variations are allowed while others are not: If some principles can be dropped from a logic, why others not?

These points can be summarized as the tension between our informal notions and the technical developments, as in the case of Tarski's works: The nature of the elements of S is unessential, but our presystematic notion says us it is not unessential.

Now some authors are more sympathetic to the idea of pure logic, but the purity of logic that they propose is just a more abstract consideration of reasoning. For example, even though Béziau has defended the idea that logics are particular mathematical structures, sometimes he embraces the traditional view of logic: "The idea of universal logic [a very general theory of logics] is to deal with any type of reasoning (...)", and a few pages later: "Universal logic is not a fixed theory, it's a progressive science in which the study of particular cases is always significant for the development of abstract reasoning that, in turn, will be fruitfully applied." (cf. [9,

---

[12]Cf. also [8, p. 135]). Béziau has made similar remarks concerning other very abstract approaches to logic in Béziau [6, p. 9]). Nonetheless, it does not mean that he sustains the theses I am going to expound below.

p. 145; 147]) The distinction pure-applied logic as well as the geometric
analogy is also a central issue in the philosophy of logic of Newton da Costa
(cf. [12] and [13]). Priest has argued explicitly and in a more regular way
for a distinction between pure logics, certain mathematical structures, and
applied logics, pure logics applied to some end (see for example [20] and
[20]), has stressed the importance of distinguishing between logic as theory
and logic as object of study ([20, p. 207]) and finally has introduced the
very fruitful notion of 'canonical application' (cf. [20]).

Let me propose an organization of logic, which is intended to be the fundamentals of a reliable account of what logic is as well as of a framework in
which to understand contemporary logic, trying to take into consideration
the virtues of the points of view of the authors I have mentioned in the preceding paragraph.[13] *A pure logic* is a kind of mathematical structure, a logical structure. A *logical structure* is a structure of the form $LS = \langle S, \vdash_{LS} \rangle$,
where S is an arbitrary structure and $\vdash_{LS}$ is also an arbitrary relation on
$\wp(S) \times S$. Equivalently, it would be described as a pair $LS = \langle S, C_{LS} \rangle$, where
C_{LS} is an arbitrary mapping $C_{LS} : \wp(S) \longrightarrow \wp(S)$. *Phenomena with a logical structure* are phenomena with a specific "enriched" logical structure, for
example thinking, describing, reasoning, the flowing of electricity in a circuit, going from one city to another, transforming a proof for A into a proof
of B, etc. *Canonical phenomena with a logical structure* are specific and
traditionally sacred phenomena with logical structures, such as describing,
reasoning, thinking. Usually the term 'logic' has been used without distinction to designate a theory as well as for designating what the theory is
about. For example, 'logic' has been used to name a certain science, the
science of right reasoning, as well as for the presumed subject matter of such
science, the structure of norms that rule right reasoning. Some terminology
should mirror this distinction. 'Logic' could be used for the study of pure
logics, 'applied logic' for the study of empirical phenomena with a logical
structure, and 'canonically applied logic' for the study of canonical empirical phenomena with a logical structure. It has been a common practice to
reserve the term without adjectives ('geometry', 'arithmetic') for the pure
part, and noting a distinction between object and theory prevents most of
the problems in the use of the term 'logic'. If there is some reticence in
using the word 'logic' for any of the parts, the study of pure logics might
be called "universal logic".[14]

On the alternative view I have just sketched, what would be subject to

---

[13]I take those authors as representative of the points of view I want to build on. I am
not claiming that nobody else has held similar ideas.

[14]I thank Prof. Priest and Mauricio Torres-Villa for urging me to discuss these terminological issues.

continuous refutation is not necessarily our concept of logic, but most of times canonically applied logics, i.e. our theories about the structure of norms that rules right reasoning, and this is no more problematic than the refutation of any other theory. Given the pure nature of logic, structural variations are not problematic: A logic without reflexivity can be still a logical structure, even though the structure of norms that rules right reasoning be indeed reflexive. For example, when Beall and Restall say "Non-transitive or non-reflexive systems of entailment may well model interesting phenomena, but they are not accounts of *logical consequence*" the answer should be, I think, that those systems are modeling logical phenomena, but probably they are not modeling canonical logical phenomena.

6   Conclusions

I have discussed Rescher's arguments against an important analogy between logic and geometry beyond pluralism, the existence of several logics and geometries. Rescher's arguments depart from the facts that (in geometry is taken for granted that there is a distinction between "pure" and "applied" (or "physical") geometry and that it is also considered that formally all geometries are "right", or none can be formally "wrong" (because they are consistent, etc.), but only would exist one right physical geometry. One of his arguments is that there is no distinction between pure and applied in the case of logic: Logic is essentially applied; it is the study of the validity of arguments. His second argument is that if there were something like a distinction between pure and applied logics, there would be many right logics, unlike applied geometry. His third argument is that what prevents an important analogy between logic and geometry is that in order to develop a system of logic one needs a "presystematic logic machinery", the preexisting idea of what logic is, while for developing a geometry no "presystematic geometry" is needed (and perhaps such presystematic geometry does not even exist). This fact, Rescher says, implies a very important difference in how do we choose between alternative proposals in each case.

I have argued that Rescher's second argument does not make its point since both applied logical monism and applied geometrical monism are far more tenable than Rescher believes. I have rejected his third argument on historical grounds: Once upon a time there was a presystematic geometric machinery, it was considered an essentially applied science, namely the study of the structure of physical space. Nonetheless, such preexisting idea of what geometry is changed dramatically. A similar conceptual change seems to be going on in logic and, more importantly, there seems to be no reason to regard logic as immune to such change.

His first argument is perhaps the strongest case for a disanalogy between logic and geometry. But as I have suggested in my reply to the second argument, the conception of what geometry is changed in nineteenth century, and so the distinction pure-applied in geometry is relatively recent. Do we have reasons to think that we could be in need to juggle around with our current conception of logic? I have suggested that it is far from being obvious that logic is a theory in which reasoning has to be an essential component. Logic can be thought of as the study of certain mathematical structures, "pure logics". Describing and thinking are phenomena that have a logical structure, but they are not the only ones. Phenomena like going from one city to another or the flowing of electricity trough a circuit also have a logical structure. A characterization of the notion 'pure logic' needs, undoubtedly, more refinement, but contrary to what Rescher and other philosophers thought the idea of 'pure logic' is not an unintelligible one, but perhaps one of the most reasonable concepts to make sense of the contemporary scene of logic.

Acknowledgments

I wish to thank Daniel González-García, Cristian Alejandro Gutiérrez-Ramírez, Claudia Olmedo-García, Juan Francisco Orea-Retif, Mauricio Torres-Villa, Ignacio Vilaró-Luna, as well as Profs. Graham Priest and David K. Miller for their very helpful comments, questions and suggestions to earlier versions of this paper.

BIBLIOGRAPHY

[1] Barceló, Axel (2003): "How Mathematical is Mathematical Logic" (in Spanish), Diánoia 15, vol. XLVIII, pp. 3–28.
[2] Beall, JC and Greg Restall (2000): "Logical Pluralism", Australasian Journal of Philosophy 78, pp. 475–93.
[3] Beall, JC and Greg Restall (2006): Logical Pluralism, Oxford: Oxford University Press.
[4] Bell, John Lane (1988): Toposes and Local Set Theories, Oxford: Oxford University Press.
[5] Béziau, Jean-Yves (1997): "Logic May be Simple. Logic, Algebra, and Congruence", Logic and Logical Philosophy 5, pp. 129–147.
[6] Béziau, Jean-Yves (2001): "From Paraconsistent Logic to Universal Logic", Sorites 12, Mayo 2001, pp. 5–32.
[7] Béziau, Jean-Yves (2005): "From Consequence Operator to Universal Logic: A Survey of General Abstract Logics", in Jean-Yves Béziau (ed.) Logica Universalis. Towards a General Theory of Logic, Germany: Birkhäuser Verlag, 2005, pp. 3–17.
[8] Béziau, Jean-Yves (2006): "Les axioms de Tarski", in R. Pouivet and M. Rebuschi (eds.), La philosophie en Pologne 1918–1939, Paris: Vrin, 2006, pp. 135–149.
[9] Béziau, Jean-Yves (2006a): "13 Questions about Universal Logic", Bulletin of the Section of Logic vol. 35, pp. 133–150.
[10] Bueno, Otávio (2002): "Can a Paraconsistent Theorist be a Logical Monist?", in W. Carnielli, M. Coniglio, and I. D'Ottaviano (eds.), Paraconsistency: The Logical Way to the Inconsistent, New York: Marcel Dekker, 2002, pp. 535–552.

[11] Coffa, Alberto (1986): "From Geometry to Tolerance: Sources of Conventionalism in Nineteenth-Century Geometry", in Robert Colodny (ed.), From Quarks to Quasars: Philosophical Problems of Modern Physics, University of Pittsburgh Series, Volume 7 Pittsburgh: Pittsburgh University Press, pp. 3–70.

[12] da Costa, Newton C. A. (1997): Logiques classiques et non classiques: Essai sur les fondements de la logique, Paris: Masson.

[13] da Costa, Newton C. A. (2000): El conocimiento científico, Mexico: UNAM. Spanish translation by Andrés Bobenrieth of O conhecimento cientifico, São Paulo: Discurso Editorial, 1997.

[14] Haack, Susan (1978): Philosophy of Logics, Cambridge: Cambridge University Press.

[15] Kneale, William y Martha Kneale (1962): The Development of Logic, Oxford: Oxford University Press.

[16] McLarty, Colin (1993): "Numbers Can Be Just What They Have To", Noûs 27, no. 4, pp. 487–498.

[17] Mortensen, Chris (1989): "Anything is possible", Erkenntnis 30, pp. 319–337.

[18] Mortensen, Chris (2005): "It isn't so, but could it be?", Logique et Analyse 48, nos. 189-192, pp. 351–360.

[19] Priest, Graham (2000): "Review of N.C.A. da Costa's Logiques classiques et non classiques", Studia Logica, vol. 64, pp. 435–443.

[20] Priest, Graham (2003): "On Alternative Geometries, Arithmetics, and Logics; A Tribute to Lukasiewicz", Studia Logica 74, pp. 441–468. Reprinted with some modifications as Chapter 10 of Priest's Doubt Truth to be Liar, Oxford: Oxford Clarendon Press, 2006.

[21] Priest, Graham (2006): In Contradiction, second edition, Oxford: Oxford Clarendon Press.

[22] Quine, Willard van Orman (1970): Philosophy of Logic, first edition, Englewood Cliffs, New Jersey: Prentice Hall.

[23] Read, Stephen (2006): "Monism: The One True Logic", in A Logical Approach to Philosophy: Essays in Honour of Graham Solomon, ed. D. DeVidi and T. Kenyon, Springer 2006, pp. 193–209.

[24] Rescher, Nicholas (1969): Many-valued Logic, USA: McGraw-Hill.

[25] Routley, Richard (1980): Exploring Meinong's Jungle and Beyond. An Investigation of Noneism and the Theory of Items, Canberra: Research School of Social Sciences, Australian National University.

[26] Shapiro, Stewart (1996): "Space, Number, and Structure: A Tale of Two Debates", Philosophia Mathematica, vol. 4, pp. 148–173.

[27] Shapiro, Stewart (2005): "Categories, Structures, and the Frege-Hilbert Controversy: The Status of Meta-mathematics", Philosophia Mathematica 3, pp. 61–77.

[28] Torretti, Roberto (2007): "Nineteenth Century Geometry", The Stanford Encyclopedia of Philosophy (Spring 2007 Edition), Edward N. Zalta (ed.), URL = http://plato.stanford.edu/archives/spr2007/entries/geometry-19th/.

Luis Estrada-González
Faculty of Humanities,
State University of Morelos,
Cuernavaca, Morelos, Mexico
E-mail: loisayaxsegrob@gmail.com

Analytic Structures and Model Theoretic Compactness

José Iovino

1 Introduction

In recent years, there has been considerable interest in replicating the successful development of first-order model theory in non first-order contexts. One such context that has emerged as an important test case is that of metric structures, i.e., structures whose sorts are metric spaces. The appeal of a rich model theory of metric structures lies in that, besides opening up new fields of interaction between model theory and other areas of mathematics (e.g., functional analysis and probability), it provides a natural generalization of the first-order theory; every structure of the of the type studied in first-order model theory can be seen as a metric structure where the underlying metric is discrete.

Several frameworks have been proposed to study metric structures from the perspective of model theory, to wit:

- · Henson's logic of *positive bounded formulas*

- · Ben-Yaacov's *compact abstract theories*, or *"cats"*

- · The framework of *continuous logic* developed by Ben-Yaacov and Usvyatsov.

Predecesssors include Chang and Keisler's *continuous model theory* [CK66] and Krivine's *real-valued logic* [Kri74].

Henson's logic of positive bounded formulas was developed in the mid 1970's, initially to understand the connections between Banach spaces and their ultrapowers (or nonstandard hulls). In the early papers [Hen74, Hen75, Hen76], Henson proved, among other things, versions of the compactness and Löwenheim-Skolem theorems for positive bounded formulas. This logical framework was used in various papers by Henson, Heinrich, and Moore that appeared during the 1970's and 1980's to study problems in Banach space geometry. (See, for example, [HHM83, HM83a, HM83b, HHM83, HH86, HHM86, HHM87].)

In the 1990's, in a series of papers [Iov94, Iov96, Iov97, Iov98, Iov99a, Iov99b, Iov99c], the author developed the theory of forking and stability for this language, showing that the resulting theory is surprisingly analogous to the first-order case. The emphasis on these papers was on structures of functional analysis whose sorts are Banach spaces, although it was observed that the basic apparatus could be adapted, with relatively straightforward adjustments, for the more general context of pointed metric spaces. A detailed introduction to the language of positive bounded formulas, via Banach space ultrapowers, appeared in [HI02]. This paper also includes an extensive bibliography.

Ben-Yaacov's concept of compact abstract classes, or "cats", originated in different settings; the concept is motivated by two situations: (i) existentially closed models of a universal first-order theory which not form an elementary class, and (ii) hyperimaginaries in a strictly simple first-order theory. Ben Yaacov showed that, in cats, positive formulas are powerful enough to yield not only model theoretic compactness, but also independence and the basic elements of simplicity theory. For a survey, the reader is referred to [BY05a].

The context of cats is more general than that of metric structures. However, Yen-Yaacov has shown that every cat that satisfies reasonable assumptions (namely, compactness) is metric, i.e., it comes from metric structures. See [BY05b].

The framework of continuous logic was proposed in the last few years Ben-Yaacov and Usvyatsov [BYU]. A self-contained introduction can be found in [BYBH08]. The approach of continuous logic is similar to that of the logic of positive bounded formulas (in fact both approaches are mathematically equivalent), but its distinctive feature is that formulas are $[0,1]$-valued rather than $\{0,1\}$-valued; all continuous functions of $[0,1]^n$ into $[0,1]$ ($n \in \mathbb{N}$) are regarded as n-ary connectives, and the quantifiers of continuous logic are the inf and sup operators. An advantage of this approach over that of positive bounded formulas is that in some situations it is more natural to write functional equations (or inequations) involving compositions of real-valued continuous functions, suprema, and infima, rather than positive bounded axioms.

Continuous logic is a simplification of (and is equivalent to) the framework of *continuous model theory* proposed by Chang and Keisler in the 1960's. See [CK66].

All of these logics satisfy model theoretic compactness. Furthermore, all of them are equivalent in terms of expressive power.

In this communication we consider the following questions:

1. Is it a mere coincidence that these logics are equivalent?

2. Is here a more expressive logic that yields a powerful, compact model theory of metric structures?

We shall see that the answer to both questions is "no". In fact, as will be shown below, in essence, what makes the aforementioned logics equivalent is the fact that they all satisfy a form of model-theoretic compactness.

2   Continuous logic

The result discussed in this communication is on abstract logics, and the model-theoretic frameworks mentioned in the introduction are examples of logics to which this result applies. As we introduce the concepts involved in the statement of the result, we will use continuous logic (CL hereafter) as the motivating example. However, aside from the fact that of the four proposals mentioned in the introduction CL is the most recent, the choice of CL as leading example in our context is rather arbitrary; as will be clear, the hypothesis of the main result are weak enough to be satisfied not only by many variations of the examples mentioned above, but also by classical extensions of first-order logic, e.g., extensions of first-order by quantifiers.

In this section we provide the basics of the syntax and semantics of CL. Previous familiarity with CL or any of the logics mentioned hereto is not needed to follow the ideas presented here. The reader interested in further aspects of continuous logic is referred to [BYU] or [BYBH08]. (Either of these references can serve as self-contained introduction.)

The structures of CL are those of the form

$$\mathcal{M} = (\, M^{(s)}, d^{(s)}, R_i, F_j, a_k \mid s \in S, i \in I, j \in J, k \in K \,),$$

where

- $(\, M^{(s)}, d^{(s)} \mid s \in S \,)$ is a family of bounded metric spaces called the *sorts* of $\mathcal{M}$; the metrics $d^{(s)}$ will be called the metrics of $\mathcal{M}$

- For each $i \in I$, R_i is a uniformly continuous function of the form

$$R_i \colon M^{(s_1)} \times \cdots \times M^{(s_n)} \to \mathbb{R},$$

where n is an integer $s_1, \ldots, s_n \in S$; the functions R_i are called the *predicates* of $\mathcal{M}$

· For each $j \in J$, F_j is a uniformly continuous function of the form

$$F_j \colon M^{(s_1)} \times \cdots \times M^{(s_n)} \to M^{(s_0)},$$

where n is an integer $s_0, \ldots, s_n \in S$; the functions F_j are called the *operations* of $\mathcal{M}$

· For each $k \in K$, a_k is a distinguished element of one of the sorts of $\mathcal{M}$; the elements a_k are called the *constants* of $\mathcal{M}$.

These will be called *metric structures*. The restriction that the sorts be bounded is given to facilitate the syntax and does not limit the class of structures under consideration; indeed, if (M, d) is an unbounded metric space and $a \in M$, (M, d) may be replaced by sorts (M_n, d), where $M_n = \{ x \in M \mid d(x, a) \le n \}$.

Examples of metric structures abound in classical mathematics: metric spaces, Banach spaces, operator spaces, measure algebras, and the kinds of structures traditionally studied in model theory; in this last case, the unmentioned metrics are regarded as discrete. More examples can be found in [BYBH08], in [BYU], and in [HI02], which focuses on structures based on Banach spaces.

Let

$$\mathcal{M} = (\ M^{(s)}, d^{(s)}, R_i, F_j, a_k \mid s \in S, i \in I, j \in J, k \in K\ ),$$

be a metric structure. A *signature* for $\mathcal{M}$ includes:

· A distinct binary relation symbol for each metric of $\mathcal{M}$

· A distinct n-ary relation symbol for each n-ary predicate of $\mathcal{M}$

· A distinct n-ary function symbol for each n-ary function of $\mathcal{M}$

· A distinct constant symbol for each constant of $\mathcal{M}$

· A *modulus of uniform continuity* for each of these functions; a modulus of uniform continuity for a function is a map $\epsilon \mapsto \delta$ such that whenever two arguments are, variable by variable, within distance δ, then their images are within distance ϵ. If L is a signature for $\mathcal{M}$ we say that $\mathcal{M}$ is and L-*structure*.

Since the sorts of $\mathcal{M}$ are bounded and the predicates of $\mathcal{M}$ are uniformly continuous, the range of each predicate is a bounded subset of $\mathbb{R}$. Without loss of generality, we can assume that the diameter of each sort of $\mathcal{M}$ is at most 1, and that the range of each predicate of $\mathcal{M}$ is a subset of $[0, 1]$.

Let $\mathcal{M}$ be an L-structure. To simplify the exposition, we will assume that $\mathcal{M}$ has a single sort, (M, d). Also, to ease notation, we will use the same symbol to denote any metric, relation, function, or constant of $\mathcal{M}$ and its respective metric, relation, function, or constant symbol in L.

The terms of L are defined as in ordinary first-order logic (i.e., starting with an infinite set of variables and and iterating function symbols). The *real-valued formulas* of L are defined inductively as follows:

· All the expressions of the form $d(t_1, t_2)$, where t_1, t_2 are terms L, and all the expressions of the form $R(t_1, \ldots, t_n)$, where R is an n-ary predicate symbol and $t_1, \ldots, t_n$ are terms of L, are real-valued formulas of L.

· If $\varphi_1, \ldots, \varphi_n$ are real-valued formulas of L and $c \colon [0,1]^n \to [0,1]$ is a continuous function, then $c(\varphi_1, \ldots, \varphi_n)$ is a real-valued formula of L.

· If x is a variable of L and φ is a real-valued formula of L, the expressions $\sup_x \varphi$ and $\inf_x \varphi$ are real-valued formulas of L.

The analogy with first-order logic should be clear. In CL, formulas represent $[0,1]$-valued instead of $\{0,1\}$-valued functions; the n-ary connectives of CL are all the continuous functions of the form $c \colon [0,1]^n \to [0,1]$, and the quantifiers are the inf and sup operators.

For every real-valued formula of φ of L there exists a nonnegative integer n such φ such that for every L-structure $\mathcal{M}$ as above, φ naturally defines a function $\varphi^M \colon M^n \to [0,1]$. The formula φ is said to be a *sentence* if $n = 0$.

DEFINITION 1. Let $\mathcal{M}$ and $\mathcal{N}$ be L structures.

1. We say that $\mathcal{M}$ and $\mathcal{N}$ are *elementarily equivalent*, and write $\mathcal{M} \equiv \mathcal{N}$, if $\varphi^M = \varphi^N$ for every sentence φ of L.

2. If $\mathcal{M}$ is a substructure of $\mathcal{N}$, we say that $\mathcal{M}$ is an *elementary substructure* of $\mathcal{N}$ if the structures $(\mathcal{M}, a \mid a \in M)$ and $(\mathcal{N}, a \mid a \in M)$ are elementarily equivalent.

The following is a basic property of CL. The proof can be found in [BYBH08] and [BYU].

THEOREM 2. *Suppose that $(I, <)$ is a linearly ordered set and $(\mathcal{M}_i \mid i \in I)$ is a family of L-structures such that $\mathcal{M}_i \prec \mathcal{M}_j$ whenever $< j$. Then, for every $i \in I$, $\mathcal{M}_i \prec \bigcup_{i \in I} \mathcal{M}_i$.*

DEFINITION 3. If $\mathcal{M}$ is a metric structure, let $(\mathcal{M}, \mathbb{R}, \leq)$ denote the structure that includes, in addition to the structure already present in $\mathcal{M}$, the set $\mathbb{R}$ as a distinguished sort, and the order $\leq$ on $\mathbb{R}$. An L-*inequality* is an expression of the form $\varphi \leq r$ or $\varphi \geq r$, where φ is a real-valued formula of L and $r \in \mathbb{R}$.

Clearly, $\mathcal{M} \equiv \mathcal{N}$ if and only if $\mathcal{M}$ and $\mathcal{N}$ satisfy the same inequalities.

3   Abstract Logics and Approximations

If L and L' are multi-sorted signatures, a *renaming* is a bijection $r\colon L \to L'$ that maps sort symbols onto sort symbols, relation symbols onto relation symbols, and function symbols onto function symbols, respecting. If $r\colon L \to L'$ is a renaming and $\mathcal{M}$ is an L-structure, $\mathcal{M}^r$ denotes the structure that results from converting $\mathcal{M}$ into an L'-structure through the map r. The structure $\mathcal{M}^r$ is called a renaming of $\mathcal{M}$.

Let us recall Lindstrom's definition of abstract logic [Lin69]:

DEFINITION 4. A logic $\mathcal{L}$ consists of the following items.

1. A class of structures, called the *structures of* $\mathcal{L}$, that is closed under isomorphisms, renamings, expansion by constants, and reducts.

2. For each multi-sorted signature L, a set $\mathcal{L}[L]$ called the *L-sentences of* $\mathcal{L}$, such that $\mathcal{L}[L] \subseteq \mathcal{L}[L']$ when $L \subseteq L'$.

3. A binary relation $\overset{\mathcal{L}}{\models}$ between structures and sentences of $\mathcal{L}$ such that:

   (a) If $\mathcal{M}$ is an L-structure of $\mathcal{L}$ and $\mathcal{M} \overset{\mathcal{L}}{\models} \varphi$, then $\varphi \in \mathcal{L}[L]$.

   (b) *Isomorphism Property.* If $\mathcal{M} \overset{\mathcal{L}}{\models} \varphi$ and $\mathcal{M}$ is isomorphic to $\mathcal{N}$, then $\mathcal{N} \overset{\mathcal{L}}{\models} \varphi$;

   (c) *Reduct Property.* If $L \subseteq L'$, $\mathcal{M}$ is a L'-structure of $\mathcal{L}$ and $\varphi \in \mathcal{L}[L]$, then $\mathcal{M} \overset{\mathcal{L}}{\models} \varphi$ if and only if $\mathcal{M} \restriction L \overset{\mathcal{L}}{\models} \varphi$;

   (d) *Renaming Property.* Suppose that $r\colon L \to L'$ is a renaming. Then for each sentence $\varphi \in \mathcal{L}[L]$ there exists a sentence $\varphi^r \in \mathcal{L}[L]$ such that $\mathcal{M} \overset{\mathcal{L}}{\models} \varphi$ if and only if $\mathcal{M}^r \overset{\mathcal{L}}{\models} \varphi^r$.

A logic $\mathcal{L}$ has *conjunctions* if for every pair of sentences $\varphi, \varphi' \in \mathcal{L}[L]$ there exists a sentence $\psi \in \mathcal{L}[L]$ such that

$$\mathcal{M} \overset{\mathcal{L}}{\models} \psi \quad \text{if and only if} \quad \mathcal{M} \overset{\mathcal{L}}{\models} \varphi \quad \text{and} \quad \mathcal{M} \overset{\mathcal{L}}{\models} \varphi'.$$

The logic $\mathcal{L}$ is said to have *negations* if for every sentence $\varphi \in \mathcal{L}[L]$ there exists a sentence $\psi \in \mathcal{L}[L]$ such that

$$\mathcal{M} \overset{\mathcal{L}}{\models} \psi \quad \text{if and only if} \quad \mathcal{M} \overset{\mathcal{L}}{\not\models} \varphi.$$

We now turn focus on model theoretic compactness. All the model theoretic frameworks for metric structures mentioned in the introduction satisfy a form of the compactness theorem. However, the form compactness satisfied by these logics is not a literal translation of the classical compactness theorem of first-order logic; the reason is that those logics that do not have negations in the sense defined above. The statement of compactness for those more general contexts involves topological perturbations. The compactness of first-order logic is a particular case when the topologies involved in the perturbations are discrete. This is a peculiarity of first-order due to the fact that its space of truth-values, $\{0, 1\}$, is a discrete. In [Iov01], the author introduced a notion *approximations* sentences for abstract logics that captures the kinds of perturbations needed to state model theoretic compactness for logics with truth values that are not necessarily discrete. We now recall the concept of approximation introduced in [Iov01]:

DEFINITION 5. Let $\mathcal{L}$ be a logic. A *system of approximations* in $\mathcal{L}$ is a binary relation $\lhd$ on the sentences of $\mathcal{L}$ such that

1. $\lhd$ is transitive;

2. If $\varphi \lhd \varphi'$ and $\varphi \in \mathcal{L}[L]$, then $\varphi' \in \mathcal{L}[L]$;

3. If $\varphi \lhd \varphi'$ and $\mathcal{M} \overset{\mathcal{L}}{\models} \varphi$, then $\mathcal{M} \overset{\mathcal{L}}{\models} \varphi'$.

If $\lhd$ is a system of approximations in a logic $\mathcal{L}$, φ is a sentence of $\mathcal{L}$ and $\varphi \lhd \varphi'$, we will say that φ' is a $\lhd$-*approximation* (or simply, an "approximation", if the underlying system of approximations is clear from the context) of φ. A *logic with approximations* is a pair $(\mathcal{L}, \lhd)$, where $\mathcal{L}$ is a logic and $\lhd$ is a system of approximations in $\mathcal{L}$.

If $(\mathcal{L}, \lhd)$ is a logic with approximations, $\mathcal{M}$ is a structure of $\mathcal{L}$, and φ is a sentence of $\mathcal{L}$, we will say that $\mathcal{M}$ *approximately satisfies* φ, and write $\mathcal{M} \overset{\mathcal{L}}{\models}_{\mathcal{A}} \varphi$, if $\mathcal{M} \overset{\mathcal{L}}{\models} \varphi'$ for every $\lhd$-approximation φ' of φ.

REMARKS 6.

1. By condition (3) in Definition 5, the relation $\overset{\mathcal{L}}{\models}_{\mathcal{A}}$ is weaker than $\overset{\mathcal{L}}{\models}$.

2. Every logic $\mathcal{L}$ can be regarded naturally as a logic with approximations by defining $\lhd$ as the diagonal relation on the sentences of $\mathcal{L}$; in other words, the only approximation of each sentence is itself. We will refer to this system of approximations as the *discrete* system on $\mathcal{L}$. Notice that, relative to the discrete system, the relations $\overset{\mathcal{L}}{\models}$ and $\overset{\mathcal{L}}{\models}_{\mathcal{A}}$ are identical.

A *theory* of a logic $\mathcal{L}$ is a set of sentences of $\mathcal{L}$. Let $(\mathcal{L}, \lhd)$ be a logic with approximations. We will say a theory Σ of $\mathcal{L}$ is *consistent* if there exists a structure $\mathcal{M}$ of $\mathcal{L}$ which approximately satisfies every sentence in Σ. We will say that Σ is *finitely consistent* if every finite subset of Σ is consistent.

DEFINITION 7. Let $(\mathcal{L}, \lhd)$ be a logic with approximations. We will say that $(\mathcal{L}, \lhd)$ *satisfies the compactness theorem* if it has the property that every theory of $\mathcal{L}$ which is finitely consistent is consistent.

Let $(\mathcal{L}, \lhd)$ be a logic with approximations. If $\mathcal{M}$ is a structure of $\mathcal{L}$, we denote by $\mathrm{Th}_{\mathcal{A}}^{\mathcal{L}}(\mathcal{M})$ the set of sentences of $\mathcal{L}$ that are approximately satisfied by $\mathcal{M}$. If $\mathcal{N}$ is a structure of $\mathcal{L}$, we write $\mathcal{M} <_{\mathcal{A}}^{\mathcal{L}} \mathcal{N}$ to indicate that $M \subseteq N$ and the structure $(\mathcal{N}, a)_{a \in M}$ approximately satisfies $\mathrm{Th}_{\mathcal{A}}^{\mathcal{L}}((\mathcal{M}, a)_{a \in M})$. (Recall that the class of structures of a logic is assumed to be closed under expansions by constants.)

DEFINITION 8. Let $(\mathcal{L}, \lhd)$ be a logic with approximations. We will say that $(\mathcal{L}, \lhd)$ *satisfies the elementary chain property* if the following condition holds. Whenever

$$\mathcal{M}_0 <_{\mathcal{A}}^{\mathcal{L}} \mathcal{M}_1 <_{\mathcal{A}}^{\mathcal{L}} \ldots <_{\mathcal{A}}^{\mathcal{L}} \mathcal{M}_n <_{\mathcal{A}}^{\mathcal{L}} \ldots \qquad (n < \omega)$$

there exists a structure $\mathcal{M}$ of $\mathcal{L}$ such that $\mathcal{M}_n <_{\mathcal{A}}^{\mathcal{L}} \mathcal{M}$ for every $n < \omega$, and $\mathcal{M}$ is uniquely determined by $\bigcup_n \mathcal{M}$.

4   Continuous Logic as an Abstract Logic

In this section we state the properties of CL that, as Theorem 9 shows, characterize it.

1. The class of structures of CL is the class of all metric structures.

2. The L-sentences of CL, for a given signature L, are all the positive boolean combinations of inequalities $\varphi \leq r$ or $\varphi \geq r$, where φ is a real-valued sentence of L and $r \in \mathbb{R}$; see Definition 3.

3. The relation $\overset{\mathrm{CL}}{\vDash}$ is the obvious one: $\mathcal{M} \overset{\mathrm{CL}}{\vDash} \varphi \leq r$ iff $\varphi^{\mathcal{M}} \leq r$, and $\mathcal{M} \overset{\mathrm{CL}}{\vDash} \varphi \geq r$ iff $\varphi^{\mathcal{M}} \geq r$

4. CL has approximations: the approximations of the inequality $\varphi \leq r$ are the inequalities of the form $\varphi \leq s$, where $s > r$, and the approximations of $\varphi \geq r$ are the inequalities of the form $\varphi \geq s$, where $s < r$; if $\sigma_1, \ldots, \sigma_n$ are inequalities and $B(\sigma_1, \ldots, \sigma_n)$ is a positive boolean combination of $\sigma_1, \ldots, \sigma_n$, than the approximations of $B(\sigma_1, \ldots, \sigma_n)$ are all the expressions of the form $B(\tau_1, \ldots, \tau_n)$, where τ_i is an approximation of σ_i, for $i = 1, \ldots, n$. We will write $\sigma < \tau$ if τ is an approximation of σ.

5. The pair $(\mathrm{CL}, <)$ satisfies the compactness theorem and the elementary chain property; see Definitions 7 and 8.

6. The pair $(\mathrm{CL}, <)$ has a weak negation: define $\overset{w}{\neg}(\varphi \leq r) = \varphi \geq r$, $\overset{w}{\neg}(\varphi \geq r) = \varphi \leq r$, and, inductively, $\overset{w}{\neg}(\sigma \wedge \tau) = \overset{w}{\neg}(\sigma) \vee \overset{w}{\neg}(\tau)$ and $\overset{w}{\neg}(\sigma \vee \tau) = \overset{w}{\neg}(\sigma) \wedge \overset{w}{\neg}(\tau)$.

5 The maximality of Continuous Logic

Let $(\mathcal{L}, \lhd)$ and $(\mathcal{L}_1, \lhd_1)$ be logics with approximations such that $\mathcal{L}$ and $\mathcal{L}_1$ have the same structures. We will say that a sentence φ of $\mathcal{L}$ is *reducible* to $\mathcal{L}_1$ if the following condition holds. For every $\lhd$-approximation φ' of φ there exist two sentences $\psi[\varphi, \varphi']$ and $\psi'[\varphi, \varphi']$ of $\mathcal{L}_1$ such that:

1. $\psi[\varphi, \varphi'] \lhd_1 \psi'[\varphi, \varphi']$;

2. If $\mathcal{M}$ is a structure of $\mathcal{L}$,

$$\mathcal{M} \overset{\mathcal{L}}{\vDash} \varphi \qquad \text{implies} \qquad \mathcal{M} \overset{\mathcal{L}_1}{\vDash} \psi[\varphi, \varphi'],$$

$$\mathcal{M} \overset{\mathcal{L}_1}{\vDash} \psi'[\varphi, \varphi'] \qquad \text{implies} \qquad \mathcal{M} \overset{\mathcal{L}}{\vDash} \varphi'.$$

We will say that $(\mathcal{L}_1, \lhd_1)$ is an *extension* of $(\mathcal{L}, \lhd)$ if every sentence of $\mathcal{L}$ is reducible to $\mathcal{L}_1$. Two logics with approximations will be called *equivalent* if they are reducible to each other.

Intuitively, $(\mathcal{L}_1, \lhd_1)$ is an extension of $(\mathcal{L}, \lhd)$ if every sentence of $\mathcal{L}$ can be approximated by sentences of $\mathcal{L}'$. As a trivial but important example let us notice that if $(\mathcal{L}, \lhd)$ is a logic with approximations, $\mathcal{L}_1$ is a logic with the same structures as $\mathcal{L}$, every sentence of $\mathcal{L}$ is a sentence of $\mathcal{L}_1$, and $\overset{\mathcal{L}_1}{\vDash}$ extends $\overset{\mathcal{L}}{\vDash}$ (in the traditional mathematical sense of the word) then $\mathcal{L}_1$ paired with the discrete system of approximations (see Remark 6) is an extension of $(\mathcal{L}, \lhd)$.

let $(\mathcal{L}, \lhd)$ be a logic with approximations. A *weak negation* on $(\mathcal{L}, \lhd)$ is a monadic operation $\overset{w}{\neg}$ on the sentences of $\mathcal{L}$ such that

1. If $\varphi \in \mathcal{L}[L]$, then $\overset{w}{\neg}\varphi \in \mathcal{L}[L]$;

2. If $\varphi \in \mathcal{L}[L]$ and $\mathcal{M}$ is an L-structure of $\mathcal{L}$, then

$$\mathcal{M} \overset{\mathcal{L}}{\vDash} \varphi \qquad \text{or} \qquad \mathcal{M} \vDash \overset{w}{\neg}\varphi;$$

3. If φ' is an approximation of φ, then

$$\mathcal{M} \overset{\mathcal{L}}{\vDash}_{\mathcal{A}} \overset{w}{\neg}\varphi' \qquad \text{implies} \qquad \mathcal{M} \overset{\mathcal{L}}{\vDash}_{\mathcal{A}} \varphi.$$

Note that if $\mathcal{L}$ is a logic with negations and $\lhd$ is the discrete system of approximations of $\mathcal{L}$ (see Remark 6), then the negation of $\mathcal{L}$ is a weak negation on $(\mathcal{L}, \lhd)$.

If $(\mathcal{L}, \lhd)$ is a logic with approximations and $\mathcal{M}, \mathcal{N}$ are structures of $\mathcal{L}$, we write $\mathcal{M} \equiv_{\mathcal{A}}^{\mathcal{L}} \mathcal{N}$ if $\mathrm{Th}_{\mathcal{A}}^{\mathcal{L}}(\mathcal{M}) \subseteq \mathrm{Th}_{\mathcal{A}}^{\mathcal{L}}(\mathcal{N})$ and $\mathrm{Th}_{\mathcal{A}}^{\mathcal{L}}(\mathcal{N}) \subseteq \mathrm{Th}_{\mathcal{A}}^{\mathcal{L}}(\mathcal{M})$. Notice that if $(\mathcal{L}, \lhd)$ has a weak negation, then each of these inclusions implies the other.

The following result was proved in [Iov01]:

THEOREM 9. *Suppose that* $(\mathcal{L}, \lhd)$ *is a logic for metric structures and that* $(\mathcal{L}, \lhd)$

· *extends* $(\mathrm{CL}, <)$,

· *satisfies the compactness theorem,*

· *satisfies the elementary chain property, and*

· *has a weak negation.*

Then $(\mathcal{L}, \lhd)$ *is equivalent to* $(\mathrm{CL}, <)$.

This theorem shows, among other things, the equivalence among all the logics for metric structures mentioned here. For instance, let us show the equivalence between CL and the logic $\mathcal{L}_{\mathrm{PB}}$ of positive bounded formulas. Both logics satisfy the compactness theorem and the elementary chain property (the proofs can be found in any basic expositions available for these two logics; see, for example, [BYBH08] for the CL and [HI02] for $\mathcal{L}_{\mathrm{PB}}$), so we just have to note that CL is an extension of $\mathcal{L}_{\mathrm{PB}}$ in the sense defined at the beginning of this section. Recall from Section 2 that two metric structures $\mathcal{M}, \mathcal{N}$ are elementarily equivalent in CL if and only if the structures $(\mathcal{M}, \mathbb{R}, <)$ and $(\mathcal{N}, \mathbb{R}, <)$ satisfy the same inequalities of the form $\varphi \leq r$ and $\varphi \geq r$, where $r \in \mathbb{R}$ and φ is a $[0,1]$-valued function built up from the ($[0,1]$-valued) predicates of L by using continuous functions $c \colon [0,1]^n \to [0,1]$ as connectives and the operators sup and inf as quantifiers. The positive bounded formulas of a signature L are all the expressions that can be built up from the basic inequalities of the form $R(t_1, \ldots, t_n) \leq r$ and $R(t_1, \ldots, t_n) \geq r$, where $r \in \mathbb{R}$, $t_1, \ldots, t_n$ are terms of L and R is a predicate of L, by using the connectives $\wedge, \vee$ and the first-order the quantifiers $\exists, \forall$ (which range over the bounded sorts — hence the name "positive bounded"). If ψ is a positive bounded formula, the approximations of ψ are defined as the formulas that result from "relaxing" all the estimates that occur in ψ, i.e., replacing all the inequalities of the form $R(t_1, \ldots, t_n) \leq r$ that occur in ψ by inequalities of the form $R(t_1, \ldots, t_n) \leq s$ for some $s > r$ and, similarly, replacing all the inequalities of the form $R(t_1, \ldots, t_n) \geq r$ by $R(t_1, \ldots, t_n) \geq s$ for some $s < r$. It is easy to see, by induction on the complexity of formulas, that every positive bounded formula is reducible to CLinequalities of the form defined in : for the quantifier-free case, it suffices to observe that every formula of the form

$$\bigwedge_{1 \leq i \leq N} \Big(\bigvee_{1 \leq \alpha \leq m(i)} R_{1,\alpha}(\bar{t}) \leq r_{1,\alpha} \vee \bigvee_{1 \leq \beta \leq n(i)} R_{2,\beta}(\bar{t}) \geq r_{2,\beta} \Big),$$

where $R_{1,\alpha}, R_{2,\beta}$ are $[0,1]$-valued predicates and $r_{1,\alpha}, r_{2,\beta} \in [0,1]$ for $1 \leq \alpha \leq m(i), 1 \leq \beta \leq n(i)$, is equivalent to the inequality

$$\max_{1 \leq i \leq N} \Big[\min(\min_{1 \leq \alpha \leq m(i)} R_{1,\alpha}(\bar{t}) \mathbin{\dot{-}} r_{1,\alpha}, \min_{1 \leq \beta \leq n(i)} r_{2,\beta} \mathbin{\dot{-}} R_{2,\beta}(\bar{t})) \Big] \leq 0,$$

where $\dot{-}$ denotes the truncated difference on $[0,1]$, i.e., $x \mathbin{\dot{-}} y$ is $x - y$ if $x \geq y$ and 0 if $x < y$; the truncated difference is a continuous function from $[0,1]^2$ into $[0,1]$, and hence a binary connective of CL, so the preceding inequality is of the type defined in Definition 3. The quantifier step of the induction is given by the fact for every real-valued function f and every $\epsilon > 0$,

- $\exists x(f(x) \leq r) \Longrightarrow \inf_x f(x) \leq r \Longrightarrow \exists x(f(x) \leq r + \epsilon)$

- $\exists x(f(x) \geq r) \Longrightarrow \sup_x f(x) \geq r \Longrightarrow \exists x(f(x) \geq r - \epsilon)$

- $\forall x(f(x) \leq r) \Longleftrightarrow \sup_x f(x) \leq r$

- $\forall x(f(x) \geq r) \Longleftrightarrow \inf_x f(x) \geq r.$

The continuous model theory framework of Chang and Keisler [CK66] extends CL (the main difference is that Chang and Keisler include many more quantifiers); but then, since both logics satisfy the compactness theorem and the elementary chain property, they must be equivalent.

First-order logic extends CL, since every inequality of the type defined in Definition 3 is a first-order sentence. The extension is proper because, as a logic for metric structures, first-order does not satisfy the compactness theorem. (In fact, the expressive power of first-order logic on Banach spaces is known to be quite high [SS78].) It seems rather striking to us that there is no logic strictly between CL and first-order satisfying the conditions of Theorem 9.

BIBLIOGRAPHY

[BY05a] Itay Ben-Yaacov, *Compactness and independence in non first order frameworks*, Bull. Symbolic Logic **11** (2005), no. 1, 28–50. MR MR2125148 (2006d:03054)

[BY05b] _____, *Uncountable dense categoricity in cats*, J. Symbolic Logic **70** (2005), no. 3, 829–860. MR MR2155268 (2006g:03063)

[BYBH08] Itaï Ben Yaacov, Alexander Berenstein, and C. Ward Henson, *Model theory for metric structures*, Model theory with applications to algebra and analysis. Vol. 2, London Math. Soc. Lecture Note Ser., vol. 350, Cambridge Univ. Press, Cambridge, 2008, pp. 315–427. MR MR2436146

[BYU] Itay Ben-Yaacov and Alex Usvyatsov, *Continuous first order logic and local stability*, To appear in the Transactions of the American Mathematical Society.

[CK66] Chen-chung Chang and H. Jerome Keisler, *Continuous model theory*, Annals of Mathematics Studies, No. 58, Princeton Univ. Press, Princeton, N.J., 1966. MR MR0231708 (38 #36)

[Hen74] C. W. Henson, *The isomorphism property in nonstandard analysis and its use in the theory of Banach spaces*, J. Symbolic Logic **39** (1974), 717–731. MR 50 #12713

[Hen75] ———, *When do two Banach spaces have isometrically isomorphic nonstandard hulls?*, Israel J. Math. **22** (1975), no. 1, 57–67.

[Hen76] ———, *Nonstandard hulls of Banach spaces*, Israel J. Math. **25** (1976), no. 1-2, 108–144.

[HH86] Stefan Heinrich and C. Ward Henson, *Banach space model theory. II. Isomorphic equivalence*, Math. Nachr. **125** (1986), 301–317.

[HHM83] S. Heinrich, C. W. Henson, and L. C. Moore, Jr., *Elementary equivalence of L_1-preduals*, Banach space theory and its applications (Bucharest, 1981), Springer, Berlin, 1983, pp. 79–90. MR 84j:46021

[HHM86] ———, *Elementary equivalence of $C_\sigma(K)$ spaces for totally disconnected, compact Hausdorff K*, J. Symbolic Logic **51** (1986), no. 1, 135–146. MR 87f:03097

[HHM87] ———, *A note on elementary equivalence of $C(K)$ spaces*, J. Symbolic Logic **52** (1987), no. 2, 368–373. MR 88k:03078

[HI02] C. Ward Henson and José Iovino, *Ultraproducts in analysis*, Analysis and logic (Mons, 1997), London Math. Soc. Lecture Note Ser., vol. 262, Cambridge Univ. Press, Cambridge, 2002, pp. 1–110. MR 1 967 834

[HM83a] C. W. Henson and L. C. Moore, Jr., *The Banach spaces $l_p(n)$ for large p and n*, Manuscripta Math. **44** (1983), no. 1-3, 1–33. MR 84g:46026

[HM83b] ———, *Nonstandard analysis and the theory of Banach spaces*, Nonstandard analysis—recent developments (Victoria, B.C., 1980), Springer, Berlin, 1983, pp. 27–112. MR 85f:46033

[Iov94] José Iovino, *Stable theories in functional analysis*, Ph.D. thesis, University of Illinois at Urbana-Champaign, 1994.

[Iov96] ———, *The Morley rank of a Banach space*, J. Symbolic Logic **61** (1996), no. 3, 928–941. MR 97j:03072

[Iov97] ———, *Definability in functional analysis*, J. Symbolic Logic **62** (1997), no. 2, 493–505. MR 98i:03046

[Iov98] ———, *Types on stable Banach spaces*, Fund. Math. **157** (1998), no. 1, 85–95. MR 99d:46013

[Iov99a] ———, *Stable Banach spaces and Banach space structures. I. Fundamentals*, Models, algebras, and proofs (Bogotá, 1995), Dekker, New York, 1999, pp. 77–95. MR 2000h:03064a

[Iov99b] ———, *Stable Banach spaces and Banach space structures. II. Forking and compact topologies*, Models, algebras, and proofs (Bogotá, 1995), Dekker, New York, 1999, pp. 97–117. MR 2000h:03064b

[Iov99c] ———, *Stable models and reflexive Banach spaces*, J. Symbolic Logic **64** (1999), no. 4, 1595–1600. MR 2001e:03068

[Iov01] ———, *On the maximality of logics with approximations.*, J. Symbolic Logic **66** (2001), no. 4, 1909–1918.

[Kri74] J.-L. Krivine, *Langages à valeurs réelles et applications*, Fund. Math. **81** (1974), 213–253, Collection of articles dedicated to Andrzej Mostowski on the occasion of his sixtieth birthday, III. MR 50 #1873

[Lin69] P. Lindström, *On extensions of elementary logic*, Theoria **35** (1969), 1–11. MR 39 #5330

[SS78] S. Shelah and J. Stern, *The Hanf number of the first order theory of Banach spaces*, Trans. Amer. Math. Soc. **244** (1978), 147–171.

José Iovino
Department of Mathematics
The University of Texas at San Antonio
San Antonio, USA.
E-mail: iovino@math.utsa.edu

On Profinite Structures

HUGO LUIZ MARIANO AND FRANCISCO MIRAGLIA

ABSTRACT. We prove that first-order profinite structures, i.e., the projective limits of downward directed systems of finite discrete structures, are pure injective, that is, they possess the extension property relative to the class of pure embeddings. We present a topological characterization of the notion of profinite structure which yields closure properties of this class of structures and a description of its quotients objects, generalizing results in [11] and [14]. We discuss an interesting elementary class of first-order topological structures and introduce the notion of a saturated family of congruences, associating to each such pair of data a functor, called Profinite Hull. It is shown, by two different approaches that this functor satisfies an universal property, yielding a pair of adjoint functors. It is also established that the profinite hull functor preserves inductive limits and quotients by a saturated congruence. It is considered a "local-global principle" naturally associated to the profinite hull functor.

Introduction

If L is a first-order language with equality, write **L-mod** for the category of L-structures and L-homomorphisms, which in this in this paper will be referred to as *L-morphisms*. The language L (although arbitrary) will remain fixed in all that follows.

We shall consider profinite L-structures, i.e. the projective limits of downward directed systems of finite discrete L-structures.

This work is a sequence to [13], where it is shown that profinite L-structures are retracts of certain ultraproducts of finite L-structures and, as a consequence, any elementary class $\mathcal{A}$ of **L-mod** that is axiomatizable by sentences of the form $\forall \vec{x}(\psi_0(\vec{x}) \to \psi_1(\vec{x}))$, where $\psi_0(\vec{x}), \psi_1(\vec{x})$ are positive-existential L-formulas, is closed under profinite limits. These considerations apply, in particular, to *Special Groups* and *Reduced Special Groups*, a first-order axiomatization of the algebraic theory of quadratic forms and its reduced counterpart, respectively. The reader is referred to [7] for a presentation of this circle of ideas, including the appropriate first-order language. Most of the results proved here generalize results first obtained for reduced special groups (see [14]) and for varieties of algebras. In particular, they also apply to Boolean algebras, that, in a sense that can be made precise (see Chapter 4 in [7]), is a subcategory of the category of reduced special groups.

The paper contains four sections. In section 1 we generalize the well-known fact that complete Boolean algebras are the injective objects in the category of Boolean algebras and in the category of Heyting algebras, by showing that profinite L-structures are pure-injectives, i.e., have the extension property with respect to the class of pure embeddings.

In section 2 we present a topological characterization of the notion of profinite L-structure, obtaining closure properties of this class of L-structures and a description of its quotients objects. This generalizes results first obtained in [11] for reduced special groups.

Section 3 introduces an elementary class $\mathcal{A}$ of **L-mod** L-structures, together with a saturated family of congruences, $\mathfrak{C}$, generalizing both the well-known concept of congruence in varieties of algebras and the congruences induced by saturated subgroups of reduced special groups (see [7]). To each such pair $(\mathcal{A}, \mathfrak{C})$ we associate:

$$\begin{cases} \text{The profinite hull functor:} & \mathcal{P} : \mathcal{A}^{top} \longrightarrow \mathcal{A}_{pf}; \\ \text{A natural transformation:} & (M \xrightarrow{\eta_M} \mathcal{P}(M))_{M \in \mathcal{A}^{top}}, \end{cases}$$

where $\mathcal{A}^{top} \subseteq \mathbf{L} - \mathbf{mod}^{top}$ is the (full) subcategory of topological structures in $\mathcal{A}$ and continuous L-morphisms, while $\mathcal{A}_{pf} \subseteq \mathcal{A}^{top}$ is the full subcategory of $\mathcal{A}^{top}$ of profinite structures in $\mathcal{A}$. We show that the profinite hull functor satisfies a universal property, yielding an adjoint pair of functors. This is proven by two different approaches: a categorical one and a topological-analytical one. We also prove that the profinite hull functor preserves directed inductive limits and quotients by a saturated congruence, generalizing results in [14].

Section 4 presents constructions and questions suggested by the present work and connected to a local-global principle naturally associated to the profinite hull functor.

1   Profinites and Injectives

For the reader's convenience we register the following

REMARK 1. a) Recall that a formula φ in L is

∗ **positive existential (p.e.)** if it is equivalent to a formula constructed from the atomic formula employing only the connectives $\land$, $\lor$ and the existential quantifier $\exists$;

∗ **positive primitive (p.p.)** if it is equivalent to a formula of the form $\exists \overline{x}\varphi$, where φ is a conjunction of atomic formulas.

* **geometrical** if it is logically equivalent to one of the form $\forall \, \bar{x}(\varphi(\bar{x}, \, \bar{y})$ $\rightarrow \psi(\bar{x}, \, \bar{y}))$, where φ, ψ are p.e.-formulas, **or** to the negation of an atomic formula.

It is well-known that every p.e.-formula is equivalent to the disjunction of finite conjunctions of p.p.-formulas.

b) A map between L-structures, $f : M \longrightarrow N$, is a **pure L-morphism** if for each p.e.-formula $\varphi(\bar{x})$ and for all $\bar{a}$ in M, $\quad M \models \varphi[\bar{a}] \quad \Leftrightarrow \quad N$ $\models \varphi[f\bar{a}]$. Hence, a $\underline{L\text{-morphism}}$ $g : M \longrightarrow N$ is pure iff it reflects p.p.-formulas. Clearly, all pure L-morphisms are L-embeddings and any elementary embedding and any L-section [1] are pure L-morphisms. $\qquad \square$

We also mention the following

FACT 2.   If Σ is a set of geometrical L-sentences and $f : M \longrightarrow N$ is a pure L-morphism, then $N \models \Sigma \; \Rightarrow \; M \models \Sigma$. $\qquad \square$

PROPOSITION 3.   *a) Let $M \xrightarrow{f} N \xrightarrow{g} P$ be L-morphisms. Then:*

   (1) *f, g pure $\; \Rightarrow \; g \circ f$ pure;*

   (2) *$g \circ f$ pure $\; \Rightarrow \; f$ pure. In particular, every L-section in **L-mod** is a pure embedding.*

b) If $f_i : M_i \longrightarrow N_i$, $i \in I$, is a family of pure L-morphisms, their product, $\prod_{i \in I} f_i : \prod_{i \in I} M_i \longrightarrow \prod_{i \in I} N_i$, is a pure L-morphism.

*c) Let $\langle \, I, \leq \rangle$ be an upward directed poset and $\mathcal{M} = \langle \, M_i; \{f_{ij} : i \leq j\} \rangle$ and $\mathcal{N} = \langle \, N_i; \{g_{ij} : i \leq j\} \rangle$ be I-diagrams in **L-mod**. Let $\lim \mathcal{M} = \langle \, M; f_i \rangle$ and $\lim \mathcal{N} = \langle \, N; g_i \rangle$ their colimits in **L-mod**. Let $\overrightarrow{\langle \, h_i \rangle}_{i \in I} :$ $\mathcal{M} \longrightarrow \mathcal{N}$ be a morphism of I-diagrams and let $\lim_{\longrightarrow} h_i = h : M \longrightarrow N$ be the limit L-morphism. Then:*

   (1) *If each h_i is pure, then $h : M \longrightarrow N$ is pure;*

   (2) *If each f_{ij} is pure, $i \leq j$ in I, then $f_i : M_i \longrightarrow M$ is pure.*

d) Let $f : M \longrightarrow N$ be a L-morphism. Then the following conditions are equivalent:

   (1) *f is a pure L-embedding;*

   (2) *There are a L-structure P, a L-morphism $g : N \longrightarrow P$ and a pure L-embedding $h : M \longrightarrow P$ such that $g \circ f = h$;*

   (3) *There are a L-structure P, a L-morphism $g : N \longrightarrow P$ and a L-elementary embedding $h : M \longrightarrow P$ such that $g \circ f = h$;*

   (4) *There is an ultrafilter pair (I, U) and a L-morphism $g' : N \longrightarrow M^I/U$ such that $g' \circ f = \delta_M$, where $\delta_M : M \longrightarrow M^I/U$ is the canonical diagonal L-elementary embedding.*

Proof.   (Sketch; full proofs appear in [14]): Items (a), (b) and (c) are straightforward. For item (d), (1) $\Rightarrow$(2) is clear: take $P = N$, $g = id$

---

[1] A L-section is a L-morphism that admits a retraction that is also a L-morphism.

and $h = f$; (2) $\Rightarrow$ (1) is item (a.2) above, which also gives (3) $\Rightarrow$ (1) because elementary embeddings are pure; (1) $\Rightarrow$ (3) is an application of the Robinson's diagram method; (4) $\Rightarrow$ (3) is obvious and (3) $\Rightarrow$ (4) follows from Scott's Lemma (Lemma 8.1.3 in [2]). ∎

Since any Boolean algebra is the directed union of its (complete) finite subalgebras, it follows from Proposition 3.(c.2) and Sikorsky's Extension Theorem that any *injective* Boolean algebra morphism is a pure embedding in the natural language of Boolean algebras.

As the projective limit of a diagram of complete Boolean algebras and complete homomorphisms is a complete Boolean algebra, it follows that the profinite Boolean algebras are complete, and it is well known that complete Boolean algebras are the injective Boolean (or Heyting) algebras relatively to the class of injective homomorphisms, so profinite Boolean algebras are injectives. We obtain a generalization of this fact: we prove, from the results in [13] below, that profinite structures are pure injective structures, the structures that have the extension property relatively to the class of pure embeddings.

REMARK 4. Recall a non-empty partially ordered set (poset), $\langle I, \leq \rangle$, is *downward directed* if for each $i, j \in I$ there is a $k \in I$ such that $k \leq i, j$. A L-structure is **profinite** if it is L-isomorphic to the limit of a diagram of *finite* L-structures over a *downward directed poset*.

Let $\langle I, \leq \rangle$ be a *downward* directed poset and let $\mathcal{M} = (M_i, \{f_{ij} : i \leq j\})$ be a diagram of finite L-structures over I. Write $(P, \{p_i : i \in I\}) = \underleftarrow{lim}\ \mathcal{M}$. We have the natural L-embedding,
$$\iota : P \hookrightarrow M = \textstyle\prod_{i \in I} M_i,$$
such that for all $i \in I$, $p_i = \pi_i \circ \iota$, where $\pi_i : M \longrightarrow M_i$ is the a canonical projection. Writing $k^\leftarrow \doteq \{j \in I : j \leq k\}$, $k \in I$, then it is straightforward that

For all $\overline{x} \in P$ and all $j, k \in I$ ($j \in k^\leftarrow \Rightarrow f_{jk}(x_j) = x_k$).

Moreover, if $\mathcal{F}$ is a filter on I, for each $J \in \mathcal{F}$ there is a natural L-morphism, $\nu_J : M_{|J} = \prod_{j \in J} M_j \longrightarrow M/\mathcal{F}$, given by $x \longmapsto x/\mathcal{F}$, where $M/\mathcal{F}$ is the reduced product $\prod_{i \in I} M_i/\mathcal{F}$. □

In [13] we prove the following

THEOREM 5. *Profinite L-structures are retracts of ultraproducts of finite L-structures. More precisely, and with the notation in the Remark 4, let $\langle I, \leq \rangle$ be an downward directed poset and let $\mathcal{M} = (M_i, \{f_{ij} : i \leq j\})$ be a diagram of finite L-structures over I. If $\underleftarrow{lim}\ \mathcal{M} = (P, \{p_i : i \in I\})$ then the L-morphism given by the composition $P \xrightarrow{\iota} \prod_{i \in I} M_i \xrightarrow{\nu_I} \prod_{i \in I} M_i/\mathcal{U}$ is an L-section, where $\mathcal{U}$ is a directed ultrafilter in I, i.e. $i^\leftarrow$*

$\in \mathcal{U}$, *for each* $i \in I$.

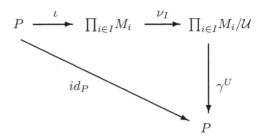

COROLLARY 6. *Let T be a L-theory axiomatized by geometrical L-sentences. Then, $\mathrm{Mod}(T)$, the full subcategory of models of T, is closed under profinite limits* [2].   □

DEFINITION 7. A L-structure T is **pure L-injective** if T is injective with respect to the class of pure L-embeddings, that is, if M, M' are L-structures, $j : M \to M'$ is a pure L-embedding and $g : M \longrightarrow T$ is a L-morphism, there is a L-morphism, $g' : M' \longrightarrow T$, extending g, i.e., $g' \circ j = g$.

It is easily established that the in class of pure L-injective structures is closed under products and retracts. We are now establish the main result of this section, namely

THEOREM 8. *Profinite L-structures are pure L-injectives.*

Proof. By Theorem 5 profinite L-structures are L-retracts of L-products of finite L-structures; hence, it is enough to prove the following:

Claim: Finite L-structures are pure L-injectives.

Proof of Claim: Let T be a finite L-structure, let $j : M \to M'$ be a pure L-embedding and let $g : M \longrightarrow T$ be a L-morphism. Proposition 3.(d) yields an ultrafilter pair, (I, U), and a L-morphism $h : M' \longrightarrow M^I/U$ such that $h \circ j = \delta_M$. Now applying the "(I, U)-ultrapower functor" to the L-morphism $g : M \longrightarrow T$ yields $g^I/U \circ \delta_M = \delta_T \circ g$. Since T is *finite*, the diagonal embedding $\delta_T : T \longrightarrow T^I/U$ is a L-*isomorphism*. Now if $g' : M' \longrightarrow T$ is the L-morphism given by the composition $M' \xrightarrow{h} M^I/U \xrightarrow{g^I/U} M'^I/U \xrightarrow{(\delta_T)^{-1}} T$, we obtain $g' \circ j = g$, as needed.

---

[2]It is straightforward that the final object of **L-mod** belongs to $\mathrm{Mod}(T)$.

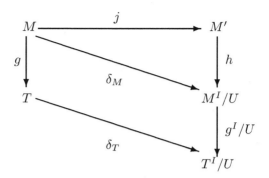

A natural question is if all structures have an "injective hull" relative to some class of embeddings. Since profinite structures are injectives, it is also natural to ask if all structures have a "profinite hull": this is indeed the case, as we shall see in the sections that follow.

2　Profinite Topological Structures

REMARK 9. We here recall basic results on congruences.

a) A *congruence* in a L-structure M is an equivalence relation $C \subseteq M \times M$ which is also a L-substructure of the L-product $M \times M$. If $f : M \longrightarrow N$ is any L-morphism and $C' \subseteq N \times N$ is a congruence in N then $f^\star(C') \doteq (f \times f)^{-1}[C'] \subseteq M \times M$ is a congruence in M, in particular, $ker(f) \doteq f^\star(\Delta_N) = \{(x, x') \in M \times M : f(x) = f(x') \in N\}$ is a congruence in M. As in the universal algebra situation, to each $C \in Cong(M) \doteq \{\text{congruences in } M\}$, there is a canonical L-structure M/C defined on the quotient set: constants and functions symbols are treated as usual in universal algebra; for a k-ary relation symbol R in L, its interpretation in the quotient is defined as follows:
$$(x_0/C, \ldots, x_{k-1}/C) \in R^{M/C} \quad \Leftrightarrow \quad \exists\, x'_0, \ldots, x'_{k-1} \in M \text{ with}$$
$x'_i/C = x_i/C$, $i < k$ and $(x'_0, \ldots, x'_{k-1}) \in R^M$.

b) A L-structure M is said to be **L-inhabited** if for each L-relational symbol R, $R^M \neq \emptyset$; in particular, $M \neq \emptyset$ because $\Delta_M = (=)^M \neq \emptyset$. If $\{M_i, i \in I\}$ is a family of L-inhabited structures and F is a filter in I then the reduced product $\prod_{i \in I} M_i/F$ is a quotient of the product structure $\prod_{i \in I} M_i$ by the L-congruence determined by the kernel of the natural map from $\prod_{i \in I} M_i$ to $\prod_{i \in I} M_i/F$.

$\square$

With notation as in Remark 9.(a), we have the fundamental theorem of L-morphisms:

PROPOSITION 10. (cf. Proposition 17.21, p. 174, [16]) *If M is a L-*

structure and $C \in Cong(M)$, the natural map $q_C : M \longrightarrow M/C$, $x \mapsto x/C$ is a L-morphism. Moreover, if $f : M \longrightarrow N$ is a L-morphism such that $C \subseteq ker(f)$, there is an unique L-morphism, $\bar{f} : M/C \longrightarrow N$ such that $f = \bar{f} \circ q_C$. □

We assume the reader is familiar with the concept of **uniform space**, as for instance in [4] or [3]. A well-known result due to André Weil guarantees that a topological space is *uniformizable* iff it is completely regular. Associated to any uniformity there is a notion of completeness and any Hausdorff uniform space has an essentially unique *completion*. Any compact Hausdorff space (X, τ) is uniformizable, by an uniquely determined uniform structure: the set of all neighborhoods of the diagonal $\Delta_X \subseteq X \times X$ in the product topology, and it is complete with this uniformity.

PROPOSITION 11. *Let $\langle I, \leq \rangle$ be a downward directed poset and $\mathcal{M}$: $I \longrightarrow$ **L-mod** a diagram of L-structures, such that for each $i \in I$, M_i is finite. As in Remark 4, write $(P, \{p_i : i \in I\}) = \lim_{\substack{\longleftarrow \\ i \in I}} M_i$ for the the projective limit of $\mathcal{M}$, $(M, \{\pi_i : i \in I\}) = \prod_{i \in I} M_i$ for the L-structure product and $\iota : P \hookrightarrow M$ for the canonical L-embedding. Suppose that for each $i \in I$, M_i is a* discrete *topological space and P is a topological subspace of M (endowed with the product topology). Then:*

a) P is a boolean *space (i.e. a Hausdorff, compact space with a basis of clopens); moreover:*

∗ The L-operations in P are continuous *functions;*

∗ the n-ary L-relations in P are closed *subsets of the product P^n, $n \in \omega$.*

Moreover, the topology in P is the coarsest such that p_i is a continuous *L-morphism, for all $i \in I$.*

b) For each $i \in I$, $ker(p_i) = \{(\vec{s}, \vec{t}) \in P \times P : p_i(\vec{s}) = p_i(\vec{t})\} = \bigcup\{(P \cap \prod_{j \neq i} M_j \times \{x_i\})^2 : x_i \in M_i\}$ is a congruence of P, of discrete finite index (i.e., the quotient topological space $P/ker(p_i)$ is discrete and finite). The set $S = \{ker(p_i) : i \in I\}$ is a fundamental system of entourages of the (unique) uniformity compatible with the topology of P.

c) If $\psi(\vec{x}, \vec{y})$ is a p.e.(L)-formula with $length(\vec{x}) = n$, $length(\vec{y}) = m$ and $\vec{a}$ is a finite sequence of elements of P with $length(\vec{a}) = n$ then $[\psi(\vec{a})]^M \doteq \{\vec{b} \in P^m : P \models \psi[\vec{a}, \vec{b}]\}$ is a closed subset of P^m. Moreover, $[\psi(\vec{a})]^M \approx \lim_{\substack{\longleftarrow \\ i \in I}} [\psi((p_i)^m \vec{b})]^{M_i}$ (as topological spaces).

Proof. a) Since $P = \bigcap\{E_{ij} \subseteq M : i \leqslant j\}$, where $E_{ij} = \{\vec{a} \in \prod_{i \in I} M_i : f_{ij}(a_i) = a_j\} \subseteq M$ is a closed subset of M (because M is Hausdorff), $P \subseteq M$ is closed and thus Boolean space. Let t a n-ary functional symbol; since $t^P : P^n \longrightarrow P$ is the unique function such that $p_i \circ t^P = t^{M_i} \circ (p_i)^n$, for each $i \in I$, t^P must be continuous. If R is a n-ary relational symbol, then R^M is a closed subset of M^n: since $(\prod_{i \in I} M_i)^n \approx \prod_{i \in I} M_i^n$, then $R^M \approx \prod_{i \in I} R^{M_i}$ and $\prod_{i \in I} R^{M_i}$ is a closed subset of $\prod_{i \in I} M_i^n$. Since P is a closed substructure of M, P^n is closed in M^n and hence $R^P = R^M \cap P^n$ is closed in P^n.

b) Fix $i \in I$; since M_i is finite and discrete the diagonal $\Delta_i {\subseteq} M_i \times M_i$ is a M_i-congruence of finite index, and so $C_i \doteq ker(p_i) = (p_i \times p_i)^{-1}[\Delta_i]$ is a P-congruence of discrete finite index, because $\overline{p_i} : P/C_i \rightarrowtail M_i$ is a *continuous injection*. For each $\vec{s} \in P$, let $C_i(\vec{s}) = \{\vec{t} \in P : (\vec{s}, \vec{t}) \in C_i\} = P \cap (\prod_{j \neq i} M_j \times \{s_i\})$. Now note that, since $(I, \leqslant)$ is downward directed, for each $I' {\subseteq}_{fin} I$ there is a $k \in I$ such that $\forall i \in I'(k \leqslant i)$ whence $C_k {\subseteq} \bigcap \{C_i : i \in I'\}$, wherefrom we conclude that $\{C_i(\vec{s}) : i \in I\}$ is a fundamental system of *open* neighborhoods of $\vec{s}$ in P. Since P is Hausdorff, we have $\{\vec{s}\} = \bigcap\{C_i(\vec{s}) : i \in I\}$; hence, if $\vec{t} \neq \vec{s}$, there is $i \in I$ such that $\vec{t} \notin C_i(\vec{s})$ (equivalently, $(\vec{s}, \vec{t}) \notin C_i$), and so $\Delta_P = \bigcap\{C_i : i \in I\}$.

Now let $W {\subseteq} P^2$ be an open neighborhood of the diagonal Δ_P; we claim that there is $i \in I$ such that $C_i {\subseteq} W$. Indeed, otherwise for all $i \in I$, $C_i \setminus W \neq \emptyset$, entailing $\bigcap\{C_i \setminus W : i \in I\} \neq \emptyset$ (an intersection of a downward directed sequence of non-empty closed subsets of the compact space P^2), which in turn yields $\Delta_P \setminus W = \bigcap\{C_i : i \in I\} \setminus W \neq \emptyset$, a contradiction.

c) By a), any L-term $t(\vec{x}, \vec{y})$ gives rise to a continuous functions on P. Let ψ' be a positive-existential formula equivalent to ψ. Again, it follows from item a) (and induction on the complexity of ψ') that $[\psi(\vec{a})]^P$ is closed in P^m. Since the "interpretation subset" of p.e.-formulas are preserved by L-morphism, we obtain a cofiltered system of (closed) subsets $\{[\psi((p_i)^m(\vec{a}))]^{M_i} {\subseteq} (M_i)^m : i \in I\}$ and a continuous function $[\psi(\vec{a})]^P \longrightarrow \varprojlim_{i \in I} [\psi((p_i)^m(\vec{a}))]^{M_i}$ between *boolean spaces* (we have $[\psi(\vec{a})]^P {\subseteq} P^m$ and $\varprojlim_{i \in I} [\psi((p_i)^m(\vec{a}))]^{M_i} {\subseteq} \varprojlim_{i \in I} (M_i)^m \approx P^m$). It is straightforward that the continuous function above is injective; hence, it suffices to prove surjectiveness to establish it to be a homeomorphism. But $\varprojlim_{i \in I} [\psi((p_i)^m(\vec{a}))]^{M_i} \approx \{\vec{b} \in P^m : \prod_{i \in I} M_i \models \psi[\vec{a}, \vec{b}]\}$ and, since $P \hookrightarrow M$ is L-pure (see Theorem 5), we get $\{\vec{b} \in P^m : \prod_{i \in I} M_i \models \psi[\vec{a}, \vec{b}]\} = \{\vec{b} \in P^m : \varprojlim_{i \in I} M_i \models \psi[\vec{a}, \vec{b}]\} = [\psi(\vec{a})]^P$, as needed. ∎

At this point, it is natural to consider the following:

DEFINITION 12. Let L be a first-order language with equality.

a) $*$ **L-mod**^{top} is the category whose objects are the *topological* L-structures (i.e., the L-structures M with a topology τ such that the interpretations of the function symbols are continuous functions) and whose morphisms are the *continuous* L-morphisms.

* **L-mod**^{disc} is the full subcategory of **L-mod**^{top} whose objects are the *discrete* L-structures.

* **L-mod**^{sep} is the full subcategory of **L-mod**^{top} whose objects are the topological L-structures such that the interpretations of the relational symbols are closed subsets (of the appropriate product).

* **L-mod**^{scomp} is the full subcategory of **L-mod**^{sep} whose objects are the *compact* L-structures.

* **L-mod**$_{fin}$ is the full subcategory of **L-mod**^{top} whose objects are the *finite* L-structures endowed with the *discrete topology*.

* **L-mod**$_{pf}$ is the full subcategory of **L-mod**^{top} whose objects are the *profinite* L-structures that are topological structures when considered with the natural *boolean topology*.

If M is a L-structure, note that:

* If $M \in$ **L-mod**^{top}, any L_M-term $t(\overline{h}, v_1, \ldots, v_n)$ yields a continuous function $t_{\overline{h}}^M : M^n \longrightarrow M$;

* If $M \in$ **L-mod**^{sep} then M is a *Hausdorff space* (Δ_M is closed in M^2) and any positive quantifier free L_M-formula, $\varphi(\overline{h}, v_1, \ldots, v_n)$, yields a closed subset of M^n, $[\varphi(\overline{h})]^M = \{ \overline{a} \in M^n : M \models \varphi[\overline{h}, \overline{a}] \}$.

* If $M \in$ **L-mod**^{scomp}, any positive existential L_M-formula , $\varphi(\overline{h}, v_1, \ldots, v_n)$ yields a closed subset of M^n, $[\varphi(\overline{h})]^M = \{ \overline{a} \in M^n : M \models \varphi[\overline{h}, \overline{a}] \}$.

* **L-mod** $\cong$ **L-mod**$^{disc} \hookrightarrow$ **L-mod**^{sep} and **L-mod**$_{fin} \hookrightarrow$ **L-mod**$_{pf} \hookrightarrow$ **L-mod**^{scomp}

b) Let $M \in$ **L-mod**^{top}. A collection S of subsets of $M \times M$ is a **pf-system in M** if :

* The elements of S are congruences in M, of discrete finite index (i.e., the quotient topological structure M/S is discrete and finite);

* S is a fundamental system of entourages of some uniformity compatible with the topology of M.

Notice that if a structure M in **L-mod**^{top} has a pf-system, then M is a completely regular space. Moreover,

* If $\mathcal{V}(M) \doteq \{C \in Cong(M) : M/C \in$ **L-mod**$_{fin}\}$ is a pf-system, then $\mathcal{V}(M)$ is the largest pf-system in M.

* If $M \in$ **L-mod**^{scomp}, then all pf-systems in M induces the *same uniformity* in M. Observe that if $C \in \mathcal{V}(M)$ then $C = (q_C \times q_C)^{-1}[\Delta_{M/C}] \subseteq M \times M$ is a *clopen* subset of $M \times M$.

THEOREM 13. *For $M \in$ **L-mod**^{top} the following are equivalent:*

(1) *M is profinite, i.e., is a projective limit of a downward directed system of discrete and finite L-structures.*

(2) *$M \in$ **L-mod**^{sep}, it has a pf-system and it is a Boolean topological space.*

Proof. (1) $\Rightarrow$ (2) follows from Proposition 11. For (2) $\Rightarrow$ (1), if S is a pf-system in M, then $(S, \subseteq)$ is a downward directed poset. For each $\Sigma \in S$, $M/\Sigma \in \mathbf{L\text{-}mod}_{fin}$, with the quotient structure. Now consider the diagram $(S, \subseteq) \longrightarrow \mathbf{L\text{-}mod}_{fin}$, where $(\Sigma \subseteq \Sigma')$ in S is taken to $q_{\Sigma, \Sigma'} : M/\Sigma \twoheadrightarrow M/\Sigma'$, the unique $\mathbf{L\text{-}mod}_{fin}$-morphism such that $(M \overset{q_{\Sigma'}}{\twoheadrightarrow} M/\Sigma') = (M \overset{q_\Sigma}{\twoheadrightarrow} M/\Sigma \overset{q_{\Sigma, \Sigma'}}{\twoheadrightarrow} M/\Sigma')$. This yields a profinite L-structure $P \doteq \underleftarrow{\lim}_{\Sigma \in S} M/\Sigma$ and a *continuous L-morphism*, $\delta_S : M \longrightarrow \underleftarrow{\lim}_{\Sigma \in S} M/\Sigma$, given by $m \mapsto (m/\Sigma)_{\Sigma \in S}$.

Fix $\Sigma \in S$ and let $\Sigma^* \doteq ker(p_\Sigma) \in Cong(P)$. Let R be a k-ary L-relational symbol; if $(\vec{m}_0, \ldots, \vec{m}_{k-1})$ is in R^P, then, since $\{\Sigma^* \in Cong(P) : \Sigma \in S\}$ is a pf-system in P (Proposition 11), it follows that $\{\prod_{i<k} \Sigma^*(\vec{m}_i) : \Sigma \in S\}$ is a fundamental system of neighborhoods of $(\vec{m}_0, \ldots, \vec{m}_{k-1}) \in P^k$. Since $p_\Sigma : P \twoheadrightarrow M/\Sigma$ is a L-morphism, $(p_\Sigma(\vec{m}_1), \ldots, p_\Sigma(\vec{m}_k)) \in R^{M/\Sigma}$ and so there are $a_0, \ldots, a_{k-1} \in M$ such that $(a_0, \ldots, a_{k-1}) \in R^M$ and $(q_\Sigma(a_0), \ldots, q_\Sigma(a_{k-1})) = (p_\Sigma(\vec{m}_0), \ldots, p_\Sigma(\vec{m}_{k-1}))$. Because $q_\Sigma = p_\Sigma \circ \delta_S$, we see that $(\delta_S(a_0), \ldots, \delta_S(a_{k-1})) \in \prod_{i<k} \Sigma^*(\vec{m}_i)$. This means that $(\delta_S)^k[R^M]$ is a *dense subset* of R^P. In particular, $\delta_S[M]$ is a dense subset of P, because $(\delta_S)^2[\Delta_M]$ is dense in Δ_P. Now we are ready to establish the following

Claim: δ_S is a $\mathbf{L\text{-}mod}^{top}$-isomorphism.

Indeed, since M is Hausdorff and S is a pf-system, if $m \neq m'$ in M, there is $\Sigma \in S$ such that $\Sigma(m) \cap \Sigma(m') = \emptyset$ and so $\Delta_M = \bigcap S$; clearly, $ker(\delta_S) = \bigcap S$, and so δ_S is *injective*. For $k \in \mathbb{N}$, since $(\delta_S)^k : M^k \longrightarrow P^k$ is a continuous injection from a compact space into a Hausdorff space, it is a *homeomorphism* onto its image. If R is a k-ary relation, because R^M and R^P are closed in M^k and P^k, respectively, and $(\delta_S)^k[R^M]$ is dense in R^P, it follows that $(\delta_S)^k[R^M] = R^P$. In particular, δ_S is *surjective*, because $(\delta_S)^2[\Delta_M] = \Delta_P$. It remains to check δ_S is a L-embedding: let R a k-ary relation and let $(m_0, \ldots, m_{k-1}) \in M^k$ be such that $(\delta_S)^k(m_0, \ldots, m_{k-1}) \in R^P$; since $(\delta_S)^k[R^M] = R^P$, there is $(m'_0, \ldots, m'_{k-1})$ in $R^M \subseteq M^k$ such that $(\delta_S)^k(m_0, \ldots, m_{k-1}) = (\delta_S)^k(m'_0, \ldots, m'_{k-1})$ and, because $(\delta_S)^k$ is injective, we obtain $(m_0, \ldots, m_{k-1}) \in R^M$, as needed. $\blacksquare$

The proof of Theorem 13 shows that condition (2) may be rewritten as:

$(2')$: M is in $\mathbf{L\text{-}mod}^{scomp}$ and has a pf-system.

Theorem 13 yields closure properties of $\mathbf{L\text{-}mod}_{pf}$ and a characterization of its quotient objects.

COROLLARY 14. *The subcategory* $\mathbf{L\text{-}mod}_{pf} \subseteq \mathbf{L\text{-}mod}^{top}$ *is closed under:*
a) Closed substructures, *i.e. if M is profinite and $M' \subseteq M$ is a substructure of M that is also a closed subset, then M', endowed with the topology induced*

by M, is profinite.
b) Isomorphism, products and general projective limits.

Proof. a) Clearly, M' is a Boolean (sub)space, with $M' \in$ **L-mod**^{sep}; and if S is a pf-system in M, then $S' \doteq \{C \cap (M' \times M') : C \in S\}$ is a pf-system in M'.

b) If $\{M_i : i \in I\}$ is a family in **L-mod**$_{pf}$, clearly $M \doteq \prod_{i\in I} M_i$ is a Boolean space. If R is a k-ary relational symbol then, since $R^M \approx \prod_{i\in I} R^{M_i} \subseteq \prod_{i\in I}(M_i)^k \approx M^k$, R^M is closed in M^k, and so $M \in$ **L-mod**^{sep}. If S_i is a pf-system in M_i, $i \in I$, then

$$S \doteq \{(\prod_{i\in I'} C_i) \times (\prod_{i\in I\setminus I'} M_i \times M_i) \subseteq (M \times M) : \text{for some } I' \subseteq_{fin} I$$

and some $C_i \in S_i, i \in I'\}$
is a pf-system in M. Now if $\mathcal{M} : \mathcal{D} \longrightarrow$ **L-mod**$_{pf}$ is any diagram based on an arbitrary small category $\mathcal{D}$, the same methods employed in the proof of Proposition 11.(a) will establish $\varprojlim_{i\in Obj(\mathcal{D})} M_i \hookrightarrow \prod_{i\in Obj(\mathcal{D})} M_i$ is a closed substructure, as needed. ∎

COROLLARY 15. *For $M \in$ **L-mod**$_{pf}$ and $\Sigma \in Cong(M)$, the following are equivalent:*
    (1) $M/\Sigma \in$ **L-mod**$_{pf}$ *with the quotient* **L-mod**^{top}*-structure;*
    (2) *There is $X \subseteq \mathcal{V}(M)$ such that $\Sigma = \bigcap X$.*

Proof. (1) $\Rightarrow$ (2) : Let S' a pf-system in M/Σ and write $q_\Sigma : M \twoheadrightarrow M/\Sigma$ for the canonical **L-mod**^{top}-morphism. If $C' \in$ S then, since $\overline{q_\Sigma} : M/(q_\Sigma)^\star(C') \longrightarrow (M/\Sigma)/C'$ is a continuous bijection (it is an **L-mod**^{top}-isomorphism), we get $(q_\Sigma)^\star(C') \in \mathcal{V}(M)$. Since M/Σ is Hausdorff, $\Delta_{M/\Sigma} = \bigcap S'$, and so $\Sigma = (q_\Sigma)^\star(\Delta_{M/\Sigma}) = \bigcap \{(q_\Sigma)^\star(C') : C' \in S'\}$.
(2) $\Rightarrow$ (1) : We first show that $M/\Sigma \in$ **L-mod**^{scomp}: clearly, it is a compact space and since $X \subseteq \mathcal{V}(M)$, Σ is an intersection of a family of *clopen* subsets of $M \times M$. Thus Σ is closed in $M \times M$ and then M/Σ is Hausdorff. It is straightforward that M/Σ is a topological L-structure, so to conclude that $M/\Sigma \in$ **L-mod**^{sep} it suffices to prove that all L-relations in M/Σ are closed. If R is a k-ary relation, then R^M is closed in M^k and, M^k being compact and $(M/\Sigma)^k$ being Hausdorff, it follows that $(q_\Sigma)^k$ is a *closed function*, and so $R^{M/\Sigma} = (q_\Sigma)^k[R_M]$ is closed in $(M/\Sigma)^k$. It remains to check that M/Σ has a pf-system. Let $W \subseteq (M/\Sigma \times M/\Sigma)$ be an open neighborhood of the diagonal ($\Delta_{M/\Sigma} \subseteq W$) and take $V = (q_\Sigma \times q_\Sigma)^{-1}[W]$; then $V \subseteq (M \times M)$ is open and $\Sigma = (q_\Sigma \times q_\Sigma)^{-1}[\Delta_{M/\Sigma}] \subseteq V$. If $K = (M \times M) \setminus V$, then K is closed in M^2.
Claim: *There is $X' \subseteq_{fin} X$ such that $K \cap \bigcap X' = \emptyset$.*
Note that the conclusion is equivalent to $\Delta_M \subseteq \bigcap X' \subseteq V$. Indeed, if

for all $X'\subseteq_{fin}X$ $K \cap \bigcap X' \neq \emptyset$, then by the compactness of $M \times M$, $K\cap\Sigma = K\cap\bigcap X \neq \emptyset$, which is impossible because $K = M^2 \setminus V \subseteq M^2 \setminus \Sigma$.

Hence, for each open W of $M/\Sigma \times M/\Sigma$, with $\Delta_{M/\Sigma}\subseteq W$, there is $X'\subseteq_{fin}X$ such that $\bigcap X'\subseteq(q_\Sigma \times q_\Sigma)^{-1}[W]$; now, Proposition 10 entails $\{(q_\Sigma \times q_\Sigma)[\bigcap X'] \ : \ X' \subseteq_{fin} X\}$ is a pf-system in M/Σ.   ∎

3   The Profinite Hull Functor

DEFINITION 16. Let $\mathcal{A}$ be a *full* subcategory of **L-mod** and let
$$\mathfrak{C} = \{\mathfrak{C}(M) : \mathfrak{C}(M) \subseteq Cong(M), M \in \mathcal{A}\}$$
be a collection of sets of congruences, parametrized by the L-structures in $\mathcal{A}$. We say that $\mathfrak{C}$ is **saturated** if it satisfies the following conditions:

[sat 1] : For all $M \in \mathcal{A}$, each element of $\mathfrak{C}(M)$ is a $\mathcal{A}$-congruence, that is, if $\Sigma \in \mathfrak{C}(M)$, then the quotient structure M/Σ is in $\mathcal{A}$; moreover, Δ_M (the identity congruence) and $M \times M$ are in $\mathfrak{C}(M)$;

[sat 2] : For all $M \in \mathcal{A}$, $\mathfrak{C}(M)$ is closed under finite intersections;

[sat 3] : $\mathfrak{C}$ is stable under inverse images, i.e., if $f : M\longrightarrow M'$ is a L-morphism in $\mathcal{A}$ and $\Sigma' \in \mathfrak{C}(M')$, then $f^\star(\Sigma') \doteq (f \times f)^{-1}[\Sigma']$ is in $\mathfrak{C}(M)$.

1. Remarks and Examples. a) With notation as in Definition 16, a saturated family of congruences, $\mathfrak{C}$, induces a contravariant functor from the category $\mathcal{A}$ into the category Ω of downward directed posets and increasing functions, as follows:
$$(M \xrightarrow{f} M') \mapsto (\mathfrak{C}(M') \xrightarrow{f^\star} \mathfrak{C}(M)),$$
i.e., $(id_{M'})^\star = id_{\mathfrak{C}(M')}$ and if $f' : M'\longrightarrow M''$ is an $\mathcal{A}$-morphism then $(f' \circ f)^\star = f^\star \circ f'^\star$. Moreover, each $\Sigma' \in \mathfrak{C}(M')$ yields a *derived $\mathcal{A}$-morphism*, $f_{\Sigma'} : M/f^\star(\Sigma') \rightarrowtail M'/\Sigma'$, the *unique* $\mathcal{A}$-morphism such that $f_{\Sigma'} \circ q_{f^\star(\Sigma')} = q'_{\Sigma'} \circ f$.

b) If L does not contains relational symbols and $\mathcal{A}$ is a variety of algebras, then $\mathcal{A}$ is equationally axiomatizable (Birkhoff's Theorem) and the full family $\{Cong(M) : M \in \mathcal{A}\}$ is saturated. In fact, it is well-known that in this case, for each $M \in \mathcal{A}$, $Cong(M)$ is closed under arbitrary intersections and directed unions, constituting an algebraic lattice under inclusion.

c) If $\mathcal{A}$ is the category of reduced special groups (RSG), then $\mathcal{A}$ is axiomatizable by geometrical sentences and it follows from results in Chapter 2 of [7] that the class of *saturated subgroups* of each RSG yields a saturated family of congruences in $\mathcal{A}$, *Sat*. As in the case of Example (b), for each RSG G, $Sat(G)$ is closed under arbitrary intersections and directed unions, also constituting an algebraic lattice under inclusion: its compact elements are the saturated subgroups that are the set of represented elements of a Pfister form over G.

d) If a saturated family of congruences in $\mathcal{A}$, $\mathfrak{C}$, is closed under arbitrary intersections and directed unions and is an algebraic lattice under inclusion (as is the case of Examples (b) and (c), above), then:

* An $\mathcal{A}$-morphism $f : M \longrightarrow M'$ yields an increasing function
$$f_\star : (\mathfrak{C}(M), \subseteq) \longrightarrow (\mathfrak{C}(M'), \subseteq) :$$
$\Sigma \in \mathfrak{C}(M) \mapsto \bigcap\{\Gamma' \in \mathfrak{C}(M') : (f \times f)[\Sigma] \subseteq \Gamma'\}$.

Moreover, we have the following *adjunction* :

for each $\Sigma \in \mathfrak{C}(M)$ and each $\Gamma' \in \mathfrak{C}(M')$, $f_\star(\Sigma) \subseteq \Gamma' \Leftrightarrow \Sigma \subseteq f^\star(\Gamma')$.

* The map $(M \xrightarrow{f} M') \mapsto (\mathfrak{C}(M) \xrightarrow{f_\star} \mathfrak{C}(M'))$ yields a *covariant* functor from $\mathcal{A}$ to the category Ω of downward directed posets and increasing functions (i.e. $(id_{M'})_\star = id_{\mathfrak{C}(M')}$ and if $f' : M' \longrightarrow M''$ is an $\mathcal{A}$-morphism then $(f' \circ f)_\star = f'_\star \circ f_\star$). □

Henceforth, assume we have a pair $(\mathcal{A}, \mathfrak{C})$ *where:*

(i) $\mathcal{A} = \mathrm{Mod}(T)$, *where T is a theory axiomatized by geometrical L-sentences* (*cf. Remark 1*). *In particular $\mathcal{A}$ is closed in* **L-mod** *under profinite limits* (*Corollary 6*);

(ii) $\mathfrak{C}$ *is a saturated family of $\mathcal{A}$-congruences, as in Definition 16.*

Write $\mathcal{A}^{top} \subseteq$ **L-mod**^{top} for the full subcategory of topological L-structures in $\mathcal{A}$ and continuous L-morphisms. Analogously, we define the subcategories $\mathcal{A}^{sep} \subseteq$ **L-mod**^{sep}, $\mathcal{A}^{scomp} \subseteq$ **L-mod**^{scomp}, $\mathcal{A}^{disc} \subseteq$ **L-mod**^{disc}, $\mathcal{A}_{fin} \subseteq$ **L-mod**$_{fin}$ and $\mathcal{A}_{pf} \subseteq$ **L-mod**$_{pf}$.

2. The Profinite Hull. For $M \in \mathcal{A}^{top}$, let
$$\mathcal{V}(M) \doteq \{C \in \mathfrak{C}(M) : \text{the quotient } \mathcal{A}^{top}\text{-object } M/C \text{ is in } \textbf{L-mod}_{fin}\}.$$
If $\Sigma, \Sigma_1, \Sigma_2 \in \mathcal{V}(M)$ and $\Sigma' \in \mathfrak{C}(M)$ is such that $\Sigma \subseteq \Sigma'$, then we have the canonical $\mathcal{A}^{top}$-arrows

$$\begin{cases} M/\Sigma \twoheadrightarrow M/\Sigma' & \text{given by} \quad m/\Sigma \mapsto m/\Sigma'; \\ M/(\Sigma_1 \cap \Sigma_2) \rightarrowtail M/\Sigma_1 \times M/\Sigma_2 & \text{given by} \quad m/\Sigma_1 \cap \Sigma_2 \mapsto (m/\Sigma_1, m/\Sigma_2) \end{cases}$$

and so $\mathcal{V}(M) \subseteq \mathfrak{C}(M)$ is a filter in $(\mathfrak{C}(M), \subseteq)$. In particular, $(\mathcal{V}(M), \subseteq)$ is a *downward directed poset*. We then obtain the **canonical diagram of** M, $D(M) : (\mathcal{V}(M), \subseteq) \longrightarrow \mathcal{A}_{fin}$, where $(\Sigma \subseteq \Sigma')$ in $\mathcal{V}(M)$ is taken to $q_{\Sigma,\Sigma'} : M/\Sigma \twoheadrightarrow M/\Sigma'$, the unique $\mathcal{A}_{fin}$-morphism such that
$$(M \xrightarrow{q_{\Sigma'}} M/\Sigma') = (M \xrightarrow{q_\Sigma} M/\Sigma \xrightarrow{q_{\Sigma,\Sigma'}} M/\Sigma').$$

The limit of this diagram yields the $\mathcal{A}_{pf}$-object **profinite hull of** M, $\mathcal{P}(M) \doteq \varprojlim_{\Sigma \in \mathcal{V}(M)} M/\Sigma$, together with a canonical $\mathcal{A}^{top}$-morphism, $\eta_M :$ $M \longrightarrow \mathcal{P}(M)$, given by $m \mapsto (m/\Sigma)_{\Sigma \in \mathcal{V}(M)}$. In more detail, given $\Sigma \in \mathcal{V}(M)$, we have the "projections on quotients", $q_\Sigma : M \twoheadrightarrow M/\Sigma$, $m \mapsto m/\Sigma$ and the "projections of the limit", $p_\Sigma : \mathcal{P}(M) \twoheadrightarrow M/\Sigma$, $(m_C/C)_{C \in \mathcal{V}(M)} \mapsto$

m_Σ/Σ, yielding a commutative cone over the diagram $D(M)$, $(q_\Sigma : M \twoheadrightarrow M/\Sigma)_{\Sigma \in \mathcal{V}(M)}$; then, η_M is the unique arrow such that $p_\Sigma \circ \eta_M = q_\Sigma$, for each $\Sigma \in \mathcal{V}(M)$. $\qquad\qquad\square$

REMARK 17. The same argument used in the paragraph preceding the statement of the Claim in the proof of Theorem 13 shows that if R is a k-ary relation in L, $(\eta_M)^k[R^M]$ is dense in $R^{\mathcal{P}(M)}$, $k \in \mathbb{N}$. In particular, $\eta_M[M]$ is dense in $\mathcal{P}(M)$, because $(\eta_M)^2[\Delta_M]$ is dense in $\Delta_{\mathcal{P}(M)}$. $\qquad\square$

To justify the adjective *canonical* employed above, we shall now show that the associations $M \mapsto D(M)$ and $M \mapsto \mathcal{P}(M)$ are functorial, and that the family $\eta = \{\eta_M : M \in \mathcal{A}^{top}\}$ is a natural transformation from the functor $id_{\mathcal{A}^{top}}$ to the functor $\iota \circ \mathcal{P}$, where $\iota : \mathcal{A}_{pf} \hookrightarrow \mathcal{A}^{top}$ is the inclusion functor. As a preliminary to this discussion, we recall the definition of morphism of diagrams over distinct bases, which should be compared with Definition 29.3, p. 349 of [16]. In Part 6 of the latter reference the reader will find a discussion of change of base in a general setting.

DEFINITION 18. Let $I = \langle I, \leq \rangle$ and $L = \langle L, \leq \rangle$ be downward directed posets and let $\mathcal{D}$ be a category. Let $\mathcal{G} = \langle G_i, g_{ji} \rangle$ and $\mathcal{H} = \langle H_l, h_{ml} \rangle$ be diagrams in $\mathcal{D}$ over I and L, respectively. A **morphism**, $\alpha : \mathcal{G} \longrightarrow \mathcal{H}$, consists of a pair, $\alpha = \langle \mathfrak{a}, \mathfrak{u} \rangle$, where $\mathfrak{a} : L \longrightarrow I$ is an increasing map and $\mathfrak{u} = \{\mathfrak{u}(l) : l \in L\}$ is a set of $\mathcal{D}$-morphisms, $\mathfrak{u}(l) : G_{\mathfrak{a}(l)} \longrightarrow H_l$, such that for all $l \leq k$ in L, $h_{lk} \circ \mathfrak{u}(l) = \mathfrak{u}(k) \circ g_{\mathfrak{a}(l),\mathfrak{a}(k)}$, i.e., the diagram below right is commutative:

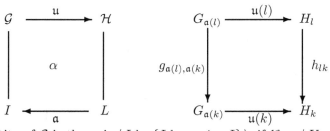

The identity of $\mathcal{G}$ is the pair $\langle Id_I, \{Id_{G_i} : i \in I\} \rangle$; if $\mathcal{K} = \langle K_p, k_{qp} \rangle$ is a diagram in $\mathcal{D}$ over the downward directed poset $\langle P, \leq \rangle$ and $\beta = \langle \mathfrak{b}, \mathfrak{v} \rangle : \mathcal{H} \longrightarrow \mathcal{K}$ is a morphism, then $\beta \circ \alpha \doteq \langle \mathfrak{a} \circ \mathfrak{b}, \mathfrak{v} \odot \mathfrak{u} \rangle$, where for each $p \in P$, $(\mathfrak{v} \odot \mathfrak{u})(p) = \mathfrak{v}(\mathfrak{b}(p)) \circ \mathfrak{u}(\mathfrak{a}(\mathfrak{b}(p))) : G_{\mathfrak{a}(\mathfrak{b}(p))} \longrightarrow K_p$. It is straightforward that the usual rules for composition are satisfied and we obtain the **category of all downward directed $\mathcal{D}$-diagrams**, $Diag_\Omega(\mathcal{D})$.

3. The "canonical diagram" Functor. a) Let $f : M \longrightarrow M'$ be a $\mathcal{A}^{top}$-morphism. By hypothesis, the (increasing) function $f^\star : \mathfrak{C}(M') \longrightarrow \mathfrak{C}(M) :$ $\Sigma' \mapsto (f \times f)^{-1}[\Sigma']$ is well defined. For $\Sigma' \in \mathcal{V}(M') \subseteq \mathfrak{C}(M')$, consider the *derived $\mathcal{A}^{top}$-morphism*, $f_{\Sigma'} : M/f^\star(\Sigma') \rightarrowtail M/\Sigma'$ given by $m/f^\star(\Sigma') \mapsto$

$f(m)/\Sigma'$, the unique $\mathcal{A}^{top}$-morphism such that $f_{\Sigma'} \circ q_{f^\star(\Sigma')} = q'_{\Sigma'} \circ f$. Then, $f^\star(\Sigma') \in \mathfrak{C}(M)$ and, since $f_{\Sigma'}$ is an *injective* continuous function into a finite discrete space, we must have $f^\star(\Sigma') \in \mathcal{V}(M) = \{C \in \mathfrak{C}(M) : M/C$ is finite and discrete$\}$. Hence, we get a map,

$$(M \xrightarrow{f} M') \; \mapsto \; ((\mathcal{V}(M'), \subseteq) \xrightarrow{f^\star} (\mathcal{V}(M), \subseteq)),$$

yielding a contravariant functor from $\mathcal{A}^{top}$ to the category Ω of downward direct posets and increasing functions: clearly, $(id_{M'})^\star = id_{\mathcal{V}(M')}$; and if $f' : M' \longrightarrow M''$, then $(f' \circ f)^\star = f^\star \circ f'^\star$.

b) Let $f : M \longrightarrow M'$ be a $\mathcal{A}^{top}$-morphism. We have the canonical diagrams $D(M) : (\mathcal{V}(M), \subseteq) \longrightarrow \mathcal{A}_{fin}$, $D(M') : (\mathcal{V}(M'), \subseteq) \longrightarrow \mathcal{A}_{fin}$ and an increasing function $f^\star : (\mathcal{V}(M'), \subseteq) \longrightarrow (\mathcal{V}(M), \subseteq)$. There is a natural way to relate the "parallel" diagrams $D(M) \circ f^\star, D(M') : (\mathcal{V}(M'), \subseteq) \longrightarrow \mathcal{A}_{fin}$: for each $\Sigma' \in \mathcal{V}(M')$ we have $D(M) \circ f^\star(\Sigma') = M/f^\star(\Sigma')$ and the derived morphism $f_{\Sigma'} : M/f^\star(\Sigma') \rightarrowtail M'/\Sigma'$. Therefore, the family $\Phi(f) \doteq (f_{\Sigma'})_{\Sigma' \in \mathcal{V}(M')}$ is a natural transformation from the diagram $D(M) \circ f^\star$ to the diagram $D(M')$: indeed, if $(\Gamma' \subseteq \Sigma') \in \mathcal{V}(M')$, then clearly $(f^\star(\Gamma') \subseteq f^\star(\Sigma')) \in \mathcal{V}(M)$ and $f_{\Sigma'} \circ q_{f^\star(\Gamma')f^\star(\Sigma')} = q'_{\Gamma'\Sigma'} \circ f_{\Gamma'}$.

In fact, we get a covariant functor Υ, from $\mathcal{A}^{top}$ to the category $Diag_\Omega(\mathcal{A}_{fin})$ (as in Definition 18), given by

$$(M \xrightarrow{f} M') \; \xmapsto{\Upsilon} \; (D(M) \xrightarrow{\langle f^\star, \Phi(f) \rangle} D(M')).$$

For functoriality, note that $(id_M^\star, \Phi(id_M)) = (id_{\mathcal{V}(M)}, id_{D(M)})$ and if $f' : M' \longrightarrow M''$, then $(f' \circ f)^\star = f^\star \circ f'^\star$ and $\Phi(f' \circ f) = \Phi(f') \odot \Phi(f)$ holds because for each $\Sigma'' \in \mathcal{V}(M'')$ the diagram below is commutative:

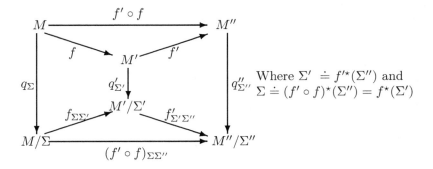

The above construction is schematically described as follows:

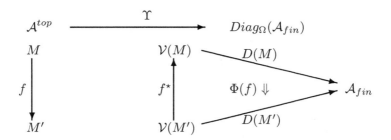

4. The Profinite Hull Functor. We saw above that there is a "canonical diagram" functor, $\Upsilon : \mathcal{A}^{top} \longrightarrow Diag_\Omega(\mathcal{A}_{fin})$, given by $(M \xrightarrow{f} M') \overset{\Upsilon}{\mapsto} (D(M) \overset{(f^\star, \Phi(f))}{\longrightarrow} D(M'))$.

Furthermore, each $M \in \mathcal{A}^{top}$ has a "profinite hull", $\mathcal{P}(M) \doteq \underleftarrow{lim}\ D(M) \in \mathcal{A}_{pf}$. To show that object-map $M \mapsto \mathcal{P}(M)$ extends to a functor, $\mathcal{P} : \mathcal{A}^{top} \longrightarrow \mathcal{A}_{pf}$, it suffices to prove there is a well-defined *functor limit*, $\underleftarrow{lim}: Diag_\Omega(\mathcal{A}_{fin}) \longrightarrow \mathcal{A}_{pf}$. The **profinite hull functor** $\mathcal{P}$ will then be the composition of the functors "limit" and "canonical diagram". With notation as in Definition 18, the existence of the "functor limit", $\underleftarrow{lim}: Diag_\Omega(\mathcal{A}_{fin}) \longrightarrow \mathcal{A}_{pf}$, is guaranteed by the following general

FACT 19. Let $\langle I, \leq \rangle$, $\langle L, \leq \rangle$ and $\langle P, \leq \rangle$ be downward directed posets. Let $\mathcal{G} = \langle G_i; g_{ji} \rangle$, $\mathcal{H} = \langle H_l; h_{ml} \rangle$ and $\mathcal{K} = \langle K_p; k_{pq} \rangle$ be diagrams in $\mathcal{A}_{fin}$ over I, L and P, respectively. Let $\alpha = \langle \mathfrak{a}, \mathfrak{u} \rangle : \mathcal{G} \longrightarrow \mathcal{H}$ and $\beta : \mathcal{H} \longrightarrow \mathcal{K}$ be change of base morphisms. Let $\langle \widehat{G}, g_i \rangle = \underleftarrow{lim}\ \mathcal{G}$, $\langle \widehat{H}, h_l \rangle = \underleftarrow{lim}\ \mathcal{H}$ and $\langle \widehat{K}, k_p \rangle = \underleftarrow{lim}\ \mathcal{K}$ be the corresponding projective limits. Then:

(1) The family $(\widehat{G} \xrightarrow{g_{\mathfrak{a}(l)}} G_{\mathfrak{a}(l)} \xrightarrow{\mathfrak{u}(l)} H_l)_{l \in L}$ is a *commutative cone* over the diagram $\mathcal{H}$. Hence, there is an *unique* $\mathcal{A}_{pf}$-morphism, $\widehat{\alpha} : \widehat{G} \longrightarrow \widehat{H}$, such that for each $l \in L$, $h_l \circ \widehat{\alpha} = \mathfrak{u}(l) \circ g_{\mathfrak{a}(l)}$.

(2) $\widehat{(\beta \circ \alpha)} = \widehat{\beta} \circ \widehat{\alpha}$ and if $\alpha = id_{\mathcal{G}}$ then $\widehat{\alpha} = id_{\widehat{G}}$. $\square$

Hence, the **Profinite Hull Functor**, $\mathcal{P} : \mathcal{A}^{top} \longrightarrow \mathcal{A}_{pf}$, is given by:

<u>Objects :</u> $M \in \mathcal{A}^{top} \mapsto \mathcal{P}(M) \doteq (\underleftarrow{lim}_{\Sigma \in \mathcal{V}(M)}\ M/\Sigma) \in \mathcal{A}_{pf}$ and $(\mathcal{P}(M) \xrightarrow{p_\Sigma} M/\Sigma)_{\Sigma \in \mathcal{V}(M)}$ is the limit cone.

<u>Morphisms :</u> $(M \xrightarrow{f} M') \in \mathcal{A}^{top} \mapsto (\mathcal{P}(M) \xrightarrow{\mathcal{P}(f)} \mathcal{P}(M')) \in \mathcal{A}_{pf}$, where $\mathcal{P}(f) : \mathcal{P}(M) \longrightarrow \mathcal{P}(M')$ is the unique $\mathcal{A}_{pf}$-morphism such that for each $\Sigma' \in \mathcal{V}(M')$, $p'_{\Sigma'} \circ \mathcal{P}(f) = f_{\Sigma'} \circ p_{f^\star(\Sigma')}$. $\square$

5. The natural transformation η. The family $(M \xrightarrow{\eta_M} \mathcal{P}(M))_{M \in \mathcal{A}^{top}}$

is a natural transformation from the functor $id_{\mathcal{A}^{top}}$ to the functor $\iota \circ \mathcal{P}$, where $\iota : \mathcal{A}_{pf} \hookrightarrow \mathcal{A}^{top}$ is the inclusion functor. It suffices to check that if $f : M \longrightarrow M'$ is a morphism in $\mathcal{A}^{top}$, then $\mathcal{P}(f) \circ \eta_M = \eta_{M'} \circ f$. Equivalently, by the universal property of $\mathcal{P}(M') = \varprojlim_{\Sigma' \in \mathcal{V}(M')} M'/\Sigma'$, it must be verified that for each $\Sigma' \in \mathcal{V}(M')$, $p'_{\Sigma'} \circ \mathcal{P}(f) \circ \eta_M = p'_{\Sigma'} \circ \eta_{M'} \circ f$. But this follows directly from the definitions and a straightforward diagram chase:

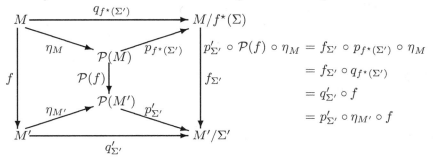

$$p'_{\Sigma'} \circ \mathcal{P}(f) \circ \eta_M = f_{\Sigma'} \circ p_{f^*(\Sigma')} \circ \eta_M$$
$$= f_{\Sigma'} \circ q_{f^*(\Sigma')}$$
$$= q'_{\Sigma'} \circ f$$
$$= p'_{\Sigma'} \circ \eta_{M'} \circ f$$

In [14] it is shown that:

∗ For each Boolean algebra B, the BA-morphism $\eta_B : B \longrightarrow \mathcal{P}(B)$ may be identified with the natural (injective) BA-homomorphism $B \rightarrowtail Parts(Stone(B)) : b \mapsto \{U \in Stone(B) : b \in U\}$;

∗ For each reduced special group G, the RSG-morphism $\eta_G : G \longrightarrow \mathcal{P}(G)$ reflects subforms; in particular, it is a complete embedding and reflects isotropy of forms over G.

We now show that the functor $\mathcal{P} : \mathcal{A}^{top} \longrightarrow \mathcal{A}_{pf}$ is a profinite hull: for each M in $\mathcal{A}^{top}$, every morphism from M to an object in $\mathcal{A}_{pf}$ factors uniquely through the natural arrow $\eta_M : M \longrightarrow \mathcal{P}(M)$; and then prove that $\mathcal{P}$ preserves inductive limits and quotients.

6. A universal property. Let $f : M \longrightarrow P$ be a $\mathcal{A}^{top}$-morphism, where $M \in \mathcal{A}^{top}$ and $P \in \mathcal{A}_{pf}$. Since $\mathcal{V}(P)$ is a pf-system in P, $\eta_P : P \longrightarrow \mathcal{P}(P)$ is an $\mathcal{A}_{pf}$-isomorphism (see the proof of Theorem 13). Hence, we may obtain an extension $\widetilde{f} \in \mathcal{A}_{pf}(\mathcal{P}(M), P)$ of f to $\mathcal{P}(M)$ (i.e. $f = \widetilde{f} \circ \eta_M$), given by $\widetilde{f} \doteq (\eta_P)^{-1} \circ \mathcal{P}(f)$. Moreover, this continuous extension is unique, since $\eta_M[M]$ is dense in $\mathcal{P}(M)$ (cf. Remark 17) and P is a Hausdorff space. Therefore, each $\mathcal{A}^{top}$-morphism, $f : M \longrightarrow P$, with $M \in \mathcal{A}^{top}$ and $P \in \mathcal{A}_{pf}$, has a unique extension, $\widetilde{f} \in \mathcal{A}_{pf}(\mathcal{P}(M), P)$, such that $f = \widetilde{f} \circ \eta_M$. In particular, the functor $\mathcal{P} : \mathcal{A}^{top} \longrightarrow \mathcal{A}_{pf}$ is left adjoint to the inclusion functor $\iota : \mathcal{A}_{pf} \hookrightarrow \mathcal{A}^{top}$ and the natural transformation η is the unit of this adjunction. □

The Theorem that follows uses a topological-analytical technique to obtain a generalization of the above result, also providing a characterization of the universal arrow η_M.

THEOREM 20. *Let* $M \in \mathcal{A}^{top}$.

a) Let $K \in \mathcal{A}^{sep}$ *that has a pf-system with respect to which it is a complete Hausdorff uniform space. Then for each* $f \in \mathcal{A}^{top}(M, K)$ *such that* $ker(f) \supseteq ker(\eta_M) = \bigcap\{ker(p_\Sigma) : \Sigma \in \mathcal{V}(M)\}$, *there is an unique* $\mathcal{A}^{top}$-*morphism,* $\tilde{f}$: $\mathcal{P}(M) \longrightarrow K$, *such that* $\tilde{f} \circ \eta_M = f$. *In particular, these conditions are satisfied if* $K \in \mathcal{A}_{pf}$ *and thus each* $f \in \mathcal{A}^{top}(M, K)$ *has an unique extension* $\tilde{f}$ *to* $\mathcal{P}(M)$, *satisfying* $\tilde{f} = \eta_K^{-1} \circ \mathcal{P}(f)$.

b) Let $T \in \mathcal{A}_{pf}$, *let* $j : M \longrightarrow T$ *be an* $\mathcal{A}^{top}$-*morphism and let* $M^+ \doteq j[M] \subseteq T$ *be the image structure. Assume that:*

$*$ $ker(j) \subseteq ker(\eta_M)$ *(hence,* η_M *factors uniquely through* M^+, *i.e.* $M^+ \cong M/ker(j)$ *and* $\eta_M = (\eta_M)^+ \circ j$);

$*$ *The L-morphism* $(\eta_M)^+ : M^+ \longrightarrow \mathcal{P}(M)$ *is uniformly continuous in the uniformity induced by* T *on* M^+ *(i.e., for each* $\Sigma \in \mathcal{V}(M)$, *there is* $C \in \mathcal{V}(T)$ *such that* $(\eta_M^+ \times \eta_M^+)[C \cap (M^+ \times M^+)] \subseteq ker(p_\Sigma))$.

$*$ *The image of each L-relation in* M *(including equality) is dense in the corresponding L-relation in* T *in the product topology (i.e., if* R *is a n-ary relation,* $C \in \mathcal{V}(T)$ *and* $(x_0, \ldots, x_{n-1}) \in R^T \subseteq T^n$, *there are* $a_0, \ldots, a_{n-1}$ *in* M *such that* $(a_0, \ldots, a_{n-1}) \in R^M$ *and* $(j(a_i), x_i) \in C$).

Then, with the notation as in (a), $\tilde{j} : \mathcal{P}(M) \longrightarrow T$ *is the unique* $\mathcal{A}_{pf}$-*isomorphism such that* $\tilde{j} \circ \eta_M = j$. *In particular,* η_M *satisfies above conditions.*

Proof. a) As $\eta_M[M]$ is a dense subset of $\mathcal{P}(M)$ and K is an Hausdorff space there is *at most one* continuous "extension" of f to $\mathcal{P}(M)$. We will construct an $\mathcal{A}^{top}$-extension f to $\mathcal{P}(M)$. Let $M' = \eta_M[M]$ be the image L-structure by η_M (so $M' \cong M/ker(\eta_M)$ in **L-mod**), endowed with the uniformity induced by $\mathcal{P}(M)$. By Proposition 10, there is an unique L-morphism, $f' : M' \longrightarrow K$, such that $f' \circ (\eta_M)_\upharpoonright = f$. We will show that f' is uniformly continuous for some fixed pf-system S in K as in the hypothesis and let $S' \doteq \{ker(p_\Sigma) \cap (M' \times M') : \Sigma \in \mathcal{V}(M)\}$ be a fundamental system of entourages of M'. Given $C \in S$, set $\Sigma \doteq (f \times f)^{-1}[C]$ and $C' \doteq ker(p_\Sigma) \cap (M' \times M')$; then $\Sigma \in \mathcal{V}(M)$ and $C' = \{(a', b') \in M' \times M' : \exists x, y \in M\ (a', b') = (\eta_M(x), \eta_M(y))$ and $p_\Sigma(a') = p_\Sigma(b')\}$, hence, $(f' \times f')[C'] = \{(f(x), f(y)) : x, y \in M\ ,\ q_\Sigma(x) = q_\Sigma(y)\} = (f \times f)[\Sigma] \subseteq C$, showing that f' is indeed uniformly continuous. Since M' is dense in $\mathcal{P}(M)$ and both $\mathcal{P}(M)$ and K are complete uniform spaces, there is an *unique* uniformly continuous map, $\tilde{f} : \mathcal{P}(M) \longrightarrow K$ such that $\tilde{f} \circ \iota = f'$, where $\iota : M' \hookrightarrow \mathcal{P}(M)$. Hence, $\tilde{f} : \mathcal{P}(M) \longrightarrow K$ is a *continuous function*, satisfying $\widetilde{f} \circ \eta_M = \widetilde{f} \circ \iota \circ (\eta_M)_{\restriction} = f' \circ (\eta_M)_{\restriction} = f$, and $\widetilde{f}$ is the *unique* continuous extension of f along η_M, as needed.

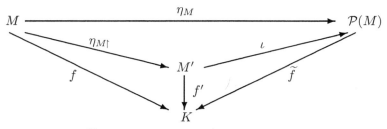

It remains to check that $\widetilde{f}$ is a L-morphism. Note that M' is a L-subalgebra of $\mathcal{P}(M)$, i.e. it contains the interpretations in $\mathcal{P}(M)$ of the constants in L, and is closed under the interpretations of all L-operations in $\mathcal{P}(M)$. If c is a constant in L, then since f' is a L-morphism, we get $c^K = f'(c^{M'}) = \widetilde{f}(c^{\mathcal{P}(M)})$, as needed. If t is a n-ary operation in L, let $h_1, h_2 : (\mathcal{P}(M))^n \longrightarrow K$ be the continuous maps

$$((\mathcal{P}(M))^n \xrightarrow{(\widetilde{f})^n} K^n \xrightarrow{t^K} K) \quad \text{and} \quad ((\mathcal{P}(M))^n \xrightarrow{t^{\mathcal{P}(M)}} \mathcal{P}(M) \xrightarrow{\widetilde{f}} K),$$

respectively and set $h : (\mathcal{P}(M))^n \longrightarrow K \times K : (\vec{m}_0, \ldots, \vec{m}_{n-1}) \mapsto (h_1(\vec{m}_0, \ldots, \vec{m}_{n-1}), h_2(\vec{m}_0, \ldots, \vec{m}_{n-1}))$. Since the diagonal Δ_K is closed in K^2 (K is Hausdorff) and h is a continuous map it follows that $h^{-1}[\Delta_K]$ is closed in $(\mathcal{P}(M))^n$. Since M' is a L-subalgebra of $\mathcal{P}(M)$, $\widetilde{f} \circ \iota = f'$ and f' is an L-morphism, we obtain
$(M')^n = \{(\vec{m}_0, \ldots, \vec{m}_{n-1}) \in (M')^n : f'(t^{M'}((\vec{m}_0, \ldots, \vec{m}_{n-1})) = t^K((f')^n(\vec{m}_0, \ldots, \vec{m}_{n-1})\}$

$\subseteq \{(\vec{m}_0, \ldots, \vec{m}_{n-1}) \in (\mathcal{P}(M))^n : \widetilde{f}(t^{\mathcal{P}(M)}((\vec{m}_0, \ldots, \vec{m}_{n-1})) = t^K((\widetilde{f})^n(\vec{m}_0, \ldots, \vec{m}_{n-1})\} = h^{-1}[\Delta_K]$.
Since M' is dense in $\mathcal{P}(M)$, the same is true of $(M')^n$ in $(\mathcal{P}(M))^n$, and so $h^{-1}[\Delta_K]$ is a dense closed subset of $(\mathcal{P}(M))^n$. Thus, $h^{-1}[\Delta_K] = (\mathcal{P}(M))^n$, i.e. $\widetilde{f}$ preserves the operation t. Let R an n-ary L-relation and let $(\vec{m}_0, \ldots, \vec{m}_{n-1}) \in R^{\mathcal{P}(M)} \subseteq (\mathcal{P}(M))^n$. Since R^K is closed in K^n, to prove $(\widetilde{f})^n(\vec{m}_0, \ldots, \vec{m}_{n-1}) \in R^K$ it suffices to show that for each neighborhood W of $(\widetilde{f})^n(\vec{m}_0, \ldots, \vec{m}_{n-1}) \in K^n$ we have $W \cap R^K \neq \emptyset$. Since $\widetilde{f}$ is continuous and $\{\prod_{i<n}(ker(p_\Sigma))(\vec{m}_i) : \Sigma \in \mathcal{V}(M)\}$ is a fundamental system of neighborhoods of $(\vec{m}_0, \ldots, \vec{m}_{n-1}) \in (\mathcal{P}(M))^n$, there is $\Sigma \in \mathcal{V}(M)$ such that $(\widetilde{f})^n[\prod_{i<n}(ker(p_\Sigma))(\vec{m}_i)] \subseteq W$. Since $(\eta_M)^n[R^M]$ is dense in $R^{\mathcal{P}(M)}$ (see Remark 17), there is $(a_0, \ldots, a_{n-1}) \in R^M$ such that $(\eta_M)^n(a_0, \ldots, a_{n-1}) \in R^{\mathcal{P}(M)} \cap \prod_{i<n}(ker(p_\Sigma))(\vec{m}_i)$; from $\widetilde{f} \circ \eta_M = f$ we get $(\widetilde{f})^n((\eta_M)^n(a_0, \ldots, a_{n-1})) = f^n(a_0, \ldots, a_{n-1}) \in R^K$, showing that

$W \cap R^K \neq \emptyset$, completing the proof that $\widetilde{f}$ is a L-morphism.

It is clear that if K is in $\mathcal{A}_{pf}$, then it satisfies the properties in statement of (a) and, since η is a natural transformation and $\eta_K : K \xrightarrow{\cong} \mathcal{P}(K)$, we have $ker(\eta_M) \subseteq ker(\mathcal{P}(f) \circ \eta_M) = ker(\eta_K \circ f) = ker(f)$, as well as $(\eta_K)^{-1} \circ \mathcal{P}(f) = \widetilde{f}$, for each $f : M \longrightarrow K$, as needed.

b) The conditions on T and $j : M \longrightarrow T$ are such that the proof of item (a) yields a *unique* $\mathcal{A}_{pf}$-morphism $\widehat{\eta}_M : T \longrightarrow \mathcal{P}(M)$, such that $\widehat{\eta}_M \circ j = \eta_M$. Then, $\widehat{\eta}_M \circ \widetilde{j} \circ \eta_M = \eta_M$ and it follows from (a) that $\widehat{\eta}_M \circ \widetilde{j} = id_{\mathcal{P}(G)}$. Since $\widetilde{j} \circ \widehat{\eta}_M \circ j = j$, $j[G]$ is dense in T and T is a Hausdorff, we obtain $\widetilde{j} \circ \widehat{\eta}_M = id_T$. Hence, $\widetilde{j}$ and $\widehat{\eta}_M$ are the unique (inverse) $\mathcal{A}_{pf}$-isomorphism between the arrows $\eta_M : M \longrightarrow \mathcal{P}(M)$ and $j : M \longrightarrow T$. ∎

A well-known general categorial result on adjoint pairs of functors yields

COROLLARY 21. *The inclusion functor* $\iota : \mathcal{A}_{pf} \hookrightarrow \mathcal{A}^{top}$ *preserves projective limits and* $\mathcal{P} : \mathcal{A}^{top} \longrightarrow \mathcal{A}_{pf}$ *preserves inductive limits. In particular, since* $\mathcal{A}$ *is a* $\forall\exists$*-axiomatizable elementary class,* $\mathcal{A}^{top} \hookrightarrow \mathbf{L\text{-}mod}^{top}$ *creates upward directed* limits, *i.e. if* $(I, \leqslant)$ *is an upward directed poset and* $D : (I, \leqslant) \longrightarrow \mathcal{A}^{top}$ *is a diagram, then the inductive limit in the category* $\mathcal{A}^{top}$ *of the composition* $(I, \leqslant) \xrightarrow{D} \mathcal{A}^{top} \hookrightarrow \mathbf{L\text{-}mod}^{top}$, $(M, (D(i) \xrightarrow{\alpha_i} M)_{i \in I})$, *is also the inductive limit of the diagram* D *in the category* $\mathcal{A}^{top}$, *thus* $(\mathcal{P}(M), (\mathcal{P}(D(i)) \xrightarrow{\mathcal{P}(\alpha_i)} \mathcal{P}(M))_{i \in I})$ *is the inductive limit in the category* $\mathcal{A}_{pf}$ *of the diagram* $\mathcal{P} \circ D : (I, \leqslant) \longrightarrow \mathcal{A}_{pf}$. □

Before stating the pertinent result for quotients we register the following

REMARK 22. Let $f : M \longrightarrow N$ be a $\mathcal{A}^{top}$-morphism with *dense image*. Since η_N is also a $\mathcal{A}^{top}$-morphism with dense image and $\mathcal{P}(f) \circ \eta_M = \eta_N \circ f$, we conclude $\mathcal{P}(f) : \mathcal{P}(M) \longrightarrow \mathcal{P}(N)$ has dense image; since it is continuous and the spaces involved are compact Hausdorff, $\mathcal{P}(f)$ is a closed surjective map. Hence:

(1) $ker(\mathcal{P}(f))$ is a *closed* congruence in $\mathfrak{C}(\mathcal{P}(M))$ and $\mathcal{P}(M)/ker(\mathcal{P}(f))$ is a Boolean space;

(2) The derived $\mathcal{A}^{top}$-arrow from $\mathcal{P}(f)$ (via Proposition 10), $g : \mathcal{P}(M)/ker(\mathcal{P}(f)) \longrightarrow \mathcal{P}(N)$, is a bijective $\mathcal{A}^{top}$-morphism and a homeomorphism of Boolean spaces.

A natural question is to know when is g^{-1} a $\mathcal{A}^{top}$-*morphism*. Note that if g is an $\mathcal{A}^{top}$-*isomorphism* then, by Corollary 15, $ker(\mathcal{P}(f))$ is the intersection of some subfamily of $\mathcal{V}(\mathcal{P}(M))$; in the course of the proof of Theorem 23 below we shall show that for a congruence Θ in $\mathfrak{C}$, this condition is also sufficient. □

Regarding quotients, we now state

THEOREM 23. *The functor* $\mathcal{P} : \mathcal{A}^{top} \longrightarrow \mathcal{A}_{pf}$ *preserves quotients. More precisely, for* $M \in \mathcal{A}^{top}$ *and* $\Theta \in \mathfrak{C}(M)$, *let* $q_{\Theta} : M \twoheadrightarrow M/\Theta$ *be the quotient* $\mathcal{A}^{top}$*-morphism and let* $\Sigma_{\Theta} = ker(\mathcal{P}(q_{\Theta})) \in \mathfrak{C}(\mathcal{P}(M))$. *Then:*
a) $\mathcal{P}(q_{\Theta}) : \mathcal{P}(M) \longrightarrow \mathcal{P}(M/\Theta)$ *is a surjective* $\mathcal{A}_{pf}$*-morphism.*
b) If $g_{\Theta} : \mathcal{P}(M)/\Sigma_{\Theta} \longrightarrow \mathcal{P}(M/\Theta)$ *is the derived* $\mathcal{A}^{top}$*-morphism from* $\mathcal{P}(q_{\Theta})$
(via Proposition 10), g_{Θ} *is a* $\mathcal{A}_{pf}$*-isomorphism.*

Proof. Item (a) follows immediately from Remark 22.

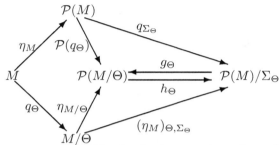

Fact. With notation as above, Σ_{Θ} is the intersection of a subfamily of $\mathcal{V}(\mathcal{P}(M))$.

Proof. Consider the subposet $\mathcal{V}_{\Theta}(M) = \{\Omega \in \mathcal{V}(M) : \Theta \subseteq \Omega\}$; then, Proposition 10 entails yields:

∗ A pair of inverse bijective increasing functions

$$
\begin{cases}
(i) & (q_{\Theta})^{\star} \restriction : \mathcal{V}(M/\Theta) \xrightarrow{\cong} \mathcal{V}_{\Theta}(M) \text{ given by} \\
& \Gamma \in \mathcal{V}(G/\Theta) \mapsto (q_{\Theta} \times q_{\Theta})^{-1}[\Gamma] \in \mathcal{V}_{\Theta}(G); \\
\\
(ii) & (q_{\Theta})_{\star} \restriction : \mathcal{V}_{\Theta}(M) \xrightarrow{\cong} \mathcal{V}(M/\Theta), \text{ given by} \\
& \Omega \in \mathcal{V}_{\Theta}(M) \mapsto (q_{\Theta} \times q_{\Theta})[\Omega] = \Omega/\Theta \in \mathcal{V}_{\Theta}(M).
\end{cases}
$$

∗ A canonical $\mathcal{A}_{pf}$-isomorphism, α_{Θ} : $\varprojlim_{\Gamma \in \mathcal{V}(M/\Theta)} (M/\Theta)/\Gamma \xrightarrow{\cong}$ $\varprojlim_{\Omega \in \mathcal{V}_{\Theta}(M)} M/\Omega$.

Now, the definition of the functor $\mathcal{P}$ guarantees that $\mathcal{P}(q_{\Theta})$: $\varprojlim_{\Omega \in \mathcal{V}(M)} M/\Omega \twoheadrightarrow \varprojlim_{\Gamma \in \mathcal{V}(M/\Theta)} (M/\Theta)/\Gamma$ is the surjective $\mathcal{A}_{pf}$-morphism such that $(m_{\Omega}/\Omega)_{\Omega \in \mathcal{V}(M)} \in \mathcal{P}(M) \mapsto ((m_{\Omega}/\Theta) / (\Omega/\Theta))_{\Omega/\Theta \in \mathcal{V}(M/\Theta)} \in \mathcal{P}(M/\Theta)$. It is straightforward that the composition $\varprojlim_{\Omega \in \mathcal{V}(M)} M/\Omega \xrightarrow{\mathcal{P}(q_{\Theta})} \varprojlim_{\Gamma \in \mathcal{V}(M/\Theta)} (M/\Theta)/\Gamma \xrightarrow{\alpha_{\Theta}} \varprojlim_{\Omega \in \mathcal{V}_{\Theta}(M)} M/\Omega$ is the "projection" $\mathcal{A}_{pf}$-morphism ρ_{Θ} : $\varprojlim_{\Omega \in \mathcal{V}(M)} M/\Omega \longrightarrow \varprojlim_{\Omega \in \mathcal{V}_{\Theta}(M)} M/\Omega$, given by $(m_\Omega/\Omega)_{\Omega \in \mathcal{V}(M)} \mapsto (m_\Omega/\Omega)_{\Omega \in \mathcal{V}_\Theta(M)}$.

Hence, $\Sigma_\Theta \doteq ker(\mathcal{P}(q_\Theta)) = ker(\mathcal{P}(q_\Theta) \circ \alpha_\Theta) = ker(\rho_\Theta) = \bigcap \{ker(p_\Omega) : \Omega \in \mathcal{V}_\Theta(M)\}$. But if $\Omega \in \mathcal{V}(M)$, then $\bar{p}_\Omega : \mathcal{P}(M)/ker(p_\Omega) \longrightarrow M/\Omega$ is a $\mathcal{A}_{fin}$-isomorphism, whence $ker(p_\Omega) \in \mathcal{V}(\mathcal{P}(M))$, showing that Σ_Θ is the intersection of a subfamily of $\mathcal{V}(\mathcal{P}(M))$, as needed.

Since $\eta_{M/\Theta} \circ q_\Theta = \mathcal{P}(q_\Theta) \circ \eta_M$, it follows that $\Theta \subseteq \eta_M^*(\Sigma_\Theta)$ and so there is a unique $\mathcal{A}^{top}$-morphism, $(\eta_M)_{\Theta,\Sigma_\Theta} : M/\Theta \longrightarrow \mathcal{P}(M)/\Sigma_\Theta$, such that $(\eta_M)_{\Theta,\Sigma_\Theta} \circ q_\Theta = q_{\Sigma_\Theta} \circ \eta_M$. By the Fact above, Σ_Θ is the intersection of some subfamily of $\mathcal{V}(\mathcal{P}(M))$ and so, by Corollary 15, $\mathcal{P}(M)/\Sigma_\Theta \in \mathcal{A}_{pf}$. Now, by Theorem 20, there is a unique $\mathcal{A}_{pf}$-morphism $h_\Theta : \mathcal{P}(M/\Theta) \longrightarrow \mathcal{P}(M)/\Sigma_\Theta$ such that $h_\Theta \circ \eta_{M/\Theta} = (\eta_M)_{\Theta,\Sigma_\Theta}$. We now claim that g_Θ and h_Θ are inverse $\mathcal{A}_{pf}$-isomorphisms and, in fact, the unique (iso)morphisms between the arrows $\mathcal{P}(q_\Theta) : \mathcal{P}(M) \twoheadrightarrow \mathcal{P}(M/\Theta)$ and $q_{\Sigma_\Theta} : \mathcal{P}(M) \twoheadrightarrow \mathcal{P}(M)/\Sigma_\Theta$, for we have:

* $h_\Theta \circ g_\Theta = id$: Since $h_\Theta \circ g_\Theta \circ q_{\Sigma_\Theta} \circ \eta_M = h_\Theta \circ \mathcal{P}(q_\Theta) \circ \eta_M = h_\Theta \circ \eta_{M/\Theta} \circ q_\Theta = (\eta_M)_{\Theta,\Sigma_\Theta} \circ q_\Theta = q_{\Sigma_\Theta} \circ \eta_M = id \circ q_{\Sigma_\Theta} \circ \eta_M$, and the conclusion follows from the universal property of η_M and the surjectivity of q_{Σ_Θ};

* $g_\Theta \circ h_\Theta = id$: Since $g_\Theta \circ h_\Theta \circ \eta_{M/\Theta} \circ q_\Theta = g_\Theta \circ (\eta_M)_{\Theta,\Sigma_\Theta} \circ q_\Theta = g_\Theta \circ q_{\Sigma_\Theta} \circ \eta_M = \mathcal{P}(q_\Theta) \circ \eta_M = \eta_{M/\Theta} \circ q_\Theta = id \circ \eta_{M/\Theta} \circ q_\Theta$, and the conclusion follows from the surjectivity of q_Θ and the universal property of $\eta_{M/\Theta}$, completing the proof of (b). It is clear from the calculations above that g_Θ and h_Θ are inverse isomorphisms between the arrows $\mathcal{P}(q_\Theta) : \mathcal{P}(M) \twoheadrightarrow \mathcal{P}(M/\Theta)$ and $q_{\Sigma_\Theta} : \mathcal{P}(M) \twoheadrightarrow \mathcal{P}(M)/\Sigma_\Theta$, while their uniqueness stems from the fact that $\mathcal{P}(q_\Theta)$ and q_{Σ_Θ} are both surjective. ∎

4   Concluding Remarks

Now we will suppose a bit more on the elementary class $\mathcal{A} \subseteq$ **L-mod**: that $\mathcal{A}$ is axiomatizable by sentences like $\forall \vec{x}(\varphi(\vec{x}) \to \psi(\vec{x}))$ where $\varphi(\vec{x}), \psi(\vec{x}) \in [\exists, \wedge, atom(L)]$ or are the negations of atomic L-formulas, in particular, $\mathcal{A}$ is *closed under L-products*. As seen in section 1, profinite structures are pure injective. Given such kind of class $\mathcal{A}$ of L-structures and a saturated family of $\mathcal{A}$-congruences, $\mathfrak{C}$, it is natural consider the subclass $\mathcal{A}_{LG} \subseteq \mathcal{A}$ of (discrete) structures M in $\mathcal{A}$ such that the canonical arrow $\eta_M : M \longrightarrow \mathcal{P}(M)$ is a *pure L-embedding*. This can be rephrased as a *local-global principle*, as follows: M is in $\mathcal{A}_{LG}$ if for all p.p. L-formulas, $\phi(\vec{x})$, and all $\bar{a} \in M^n$,

$$[\mathrm{LG}] \quad \begin{cases} M \models \phi[\bar{a}] & \Leftrightarrow & \mathcal{P}(M) \models \phi[\eta_M(\bar{a})] \\ & \Leftrightarrow & \text{For all } C \in \mathfrak{C}(M), \text{ such that } M/C \text{ is finite,} \\ & & M/C \models \phi[\bar{a}/C]. \end{cases}$$

The following examples illustrate the principle [LG]:

(a) Boolean Algebras. Every boolean algebra satisfies [LG]. In [14], the BA-morphism $\eta_B : B \longrightarrow \mathcal{P}(B)$ is identified with the BA-embedding of B into the clopens of its Stone space. But it follows from Proposition 3.(c) and Sikorsky's Extension Theorem that all BA-monics are pure, since any Boolean algebra is the directed limit of its *finite* subalgebras.

(b) Reduced Special Groups (RSG). By a result in [9], formulated in the dual category of abstract order spaces, there are reduced special group that do not verify [LG] (see also [15]). In [1] is shown that the subclass of RSGs that satisfy this local-global principle is also a $\forall\exists$-axiomatizable elementary class . In [14] it is shown that a weaker formulation of [LG] holds for all RSGs: the morphism η_G $G \longrightarrow \mathcal{P}(G)$ reflects subforms; in particular, it is a complete embedding and reflects isotropy of forms with coefficients in G.

From the universal property of the profinite hull functor (Theorem 20) and Proposition 3, it follows that $\mathcal{A}_{LG}$ $\hookrightarrow$ $\mathcal{A}$ is closed under L-isomorphisms, pure L-substructures and L-products, while the complementary subclass, $(\mathcal{A} \setminus \mathcal{A}_{LG})$ is closed under reduced powers. We then pose the following

Problem 1. *Is the class $\mathcal{A}_{LG}$ closed under quotients by elements of $\mathfrak{C}$?*
If L-structures in $\mathcal{A}$ are L-*inhabited* (see Remark 9.(b)) and the answer to the above problem is affirmative, then any reduced product of structures in $\mathcal{A}_{LG}$ is also in $\mathcal{A}_{LG}$. A model-theoretic consequence of this and the closure properties established above shows that $\mathcal{A}_{LG}$ is an L-elementary class *axiomatizable by Horn sentences* (see [5] Theorems 4.1.12 and 6.2.5).

Acknowledgements

The first author thanks the supervision of professor Walter Carnielli in a post-doctoral fellowship in CLE-UNICAMP with support of FAPESP and the supervision of professor Maximo Dickmann in a post-doctoral fellowship at the Logic Group of the University of Paris VII, with financial support of CAPES.

BIBLIOGRAPHY

[1] V. Astier, M. Tressl, *Axiomatization of local-global principles for pp-formulas in spaces of orderings*, Archive for Mathematical Logic **44** (2005), 77-95.

[2] J. L. Bell, A. B. Slomson, **Models and Ultraproducts: an Introduction**, North-Holland Publishing Company, Amsterdam, Netherlands, 1971.

[3] N. Bourbaki, **General Topology, part 1**, Elements of Mathematics, Addison-Wesley Publishing Company, Great Britain, 1966.

[4] D. Bushaw, **Elements of General Topology**, John Wiley & Sons, N. York, 1967.

[5] C. C. Chang, H. J. Keisler, **Model Theory**, North-Holland Publishing Company, Amsterdam, Netherlands, 1990.

[6] M. Dickmann, F. Miraglia, *On Quadratic Forms whose total signature is zero mod* 2^n. *Solution to a problem of M. Marshall*, Inventiones mathematicae **133** (1998), 243-278.

[7] M. Dickmann, F. Miraglia, **Special Groups: Boolean-Theoretic Methods in the Theory of Quadratic Forms**, Memoirs of the AMS **689**, American Mathematical Society, Providence, USA, 2000.

[8] D. P. Ellerman, *Sheaves of structures and generalized ultraproducts*, Annals of Mathematical Logic **7** (1974), 163-195.

[9] P. Gladki, M. A. Marshall, *The pp conjecture for spaces of orderings of rational conics*, Algebra and its Applications, **6** (2007), 245-257.

[10] A. Grothendieck, J. L. Verdier, *Préfaisceaux*, Exposé I in **SGA 4**, Lecture Notes in Mathematics **269**, Springer-Verlag, Berlin, Germany, 1972, 1-217.

[11] A. L. de Lima, *Les groupes spéciaux. Aspects Algébriques et Combinatoires de la Théorie des Espaces d'Ordres Abstraits*, Thèse de doctorat, Université Paris VII, Paris, France, 1996.

[12] H. L. Mariano, F. Miraglia, *Logic, Partial Orders and Topology*, Manuscrito **28** n.2, (2005), 449-546.

[13] H. L. Mariano, F. Miraglia, *Profinite Structures are Retracts of Ultraproducts of Finite Structures*, Reports on Mathematical Logic, **42** (2007), 169-181.

[14] H. L. Mariano, *Contribuições à teoria dos Grupos Especiais*, Phd Thesis, University of São Paulo, São Paulo, Brazil, 2003.

[15] M. A. Marshall, *Open questions in the theory of spaces of orderings*, Journal of Symbolic Logic **67** (2002), 341-352.

[16] F. Miraglia, **An Introduction to Partially Ordered Structures and Sheaves**, Contemporary Logic Series, vol 1, Polimetrica Scientific Publisher, Milan Italy, 2006.

[17] L. Ribes, **Introduction to Profinite Groups and Galois Cohomology**, Queen's Papers in Pure and Applied Mathematics **24**, Queen's University, Ontario, Canada, 1970.

Hugo Luiz Mariano and Francisco Miraglia
Instituto de Matemática e Estatística
University of São Paulo
São Paulo, Brazil
E-mails: {hugomar,miraglia}@ime.usp.br

A Structural Perspective on
Mathematics (and Mathematical Proofs)

JAIRO JOSÉ DA SILVA

ABSTRACT. Often in mathematics theorems of *interpreted* theories are proved in the context of other theories, governing differently conceived mathematical domains. This change of context, however, involves a re-conceptualization of the statement proven, which implies a change of meaning. But mathematicians do not take this into consideration, which poses an important question for the philosophy of mathematics: why is the change of meaning that comes together with the re-conceptualization of mathematical statements mathematically irrelevant?

The introduction of *purely formal* objects in the domains of interpreted theories (for instance, imaginary "numbers" into numerical domains) for solving problems originally circumscribed to these domains by strictly symbolic means also raises a question: how can what has *no* meaning have any bearing on what has a definite meaning?

The answer to these questions lies in the fact that mathematical assertions are purely formal, that is, their truth-value does not depend on their material content. Since mathematics gives voice only to formal truths, mathematical statements can be freely reinterpreted and still preserve their identity and mathematical domains, no matter the particular meaning attributed to them, embedded one into another or extended by the adjunction of formal entities without absurdity. By so doing we may be able to obtain more convenient contexts where formal properties of the original domains can be established in easier or more enlightening ways.

The formal character of mathematics accounts for its applicability to science independently of its being literally true — i.e. true of independently existing things, either objective domains or structures (against the so-called indispensability argument). The utility of mathematics in scientific explanation, I claim, depends solely on its ability to provide formal contexts where formal aspects of experience can be conveniently represented by means of isomorphic embeddings.

Let us suppose for the sake of argumentation that arithmetic is concerned with a particular kind of abstracts objects — cardinal numbers — subsumed under a particular concept — that of cardinality, number or quantity —. Obviously, under such a presupposition the meaning of numerical statements (whatever "meaning" happens to mean) must involve somehow the meaning attached to the concept of number. Arithmetical statements must express some aspect of the concept under which its objects are assembled; or so it seems.

Now, suppose that in order to prove an arithmetical statement we resort to another mathematical theory, say, complex analysis. Of course, complex numbers constitute only a formal expansion of the domain of cardinal numbers, since they do not fall under the ruling concept of arithmetic (complex numbers do not measure quantity). In what sense, then, the statement proven is the *same* statement we had before complex numbers came into the picture? It is obvious, I think, that the statement considered as one about complex numbers does not, ipso facto, preserve the meaning of the original statement, which involved, as presupposed, the idea of number as a measure of quantity.

Now, mathematicians have made this moving about freely of mathematical statements from one to other, more convenient contexts into a powerful method of proof, and, moreover, do not think they change, by so doing, *that* which they prove. The conclusion then presses itself that, from a strictly mathematical perspective, the meaning of mathematical statements, at least insofar as this is related to how the objects to which the statements refer are conceived, *has no mathematical relevance*.

Let us for the moment accept uncritically the familiar distinction between what is conventionally called interpreted (or contentual) theories, i.e. theories of well-determinate concepts or objectual domains, and non-interpreted (or purely symbolic) mathematical theories, i.e. rule-governed systems of symbolic manipulations[1]. On the one hand, theories supposedly focused on objects of a certain type (objects falling under a certain concept or sharing a common meaning); on the other, "mere" manipulations of signs without meaning other than the purely operational.

Unlike the nominal terms of well-formed sentences of the language of non-interpreted mathematical theories, those of interpreted theories are supposed to refer to determinate elements of the domains of these theories (the domains they describe), and these sentences themselves to possible states-of-affairs in these domains (to say that any well-formed sentence of the language of a theory represents a possible state-of-affairs of the domain of

---

[1]One of the conclusions of this paper is that this distinction is mathematically irrelevant.

this theory is equivalent — by stipulation — to saying that any such sentence has a determinate, although possibly unknown truth-value; but neither assertion expresses a *fact* or a *factual* hypothesis[2]). If we define meaning in terms of truth conditions, the meaning of any assertion of an interpreted theory is determined by the situation (i.e. the possible state-of-affairs) that would make it true — its truth-value depending on what happens to be the case in the domain of the theory. Assertions of non-interpreted theories are, on the contrary, supposed to be devoid of meaning, except the purely formal one that grammatical correction suffices to guarantee.[3]

It is however a common practice in mathematics to de-contextualize and re-contextualize assertions (by this I mean proving assertions referring to objects of a *particular* type in *another* theory, governing objects of a *different* type) in order to give them better, easier, more interesting or just different proofs. But, obviously, this reorientation redirects reference; the assertions no longer refer to the domains to which they originally referred. And since they no longer express the same situations they no longer have the same meaning, even if syntactically nothing changes. In short, in proving a theorem in an interpreted theory that differs from that in which the theorem was originally stated we are in fact proving a different theorem, if the identity of (material) meaning is the criterion for the identity of theorems[4].

Let us go back to the example I chose to open this paper, and suppose we manage to prove an arithmetical statement using complex analysis. This poses no problem if we accept the shift of meaning, taking the statement in cause as referring to complex numbers and the theorem proven as establishing a fact in the complex domain. But the situation becomes paradoxical, apparently, if we insist in preserving the original meaning of the displaced theorem. To what extent are we entitled to say that the fact we proved involves natural numbers as originally conceived, viz., as particularizations of the notion of quantity?

It is in general taken for granted that natural numbers are *in fact* particular complex numbers, and that no problem arises in proving theorems

---

[2]Cf. my "Husserl on the Principle of Excluded Middle", in G. Banham (ed.) *Husserl and the Logic of Experience*, Houndmills, Basingstoke, Hampshire: Palgrave Macmillan, 2005.

[3]We can draw a distinction between (semantic) *material* meaning and (syntactic) *formal* meaning of assertions in general; the former being involved in the representation of particular situations in particular domains, the latter with abstract formal situations capable of being embodied in materially distinct domains.

[4]My point is that it is *not*. A theorem can remain invariant (the same theorem) even if the reference of its non-logical terms, and then its material meaning, changes. What mathematical assertions really express — via their formal meanings — are formal situations (for instance, that a particular materially indeterminate binary relation is reflexive, and things of the sort).

of the arithmetic of natural numbers in any theory thematically focused on the complex numbers. But this of course involves giving the natural numbers another meaning, no longer that which they have as objects of their original theory. Accepting the analytic proof as a proof of the arithmetical statement as originally meant presupposes that the natural numbers can be taken alternatively as either quantitative forms (or in any case entities related to the concept of quantity) or complex numbers that have *nothing* to do with this concept. It is not clear, at least not to me, that retaining the original meaning of an assertion in a context of proof where it receives a different interpretation is a coherent procedure — *if, of course, the (material) meaning of mathematical assertions is indeed mathematically relevant.*

But things can get worse. It is also a common mathematical practice to divest contexts of proof of any intrinsic meaning, thus reducing them to domains of "mere symbolic manipulation". The puzzle becomes more intriguing now: how can what has *no meaning* (instead of just another meaning, as above) be useful in proving what has a fixed meaning? If for instance we accept that the arithmetic of cardinal numbers is a theory of objects of a determinate type — numbers, which we associate with the notion of quantity — how is it possible that in order to prove facts concerning these entities — and supposedly involving and maybe already pre-determined by the meaning attributed to them — it is profitable, and often unavoidable to move to a new context, containing entities that are not numbers in the original sense and only behave *formally* like numbers (remember that complex numbers were useful in arithmetic and algebra *well before* they were given a meaning in terms of vectors in the plane)? How can extensions of the domain of numbers proper by the adjunction of entities that only *behave* like numbers from a purely formal perspective be of any help in proving facts that should follow exclusively from the meaning attached to the concept of number? Moreover — what is more intriguing — how can a fact proved using a formal extension of the concept of number still be accepted as involving the *meaning* originally attached to this concept (when this proof could be carried through only because this meaning was explicitly cancelled)?

Whether arithmetic is an *objectual* theory, i.e. a theory of objects — the numbers —, understood indifferently as attributes of concepts or classes, abstract forms of equinumerous collections, objects of intuition or reason or what have you, or a *conceptual* theory, whose thematic concept can be thought as given in some original insight, derived from experience or freely invented; no matter what your favorite philosophy of arithmetic is, if arithmetic is a science of something, a contentual or interpreted theory, it should, or so it seems, be responsible for its domain, whose facts should not be left for the care of others to establish, much less so if the helping hand comes from blind rule-governed symbolic systems.

In short, extending a mathematical domain with the introduction of "alien" objects involves either reconceptualization, that is, the attribution of another meaning to this domain, or formal abstraction, i.e. the reduction of the elements of the domain to the most general form of objects devoid of any intrinsic meaning upon which we simply operate (a domain of pure symbolic manipulation). In any case, we are no longer dealing with the original domain with its original meaning.

That we can prove mathematical assertions by resorting to extensions of its theory in which the meaning originally attached to it is lost did not, surprisingly, elicit much discussion in the philosophical literature, even among realists, who believe in the existence of mathematical objects and, moreover, that mathematical theories are theories *of* these objects (shouldn't they explain how and why non-numbers can be useful, even indispensable in proving facts about numbers?). There are, however, exceptions. The philosopher Edmund Husserl was much concerned with this problem and gave an answer to it around 1890 (and more explicitly in 1901 in a couple of conferences given at Göttingen[5]). According to him, provided the original theories are complete and their extensions consistent[6], no logical or epistemological problems arise from moving to extensions in order to prove facts under the jurisdiction of the original theories. In this case, he thought, these extensions could be used, if useful, but would be essentially unnecessary and would have no real impact on the original theories. Non-sense was admitted provided it is useful and, in the end, dispensable.

In my opinion this answer is not satisfactory for mainly one reason: although it safeguards the original theories from inconsistencies and "epistemic intromission" by other theories, it does not explain *why* extensions are so often useful. In this paper I will try and offer an answer to this question[7].

---

[5]See "Review of Schröder's *Lectures on the Algebra of Logic*", *Göttinger Gelehrte Anzeigen*, nº 7, pp. 243-278, 1891 and *Husserliana XII*, pp. 430-451, The Hague: Martinus Nijhoff, 1970.

[6]Husserl original term for "complete" is "*definite*". Of course, extensions of definite theories are expressed in richer languages, the original theories being complete only with respect to the sentences of their languages (Husserl distinguishes between a restricted and a general notion of definiteness; in his answer to the problem of "imaginaries" he only required restricted definiteness).

[7]Although Husserl had by the time he published his *Logical Investigations* (1900-01) the conception of (purely) *formal* mathematical theories as theories of materially indeterminate objects considered from an exclusively formal perspective (according to him, these theories belong to *formal ontology*, a province of formal logic), he failed to noticed that even contentual theories are in a sense purely formal. Husserl always showed some mistrust of purely formal theories (and formalization in general) for the sake of themselves (with respect to Husserl attitude towards formalism see, for instance, his last work:

In general we try to capture our understanding of a mathematical domain in a set of basic truths (the axioms) from which all the truths concerning the domain could (and should) follow logically. Usually axiomatization comes only at a later stage of the investigation of a mathematical domain or, alternatively, the concept that governs it; we often try to carry out proofs on an intuitive basis − that is, on the basis of our intuitive understanding of the meaning attached to the domain in question − and then eventually select some sufficient set of basic premises that fully articulate this understanding. But once we have succeeded in axiomatizing a mathematical domain, proving theorems about it should ideally be simply a matter of deriving them from the axioms by strictly logical means; to allow new "insights" to come in is tantamount to giving the domain new shades of meaning, which may or may not be consistent with the original meaning.

That new insights are needed in carrying out proofs indicates that the axioms already selected do not exhaust the meaning associated with the domain, and turning these "insights" into new axioms is a way of better articulating our intuitive grasp of the domain and complete the axiomatization, provided, of course, the new axioms do not contradict those already selected. Eventually we may end up with a complete set of axioms, in which case the theory cannot be extended on pain of contradiction. Introducing, however, axioms that explicitly contradict some of those already in place, as is often the case when "extending" a domain[8], requires that some conflicting axioms be removed from the original theory, which amounts to changing its meaning. Therefore, any statement proved in the extended theory, even if the language remained unchanged, differs in meaning from the syntactically identical statement proved in the original theory. To prove something with an unaltered meaning by giving it another meaning is, or so it seems, simply blatantly inconsistent. The embarrassment caused by the use of "imaginary numbers" in the arithmetic of real numbers, or infinitesimals in real analysis are historical examples of recognized inconsistencies, which nonetheless did not cause mathematicians to abandon these methods.

The role of the philosopher of mathematics is not, however, to criticize mathematical practices, but to explain them. We must then accept the fact that mathematical proofs often ignore the meaning the assertion being

---

The Crisis of European Science and Transcendental Phenomenology: an Introduction to Phenomenological Philosophy, Evanston, Illinois: Northwestern University Press, 1970). According to Husserl, formal theories only become *science* when applied, i.e. when becoming theories of something in particular.

[8]For instance, simply introducing the imaginary unit into the field of real numbers by means of a new axiom (there is an x such that $x^2 = -1$) flagrantly contradicts the axioms that characterize the field of real numbers (from which we can *prove* that the equation $x^2 + 1 = 0$ has *no* solution).

proved has, inserting it in contexts in which this meaning cannot be consistently preserved; and that a statement can be proved even if its proof is carried out in a context whose meaning is incompatible with the meaning originally attached to it. Now, how to understand the nature of mathematical theories so that this mingling of theories and domains causes no scandal?

In mathematics, there is, and there should be no a priori restriction on where to look for help in order to prove whatever we may want to prove[9], and one of the standard mathematical techniques of proof is the immersion of the domain of interest (or a re-conceptualized version of it) into another so that the structuring operations and relations of the domain of primary interest appear as restrictions of (all or some of) the operations and relations of the larger domain into which it is embedded. By so doing we may be able to prove truths in the larger domain whose restriction to the original domain are the facts we wanted to prove.

But, as I have been insisting, with each new, differently contextualized proof the statement being proved has a different meaning. Since in general there cannot be *a priori* eliminated the possibility of finding new proofs of well-established theorems, in completely different contexts, the conclusion imposes itself that the meaning of a mathematical statement is not determined once and for all, but is open to changes as new proofs of the statement are found[10].

But, as I have been arguing, this all too common mathematical practice seems to harbor a contradiction. So, if this practice is acceptable, as is obviously the case, the meaning attached to contentual theories cannot have any import on purely mathematical facts concerning their domains. This is, in a nutshell, the *formalist* conception of mathematics, usually badly misconstrued as the claim that mathematics is only merely a rule-abiding game, like chess. Correctly formulated formalism only says that mathematical truths (and a science providing us with truths cannot be a mere game) are formal. This, to my view, is close to being a self-evident truth (often

---

[9]For example, the fundamental theorem of algebra, albeit being a theorem of the *algebra* of complex numbers can also be proved by analytic or topological methods. This phenomenon is so utterly common in mathematics that we tend to overlook its relevance.

[10]Let me at this point advance the main thesis of this paper: although a different proof can indeed give a theorem a new material meaning, it is not the material, but only the formal meaning of mathematical statements, which differently contextualized proofs do not alter, that is mathematically relevant. New proofs of a mathematical statement, carried out in different theories, show that the domains of these theories have some identical formal aspects (in particular those expressed by the theorem in question). What in the end drives mathematicians to search, in different domains, for new proofs of known theorems is the desire to establish formal connections among different mathematical domains, which can be very rich in mathematical consequences.

clouded by philosophical *parti pris*, whose analysis would take us far afield).

In other words, mathematical theories, even interpreted mathematical theories, do not really care for *what* the objects of their domains are, but only for *how* they relate to each other, irrespectively of the particular nature of the relations involved. Regardless of the fact that its intended interpretation may have been effective for the establishment of a theory (the formal description provided by the theory may have been originally established as the description of the formal properties of one particular domain), the theory is not in a strong sense a theory of its intended interpretation. Or better, it is, but in a limited sense: the theory describes only purely formal aspects of the intended model, which any other model of the theory also exhibits. The difference between an interpreted and a purely formal theory is that whereas the former provides a formal description *in concreto*, the latter does so *in abstracto*.

It is irrelevant whether a mathematical theory is designed to describe a *determinate* domain of objects, all it describes are its formal properties, that is, how its objects relate to each other irrespectively of their particular nature, or the particular nature of these relations. The axioms of contentual arithmetic, for instance, are not, in a sense, about and do not describe the properties of numbers *as such*, but only the formal properties of their arrangement (or, equivalently, the formal structure underlying the domain of numbers).

Another word for form is structure (which better conveys the idea of an articulation among elements); so, any mathematical theory — contentual or not — is in this sense only a set of truths concerning structural arrangement. So, mathematical theories are formal or structural descriptions; structural or formal facts are all mathematics really cares about. This is precisely what we mean when we say that mathematics is a formal science (whether or not formalized as a calculus)[11].

It is because the nature, i.e. the particular meaning attached to the domain of the intended interpretation (or the concept that governs it) plays no role in mathematical proofs that we can change the domain, or its meaning, in order to carry out proofs. In proving, for instance, a theorem of the arithmetic of natural numbers in complex analysis, we are in fact proving a structural fact that holds in a substructure of the complex domain isomorphic to the natural numbers; so, it is also a structural fact of the

---

[11]I want to avoid the trap of trying to define *mathematically* what a structure is; the idea of course is that it is *that* which identically structured domains have in common (i.e. structures are abstract entities). Since identically structured domains are those that are *isomorphic*, the idea then is that structures are isomorphism types. Structures are then characterized syntactically by categorical theories, but any formal assertion can be seen as a structural description.

domain of cardinal numbers, since structural truths are preserved under isomorphisms. Since formal arithmetical truths concerning ω-sequences is *all* that arithmetic gives us, even if we think of it as describing properties of a *determinate* ω-sequence — the cardinal numbers, whatever they are — we can accept the fact proved with the help of complex analysis as a bona fide theorem of contentual, interpreted arithmetic. In few words, contrary to what Husserl thought (that the use of imaginary entities, being contradictory with the *meaning* of the theory from whose point of view these elements are, precisely, imaginary, can be vindicated only if essentially superfluous), the transition through "the imaginary" works because mathematics — even contentual mathematics — is purely formal; the particular interpretation attributed to the non-logical terms of the mathematical theories is not mathematically relevant.

Hence, all of mathematics, not only non-interpreted, purely symbolic theories belongs to what Husserl called *formal ontology* (a term he introduced to denote the a priori science of objects thought simply as bearers of attributes, which, according to him, included purely symbolic mathematical theories, but not physical geometry and other mathematical sciences that — he thought — had a fixed domain, the perceptual space in the case of geometry).

We usually reserve the word "formal" (as Husserl did) to non-interpreted, i.e. materially indeterminate theories. But, as we have seen, it is really immaterial whether mathematical theories are or are not given a material interpretation; the difference between contentual and "purely formal", or better, interpreted and non-interpreted theories is that whereas the former direct their attention to structures instantiated by privileged objectual domains, the latter establish structures by fiat (therefore the need for proofs of consistency in order to establish that these structures are structures of possible domains of objects[12]).

Extending the field of the real numbers by the adjunction of imaginary numbers for a better treatment of algebraic equations, or space by the adjunction of points and lines at the infinite for the study of its projective prop-

---

[12]If consistency cannot be proved, we settle for the second best, i.e. to use the theory until a contradiction comes out into the open; in this case we reform the theory — as a rule as little as needed — in order to eliminate the problem, and proceed. But in truth not even evident inconsistencies have forced mathematicians to abandon useful formal methods. After all, Berkeley's well-founded criticism of the infinitesimal calculus did not disturb mathematicians enough for them to drop so useful a technique. What the so-called arithmetization of analysis accomplished was not, as sometimes argued, to give calculus a decent subject matter (as part of arithmetic), but to show its relative consistency with arithmetic. (Analogously for imaginary numbers; their geometrical representation provided a much-needed proof of relative consistency — with respect to geometry —, not a "meaning" they never needed).

erties, are only two already classic examples of the utility of "imaginaries" in mathematics. The reason why this works is not difficult to understand now from our point of view that all that mathematics does is provide formal descriptions: to extend a domain of objects (or, equivalently, to reconceptualize its ruling concept so as to immerse it in a larger domain falling under a different concept) is tantamount to isomorphically embedding the original structure into a richer one, where certain formal properties of the original structure belong more naturally, stand out more conspicuously or are maybe more easily perceived[13].

In particular, this way of interpreting the nature of mathematics (and mathematical proofs) avoids uncomfortable ontological commitments. It is not because theories supposedly of certain types of objects are useful or maybe unavoidable, that these objects exist in any pregnant sense (for instance, as entities independent of the theories that posit them). Even purely intentional (in a way, fictional) objects, whose existence is strictly internal to their theories (i.e. that exist merely as intentional correlates of their theories[14]), can be good for instantiating structures in which other structures can be immersed so as to have their (formal) properties highlighted.

---

[13] Husserl realized that formally abstracting a theory and extending it to a larger purely formal theory is a powerful method of proof in mathematics, but, since he believed that the original theory − previous to being abstracted into an empty formal theory − *and the meaning associated with it* should be sufficient for proving whatever was true of its domain, he insisted on limiting the applicability of the method to complete theories (i.e. those that could in principle prove all the facts of their domains).

[14] Mathematical theories are artifacts, and as such they are objectively existing things; the objects they posit inherit this objectivity: anyone who masters a theory can refer to its objects and investigate the truth of assumptions about them by simply verifying whether they are justified by the theory − in case no final decision can be reached, the conjecture in question is undecidable (in the theory). Mathematical theories and their objects are as objective − although as unreal, i.e. not real as space-temporal objects − as Mahler's fifth symphony (do not confuse Mahler's symphony with its multiple materializations, i.e. its written representations or performances). Frege is not far from saying something analogous in the *Grundlagen*; for him, numbers are as objective as the line of the Equator, which although not existing in physical space can be referred to and perfectly located by means of an objectively accessible system of coordinates. Something similar can, I think, be said about numbers vis-à-vis arithmetic: numbers are materially indeterminate formal objects whose *only* properties are those formal properties they possess as elements of a structure defined by the axioms of arithmetic. So, there are questions about numbers that are meaningless, for instance: does 2 belong to 3? Even if we interpret arithmetic in set theory, thus seeing numbers as sets, this question cannot be answered, for it does not involve a *formal* property of numbers − it does not transfer to other interpretations of arithmetic. In short, the relation of belonging is not an *arithmetical* relation. In formal theories we do not know *what* we are referring to (the best answer is that we refer to nothing in particular or, if we insist in giving theories determinate domains, we can say that formal theories refer to formal objects), only what formal properties these things have (whatever they are).

In order to investigate a structure it is perfectly all right to embed it into purely "fictional" – but "richer" – structures.

Suppose, for the sake of argumentation that Kronecker was right, that good old God created the natural numbers, which as a consequence exist as independent entities (independently of us, at least), and that the rest of the so-called mathematical objects are of our creation. Since from the mathematical point of view these objects are only points of structural articulation, like the joints of a mechanism, all we have really created were structures (or mere structural descriptions of logically possible domains). Now, if our creations admitted substructures isomorphic to that of the natural numbers, the structure of ω-sequences (imagine that we simply invented structures with substructures identical – i.e. isomorphic – to the structure of God's creation, supposing this is not sacrilegious), we can discover structural properties of the divine creation by showing that the relevant segment of our creatures have these properties, despite the fact that our structures do not "really" exist. Nothing that we invented needs to exist in the same sense we are supposing the natural numbers to exist, neither the objects nor their structures; structural similarity is all that matters, and it can very well be established between reality and fantasy.

In a strictly analogous way we can understand the application of mathematics to natural sciences or daily life. Physical reality itself, or our experience of it is obviously structured; it requires deeply rooted empiricist prejudices to think that our direct experience of reality consists of inarticulate sense data. However, our experience is not usually structured in a rich mathematical manner – not even our experience of space –; perceptual structures (as we may call them) are in general much poorer than mathematical structures (and this can count against a certain structuralist account of the applicability of mathematics, according to which the application of mathematical theories to reality boils down to the "filling" of empty structures with real stuff).

We usually say, for instance, that geometrical "objects" – points, lines, figures, in short, the entities geometry "is about" –, or at least some of them, are idealized abstractions of experience, but this is not an adequate manner of expressing the facts. Geometrical objects are inventions of ours that do not exist in reality (we "invent" them by means of definitions, i.e. formal stipulations); these objects, of course, can be *suggested* by experience[15], but are nonetheless creatures of intelligence that are not given to

---

[15]Not pure inarticulate sensorial experience, though, but structured experience conforming to certain presuppositions (like, for instance, those that according to Helmholtz sustain our classical – Euclidian – conception of perceptual space. See, for instance, "On the Factual Foundations of Geometry", in P. Pesic (ed.), *Beyond Geometry: Classic Papers from Riemann to Einstein*. New York: Dover, 2007).

the senses.

To apply geometry to our experience of space means to find adequate formal contexts (described by mathematical theories of space, i.e. geometries) where to represent our spatial experience by means of isomorphic embeddings. Notice in passing that it is not intrinsically necessary that the geometrical models of our spatial experience be Euclidian; it is a matter of fact, not right (just like Poincaré and Gauss wanted), which geometry best fits the experience. The roughness of our perception of spatial structure, in particular, can accommodate different geometrical representations of perceptual space. If our perception of spatial structure, or our scientific reconstruction of it, as seems to be in fact the case, is dependent of physical phenomena and what happens to be located in space, we can hardly expect geometrical representations to be a priori.

The explanation of the application of mathematics to science or to itself follows invariably the same pattern (such uniformity, I think, counts as an argument in favor of this view): mathematics provides us with logically articulated formal (equivalently, structural) descriptions (i.e., formal theories of logically possible objectual domains that have no necessary existence outside mathematics) we can use to investigate the formal properties of domains of our interest (in science, these are typically sets of experimental data displaying different degrees of structural articulation).

To conclude I want, based on the characterization of mathematics as a formal science that I presented here, to establish a distinction between mathematical and analytic assertions, and thus between mathematics and logic. Husserl, following Bolzano, defined *analyticity* as the character of truths that remain true by arbitrary reinterpretations of the non-logical terms occurring in their linguistic expressions. It is the idea of analyticity as the character of what is true due only to form, not content, which corresponds to our modern conception of *validity*.

So, a true assertion − i.e. one that is true in some interpretation − is *analytic* provided it is true in *any* interpretation for its language. An analytic truth is one that is true in all domains of relevant signature. This definition brings forth the close relationship, but also the distinction between analytic and mathematical truths. Whereas the former remain true under reinterpretations within the class of all the interpretations of their languages, the latter remain necessarily true only under reinterpretations that preserve the truth of the formal stipulations on which their truth logically depends. Both mathematical and analytic truths, however, share the property of being *formally* true; unconditionally true the former, conditionally true the latter. In this lies the logical character of mathematics.

From the point of view I defended here, which one may call, to be on the safe side, *ontologically non-committed formalism or structuralism*[16], I believe I was able to show that we can make sense of the common mathematical practice of embedding mathematical domains into one another and extending theories in order to prove theorems. As I have argued, no matter the meaning attached to mathematical domains, all that matters is their formal properties, which are preserved under isomorphic embeddings. It is also immaterial whether the extended domains exist or not; all that is demanded of extensions is consistency.

It seems then that Hilbert was right after all; mathematicians are free to create whatever they like, provided their creations are consistent. The history of mathematics shows that inventing theories, no matter how little intuitive, how far from being literally true (i.e. true *of* something that exists independently of them, or true in the sense of the correspondence theory of truth), or even how *meaningless* they happen to be, provided they solve problems, is precisely what mathematics is all about. We could see this simply as sheer irresponsibility, which sometimes mysteriously pays off by being useful, but we need not.

I think even a quick glance at some real mathematical proof procedures forces upon us a structural view of mathematics, one in which the "mystery" of the applicability of mathematics, to science and to itself, receives a most natural answer[17]. In particular, this perspective allows us to see the error of the so-called *indispensability argument*. According to this argument mathematical objects *must* exist, since the applicability of mathematics to science requires mathematics to be true, and no true theory can be about non-existing entities. The error of this line of thought, of course, is that mathematics is not "about" objects, but forms or structures; and forms can be useful for investigating other forms even if one, some or all of them are not forms of anything that really exists.

---

[16]Whether mathematical objects, manifolds or structures really exist, in the strong sense realists believe they exist, or are only fictional creations, does not really matter; nothing would change one way or the other in the structural account of mathematics I presented here. In fact, I do not believe the so-called ontological problem can be definitively "solved", and even if it could, I do not think the "solution" would matter so much to our understanding of the nature of mathematics and the vastly more important problem of the philosophy of mathematics: how to account for the applicability of mathematics?.

[17]Of course, I have in mind mostly the "reasonable" effectiveness of mathematics. What Wigner called its "unreasonable" effectiveness ("The Unreasonable Effectiveness of Mathematics in the Natural Sciences", *Communications in Pure and Applied Mathematics*, vol. 13, n° 1, February 1960) may require a more elaborate explanation (which, however, I think can still be provided along the same, or analogous lines).

Jairo José da Silva
Mathematics Department
UNESP
Rio Claro, Brazil
E-mail: jairomat@linkway.com.br.

SECTION 4

LOGIC, PROOFS AND GAMES

Goal-Directed Tableaux

JOKE MEHEUS AND KRISTOF DE CLERCQ

ABSTRACT. This paper contains a new format for analytic tableaux, called goal-directed tableaux. Their main interest lies in the fact that the search for a closed tableau proceeds in a highly constrained way. The goal-directed tableaux do not form a complete decision method for propositional classical logic (because they do not sustain *Ex Falso Quodlibet*). For consistent sets of premises, however, they lead to the same results as the usual analytic tableaux for classical logic.

1 Introduction

The aim of this paper is to present a new kind of analytic tableaux for propositional logic that we shall call goal-directed tableaux. Although originally intended for a new approach to tableaux-based abduction (see [11]), the goal-directed tableaux turn out to be interesting in themselves. Their main interest lies in the fact that the search for a closed tableau proceeds in a goal-directed manner: the format ensures that no formula is decomposed in a branch unless this formula is (in a specific way) related to the top formula of that branch. It is moreover warranted that the search space is highly constrained: at each stage of a tableau construction, the last formula of an open branch uniquely determines the next step for that branch.

The method presented here is inspired by the method for *parent clash restricted* (PCR) tableaux (see [15]), which are called tightly connected tableaux in [5] and in [8].[1] Tightly connected tableaux are effective in avoiding redundant steps—steps that do not contribute in finding a closed tableau. This is because in a tightly connected tableau, no branching rule is applied to a formula, unless this leads (in one or more steps) to the closure of some branch.

The existing methods for tightly connected tableaux are complete for classical logic (see [8, pp. 176-180] for proofs).[2] They exhibit, however, a serious shortcoming: they are not confluent. A tableau method is said to be *confluent* if it has the following property: if $\Gamma \vDash A$, then *every* tableau for $\Gamma \vDash A$ can be extended to a closed tableau. What this comes to is that the *order* in which the steps are performed does not influence the final result:

---

[1]Both Bibel's connection method (see [3]) and Loveland's model elimination calculus (see [9]) are related to the method of tightly connected tableaux—see [8, p. 182].

[2]A tableau method is said to be complete if it satisfies the property that, whenever $\Gamma \vDash A$, there exists a closed tableau for $\Gamma \vDash A$.

no matter how one proceeds, one will be able to find a closed tableau, if there is one.

As a side-effect of its goal-directedness, the method presented in this paper is neither complete nor confluent for classical propositional logic. However, unlike the existing methods for tightly connected tableaux, it is complete as well as confluent for the paraconsistent logic **CL⁻** from [1]—the system **CL⁻** validates all rules of classical propositional logic, except for *Ex Falso Quodlibet*. Another way to put is that the method is complete and confluent, whenever the set of premises is consistent.

In the next section, we introduce the idea of a tightly connected tableau and show how the confluence problem can be solved.

2　Some Preliminary Observations

Suppose that we are asked to find out whether f is a semantic consequence of $\{a \supset \neg b,\ a \vee \neg c,\ b \vee ((d \vee e) \vee c),\ a \vee \neg b,\ \neg a \vee \neg c,\ f \vee \neg d,\ g \vee h,\ \neg e\}$, and that we make an (unsigned) analytic tableau for this, as is shown in Figure 1. The numbers in the tableau refer to the stage at which the formulas are written down and are added for convenience only.

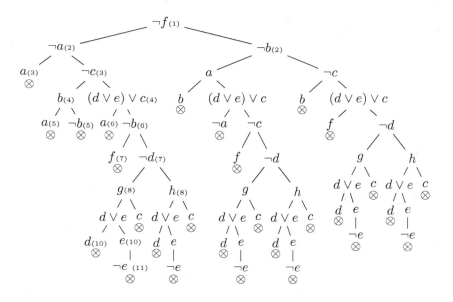

Figure 1. An analytic tableau for $a \supset \neg b,\ a \vee \neg c,\ b \vee ((d \vee e) \vee c),\ a \vee \neg b,\ \neg a \vee \neg c,\ f \vee \neg d,\ g \vee h,\ \neg e \vDash f$

The tableau in Figure 1 satisfies the only condition that Smullyan [14, p. 18] offers for efficient tableaux: formulas that do not cause branching (if any) are decomposed before those that cause branching. It is easily

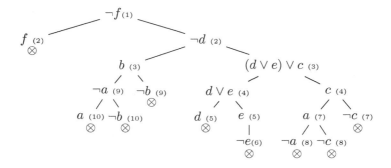

Figure 2. A more concise tableau for $a \supset \neg b$, $a \vee \neg c$, $b \vee ((d \vee e) \vee c)$, $a \vee \neg b$, $\neg a \vee \neg c$, $f \vee \neg d$, $g \vee h$, $\neg e \models f$

observed, however, that the tableau includes several redundant steps. A more concise tableau for the same problem is presented in Figure 2.

In order to explain what the difference is between the two tableaux, we first need some terminology. We shall say that a premise A is *introduced* in a branch θ at stage s of the tableau if, at stage s, A is appended to θ or θ is extended by applying a tableau rule to A. Thus, in the tableau in Figure 1, the premise $a \supset \neg b$ is introduced in both branches at stage 2 of the tableau and $\neg e$ is introduced in the seventh branch from the left at stage 11.

We shall say that a formula A is a *positive part* of a formula B iff A can be obtained from B by zero or more applications of the tableau rules. Thus, $\neg a$ is a positive part of $\neg a$ as well as of $a \supset b$ and $\neg a \vee \neg b$. Where $*A$ denotes the complement of A (that is, B if A has the form $\neg B$ and $\neg A$ otherwise), we shall say that a formula A is *connected* to a formula B iff $*B$ is a positive part of A. Thus, for instance, $c \vee d$ is connected to $\neg c$ and $a \equiv b$ is connected to a, $\neg a$, b and $\neg b$.

The difference between the two tableaux is that, in the second one, no premise A is introduced in a branch θ *unless* A is connected to some formula B that occurs in θ. This warrants that applying the tableau rules to A will result in at least one extension of θ that is *closed*. By thus restricting the application of the tableau rules, it is avoided that redundant premises cause unnecessary branching.

When the tableau in Figure 2 is at stage 1, there is only one premise that satisfies the criterion from the previous paragraph, namely $f \vee \neg d$. Applying the appropriate tableau rule to $f \vee \neg d$ results in two extensions and the left one is immediately closed. At stage 2, there is again only one premise that satisfies the criterion, namely $b \vee ((d \vee e) \vee c)$. Applying the tableau rules to this formula as well as to one of its parts, namely $(d \vee e) \vee c$, leads to four new extensions at stage 5 of the tableau and one of them is closed.

A closer inspection of the tableau in Figure 2 reveals that an even stronger property holds for it: whenever a premise A is introduced in a branch θ, A is connected to the *last* formula that occurs in θ. A tableau that satisfies this property will be called *tightly connected*. It will be called *connected* if it holds true that whenever a premise A is introduced in a branch θ, A is connected to *some* formula that occurs in θ. The difference is illustrated in Figure 3: both tableaux are connected but only the right one is tightly connected.

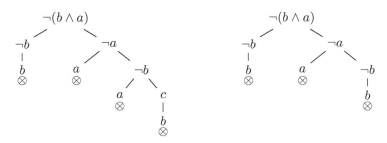

Figure 3. A connected tableau and a tightly connected tableau for $a \vee \neg b$, $a \vee c$, $b \vDash b \wedge a$

It is easily observed that the tableau in Figure 1 is neither tightly connected nor connected. For instance, the premise $a \supset \neg b$ is introduced in both branches at stage 2 of the tableau although it is not connected to any formula in the tableau at that stage. Also, the premise $g \vee h$ is introduced several times, but is not connected to any formula in the tableau. Note that, in the tableau in Figure 2, the introduction of $a \supset \neg b$ is postponed until stage 9 of the tableau and $g \vee h$ is not introduced in any of the branches.

The question is now whether it is possible to restrict our attention to tableaux that are *tightly* connected. The advantage of a tightly connected tableau is that, in order to decide how the tableau should be expanded, only the last formula of each branch should be considered.

At first sight, it seems that the answer to this question must be negative. Consider, for instance, the tableau in Figure 4. At stage 2 of the tableau $a \supset c$ and $b \supset c$ are added to the tableau in view of the premise $(a \supset c) \wedge (b \supset c)$ and the fact that $(a \supset c) \wedge (b \supset c)$ is connected to $\neg c$. At stage 3, the formula $b \supset c$ is decomposed. If we would require that the tableau be tightly connected, the procedure would stop here and the tableau would not be closed. Still, it is clear that the tableau can be closed by first decomposing $a \supset c$ and next introducing $d \supset a$ and d.

Another example is presented in Figure 5. Also here, a closed tableau exists, but the tableau in Figure 5 cannot be expanded into a closed tableau that is tightly connected.

What both examples illustrate is that simply imposing the restriction concerning tight connectedness on the usual tableaux results in a method that is not confluent.

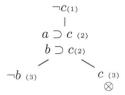

Figure 4. An unsuccessful attempt to construct a tightly connected tableau for $(a \supset c) \wedge (b \supset c)$, $d \supset a$, $d \vDash c$

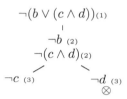

Figure 5. An unsuccessful attempt to construct a tightly connected tableau for a, $a \supset b \vDash b \vee (c \wedge d)$

To overcome this difficulty, we shall make two changes to the usual format of analytic tableau. The first is that we shall work with *labelled* formulas. In order to construct a tableau for $\Gamma \vDash A$, we first define the set $(\Gamma \cup \{\neg A\})^{\lambda}$. The idea is that $(\Gamma \cup \{\neg A\})^{\lambda}$ is the set of all formulas that are obtained from members of $\Gamma \cup \{\neg A\}$ by underlining exactly one occurrence of a schematic letter in them. For instance, where $\Gamma \cup \{\neg A\} = \{a \supset b,\ a \wedge (b \supset c),\ \neg(b \wedge c)\}$, $(\Gamma \cup \{\neg A\})^{\lambda} = \{\underline{a} \supset b,\ a \supset \underline{b},\ \underline{a} \wedge (b \supset c),\ a \wedge (\underline{b} \supset c),\ a \wedge (b \supset \underline{c}),\ \neg(\underline{b} \wedge c),\ \neg(b \wedge \underline{c})\}$.

The second change is that we shall work with tableaux in which the different search paths are *separated* from each other. This requires some more explanation. Consider again the question whether c is a semantic consequence of $\{(a \supset c) \wedge (b \supset c),\ d \supset a,\ d\}$. One search path is presented in Figure 4. This search path is unsuccessful—it does not lead to the closure of all branches of the tableau. There is, however, a different search path that is successful. This second search path is presented in Figure 6. Note especially that the tableau in Figure 6 is tightly connected.

In an ordinary analytic tableau, different search paths are simply concatenated. Thus, if one starts the tableau for $(a \supset c) \wedge (b \supset c)$, $d \supset a$, $d \vDash c$ as in Figure 4, one extends the open branch of the tableau with the tree below $a \supset c$ from Figure 6. The result is a closed tableau that is connected, but not tightly connected. The idea behind the goal-directed tableaux is that they enable the *parallel* exploration of different search paths in such a way that each search path in the tableau is tightly connected.

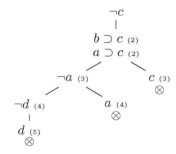

Figure 6. A successful attempt to construct a tightly connected tableau for $(p \supset r) \wedge (q \supset r)$, $s \supset p$, $s \vDash r$

To realize this technically, we distinguish between two kinds of junctions: *and-junctions* and *or-junctions*. The or-junctions are the usual junctions of analytic tableaux and are represented as usual. The and-junctions are represented by downwards forks and are used for the introduction of premises as well as for the decomposition of those formulas that in a Smullyan-tableau do not cause branching.

With respect to this new format, we shall say that a tableau is *tightly connected* if every formula in the tableau, except for the top formula, is either connected to its parent or is a positive part of its parent. The negation of the conclusion by which we start the tableau will be called the *main goal* of the tableau and, at each stage of the tableau, the last formula of an open branch will be called the *current goal* of that branch.

In Figure 7, we present a goal-directed tableau for $(a \supset c) \wedge (b \supset c)$, $d \supset a$, $d \vDash c$. The main goal of the search tree is $\neg c$. As c occurs twice in the premise $(a \supset c) \wedge (b \supset c)$, this premise is connected in two different ways to the main goal. This is why in the search tree it is entered twice, but with a different label. At stage 3 of the search tree, we begin the search path to the right by adding the labelled conjunct $b \supset \underline{c}$. At stage 4, we add $\neg b$ and $\underline{c}$. At this stage, the search path to the right stops: there is no labelled premise that is connected to $\neg b$. This is why we now start working on the search path to the left. Again, we first add the labelled conjunct. Decomposing this formula, we obtain $\neg a$ and $\underline{c}$. The difference with the search path to the right is that now there is a labelled premise that is connected to the last formula of the open branch, namely $d \supset \underline{a}$. As there is only one labelled premise that satisfies this criterion, we add $d \supset \underline{a}$ as the sole and-successor of $\neg a$ (stage 7). Two stages later the search path to the left is closed—both its branches are closed. At this moment, we know that, if only we give up the constraint on tight connectedness, also the search path on the right can be closed. It suffices that we copy the formulas added at stages 5-9 in the open branch in the search path to the right. This is why we mark the open branch in the right search path with a "C".

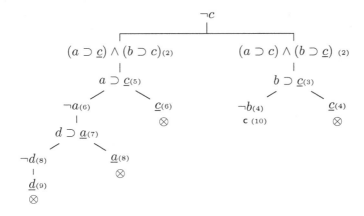

Figure 7. A closed goal-directed tableau for $(a \supset c) \wedge (b \supset c)$, $d \supset a$, $d \vDash c$

The goal-directed tableau for a, $a \supset b \vDash b \vee (c \wedge d)$ is presented in Figure 8. As applying the appropriate tableau rule to the main goal leads to two conclusions, we use an and-junction for its decomposition (stage 2). At stage 3, we decompose $\neg(c \wedge d)$ in the right branch. As no premise is connected to the resulting literals, the search path to the right stops and we continue the search path to the left. The only premise that is connected to $\neg b$ is $a \supset b$. Hence, this premise is entered as the sole and-successor of $\neg b$ (stage 4). After decomposing $a \supset b$ (stage 5) and introducing the premise a (stage 6), both branches to the left are closed. In view of this, both branches to the right are marked with a "C" at stage 7. Remark that also this tableau is tightly connected: apart from the main goal, every formula in the tableau is connected to its parent or is a positive part of its parent.

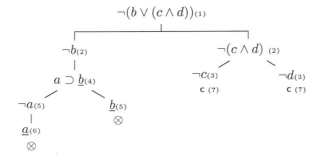

Figure 8. A closed goal-directed tableau for a, $a \supset b \vDash b \vee (c \wedge d)$

The tableau for $\models \neg a \lor a$ (see Figure 9) illustrates why also the negation of the conclusion has to be included in the set of labelled formulas. At stage 2, an and-junction is used for the decomposition of the main goal. This is why we need to reintroduce (a labelled version of) the main goal in the tableau in order to obtain a closed branch. As soon as the branch to the left is closed, the one to the right may be C-marked.

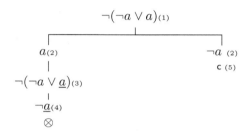

Figure 9. A closed goal-directed tableau for $\models \neg a \lor a$

3 The Goal-Directed Tableau Method

In this section, we present the definitions and rules for the goal-directed method and prove that it is adequate. In order to define the tableau rules, we make use of $\mathfrak{a}$- and $\mathfrak{b}$-formulas (as in [14]). This distinction also allows one to define the positive part relation in a concise way.

$\mathfrak{a}$	$\mathfrak{a}_1$	$\mathfrak{a}_2$	$\mathfrak{b}$	$\mathfrak{b}_1$	$\mathfrak{b}_2$
$(A \land B)$	A	B	$\neg(A \land B)$	$*A$	$*B$
$(A \equiv B)$	$(A \supset B)$	$(B \supset A)$	$\neg(A \equiv B)$	$\neg(A \supset B)$	$\neg(B \supset A)$
$\neg(A \lor B)$	$*A$	$*B$	$(A \lor B)$	A	B
$\neg(A \supset B)$	A	$*B$	$(A \supset B)$	$*A$	B
$\neg\neg A$	A	A			

The following clauses constitute a recursive definition of the *positive part* relation for propositional logic. An expression of the form $\mathrm{pp}(A, B)$ will be read as "A is a positive part of B".

1. $\mathrm{pp}(A, A)$.

2. $\mathrm{pp}(A, \mathfrak{a})$ if $\mathrm{pp}(A, \mathfrak{a}_1)$ or $\mathrm{pp}(A, \mathfrak{a}_2)$.

3. $\mathrm{pp}(A, \mathfrak{b})$ if $\mathrm{pp}(A, \mathfrak{b}_1)$ or $\mathrm{pp}(A, \mathfrak{b}_2)$.

4. If $\mathrm{pp}(A, B)$ and $\mathrm{pp}(B, C)$, then $\mathrm{pp}(A, C)$.

We shall say that a formula A is *connected* to a formula B iff pp($*B, A$). Where A^λ is $\{B \mid B$ is obtained from A by underlining *exactly one* occurrence of a schematic letter in $A\}$, the set Δ^λ will stand for $\bigcup_{A \in \Delta} A^\lambda$. A formula will be called labelled if it contains a schematic letter that is underlined. Where A is a literal, an expression of the form $\sigma(A)$ will refer to the schematic letter that occurs in A. We shall say that a formula A is *fulfilled* in a branch θ if, where A is an $\mathfrak{a}$-formula both $\mathfrak{a}_1$ and $\mathfrak{a}_2$ occur in θ, and where A is a $\mathfrak{b}$-formula, $\mathfrak{b}_1$ or $\mathfrak{b}_2$ occurs in θ.

In addition to the starting rule, the procedure consists of a premise rule and three rules for extending the tableau.

SR Start the tableau for $\Gamma \vDash A$ by writing down $*A$.

PR If θ is an open branch of a tableau for $\Gamma \vDash A$, the last formula C of θ is a *literal*, and $B_1, \ldots, B_n$ ($n \geq 1$) are the members of $(\Gamma \cup \{\neg A\})^\lambda$ that are connected to C and in which $\sigma(C)$ occurs underlined, then $B_1, \ldots, B_n$ may be adjoined to θ as the n and-successors of C.

EXT1 If some *labelled* $\mathfrak{a}$ is the current goal of some branch θ and $\mathfrak{a}_1$ (respectively $\mathfrak{a}_2$) is labelled in $\mathfrak{a}$, then $\mathfrak{a}_1$ (respectively $\mathfrak{a}_2$) may be adjoined to θ as the sole and-successor of $\mathfrak{a}$.

EXT2 If some *non-labelled* $\mathfrak{a}$ is the current goal of some θ, then $\mathfrak{a}_1$ and $\mathfrak{a}_2$ may be adjoined to θ as the two and-successors of $\mathfrak{a}$.

EXT3 If some (labelled or non-labelled) $\mathfrak{b}$ is the current goal of some branch θ, then $\mathfrak{b}_1$ and $\mathfrak{b}_2$ may be adjoined to θ as the two or-successors of $\mathfrak{b}$.

In addition to the rules for extending the tableau, we also need a marking definition. Where $B_1, \ldots, B_n$ are the n and-successors of some formula A, we shall say that $B_1, \ldots, B_n$ are each others *and-siblings*.

DEFINITION 1. A branch θ is C-marked in a tableau at stage s iff it contains an and-successor B of some A and, for some and-sibling C of B, all branches that go through C are closed.

Thus, at stage 10 of the tableau in Figure 7, one of the branches is C-marked: the branch contains an and-successor of the main goal $\neg c$ (namely, $(a \supset c) \wedge (b \supset \underline{c})$), and there is an and-sibling of $(a \supset c) \wedge (b \supset \underline{c})$ (namely $(a \supset \underline{c}) \wedge (b \supset c)$) such that all branches that go through $(a \supset \underline{c}) \wedge (b \supset c)$ are closed.

The definitions for open and closed branches are as usual, except that the marking of branches has to be taken into account.

DEFINITION 2. A branch is *finished* iff no tableau rule can be applied to it.

DEFINITION 3. A branch is *closed* iff it contains some formula and its negation.

DEFINITION 4. A branch is *open* iff it is finished, it is not closed *and* it is not C-marked.

In order to define the notions of closed and open tableaux and in view of the adequacy proofs, we need to make a distinction between *stopped* tableaux and *completed* tableaux. This requires that we first define the conditions under which a formula is *complete*:

DEFINITION 5. A branch θ of a tableau for $\Gamma \vDash A$ is *complete* iff it is finished and for every $C \in (\Gamma \cup \{\neg A\})^\lambda$ that does not occur in θ, there is some branch θ' such that C occurs in θ' and C is an and-sibling of some formula B that occurs in θ.

The definitions for *completed* tableaux and for *stopped* tableaux are as follows:

DEFINITION 6. A tableau is *completed* iff every branch in the tree is complete or closed or C-marked.

DEFINITION 7. A tableau is *stopped* iff no tableau rule can be applied to it, but it is not completed.

We shall say that a *tableau* is *finished* iff it is completed or it is stopped. In view of the above, the notions of closed and open tableaux are defined as follows:

DEFINITION 8. A tableau is *closed* iff every branch in the tree is closed or C-marked.

DEFINITION 9. A tableau is *open* iff it is finished and it is not closed.

We now present the adequacy proofs for the goal-directed procedure.

THEOREM 10. *If the tableau for* $A_1, \ldots, A_n \vDash B$ *is closed, then* $A_1, \ldots, A_n \vDash B$.

Proof. Suppose that $A_1, \ldots, A_n \nvDash B$. It follows that there is a valuation function v that assigns the value 1 to $A_1, \ldots, A_n$ as well as to $\neg B$. It is easily shown by induction that, at each stage of the tableau, there is a branch θ in the tableau such that (i) v *agrees with* θ (that is, v assigns the value 1 to all formulas that occur in θ) and (ii) θ is not C-marked.

After applying the starting rule SR, it is obvious that v agrees with the sole branch in the tableau and that the latter is not C-marked (basis). Suppose that v agrees with θ at some stage s, that θ is not C-marked at stage s, and that we apply a rule to it. We have to distinguish four cases. The applied rule is:

PR.   In view of the induction hypothesis and the fact that v assigns the value 1 to all premises, v obviously agrees with all extensions of θ at stage $s + 1$. But then, none of the extensions is closed at stage

$s + 1$, and hence, in view of the marking definition, none of them is C-marked.

EXT1. As v obviously agrees with the single extension of θ at stage $s + 1$, the latter is not C-marked at stage $s + 1$.

EXT2. As v obviously agrees with both resulting extensions, neither of them is C-marked at stage $s + 1$.

EXT3. As v agrees with θ at stage s, v obviously agrees with at least one of the resulting extensions. Moreover, as θ is not C-marked at stage s, the marking definition warrants that neither extension is C-marked at stage $s + 1$.

It follows that, at each stage of the tableau, there is at least one branch that is not closed (as v agrees with it) and that is not C-marked. Hence, the finished tableau is not closed. ■

THEOREM 11. *If the tableau for $A_1, \ldots, A_n \models B$ is open and completed, then $A_1, \ldots, A_n \nvDash B$*

Proof. Consider a branch θ in the completed tableau for $A_1, \ldots, A_n \models B$ that is open and extend it in the following way. Look for the topmost formula C (if there is any) that is an and-sibling of some $C_1, \ldots, C_m$ and extend θ first with C_1 and the complete tree below C_1. Next, extend all the thus obtained extensions with C_2 and the tree below C_2 and so on. Repeat the procedure for all the and-junctions that occur in the extensions. It is easily observed (in view of the definition of a completed tableau) that in the final result all extensions of θ will contain all premises. Moreover, as θ was not C-marked, at least one of the extensions of θ will be open.

Consider a branch θ' in the extended tableau that is open and that contains all premises. As the tableau was completed, any b-formula that occurs in θ' is fulfilled in θ'. Moreover, in view of the way in which θ'' was obtained, it is obvious that, for any a-formula that is not fulfilled in θ'', $\mathfrak{a}_1$ (respectively $\mathfrak{a}_2$) is contained in θ'' and $\mathfrak{a}_2$ (respectively $\mathfrak{a}_1$) is not connected to any literal that occurs in θ'. Hence, if θ' is further extended until all formulas are fulfilled (by means of the usual rules for analytic tableaux), all thus obtained extensions will be open. Let θ'' be one of the resulting extensions and consider a valuation function v that assigns the value 1 to all literals that occur in θ''.

As all formulas that occur in θ'' are fulfilled in θ'', it follows, by an obvious induction on the complexity of formulas (their length), that v assigns the value 1 to all formulas that occur in θ''. Hence, as θ'' contains all premises, it follows that $A_1, \ldots, A_n \nvDash B$. ■

As was mentioned above, the tableau method does not sustain *Ex Falso Quodlibet*. This is directly related to the fact that the premise rule is restricted to formulas that are connected to formulas that already occur in a branch. Because of this, the tableau method delivers all the classical consequences of an inconsistent set of premises, except for those that are trivial.

For instance, whereas it is neither possible to obtain a closed tableau for a, $\neg a$, $\neg a \vee b \vDash c$ (none of the premises will be introduced in a goal-directed tableau) nor for a, $\neg a$, $\neg a \vee b \vDash \neg b$, one does obtain closed tableaux for a, $\neg a$, $\neg a \vee b \vDash b$ as well as for (i) a, $\neg a$, $\neg a \vee b \vDash a \vee b$, (ii) a, $\neg a$, $\neg a \vee b \vDash a \vee \neg b$, (iii) a, $\neg a$, $\neg a \vee b \vDash \neg a \vee b$, and (iv) a, $\neg a$, $\neg a \vee b \vDash \neg a \vee \neg b$.

It can be shown, however, that, if Γ is consistent and the tableau for $\Gamma \vDash B$ is open and stopped, then $\Gamma \nvDash B$.

THEOREM 12. *If the tableau for* $\Gamma \vDash A$ *is open and stopped, then* $\Gamma \nvDash A$ *or* Γ *is inconsistent.*

Proof. If the tableau for $\Gamma \vDash A$ is open and stopped, there is a branch θ in it that is open but not complete. Hence, there are one or more premises B that do not occur in θ and for which there is no formula C such that C occurs in θ and C is an and-sibling of B. In view of the tableau rules, it is easily observed that these premises are not connected to any formula in θ. Hence, the only way in which completing θ may lead to closure is when the premises themselves are inconsistent. The same reasoning applies to any other branch in the tableau that is open but not complete. ∎

4   Comparisons and Possible Extensions

As will be clear from the previous sections, the goal-directed method is especially suited for the refutation of sets that are not minimally inconsistent. In cases like this, the goal-directed method avoids, in a very natural way, that redundant premises are introduced in the search tree. This is an important difference with other methods that were primarily designed to increase the computational efficiency of analytic tableaux, such as the **KE** system (see [7] and [4]). Apart from the starting rule, the **KE** system consists of two elimination rules and one branching rule:

SR*   Start the tableau for $\Gamma \vDash A$ by writing down all members of Γ as well as $\neg A$.

E1   Where θ is an open branch in which some $\mathfrak{a}$ occurs, both $\mathfrak{a}_1$ and $\mathfrak{a}_2$ may be adjoined to θ.

E2   Where θ is an open branch in which some $\mathfrak{b}$ occurs, $\mathfrak{b}_1$ (respectively $\mathfrak{b}_2$) may be adjoined to θ, *provided* that $*\mathfrak{b}_2$ (respectively $*\mathfrak{b}_1$) occurs in θ.

PB   Where θ is an open branch and A is its last formula, both B and $\neg B$ may be adjoined to θ as the two or-successors of A.

In [5], a simple refutation procedure is presented for **KE** (called the *canonical procedure*) and is proven to be complete. The canonical procedure warrants that the rule PB is only applied when none of the elimination rules is applicable and moreover that any application of the rule PB results in the decomposition of at least one formula.

From a computational point of view, the canonical procedure for **KE** is more efficient than the goal-directed method. For instance, there are cases where the goal-directed method necessarily leads to repeated subtrees whereas the **KE** procedure does not. By way of example, compare the tableaux in Figures 10 and 11. In the goal-directed tableau, the open branches that end with $\neg a$ can only be closed by repeating the subtree below $\neg a$ from the left branch. More importantly, it has been shown that the canonical procedure for **KE** can polynomially simulate (p-simulate)[3] the Smullyan tableau method, whereas the latter cannot p-simulate the former (see [5, p. 108–110]). A similar complexity result cannot be proven for the goal-directed method presented here.

$$a \vee b$$
$$a \vee \neg b$$
$$\neg a \vee (c \wedge d)$$
$$\neg(a \wedge (c \wedge d))$$

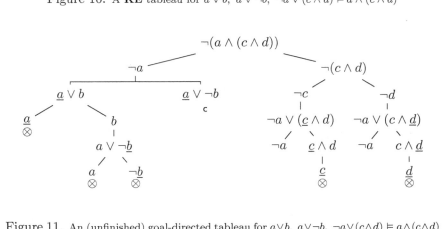

Figure 10. A **KE** tableau for $a \vee b$, $a \vee \neg b$, $\neg a \vee (c \wedge d) \vDash a \wedge (c \wedge d)$

Figure 11. An (unfinished) goal-directed tableau for $a \vee b$, $a \vee \neg b$, $\neg a \vee (c \wedge d) \vDash a \wedge (c \wedge d)$

Still, the **KE** method does not prevent one from decomposing redundant formulas and hence may lead to unnecessary branchings. For instance, the tableau in Figure 12 is constructed in accordance with the canonical procedure for **KE**, but clearly contains an unnecessary branching. No tableau

---

[3]See [5, p. 56] for a definition of p-simulation.

that is based on the idea of connectedness will contain this branching (see also Figure 2).

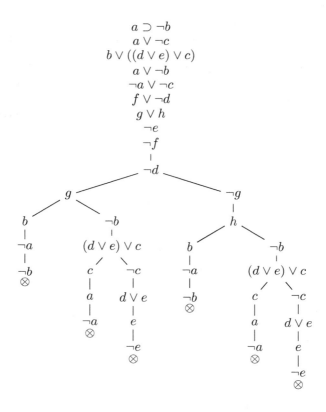

Figure 12. An inefficient **KE** tableau for $a \supset \neg b$, $a \vee \neg c$, $b \vee ((d \vee e) \vee c)$, $a \vee \neg b$, $\neg a \vee \neg c$, $f \vee \neg d$, $g \vee h$, $\neg e \vDash f$

Although this was not its primary aim, the efficiency of the goal-directed method may be increased in a number of ways. An obvious extension is that a derived rule is introduced for or-junctions:

EXT3*   If some (labelled or non-labelled) $\mathfrak{b}$ is the current goal of some branch θ and $*\mathfrak{b}_1$ (respectively $*\mathfrak{b}_2$) occurs in θ, then $\mathfrak{b}_2$ (respectively $\mathfrak{b}_1$) may be adjoined to θ as the sole or-successor of $\mathfrak{b}$.

In addition to this, one may introduce a marking definition that avoids the repetition of closed subtrees:[4]

---

[4]This is similar to the *checking* procedure in *improved analytic tableaux* (see [15, p. 25–26]).

DEFINITION 13. A branch θ is R-marked iff its last formula is A and there is an ancestor of A that has a child that is also A and that is at the top of a *closed* subtree.

A final extension concerns the labelling procedure. Instead of labelling schematic letters, one may opt for labelling subformulas. This may be realized by defining A^λ as $\{B \mid B$ is obtained from B by underlining exactly one *subformula* in $A\}$ and by reformulating the premise rule:

PR   If θ is an open branch of a tableau for $\Gamma \vDash A$, C is the last formula of θ, and $B_1, \ldots, B_n$ ($n \geq 1$) are the members of $(\Gamma \cup \{\neg A\})^\lambda$ that are connected to C and in which $*C$ occurs labelled, then $B_1, \ldots, B_n$ may be adjoined to θ as the n and-successors of C.

We shall use the term *improved* goal-directed tableaux for tableaux that are obtained by means of these extensions. All three extensions are illustrated in the tableau in Figure 13 (compare with the tableau from Figure 11).

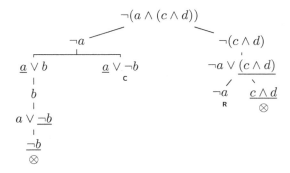

Figure 13. An improved goal-directed tableau for $a \vee b$, $a \vee \neg b$, $\neg a \vee (c \wedge d) \vDash a \wedge (c \wedge d)$

5   In Conclusion

The goal-directed tableaux presented in this paper bear some resemblances with the goal-directed proofs from [2] (and are actually inspired by them). One of the main differences, however, is that the extension to the predicative level is quite hard in the case of goal-directed proofs but rather straightforward in the case of goal-directed tableaux. Another advantage is that for certain applications (such as identifying the formulas that can be 'abduced' from a given set of premises), goal-directed tableaux lead to more elegant solutions than goal-directed proofs (see [11] for an approach to abduction in terms of goal-directed tableaux and [10] for one in terms of goal-directed proofs).

Acknowledgements

Research for this paper was supported by subventions from Ghent University and from the Research Foundation – Flanders (FWO - Vlaanderen). The second author is a Postdoctoral Fellow of the Research Foundation – Flanders. We are indebted to Dagmar Provijn for comments on a previous version.

BIBLIOGRAPHY

[1] Diderik Batens. A paraconsistent proof procedure based on classical logic. See http://www.cle.unicamp.br/wcp3/ for an abstract.

[2] Diderik Batens and Dagmar Provijn. Pushing the search paths in the proofs. A study in proof heuristics. *Logique et Analyse*, 173–175:113–134, 2001. Appeared 2003.

[3] Wolfgang Bibel. *Automated Theorem Proving*. Vieweg, Braunschweig, 1982.

[4] Marcello D'Agostino. Are tableaux an improvement on truth-tables? Cut-free proofs and bivalence. *Journal of Logic, Language and Information*, 1:235–252, 1992.

[5] Marcello D'Agostino. Tableau methods for classical propositional logic. In D'Agostino et al. [6], pages 45–123.

[6] Marcello D'Agostino, Dov Gabbay, Reiner Hähnle, and Joachim Posegga, editors. *Handbook of Tableau Methods*. Kluwer, Dordrecht, 1999.

[7] Marcello D'Agostino and Marco Mondadori. The taming of the cut. Classical refutations with analytic cut. *Journal of Logic and Computation*, 4:285–319, 1994.

[8] Reinhold Letz. First-order tableau methods. In D'Agostino et al. [6], pages 125–196.

[9] Donald W. Loveland. *Automated Theorem Proving: A Logical Basis*. North Holland, Amsterdam, 1978.

[10] Joke Meheus and Dagmar Provijn. Abduction through semantic tableaux versus abduction through goal-directed proofs. *Theoria*, 22/3(60):295–304, 2007.

[11] Joke Meheus and Dagmar Provijn. A goal-directed procedure for tableaux-based abduction. To appear.

[12] Marco Mondadori. Classical analytical deduction. Annali dell'Università di Ferrara, Sez. III. Discussion paper 1, Università di Ferrara, 1988.

[13] Marco Mondadori. Classical analytical deduction, Part II. Annali dell'Università di Ferrara, Sez. III. Discussion paper 5, Università di Ferrara, 1988.

[14] Raymond M. Smullyan. *First Order Logic*. Dover, New York, 1995. Original edition: Springer, 1968.

[15] André Vellino. *The Complexity of Automated Reasoning*. PhD thesis, University of Toronto, 1989.

Joke Meheus and Kristof De Clercq
Centre for Logic and Philosophy of Science
University of Ghent
Ghent, Belgium
E-mails: {Joke.Meheus,Kristof.DeClercq}@UGent.be

Intuitionistic Sequent System with Multiple Succedent

María Fernanda Pallares

ABSTRACT. Sequent systems were presented by Gerhard Gentzen in the thirties. The main difference between the classical and intuitionistic versions was the cardinality of the succedent: intuitionistic sequents have only one formula in the succedent. In 1990, studies on how dependency relations between formulas could guarantee the intuitionistic character of a system, started to appear. This line of research also enabled the development of proof theory of other nonclassical logics. In this text we present the intuitionistic system FIL1, an extension to first order logic of the system FIL developed by De Paiva and Pereira for propositional logic. This extension is sound, complete and satisfies cut-elimination.

1 Introduction.

Gerhard Gentzen developed the *Sequent Calculus* in order to obtain normal derivations. The main difference between Gentzen's classical (LK) and intuitionistic (LJ) calculi lay in the concept of sequent: succedents of intuitionistic sequents have at most one formula. Multiple succedents allow, for instance, the derivation of the Principle of Excluded Middle and Peirce Law which represent the power of classical logic and it is widely believed that multiple succedents are responsible for granting the classical strength to a system. After the introduction of Gentzen's work, intuitionistic systems were developed in which the cardinality restriction was not based on the concept of sequent (usually known as a *global* restriction), but on some rules of inference (*local* restriction). The first multiple succedent intuitionistic systems were Maehara's LJ' (1954) and Dragalin's GHPC (1979). Maehara's system differs from LK in the following rules:

$$\frac{A, \Gamma \Rightarrow}{\Gamma \Rightarrow \neg A} \ (\neg_R) \qquad \frac{A, \Gamma \Rightarrow B}{\Gamma \Rightarrow A \rightarrow B} \ (\rightarrow_R) \qquad \frac{\Gamma \Rightarrow A(a)}{\Gamma \Rightarrow \forall x A(x)} \ (\forall_R)$$

In the propositional fragment, all the restrictions can be concentrated on the conditional rule with the negation as a defined symbol. This was how Schellinx's $IL^>$ system (see appendix) was presented. Schellinx proves that $IL^>$ is an intuitionistic system. It fulfills, in relation with IL[1], the following

---

[1]IL is the version of LJ with the bottom as a primitive symbol.

theorems:

THEOREM 1. $\vdash_{IL>} \Gamma \Rightarrow \Delta$ *if and only if* $\vdash_{IL} \Gamma \Rightarrow \bigvee \Delta$.

COROLLARY 2. *If* $\vdash_{IL>} \Rightarrow A$, *then* $\vdash_{IL} \Rightarrow A$.

Regarding the reduction steps designed by Gentzen to eliminate cuts, it was widely believed [2] that a proof of the cut-elimination theorem for LJ' (or $IL^>$) could be obtained by means of a simple adaptation of Gentzen's techniques. However, in the beginning of the nineties, two examples were presented indicating the contrary. In "Some Syntactical Observations on Linear Logic", Schellinx shows what happens if we try to eliminate cuts with rules that have cardinality restrictions:

$$
\cfrac{\cfrac{\Gamma \Rightarrow \Delta, A}{\Gamma \Rightarrow \Delta, (A \vee B)} \quad \cfrac{(A \vee B), \Theta, C \Rightarrow D}{(A \vee B), \Theta \Rightarrow (C \to D)}(*)}{\Gamma, \Theta \Rightarrow \Delta, (C \to D)} \quad \Rightarrow \quad \cfrac{\cfrac{\Gamma \Rightarrow \Delta, A}{\Gamma \Rightarrow \Delta, (A \vee B)} \quad (A \vee B), \Theta, C \Rightarrow D}{\cfrac{\Gamma, \Theta, C \Rightarrow \Delta, D}{\Gamma, \Theta \Rightarrow \Delta, (C \to D)}(**)}
$$

where (*) is an application of $(\to_R)$ in LJ' and (**) is not an application of the same rule. In the attempt to find an adequate formulation for intuitionistic linear logic, Pereira discovered the second (counter)example:

$$
\cfrac{\cfrac{p \Rightarrow p}{p \Rightarrow p, \perp} \quad \cfrac{\perp, 0 \Rightarrow q}{\perp \Rightarrow (0 \multimap q)}}{p \Rightarrow (0 \multimap q), p}
$$

The limitation which the cardinality of the succedent meant for intuitionistic logic, was also an obstacle for proof theory of other non-classical logics. Before the nineties, there was extensive work on model theory of intermediate (between classical and intuitionistic) logics also incorporating Maehara's formulation. But it was only in the nineties that new lines of research enabled new frames for other logics. One of them was the formulation of *Hypersequents* by Avron. *Hypersequents* are sets of sequents that increase the applications of Gentzen-style systems. Avron's paper "The Method of Hypersequents in the Proof Theory of Propositional Non-Classical Logics" is a very clear diagnosis of the limits of Gentzen's system and why we need new frameworks.

We intend to work along another line of research. Motivated by the limits of the original sequent system, others systems included information about the dependency relations between the formulas of the antecedent and the succedent. We would like to highlight the research carried out by Hyland and De Paiva (and later on, with Braüner's cooperation) for Intuitionistic Linear Logic (FILL), Kashima and Shimura's system (the first cut-free

---

[2]See, for example, Takeuti. *Proof Theory*, p.60.

system for Constant Domains Logic), Braüner's cut-free system for Modal
Logic S5, De Paiva and Pereira's intuitionistic system (FIL) and NFIL, a
version of the latter for Natural Deduction accomplished by Franklin.

These systems, not only show us how the multiple succedent stimulate
many results, but also offer a new way of understanding the particulari-
ties of the different logics. Many of these systems were presented independently. Therefore, a systematic approach for these logics is yet pending.
This work on first-order aims to contribute to this task. Systems with the
concept of *dependency relations* have the same good properties we recognize
in Gentzen-systems. The system we present here illustrates how, just by
adding a simple notation, we can modify the field of application of sequent
systems. The significance of this, is not merely the presentation of a new
system for a specific logic, but the possibility it presents of handling many
logics in proof theory.

Let us consider the system FIL (designed by De Paiva and Pereira) which
is the main reference of our work. In "Uma Breve Nota sobre o Condicional
e Múltiplas Conclusões", Pereira emphasizes the role of the conditional: the
implicational fragment is enough to distinguish classical from intuitionistic
logic. This does not happen with the fragment $\{\vee, \wedge\}$ nor with $\{\wedge, \neg\}$, which
is a curious fact considering that the latter is complete from the classical
point of view.

De Paiva and Pereira define a new kind of sequent. A *decorated sequent*
has the following form:

$$A_1(n_1), ..., A_k(n_k) \Rightarrow B_1/S_1, ..., B_m/S_m$$

where:

(a) A_i $(1 \le i \le k)$ and B_j $(1 \le j \le m)$;

(b) n_i $(1 \le i \le k)$ is a natural number which is the *index* of the formula
A_i;

(c) S_j $(1 \le j \le m)$ is the set of indexes which the formula B_j depends
on.

The intuitionistic character focuses on the restriction of the $(\rightarrow_R)$ rule:

$$\frac{\Gamma, A(n) \Rightarrow \Delta, B/S}{\Gamma \Rightarrow \Delta, (A \rightarrow B)/S^\circ} (\rightarrow_R)$$

(i) For each S' in Δ, the index n does not belong to S'.
(ii) If n ϵ S, then $S^\circ = S - \{n\}$. In the other cases, $S^\circ = $ S.

Here, the multiple succedent does not make the system collapse back into
classical logic and it is a $(\rightarrow_R)$ rule without any restriction that makes this
possible. If we consider that the informal meaning of a sequent $A_1, ..., A_n \Rightarrow$

$B_1, ..., B_m$ is the same as the meaning of the formula $(A_1 \wedge ... \wedge A_n \rightarrow B_1 \vee ... \vee B_m)$, how should we interpret the commas of a multiple succedent sequent that is also intuitionistic? We can see that the succedents are not completely explained by the properties of the disjunction. To do this, we also need to know the disjunction behaves with respect to other logical constants. Therefore, the difference between classical and intuitionistic multiple succedent systems can be characterized by the way the conditional interacts with the disjunction represented by the commas in the succedent. While this internalization can be performed without any restrictions in classical logic, in FIL, the conditional is internalized in a disjunction with special restrictions concentrated on the conditional rule: not all the hypotheses are available in order to introduce an intuitionistic conditional. This is how FIL ensures its intuitionistic character.

In the text mentioned above, De Paiva and Pereira prove cut-elimination and suggest as a model, the proof of soundness presented by Kashima and Shimura. In the cut-elimination theorem with Gentzen's method, the use of mix can block out the application of the $(\rightarrow_R)$ rule. In this case, they present a new version of the cut named *indexed cut*:

$$\frac{\Gamma \Rightarrow \Delta \quad \Gamma' \Rightarrow \Delta'}{\Gamma, \Gamma' \Rightarrow \Delta, \Delta'} \langle A; \alpha_1, ..., \alpha_k; \beta_1, ..., \beta_j \rangle$$

where $< A; \alpha_1, ..., \alpha_k; \beta_1, ..., \beta_j >$ means that the formula 'A' must be deleted only in the positions $\alpha_1, ..., \alpha_k$ of the succedent in the left premise and in the positions $\beta_1, ..., \beta_j$ of the antecedent of the right premise. The indexed cut is a particular case of the mix and it operates as follows:

$$\frac{\dfrac{\pi_1}{\dfrac{\Gamma \Rightarrow \Delta, A/S^1, B/S^2}{\Gamma \Rightarrow \Delta, (A \vee B)/(S^1 \cup S^2)} (\vee_R)} \quad \dfrac{\pi_2}{\dfrac{(A \vee B)(n), \Gamma', C(m) \Rightarrow \Delta', D/(S^3 \cup \{n, m\})}{(A \vee B)(n), \Gamma' \Rightarrow \Delta, (C \rightarrow D)/(S^3 \cup \{n\})} (\rightarrow_R)}}{\Gamma, \Gamma' \Rightarrow \Delta, \Delta'[n|(S^1 \cup S^2)], (C \rightarrow D)/S^3[n|(S^1 \cup S^2)]} \langle (A \vee B); 1; 1 \rangle$$

$$\longmapsto$$

$$\frac{\dfrac{\dfrac{\pi_1}{\dfrac{\Gamma \Rightarrow \Delta, A/S^1, B/S^2}{\Gamma \Rightarrow \Delta, (A \vee B)/(S^1 \cup S^2)} (\vee_R)} \quad \dfrac{\pi_2}{(A \vee B)(n), \Gamma', C(m) \Rightarrow \Delta', D/(S^3 \cup \{m, n\})} \langle (A \vee B); 1; 1 \rangle}{\Gamma, \Gamma'^{\circ}, C(m) \Rightarrow \Delta, \Delta'[n|(S^1 \cup S^2)], D/(S^3 \cup \{m\})[n|(S^1 \cup S^2)]}}{\Gamma, \Gamma' \Rightarrow \Delta, \Delta'[n|(S^1 \cup S^2)], (C \rightarrow D)/S^3[n|(S^1 \cup S^2)]} (\rightarrow_R)$$

Note that, if dependency relations are taken care of, there are no problems with the succedent of the $(\rightarrow_R)$ rule as it was the case with LJ' or $IL^>$. So, in the propositional case, a Gentzen's style proof can be performed fluently.

2   FIL1: extension of FIL to the first-order logic.

We present the system FIL1 (multiplicative version).

NOTATION 3.

- Γ, Γ', Δ, Δ' are sequences of formulas.

- Γ and Δ represent the formulas with the indexes or sets, that is,
  $\Gamma = A_1(n_1), ..., A_k(n_k)$ and $\Delta = B_1/S_1, ..., B_m/S_m$.

- In the $\perp$-axiom, k $\geq$ 1.

- The rules with two premises do not have common indexes.

- $\Delta[k|S]$ is the result of replacing each S in Δ by k ϵ S' by ((S'- {k})$\cup S$).

- $\Delta[k, S]$ is the result of replacing each S' in Δ where k ϵ S' by $(S' \cup S)$.

Axioms

$$A(n) \Rightarrow A/\{n\} \qquad \perp(n) \Rightarrow A_1/\{n\}, ..., A_k/\{n\}$$

Structural rules

$$\frac{\Gamma \Rightarrow A/S, \Delta \qquad A(n), \Gamma' \Rightarrow \Delta'}{\Gamma, \Gamma' \Rightarrow \Delta, \Delta'[n|S]} \ (Cut)$$

$$\frac{\Gamma, A(n), B(m), \Gamma' \Rightarrow \Delta}{\Gamma, B(m), A(n), \Gamma' \Rightarrow \Delta} \ (E_L)$$

$$\frac{\Gamma \Rightarrow \Delta, A/S, B/S', \Delta'}{\Gamma \Rightarrow \Delta, B/S', A/S, \Delta'} \ (E_R)$$

$$\frac{\Gamma \Rightarrow \Delta}{A(n), \Gamma \Rightarrow \Delta} \ (W_L)$$
where n is a new index
introduced in some or many
dependency sets of Δ.

$$\frac{\Gamma \Rightarrow \Delta}{\Gamma \Rightarrow \Delta, A/\{\}} \ (W_R)$$

$$\frac{\Gamma, A(n), A(m) \Rightarrow \Delta}{\Gamma, A(k) \Rightarrow \Delta[max(n,m)|\{k\}]} \ (C_L)$$
where k = min(n,m).

$$\frac{\Gamma \Rightarrow \Delta, A/S, A/S'}{\Gamma \Rightarrow \Delta, A/S \cup S'} \ (C_R)$$

Operational rules

$$\frac{\Gamma, A(n), B(m) \Rightarrow \Delta}{\Gamma, (A \wedge B)(k) \Rightarrow \Delta[max(n,m)|\{k\}]} \ (\wedge_L)$$
where k = min(n,m).

$$\frac{\Gamma \Rightarrow \Delta, A/S \quad \Gamma' \Rightarrow \Delta', B/S'}{\Gamma, \Gamma' \Rightarrow \Delta, \Delta', (A \wedge B)/S \cup S'} \ (\wedge_R)$$

$$\frac{\Gamma, A(n) \Rightarrow \Delta \quad \Gamma', B(m) \Rightarrow \Delta'}{\Gamma, \Gamma'(A \vee B)(k) \Rightarrow \Delta[i|\{k\}], \Delta'[i|\{k\}]} \ (\vee_L)$$
where k = min(n,m) and i= max(n,m).

$$\frac{\Gamma \Rightarrow \Delta, A/S, B/S'}{\Gamma \Rightarrow \Delta, (A \vee B)/S \cup S'} \ (\vee_R)$$

$$\frac{\Gamma \Rightarrow \Delta, A/S \quad B(n), \Gamma' \Rightarrow \Delta'}{(A \rightarrow B)(n), \Gamma, \Gamma' \Rightarrow \Delta, \Delta'[n, S]} \ (\rightarrow_L)$$

$$\frac{\Gamma, A(n) \Rightarrow \Delta, B/S}{\Gamma \Rightarrow \Delta, (A \rightarrow B)/S^\circ} \ (\rightarrow_R)$$
(i) For each S' in Δ
n does not belong to S'.
(ii) if n ϵ S,
then $S^\circ = S - \{n\}$.
In other cases, S°= S.

$$\frac{\Gamma, A(a)(n) \Rightarrow \Delta}{\Gamma, \exists x A(x)(n) \Rightarrow \Delta} \ (\exists_L)$$
where 'a' does not occur in the lower sequent.

$$\frac{\Gamma \Rightarrow \Delta, A(t)/S}{\Gamma \Rightarrow \Delta, \exists x A(x)/S} \ (\exists_R)$$

$$\frac{\Gamma, A(t)(n) \Rightarrow \Delta}{\Gamma, \forall x A(x)(n) \Rightarrow \Delta} \ (\forall_L)$$

$$\frac{\Gamma \Rightarrow \Delta, A(a)/S}{\Gamma \Rightarrow \Delta, \forall x A(x)/S} \ (\forall_R)$$
(i)'a' does not occur in
the lower sequent.
(ii)for each S' in Δ,
S' $\cap$ S = $\emptyset$.

2.1 Completeness.

The completeness of FIL1 can be proved syntactically via the intuitionistic system $IL^>$. Let us bear in mind that the formulas in FIL1 are indexed, so they are different sequents. According to this, we mark this difference with Γ' and Δ' Therefore, every derivation in $IL^>$ can be easily transformed in a derivation of FIL1 just by adding the notation on the dependency relations.

THEOREM 4. *If* $\vdash_{IL>} \Gamma \Rightarrow \Delta$, *then* $\vdash_{FIL1} \Gamma' \Rightarrow \Delta'$.

Proof. By induction on derivations. ■

THEOREM 5. *If* $\vdash_{IL} \Gamma \Rightarrow \bigvee \Delta$, *then* $\vdash_{FIL1} \Gamma' \Rightarrow \Delta'$.

Proof. By transitivity between theorems 1 and 4. ■

2.2 Soundness.

THEOREM 6. *If* $\vdash_{FIL1} \Gamma' \Rightarrow \Delta'$, *then* $\vdash_{IL>} \Gamma \Rightarrow \Delta$.

THEOREM 7. *If* $\vdash_{FIL1} \Gamma' \Rightarrow \Delta'$, *then* $\vdash_{IL} \Gamma \Rightarrow \bigvee \Delta$.

Proof. By transitivity between theorems 1 and 6.   ∎

The differences between the systems FIL1 and $IL^>$ are $(\Rightarrow_R)$ and the $(\forall_R)$ rules. Thus, in order to prove the soundness of FIL1, it is enough to show that derivations of FIL1 can be transformed into derivations where the application of $(\to_R)$ and $(\forall_R)$ have only one formula in the succedent of the premise. However, both cases need to be analyzed in different forms. De Paiva and Pereira justify the soundness of the propositional fragment using Kashima and Shimura's strategy. Therefore, our main task will be to explain the $(\forall_R)$ rule. Note that one restriction in the $(\forall_R)$ rule of FIL1 indicates that the active formula A(a) has an exclusive set of hypotheses. So, with the information of the dependency relations, we can fragment this derivation into (possibly) many others. Now, one of the fragments obtained will be a derivation of the sequent $\Gamma_i \Rightarrow A(a)$ where Γ_i is a submultiset of Γ containing the hypotheses which A(a) depends on. By isolating this derivation, we can introduce the universal from a premise with only one formula in the succedent and restore the rest of the formulas by weakening. Thus, the introduction of universal formulas in FIL1 is equal to the introduction in $IL^>$.

Our main problem will be how we deal with the formulas introduced by weakening: both the theorems and the dependency sets of the formulas introduced by (W_R), are empty. If we split the succedent according to the hypotheses which they depend on, we will need to express this difference in a notational way to avoid the case $\vdash_{FIL1} A/\{\}$ where A is any kind of formula. Hence, we will add a star both to the dependency set of the formulas introduced by (W_R): $A/\{\}^*$ and to the indexes of the formulas introduced by (W_L): $A(n^*)$. We will call the formulas, sets and indexes with stars *W-formulas*, *W-indexes* and *W-sets* and the rest of the formulas will be called *genuine*.

We will need to address a further complexity that rises when using the other rules and when the formulas introduced by weakening are combined with other formulas. We define these operations[3] in order to prove the soundness. In the first four, the combination of expressions with and without "*" result in the elimination of the "*":

(C_L): $[S^* \cup S'] = [S \cup S']$

(C_R): $[min(n^*, m)] = [min(n, m)]$

$(\vee_R)$: $[S^* \cup S'] = [S \cup S']$

$(\wedge_L)$: $[min(n^*, m)] = [min(n, m)]$

$(\vee_L)$: $[min(n^*, m)] = [min(n, m)]^*$ and for every index $j \in S$ such as $n \in S$ and/or $m \in S$, then j is replaced by j^* in the lower sequent.

---

[3]We think the role and relation of the W-formulas in a dependency relation system deserve a deeper study.

$(\wedge_R)$: $[S^* \cup S'] = [S \cup S']^*$ and for every index j $\in$ S $\cup$ S', then j is replaced by j^* in the lower sequent.

(Cut): when $\Delta'[n|S^*]$ for some S' in Δ such that n $\in \Delta$, $((S' - \{n\}) \cup S^*) = [((S' - \{n\}) \cup S)]^*$ and for every index j $\in [((S' - \{n\}) \cup S)]^*$, then j is replaced by j^* in the lower sequent.

(Cut): when $\Delta'[n|S]$ for some $S^{*'}$ in Δ such as n $\in \Delta$, $((S^{*'}-\{n\}) \cup S) = [((S' - \{n\}) \cup S)]^*$ and for every index j $\in [((S' - \{n\}) \cup S)]^*$, then j is replaced by j^* in the lower sequent.

$(\rightarrow_L)$: when $\Delta'[n, S^*]$ for some S' in Δ such as n $\in \Delta$, $(S' \cup S^*) = [(S' \cup S)]^*$ and for every index j $\in [(S' \cup S)]^*$, then j is replaced by j^* in the lower sequent.

$(\rightarrow_L)$: when $\Delta'[n, S]$ for some $S^{*'}$ in Δ such that n $\in \Delta$, $(S^{*'} \cup S) = [(S' \cup S)]^*$ and for every index j $\in [(S' \cup S)]^*$, then j is replaced by j^* in the lower sequent .

The universal quantification will be justified along the following lines: firstly, we will prove that if the active formula is a W-formula, it can be deleted and the result will be a derivable sequent (Lemma 10). Then, we will show that the resulting derivable sequent can be restored by an already quantified weakening (Lemma 12). The quantification of a genuine formula is justified in Lemma 11, where we will split each derivation isolating the formula to be quantified and prove that the quantification is done by applying the traditional intuitionistic rule. We will also prove that, in both cases, the same dependency relations are always restored.

Thus, the proof of soundness is a process by which for every derivation of FIL1, we obtain a derivation where every application of the $(\forall_R)$rule has only one formula in the premise. Finally, we will consider these new derivations and use the lemma of Kashima and Shimura in order to obtain derivations where the $(\rightarrow_R)$ rule has one formula in the premise. For the sake of brevity, we will present the lemmas and examples as briefly as possible.

NOTATION 8.

- For this proof, the antecedents and the succedents are multisets of formulas.

- To express variations in the contexts, formulas or sets we add the symbol 'o'. For instance, Γ° is a variations of Γ, etc.

- $[A]\Gamma$ will express that $[A]$ is a detached formula that belongs to Γ.

DEFINITION 9 (Partition). Let $\Gamma \Rightarrow \Delta$ be a sequent where Γ and Δ are multisets. $\Gamma_1; ...; \Gamma_m \Rightarrow \Delta_1; ...\Delta_n$ is a *partition*($\wp$) of $\Gamma \Rightarrow \Delta$ where Γ_i $(1 \leq i \leq m)$ is a submultiset of Γ and Δ_i $(1 \leq j \leq n)$ is a submultiset of Δ.

We can refer to the sections of a partition as 'part 1','part 2', etc. in the antecedent (succedent).

Following Kashima and Shimura, we will say that a *partition* allows to organize the formulas in a sequent according to their dependency relations.

LEMMA 10. *Let* $\Gamma \Rightarrow \Delta$ *be a derivable sequent in FIL1. Then, the sequent* $\Gamma^{-w} \Rightarrow \Delta^{-w}$ *is derivable in FIL1 (where* Γ^{-w} *is the result of excluding the W-formulas of* Γ *and* Δ^{-w} *is the result of excluding the W-formulas of* Δ *).*

Proof. By induction over derivations of FIL1. For instance:
R = $(\wedge_R)$.
(a) If both active formulas are W-formulas:

$$\frac{\overset{\pi_1}{\Gamma \Rightarrow \Delta, C/S^{1*}} \qquad \overset{\pi_2}{\Gamma' \Rightarrow \Delta', D/S^{2*}}}{\Gamma, \Gamma' \Rightarrow \Delta, \Delta', (C \wedge D)/(S^1 \cup S^2)^*} \ (\wedge_R)$$

By I.H., we have π_1' of $\Gamma^{-w} \Rightarrow \Delta^{-w}$ and π_2' of $\Gamma'^{-w} \Rightarrow \Delta'^{-w}$. We can obtain a derivation of the conclusion without W-formulas in the following form:

$$\frac{\dfrac{\overset{\pi_1'}{\Gamma^{-w} \Rightarrow \Delta^{-w}}}{\Gamma^{-w} \Rightarrow \Delta^{-w}, A/\{\}^*} \ (W_R) \qquad \dfrac{\overset{\pi_2'}{\Gamma'^{-w} \Rightarrow \Delta'^{-w}}}{A(n^*), \Gamma'^{-w} \Rightarrow \Delta'^{-w}} \ (W_L)}{\Gamma^{-w}, \Gamma'^{-w} \Rightarrow \Delta^{-w}, \Delta'^{-w}} \ (cut)$$

(b) If only one of the active formulas is a W-formula:

$$\frac{\overset{\pi_1}{\Gamma \Rightarrow \Delta, C/S^{1*}} \qquad \overset{\pi_2}{\Gamma' \Rightarrow \Delta', D/S^2}}{\Gamma, \Gamma' \Rightarrow \Delta, \Delta', (C \wedge D)/(S^1 \cup S^2)^*} \ (\wedge_R)$$

By I.H., we have π_1' of $\Gamma^{-w} \Rightarrow \Delta^{-w}$ and π_2' of $\Gamma'^{-w} \Rightarrow \Delta'^{-w}, D/S^{2-w}$.
(We see that if one of the components is genuine by I.H., we have a derivation of $\Gamma^{-w} \Rightarrow \Delta^{-w}$ from the premise with the other component). ■

LEMMA 11. *Let* $\Gamma \Rightarrow \Delta_1/S_1; ...; \Delta_n/S_n$ *be a derivable sequent in FIL1 where* Γ, Δ_i *and* $\Delta_j (1 \le i \le n)$ *are multisets of formulas,* S_i *and* $S_j (1 \le i \le n)$ *is the generalized union of the dependency sets of the formulas of* Δ_i *and* Δ_j *respectively and:*
  (i) Each Δ_i *has, at least, one formula that is not a W-formula.*
  (ii) For every S_i *and* S_j *in the succedent,* $S_i \cap S_j = \emptyset$.
Then, for each Δ_i/S_i *there is a* $\Gamma_i \subseteq \Gamma$ *such as* $\vdash_{FIL} \Gamma_i \Rightarrow \Delta_i/S_i$ *and* Γ_i *has the hypotheses which the formulas of* Δ_i *depend on.*

Proof. By induction over the length of derivations.

We consider a derivation π of $\Gamma \Rightarrow \Delta_1/S_1; ...; \Delta_n/S_n$ where the sets S_i are disjoint and the last rule applied is 'R'.

R= ($\wedge_L$)

$$\pi$$

$$\frac{\Gamma, A(m), B(n) \Rightarrow \Delta/S}{\Gamma, (A \wedge B)(k) \Rightarrow \Delta_1/S_1^\circ; ...; \Delta_n/S_n^\circ} \ (\wedge_L)$$

where the formula $(A \wedge B)(\mathrm{k})$, $\mathrm{k} = min(m, n)$ and the dependency sets S^j in the premise such that n or m belong to them, they can be replaced in the conclusion by $[(S^j - max(m, n))|k]$.

Due to the conditions of the lemma, the index k occurs only in one of the components of the intersection. However, k is the result of the function min and we could have in this premise a group of sets with index m that could be disjoint from sets with index n. In this case we would have in the conclusion a family that is the union of all these sets. Note that the relations of intersection between the dependency sets in the conclusion can be different from those of the premise. So, we study the possibilities:

(a) If n and m belong to sets that are not disjoint in the premise, then we have by I.H. $\vdash_{FIL1} \Gamma_i, [A(m), B(n)] \Rightarrow \Delta_i/(S_i \cup \{m, n\})$ and by $(\wedge_L)$, $\vdash_{FIL1} \Gamma_i, [(A \wedge B)(k)] \Rightarrow \Delta_i/(S_i \cup \{k\})$.

(b) If n and m belong to sets that are disjoint in the premise, by I.H., we have $\Gamma_i, [A(m)] \Rightarrow \Delta_i/(S_i \cup \{m\})$ and $\Gamma_j, [B(n)] \Rightarrow \Delta_j/S_j \cup \{n\}$. From each one, we can obtain other derivations like this example:

$$\pi'$$

$$\frac{\dfrac{\Gamma_i, [A(m)] \Rightarrow \Delta_i/(S_i \cup \{m\})}{\Gamma_i, [A(m), B(n^*)] \Rightarrow \Delta_i/(S_i \cup \{m\})} \ (W_L)}{\Gamma_i, [(A \wedge B)(k)] \Rightarrow \Delta_j/(S_i \cup \{k\})} \ (\wedge_L)$$

(c) We consider a submultiset Δ_i that is disjoint from the rest of the conclusion such that k does not belong to S_i. Then, the formulas of Δ_i in the premise do not depend on the formula with index n nor on the formula with index m. By I.H. ,$\vdash_{FIL1} \Gamma_i \Rightarrow \Delta_i/S_i$. ∎

LEMMA 12. *Let π be a derivation in FIL1 of $\Gamma \Rightarrow \Delta$. Then, there is a derivation π' of $\Gamma \Rightarrow \Delta$ in FIL1 where every application of the $(\forall_R)$ rule in π' has only one formula in the succedent of the premise.*

Proof. We offer a procedure that goes through each derivation of FIL1 transforming every application of the $(\forall_R)$ rule. Let us consider a derivation π where R is the last rule applied and R = $(\forall_R)$. We have three possibilities:

(a) A(a) is a W-formula in the sequent $\Gamma \Rightarrow A(a)/S^*, \Delta/S'$. By lemma 1, there is a derivation π' of the sequent $\Gamma^{-w} \Rightarrow \Delta^{-w}/S^{-w}$ from which we can recuperate the last sequent of π with the formula A(a) quantified and the same dependency relations. Note that we are using the (W_L) rule introducing the new indexes in (possibly) some sets of the succedent:

$$\pi'$$

$$\cfrac{\cfrac{\Gamma^{-w} \Rightarrow \Delta^{-w}/S'^{-w}}{\Gamma^{-w} \Rightarrow \forall x A(x)/\{\}^*, \Delta/S'^{-w}} \ (W_R)}{\Gamma \Rightarrow \forall x A(x)/S^*, \Delta/S'} \ (W_L)$$

(b) A(a) is not a W-formula in the sequent $\Gamma \Rightarrow A(a)/S; \Delta/S'$. Since the last rule of inference in π is $(\forall_R)$ we know that $(S \cap S') = \emptyset$. By Lemma 11, we have a derivation π' of the sequent $\Gamma_i \Rightarrow A(a)/S$ where Γ_i has the hypotheses which A(a) depends on:

$$\pi'$$

$$\cfrac{\cfrac{\cfrac{\Gamma_i \Rightarrow A(a)/S_i}{\Gamma_i \Rightarrow \forall x A(x)/S_i} \ (\forall_R)}{\Gamma_i \Rightarrow \forall x A(x)/S_i, \Delta/\{\}^*} \ (W_R)}{(\Gamma - \Gamma_i), \Gamma_i \Rightarrow \forall x A(x)/S_i, \Delta/S'^*} \ (W_{L,R})$$

where, by arranging the indexes of the formulas of $\{\Gamma - \Gamma_i\}$ in a suitable way, we recuperate the same dependency relations and the premises of $(\forall_R)$ only have one formula.

(c) There is one last case we need to consider: the case where the succedent has the form $\Gamma \Rightarrow A(a)/S, B/S^{*'}$. In this case, if we want to quantify the formula A(a), we have to appeal to Lemma 10 and delete $B/S^{*'}$, obtaining $\Gamma^{-w} \Rightarrow A(a)/S$ allowing to quantify and restore $B/S^{*'}$ and $\Gamma - \Gamma^{-w}$ by weakening. ∎

The (W_L) rule in FIL1 allows us to introduce the new index in the succedent. We do not need this feature for an intuitionistic system because it does not affect the deductive power of the system. We use it only in this lemma in order to recuperate dependency relations.

2.3 Justifying the conditional.

As we mentioned, in order to justify the rule of the conditional with a multiple-succedent premise, we use the model provided by Kashima and Shimura with the following difference: we apply the lemma to derivations of FIL1 that already have the applications of the $(\forall_R)$ rule with only one formula in the succedent of the premise. The result will be a derivation of $IL^>$ (with the notation of the dependency relations).

LEMMA 13. *Let π be a proof of a sequent $S = \Gamma \Rightarrow \Delta$ in FIL1 where the premises of all the applications of $(\forall_R)$ have only one formula in the succedent. Let $\Gamma_1; \Gamma_2 \Rightarrow \Delta_1; \Delta_2$ be a partition of S such as the formulas* *of Δ_1 do not depend on the formulas of Γ_2. Then, we can construct a derivation π' of the sequent $S' \equiv \Gamma_1 \Rightarrow \Delta_1, (\bigwedge(\Gamma_2) \to \bigvee(\Delta_2))$ in FIL1 where every application of $(\to_R)$ in π' has only one formula in the succedent.*

Proof. The proof is an induction over derivations, where two subcases are taken into consideration: first, the main formula of each rule is in 'part 1' of the multiset and second, the main formula of each rule is in 'part 2'. When we use I.H., for instance, in $(\wedge_R)$ and $(\vee_R)$ we have more than one formula in each premise. That is why 'part 1'and 'part 2' may have more than one formula:

R $= (\vee_R)$.

(a) $\wp = \Gamma_1, \Gamma_2 \Rightarrow \Delta_1, (A \vee B); \Delta_2$. Since the $(\vee_R)$ rule joins two sets, if the disjunction is in part 1, both components are in part 1 of the premise. So, by I.H., we have a derivation π of $\Gamma_1 \Rightarrow \Delta_1, A, B; (\bigwedge \Gamma_2 \to \bigvee \Delta_2)$ and by the $(\vee_R)$ rule, we obtain $\Gamma_1 \Rightarrow \Delta_1, (A \vee B); (\bigwedge \Gamma_2 \to \bigvee \Delta_2)$.

(b) $\wp = \Gamma_1, \Gamma_2 \Rightarrow \Delta_1; \Delta_2, (A \vee B)$.

If the main formula and both active formulas are in part 2, then the partition of the premise is $\Gamma_1, \Gamma_2 \Rightarrow \Delta_1; \Delta_2, A, B$. By I.H. ,we have the derivation of $\Gamma_1 \Rightarrow \Delta_1, (\bigwedge \Gamma_2 \to (\bigvee \Delta_2 \vee (A \vee B)))$.

(c) $\wp = \Gamma_1, \Gamma_2 \Rightarrow \Delta_1; \Delta_2, (A \vee B)$. Let us consider the case where one component is in Δ_1 and the other one is in Δ_2.

Note that it is enough that one of the active formulas is in part 2 of Δ in order for the disjunction to stay in the part 2, that is, depending on Γ_2. For instance, by I.H., we can have a derivation π' of $\Gamma_1 \Rightarrow \Delta_1, A; (\bigwedge \Gamma_2 \to (\bigvee \Delta_2 \vee B))$. From π' we can continue as follows:

$$
\cfrac{
  \cfrac{
    \Gamma_1 \Rightarrow \Delta_1, A; (\bigwedge(\Gamma_2 \to \bigvee \Delta_2 \vee B)}{\Gamma_1 \Rightarrow \bigvee \Delta_1, A \vee (\bigwedge \Gamma_2 \to (\bigvee \Delta_2 \vee B))}(\vee_R) \quad
    \cfrac{
      \cfrac{
        \cfrac{
          \cfrac{
            A \Rightarrow A \quad
            \cfrac{
              \bigwedge \Gamma_2 \Rightarrow \bigwedge \Gamma_2 \quad
              \cfrac{\bigvee \Delta_2 \Rightarrow \bigvee \Delta_2 \quad B \Rightarrow B}{\bigvee \Delta_2 \vee B \Rightarrow \bigvee \Delta_2, B}(\vee_L)
            }{\bigwedge \Gamma_2, (\bigwedge \Gamma_2 \to (\bigvee \Delta_2 \vee B)) \Rightarrow \bigvee \Delta_2, B}(\to_L)
          }{\bigwedge \Gamma_2, A \vee (\bigwedge \Gamma_2 \to (\bigvee \Delta_2 \vee B)) \Rightarrow \bigvee \Delta_2, A, B}(\vee_L)
        }{\bigwedge \Gamma_2, A \vee (\bigwedge \Gamma_2 \to (\bigvee \Delta_2 \vee B)) \Rightarrow \bigvee \Delta_2, (A \vee B)}(\vee_R)
      }{\bigwedge \Gamma_2, A \vee (\bigwedge \Gamma_2 \to (\bigvee \Delta_2 \vee B)) \Rightarrow (\bigvee \Delta_2 \vee (A \vee B))}(\vee_R)
    }{A \vee (\bigwedge \Gamma_2 \to (\bigvee \Delta_2 \vee B)) \Rightarrow \bigwedge \Gamma_2 \to (\bigvee \Delta_2 \vee (A \vee B))}(\to_R)
}{\Gamma_1 \Rightarrow \Delta_1; (\bigwedge \Gamma_2 \to \bigvee(\Delta_2 \vee (A \vee B)))}(cut)
$$

■

2.4   Cut-elimination for FIL1.

For the propositional case, as we have seen, cuts can be eliminated in Gentzen's style using the indexed cut. The same idea cannot be applied to the $(\forall_R)$ rule. Therefore, it is necessary to use Lemma 11.

LEMMA 14. *Let π be a derivation of the sequent $\Gamma \Rightarrow \Delta$ where:*
(i) the last rule applied in π is the indexed cut;

(ii) there is not another application of the indexed cut in π.
Then, π can be transformed into a free-(indexed)cut derivation π' of the
same sequent $\Gamma \Rightarrow \Delta$.

Proof. By induction on the pair (φ, ψ) where φ is the *degree*[4] of the cut
formula and ψ is the *rank* of the application of the rule.

Let us consider the universal case. By Lemma 11, we can split the premise
of the cut and use the sequent that does not depend on the cut formula
restoring the rest of the formulas of the conclusion by weakening. We would
like to point out that, if the dependency relations are suitable, this idea can
be applied to other cases speeding up the elimination of cuts. If, by means
of lemma 2, we can isolate a sequent from one of the premises that does not
have any occurrences of the cut formula to be deleted, we can obtain the
conclusion immediately by weakening.

So, let us consider the premise $(A \vee B)(n), \Gamma' \Rightarrow \Delta', C(a)$ as an example.
We observe which of the components of the succedent depends on $(A \vee B)(n)$.
In this case, we know that it is always possible to apply LemmaL2 and
isolate one formula because of the conditions required by the $(\forall_R)$ rule (the
set of the active formula must be disjoint from the rest). Consider that the
formula C(a) depends on $(A \vee B)(n)$. Thus, using the lemma we have a
sequent $\Gamma'_i \Rightarrow \Delta'_i/S_i$ where the index n does not occur in S_i. By (W_R) and
(W_L), we obtain the conclusion of the cut $\Gamma, \Gamma' \Rightarrow \Delta, \Delta', \forall x C(x)/S^*$. We
can carry out a similar procedure in the case that Δ_i depends on $(A \vee B)(n)$.
∎

THEOREM 15. *Let π be a derivation $\Gamma \Rightarrow \Delta$ in FIL1. Then, it can be
transformed in a cut-free derivation π'.*

Proof. Directly from Lemma 14.   ∎

3   Conclusion.

As we have just seen, in the case of the conditional, the main difference for
First Order Logic lies in the conditions for introducing the universal quantifier in the disjunction expressed by the commas in the multiple succedent.
We demonstrated that, if we know the dependency relations and if we have
an introduction of the universal quantifier from a premise $\Gamma \Rightarrow A(a), \Delta$ in
a derivation of FIL1, this derivation can be split up in several intuitionistic
derivations, one of which, contains the active formula A(a) and the universal
can be introduced in the intuitionistic traditional way. We also took care
of the cases that were affected by the weakening rule.

As a final note, we would like to refer to some issues concerning FIL1
which need to be addressed. The first one is to perform a version of the
first-order logic applied to the Natural Deduction system (NFILL) developed by Franklin. A second one is to take into account Prawitz's work

---

[4]We use *degree* and *rank* in the standard way.

on the weakness of the requirements for the variables in the universal introduction and existential elimination in *Natural Deduction*. It would be only natural to accomplish the analogous treatment for sequent calculus. Finally, we would like to point out again that this approach can be used to develop systems for non-classical logics. Another interesting issue to address is the study of possible non-classical arithmetics. This rises from the curious result presented by Troelstra[5] stating that there is no intermediate arithmetic working with Constant Domain Logic that satisfies, for instance, the disjunction property. It becomes Peano Arithmetic.

Acknowledgements

This work is part of my thesis at PUC-Rio completed under the supervision of Prof. Luiz Carlos Pereira and with the financial support of CAPES (Brazil). I would like to thank Prof. Maria da Paz N.de Medeiros, Wagner Sanz for the motivating discussion on Gentzen-type systems, the UDELAR (Uruguay), CLE (UNICAMP) and the Organizing Committee for the support during the *CLE 30 years-XV EBL-XVI SLALM* events.

Apendix. System $IL^>$. H. Schellinx. 1991.

Axioms

$$A \Rightarrow A \qquad \Gamma, \bot \Rightarrow \Delta$$

Structural rules

$$\frac{\Gamma \Rightarrow A, \Delta' \qquad A, \Gamma' \Rightarrow \Delta}{\Gamma, \Gamma' \Rightarrow \Delta', \Delta} \ (Cut)$$

$$\frac{\Gamma, A, B, \Gamma' \Rightarrow \Delta}{\Gamma, B, A, \Gamma' \Rightarrow \Delta} \ (E_L) \qquad \frac{\Gamma \Rightarrow A, B, \Delta}{\Gamma \Rightarrow B, A, \Delta} \ (E_R) \qquad \frac{\Gamma \Rightarrow \Delta}{\Gamma, A \Rightarrow \Delta} \ (W_L) \qquad \frac{\Gamma \Rightarrow \Delta}{\Gamma \Rightarrow A, \Delta} \ (W_R)$$

$$\frac{\Gamma, A, A \Rightarrow \Delta}{\Gamma, A \Rightarrow \Delta} \ (C_L) \qquad \frac{\Gamma \Rightarrow A, A, \Delta}{\Gamma \Rightarrow A, \Delta} \ (C_R)$$

Operational rules

$$\frac{\Gamma, A \Rightarrow \Delta \qquad \Gamma, B \Rightarrow \Delta}{\Gamma, (A \vee B) \Rightarrow \Delta} \ (\vee_L) \qquad \frac{\Gamma \Rightarrow A, \Delta}{\Gamma \Rightarrow (A \vee B), \Delta} \ (\vee_R^1) \qquad \frac{\Gamma \Rightarrow B, \Delta}{\Gamma \Rightarrow (A \vee B), \Delta} \ (\vee_R^2)$$

$$\frac{\Gamma, A \Rightarrow \Delta}{\Gamma, (A \wedge B) \Rightarrow \Delta} \ (\wedge_L^1) \qquad \frac{\Gamma, B \Rightarrow \Delta}{\Gamma, (A \wedge B) \Rightarrow \Delta} \ (\wedge_L^2) \qquad \frac{\Gamma \Rightarrow A, \Delta \qquad \Gamma \Rightarrow B, \Delta}{\Gamma \Rightarrow (A \wedge B), \Delta} \ (\wedge_R)$$

$$\frac{\Gamma \Rightarrow A, \Delta \qquad \Gamma, B \Rightarrow \Delta}{\Gamma, (A \rightarrow B) \Rightarrow \Delta} \ (\rightarrow_L) \qquad \frac{\Gamma, A \Rightarrow B}{\Gamma \Rightarrow (A \rightarrow B)} \ (\rightarrow_R)$$

$$\frac{A(t), \Gamma \Rightarrow C}{\forall x A(x)\Gamma \Rightarrow \Delta} \ (\forall_L) \qquad \frac{\Gamma \Rightarrow A(a)}{\Gamma \Rightarrow \forall x A(x)} \ (\forall_R)$$

$$\frac{A(a), \Gamma \Rightarrow \Delta}{\exists x A(x), \Gamma \Rightarrow \Delta} \ (\exists_L) \qquad \frac{\Gamma \Rightarrow \Delta, A(t)}{\Gamma \Rightarrow \Delta, \exists x(A(x))} \ (\exists_R)$$

(The variable a cannot occur in the lower sequent
in the $(\exists_L)$ and the $(\forall_R)$ rules.)

---

[5]See: Troelstra. *Metamathematical Investigation of Intuitionistic Arithmetic and Analysis.*

BIBLIOGRAPHY

[1] Avron, A. The Method of Hypersequents in the Proof Theory of Propositional Non-Classical Logics. In W. Hodges, M. Hyland, C. Steinhorn and J. Truss, editors, *Logic: Foundations to Applications*, pages 1–32. Oxford Science Publication, 1996.

[2] Braüner, T. and De Paiva, V. Cut-Elimination for Full Intuitionistic Linear Logic. BRICS Report Series, RS-96-10, 1996.

[3] De Paiva, V. and Pereira, L. C. A new proof system for intuitionistic logic. *The Bulletin of SL*, 1(1):101, 1995.

[4] De Paiva, V. and Pereira, L. C. A Short Note on Intuitionistic Propositional Logic with Multiple Conclusions. Available at:
http://www.cs.bham.ac.uk/ vdp/publications/fil.pdf.

[5] Franklin, L. Sistemas com múltiplas conclusões para a lógica proposicional intuicionista. Master thesis, PUC-Rio, 2000.

[6] Gentzen, G The collected papers of Gerhard Gentzen. Szabo, M. E., editor, North-Holland, 1969.

[7] Kashima, R. and Shimura, T. Cut-Elimination Theorem for the Logic of Constant Domain's. *MLQ* 40:153-172, 1994.

[8] Pereira, L. C. Uma Breve Nota sobre o Condicional e Múltiplas Conclusões. In Evora, F. et al., editors, *Lógica e ontología. Ensaios em homenagem a Balthasar Barbosa Filho*, São Paulo, 2004.

[9] Schellinx, H. Some Syntactical Observations on Linear Logic. *J. Logic Computation*, 1(4):537-559, 1991.

[10] Takeuti, G *Proof Theory*, North-Holland, 1975.

[11] Troelstra, A. S. Metamathematical Investigation of Intuitionistic Arithmetic and Analysis. LNinM, 344, 1973.

María Fernanda Pallares
Philosophy Institute
University of the Republica
Montevideo, Uruguay
E-mail: pallaresmf@gmail.com

On the Proof Theory of $\mathcal{ALC}$

Alexandre Rademaker, Edward H. Haeusler and Luiz
C. Pereira

ABSTRACT. Description Logic (DL) refer to a family of logics widely
used in Knowledge Representation. By means of DL, KR systems
provide various inference capabilities to deduce implicit knowledge
from the explicitly one. Most of the inference engines provided to DL
are based on Tableaux. They mostly rely on emulating DL reasoning by means of its natural first-order interpretation. This first-order
feature are present even when the DLs involved is a propositional
one, as $\mathcal{ALC}$, a basic Decription Logic. In a previous research report, we introduce a sequent calculus for $\mathcal{ALC}$, a basic Description
Logic, named $S_{\mathcal{ALC}}$. This sequent calculus is purely propositional.
Moreover, it follows a labelled deductive system tradition.

In this paper we present $S_{\mathcal{ALC}}$ and the proof of cut-elimination for
it. $S_{\mathcal{ALC}}$ was designed with the main motivation of trying to provide
a way to extract computational content of $\mathcal{ALC}$ proofs, for it can be
seen as an intermediate step towards a Natural Deduction System for
$\mathcal{ALC}$. Obtaining a Natural Deduction system for $\mathcal{ALC}$ is the natural
development of the investigation reported here.

1 Introduction

Description Logics is a family of formalisms used to represent knowledge of a
domain. In contrast with others knowledge representation systems, Description Logics are equipped with a formal, logic-based semantics. Knowledge
representation systems based on description logics provide various inference
capabilities that deduce implicit knowledge from the explicitly represented
knowledge [1].

The use of Description Logics by regular users, that is, non-technical
users, would be wider if the computed inferences could be presented as a
natural language text – or any other presentation format at the domain's
specification level of abstraction – without requiring any knowledge on logic
to be understandable [8].

In [9] we presented the system $S_{\mathcal{ALC}}$ a sequent calculus for $\mathcal{ALC}$ [1] and
proved that it is sound and complete. The first motivation for developing
such system is the extraction of computational content of $\mathcal{ALC}$ proofs. More
precisely, this system was developed to allow the use of natural language

---

[1] $\mathcal{ALC}$ means Attributive Language with Complements, a basic Description Logic.

to rendering of a Natural Deduction System proof. So that, the present calculus for $\mathcal{ALC}$ is an intermediate step towards a Natural Deduction System for $\mathcal{ALC}$ [4]. For instance, a natural language rendering of a Natural Deduction System is worthwhile in a context like proof of conformance in security standards [5].

Our Sequent Calculus, compared with other approaches like Tableaux [11] and the Sequent Calculus for $\mathcal{ALC}$ [6, 8, 2] based on this very Tableaux, does not use individual variables (first-order ones) at all. The main mechanism in our system is based on labeled formulas. The labeling of formulas is among one of the most successful artifacts for keeping control of the context in the many existent quantification in Logical system and modalities. For a detailed reading on this approach, we point out [10, 7].

In this paper we prove that $S_{\mathcal{ALC}}$ has the desirable property of allowing the deterministic constructions of proofs. That is, we prove that the *cut rule* can be eliminated from the system $S_{\mathcal{ALC}}$ without lost the completeness and soundness. The result present here is an important step toward the implementation of a theorem prover for $S_{\mathcal{ALC}}$.

This paper is structured as follows. Section 2 presents our Sequent Calculus for $\mathcal{ALC}$. Section 3 presents the proof of cut-elimination for $S_{\mathcal{ALC}}$. Finally, in Section 4 we point out further works and present some conclusions.

2   The $\mathcal{ALC}$ sequent calculus

$\mathcal{ALC}$ is a basic description language [1] in which elementary descriptions are *atomic concepts*, which denote sets of individuals, and *atomic roles*, which denote binary relationships between individuals. Complex descriptions can be built from them inductively with concept constructors:

$$\phi_c ::= \bot \mid A \mid \neg\phi_c \mid \phi_c \sqcap \phi_c \mid \phi_c \sqcup \phi_c \mid \exists R.\phi_c \mid \forall R.\phi_c$$

where A stands for atomic concepts and R for atomic roles.

The Sequent Calculus for $\mathcal{ALC}$ that it is shown in Figure 1 was first presented in [9] where it was proved to be sound and complete. It is based on the extension of the language ϕ_c. Labels are lists of (possibly skolemized) role symbols. Its syntax is as follows:

$$L ::= R, L \mid R(L), L \mid \emptyset$$
$$\phi_{lc} ::= {}^L\phi_c{}^L$$

where R stands for atomic role names, L for list of atomic role names and $R(L)$ is an skolemized role expression.

Each *consistent* labeled $\mathcal{ALC}$ concept has an $\mathcal{ALC}$ concept equivalent. For instance, ${}^{Q_2,Q_1}\alpha^{R_1(Q_2),R_2}$ is equivalent to $\exists R_2.\forall Q_2.\exists R_1.\forall Q_1.\alpha$.

Let α be a ϕ_{lc} formula; the function $\sigma : \phi_{lc} \to \phi_c$ transforms a labeled $\mathcal{ALC}$ concept into an $\mathcal{ALC}$ concept. $\mathcal{ALC}$ sequents are expressions of the form $\Delta \Rightarrow \Gamma$ as a *sequent* where Δ and Γ are finite sequences of labeled concepts. The natural interpretation of the sequent $\Delta \Rightarrow \Gamma$ is the $\mathcal{ALC}$ formula $\prod_{\delta \in \Delta} \sigma(\delta) \sqsubseteq \bigsqcup_{\gamma \in \Gamma} \sigma(\gamma)$.

The lists of labels will be omitted whenever it is clear that a rule does not take into account their specific form. This is the case for the structural rules. In all rules α, β stands for $\mathcal{ALC}$ concepts (formulas without labels), γ, δ stands for labeled concepts, Γ, Δ for list of labeled concepts. L, M, N stands for list of roles. All of this letters may have indexes whenever necessary for distinction. If $^{L_1}\alpha^{L_2}$ is a consistently labeled formula then $\mathcal{D}(L_2)$ is the set of role symbols that occur inside the *skolemized role expressions* in L_2. Note that $\mathcal{D}(L_2) \subseteq L_1$ always holds.

Let us consider a labeled formula $^{L_1}\alpha^{L_2}$; the notation $^{\frac{L_2}{L_1}}\alpha^{\frac{L_1}{L_2}}$ denotes the exchanging of the universal roles occurring in L_1 for the existential roles occurring in L_2 in a consistent way such that the skolemization is dually placed. It is used to express the negation of labeled concepts. If $\beta \equiv \neg\alpha$ the formula $^{\frac{Q}{R}}\beta^{\frac{R}{Q(R)}}$ is the negation of $^R\alpha^{Q(R)}$.

The restrictions in the rules ($\forall$-r) and ($\forall$-l) means that the role R can only be removed from the left list of labels if none of the skolemized role expressions in the right list depends on it.

3 The cut-elimination theorem

In this section we adopt the usual terminology of proof theory for sequent calculus presented in [3, 12]. We follow Gentzen's original proof for cut elimination with the introduction of the *mix* rule.

Let δ be a labeled formula. An inference of the following form is called *mix* with respect to δ:

$$\frac{\Delta_1 \Rightarrow \Gamma_1 \quad \Delta_2 \Rightarrow \Gamma_2}{\Delta_1, \Delta_2^* \Rightarrow \Gamma_1^*, \Gamma_2} \ (\delta)$$

where both Γ_1 and Δ_2 contain the formula δ, and Γ_1^* and Δ_2^* are obtained from Γ_1 and Δ_2 respectively by deleting all the occurrences of δ in them.

In order to obtain an easier presentation of our cut elimination we introduce four additional rules of inference called *quasi-mix* rules. One may note that those rules are related to combinations of possible applications of *prom-1* and *prom-2* rules followed by *cut* rule. This explain why δ and γ are used below, that is, to respect the restrictions of *prom-1* (resp. *prom-2*) regarding the number of formulas in the antecedent (resp. consequent).

$$\frac{\delta \Rightarrow \Gamma_1 \quad \Delta_2 \Rightarrow \Gamma_2}{\delta^{+R}, \Delta_2^* \Rightarrow \Gamma_1^{*+R}, \Gamma_2} \ (^{L_1}\alpha^{L_2}, {}^{L_1}\alpha^{L_2, R}) \qquad \frac{\Delta_1 \Rightarrow \Gamma_1 \quad {}^{L_1}\alpha^{L_2} \Rightarrow \Gamma_2}{\Delta_1, \Rightarrow \Gamma_1^*, \Gamma_2{}^{+R}} \ (^{L_1}\alpha^{L_2, R}, {}^{L_1}\alpha^{L_2})$$

$$\frac{\Delta_1 \Rightarrow {}^{L_1}\alpha^{L_2} \quad \Delta_2 \Rightarrow \Gamma_2}{{}^{+R}\Delta_1, \Delta_2^* \Rightarrow \Gamma_2} \ (^{L_1}\alpha^{L_2}, {}^{R, L_1}\alpha^{L_2}) \qquad \frac{\Delta_1 \Rightarrow \Gamma_1 \quad \Delta_2 \Rightarrow \gamma}{\Delta_1, {}^{+R}\Delta_2^* \Rightarrow \Gamma_1^*, {}^{+R}\gamma} \ (^{R, L_1}\alpha^{L_2}, {}^{L_1}\alpha^{L_2})$$

where in the tuple on right indicates the two mix formulas of this inference. Γ_1 contains the first projection of the tuple, Δ_2 contains the second projection and Γ_1^* and Δ_2^* are obtained from Γ_1 and Δ_2 by deleting all occurrences

$$\overline{\alpha \Rightarrow \alpha} \qquad\qquad \overline{\bot \Rightarrow \alpha}$$

$$\frac{\Delta \Rightarrow \Gamma}{\Delta, \delta \Rightarrow \Gamma} \text{ weak-l} \qquad\qquad \frac{\Delta \Rightarrow \Gamma}{\Delta \Rightarrow \Gamma, \gamma} \text{ weak-r}$$

$$\frac{\Delta, \delta, \delta \Rightarrow \Gamma}{\Delta, \delta \Rightarrow \Gamma} \text{ contraction-l} \qquad\qquad \frac{\Delta \Rightarrow \Gamma, \gamma, \gamma}{\Delta \Rightarrow \Gamma, \gamma} \text{ contraction-r}$$

$$\frac{\Delta_1, \delta_1, \delta_2, \Delta_2 \Rightarrow \Gamma}{\Delta_1, \delta_2, \delta_1, \Delta_2 \Rightarrow \Gamma} \text{ perm-l} \qquad\qquad \frac{\Delta \Rightarrow \Gamma_1, \gamma_1, \gamma_2, \Gamma_2}{\Delta \Rightarrow \Gamma_1, \gamma_2, \gamma_1, \Gamma_2} \text{ perm-r}$$

$$\frac{\Delta, {}^{L_1, R}\alpha^{L_2} \Rightarrow \Gamma}{\Delta, {}^{L_1}(\forall R.\alpha)^{L_2'} \Rightarrow \Gamma} \;\forall\text{-l} \qquad\qquad \frac{\Delta \Rightarrow \Gamma, {}^{L_1, R}\alpha^{L_2}}{\Delta \Rightarrow \Gamma, {}^{L_1}(\forall R.\alpha)^{L_2'}} \;\forall\text{-r}$$

$$\frac{\Delta, {}^{L_1}\alpha^{R(L_1), L_2} \Rightarrow \Gamma}{\Delta, {}^{L_1}(\exists R.\alpha)^{L_2} \Rightarrow \Gamma} \;\exists\text{-l} \qquad\qquad \frac{\Delta \Rightarrow \Gamma, {}^{L_1}\alpha^{R(L_1), L_2}}{\Delta \Rightarrow \Gamma, {}^{L_1}(\exists R.\alpha)^{L_2}} \;\exists\text{-r}$$

$$\frac{\Delta, {}^{L}\alpha^{\emptyset}, {}^{L}\beta^{\emptyset} \Rightarrow \Gamma}{\Delta, {}^{L}(\alpha \sqcap \beta)^{\emptyset} \Rightarrow \Gamma} \;\sqcap\text{-l} \qquad\qquad \frac{\Delta \Rightarrow \Gamma, {}^{L}\alpha^{\emptyset} \quad \Delta \Rightarrow \Gamma, {}^{L}\beta^{\emptyset}}{\Delta \Rightarrow \Gamma, {}^{L}(\alpha \sqcap \beta)^{\emptyset}} \;\sqcap\text{-r}$$

$$\frac{\Delta, {}^{\emptyset}\alpha^{L} \Rightarrow \Gamma \quad \Delta, {}^{\emptyset}\beta^{L} \Rightarrow \Gamma}{\Delta, {}^{\emptyset}(\alpha \sqcup \beta)^{L} \Rightarrow \Gamma} \;\sqcup\text{-l} \qquad\qquad \frac{\Delta \Rightarrow \Gamma, {}^{\emptyset}\alpha^{L}, {}^{\emptyset}\beta^{L}}{\Delta \Rightarrow \Gamma, {}^{\emptyset}(\alpha \sqcup \beta)^{L}} \;\sqcup\text{-r}$$

$$\frac{\Delta \Rightarrow \Gamma, {}^{L_1}\alpha^{L_2}}{\Delta, {}^{L_2}\neg\alpha^{L_1} \Rightarrow \Gamma} \;\neg\text{-l} \qquad\qquad \frac{\Delta, {}^{L_1}\alpha^{L_2} \Rightarrow \Gamma}{\Delta \Rightarrow \Gamma, {}^{L_2}\neg\alpha^{L_1}} \;\neg\text{-r}$$

$$\frac{{}^{L_1}\alpha^{L_2} \Rightarrow {}^{M_1}\beta_1{}^{N_1}, \ldots, {}^{M_n}\beta_n{}^{N_n}}{{}^{L_1}\alpha^{L_2, R} \Rightarrow {}^{M_1}\beta_1{}^{N_1, R}, \ldots, {}^{M_n}\beta_n{}^{N_n, R}} \text{ prom-1}$$

$$\frac{{}^{M_1}\beta_1{}^{N_1}, \ldots, {}^{M_n}\beta_n{}^{N_n} \Rightarrow {}^{L_1}\alpha^{L_2}}{{}^{R, M_1}\beta_1{}^{N_1'}, \ldots, {}^{R, M_n}\beta_n{}^{N_n'} \Rightarrow {}^{R, L_1}\alpha^{L_2'}} \text{ prom-2}$$

$$\frac{\Delta_1 \Rightarrow \Gamma_1, {}^{L_1}\alpha^{L_2} \quad {}^{L_1}\alpha^{L_2}, \Delta_2 \Rightarrow \Gamma_2}{\Delta_1, \Delta_2 \Rightarrow \Gamma_1, \Gamma_2} \text{ cut}$$

In the rules ($\forall$-r) and ($\forall$-l) $R \notin \mathcal{D}(L_2)$ must hold. In rules (prom-2), ($\forall$-r) and ($\forall$-l), the notation L_2', N_i' means the reconstructions of the skolemized expressions on those lists regarding the modification of the lists L_1 and M_i, respectively.

Figure 1. The System $S_{\mathcal{ALC}}$

of the first and second projection, respectively. The notation Δ^{+R} (resp. $^{+R}\Delta$) means the addition of R on the right (resp left) list of labels of all $\delta \in \Delta$. In what follows, we will consider the *mix* rule as a special case of *quasi-mix* rule.

DEFINITION 1 (The $S^*_{\mathcal{ALC}}$ system). We call $S^*_{\mathcal{ALC}}$ the new system obtained from $S_{\mathcal{ALC}}$ by replacing the cut rule by the *quasi-mix* rules.

LEMMA 2. *The systems $S_{\mathcal{ALC}}$ and $S^*_{\mathcal{ALC}}$ are equivalent, that is, a sequent is $S_{\mathcal{ALC}}$-provable if and only if that sequent is also $S^*_{\mathcal{ALC}}$-provable.*

Proof. All applications of *mix* rule can be replaced by applications of cut rule provide that all the repetitions of the cut formula in the upper sequents being first reduced to just one occurrence on each sequent. This is easily done by one or more application of the contraction and permutations rules. The remaining four *quasi-mix* rules are derived from inferences where the promotional rules (*prom-1* and *prom-2*) are applied just before a *mix* rule. In that way, one can reduce all the applications of *quasi-mix* rule by sequence of *prom-1* or *prom-2* followed by *mix* rule. ∎

By Lemma 2, it is sufficient to show that the *quasi-mix* rules is redundant in $S^*_{\mathcal{ALC}}$, since a proof in $S^*_{\mathcal{ALC}}$ without *quasi-mix* is at the same time a proof in $S_{\mathcal{ALC}}$ without cut.

DEFINITION 3 ($\mathcal{T}$-$S_{\mathcal{ALC}}$ system). $S_{\mathcal{ALC}}$ was defined with initial sequents of the form $\alpha \Rightarrow \alpha$ with α a $\mathcal{ALC}$ concept definition (logical axiom). However, it is often convenient to allow for other initial sequents. So if $\mathcal{T}$ is a set of sequents of the form $\Delta \Rightarrow \Gamma$, where Δ and Γ are sequences of $\mathcal{ALC}$ concept descriptions (non-logical axioms), we define $\mathcal{T}$-$S_{\mathcal{ALC}}$ to be the proof system defined like $S_{\mathcal{ALC}}$ but allowing initial sequents to be from $\mathcal{T}$ too.

The Definition 3 can be extend to the system $S^*_{\mathcal{ALC}}$ in the same way, obtaining the system $\mathcal{T}$-$S^*_{\mathcal{ALC}}$.

DEFINITION 4 (Free-quasi-mix free proof). Let P be a proof in $\mathcal{T}$-$S^*_{\mathcal{ALC}}$. A formula occurring in P is anchored (by a $\mathcal{T}$-sequent) if it is a direct descendent of a formula occurring in an initial sequent in $\mathcal{T}$. A *quasi-mix* inference in P is *anchored* if either:

(i) the mix formulas are not atomic and at least one of the occurrences of the mix formulas in the upper sequents is anchored, or

(ii) the mix formulas are atomic and both of the occurrences of the mix formulas in the upper sequents are anchored.

A *quasi-mix* inference which is not anchored is said to be *free*. A proof P is *free-quasi-mix free* if it contains no free *quasi-mixes*.

THEOREM 5 (Free-mix Elimination). *Let $\mathcal{T}$ be a set of sequents. If $\mathcal{T}$-$S^*_{\mathcal{ALC}} \vdash \Delta \Rightarrow \Gamma$ then there is a* free-quasi-mix free *proof in $\mathcal{T}$-$S^*_{\mathcal{ALC}}$ of $\Delta \Rightarrow \Gamma$.*

This means that any theorem in the $(S_{\mathcal{ALC}})$ system can be deterministically provable. Theorem 5 is a consequence of the following lemma.

LEMMA 6. *If P is a proof of S (in $\mathcal{T}$-$S^*_{\mathcal{ALC}}$) which contains only one* freequasi-mix, *occurring as the last inference, then S is provable without any free-mix.*

Theorem 5 is obtained from Lemma 6 by simple induction over the number of *quasi-free-mix* occurring in a proof P.

We can now concentrate our attention on Lemma 6. First we define three scalars as a measure of the complexity of the proof. The *grade* of a formula $^{L_1}\alpha^{L_2}$ is defined as the number of logical symbols of α (denoted by $g(^{L_1}\alpha^{L_2})$). The *label-degree* of a formula $^{L_1}\alpha^{L_2}$ is defined as $ldegree(^{L_1}\alpha^{L_2}) = |L_1| + |L_2|$ where $|L|$ means the length of the list L. The grade of a mix is the grade of the *mix* formula. The grade of a *quasi-mix* is $g(\gamma, \gamma') = g(\gamma) + g(\gamma')$ In a similar way, the label-degree of a mix is the label-degree of the *mix* formula. The label-degree of a *quasi-mix* is $ldegree(\gamma, \gamma') = ldegree(\gamma) + ldegree(\gamma')$. Given a proof P containing only one application of *quasi-mix* rule as the last rule of inference, we say that the grade of P (denote by $g(P)$) and the label-degree of P (denoted by $ldegree(P)$) as the grade and label-degree of that *quasi-mix*.

Let P be a proof containing only one *quasi-mix* as its last inference:

$$J \, \frac{\Delta_1 \Rightarrow \Gamma_1 \qquad \Delta_2 \Rightarrow \Gamma_2}{\Delta_1, \Delta_2^* \Rightarrow \Gamma_1^*, \Gamma_2} \, (\gamma, \gamma')$$

We refer to the left and right sequents as S_1 and S_2 respectively, and to the lower sequent as S. We call a thread in P a left (or right) thread if it contains the left (or right) upper sequent of the *quasi-mix* J. The *rank* of the thread $\mathcal{F}$ in P is defined as the number of consecutive sequents, counting upward form the left (right) upper sequent of J, that contains γ (γ') in its succedent (antecedent). Since the left (right) upper sequent always contains the mix formulas, the rank of a thread in P is at least 1. The rank of a thread $\mathcal{F}$ in P is denoted by $rank(\mathcal{F}; P)$ and is defined as follows:

$$rank_l(P) = \max_{\mathcal{F}}(rank(\mathcal{F}; P)),$$

where $\mathcal{F}$ ranges over all the left threads in P, and

$$rank_r(P) = \max_{\mathcal{F}}(rank(\mathcal{F}; P)),$$

where $\mathcal{F}$ ranges over all the right threads in P. The rank of P is defined as

$$rank(P) = rank_l(P) + rank_r(P),$$

where $rank(P) \geq 2$.

Proof. We prove Lemma 6 by lexicographically induction on the ordered triple (*grade*,*ldegree*,*rank*) of the proof P. We divide the proof into two main cases, namely *rank* = 2 and *rank* > 2 (regardless of *grade* and *ldegree*).[2]

Case 1: *rank* = 2 We shall consider several cases according to the form of the proofs of the upper sequents of the *quasi-mix*.

1.1) The left upper sequent S_1 is a logical initial sequent. There are several cases to be examined.

    1.1.1) P has the form on the left. We can easily obtain the same end-sequent without using the *quasi-mix* as follows on the right.

$$J \; \frac{\alpha \Rightarrow \alpha \qquad \Delta_2 \Rightarrow \Gamma_2}{\alpha^R, \Delta_2^* \Rightarrow \Gamma_2} (\alpha, \alpha^R)$$

$$\cfrac{\cfrac{P_1}{\cfrac{\Delta_2 \Rightarrow \Gamma_2}{\cfrac{\alpha^R, \ldots, \alpha^R, \Delta_2^* \Rightarrow \Gamma_2}{\alpha^R, \Delta_2^* \Rightarrow \Gamma_2} \text{ some contractions}}\text{ some permutations}}}{}$$

All other cases are treated in a similar way.

1.2) The right upper sequent S_2 is a logical initial sequent. Similar as cases 1.1.

1.3) S_1 or S_2 (or both) are non-logical initial sequents. In this case, it is obvious that the *quasi-mix* is not a *free* and it does not need to be eliminated.

1.4) Neither S_1 nor S_2 are an initial sequents, and S_1 is the lower sequent of a structural inference J_1. Since $rank_l(P) = 1$, the first mix formula $^{L_1}\alpha^{L_2}$ cannot appear in the succedent of the upper sequent of J_1, that is, J_1 must be *weak-r*, whose introduced formula is $^{L_1}\alpha^{L_2}$. Again there are several cases to be examined.

    1.4.1) Let us consider the *quasi-mix* case $(^{L_1}\alpha^{L_2}, {}^{L_1}\alpha^{L_2,R})$:

$$J \; \cfrac{J_1 \cfrac{\delta \Rightarrow \Gamma_1}{\delta \Rightarrow \Gamma_1, {}^{L_1}\alpha^{L_2}} \qquad \Delta_2 \Rightarrow \Gamma_2}{\delta^{+R}, \Delta_2^* \Rightarrow \Gamma_1^{+R}, \Gamma_2} (^{L_1}\alpha^{L_2}, {}^{L_1}\alpha^{L_2,R})$$

    where Γ_1 does not contain $^{L_1}\alpha^{L_2}$. We can eliminate the *quasi-mix* as follows:

---

[2]The complete proof can be found at http://www.tecmf.inf.puc-rio.br/ AlexandreRademaker/salc/.

$$\frac{\delta \Rightarrow \Gamma_1}{\delta^{+R} \Rightarrow \Gamma_1{}^{+R}} \text{ prom-1}$$

$$\frac{}{\text{some weakenings}}$$

$$\frac{\Delta_2^*, \delta^{+R} \Rightarrow \Gamma_1{}^{+R}, \Gamma_2}{\text{some permutations}}$$

$$\delta^{+R}, \Delta_2^* \Rightarrow \Gamma_1{}^{+R}, \Gamma_2$$

All other cases are treated in a similar way.

1.5) The same conditions that hold for 1.4 but with S_2 as the lower sequent of structural inference instead of S_1. As in 1.4.

1.6) Neither S_1 nor S_2 are an initial sequents and S_1 is the lower sequent of a *prom-1* rule application and J is a *mix* rule application.

$$J \quad \cfrac{\cfrac{\begin{array}{c}P_1\\[2pt]\delta \Rightarrow \Gamma_1\end{array}}{\delta^{+R} \Rightarrow \Gamma_1{}^{+R}}\text{ prom-1} \qquad \cfrac{P_2}{\Delta_2 \Rightarrow \Gamma_2}}{\delta^{+R}, \Delta_2^* \Rightarrow \Gamma_1^{*+R}, \Gamma_2} \ (\gamma^{+R})$$

where by assumption none of the proofs P_n for $n \in \{1,2\}$ contain a mix or *quasi-mix*. Moreover, Γ_1 does not contain γ^{+R} since $rank_l(P) = 1$. That is, the *prom-1* rule introduced the mix formula of J. We can replace the application of the *mix* rule by an application of *quasi-mix* rule as follows:

$$\cfrac{\cfrac{P_1}{\delta \Rightarrow \Gamma_1} \qquad \cfrac{P_2}{\Delta_2 \Rightarrow \Gamma_2}}{\delta^{+R}, \Delta_2^* \Rightarrow \Gamma_1^{*+R}, \Gamma_2} \ (\gamma, \gamma^{+R})$$

The new *quasi-mix* rule has label-degree less than $ldegree(\gamma^{+R}, \gamma^{+R})$. So by the induction hypothesis, we can obtain a proof which contains no mixes.

1.7) Similar case as above with S_1 being lower sequent of a *prom-2* or S_2 being lower sequent of *prom-1* or *prom-2*. We apply similar reductions transforming *mix* application into *quasi-mix* rules applications. Always "moving" the *mix* upward into the direction of the *prom-1* or *prom-2* inference.

1.8) Both S_1 and S_2 are lower sequents of logical inferences and $rank_l(P) = rank_r(P) = 1$, J being a *mix* with the mix formula γ of each side being the principal formula of the logical inference. We use induction on the grade, distinguishing several cases according to the outermost logical symbol of γ:

i) The outermost logical symbol is $\sqcap$. P has the form:

$$\cfrac{\cfrac{\begin{array}{cc} P_1 & P_2 \\ \Delta_1 \Rightarrow \Gamma_1, {}^L\alpha^\emptyset & \Delta_1 \Rightarrow \Gamma_1, {}^L\beta^\emptyset \end{array}}{\Delta_1 \Rightarrow \Gamma_1, {}^L(\alpha \sqcap \beta)^\emptyset} \sqcap\text{-r} \quad \cfrac{\begin{array}{c} P_3 \\ \Delta_2, {}^L\alpha^\emptyset, {}^L\beta^\emptyset \Rightarrow \Gamma_2 \end{array}}{\Delta_2, {}^L(\alpha \sqcap \beta)^\emptyset \Rightarrow \Gamma_2} \sqcap\text{-l}}{\Delta_1, \Delta_2 \Rightarrow \Gamma_1, \Gamma_2} ({}^L(\alpha \sqcap \beta)^\emptyset)$$

where by assumption none of the proofs P_n for $n \in \{1,2,3\}$ contain a *quasi-mix*. We transform P into:

$$\cfrac{\cfrac{\begin{array}{cc} & P_1 \qquad\qquad P_3 \\ P_2 & \cfrac{\Delta_1 \Rightarrow \Gamma_1, {}^L\alpha^\emptyset \quad \Delta_2, {}^L\alpha^\emptyset, {}^L\beta^\emptyset \Rightarrow \Gamma_2}{\Delta_1, \Delta_2, {}^L\beta^\emptyset \Rightarrow \Gamma_1, \Gamma_2} ({}^L\alpha^\emptyset) \\ \Delta_1 \Rightarrow \Gamma_1, {}^L\beta^\emptyset & \end{array}}{\cfrac{\Delta_1, \Delta_1, \Delta_2 \Rightarrow \Gamma_1, \Gamma_1, \Gamma_2}{\text{some permutations and contractions}} ({}^L\beta^\emptyset)}}{\Delta_1, \Delta_2 \Rightarrow \Gamma_1, \Gamma_2}$$

which contains two mix but both with grade less than $g({}^L(\alpha \sqcap \beta)^\emptyset)$. So by induction hypothesis, we can obtain a proof which contains no mixes. Note that the mix $({}^L\alpha^\emptyset)$ is now the last inference rule of a proof which contains no mix. Given that, this mix can be omitted using the transformations defined above.

ii) The outermost logical symbol is $\forall$. In this case S_1 and S_2 must be lower sequents of $\forall$-r and $\forall$-l rule, respectively. P is:

$$\cfrac{\cfrac{\begin{array}{c} P_1 \\ \Delta_1 \Rightarrow \Gamma_1, {}^{L_1,R}\alpha^{L_2} \end{array}}{\Delta_1 \Rightarrow \Gamma_1, {}^{L_1}\forall R.\alpha^{L_2}} \forall\text{-r} \quad \cfrac{\begin{array}{c} P_2 \\ \Delta_2, {}^{L_1,R}\alpha^{L_2} \Rightarrow \Gamma_2 \end{array}}{\Delta_2, {}^{L_1}\forall R.\alpha^{L_2} \Rightarrow \Gamma_2} \forall\text{-l}}{\Delta_1, \Delta_2 \Rightarrow \Gamma_1, \Gamma_2} ({}^{L_1}\forall R.\alpha^{L_2})$$

which again by hypothesis, none of the proofs P_n for $n \in \{1,2\}$ contain a mix. These proof can be transformed into:

$$\cfrac{\begin{array}{cc} P_1 & P_2 \\ \Delta_1 \Rightarrow \Gamma_1, {}^{L_1,R}\alpha^{L_2} & \Delta_2, {}^{L_1,R}\alpha^{L_2} \Rightarrow \Gamma_2 \end{array}}{\Delta_1, \Delta_2 \Rightarrow \Gamma_1, \Gamma_2} ({}^{L_1,R}\alpha^{L_2})$$

which contains one mix with grade less than $g({}^{L_1}\forall R.\alpha^{L_2})$. So by induction hypothesis, we can obtain a proof which contains no mixes.

iii) The outermost logical symbol is $\neg$. P is:

$$\cfrac{\cfrac{\begin{array}{c} P_1 \\ \Delta_1, {}^{L_1}\alpha^{L_2} \Rightarrow \Gamma_1 \end{array}}{\Delta_1 \Rightarrow \Gamma_1, {}^{L_2}_{L_1}\neg\alpha^{L_1}_{L_2}} \neg\text{-r} \quad \cfrac{\begin{array}{c} P_2 \\ \Delta_2 \Rightarrow \Gamma_2, {}^{L_1}\alpha^{L_2} \end{array}}{\Delta_2, {}^{L_2}_{L_1}\neg\alpha^{L_1}_{L_2} \Rightarrow \Gamma_2} \neg\text{-l}}{\Delta_1, \Delta_2 \Rightarrow \Gamma_1, \Gamma_2} ({}^{L_2}_{L_1}\neg\alpha^{L_1}_{L_2})$$

This proof can be transformed into:

$$
\dfrac{
\dfrac{
\dfrac{
\dfrac{\overset{\displaystyle P_2}{\Delta_2 \Rightarrow \Gamma_2,\,{}^{L_1}\alpha^{L_2}} \qquad \overset{\displaystyle P_1}{\Delta_1,\,{}^{L_1}\alpha^{L_2} \Rightarrow \Gamma_1}}{\Delta_2, \Delta_1 \Rightarrow \Gamma_2, \Gamma_1}\;({}^{L_1}\alpha^{L_2})
}{\text{some permutations}}
}{\Delta_1, \Delta_2 \Rightarrow \Gamma_1, \Gamma_2}
}{}
$$

which contains one mix with grade less than $g(\overset{L_2}{\overset{}{L_1}}(\neg\alpha)\overset{L_1}{\overset{}{L_2}})$. So by the induction hypothesis, we can obtain a proof which contains no mixes.

The remaining cases where the outermost logical symbol of δ is $\sqcup$ and $\exists$ can be treated in a similar way.

1.9) Both S_1 and S_2 are lower sequents of logical inferences, $rank_l(P) = rank_r(P) = 1$ and J being a *quasi-mix* (γ, γ^{+R}) where the mix formulas on each side is the principal formula of the logical inferences. Let us here present just the case $\sqcup$. In this case S_1 and S_2 must be lower sequents of $\sqcup$-r and $\sqcup$-l rule, respectively:

$$
\dfrac{
\dfrac{\overset{\displaystyle P_1}{\delta \Rightarrow \Gamma_1,\,{}^{\emptyset}\alpha^{L},\,{}^{\emptyset}\beta^{L}}}{\delta \Rightarrow \Gamma_1,\,{}^{\emptyset}(\alpha \sqcup \beta)^{L}}\sqcup\text{-r}
\qquad
\dfrac{\overset{\displaystyle P_2}{\Delta_2,\,{}^{\emptyset}\alpha^{L,R} \Rightarrow \Gamma_2} \qquad \overset{\displaystyle P_3}{\Delta_2,\,{}^{\emptyset}\beta^{L,R} \Rightarrow \Gamma_2}}{\Delta_2,\,{}^{\emptyset}(\alpha \sqcup \beta)^{L,R} \Rightarrow \Gamma_2}\sqcup\text{-l}
}{\delta^{+R}, \Delta_2 \Rightarrow \Gamma_1{}^{+R}, \Gamma_2}\;({}^{\emptyset}(\alpha \sqcup \beta)^{L},\,{}^{\emptyset}(\alpha \sqcup \beta)^{L,R})
$$

This proof can be transformed into:

$$
\dfrac{
\dfrac{
\dfrac{\overset{\displaystyle P_1}{\delta \Rightarrow \Gamma_1,\,{}^{\emptyset}\alpha^{L},\,{}^{\emptyset}\beta^{L}} \qquad \overset{\displaystyle P_2}{\Delta_2,\,{}^{\emptyset}\alpha^{L,R} \Rightarrow \Gamma_2}}{\delta^{+R}, \Delta_2 \Rightarrow \Gamma_1{}^{+R},\,{}^{\emptyset}\beta^{L,R}, \Gamma_2}\;({}^{\emptyset}\alpha^{L},\,{}^{\emptyset}\alpha^{L,R}) \qquad \dfrac{\overset{\displaystyle P_3}{\Delta_2,\,{}^{\emptyset}\beta^{L,R} \Rightarrow \Gamma_2}}{}}{\delta^{+R}, \Delta_2, \Delta_2 \Rightarrow \Gamma_1{}^{+R}, \Gamma_2, \Gamma_2}\;({}^{\emptyset}\beta^{L,R})
}{\text{some permutations and contractions}}
}{\delta^{+R}, \Delta_2 \Rightarrow \Gamma_1{}^{+R}, \Gamma_2}
$$

which again contains one *mix* and one *quasi-mix*, but both with grade less than the grade of *quasi-mix* on P. So by the induction hypothesis, we can obtain a proof which contains no *quasi-mixes* at all. All other cases of outermost logical symbol in *quasi-mix* inferences can be obtained in a similar way.

Case 2: $r > 2$, i.e., $rank_l(P) > 1$ **and/or** $rank_r(P) > 1$ The induction hypothesis is that from every proof Q which contains a *quasi-mix* only as the last inference, and which satisfies either $g(Q) < g(P)$, or $g(Q) = g(P)$ and $rank(Q) < rank(P)$, we can eliminate the application of the *quasi-mix*.

2.1) $rank_r(P) > 1$

2.1.1) Let us consider a *quasi-mix* of the form (δ, δ^{+R}) in which Δ_1 contains δ^{+R}. In this case, we construct a new proof as follows.

$$\frac{\dfrac{\Delta_2 \Rightarrow \Gamma_2}{\text{some permutations and contractions}}}{\dfrac{\delta^{+R}, \Delta_2^* \Rightarrow \Gamma_2}{\Delta_1, \Delta_2^* \Rightarrow \Gamma_1^*, \Gamma_2}\text{ some weakenings and permutations}}$$

where the assumption Δ_1 contains δ^{+R} were used in the last inference to construct Δ_1. When Γ_2 contains δ and in the other *quasi-mix* cases the transformation is similar.

2.1.2) S_2 is the lower sequent of a inference J_2, where J_2 is not a logical inference whose principal formula is δ. We will consider just the case where the *quasi-mix* is of the form (δ, δ^{+R}), the other cases of quasi-mix can be treated in a similar way. P has the form:

$$\frac{\dfrac{P_1}{\Delta_1 \Rightarrow \Gamma_1} \quad J_2 \dfrac{\dfrac{P_2}{\Phi \Rightarrow \Psi}}{\Delta_2 \Rightarrow \Gamma_2}}{\Delta_1, \Delta_2^* \Rightarrow \Gamma_1^*, \Gamma_2} (\delta, \delta^{+R})$$

where P_1 and P_2 contain no *quasi-mixes* and Φ contains at least one occurrence of δ^{+R}. We first consider the proof P':

$$\frac{\dfrac{P_1}{\Delta_1 \Rightarrow \Gamma_1} \quad \dfrac{P_2}{\Phi \Rightarrow \Psi}}{\Delta_1, \Phi^* \Rightarrow \Gamma_1^*, \Psi} (\delta, \delta^{+R})$$

$g(P) = g(P')$, $rank_l(P') = rank_l(P)$ and $rank_r(P') = rank_r(P)$ -1. Thus, by the induction hypothesis, the final sequent in P' is provable without *quasi-mix*. Given that, we can now construct a proof:

$$\frac{\dfrac{\dfrac{P'}{\Delta_1, \Phi^* \Rightarrow \Gamma_1^*, \Psi}}{\Phi^*, \Delta_1 \Rightarrow \Gamma_1^*, \Psi}\text{ some permutations}}{J_2 \dfrac{}{\Delta_2^*, \Delta_1 \Rightarrow \Gamma_1^*, \Gamma_2}}$$

In the case that the auxiliary formula in J_2 in P is a mix in Φ, we need an additional weakening before J_2 in the last proof.

2.1.3) Δ_1 contains no δ's, S_2 is the lower sequent of a logical inference whose principal formula is δ and J is a *mix* rule inference. We have to consider several cases according to the outermost logical symbol of δ:

i) The outermost logical symbol of δ is $\sqcup$. Let us consider a proof P whose last part is of the form:

$$
\cfrac{
\cfrac{}{\Delta_1 \Rightarrow \Gamma_1}\ (P_1)
\qquad
\cfrac{
\cfrac{\Delta_2, {}^\emptyset\alpha^L \Rightarrow \Gamma_2}{}\ (P_2)
\qquad
\cfrac{\Delta_2, {}^\emptyset\beta^L \Rightarrow \Gamma_2}{}\ (P_3)
}{\Delta_2, {}^\emptyset(\alpha \sqcup \beta)^L \Rightarrow \Gamma_2}
}{\Delta_1, \Delta_2^* \Rightarrow \Gamma_1^*, \Gamma_2}\ ({}^\emptyset(\alpha \sqcup \beta)^L)
$$

Assuming that $\delta = {}^\emptyset(\alpha \sqcup \beta)^L$ is in Δ_2, consider the proof Q_1 (left) and Q_2 (right):

$$
\cfrac{
\cfrac{}{\Delta_1 \Rightarrow \Gamma_1}\ (P_1)
\qquad
\cfrac{}{\Delta_2, {}^\emptyset\alpha^L \Rightarrow \Gamma_2}\ (P_2)
}{\Delta_1, \Delta_2^*, {}^\emptyset\alpha^L \Rightarrow \Gamma_1^*, \Gamma_2}\ (\delta)
\qquad\qquad
\cfrac{
\cfrac{}{\Delta_1 \Rightarrow \Gamma_1}\ (P_1)
\qquad
\cfrac{}{\Delta_2, {}^\emptyset\beta^L \Rightarrow \Gamma_2}\ (P_3)
}{\Delta_1, \Delta_2^*, {}^\emptyset\beta^L \Rightarrow \Gamma_1^*, \Gamma_2}\ (\delta)
$$

We have $g(Q_1) = g(Q_2) = g(P)$, $rank_l(Q_1) = rank_l(Q_2) = rank_l(P)$ and $rank_r(Q_1) = rank_r(Q_2) < rank_r(P)$. Hence, by the induction hypothesis, the end-sequents of P_1 and P_2 are provable without a *mix*. Let us consider new proofs without *mix* Q_1' and Q_2' in the construction of P':

$$
\cfrac{
\cfrac{}{\Delta_1 \Rightarrow \Gamma_1}\ (P_1)
\qquad
\cfrac{
\cfrac{}{\Delta_1, \Delta_2^*, {}^\emptyset\alpha^L \Rightarrow \Gamma_1^*, \Gamma_2}\ (Q_1')
\qquad
\cfrac{}{\Delta_1, \Delta_2^*, {}^\emptyset\beta^L \Rightarrow \Gamma_1^*, \Gamma_2}\ (Q_2')
}{\Delta_1, \Delta_2^*, {}^\emptyset(\alpha \sqcup \beta)^L \Rightarrow \Gamma_1^*, \Gamma_2}\ ({}^\emptyset(\alpha \sqcup \beta)^L)\ \sqcup\text{-l}
}{\Delta_1, \Delta_1, \Delta_2^* \Rightarrow \Gamma_1^*, \Gamma_1^*, \Gamma_2}
$$

Then, $g(P') = g(P)$, $rank_l(P') = rank_l(P)$ and $rank_r(P') = 1$, since Δ_1 and Δ_2^* do not contain ${}^\emptyset(\alpha \sqcup \beta)^L$. By the induction hypothesis the end-sequent of P' is provable without a *mix*.

The remaining cases where the outermost logical symbol of δ is $\forall$, $\exists$, $\sqcap$ and $\neg$ are treated in a similar way.

2.1.4) The same conditions that hold for 2.1.3 but J is a *quasi-mix* rule inference. We have to consider several cases according to the outermost logical symbol of δ. All the cases are treated in a similar way of cases 2.1.3.

2.2) $rank_r(P) = 1$ and $rank_l(P) > 1$. This case is proved as in case 2.1 above.

∎

4   Conclusion and further works

Future investigation must also include: (1) the extension of this calculus in order to deal with stronger Description Logics, mainly, $\mathcal{SHIQ}$ [1]; (2) the development of a Natural Deduction System based on $S_{\mathcal{ALC}}$.

Another interesting thing to be investigated is a comparison with others inference algorithms – like the structural subsumption algorithms and Tableaux [1]. Furthermore, we have also start the development of a prototype theorem prover for $S_{\mathcal{ALC}}$. For example, it seems that when restricted to the kind of formulas that the structural subsumption algorithm can handle, a cut-free proof of a subsumption correspond stepwise to the procedure followed by this very algorithm. This would be an advantage of our sequent calculus regarding to the usual Tableaux for description logics.

It is interesting to note that the labels in our sequent calculus represent the relationship between the worlds instead worlds themselves as it is usual in labeled deductive systems for modal logics.

BIBLIOGRAPHY

[1] F. Baader. *The Description Logic Handbook: theory, implementation, and applications.* Cambridge University Press, 2003.

[2] A. Borgida, E. Franconi, I. Horrocks, D. McGuinness, and P. Patel-Schneider. Explaining $\mathcal{ALC}$ subsumption. In *Proceddings of the International Workshop on Description Logics*, pages 33–36, 1999.

[3] S. R. Buss. An introduction to proof theory. In S. R. Buss, editor, *Handbook of Proof Theory, Studies in Logic and the Foundations of Mathematics*, page 811. Elsevier, Amsterdam, 1998.

[4] D. de Oliveira, C. S. de Souza, and E. Haeusler. Structured argument generation in a logic based kb-system. In L. S. Moss, J. Ginzburg, and M. de Rijke, editors, *Logic Language and Computation*, number 96 in CSLI Lecture Notes, pages 237–265. CSLI, Stanford, California, 1 edition, 1999.

[5] F. do Amaral, C. Bazílio, G. M. H. da Silva, A. Rademaker, and E. Haeusler. An Ontology-based Approach to the Formalization of Information Security Policies. *EDOCW*, 0:1, 2006.

[6] M. Fitting. *Proof methods for modal and intuitionistic logics.* Reidel, 1983.

[7] D. M. Gabbay. *Labelled deductive systems*, volume 1. Oxford University Press, 1996.

[8] D. McGuinness. *Explaining Reasoning in Description Logics.* PhD thesis, Rutgers University, 1996.

[9] A. Rademaker, F. do Amaral, and E. Haeusler. A Sequent Calculus for $\mathcal{ALC}$. Monografias em Ciência da Computação 25/07, Departamento de Informática, PUC-Rio, 2007.

[10] C. Renteria and E. Haeusler. A natural deduction system for keisler logic. *Eletronic Notes in Theoretical Computer Science*, 123:229–240, 2005.

[11] M. Schmidt-Schau and G. Smolka. Attributive concept descriptions with complements. *Artificial Intelligence*, 48(1):1–26, 1991.

[12] G. Takeuti. *Proof Theory. Number 81 in Studies in Logic and the Foundations of Mathematics.* North-Holland, 1975.

Alexandre Rademaker and Edward Hermann Haeusler
Informatic Department
Pontifical Catholic University of Rio de Janeiro, Brazil
E-mail: {arademaker,hermann}@inf.puc-rio.br

Luiz Carlos Pereira
Philosophy Department
Pontifical Catholic University of Rio de Janeiro, Brazil
E-mail: luiz@inf.puc-rio.br

Strategies. What's in a Name?

Dagmar Provijn

ABSTRACT. In this paper, I will show that Hintikka's notion of 'strategy' can refer to proof-heuristic reasoning as well as to methodological reasoning forms. Formulating this distinction allows for a better understanding of the notion and for an easier way to tackle the problem of formalization. Contrary to Hintikka's opinion, heuristic reasoning can be implemented in formal proofs by means of goal-directed procedures. A goal-directed proof procedure for propositional classical logic will be presented to show how the incorporation of search steps in a proof allows for a perspicuous formulation of heuristic principles that lead the proof search. I will also refer to the fact that methodological reasoning forms, on the other hand, can be formally represented by means of adaptive logics.

1 Introduction

Throughout many of his papers Hintikka has inveighed bitterly against the stepmotherly way in which 'strategic' principles are treated in both the development and teaching of formal deductive logic and reasoning in general [9, 11]. In broad terms this criticism is based on a distinction between 'definitory' rules – the rules of inference of a formal system spelling out which steps are allowed at a certain stage of a proof – and the 'strategic' principles determining which of these possible inference steps are the most valuable with reference to the goal to be obtained. The rules refer to correctness, the principles to excellence in reasoning.

To make things clear, Hintikka draws an analogy between constructing logical proofs and playing chess. In a game of chess the 'definitory' rules tell the player which moves are permitted. However, even knowing these rules by heart won't help much to checkmate an experienced opponent. In order to obtain this goal – viz. winning the game – the player needs 'strategic' principles, well-thought-out assessments of the consecutive steps that are needed to checkmate the opponent. A same division of labor applies to the construction of logical proofs. The quasi-random application of inference rules may be instructive in a first rendezvous with logic, but once more complex proofs are intended this approach will show highly inadequate. From that moment on, heuristic principles or 'strategies' will be required to obtain the goal or subgoals in an efficient way. Contrary to 'definitory' rules, 'strategic' principles cannot refer to single steps in a reasoning process. At the very least these 'strategic' principles will refer to series of steps. Hence, even if the formulation of optimal 'strategies' is feasible, it will never be in a recursive way [10, pp. 4–5].

However, interpreting strategies as heuristic principles as such does not fully cover the overtones. Though the 'definitory' rules and 'strategic' principles govern the same kinds of inferential steps, the former are validated according to their ability to carry over truth or high probability to the conclusion, while the latter enable us to acquire new information and should rather lead to truth in the long run [12, pp. 98–100]. So, 'strategic' principles do not only guide 'definitory' rules for being applied in an efficient way, they also lead to new information and as such they clear the way for ampliative forms of reasoning. According to Hintikka, all reasoning and argumentation is to be considered as a question-answer sequence containing both deductive and interrogative steps. This process of questioning and deducing can be codified within his *interrogative model of reasoning and inquiry* that should be taken as an *interrogative game* to put emphasis on the significance of the strategies that are involved. This game is played by an *Inquirer* and an *Oracle/Nature* being neither more nor less than an external source of information rendering only true answers.[1] In this game, the Inquirer always has the choice to make some deductive move on the basis of what has already been established or to ask a question – the presupposition of which should also occur in what is established – to the *Oracle* in order to obtain some new information if the *Oracle* provides an answer. However, knowing what deductive steps can be made and what questions can be asked on the basis of the given or generated set of information does not answer the main question the Inquirer finds himself confronted with: 'Which proposition(s) should be used first to make a deductive move or should be used as a presupposition of a question?'. An answer to this question can only be established by 'strategic' means; and for Hintikka the principles underlying the choice of a specific deductive move and those underlying the choice of a particular question are roughly the same.

To show that 'strategic' thinking by means of the *interrogative model of inquiry* is valuable for argumentation in general, Hintikka refers to the magnificent example of 'the curious incident of the dog in the night-time' from Sir Arthur Conan Doyle's detective story 'Silver Blaze' featuring Sherlock Holmes [11, pp. 318–319]. Hintikka points out the mysterious fact that the watchdog did nothing during the night-time, even at the moment when the horse Silver Blaze was stolen. In fact, Hintikka refers to the following excerpt: "Is there any point to which you would wish to draw my attention?" "To the curious incident of the dog in the night-time." "The dog did nothing in the night-time." "That was the curious incident", remarked Sherlock

---

[1]The more sophisticated version of the interrogative game, introducing bracketing to deal with an Oracle that is not compelled to render true answers – see e.g.[13], is left out as it will not influence the point I want to make here.

Holmes.[2] According to Hintikka the utterances of Holmes in fact replace three questions: 'Was there a watchdog in the stables?', 'Did it bark during the night, including the time when the horse was stolen?' and 'Now who is it that a trained watchdog does not bark at in the middle of the night?'[11, p. 319].

As the presence of the dog and its silence – the stable boys were not alerted – are known facts, Hintikka concludes: 'Hence it must have been the stable master himself who stole the horse...'[11, p. 319]. In terms of the 'interrogative game' it is the introduction of the dog as a new individual in the inquiry that deserves all credentials. Besides, if this 'interrogative argument' would have been evaluated as a single reasoning step, it would lose all of its splendor. In game-theoretical terms, this troublesome line of argument obtains its value by supporting and being supported by other arguments to line up in the desired payoff of finding the truth. As such, it is only in view of Holmes' whole 'strategy' that all the marvel of the argument pops up. Again this supports a holistic view concerning 'strategies'.

2   On strategies

In [23], Wiśniewski refers to two features of the *interrogative model of inquiry* that reveal much of the tenuousness of the concept 'strategy' as it is presented by Hintikka. First, the *interrogative model of inquiry* does not presuppose a logic of question-handling; the inferences that take place are deductive and questions are only means to obtain new information. Wiśniewski however argues for a logic in which questions serve as premises and conclusions in order to obtain a realistic theory of inquiry that does not rely on 'external' strategic considerations – see [23, pp. 390–391]. Second, even when *Nature* gives a true answer to the *Inquirer*, it can nevertheless be completely useless when not contributing to any possible deductive inference that supports the finding of the main aim or goal of the inquiry – see [23, pp. 391].

In fact, it is even hard to find concrete formulations of the 'strategies' that are so greatly appreciated and praised by Hintikka.[3] This lack of extensive elaboration of the 'strategies' so often referred to becomes clear in [12] where he offers a general account of what 'strategic' thinking is about in view of deductive rules and abductive questions; the latter in fact dealing with ampliative reasoning in general – see [12, p. 108] and [19, p. 270]:

> From the strategic viewpoint, the crucial question about abductive questions is: Which one to ask first? [...] Probably the only general answer, which unfortunately does not yield any directly applicable recipes, is to say that the choice of the right questions

---

[2]See   http://ebooks.adelaide.edu.au/d/doyle/arthur_conan/d75me/silver.blaze.html for an electronic version of the story.

[3]In this respect, see also [14] from page 97 on.

depends on one's ability to anticipate their answers. Strategi-
cally, there is little difference between selecting a question to
ask in preference to others and guessing what its answer will be
— and guessing how it compares with the expected answers to
other questions that could be asked. And even in the case of
deductive rules, the secret of a good strategist is to be able to
anticipate where the inferences lead. [12, p. 109]

Another lightening excerpt stems from [8]:

The closer examination of the payoffs and different strategies
would take us too far. Suffice it to mention here one interesting fact. Most of the deductive moves – including some of the
most interesting ones – can be replaced by a suitable question,
assuming that an answer to it is forthcoming.

For instance, assume that $(F_1 \lor F_2)$ occurs in the left column
of some *subtableaux* σ_j. A deductive move might involve splitting σ_j in two, with F_1 and F_2, respectively, added to their left
columns. However, instead of doing so one could ask "Is it the
case that F_1 or is it the case that F_2?" This can be done because
the presupposition of this question is $(F_1 \lor F_2)$. Whichever the
answer is, one is saved the trouble of continuing the construction of one of the two *subtableaux* into which the deductive move
would have split σ_j. [8, p. 168]

The main problem with 'strategic' thinking as presented and defended
by Hintikka is that the determination of its meaning heavily depends on
the specific construction of the *interrogative model of inquiry* and that its
possible specifications are way too diverse to fit into this model. First of
all, Hintikka puts too much stress on the importance of questions, while in
fact it is the determination of the right presuppositions that makes you ask
the right questions. Secondly, the *interrogative model of inquiry* contains
no elements that facilitate the anticipation of the right formulas or presuppositions that should be derived in view of deriving the main goal. Above
that, too much of the success that is promised by the *interrogative model of
inquiry* depends on the selection and the specific content of the *Oracle*.

I shall distinguish the two interpretations of 'strategic' thinking that are
present in Hintikka's notion of 'strategy'. First of all, there are the heuristics
that govern the 'definitory' rules of formal deductive inference systems and
secondly there are the holistic 'strategies' that refer to ampliative methodological reasoning forms such as abduction, induction, compatibility, etc.
Stating this distinction explicitly provides a better understanding of the
notion and allows for a specific handling of all the different reasoning forms
that are incorporated in it. This diversification in the meaning of 'strategies' is not a criticism on Hintikka's conception, it rather simplifies their
understanding and respective formalizations.

In sections 3 and 5 I will show that it is possible to develop proof procedures that allow to construct proofs in a goal-directed way and that explicate the heuristic reasoning involved in this process. In these sections, it will become clear how the proof procedure allows to anticipate where the inferences lead and in fact should lead in order to obtain the main goal of a search process, which is in line with Hintikka's claim that 'the secret of a good strategist is to be able to anticipate where the inferences lead' [12, p. 109]. I will also refer to the fact that these proofs can render the presuppositions for questions that serve the search for the main goal and that they allow for the selection of 'potential abductive explanations' – see sections 3 and 4. Roughly speaking, one may think of these goal-directed proofs as the proof theoretic counterpart of Hintikka's Beth-tableaux used in the *interrogative model of inquiry*, including part of the 'strategies'.

Though the goal-directed proof theories may solve the second problem that was indicated by Wiśniewski, I will not go as far as to say that it also solves the first. Even if the goal-directed proofs may render presuppositions for the generation of questions, it cannot be considered as a real theory of inquiry. Therefore logics of inquiry should be studied as a specific brand of logics of which one can also study the proof-heuristics once the 'definitory' rules are determined.

Also the fact that goal-directed proofs may render 'potential abductive explanations' does not solve the topic of ampliative reasoning in general. Hintikka in fact refrains from making a distinction between new information obtained through specific methodological reasoning that can be formalized, and new information obtained through directing a question to an external source (of which the choice can also be led by specific methodological reasoning). Above that, abduction itself seems a fair way to stipulate 'potential abductive explanations' for a particular fact, given a background theory from which they are generated – which is in line with [1] and [17]. If these potential explanations are multiple or whenever the background theory is inadequate to render any of these explanations, the abductive process may serve as a clue for further investigations that lead to novel results. On the other hand, it seems quite unrealistic to believe that the process of generating '(potential) abductive explanations' will render novel results itself. For that we will have to study the role abductive reasoning can play together with other kinds of ampliative (and in fact also corrective) reasoning. At least the formal representation of ampliative and corrective reasoning forms points in this direction – see for example [2] on inductive generalizations, [6] on compatibility, [16] on the use of analogical reasoning and [18] on abductive reasoning. The problem of ampliative reasoning forms and the specific function of external sources of information will be discussed in section 4.

3   On proof-heuristics

Hintikka's claim that the study of heuristics has played second fiddle to the elaboration of 'definitory' rules certainly makes sense. Concerning 'strategic' reasoning, one should however pay attention to the distinction between the possibility to push formal elements of search paths in the proof itself and the ideal of directly obtaining the most efficient proofs: the former might be a possibly unsuccessful but insightful and relevant practice that allows for the development of perspicuous heuristics, the latter is aimed at convincing.

In line with Hintikka, Suppes stresses that a great deal of the reasoning about a certain domain stems from nonverbal learning and that an excessive formalization of these reasoning processes should be avoided. In this respect, Suppes supports Hintikka's chess-analogy and even adds that at higher levels of mastery the strategies are only partly explainable by the players themselves [22, p. 744–745].

In [7], a proof format is elaborated allowing to push formal elements of search steps into formal proofs for $\Gamma \vdash G$ in which Γ is a premise set and G the goal of the search process. This proof format allows for the development of goal-directed proof procedures that are composed of (i) a *set of inference rules* based on the new proof format – i.e. for classical propositional logic in this paper –, (ii) a *positive part relation* – see Definition 1 and table 1 – that allows for the selection of 'relevant premises' by determining whether a goal-formula is derivable from them, (iii) a *set of marking definitions* – see Definitions 2, 3 and 4 – that determine which goal-formulas are still to be searched for in order to obtain the main goal G and (iv) a *set of instructions* that form restrictions on the application of inference rules depending on the formulas already occurring in the proof for $\Gamma \vdash G$.

The proof-heuristics that results from the set of instructions can be restrictive in different degrees but it will in most of its formulations allow for goal-directed and efficient proofs. In the goal-directed proofs, formulas are derived that have the following form

$$[B_1, \ldots, B_n]A$$

meaning that A is derived on the condition $[B_1, \ldots, B_n]$.[4] The rules of the inference system are such that:

'if $[\Delta]A$ is derivable from a set of premises Γ then $\Gamma \cup \Delta \vdash A$.'

The condition $[B_1, \ldots, B_n]$ on which a formula A is derived guides the search process as it reminds us that each B_i $(1 \leq i \leq n)$ has not been derived yet, but should be in order to derive A. The goal-directed proof procedure provides for classical propositional logic a decision method for $A_1, \ldots A_n \vdash B$ and a positive test for $\Gamma \vdash A$. However, the fact that part of the reasoning process for G is implemented in the proof is the most promising in view of defining further heuristic principles.[5]

---

[4]If the condition is empty (notation: $[\emptyset]A$ or simply A), A is said to be derived unconditionally.

[5]The formal aspects of this goal-directed proof procedure will be formulated in section 5.

Let us consider an example of a goal-directed proof for the following Γ and G:

$$\Gamma = \{(p \wedge q) \supset r, \sim s \vee p, t \wedge s, t \supset u\} \text{ and } G = \{r\}$$

The procedure instructs to start the proof with the introduction of $[r]r$ by means of the Goal rule, which might seem redundant but in fact guides the initiation of the search process (and it is even necessary to derive theorems of the logic at hand). The goal-directed proof proceeds as follows. First of all, the procedure will instruct to search for a premise from which r may be derived. If r is a *positive part* of a premise – viz. it is derivable from it, see Definition 1 – then this premise is introduced in the proof; in this case only $(p \wedge q) \supset r$ satisfies this requirement. Once this premise is introduced, it can be analyzed in such a way as to show which further steps have to be made for deriving r.

1	$[r]r$		Goal
2	$(p \wedge q) \supset r$		Prem
3	$[p \wedge q]r$	2	$\supset$E
4	$[p, q]r$	3	C$\wedge$E

Line 3 states that r is derivable if $p \wedge q$ is, and as $p \wedge q$ is not a *positive part* of any of the premises (cannot be derived from any of the premises as such), the condition is further analyzed at the next stage of the proof, as the separate derivation of p and q may also lead to the derivation of r. The search process can be stopped after the introduction of line 4 if we only stick to the premise set Γ as q is not a *positive part* of a premise. Nevertheless, there are good reasons to continue the proof here in order to show the versatility of the proof format and to formulate some more general claims about heuristics.

5	$\sim s \vee p$		Prem
6	$[s]p$	5	$\vee$E
7	$t \wedge s$		Prem
8	s	7	$\wedge$E
9	p	6,8	Trans

From lines 5 to 9, p is derived through a search process that first leads to the derivation of s. The final version of the proof looks as follows:

1	$[r]r$		Goal	
2	$(p \wedge q) \supset r$		Prem	
3	$[p \wedge q]r$	2	$\supset$E	
4	$[p, q]r$	3	C$\wedge$E	R10

5	$\sim s \vee p$		Prem	
6	$[s]p$	5	$\vee$E	R9
7	$t \wedge s$		Prem	
8	s	7	$\wedge$E	
9	p	6,8	Trans	
10	$[q]r$	4,9	Trans	

The derivation of p at line 9 allows to mark line 6 as redundant because it is no longer useful to search for s – see Definition 2. It also leads to the elimination of p from the condition at line 4 by means of the Transitivity rule, which results in line 10 and the redundancy marking of line 4.

Suppose Γ also contains the premises $v \supset q$ and $u \supset (q \wedge w)$. The previous proof then continues as follows:

11	$v \supset q$		Prem	
12	$[v]q$	11	$\supset$E	R20
13	$u \supset (q \wedge w)$		Prem	
14	$[u]q \wedge w$	13	$\supset$E	
15	$[u]q$	14	$\wedge$E	R20
16	$t \supset u$		Prem	
17	$[t]u$	16	$\supset$E	R19
18	t	7	$\wedge$E	
19	u	17,18	Trans	
20	q	15,19	Trans	
21	r	10,20	Trans	

In this new situation there is no reason at all to stop the search procedure after line 4, even if Γ would only be extended with $v \supset q$. In view of the final result one might put forward that steps 11 and 12 are redundant, since they represent an unsuccessful search path. One should however keep in mind that the usefulness of these steps for the final goal of the search procedure can only be established once one has gone through the search path itself. At stage 10 of the proof it is heuristically valuable to start this search path as it may lead to the derivation of q. Even though one might imagine the perfect heuristics that instantly results in the most elegant and efficient proof for $\Gamma \vdash G$ – which I presume to be impossible –, this should not lead to the depreciation of a formal representation of goal-directed reasoning in which, along with the relevant steps that are made in view of the main goal, only the given premise set Γ and the already performed search steps of the reasoning process are present. Besides, the formal representation of the search procedure contains the elements that are needed to construct the shortest and most convincing proof for $\Gamma \vdash G$. It is a fact that the proof of the pudding is in the eating, but knowing that the right ingredients and the right method have been used to prepare it already reveals a lot of the quality it will show. Each step in the goal-directed proofs is heuristically valuable in view of the main goal and it is supported by the insight in the premises at that stage of the proof and by the search steps that are already made. Looking at the proof with the original Γ one should notice that t is not derived because it has no heuristic value relative to the derivation of r. This in contrast to the extended Γ where the derivation of t plays a pivotal role in the derivation of the same main goal. Notwithstanding each inference step can only be evaluated in the whole search process, it is possible to formulate a heuristic procedure that renders a formal description of the search process step-by-step, at least if one is able to agree with a goal-directed proof procedure to construct a proof for $\Gamma \vdash G$.

Hence the chess-analogy falls short in that a game of chess always begins with an identical starting situation and with an exclusive goal such that the flexibility of assimilating and anticipating the moves of the opponent combined with an initial strategy is primordial to play a successful game. The search for a proof for $\Gamma \vdash G$ on the other hand starts differently as there is no stubborn opponent and as the goal already contains specific information leading to a sensible selection and analysis of the premises and a heuristically relevant application of 'definitory' rules to obtain the main goal or subgoals that are derived in the formal search process [20, 4].

However, one may not conclude that the study of proof-heuristics has become trivial now. In fact, what is presented here is a rather general mechanism through which search steps can be introduced in formal proofs in order to formulate perspicuous proof heuristics for particular inference systems. This means that each particular logic will have its own specifications concerning inference rules, positive part relation, marking definitions and heuristic instructions that have to be fixed. Above that, for each logic a multiplicity of specific 'strategies' can be developed according to different deterministic configurations of the heuristic instructions – think for example of the elementary choice between breadth-first and depth-first search path construction. The choice for specific 'strategies' – as specific configurations of instructions – in particular situations can form a study on its own. Though, I rather opt for a more liberal formulation of the proof-heuristics that can be further determined according to the insights that are gained during a specific search process. Again, this fits with the idea that the proof of the pudding is in the eating, though without having to start from scratch.

One may certainly support Suppes' claim that much of the acquisition of formal reasoning runs unconsciously. However, this should not restrain us from formulating formal representations of how we learn and how we actually perform goal-directed reasoning processes via more or less restrictive heuristics. This at least enables us to evaluate and discuss a more precise formulation of what this heuristic thinking may be.

4   On methodological reasoning forms

Another reason why the proof according to the original Γ, presented in the previous section, was not stopped is that the derivation of line 10 forms the basis for the determination of a 'potential abductive explanation'. The elaboration of this application of the goal-directed proof procedure has started in [17] and should form an alternative for the tableaux-based determination of abductive reasoning as presented in [1]. However, these dead-ends in the goal-directed search process – viz. line 10 in the original proof and line 12 in the extended proof – not only serve as a basis for 'abductive explanations'; they can also function as triggers that ask for an extension of our background knowledge in the given problem solving situation. As such they can serve as clues for the initiation of other ampliative reasoning processes. The latter application of the goal-directed proof format is extensively elaborated by Batens in [4].

Actually, Batens' text deals with the formal representation of methodological reasoning forms that can be corrective (handling inconsistencies and ambiguity) or ampliative (abduction, induction, compatibility, etc.) by means of formal problem solving processes. Even more, it combines both the heuristics involved in formal deductive inference systems and the formalization of methodological reasoning forms by means of a formal framework for problem solving processes. For that, Batens combines (i) adaptive logics – rendering formal explications of methodological reasoning forms for which there is no positive test[6] – necessarily generating dynamic proofs in which specific reasoning steps can in turn be considered IN and OUT according to the insights in the reasoning process at a certain stage; (ii) the prospective dynamics of the goal-directed proof format that allows to introduce formal elements of the dynamic search process into the proofs[7]; and (iii) erotetic logics to introduce problems as sets of questions on the basis of the conditions that are generated by the goal-directed proof procedure. This formal representation of problem solving processes shows the possibility for the determination of the presuppositions that are needed to introduce a question by means of a step-by-step procedure. This procedure already contains some elements of the goal-directed reasoning that belongs to particular methodological reasoning forms and it facilitates the development of a more or less restrictive heuristics.

However, one should keep in mind that the formalized problem solving processes are content guided in the sense that during the reasoning process parts of the original setup may change according to the insights that were gained along its different search paths. The fact that we do not know in

---

[6]See [5] for a general framework of adaptive logics and http://logica.ugent.be/centrum/writings/pubs.php for an extensive list of publications on adaptive logics and their applications on abduction, induction, compatibility, etc.

[7]In [3] it is shown that goal-directed proof procedures can be formulated for the dynamic proofs of adaptive logics.

advance which of the possible setups is the one that will be successful in the end, should not restrain us from the study of the heuristics and the methodological reasoning processes that fit best to each specific stage of the problem solving process.

If we compare this position with the following ideas that Hintikka has about the matter, we can make some further comments:

> There is nevertheless no such thing as *the* logic of discovery. For one important thing, the structure of an interrogative process depends crucially on the class of answers Φ which the oracle(s) will give. (It is assumed to be constant throughout an interrogative inquiry.) How is Φ determined? In applications, there are usually various *ad hoc* restrictions on Φ which are hard to reach any overview on. For instance, in the case of scientific reasoning their study involves questions of observational and experimental methodology.
>
> What can be studied here by means of logical methods is how our interrogative logic depends on the structural restrictions on available answers.[13, p. 60]

The point on there not being *the* logic of discovery is hardly questionable, even for supporters of the rationality of the process of discovery. However, I want to comment on Hintikka's further claims in view of the heuristic and methodological formalizations I mentioned before. In [15], Meheus argues that problems giving rise to creative steps are always ill-defined in the sense that the background information used to solve the problem is incomplete or inconsistent. The kinds of specific violations of logical properties presupposed by a problem solver are in great number. However, adaptive logics are formal tools that adapt themselves to situations in which these specific violations occur. They do not invalidate specific sets of inference rules to avoid triviality, but invalidate specific *applications* of inference rules – i.e. those that would lead to triviality when a certain presupposition is violated.

This application of adaptive logics to formalize many of the reasoning processes involved in ampliative and corrective (and possibly creative) problem solving situations as defended in [4, 15] contrasts strongly with Hintikka's inability or omission to specify the 'strategies' that could replace these formalizations and that make the 'Inquirer' ask the right questions and select the appropriate 'Oracle(s)'. The fact that Hintikka does not really specify what exactly can be represented by means of 'strategies' and which of the 'Oracles' could possibly be replaced by ampliative or corrective reasoning strongly supports this lack of detail in Hintikka's *interrogative model of inquiry*. In section 4 of [4] Batens even shows that the notion 'Oracle' is less trivial than how it is applied by Hintikka, as its content has to vary according to changes in the situation the problem solver is confronted with. Batens also formulates an alternative for external resources that can be used as epistemic devices in formal problem solving processes.

Let me return now to the 'Silver Blaze' case. First of all, the excerpt chosen by Hintikka seems to function as a cunning trigger for the reader to step into the reasoning process and the deliberation that Holmes has already performed at that stage of the story. The full argumentation in which its merits are revealed is spelled out when Holmes defends his case in the end. When the excerpt is introduced, the three questions are already answered for Holmes, and as such do not serve a heuristic function for him. For the reader, it can only have this function if he or she already made the same reasoning steps as Holmes, otherwise the other possibilities – the dog being drugged or the positive behavior of the dog towards other familiar people – are still open for investigation.[8] Secondly, even if these questions are useful to establish the right conclusions, they are not important as such, it is rather the goal-directed search process to their preconditions that needs full attention. As holds for every reasoning step, these preconditions receive their heuristic value in view of the other reasoning steps that are already performed and, as I have shown before, this can be represented by a step-by-step procedure. Even though the reasoning in Holmes' case is based on ampliative reasoning forms – not in the least on abduction –, this should not be an argument against a stepwise formalization of it, as the formal problem solving processes from [4] bear at least the potential to formalize these kinds of ampliative reasoning. Finally and foremost, the references to detective stories as examples of excellence in reasoning have a perfidious side. These stories are representative for ideal situations in which the most elegant and efficient proofs are presented to make the star of the detective shine brighter than ever before. For these ideal situations it is in fact impossible to describe a step-by-step heuristic search method. On the other hand, it are the goal-directed problem solving reasoning processes leading to the wonderful proofs that deserve our attention and we have good reasons to believe that it is possible to develop goal-directed proof procedures to formally represent them.

5   A Goal-directed proof procedure for propositional Classical Logic

In this section I will present a goal-directed proof procedure for propositional classical logic (henceforth **CL**) to support part of the previous argumentation and to clarify how this kind of procedures function. For reasons of space, I present a version in which the instructions are immediately linked to the inference rules as in [4, 17]. This version is less deterministic than the one presented in [7] where a separate list of instructions is formulated. As mentioned in section 3 this leaves ample space for a further determination of the proof-heuristics according to the insights that are gained during specific

---

[8]In this sense the third question is rather premature and should be preceded by the question: Why didn't the dog bark during the night, including the time when the horse was stolen?

search processes.

One should remind that the procedure is developed to show that a main goal G is derivable from a set of premises Γ. Formulas in the proofs of the form $[\Delta]A$ indicate that the formula A is derived on the condition Δ which is always a finite set of formulas. The rules of the procedure are such that

(†)'if $[\Delta]A$ is derivable from a set of premises Γ then $\Gamma \cup \Delta \vdash A$.'

Consequently, every line in the proof will express information about derivability from the premise set Γ. As we search for the main goal G, it is obvious that the formula of the first line of a goal-directed proof is $[G]G$ – see the Goal-rule below in this section and the example from section 3. The G in the condition reminds us that this is the only formula we search for at the start of the proof. Through the introduction of the formulas in the condition, we obtain a heuristic tool that tells us which formulas we have to search for in order to obtain the main goal. Hence, at every stage of the proof one should search for a formula that occurs in a condition. Two problems arise in this respect: (i) what steps should one take in order to derive a formula in a condition, and (ii) should one always try to obtain all members of all conditions in a proof?

The second problem is easiest to handle by means of marking definitions. Lines containing superfluous conditions will be marked and as such the formulas in their condition will be no longer searched for. The reasons for marking are redundancy, pure inconsistency and loops – see Definitions 2, 3 and 4 below.

In view of (i), whenever we search for a certain formula A, we will check whether this formula can be obtained from a premise or a formula already derived in the proof. Claiming that it should be obtainable from these is stronger than just demanding that it occurs in them (p occurs in $p \supset q$ but is not obtainable from it, at least not in an efficient way; on the other hand, p is obtainable from $p \vee q$, $p \wedge q$, etc.). Before a premise will be introduced and analyzed if needed, some formula in a condition has to be an obtainable part of it. The notion 'obtainable part' is defined more precisely as a *positive part* relation between certain formulas.

For a concise formulation of the *positive part* relation and the inference rules we distinguish between $\mathfrak{a}$- and $\mathfrak{b}$-formulas, based on a theme from [21]. Let $*A$ denote the 'complement' of A, viz. B if A has the form $\sim B$ and $\sim A$ otherwise.

The following clauses constitute a recursive definition of the *positive part* relation for **CL**:

DEFINITION 1.

1. $\mathrm{pp}(A, A)$.

2. $\mathrm{pp}(A, \mathfrak{a})$ if $\mathrm{pp}(A, \mathfrak{a}_1)$ or $\mathrm{pp}(A, \mathfrak{a}_2)$.

$\mathfrak{a}$	$\mathfrak{a}_1$	$\mathfrak{a}_2$	$\mathfrak{b}$	$\mathfrak{b}_1$	$\mathfrak{b}_2$
$A \wedge B$	A	B	$\sim(A \wedge B)$	$*A$	$*B$
$A \equiv B$	$A \supset B$	$B \supset A$	$\sim(A \equiv B)$	$\sim(A \supset B)$	$\sim(B \supset A)$
$\sim(A \vee B)$	$*A$	$*B$	$A \vee B$	A	B
$\sim(A \supset B)$	A	$*B$	$A \supset B$	$*A$	B
$\sim\sim A$	A	A			

Table 1. $\mathfrak{a}$- and $\mathfrak{b}$-formulas as variation on a theme from [21]

3. $\mathrm{pp}(A, \mathfrak{b})$ if $\mathrm{pp}(A, \mathfrak{b}_1)$ or $\mathrm{pp}(A, \mathfrak{b}_2)$.

4. If $\mathrm{pp}(A, B)$ and $\mathrm{pp}(B, C)$, then $\mathrm{pp}(A, C)$.

For example: $\mathrm{pp}(\sim p, (p \vee q) \supset r)$ as $\mathrm{pp}(\sim p, \sim(p \vee q))$ and $\mathrm{pp}(\sim(p \vee q), (p \vee q) \supset r)$.

It is obvious that if we search for an $\mathfrak{a}$-formula (viz. one that occurs in a condition) then we should try to derive both $\mathfrak{a}_1$ and $\mathfrak{a}_2$, and that in the case of a $\mathfrak{b}$-formula it is sufficient to derive $\mathfrak{b}_1$ or $\mathfrak{b}_2$. On the other hand, if an $\mathfrak{a}$-formula is derived in a proof, then we can also derive $\mathfrak{a}_1$ and $\mathfrak{a}_2$ in the proof. If a $\mathfrak{b}$-formula is derived in a proof, then we can also derive $\mathfrak{a}_1$ or $\mathfrak{a}_2$ on the condition that $*\mathfrak{a}_2$ or $*\mathfrak{a}_1$ can be derived in the proof.

Suppose we search a proof for the following Γ and G:
$\Gamma = \{(p \wedge q) \supset r, s \wedge p, s \supset t\}$ and $G = \{r\}$

1	$[r]r$		Goal
2	$(p \wedge q) \supset r$		Prem

The proof is started with the introduction of $[r]r$ on the first line. As the main goal r is a positive part of the premise $(p \wedge q) \supset r$, the latter is added to the proof.

3	$[p \wedge q]r$	2	$\supset$E
4	$[p, q]r$	3	C$\wedge$E

As r is not a positive part of any of the other premises, the premise introduced on line 2 is analyzed by a formula analyzing rule (see below) in such a way that r can be derived whenever $p \wedge q$ is. As the latter is not a positive part of any of the premises and as there are no other goals to search for, the only option is to analyze the condition of line 3 by a condition analyzing rule which results in the separate goals p and q.

1	$[r]r$		Goal	
2	$(p \wedge q) \supset r$		Prem	
3	$[p \wedge q]r$	2	$\supset$E	
4	$[p, q]r$	3	C$\wedge$E	R7
5	$p \wedge s$		Prem	
6	p	5	$\wedge$E	
7	$[q]r$	4,6	Trans	

As p can be derived from premise $p \wedge s$, the latter is introduced and analyzed. Once p is derived in the proof, it can be removed from the condition on line 4 and as such line 4 can be R-marked, meaning that the condition on this line no longer serves as a set of goals that direct the goal-directed search process and consequently loses its heuristic role in the proof. Remark that premise $s \supset t$ is not introduced and consequently not analyzed in the proof as it cannot play a role in the derivation of r given this specific Γ.

The anticipatory character of the procedure – in view of the excerpts of section 2 – can easily be shown by way of an example. If we search for a proof of $G = \{r\}$ from $\Gamma = \{p \vee q, p \supset r, s \supset q\}$ and interpret the lines in the goal-directed proof as questions to an external source of information (that can also be replaced by a specific mode of methodological reasoning – see section 4), we obtain the following search process. The first question will be "Is r the case?". If it is not answered by the source, the occurrence of $[p]r$ in the proof allows for the question "Is p the case?", which will lead to the question "Is $\sim q$ the case?" whenever the second question stays unanswered and $[\sim q]p$ is derived in the proof. Obviously, each question is in function of obtaining the main goal. Besides, no question will be generated on the basis of $s \supset q$ as this would be useless in view of the search process. All this is in strong contrast with the tableaux-construction that Hintikka will get – in view of the second excerpt from section 2. When the *Oracle* is unable to give an answer, one will obtain a regular tableaux with three splits and questions that do not directly relate to a search process for r.

We now move to the instructions for the procedure **CLc** based on **CL**. Two general restrictions have to be taken into account:

R1 Formula analyzing rules are not applied on formulas introduced by the Goal rule.

R2 No rule is applied to repeat a marked or unmarked line.

A third restriction needs to be imposed for infinite Γ. Let $\Gamma_1, \Gamma_2, \ldots$ be a limiting sequence of Γ iff $\Gamma_1 \subset \Gamma_2 \subset \ldots$ and $\Gamma = \Gamma_1 \cup \Gamma_2 \cup \ldots$.

R3 If Γ is infinite, a proof for $\Gamma \vdash G$ is defined in terms of a limiting sequence $\Gamma_1, \Gamma_2, \ldots$ of Γ as follows: first the instructions are applied to Γ_1; if the procedure stops, the proof is continued by applying the instructions to Γ_2; etc.[9]

The instruction *Goal* introduces the main goal in the proof:

Goal   Start a goal-directed proof with:

1      $[G]G$ Goal

Premises are only introduced in the proof if a goal of an unmarked line is a positive part of it:

Prem   If A is a goal of an unmarked line, $B \in \Gamma$, and $pp(A, B)$, then one may add:

k B Prem

If a goal A is a *positive part* of a formula B that was introduced by Prem, the formula analyzing rules allow one to analyze B until $[\Delta]A$ is derived on a line in the proof. The formula analyzing rules are summarized as follows:[10]

$$\frac{[\Delta]\mathfrak{a}}{[\Delta]\mathfrak{a}_1 \quad [\Delta]\mathfrak{a}_2} \qquad \frac{[\Delta]\mathfrak{b}}{[\Delta \cup \{*\mathfrak{b}_2\}]\mathfrak{b}_1 \quad [\Delta \cup \{*\mathfrak{b}_1\}]\mathfrak{b}_2}$$

The general form of the rules is $[\Delta]A/[\Delta \cup \Delta']B$. Their application is governed by the following instruction (in which R refers to the name of the analyzing rule):

FAR   If C is a goal of an unmarked line, $[\Delta]A$ is the formula of an unmarked line i, $[\Delta]A/[\Delta \cup \Delta']B$ is a formula analyzing rule, and $pp(C, B)$, then one may add:

k $[\Delta \cup \Delta']B$ i R

The names of the formula analyzing rules are †E, with $\dagger \in \{\wedge, \vee, \supset, \sim, \equiv\}$ or $\sim$‡E with $\ddagger \in \{\wedge, \vee, \supset, \equiv\}$.

The condition analyzing rules are summarized as follows:

$$\frac{[\Delta \cup \{\mathfrak{a}\}]A}{[\Delta \cup \{\mathfrak{a}_1, \mathfrak{a}_2\}]A} \qquad \frac{[\Delta \cup \{\mathfrak{b}\}]A}{[\Delta \cup \{\mathfrak{b}_1\}]A \quad [\Delta \cup \{\mathfrak{b}_2\}]A}$$

The general form of the rules is $[\Delta \cup \{B\}]A/[\Delta \cup \Delta']A$. Their application is governed by:

---

[9]To see the need for R3, let $\Gamma = \{p \wedge \sim p\} \cup \{q_i \supset q_{i+1} \mid i \in \{0, 1, \ldots\}\}$ and consider a **CL**-proof for $\Gamma \vdash \sim q_0$. Without R3, the proof would never stop—EFQ would never be applied and $\sim q_0$ would never be derived.

[10]If two formulas occur at the bottom line of a rule, both variants may be derived (separately) in the proof.

CAR   If A is a goal of an unmarked line, $[\Delta \cup \{B\}]A$ is the formula
           of an unmarked line i, $[\Delta \cup \{B\}]A/[\Delta \cup \Delta']A$ is a condition
           analyzing rule, then one may add:

k $[\Delta \cup \Delta']A$ i R

The names of the condition analyzing rules are equal to the names of the
formula analyzing rules preceded by a C.

As $A \lor \sim A$ is valid in **CL**, *Excluded Middle* allows for the elimination of
certain goals by the following instruction:

EM   If A is a goal of an unmarked line, $[\Delta \cup \{B\}]A$ and $[\Delta' \cup \{\sim B\}]A$
           are the respective formulas of the unmarked lines i and j, and
           $\Delta \subseteq \Delta'$ or $\Delta' \subseteq \Delta$, then one may add:

k $[\Delta \cup \Delta']A$ i, j EM

The instruction *Transitivity* allows both for the elimination of goals that
are derived unconditionally and for the generation of alternative conditions
(if the goals of a certain condition are themselves conditionally derived in
the proof):

Trans   If A is a goal of an unmarked line, and $[\Delta \cup \{B\}]A$ and $[\Delta']B$
           are the respective formulas of the unmarked lines i and j, then
           one may add:

k $[\Delta \cup \Delta']A$ i, j Trans

If no further steps can be made in view of the previous rules, and as
CL validates Ex Falso Quodlibet, one should apply the following somewhat
unusual version of Ex Falso Quodlibet:[11]

EFQ   Where $A \in \Gamma$, G may be introduced on the condition $\{\sim A\}$.

The last elements of the procedure are the *marking definitions*:

A line on which $[\Delta \cup \Delta']A$ has been derived is redundant and hence *R-
marked* if $[\Delta]A$ has been derived in the proof. Evidently, searching for the
members of Δ' is useless to obtain A.

DEFINITION 2. Line i on which $[\Delta]A$ is derived, is R-marked on a stage
of a proof if on that stage $[\Delta']A$ is derived and $\Delta' \subset \Delta$.

A condition contains a flat inconsistency if both A and $\sim A$ occur in
it. If the derivation of G relies on the derivation of an inconsistency, the
procedure takes care of this by means of EFQ. Lines of which the condition
contains a flat inconsistency are *I-marked*.

---

[11]This rule tells us that, if we are able to derive the negation of the premise A from
the premises, then the latter are inconsistent and hence G is **CL**-derivable from them.

DEFINITION 3. Line i on which $[\Delta]A$ is derived, is I-marked if Δ is flatly inconsistent.

Lines on which $[\Delta \cup \{A\}]A$ is derived can only lead to loops in the search process and are L-marked. This also indicates that the search process for A should be led by other conditions, if possible.

DEFINITION 4. Line i on which $[\Delta]A$ is derived, is L-marked if $A \in \Delta$, unless line i was introduced by means of the Goal rule.

A proof is *finished* whenever G is derived. A proof is *stopped* if it is finished or if no further instructions can be applied.

The following theorems and corollaries have been proved in [7]:

THEOREM 5. *If* $\Gamma \vdash_{\mathbf{CLc}} [\Delta]A$, *then* $\Gamma \cup \Delta \vdash_{\mathbf{CL}} A$.

This shows that **CLc** is adequate in view of the requirement (†) we mentioned at the start of this section.

COROLLARY 6. *If* $\Gamma \vdash_{\mathbf{CLc}} G$, *then* $\Gamma \vdash_{\mathbf{CL}} G$. *(Soundness.)*

THEOREM 7. *If* $\Gamma \vdash_{\mathbf{CL}} G$ *then* $\Gamma \vdash_{\mathbf{CLc}} G$. *(Completeness.)*

COROLLARY 8. **CLc**-*proofs are a decision method for* $A_1, \ldots A_n \vdash_{\mathbf{CL}} B$.

6   Conclusion

I have argued that Hintikka's notion of 'strategy' covers too much in order to be perspicuously definable. Part of this vagueness is caused by Hintikka's obstinate belief that the *interrogative model of inquiry* is *the* general model of inquiry and reasoning, whereas most of its structure is in fact 'definitory'. However, the 'strategies' that should warrant part of the diversity of the model are rarely found in Hintikka's extensive work and it is even hard to see how they could be fit into the model. Another part of the problem is caused by the fact that it is not clearly specified how the 'Oracles' are composed and selected, notwithstanding their content is primordial for the value of the results that the model should render.

An alternative approach based on a distinction between 'strategies' as proof-heuristic methods and 'strategies' as methodological reasoning forms is formulated. The former can be formally represented as goal-directed proof procedures as shown in section 5 and explained in section 3. The latter can be formalized by means of corrective or ampliative adaptive logics as described in 4. I even claim that part of the 'Oracles' can be replaced by ampliative reasoning and as such can be formalized as well. Both formalizations of proof-heuristic and methodological reasoning forms can be combined in the formal problem solving processes as presented in [4]. As such, a multiplicity of goal-directed methodological reasoning forms can be formally elaborated and investigated.

Further research should be directed at the interaction between different methodological reasoning forms and how this interaction can be guided by information that is obtained from the ongoing formal problem solving process. Similar research should be performed in order to find out if further restrictions can be added to a rather permissive initial proof-heuristics in the course of the problem solving process. Also the treatment and selection of external sources of information – i.e. the remaining 'Oracles' that cannot be replaced by methodological reasoning – still requires some further investigations.

Acknowledgements

The author is Postdoctoral Fellow of the Research Foundation – Flanders (FWO - Vlaanderen). He wishes to thank Jean Paul Van Bendegem, Diderik Batens and Joke Meheus for comments on a previous draft of the paper. He also wishes to thank the referees for many interesting comments.

BIBLIOGRAPHY

[1] Atocha Aliseda. *Abductive Reasoning. Logical Investigations into Discovery and Explanation.* Springer, Dordrecht, 2006.

[2] Diderik Batens. The basic inductive schema, inductive truisms, and the research-guiding capacities of the logic of inductive generalization. *Logique et Analyse*, 185–188:53–84, 2004. Appeared 2005.

[3] Diderik Batens. A procedural criterion for final derivability in inconsistency-adaptive logics. *Journal of Applied Logic*, 3:221–250, 2005.

[4] Diderik Batens. Content guidance in formal problem solving processes. In Olga Pombo and Alexander Gerner, editors, *Abduction and the Process of Scientific Discovery*, pages 121–156. Centro de Filosofia das Ciências da U. de Lisboa, Lisboa, 2007.

[5] Diderik Batens. A universal logic approach to adaptive logics. *Logica Universalis*, 1:221–242, 2007.

[6] Diderik Batens and Joke Meheus. The adaptive logic of compatibility. *Studia Logica*, 66:327–348, 2000.

[7] Diderik Batens and Dagmar Provijn. Pushing the search paths in the proofs. A study in proof heuristics. *Logique et Analyse*, 173–175:113–134, 2001. Appeared 2003.

[8] Jaakko Hintikka. Sherlock Holmes Confronts Modern Logic: Towards a Theory of Information-Seeking through Questioning. In Umberto Eco and Thomas A. Sebeok, editors, *The Sign of Three. Dupin, Holmes, Peirce*, chapter 7, pages 154–169. Indiana University Press, Bloomington, 1983.

[9] Jaakko Hintikka. *Inquiry as Inquiry: A Logic of Scientific Discovery.* Kluwer, Dordrecht, 1999.

[10] Jaakko Hintikka. Is logic the key to all good reasoning? In *Inquiry as Inquiry: A Logic of Scientific Discovery* [9], chapter 1, pages 1–24.

[11] Jaakko Hintikka. The role of logic in argumentation. In *Inquiry as Inquiry: A Logic of Scientific Discovery* [9], chapter 2, pages 25–46.

[12] Jaakko Hintikka. What is abduction? The fundamental problem of contemporary epistemology. In *Inquiry as Inquiry: A Logic of Scientific Discovery* [9], chapter 4, pages 91–113.

[13] Jaakko Hintikka, Ilpo Halonen, and Arto Mutanen. Interrogitive logic as a general theory of reasoning. In *Inquiry as Inquiry: A Logic of Scientific Discovery* [9], chapter 3, pages 47–90.

[14] Sangmo Jung. *The Logic of Discovery. An Interrogative Approach to Scientific Inquiry.* Peter Lang, New York, 1996.

[15] Joke Meheus. Deductive and ampliative adaptive logics as tools in the study of creativity. *Foundations of Science*, 4:325–336, 1999.

[16] Joke Meheus. Analogical reasoning in creative problem solving processes: Logico-philosophical perspectives. In Fernand Hallyn, editor, *Metaphor and Analogy in the Sciences*, pages 17–34. Kluwer, Dordrecht, 2000.

[17] Joke Meheus and Dagmar Provijn. Abduction through semantic tableaux versus abduction through goal-directed proofs. *Theoria*, 22–23(60):295–304, 2007.

[18] Joke Meheus, Liza Verhoeven, Maarten Van Dyck, and Dagmar Provijn. Ampliative adaptive logics and the foundation of logic-based approaches to abduction. In Lorenzo Magnani, Nancy J. Nersessian, and Claudio Pizzi, editors, *Logical and Computational Aspects of Model-Based Reasoning*, pages 39–71. Kluwer, Dordrecht, 2002.

[19] Sami Paavola. Abduction as a logic and methodology of discovery: the importance of strategies. *Foundations of Science*, 9:267–283, 2004.

[20] Dagmar Provijn. *Prospectieve dynamiek. Filosofische en technische onderbouwing van doelgerichte bewijzen en bewijsheuristieken*. PhD thesis, Ghent University (Belgium), 2005. Unpublished PhD thesis.

[21] Raymond M. Smullyan. *First Order Logic*. Dover, New York, 1995. Original edition: Springer, 1968.

[22] Patrick Suppes. Hintikka's generalizations of logic and their relation to science. In Randall E. Auxier and Lewis E. Hahn, editors, *The philosophy of Jaakko Hintikka*, pages 737–756. Open Court Publishing Company, Chicago, 2006.

[23] Andrzej Wiśniewski. Erotetic search scenarios. *Synthese*, 134:389–427, 2003.

Dagmar Provijn
Centre for Logic and Philosophy of Science
Ghent University, Belgium
E-mail: Dagmar.Provijn@UGent.be

Diagrams, Visualization and Operational Constraints

Sérgio Schultz

ABSTRACT. In the present paper, we will question the thesis according to which proofs with diagrams are heterogeneous, i.e. they involve cases of visual reasoning. In the first section, we will examine the informal use of Venn's diagrams in set-theoretical proofs attempting do determine in which sense proofs with diagrams could be heterogeneous. According to our explanation, such proofs would supposedly involve visual reasoning in the sense that the validity of inferences would depend on visual characteristics of the diagrams and not on derivation rules. In a second section, we will investigate which aspects would be relevant for the validity of diagrammatic inferences. These aspects would be operational constraints observed by diagrams, and these constraints would be the reflex, in the drawing rules of diagrams, of geometrical constraints that they obey while being visualized objects. In the third section, we will argue that the conception of diagrammatic proofs involving essentially a visual element presents serious problems. These problems are related to the fact that essential properties of diagrams do not depend on which visualized properties diagrams really have.

1 Introduction

Along the last decades, proofs with diagrams, traditionally perceived as prone to being under suspicion, started to have their legitimacy re-evaluated and defended. Although diagrams are seen by their defenders as signs, proofs with diagrams would be cases of heterogeneous inferences. In other words, such proofs would involve not only symbolic manipulation but also reasoning based on the visualization of diagrams as well, and, thus, would differ from sentential proofs and calculus procedures such, as for example, truth-tables.[1] We will question precisely this thesis in the present paper, questioning the role performed by visualization in proofs with diagrams.

---

[1] We refer here, basically, to BARWISE and ETCHEMENDY 1996a and 1996b, BARWISE and HANNER 1996 and LEMON and SHIN 2003. GREAVES also seems to support this position when he argues that one of the principle factors for the discredit given to proofs with diagrams after the XVIII century is related to the fact that these would involve, in an essential way, spatial intuition which became then to be seen as poorly trustful. Concerning this, see GREAVES 2002, p. 2.

In the first section, we will analyze an example of informal diagrammatic proof attempting to clarify on which basis they would be heterogeneous. As it will be explained, diagrammatic proofs heterogeneity takes place because they involve association between the validity of inferences and diagrams visualized characteristics. In the second section, we will examine which aspects supposedly visualized guarantee the validity of inferences. Such aspects would be the operational constraints observed by diagrams, constraints that would reflect, in drawing rules, geometric constraints that diagrams observe while being visualized objects. And third, we will argue that essential properties of diagrams, those that can be truly employed in proofs, do not depend on which visualized properties diagrams really do have.

We will limit our research in the present paper to informal diagrammatic proofs and we will deal with them as arguments with which we prove things for us and for others[2] and not as mathematical structures, as it happens in proof theory. Besides, even though we only employ examples of Venn's and geometrical diagrams, as far as they are paradigmatic cases in the literature concerning this subject, we intend that our arguments be applied to every diagrammatic proof.

2   Inference and visualization

Diagrams are usually conceptualized as signs that represent notions or structures through homomorphism relations[3]. Thus, that a circle A with a x marked in its center represents the pertinence of x to set A is something as highly visualized as the expression "$x \in A$" represents that relation. In both cases, we have a visualized object – an inscription – that according to determined interpretation has meaning or represents a given concept. Given another interpretation, that very same diagram can represent that A is the unitarian set of x or still, in the absence of any other interpretation, that it represents absolutely nothing.

Even though they are not signs with similarities or visualized instances of concepts such as, for example, geometrical diagrams, authors like Barwise, Etchemendy and Hammer[4] defend that diagrammatic proofs of logic and mathematical propositions involve heterogeneous inferences. However, how visualization would be involved in proofs? In order to answer this issue, let us consider the following example[5]: we would like to prove that ($C \subseteq$

---

[2]Regarding the difference between idealized proofs (conceived as mathematical structures) and proofs as those which we use for proving (provings), see CHATEAUBRIAND 2005, chapter 20.

[3]Besides the works mentioned in note 1, this stance is also present in SHIN 1996 and HAMMER 2002.

[4]This refers to what these authors have stated in their texts as well as the other texts mentioned in footnote 1 above.

[5]The example, as well as the diagram description resulting from the use of a set of operations is adapted from SHIMOJIMA 1996, p. 29-30. The original example presents the syllogism "all C is B, no B is A, therefore no C is A", and we, instead of considering

$B \wedge B \cap A = \varnothing) \to C \cap A = \varnothing$. Therefore, we represent the antecedent drawing three circles partially overlapping each other, named A, B and C, shading the complement of B in relation to C and shading the intersection or circles A and B. Thus, we obtain the following diagram:

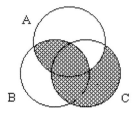

Figure 1. $C \subseteq B \wedge B \cap A = \varnothing$

From the observation that the intersection of circles C and A is shaded, we can read (or infer) from the diagram the truth of the consequent – "$C \cap A = \varnothing$" – and, thus, we can end this demonstration.

The construction of the above diagram can be thought as the result of the employment of the following rules or drawing operations:

(ω_1) Draw three circles partially overlapping each other, A, B, and C, so that there is only one area which is common to A and B, to B and C, to A and C, and to A, B and C, and that each circle has a region which is not common to the other two;

(ω_2) Shade the complement of C in relation to B; and

(ω_3) Shade the intersection of B and A.

When drawing the diagram in accordance to those rules – which give us the diagram that corresponds to $C \subseteq B \wedge B \cap A = \varnothing$ – we draw a diagram that presents the following characteristic:

(σ_1) The intersection (the common area) of C and A is shaded, and can be read as "$C \cap A = \varnothing$".

Therefore, we have an inference of the consequent from the antecedent, which proves the conditional.

A fundamental difference between diagrammatic proofs as the exemplified above and the corresponding sentential proofs lies, as affirms Seoane, in the fact that logical terms perform, from an inferential point of view, a fundamental role in the first ones and highly reduced, if not nil, in the last ones.[6] Therefore, in the above diagrammatic proof, it was only needful to draw the diagram corresponding to "$(C \subseteq B \wedge B \cap A = \varnothing)$" and then obtain the consequent freely. On the other hand, in a sentential proof of the same result, we would have to employ a set of rules of logical inference or base our understanding of the meaning of the logical terms to establish inferences.

---

the syllogism, considered the corresponding set-theoretical statement.

[6] See SEOANE 2006.

Besides, according to SHIMOJIMA 1996, the rules (ω_1) – $(\omega 3)$ do not seem to have as a logical consequence that the intersection of A and C should be shaded. That we cannot draw diagrams observing rules (ω_1) – $(\omega 3)$ which do not present property (σ_1) would place a constraint on diagrams mainly because they are objects capable of being visualized.[7] As such, these characteristics or constraints would also be able to be visualized and could be used, in diagrammatic proofs, instead of rules of logical inference employed in sentential proofs.

According to this conception, the diagrammatic deduction of "$C \cap A = \varnothing$" taken from "$C \subseteq B \wedge B \cap A = \varnothing$" could be described as follows: First, we draw, according to rules, the diagram that corresponds to $C \subseteq B \wedge B \cap A = \varnothing$. Next, we *visualize* the diagram and notice that the intersection between A and C is (and should be) shaded – i.e. it gives us the reading $C \cap A = \varnothing$. Finally, we associate visualization of this fact with the validity of the inference of "$C \cap A = \varnothing$" from "$C \subseteq B \wedge B \cap A = \varnothing$" concluding that the corresponding conditional is true.[8]

Proofs with diagrams would also differ from calculus procedures as truth-tables and the common procedures of addition, subtraction, multiplication and division. Employing these procedures, we only manipulate signs according to rules, and then we read the result. Visualization here is involved only in a trivial meaning which means that to read the signs we need to visualize them. On the other hand, a diagrammatic proof would be analogous, for example, to the inference starting from the information that Anna is the woman who is in the room talking to a man with a beard and from the visualization that the only two bearded men are talking with the same woman, thus concluding that that woman is Anna. Commenting this inference, Barwise and Etchemendy affirm:

---

[7]See SHIMOJIMA 2006, p. 43-44.

[8]According to Barwise and Etchemendy (1996a, p. 9-10), what makes this supposed proof a real proof is the existence of a homomorphism between the drawing of diagrams and set-theoretical operations. As a consequence of this homomorphism, that the diagram built according to rules (ω_1) – $(\omega 3)$ has the characteristic (σ_1) means that the sentence represented by a diagram drawn according to those rules implies the sentence that we can read a diagram with the mentioned characteristic. Thus, as Shin and Lemon affirm (2003, section 3), "if a certain relationship between diagram objects must be drawn, then the corresponding relation can be inferred as logically valid".

[...], the crucial feature of this example is that the conclusion associates a name with a person in a way that transcends each domain individually, both the linguistic and the visual. Because of this, the reasoning cannot be accurately modeled by deductions in a standard formal language. The nearest sentential analog to this conclusion might associate a name with some description ("Anna is the woman who...") , rather than with Anna herself. Alternatively, we might employ some deictic, demonstrative, or indexical element ("That woman is Anna"), but of course it is not a sentence, in isolation, that is the conclusion of your reasoning. Only when we interpret the demonstrative as referring to Anna have we captured the genuine content of your conclusion (BARWISE and ETCHEMENDY 1996a, p. 6)

According to this quotation, what makes inference heterogeneous is the fact that the conclusion consists not in a sentence but in the association of an object visualized with the sign "Anna". Similarly, what would make a diagrammatic proof heterogeneous would be the fact that it involves essentially the association of a visualized aspect of the diagram with the validity of the inference.

3   Diagrammatic proofs and operational constraints

Even though proofs with diagrams, according to the concept set above, are analogous to arguments as that concerning Anna, there are two aspects that establish a difference. The first is related to the inference concerning Anna, the situation of visualized aspect is the very same Anna. In the case of a diagrammatic proof, it is the visualization of diagrams that represent sets which is at stake and not the visualization of the said sets. Besides, and of paramount importance, Anna's visual character example is related to the content of the conclusion, which consists in the association between a visualized object and a sign. In a proof of logical or set-theoretical results employing Venn's diagrams, however, the visualization of a determined aspect of the diagram is associated to the very validity of the inference.

In other words, if we wanted to alter the inference concerning Anna in order to eliminate the need to visualize, we could modify the conclusion in such a way that it could be expressed in the sentence "That woman is Anna", as suggested beforehand. To eliminate a supposedly need to visualize in a proof with diagrams, on the other hand, it would be necessary to include an inference rule that would allow us to go from a diagram that corresponds to a proposition to a diagram that corresponds to the other.[9] Thus, we would

---

[9]With the introduction of this rule, the resulting proof would consist in, similarly to a

have visual reasoning in diagrammatic proofs meaning that the validity of inference would depend on visualized aspects of diagrams.

We could ask ourselves now which aspects supposedly visualized are those that associate themselves with the validity/non-validity of diagrammatic inferences. In the specific literature regarding this subject, for example, Barwise and Etchemendy 1996a, and mainly Shimojima 1996 – such aspects are identified as being the restrictions that govern diagrams. So we find in BARWISE and ETCHEMENDY 1996a that

> *Diagrams are physical situations. They must be, since we can see them. As such they obey their own set of constraints. [...] By choosing a representational scheme appropriately, so that the constraints on the diagrams have a good match with the constraints on the described situation, the diagram can generate a lot of information that the user never need infer. Rather, the user can simply read off facts from the diagram as needed. (BARWISE and ETCHEMENDY 1996a, p. 23).*

In the same text, we also find the following passage relating visual inferences, as the ones we would do with diagrams, to constraints:

> *[...] the perceptual system is an enormously powerful system and carries out a great deal of what one would want to call inference, and which has indeed been called perceptual inference. It is not surprising that this is so, given the fact that visual situations satisfy their own family of constraints. [...] And so is not surprising that people use the tools this system provides in reasoning. (BARWISE and ETCHEMENDY 1996a, p. 25).*

The notion of restriction regarding diagrams employed by Barwise and Etchemendy is developed by Shimojima (1996) in terms of operational constraints, i.e. restrictions concerning the effect of the operations through which we drew the diagram are of such a nature that the said diagram cannot avoid having determined characteristics. Let us consider the example of the diagrammatic proof given in section 1. There, we drew a diagram according to rules $(\omega_1) - (\omega_3)$:

(ω_1) Draw three circles partially overlapping each other, A, B, and C, so that there is only one area which is common to A and B, to B and C, to A

---

formal sentential proof, the use of rules of construction and transformation of signs, and visualization would be involved only by the fact that diagrams, because they are written signs, would have to be visualized when read.

and C, and to A, B and C, and that each circle has a region which is not common to the other two;

(ω_2) Shade the complement of C in relation to B; and

(ω_3) Shade the intersection of B and A.

As a result of these operations, we obtain a diagram that presents characteristic (σ_1):

(σ_1) the intersection of A and C is also shaded.

Besides, (σ_1) is not an accidental characteristic of the diagram, in other words, it is governed by an operational constraint meaning that when the diagram is drawn according to those rules, it *should* present the characteristic. However, Shimojima (1996) argues that (σ_1) is not a logical consequence of (ω_1) – (ω_3) since:

> There are some exceptional circumstances in which it does not hold. Think of drawing the diagram with the toy pen whose "magic" ink fades away a few seconds after it is put on the paper, or imagine drawing diagrams into a computer which automatically distorts, moves, and sometimes erase what you draw. In these circumstances, even if you execute all the operations described above, there is no guarantee that the intersection of a circle labeled "As" and a circle labeled "Cs" gets shaded. Nevertheless, the operational constraint from $\omega_1 \, {}^\circ\omega_2 \, {}^\circ\omega_3$ to out(σ_1) is fairly reliable – reliable enough for the operation method Venn to depend on it in Scenario 1 [the proof of diagram 1 above] (SHIMOJIMA 1996, p. 44-45.)

Operational constraints would not be related to logical consequences of operations or drawing rules, but to diagrams while visualized objects. Thus, Shimojima affirms:

> Even if we bar the above tricky circumstances and assume that everything is preserved until the end of the derivation, the constraint from $\omega_1 \, {}^\circ\omega_2 \, {}^\circ\omega_3$ to out(σ_1) is still not a logical necessity. The constraint holds because the following geometrical (or topological) constraints on the diagram site: if both the complement of the B-circle with respect to the C-circle and the intersection of the B-circle and the A-circle are shaded in a normal diagram site, the intersection of C-circle and the A-circle is also shaded in the diagram site. This is a case in which a local constraint on diagrams is reflected in a local constraint on operations [...]. (SHIMOJIMA 1996, p. 45)

According to the concept mentioned above, operational constraints would
be the counterparts, in drawing rules, of geometrical or topological constraints. Besides, given the examples presented by Shimojima in the first
quotation, they would neither be related to logical consequences nor to material consequences of drawing rules. In other words, unless we took for
granted that geometry is synthetic a *priori*, constraints would not be the
reflection of properties or geometrical constraints that we know a priori belong to diagrams. In fact, relevant properties for constraint would be such
that we would know them a *posteriori*, through visualization, if a diagram
has them or not. According to this analysis, we would have visual reasoning
in diagrammatic proofs in the sense that validity of inference would depend
on operational constraints which govern diagrams and these would be the
reflex, in drawing operations, of geometrical properties that diagrams have
when visualized.[10]

4   Visualization and essential properties

While the explanation of diagrammatic proofs in terms of operational constraints sounds correct, the explanation of the last ones mentioned above
presents serious problems. First, the argument that intends to demonstrate
that operational constraints cannot be understood in terms of logical consequence is, we believe, fallacious. The supposed counter-examples regarding
constraints concerning the relation of consequence involve actions that take
place throughout time: the pen whose magical ink fades after some seconds,
and the computer that moves, distorts or erases that that we had drawn.
They are, in fact, counter-examples only if we take for granted that operational constraints are related to a temporal relation among rules and to
the fact that the diagram presents certain characteristics. We have, then, a
principle petition, since we presuppose that operational constraints are not
related to logical consequences of drawing rules.

Besides, it does not sound true that geometrical constraints reflected
by operations are visualized properties of diagrams. Let us consider the
example given by Shimojima in the quotation above:

---

[10]Shimojima does not explicitly subscribe to this thesis, but it is formulated and used
explicitly by Barwise and Etchemendy, and the part in which the authors formulate it
(1996a, p. 23) is taken as the starting point for Shimojima 1996, (see p. 28). GREAVES
2002 points out the historical importance of this thesis when he affirms that "[. . .] our
ability to manipulate and interpret diagrams in formal proofs has historically been closely
linked to the functioning of our spatial intuition, and so the validity of proofs which rely
on diagrams has been thought of as at least partially contingent on the reliability of
these intuitions" (p. 6) Throughout his book, Greaves contests this thesis at no moment,
stating only that to justify the use of diagrams in the Fregean philosophy of logic would be
necessary to argue for the analytical character of our knowledge of space (p. 205). On the
other hand, he subscribes to the importance of geometrical properties of diagrams – be
them visualized or not – when he affirms: diagrammatic representations can be recognized
by the extent to which the geometric properties of the representation are relevant to their
interpretation, and the ways in which these properties impact the reasoning which are
licensed by the overall theory (p. 2).

(r) if both the complement of circle B in relation to C and the intersection of circles B and A are shaded on a normal diagram, the intersection of circles C and A on this diagram is also shaded.[11]

We can know, through visualization, that a diagram drawn under a normal situation has both characteristics (σ_1) and (σ_2) below[12]:

(σ_1) the intersection of C and A is shaded; and

(σ_2) the complement of B in relation to C and the intersection of A and B are shaded.

As a consequence, we can notice whether or not the diagram has the first property and not the second one, and so whether it follows "if (σ_2) then (σ_1)". However, in order that the geometrical constraint (r), and also the corresponding operational constraint, can be sufficiently reliable to be used in a proof, it has to be worth – at least for normal diagrammatical sites. In other words, to legitimize the association of the constraint to the validity of the inference of "$C \cap A = \varnothing$" from $C \subseteq B \wedge B \cap A = \varnothing$ – respectively, the statements corresponding to (σ_1) and (σ_2) – the constraint should have the form "if (σ_2) then necessarily (σ_1)". However, we cannot know, through visualization, whether the diagram complies with this norm[13] Therefore, either operational constraints are not the reflection of constraints that diagrams possess while visualized objects or the notions of operational constraint and of geometrical constraint are useless to clarify proofs with diagrams, since they cannot differentiate legitimized proofs from the non-legitimized ones.

The problem that we have just pointed out deals with the validity of reasoning with diagrams in logic and mathematics which is under the condition that all characteristics of the diagram on which we base our reasoning be essential. Thus, if the call for a constraint is valid, then the constraint has the form "if p then necessarily q", where the consequent presents the notion of one property which should be present in all diagrams that match p. Visualizing a diagram, however, we can only know whether it has or not determined characteristics, we do not know which ones are essential and which ones are accidental ones. Therefore, for example, considering triangle ABC below only as a visualized object, we do not know whether it is an essential or accidental property that it presents acute angles.

---

[11]This would be the geometrical constraint operating in diagram 1 above.

[12]By normal situation we mean those situations when the diagram is not drawn, for example, employing magical ink or an evil computer (p. 8).

[13]We also cannot know, visualizing the diagram, that "if (σ_2) then (σ_1)" is valid for every normal site of the diagram, which is essential for this case. The non-universality of the constraint, supposing that the diagrams are reliable, would imply in the existence of counter-examples in order to prove the result, in the case, "$(C \subseteq B \wedge B \cap A = \varnothing) \rightarrow C \cap A = \varnothing$".

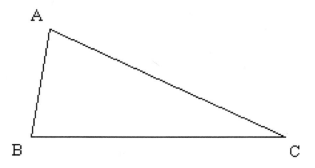

Figure 2. Triangle ABC

Within the context of an Euclidian theorem which affirms "be ABC a triangle, the sum of the internal angles of the triangle equal to the sum of two right angles", given the fact that ABC is an acute triangle is an accidental characteristic of the diagram. However, within the context of a theorem that affirms "be ABC an acute triangle, ...", that property will be essential for that diagram. What determines, in this case, whether the property is essential or not is the statement of the theorem – "be ABC an acute triangle ..." – and, within other contexts, this could be determined by the procedure employed to the building of the diagram.

Besides, if we draw diagram 2 above drawing a segment AC perpendicular to segment AB, and then drawing a third segment BC from points B and C, the resulting figure – triangle ABC – will be a right angle triangle even though, visually, it does not have this property. The same occurs if the diagram is given within the context of a proof of a statement that affirms "be ABC a rectangle triangle, ...". In these cases, someone could ask, to avoid misunderstandings, that the triangle should be corrected so that it would look like a rectangle triangle.[14] However, if no step of the proof involves the supposition that the ABC triangle is not a rectangle one, the proof does not become non-valid because it used a diagram which in fact is not a rectangle triangle. By the way, to infer that the CÂB angle is not a right angle, in this situation, would involve making an error, and not the opposite.

That diagram 2 above, within the described context in the previous paragraph, does not have the appearance of a rectangle triangle consists in a similar error that takes place when calculating "$(16 + 35) \times 3$" we write numeral '5' in longhand in such a way that it may look like '6'. Once that we manipulate numeral '5' written according to manipulation rules of numeral '5', the graphic error does not interfere in the acuity of the calculation. Sim-

---

[14]We should bear in mind that it is practically and physically impossible to draw perfect geometric figures; thus, the most we can accomplish is a diagram that looks like a rectangle triangle.

ilarly, if all the rules and procedures employed on diagram 2 can be truly applied to rectangle triangles, the fact that the diagram does not have the appearance of a rectangle triangle does not interfere in the acuity of the proof.

Under certain circumstances, however, mistakes in the drawing of the diagram can affect the proof. So let us consider the solution of the first problem of Euclid' Elements: given a segment AB, build an equilateral triangle. If we draw two circles C and C' with a radius r common to both in such a way that the lines in the diagram which represent the circles' perimeters do not intersect, we will not have a third point from which to draw the triangle. The problem in this case, however, concerns the mistake in the drawing of the diagrams which has made it impossible to apply the rules to the triangle construction, and not simply that the diagram does not present a certain visualized property. The same problem surfaces when, throughout a proof, we forget to write a negation – instead of writing "$\neg p \lor q$" we write "$p \lor q$" – and, then, attempt to apply the rule of eliminating the negation. In both cases, we have a writing mistake and not a logical mistake, i.e. a fallacious step in the proof.[15]

Taking into account cases of graphic writing mistakes in diagrams suggests us not only that we may not know, through visualization, which are the properties of the diagram that can be truly employed in a proof, i.e. which are its essential properties. It also points out the fact that visualized properties of diagrams are irrelevant to determine which its essential properties are. In other words, we can neither infer from the fact that the diagram presents property, that it is an essential property to the diagram, nor infer, because the diagram does not present a determined property, that this property cannot be validly employed in the proof. Therefore, we can, for example, within certain contexts, ask for the property of being a rectangle triangle regarding diagram 2 above, even if the drawn figure does not have this property, and, thus, we cannot visualize it in the diagram.

5 Conclusion

Since visualized properties of diagrams are irrelevant to determine which are its essential properties and which ones are accidental, they are also irrelevant to validate or not inferences regarding diagrams. This is a consequence of the fact that diagrams are, as Leibniz thought, signs and not instances of concepts, as Kant regarded geometrical diagrams.1 The moment we regard diagrams as physical objects that, by chance, promote determined concepts,

---

[15]So whoever draws the circles in a way that lines which represent their perimeter do not cross each other, draws the diagram wrongly. As a consequence, since we clearly understand the drawing procedures of diagrams, we notice that, for example, the Euclidean solution for problem 1.1 – given a segment of line AB, build an equilateral triangle – does not present a gap. This seems to be Shabel's answer to the objections raised against Euclid concerning continuity axioms, SHABEL 2003, Part 1, especially p. 27-29. A similar position regarding errors in drawing diagrams is also defended by Kenneth Manders, in MANDERS 1995.

we can only discuss essential and accidental properties within the metaphysical meaning of the term.[16] However, this is not the meaning we refer to when we mention, for example, that to have a 'x' written in the area of figure A is an essential property to a diagram that represents "$x \in A \cap B$"

If we consider diagrams as signs, however, visualization helps only to check whether the diagram was drawn and manipulated correctly. Similarly, it is necessary to visualize and compare the written expressions "$q \vee r$" and "$p \vee (q \vee r)$" to conclude that both were written correctly and that one can be inferred from the other through the rule of conjunction elimination. Either in the case of diagrammatic proofs and or in the case of written sentential proofs, visualization is involved only under a trivial meaning when if we do not visualize and identify which formulae were written down, we cannot know what had been written.

We should point out, finally, that from this does not mean that proofs with diagrams are non-valid, but that their validity is incompatible with their elucidation as cases of heterogeneous reasoning. This suggests the need for a re-evaluation of comparison between proofs with diagrams and calculation procedures.

Acknowledgements

I would like to thank professors, Frank Thomas Sautter (UFSM – Brazil), José Seoane (Udelar – Uruguay), Marco Ruffino (UFRJ – Brazil), Luiz Carlos Pereira (Pontifical Catholic University of Rio de Janeiro – Brazil), Oswaldo Chateaubriand (Pontifical Catholic University of Rio de Janeiro – Brazil) and specially Abel Lassalle Casanave (UFSM – Brazil) for their comments, critiques and suggestions for preliminary versions of the present paper. I would also like to thank Ms Viviane Horbach for her help in preparing the English version for this text.

BIBLIOGRAPHY

[1] ALLWEIN, G., AND BARWISE, J. *Logical Reasoning with Diagrams*. Oxford University Press, 1996.

[2] BARWISE, J., AND ETCHEMENDY, J. *Visual Information and Valid Reasoning IN: ALLWEIN, G e BARWISE, J. (eds): Logical Reasoning with Diagrams*. Oxford University Press, 1996.

[3] BARWISE, J., AND HAMMER, E. *Diagrams and the Concept of Logical System*. Oxford University Press, 1996.

---

[16]As Lassalle Casanave states, in LASSALLE CASANAVE 2007, there would be two other difficulties related to the conception that figures are stances: the first one refers to the very character of the stances of the figures, which empirically is always approximated. This difficulty can lead to the corroboration or not of an "exact" true nature of geometry or even state that geometrical propositions are literally false, or even present notions as abstraction, idealization, etc. The second difficulty concerns the role played by figures in a demonstration: how can one conclude from one instance a universal proposition? When conceived as a sign, it is irrelevant that a figure should be exactly a stance of a concept; besides, universality is the result of correct manipulation of signs in accordance to rules (p. 59, my translation).

[4] CASANAVE, A. L. *Conhecimento Simbólico na Investigação de 1764.* Analitica, 2007.

[5] CHATEAUBRIAND, O. *Logical Forms. Part II. Logic, Language and Knowledge.* Centro de Lógica, Epistemologia e História da Ciência / UNICAMP, 2005.

[6] GREAVES, M. *The Philosophical Status of Diagrams.* Stanford: CSLI Publications., 2002.

[7] HAMMER, E. *Diagrammatic Logic IN: GABBAY, D. M. and GUENTHNER, F. (eds) Handbook of Philosophical Logic, Vol. 4, 2ed.* Kluwer Academic Publishers, 2002.

[8] LEMON, O., AND SHIN, S. J. *Diagrams. The Stanford Encyclopedia of Philosophy (Winter 2003 Edition), Edward N. Zalta (ed.), IN: http://plato.stanford.edu/archives/win2003/entries/diagrams/. Access in 2007-12.* 2007.

[9] MANDERS, K. *The Euclidean Diagram. IN: MANCOSU, P. (ed).: The Philosophy of Mathematical Practice.* Clarendon Press, 1995.

[10] SEOANE, J. *Representar y demostrar. Observaciones preliminares sobre diagramas.* Representaciones, 2006.

[11] SHABEL, L. A. *Mathematics in Kant's Critical Philosophy.* New York & London: Routledge, 2003.

[12] SHIMOJIMA, A. *Operational Constraints in Diagrammatic Reasoning IN: ALLWEIN, G e BARWISE, J. (eds): Logical Reasoning with Diagrams.* Oxford University Press, 1996.

[13] SHIN, S. J. *Situation-Theoretic Account of Valid Reasoning with Diagrams IN: ALLWEIN, G e BARWISE, J. (eds): Logical Reasoning with Diagrams.* Oxford University Press, 1996.

Sérgio Schultz
Ph.D. Program in Philosophy
Pontifical Catholic University of Rio de Janeiro
Rio de Janeiro, Brazil
E-mail: sergioschultz@yahoo.com.br

Quantifying in Extensive Games

DAVI ROMERO DE VASCONCELOS AND EDWARD HERMANN HAEUSLER

ABSTRACT. Recently, there have been a lot of approaches that uses a modal logic to model and reason about games. It seems natural to study quantification whenever games are modeled by means of the modal approach. In this work, we use a first-order modal logic, namely Game Analysis Logic (GAL), in order to study the alternatives of quantification for the most used solution concepts, namely Nash Equilibrium (NE) and subgame perfect equilibrium (SPE), for extensive games. We also characterize these concepts by means of the structure of an extensive game according to the different ways of quantification. Despite the fact that quantifying in a modal context might be troublesome, we show that for NE and SPE the alternatives are equivalent each other.

1 Introduction

Games are abstract models of decision-making in which decision-makers (players) interact in a shared environment to accomplish their goals. Several models have been proposed to analyze a wide variety of applications in many disciplines such as Mathematics, Computer Science and even political and social sciences among others.

Game Theory [8] has its roots in the work of von Neumann and Morgenstern [7] that used mathematics in order to model and analyze games. Games are models in which the decision-makers pursue rational behavior. Rationality means that the players choose their actions after some process of optimization and take into account their knowledge or expectations of the other players' behavior. Game Theory provides general game definitions as well as reasonable solution concepts for many kinds of situations in games. Typical examples of this kind of research come from phenomena emerging from Markets, Auctions and Elections.

Although historically Game Theory has been considered more suitable to perform quantitative analysis than qualitative ones, there have been a lot of approaches that emphasizes Game Analysis on a qualitative basis, by using an adequate logic in order to express games as well as their solution concepts. Some of the most representatives of these logics are: Coalitional Logic [9]; Game Logic [10]; Game Logic with Preferences [14]; Alternating-time Temporal Logic (ATL) [2] and its variation Counter-factual ATL (CATL) [13], a good progress report of the use of ATL and its extensions is provided in [6]; Coalitional Game Logic (CGL) [1] that reasons about coalitional games. To see more details about the connections and open problems between logic and games, we point out [12].

It is well-known that quantifying in modal logics is troublesome [4]. It seems natural to study quantification whenever games are modeled by means of the modal approach. In order to observe how this problem may happen in games as simple as extensive games with perfect information [8], we use a first-order modal logic which is based on the standard logic CTL [3], namely *Game Analysis Logic* (GAL) [15]. In [16], we represent these games by means of models of GAL and their main solution concepts that are Nash equilibrium (NE) and subgame perfect equilibrium (SPE) by means of modal formulas; however, we do not take into account the alternatives of quantification in the context of games.

The main difference to the logics mentioned above is that GAL has a first-order apparatus. As a consequence, we are able to define many concepts, such as utility, in an easier way; moreover, we can study quantification in the context of games that in the main guideline of this work. It is worth mentioning that the ATL logic, in which the operators of CTL are parameterized by sets of players, can be seen as a fragment of GAL using the first-order feature of GAL; thus, there is no need for such a parameterization in GAL.

The solution concept of NE requires that each player's strategy be optimal, given the other players' strategies. And, the solution concept of SPE requires that the action prescribed by each player's strategy be optimal, given the other players' strategies, after every history. In SPE concept, the structure of the extensive game is taken into account explicitly, while, in the solution concept of NE, the structure is taken into account only implicitly in the definition of the strategies. Thus, the usual definitions of NE for extensive games does not regard to the sequential structure of the game. We can see this clearly in this quotation:

"*The first solution concept [Nash Equilibrium] we define for an extensive game ignores the sequential structure of the games; it treats the strategies as choices that are made once and for all before play begins.*"[8, pages 93]

Other authors discuss that NE is related to the structure of the extensive game; however, their formal definitions are presented in the usual way. See the quotation below:

"*We also saw that some of these Nash Equilibria may rely on "empty threats" of suboptimal play at histories that are not expected to occur - that is, at histories off the path of the equilibrium.*"[5, pages 72]

Extensive games are quite related to Kripke frames. The structure of an extensive game resembles a tree where nodes correspond to each player's position in the game, according an adequate labeling, while links, or relation between nodes, correspond to each player's action (move) considering its profile. In this setting, a first-order quantified modal logic, as GAL, can be used to express properties on these structures whenever they are taken as first-order kripke frames. When expressing such complex concepts as Nash Equilibrium and Subgame Perfect Equilibrium[1] in a first-order modal language, we are faced to distinct ways of quantifications. So to say, the concept is "necessary for each state", or, "for each state is necessary that...". Even expressing NE and SPE in natural language may raise some interesting consequences, mainly, if one consider the extensive representation of a game.

In this article, we characterize Nash Equilibrium by means of the rationales of the players on the equilibrium's path. Moreover, we aim to study the different ways of quantification for NE and SPE solution concepts in a first-order modal logic (GAL) regarding to the paths of the game.

This work is divided into 4 parts: Section 2 introduces Game Analysis Logic; Section 3 presents extensive games as well their correspondence in the GAL logic; and, finally, Section 4 concludes this work.

2   Game Analysis Logic (GAL)

We model and analyze games using a many-sorted modal first-order logic language, called *Game Analysis Logic* (GAL), that is a logic based on the standard Computation Tree Logic (CTL) [3]. A game is a model of GAL, called game analysis logic structure, and an analysis is a formula of GAL.

The *games* that we model are represented by a set of states $\mathcal{S}E$ and a set of actions $\mathcal{C}\mathcal{A}$.

A *state* $e \in \mathcal{S}E$ is defined by both a first-order interpretation and a set of players, where: 1- The first-order interpretation is used to represent the choices and the consequences of the players' decisions. For example, we can use a list to represent the history of the players' choices until certain state; 2- The set of players represents the players that have to decide simultaneously at a state. This set must be a subset of the players' set of the game. The other players cannot make a choice at this state. For instance, we can model games such as auction games, where all players are in all states, or even games as Chess or turn-based synchronous game structure, where only a single player has to make a choice at each state. Notice that we may even have some states where none of the players can make a decision that can be seen as states of the nature.

An *action* is a relation between two states e_1 and e_2, where all players in the state e_1 have commit themselves to move to the state e_2. Note that this is an extensional view of how the players commit themselves to take a joint action. Of course, this can have an intentional view and be expressed in a formal language.

We refer to $(A_k)_{k \in K}$ as a sequence of A_k's with the index $k \in K$. Sometimes we will use more than one index as in the example $(A_{k,l})_{k,l \in K \times L}$. We can also use $(A_k, B_l)_{k \in K, l \in L}$ to denote the sequence of $(A_k)_{k \in K}$ followed by the sequence $(B_l)_{l \in L}$. Throughout of this article, when the sets of indexes

---

[1]These are among the most important solution concepts in game theory

are clear in the context, we will omit them.

A *path* $\pi(e)$ is a sequence of states (finite or infinite) that could be reached through the set of actions from a given state e that has the following properties: 1- The first element of the sequence is e; 2- If the sequence is infinite $\pi(e) = (e_k)_{k \in \mathbb{N}}$, then $\forall k \geq 0$ we have $\langle e_k, e_{k+1} \rangle \in \mathcal{CA}$; 3- If the sequence is finite $\pi(e) = (e_0, \ldots, e_l)$, then $\forall k$ such that $0 \leq k < l$ we have $\langle e_k, e_{k+1} \rangle \in \mathcal{CA}$ and there is no e' such that $\langle e_l, e' \rangle \in \mathcal{CA}$. The game behavior is characterized by its paths that can be finite or infinite. Finite paths end in a state where the game is over, while infinite ones represent a game that will never end.

Below we present the formal syntax and semantics of GAL. As usual, we call the sets of sorts S, predicate symbols P, function symbols F and players N as a non-logic language in contrast to the logic language that contains the quantifiers and the connectives. We use the notation Υ_s, where s is a sort, to denote the set of terms of sort s defined in a standard way. The modalities can be read as follows.

- $[EX]\alpha$ - 'exists a path α in the next state'
- $[EF]\alpha$ - 'exists a path α in the future'
- $[EG]\alpha$ - 'exists a path α globally'
- $E(\alpha \mathcal{U} \beta)$ - 'exists a path α until β'

- $[AX]\alpha$ - 'for all paths α in the next state'
- $[AF]\alpha$ - 'for all paths α in the future'
- $[AG]\alpha$ - 'for all paths α globally'
- $A(\alpha \mathcal{U} \beta)$ - 'for all paths α until β'

DEFINITION 1 (Syntax of GAL). Let $\langle S, F, P, N \rangle$ be a non-logic language, and $t_1, \ldots, t_n$ be terms of sorts $s_1, \ldots, s_n$, and t_1' be term of sort s_1, and $P : s_1 \ldots s_n$ be a predicate symbol, and i be a player, and x_s be a variable of sort s. The **logic language of GAL** is generated by the following BNF definition:

$$\Phi ::= \top \mid i \mid P(t_1, \ldots, t_n) \mid (t_1 \approx t_1') \mid (\neg\Phi) \mid (\Phi \wedge \Phi) \mid (\Phi \vee \Phi) \mid (\Phi \rightarrow \Phi) \mid$$
$$\exists x_s \Phi \mid \forall x_s \Phi \mid [EX]\Phi \mid [AX]\Phi \mid [EF]\Phi \mid [AF]\Phi \mid [EG]\Phi \mid [AG]\Phi \mid$$
$$E(\Phi \, \mathcal{U} \, \Phi) \mid A(\Phi \, \mathcal{U} \, \Phi)$$

It is well-known that the operators $\vee, \wedge, \bot, \forall x, [AF], [EF], [AG], [EG]$ and $[EX]$ can be given by the following usual abbreviations.

- $\alpha \wedge \beta \iff \neg(\alpha \rightarrow \neg\beta)$
- $[EX]\alpha \iff \neg[AX]\neg\alpha$
- $[AG]\alpha \iff \neg E(\top \, \mathcal{U} \, \neg\alpha)$

- $\alpha \vee \beta \iff (\neg\alpha \rightarrow \beta)$
- $[AF]\alpha \iff A(\top \, \mathcal{U} \, \alpha)$
- $[EG]\alpha \iff \neg A(\top \, \mathcal{U} \, \neg\alpha)$

- $\bot \iff \neg\top$
- $[EF]\alpha \iff E(\top \, \mathcal{U} \, \alpha)$
- $\forall x \alpha(x) \iff \neg \exists x \neg \alpha(x)$

DEFINITION 2 (Structure of GAL). Let $\langle S, F, P, N \rangle$ be a non-logic language of GAL. A **Game Analysis Logic Structure** for this non-logic language is a tuple $\mathcal{G} = \langle \mathcal{SE}, \mathcal{SE}_o, \mathcal{CA}, (\mathcal{D}_s), (\mathcal{F}_{f,e}), (\mathcal{P}_{p,e}), (N_e) \rangle$ such that:

- $\mathcal{SE}$ is a non-empty set, called the set of states.

- $\mathcal{SE}_o$ is a set of initial states, where $\mathcal{SE}_o \subseteq \mathcal{SE}$.

- For each state $e \in \mathcal{SE}$, N_e is a subset of N.

- $\mathcal{CA} \subseteq \mathcal{SE} \times \mathcal{SE}$, called the set of actions of the game[2], in which if there is at least one player in the state e_1, then exists a state e_2 such that $\langle e_1, e_2 \rangle \in \mathcal{CA}$.

- For each sort $s \in S$, $\mathcal{D}_s$ is a non-empty set, called the domain of sort s[3].

- For each function symbol $f : w \to s$ of F and each state $e \in \mathcal{SE}$, $\mathcal{F}_{f,e}$ is a function such that $\mathcal{F}_{f,e} : \left(\prod_{s_k \in w} \mathcal{D}_{s_k} \right) \to \mathcal{D}_s$.

- For each predicate symbol $p : w$ of P and state $e \in \mathcal{SE}$, $\mathcal{P}_{p,e}$ is a relation such that
$$\mathcal{P}_{p,e} \subseteq \left(\prod_{s_k \in w} \mathcal{D}_{s_k} \right).$$

A GAL-structure is of finite model if the set of states $\mathcal{SE}$ and each set of domains $\mathcal{D}_s$ are finite. Otherwise, it is of infinite model. Note that even when a GAL-structure is finite we might have infinite paths.

The decision on having constant domains for GAL semantics relies on the fact that we are mainly interested in modeling perfect information games. In this case, every component of a game is known to every player and remains the same. There is no need to consider different domains for each state of a Perfect Information Game, since, nothing will be added or excluded. The situation might be different if we would modelling the rather wider class of Imperfect Information Games. This is, however, subject of a further research.

In order to provide the semantics of GAL, we define a valuation function as a mapping σ_s that assigns to each free variable x_s of sort s some member $\sigma_s(x_s)$ of domain $\mathcal{D}_s$. As we use terms, we extend every function σ_s to a function $\bar{\sigma}_s$ from state and term to element of sort s that is done in a standard way. When the valuation functions are not necessary, we will omit them.

DEFINITION 3 (Semantics of GAL). Let $\mathcal{G} = \langle \mathcal{SE}, \mathcal{SE}_o, \mathcal{CA}, (\mathcal{D}_s), (\mathcal{F}_{f,e}),$ $(\mathcal{P}_{p,e}), (N_e) \rangle$ be a GAL-structure, and (σ_s) be valuation functions, and α be a GAL-formula, where $s \in S, f \in F, p \in P$ and $e \in \mathcal{SE}$. **We write $\mathcal{G}, (\sigma_s) \models_e \alpha$ to indicate that the state e satisfies the formula α in the structure $\mathcal{G}$ with valuation functions (σ_s).** The formal definition of satisfaction $\models$ proceeds as follows:

- $\mathcal{G}, (\sigma_s) \models_e \top$.

- $\mathcal{G}, (\sigma_s) \models_e i \iff i \in N_e$

---

[2] This relation is not required to be total as in the CTL case. The idea is because we have finite games.

[3] In algebraic terminology $\mathcal{D}_s$ is a carrier for the sort s.

- $\mathcal{G},(\sigma_s) \models_e p(t_1^{s_1}, ..., t_n^{s_n}) \iff \langle \bar{\sigma}_{s_1}(e, t_1^{s_1}), ..., \bar{\sigma}_{s_n}(e, t_n^{s_n}) \rangle \in \mathcal{P}_{p,e}$

- $\mathcal{G},(\sigma_s) \models_e (t_1^s \approx t_2^s) \iff \bar{\sigma}_s(e, t_1^s) = \bar{\sigma}_s(e, t_2^s)$

- $\mathcal{G},(\sigma_s) \models_e \neg\alpha \iff \text{NOT } \mathcal{G},(\sigma_s) \models_e \alpha$

- $\mathcal{G},(\sigma_s) \models_e (\alpha \rightarrow \beta) \iff \text{IF } \mathcal{G},(\sigma_s) \models_e \alpha \text{ THEN } \mathcal{G},(\sigma_s) \models_e \beta$

- $\mathcal{G},(\sigma_s) \models_e [AX]\alpha \iff \forall e' \in \mathcal{SE}$ such that $\langle e, e' \rangle \in \mathcal{CA}$ we have $\mathcal{G},(\sigma_s) \models_{e'} \alpha$ (see Figure 1.a).

- $\mathcal{G},(\sigma_s) \models_e E(\alpha \, \mathcal{U} \, \beta) \iff$ exists a finite (or infinite) path $\pi(e) = (e_0 e_1 e_2 ... e_i)$, such that exists a k where $k \geq 0$, and $\mathcal{G},(\sigma_s) \models_{e_k} \beta$, and for all j where $0 \leq j < k$, and $\mathcal{G},(\sigma_s) \models_{e_j} \alpha$ (see Figure 1.b).

- $\mathcal{G},(\sigma_s) \models_e A(\alpha \, \mathcal{U} \, \beta) \iff$ for all finite (and infinite) paths such that $\pi(e) = (e_0 e_1 e_2 ... e_i)$, exists a k where $k \geq 0$, and $\mathcal{G},(\sigma_s) \models_{e_k} \beta$, and for all j where $0 \leq j < k$, and $\mathcal{G},(\sigma_s) \models_{e_j} \alpha$ (see Figure 1.c).

- $\mathcal{G},(\sigma_s, \sigma_{s_k}) \models_e \exists x_{s_k}\alpha \iff$ exists $d \in \mathcal{D}_{s_k}$ such that $\mathcal{G},(\sigma_s, \sigma_{s_k}(x_{s_k}|d)) \models_e \alpha$, where $\sigma_{s_k}(x_{s_k}|d)$ is the function which is exactly like σ_{s_k} except for one thing: At the variable x_{s_k} it assumes the value d. This can be expressed by the equation:

$$\sigma_s(x_{s_k}|d)(y) = \begin{cases} \sigma_s(y), & \text{if } y \neq x_{s_k} \\ d, & \text{if } y = x_{s_k} \end{cases}$$

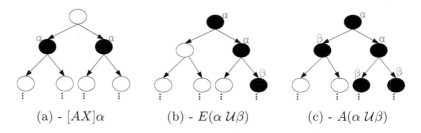

(a) - $[AX]\alpha$ (b) - $E(\alpha \, \mathcal{U} \beta)$ (c) - $A(\alpha \, \mathcal{U} \beta)$

Figure 1. Modal Connectives of GAL.

It is well-known that there is no complete and sound Deductive System for a first-order CTL [11]. Thus, GAL is also non-axiomatizable. However, we argue that we can reason about games using a model checking approach for GAL. In [16], we present a prototype of a model-checker for GAL, namely GALV, that has been developed according to the main intentions of the approach advocated here. GALV is available for download at www.tecmf.inf.puc-rio.br/DaviRomero.

One of the main problematic issues in the first-order modal context is the interaction between the modal operators and the quantifiers. In order to see that, consider the following two formulas.

$$[EF]\forall x Free(x) \tag{1}$$

$$\forall x[EF]Free(x) \tag{2}$$

Formula 1 asserts that at some day everybody will simultaneously be free. On the other hand, formula 2 asserts that everybody will be free at some day, but it does not imply that everybody will simultaneously be free. It should be clear that formula 1 implies formula 2, but the converse dos not hold.

When the quantifier appears before the modal operator, we say that the quantification is *de re*. On the other hand, when the quantifier appears after the modal operator, we say that the quantification is *de dicto*.

In order to study the alternatives of quantification in the context of extensive games, we are concerned with the relationship between the universal quantification and the operators $[EG]$ and $[AG]$. In the first case, *de dicto* implies *de re*, but it does not hold the converse. On the other hand, for the operator $[AG]$ we have an equivalence between the two alternatives of quantification. Thus, the below schemes are valid in GAL. Similar relations can be obtained for other operators.

- $[EG]\forall x\alpha(x) \rightarrow \forall x[EG]\alpha(x)$

- $[AG]\forall x\alpha(x) \leftrightarrow \forall x[AG]\alpha(x)$

3   Game Theory in Game Analysis Logic

In this section we present the correspondence between the extensive games and the GAL-structures as well as the solution concepts of NE and SPE and the formulas of GAL.

An extensive game is a model in which each player can consider his plan of action at every time of the game at which he or she to make a choice. There are two kinds of models: game with perfect information; and games with imperfect information. For the sake of simplicity we restrict the games to models of perfect information. A general model that allows imperfect information is straightforward. Below we present the formal definition and the example depicted in Figure 2.a.

DEFINITION 4. An **extensive game with perfect information** is a tuple $\langle \mathbf{N}, \mathbf{H}, \mathbf{P}, (\mathbf{u_i}) \rangle$, where

- **N** is a set, called the set of players.

- **H** is a set of sequences of actions (finite or infinite), called the set of histories, that satisfies the following properties

  – the empty sequence is a history, i.e. $\emptyset \in H$.

- if $(a_k)_{k \in K} \in H$ where $K \subseteq \mathbb{N}$ and for all $l \leq |K|$, then $(a_k)_{k=0,...,l} \in H$.

- if $(a_0 \ldots a_k) \in H$ for all $k \in \mathbb{N}$, then the infinite sequence $(a_0 a_1 \ldots) \in H$.
A history h is **terminal** if it is infinite or it has no action a such that $(h, a) \in H$. We refer to $\mathbf{T}$ as the set of terminals.

- **P** is a function that assigns to each non-terminal history a player.

- For each player $i \in N$, a utility function $\mathbf{u_i}$ on T.

P is used to indicates the player's turn. Since it has no further restrictions, we are not dealing only with alternating games. u_i measures the profit of each player at the end of a possible run.

EXAMPLE 5. An example of a two-player extensive game $\langle \mathbf{N}, \mathbf{H}, \mathbf{P}, (\mathbf{u_i}) \rangle$, where:

- $\mathbf{N} = \{1, 2\}$;

- $\mathbf{H} = \{\emptyset, (A), (B), (A, L), (A, R)\}$;

- $\mathbf{P}(\emptyset) = 1$ and $\mathbf{P}((A)) = 2$;

- $\mathbf{u_1}((B)) = 1$, $\mathbf{u_1}((A, L)) = 0$, $\mathbf{u_1}((A, R)) = 2$, $\mathbf{u_2}((B)) = 2$, $\mathbf{u_2}((A, L)) = 0$, $\mathbf{u_2}((A, R)) = 1$.

A *strategy of player* i is a function that assigns an action for each non-terminal history for each $P(h) = i$. For the purpose of this article, we represent a strategy as a tuple. In order to avoid confusing when we refer to the strategies or the histories, we use '$\langle$' and '$\rangle$' to the strategies and '(' and ')' to the histories. In Example 5, Player 1 has to make a decision only at the initial state and he or she has two strategies $\langle A \rangle$ and $\langle B \rangle$. Player 2 has to make a decision after the history (A) and he or she has two strategies $\langle L \rangle$ and $\langle R \rangle$. We denote $\mathbf{S_i}$ as the set of player i's strategies. We denote $s = (s_i)$ as a strategy profile. We refer to $\mathbf{O}(\mathbf{s_1}, \ldots, \mathbf{s_n})$ as an outcome that is the terminal history when each player follows his or her strategy s_i. In Example 5, $\langle \langle B \rangle, \langle L \rangle \rangle$ is a strategy profile in which Player 1 chooses B after the initial state and Player 2 chooses L after the history (A), and $O(\langle B \rangle, \langle L \rangle)$ is the outcome (B). In a similar way, we refer to $\mathbf{O_h}(\mathbf{h}, \mathbf{s_1}, \ldots, \mathbf{s_n})$ as the outcome when each player follows his or her strategy s_i from history $\mathbf{h}$. In Example 5, $O_h((A), \langle B \rangle, \langle L \rangle)$ is the outcome (A, L) and $u_1(O_h((A), \langle B \rangle, \langle L \rangle)) = u_1((A, L)) = 0$.

We can model an extensive game $\Gamma = \langle N, H, P, (u_i) \rangle$ as a GAL-structure in the following way. Each history $h \in H$ (from the extensive game) is represented by a state, in which a 0-ary symbol h designates a history of Γ (the one that the state is coming from), so h is a non-rigid designator. The set of the actions of the GAL-structure is determined by the set of

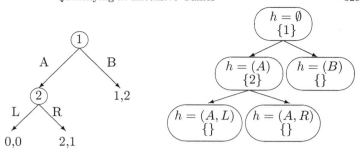

(a) - Extensive form representation   (b) - A GAL representation

Figure 2. *Mapping an extensive game into a GAL model.*

actions of each history, i.e., given a history $h \in H$ and an action a such that $(h, a) \in H$, then the states namely h and (h, a) are in the set of actions of the GAL-structure, i.e. $\langle h, (h, a) \rangle \in \mathcal{CA}$. Function P determines the player that has to make a choice at every state, i.e. $N_h = \{P(h)\}$. The utilities functions are rigidly defined as in the extensive game. The initial state is the state represented by the initial history of the extensive game, i.e. $H_o = \{\emptyset\}$. Sorts H and T are interpreted as the histories and terminal histories of the extensive game, respectively, i.e., $\mathcal{D}_H = H$ and $\mathcal{D}_T = T$. Sort U represents the utility values and is interpreted as the set of all possible utility values of the extensive game[4]. In order to define the solution concept of the subgame perfect equilibrium and the Nash equilibrium, we add to this structure the sets of players' strategies $(\mathcal{D}_{S_i})$ and functions O, O_h and, in addition, $\in: H \times H$ a predicate that states whether a history h precedes a history h'. To summarize, a **GAL-structure for an extensive game with perfect information** $\Gamma = \langle N, P, H, (u_i) \rangle$ is the tuple $\langle H, H_o, \mathcal{CA}, (\mathcal{D}_H, \mathcal{D}_T, \mathcal{D}_{S_i}, \mathcal{D}_U), (u_i, h_h, O, O_h), (\in, \geq), (N_h) \rangle$ with non-logic language $\langle (H, T, S_i, U), (h :\to H, u_i : T \to U, O : S \to T, O_h : H \times S \to T), (\in: H \times H, \geq: U \times U), N \rangle$. The example below is the GAL-structure (see Figure 2.b) of Example 5 (see Figure 2.a).

EXAMPLE 6. The GAL-structure of Example 5 is $\langle H, H_o, \mathcal{CA}, (\mathcal{D}_H, \mathcal{D}_T, \mathcal{D}_{S_1}, \mathcal{D}_{S_2}, \mathcal{D}_U), (h_h, u_1, u_2, O, O_h), (\geq), (N_h) \rangle$ with non-logic language $\langle (H, T, S_1, S_2, U), (h :\to H, u_1 : T \to U, u_2 : T \to U, O : S_1 \times S_2 \to T, O_h : H \times S_1 \times S_2 \to T), (\in: H \times H, \geq: U \times U), \{1, 2\} \rangle$ where

- $H = \{\emptyset, (A), (B), (A, L), (A, R)\}$ and $H_o = \{\emptyset\}$;

- $\mathcal{CA} = \{\langle \emptyset, (A) \rangle, \langle \emptyset, (B) \rangle, \langle (A), (A, L) \rangle, \langle (A), (A, R) \rangle\}$;

- $\mathcal{D}_{S_1} = \{\langle A \rangle, \langle B \rangle\}$, $\mathcal{D}_{S_2} = \{\langle L \rangle, \langle R \rangle\}$ and $\mathcal{D}_U = \{0, 1, 2\}$;

- $\mathcal{D}_H = \{\emptyset, (A), (B), (A, L), (A, R)\}$ and $\mathcal{D}_T = \{(B), (A, L), (A, R)\}$.

---

[4]Note that this set is finite if the game is finite.

- $h_\emptyset = \emptyset$, $h_{(A)} = (A)$, $h_{(B)} = (B)$, $h_{(A,L)} = (A, L)$, $h_{(A,R)} = (A, R)$;

- $N_\emptyset = \{1\}$, $N_{(A)} = \{2\}$, $N_{(B)} = N_{(A,L)} = N_{(A,R)} = \{\}$;

- Functions O, O_h, u_1 and u_2 are rigidly defined as in the extensive game;

- Predicates $\in$ and $\geq$ are rigidly defined as in the extensive game.

The most used solution concepts for extensive games are Nash Equilibrium (NE) and Subgame Perfect Equilibrium (SPE). The solution concept of NE requires that each player's strategy be optimal, given the other players' strategies. And, the solution concept of SPE requires that the action prescribed by each player's strategy be optimal, given the other players' strategies, after every history. In SPE concept, the structure of the extensive game is taken into account explicitly, while, in the solution concept of NE, the structure is taken into account only implicity in the definition of the strategies; however, we characterize NE by means of the rationales of the players on the equilibrium's path[5].

We can consider the definitions of NE and SPE regarding the quantification alternatives of *de dicto* and *de re*. Definitions 7 and 8 as well as definitions 9 and 10 seem to be adequate, respectively, to express SPE and NE. The quantifications of the strategies take place in different contexts in these definitions. Note the emphasis (by means of boldface) on the role played by the quantification in each definition.

DEFINITION 7. A subgame perfect equilibrium of an extensive game $\Gamma = \langle N, H, P, (u_i) \rangle$ is a strategy profile $s^* = \langle s_1^*, \ldots, s_n^* \rangle$ such that for every player $i \in N$ and every history $h \in H$ for which $P(h) = i$ we have

$$u_i(O_h(h, s_1^*, \ldots, s_n^*)) \geq u_i(O_h(h, s_1^*, \ldots, s_i, \ldots, s_n^*)),$$

for every strategy $s_i \in S_i$.

DEFINITION 8. A subgame perfect equilibrium of an extensive game $\Gamma = \langle N, H, P, (u_i) \rangle$ is a strategy profile $s^* = \langle s_1^*, \ldots, s_n^* \rangle$ such that for every player $i \in N$, **every strategy** $s_i \in S_i$ and every history $h \in H$ for which $P(h) = i$ we have

$$u_i(O_h(h, s_1^*, \ldots, s_n^*)) \geq u_i(O_h(h, s_1^*, \ldots, s_i, \ldots, s_n^*))$$

DEFINITION 9. A Nash equilibrium of an extensive game $\Gamma = \langle N, H, P, (u_i) \rangle$ is a strategy profile $s^* = \langle s_1^*, \ldots, s_n^* \rangle$ such that for every player $i \in N$ and every history on the path of the strategy profile s^* (i.e. $h \in O(s^*)$) for which $P(h) = i$ we have

$$u_i(O_h(h, s_1^*, \ldots, s_n^*)) \geq u_i(O_h(h, s_1^*, \ldots, s_i, \ldots, s_n^*)),$$

---

[5]This NE definition regards to the structure of an extensive game, yet is an equivalent one to the standard. In the appendix, we prove the correctness of such definitions.

for every strategy $s_i \in S_i$.

DEFINITION 10. A Nash equilibrium of an extensive game $\Gamma = \langle N, H, P,$ $(u_i) \rangle$ is a strategy profile $s^* = \langle s_1^*, \ldots, s_n^* \rangle$ such that for every player $i \in N$, **every strategy** $s_i \in S_i$ and every history on the path of the strategy profile s^* (i.e. $h \in O(s^*)$) for which $P(h) = i$ we have

$$u_i(O_h(h, s_1^*, \ldots, s_n^*)) \geq u_i(O_h(h, s_1^*, \ldots, s_i, \ldots, s_n^*)).$$

We invite the reader to verify that the outcomes $\langle \langle A \rangle, \langle R \rangle \rangle$ and $\langle \langle B \rangle, \langle L \rangle \rangle$ are the Nash equilibria in Example 5. Game theorists can argue that the solution $\langle \langle B \rangle, \langle L \rangle \rangle$ is not reasonable when the players regard to the sequence of the actions. To see that the reader must observe that after the history (A) there is no way for Player 2 commit himself or herself to choose L instead of R since he or she will be better off choosing R (his or her utility is 1 instead of 0). Thus, Player 2 has an incentive to deviate from the equilibrium, so this solution is not a subgame perfect equilibrium. On the other hand, we invite the reader to verify that the solution $\langle \langle A \rangle, \langle R \rangle \rangle$ is the only subgame perfect equilibrium.

Consider the following formulas as expressing subgame perfect equilibrium definitions 7 and 8, respectively. A strategy profile $s^* = \langle s_1^*, \ldots, s_n^* \rangle$ is a SPE if and only if formula 3 (or formula 4) holds at the initial state $\emptyset$, where each $\sigma_{S_i}(v_{s_i}^*) = s_i^*$. This is in fact the case, if one verifies by the mapping from extensive games into GAL models. In fact, taking *de dicto* as well *de re* into account for $[AG]$ the formulas are equivalent each other.

$$[AG](\bigwedge_{i \in N} i \to \forall v_{s_i}(u_i(O_h(h, v_{s_1}^*, \ldots, v_{s_n}^*)) \geq$$
$$u_i(O_h(h, v_{s_1}^*, \ldots, v_{s_i}, \ldots, v_{s_n}^*)))) \quad (3)$$

$$\forall v_{s_1} \ldots \forall v_{s_n}[AG](\bigwedge_{i \in N} i \to (u_i(O_h(h, v_{s_1}^*, \ldots, v_{s_n}^*)) \geq$$
$$u_i(O_h(h, v_{s_1}^*, \ldots, v_{s_i}, \ldots, v_{s_n}^*)))) \quad (4)$$

On the other hand, formulas 5 and 6 expressing Nash equilibrium according to definitions 9 and 10, respectively, are not equivalent under both interpretation for quantification for $[EG]$. As this relationship might suggest, formula 5 represents NE, while formula 6 does not represent. Although, both formulas represent NE because both formulas take always the same path (the equilibrium's path). Thus, a strategy profile $s^* = \langle s_1^*, \ldots, s_n^* \rangle$ is a NE if and only if formula 5 (or formula 6) holds at the initial state $\emptyset$, where each $\sigma_{S_i}(v_{s_i}^*) = s_i^*$.

$$[EG]\left(\left(\begin{array}{c} h \in O(v_{s_1}^*, \ldots, v_{s_n}^*) \quad \wedge \\ \bigwedge_{i \in N} i \to \forall v_{s_i}(u_i(O_h(h, v_{s_1}^*, \ldots, v_{s_n}^*)) \geq u_i(O_h(h, (v_{s_1}^*, \ldots, v_{s_i}, \ldots, v_{s_n}^*)))) \end{array} \right) \right)$$
$$(5)$$

$$\forall v_{s_1} \ldots \forall v_{s_n} [EG] \left(\left(\begin{array}{c} h \in O(v_{s_1}^*, \ldots, v_{s_n}^*) \quad \wedge \\ \bigwedge_{i \in N} i \rightarrow \left(u_i(O_h(h, v_{s_1}^*, \ldots, v_{s_n}^*)) \geq u_i(O_h(h, (v_{s_1}^*, \ldots, v_{s_i}, \ldots, v_{s_n}^*))) \right) \end{array} \right) \right)$$

$$(6)$$

In order to guarantee the correctness of the representation of both sub-game perfect equilibrium and Nash equilibrium, we state the theorem below. Proof is provided in Appendix A.

THEOREM 11. *Let Γ be an extensive game, and $\mathcal{G}_\Gamma$ be a GAL-structure for Γ, and α be a subgame perfect equilibrium formula for $\mathcal{G}$ as defined in equation 3 (or in equation 4), and β be a Nash equilibrium formula as defined in equation 5 (or in equation 6), and (s_i^*) be a strategy profile, and (σ_{S_i}) be valuations functions for sorts (S_i).*

1. *A strategy profile (s_i^*) is a SPE $\Longleftrightarrow \mathcal{G}_\Gamma, (\sigma_{S_i}) \models_\emptyset \alpha$, where each $\sigma_{S_i}(v_{s_i}^*) = s_i^*$*

2. *A strategy profile (s_i^*) is a NE $\Longleftrightarrow \mathcal{G}_\Gamma, (\sigma_{S_i}) \models_\emptyset \beta$, where each $\sigma_{S_i}(v_{s_i}^*) = s_i^*$*

4   Conclusion

In this work, we have used a first-order modal logic (GAL) to model and reason about extensive games, in which a game is a model of GAL and a solution concept is a formula. As one can expect, quantifying in the context of extensive games might be troublesome. However, for the most used solution concepts, namely Nash equilibrium (NE) and subgame perfect equilibrium (SPE), the alternatives of quantification *de re* and *de dicto* are equivalente each other. For SPE, the equivalence is given by the equivalence for $[AG]$. On the other hand, the formulas for NE are not equivalent under both interpretation for quantification for $[EG]$; however, the equivalence for NE is true, since the quantification is taken by a specific path (the equilibrium's path). As a future work, we intend to characterize the solution concept of iterated elimination of weakly dominated strategies (forward induction) by means of the structure of an extensive game, and, in addition, study the alternatives of quantification.

Appendix A - Proof of Theorem 11

Here we prove the correctness of the definitions 9 and 10 of Nash equilibrium (see lemma below) as well as the correspondence in the GAL logic (see theorem below).

LEMMA 12. *The following assertions are equivalent each other*

1. *For every player $i \in N$ we have $u_i(O(s_1^*, \ldots, s_n^*)) \geq u_i(O(s_1^*, \ldots, s_i, \ldots, s_n^*))$ for every $s_i \in S_i$.*

2. *For every $h \in O(s_1^*, \ldots, s_n^*)$ in which $P(h) = i$ we have $u_i(O_h(h, s_1^*, \ldots, s_n^*)) \geq u_i(O_h(h, s_1^*, \ldots, s_i, \ldots, s_n^*))$ for every $s_i \in S_i$.*

3. *For every $s_i \in S_i$ and every $h \in O(s_1^*, \ldots, s_n^*)$ in which $P(h) = i$ we have*
   $$u_i(O_h(h, s_1^*, \ldots, s_n^*)) \geq u_i(O_h(h, s_1^*, \ldots, s_i, \ldots, s_n^*)).$$

Proof.

- *($1 \Longrightarrow 2$) Suppose by contradiction that NOT for every $h \in O(s_1^*, \ldots, s_n^*)$ in which $P(h) = i$ we have $u_i(O_h(h, s_1^*, \ldots, s_n^*)) \geq u_i(O_h(h, s_1^*, \ldots, s_i, \ldots, s_n^*))$ for every $s_i \in S_i$. Thus, there is a $h \in O(s_1^*, \ldots, s_n^*)$ in which $P(h) = i$ we have that there is a strategy $s_i \in S_i$ such that $u_i(O_h(h, s_1^*, \ldots, s_n^*)) < u_i(O_h(h, s_1^*, \ldots, s_i, \ldots, s_n^*))$. On the other hand, we have by hypothesis that $u_i(O(s_1^*, \ldots, s_n^*)) \geq u_i(O(s_1^*, \ldots, x_i, \ldots, s_n^*))$ for every $x_i \in S_i$, and, since s_i may be taken as x_i, we have that $u_i(O_h(h, s_1^*, \ldots, s_n^*)) \geq u_i(O_h(h, s_1^*, \ldots, s_i, \ldots, s_n^*))$.*

- *($2 \Longrightarrow 1$)Suppose by contradiction that NOT for every $i \in N$ we have $u_i(O(s_1^*, \ldots, s_n^*)) \geq u_i(O(s_1^*, \ldots, s_i, \ldots, s_n^*))$ for every $s_i \in S_i$. Then, there is a player $i \in N$ such that $u_i(O(s_1^*, \ldots, s_n^*)) < u_i(O(s_1^*, \ldots, s_i, \ldots, s_n^*))$ for some $s_i \in S_i$. Now consider the following cases: Player i does not make a move at a history $h \in O(s_1^*, \ldots, s_n^*)$, and, thus he or she cannot change his or her utility, i.e. $u_i(O(s_1^*, \ldots, s_n^*)) = u_i(O(s_1^*, \ldots, s_i, \ldots, s_n^*))$. Player i makes a move at a history $h \in O(s_1^*, \ldots, s_n^*)$, which means that there is a history $h \in O(s_1^*, \ldots, s_n^*)$ in which $P(h) = i$ such that $u_i(O_h(h, s_1^*, \ldots, s_n^*)) < u_i(O_h(h, s_1^*, \ldots, s_i, \ldots, s_n^*))$ (contradiction with the hypothesis of 2).*

- *Proof $1 \Longrightarrow 3$ is similar to $1 \Longrightarrow 2$.*

- *Proof $3 \Longrightarrow 1$ is similar to $2 \Longrightarrow 1$.*

■

THEOREM 13. *Let Γ be an extensive game, and $\mathcal{G}_\Gamma$ be a GAL-structure for Γ, and α be a subgame perfect equilibrium formula for $\mathcal{G}$ as defined in equation 3 (or in equation 4), and β be a Nash equilibrium formula as defined in equation 5, and (s_i^*) be a strategy profile, and (σ_{S_i}) be valuations functions for sorts (S_i).*

1. *A strategy profile (s_i^*) is a SPE $\Longleftrightarrow \mathcal{G}_\Gamma, (\sigma_{S_i}) \models_\emptyset \alpha$, where each $\sigma_{S_i}(v_{s_i}^*) = s_i^*$*

2. *A strategy profile (s_i^*) is a NE $\Longleftrightarrow \mathcal{G}_\Gamma, (\sigma_{S_i}) \models_\emptyset \beta$, where each $\sigma_{S_i}(v_{s_i}^*) = s_i^*$*

Proof.

1. *A strategy profile* (s_i^*) *is a SPE* $\Longleftrightarrow \mathcal{G}_\Gamma, (\sigma_{S_i}) \models_\emptyset \alpha$, *where each* $\sigma_{S_i}(v_{s_i}^*)$
   $= s_i^*$

   Due to the equivalence between de re *and* de dicto *of the universal quantification of the operator* $[AG]$, *we prove* de dicto *only.*

   A strategy profile (s_i^*) *is a SPE of* Γ.
   $\Longleftrightarrow_{def}$ *for every player* i *and every history* $h \in H$ *for which* $P(h) = i$ *we have* $u_i(O_h(h, s_1^*, \ldots, s_n^*)) \geq u_i(O_h(h, s_1^*, \ldots, s_i, \ldots, s_n^*))$, *for every strategy* $s_i \in S_i$.

   By the definition of $\mathcal{G}_\Gamma$ *from* Γ, *we have that every state of* $\mathcal{G}_\Gamma$, *which represents a history of* Γ, *is reached by a path from the initial state* $\emptyset$; *moreover, we have that each domain of player* i's *strategy* $\mathcal{D}_{S_i}$ *is interpreted by the set of strategies* S_i *(i.e.* $\mathcal{D}_{S_i} = S_i$), *and the player that has to take a move in a state* e_k, *which represents the history* h_k, *is defined by the function* P *(i.e.* $N_{e_k} = \{P(h_k)\}$), *and, finally, the symbol* h *is interpreted in* e_k *by the history* h_k *(i.e.* $\bar{\sigma}_H(e_k, h) = h_k$). *As a consequence of this definition, we have*

   $\Longleftrightarrow$ *for all paths* $\pi(\emptyset) = e_0, e_1, \ldots$ *and for all* $k \geq 0$, *for every player* $i \in N$ *such that IF* $i \in N_{e_k}$ *THEN we have for all* $d_i \in \mathcal{D}_{S_i}$
   $(u_i(O_h(\bar{\sigma}_H(e_k, h), s_1^*, \ldots, s_n^*) \geq u_i(O_h(\bar{\sigma}_H(e_k, h), s_1^*, \ldots, d_i, \ldots, s_n^*)))$.

   As function O_h *and utility functions* (u_i) *are rigidly interpreted as in the extensive game* Γ, *we have*

   $\Longleftrightarrow$ *for all paths* $\pi(\emptyset) = e_0, e_1, \ldots$ *and for all* $k \geq 0$, *for every player* $i \in N$ *such that IF* $\mathcal{G}_\Gamma, (\sigma_s) \models_{e_k} i$ *THEN for all* $d_i \in \mathcal{D}_{S_i}$ *we have* $\mathcal{G}_\Gamma, (\sigma_{S_i}(v_{S_i}|d_i) \models_{e_k} (u_i(O_h(h, v_{S_1}^*, \ldots, v_{S_n}^*)) \geq u_i(O_h(h, v_{S_1}^*, \ldots, v_{S_i}, \ldots, v_{S_n}^*)))$, *where* $\sigma_{S_i}(v_{S_i}^*) = s_i^*$.

   $\Longleftrightarrow$ *for all paths* $\pi(\emptyset) = e_0, e_1, \ldots$ *and for all* $k \geq 0$, *we have* $\mathcal{G}_\Gamma, (\sigma_{S_i}) \models_{e_k} (\bigwedge_{i \in N} i \rightarrow \forall v_{S_i}(u_i(O_h(h, v_{S_1}^*, \ldots, v_{S_n}^*)) \geq u_i(O_h(h, v_{S_1}^*, \ldots, v_{S_i}, \ldots, v_{S_n}^*))))$, *where each* $\sigma_{S_i}(v_{S_i}^*) = s_i^*$.

   $\Longleftrightarrow$
   $\mathcal{G}_\Gamma, (\sigma_{S_i}) \models_{e_k} [AG[\bigwedge_{i \in N} i \rightarrow \forall v_{S_i}(u_i(O_h(h, v_{S_1}^*, \ldots, v_{S_n}^*)) \geq u_i(O_h(h, v_{S_1}^*, \ldots, v_{S_i}, \ldots, v_{S_n}^*))))$, *where each* $\sigma_{S_i}(v_{S_i}^*) = s_i^*$.

2. *A strategy profile (s_i^*) is a NE $\iff \mathcal{G}_\Gamma, (\sigma_{S_i}) \models_\emptyset \beta$, where each $\sigma_{S_i}(v_{s_i}^*)$ $= s_i^*$*

Despite the fact that the alternatives of quantification for [EG] are not equivalent each other, formula 5 and 6 express NE. The proofs are similar each other and use the lemma above, which guarantees the correctness of the definitions 9 and 10. Thus, we present de dicto alternative only.

(a)

> *A strategy profile (s_i^*) is a NE of Γ.*
>
> $\iff_{def}$ *for every player i and every history $h \in O(s^*)$ in which $P(h) = i$ we have $u_i(O(s_1^*, \ldots, s_n^*)) \geq u_i(O(s_1^*, \ldots, s_i, \ldots, s_n^*))$, for every strategy $s_i \in S_i$.*
>
> *We take the path $\pi(\emptyset) = e_0, e_1, \ldots$ in $\mathcal{G}_\Gamma$ that is defined by histories $h_0, h_1, \ldots$ on the equilibrium's path $O(s_1^*, \ldots, s_n^*)$ according to definition of $\mathcal{G}_\Gamma$ from Γ. Thus, we have*
>
> $\iff$ *there is a path $\pi(\emptyset) = e_0, e_1, \ldots$ such that for all $k \geq 0$ we have $\bar{\sigma}_H(e_k, h) \in O(s_1^*, \ldots, s_n^*)$ AND for every player $i \in N$ IF $i \in N_{e_k}$ THEN for all $s_i \in S_i$ we have $(u_i(O_h(\bar{\sigma}_H(e_k, h), s_1^*, \ldots, s_n^*)) \geq u_i(O_h(\bar{\sigma}_H(e_k, h), (s_1^*, \ldots, s_i, \ldots, s_n^*))))$, where each $\sigma_{S_i}(v_{S_i}^*) = s_i^*$.*
>
> *As function O, O_h and utility functions (u_i) are rigidly interpreted as in the extensive game Γ, we have*
>
> $\iff_{def}$ *there is a path $\pi(\emptyset) = e_0, e_1, \ldots$ such that for all $k \geq 0$ we have $\mathcal{G}_\Gamma, (\sigma_{S_i}) \models_{e_k} h \in O(v_{S_1}^*, \ldots, v_{S_n}^*)$ AND $\mathcal{G}_\Gamma, (\sigma_{S_i}) \models_{e_k} \bigwedge_{i \in N} i \to \forall v_{S_i}(u_i(O_h(h, v_{S_1}^*, \ldots, v_{S_n}^*)) \geq u_i(O_h(h, (v_{S_1}^*, \ldots, v_{S_i}, \ldots, v_{S_n}^*))))$, where each $\sigma_{S_i}(v_{S_i}^*) = s_i^*$.*
>
> $\iff$
>
> $$\mathcal{G}_\Gamma, (\sigma_{S_i}) \models_\emptyset [EG]\left(\begin{array}{c} h \in O(v_{S_1}^*, \ldots, v_{S_n}^*) \quad \wedge \\ \left(\bigwedge_{i \in N} i \to \forall v_{S_i}(u_i(O_h(h, v_{S_1}^*, \ldots, v_{S_n}^*)) \geq \right. \\ \left. u_i(O_h(h, (v_{S_1}^*, \ldots, v_{S_i}, \ldots, v_{S_n}^*)))) \right) \end{array} \right)$$
>
> *where each $\sigma_{S_i}(v_{S_i}^*) = s_i^*$.*

∎

BIBLIOGRAPHY

[1] Thomas Ågotnes, Michael Wooldridge, and Wiebe van der Hoek. On the logic of coalitional games. In P. Stone and G. Weiss, editors, *AAMAS '06:Proceedings of the Fifth International Conference on Autonomous Agents and Multiagent Systems*, pages 153–160, Hakodate, Japan, May 2006. ACM Press.

[2] Rajeev Alur, Thomas A. Henzinger, and Orna Kupferman. Alternating-time temporal logic. *J. ACM*, 49(5):672–713, 2002.

[3] Edmund M. Clarke and E. Allen Emerson. Design and synthesis of synchronization skeletons using branching-time logic. In *Workshop on Logic of Programs*, pages 52–71. Springer, may 1981.

[4] Melvin Fitting and Richard L. Mendelsohn. *First-order Modal Logic*. Kluwer Academic Publishers, 1999.

[5] Drew Fudenberg and Jean Tirole. *Game Theory*. MIT Press, October 1991.

[6] Paul Dunne Michael Wooldridge, Thomas gotnes and Wiebe van der Hoek. Logic for automated mechanism design – a progress report. In *To appear in proceedings AAAI 2007*. AAAI Press, 2007.

[7] J. Von Neumann and O. Morgenstern. *Theory of Games and Economic Behavior*. John Wiley and Sons, 1944.

[8] M. J. Osborne and A. Rubinstein. *A Course in Game Theory*. MIT Press, 1994.

[9] M. Pauly. *Logic for Social Software*. Illc dissertation series 2001-10, University of Amsterdam, 2001.

[10] Marc Pauly and Rohit Parikh. Game logic - an overview. *Studia Logica*, 75(2):165–182, 2003.

[11] G. Michele Pinna, Franco Montagna, and Elisa B. P. Tiezzi. Investigations on fragments of first order branching temporal logic. *Mathematical Logic Quarterly*, 48:51–62, 2002.

[12] Johan van Benthem. Open problems in logic and games. 2005. available at http://www.illc.uva.nl/lgc/postings.html.

[13] Wiebe van der Hoek, Wojciech Jamroga, and Michael Wooldridge. A logic for strategic reasoning. In *AAMAS '05: Proceedings of the Fourth International Joint Conference on Autonomous Agents and Multiagent Systems*, pages 157–164, New York, NY, USA, 2005. ACM Press.

[14] Sieuwert van Otterloo, Wiebe van der Hoek, and Michael Wooldridge. Preferences in game logics. In *AAMAS '04: Proceedings of the Third International Joint Conference on Autonomous Agents and Multiagent Systems*, pages 152–159, Washington, DC, USA, 2004. IEEE Computer Society.

[15] D. R. Vasconcelos. *Lógica Modal de Primeira-ordem para Raciocinar sobre Jogos (in Portuguese)*. PhD thesis, PUC-Rio, April 2007.

[16] D. R. Vasconcelos and E. H. Haeusler. Reasoning about Games via a First-order Modal Model Checking Approach. In *SBMF'07 – 10th Brazilian Symposium on Formal Methods*, Ouro Preto, MG, Brazil, 2007.

Davi Romero de Vasconcelos
Federal University of Ceará
Quixadá, Brazil
E-mail: daviromero@ufc.br

Edward Hermann Haeusler
Pontifical Catholic University of Rio de Janeiro
Rio de Janeiro, Brazil
E-mail: hermann@inf.puc-rio.br

SECTION 5

LOGIC AND ITS
ALGEBRAIC SIDE

A Duality for Three-valued Łukasiewicz Δ-implication Algebras

M. ABAD, C. R. CIMADAMORE, J. P. DÍAZ VARELA

ABSTRACT. In this work we give a topological representation for three-valued Łukasiewicz Δ-implication algebras. Every three-valued Łukasiewicz Δ-implication algebra is represented as a union of a unique family of implication filters of a suitable three-valued Łukasiewicz algebra. Inspired in this representation, we introduce the notion of topological three-valued implication space, and we prove a dual equivalence. As an application, we describe the three-valued implication space of free three-valued Łukasiewicz Δ-implication algebras.

1 Introduction and preliminaries

The class of three-valued Łukasiewicz algebras was introduced by G. Moisil in 1940 [12] as the algebraic counterpart of Łukasiewicz three-valued logic, and it was A. Monteiro in the early nineteen sixties who developed a deep investigation of its structure [10]. From the logical point on view, Monteiro's result means that Łukasiewicz three-valued logic is an axiomatic extension of Nelson's constructive logic with strong negation. The class of three-valued Łukasiewicz algebras can be considered as a subvariety of the variety of MV-algebras. In fact, it is term-equivalent to the subvariety MV_3 studied by Grigolia [7] and others.

A. Monteiro was also interested in studying algebraically the three-valued Łukasiewicz implicative calculus, and so he introduced in [11] the three-valued implicative Łukasiewicz calculus. The main results of the course are contained in [9].

An extension of three-valued Łukasiewicz implicative calculus is the modal three-valued Łukasiewicz implicative calculus obtained by adding the Baaz connective Δ (see [3]) and the axioms $(\Delta x \rightarrow \Delta y) \rightarrow \Delta(\Delta x \rightarrow y)$, $\Delta(\Delta x \rightarrow y) \rightarrow (x \rightarrow \Delta y)$ and $(x \rightarrow y) \rightarrow (\Delta x \rightarrow y)$. The three-valued Δ-implication Łukasiewicz algebras are the algebraic models of the modal three-valued Łukasiewicz implicative calculus. They are the class of all $\{\rightarrow, \Delta, 1\}$-subreducts of three-valued Łukasiewicz algebras [6].

The objective of this work is to investigate the class of three-valued Δ-implication Łukasiewicz algebras. We construct for every three-valued Łukasiewicz Δ-implication algebra $\mathbf{A}$, a three-valued Łukasiewicz algebra $\mathbf{CL}_3(\mathbf{A})$ such that the implication filter $F(\mathbf{A})$ generated by $\mathbf{A}$ in $\mathbf{CL}_3(\mathbf{A})$ is maximal and $\mathbf{CL}_3(\mathbf{A})/F(\mathbf{A})$ is the 2-element chain. Moreover, $\mathbf{A}$ is represented as a union of a unique family of implication filters of $\mathbf{CL}_3(\mathbf{A})$. Inspired in this, we introduce the notion of topological three-valued implication space, and we give a topological representation for three-valued Łukasiewicz Δ-implication algebras. As an application, we describe the topological space of free algebras.

We recall that a *three-valued Łukasiewicz algebra* is an algebra $\mathbf{L} = \langle L, \vee, \wedge, \sim, \nabla, 0, 1 \rangle$ of type $(2, 2, 1, 1, 0, 0)$ such that

1. $1 \vee x = 1$,

2. $x \wedge (x \vee y) = x$,

3. $x \wedge (y \vee z) = (z \wedge x) \vee (y \wedge x)$,

4. $\sim\sim x = x$,

5. $\sim(x \wedge y) = \sim x \vee \sim y$,

6. $\sim x \vee \nabla x = 1$,

7. $\sim x \wedge x = \sim x \wedge \nabla x$,

8. $\nabla(x \wedge y) = \nabla x \wedge \nabla y$.

The equational class of all three-valued Łukasiewicz algebras will be denoted by $\mathcal{L}_3$. The three-element chain L_3, $0 < 1/2 < 1$, with the natural lattice structure and the operations $\sim$ and ∇ defined as

$$\sim 0 = 1, \ \sim 1/2 = 1/2, \ \sim 1 = 0,$$

$$\nabla 0 = 0, \ \nabla 1/2 = 1, \ \nabla 1 = 1,$$

will be denoted by $\mathbf{L}_3$. It is well known that every subdirectly irreducible algebra in $\mathcal{L}_3$ is isomorphic to $\mathbf{L}_3$ or to its subalgebra $\mathbf{L}_2$ with universe $\{0, 1\}$. Besides, these two algebras are simple.

For every $\mathbf{L} \in \mathcal{L}_3$ we can define the operations Δ and $\rightarrow$ in L by the terms

$$\Delta x = \sim\nabla\sim x, \quad x \rightarrow y = (\nabla\sim x \vee y) \wedge (\nabla y \vee \sim x).$$

A subset $F \subseteq L$ is said to be an implication filter of $\mathbf{L}$ if $1 \in F$ and if $x \in F$ and $x \rightarrow y \in F$, then $y \in F$. It is also known that F is an implication filter of $\mathbf{L}$ if and only if F is a filter of the underlying lattice and satisfies that $\Delta x \in F$, whenever $x \in F$. If $X \subseteq L$, we denote by $F(X)$ the implication filter generated by X in $\mathbf{L}$. It is easy to see that $F(X) = \{a \in L : a \geq \bigwedge_{i=1}^{n} \Delta x_i, x_i \in X, n \in \mathbb{N}\}$ and if X is increasing then $F(X) =$

$\{a \in L : a = \bigwedge_{i=1}^{n} \Delta x_i, x_i \in X, n \in \mathbb{N}\}$. Congruences in $\mathcal{L}_3$ are determined by implication filters.

An *implication algebra* is an algebra $\mathbf{A} = \langle A, \rightarrow \rangle$ of type (2) that satisfies the equations

(T1) $(x \rightarrow y) \rightarrow x = x$,

(T2) $(x \rightarrow y) \rightarrow y = (y \rightarrow x) \rightarrow x$,

(T3) $x \rightarrow (y \rightarrow z) = y \rightarrow (x \rightarrow z)$.

If $\mathbf{A}$ is an implication algebra, there is a Boolean algebra $\mathbf{B}$ having $\mathbf{A}$ as an implication subalgebra of $\mathbf{B}$ (see [2]). Let $B(\mathbf{A})$ be the Boolean subalgebra generated by A in $\mathbf{B}$, and $F(\mathbf{A})$ the filter generated by A in $B(\mathbf{A})$. The implication algebra $\mathbf{A}$ is increasing in $B(\mathbf{A})$, (see [1]), and consequently, A is isomorphic to a union of filters of $B(\mathbf{A})$.

We denote by $\mathrm{St}(\mathbf{B})$ the Stone space of a Boolean algebra $\mathbf{B}$, that is, the topological space of ultrafilters of $\mathbf{B}$ whose topology has as a base the family $\{N_a = \{U \in \mathrm{St}(\mathbf{B}) : a \in U\}\}_{a \in B}$, and we denote by $\mathrm{Clop}(\mathbf{X})$ the Boolean algebra of all *clopen* subsets of a Boolean space $\mathbf{X}$.

For each Boolean algebra $\mathbf{B}$, the correspondence $F \rightarrow C_F = \{U \in \mathrm{St}(\mathbf{B}) : F \subseteq U\}$ gives a dual order isomorphism form the set of all filters of $\mathbf{B}$ onto the set of all closed sets of $\mathrm{St}(\mathbf{B})$, ordered by inclusion.

Consider the following Boolean algebra, called the *Boolean closure* of $\mathbf{A}$,

$$\mathbf{Bo}(\mathbf{A}) = \begin{cases} B(\mathbf{A}) & \text{if } F(\mathbf{A}) \neq B(\mathbf{A}) \\ B(\mathbf{A}) \times \{0,1\} & \text{if } F(\mathbf{A}) = B(\mathbf{A}) \end{cases}.$$

It has been proved that for any implication algebra $\mathbf{A}$, $\mathbf{Bo}(\mathbf{A})$ is the least, up to isomorphism, Boolean algebra in which the filter $F(\mathbf{A})$ is an ultrafilter. As two implication algebras may have the same Boolean closure, they are distinguished by means of the family of all maximal elements in the set of all filters of $\mathbf{Bo}(\mathbf{A})$ contained in A. We will denote this family by $\mathrm{M}(\mathbf{A})$.

We recall (see [1]) that an *implication space* is a triple $\langle \mathbf{X}, u, \mathcal{C} \rangle$ such that

(E1) $\mathbf{X}$ is a Boolean space, with underlying set X,

(E2) u is a fixed element of X,

(E3) $\mathcal{C}$ is an antichain, with respect to inclusion, of closed sets of $\mathbf{X}$ such that $\bigcap \mathcal{C} = \{u\}$,

(E4) if C is a closed subset of $\mathbf{X}$ such that for every clopen N of $\mathbf{X}$, $C \subseteq N$ implies $D \in \mathcal{C}$ for some $D \subseteq N$, then there exists $D' \in \mathcal{C}$ such that $D' \subseteq C$.

A function $f \colon \mathbf{X}_1 \to \mathbf{X}_2$ between two implication spaces $\langle \mathbf{X}_1, u_1, \mathcal{C}_1 \rangle$ and $\langle \mathbf{X}_2, u_2, \mathcal{C}_2 \rangle$ is *i-continuous* if f is continuous, $f(u_1) = u_2$, and for all $C \in \mathcal{C}_2$ there is $D \in \mathcal{C}_1$ such that $D \subseteq f^{-1}[C]$.

We denote by $\mathcal{I}$ the category whose objects are implication algebras and whose arrows are implication homomorphisms, and by $\mathcal{X}$ the category with implication spaces as objects and *i-continuous* maps as arrows. If $\mathbf{A}$ is an object in $\mathcal{I}$, then

$$\mathbb{X}(\mathbf{A}) := \langle \operatorname{St}\left(\operatorname{Clop}\left(\mathbf{Bo}(\mathbf{A})\right)\right), F\left(\mathbf{A}\right), \mathcal{C}\left(\mathbf{A}\right) \rangle,$$

is in $\mathcal{X}$, where $\mathcal{C}(\mathbf{A}) = \{C_F : F \in \mathsf{M}(\mathbf{A})\}$ is the family of closed sets corresponding to $\mathsf{M}(\mathbf{A})$. For each implication homomorphism $f \colon \mathbf{A}_1 \to \mathbf{A}_2$, there is a Boolean homomorphism $\hat{f} \colon \mathbf{Bo}(\mathbf{A}_1) \to \mathbf{Bo}(\mathbf{A}_2)$ (see [1, Corollary 2.3]). Then we define $\mathbb{X}(f) := \operatorname{St}(\hat{f}) \colon \operatorname{St}(\mathbf{Bo}(\mathbf{A}_2)) \to \operatorname{St}(\mathbf{Bo}(\mathbf{A}_1))$ by $\operatorname{St}(\hat{f})(U) = \hat{f}^{-1}(U)$, which is *i-continuous*. Conversely, given an implication space $\langle \mathbf{X}, u, \mathcal{C} \rangle$, the algebra

$$\mathbb{I}(\mathbf{X}) := \{N \in \operatorname{Clop}(\mathbf{X}) : C \subseteq N \text{ for some } C \in \mathcal{C}\}$$

is an implication algebra, and for each *i-continuous* $h \colon \mathbf{X}_1 \to \mathbf{X}_2$ between two implication spaces $\langle \mathbf{X}_1, u_1, \mathcal{C}_1 \rangle$ and $\langle \mathbf{X}_2, u_2, \mathcal{C}_2 \rangle$, the map

$$\mathbb{I}(h) := \operatorname{Clop}(h) \!\restriction_{\mathbb{I}(\mathbf{X}_2)} \colon \mathbb{I}(\mathbf{X}_2) \to \mathbb{I}(\mathbf{X}_1)$$

defines an implication homomorphism. Moreover, $\mathbb{X} \colon \mathcal{I} \rightsquigarrow \mathcal{X}$ and $\mathbb{I} \colon \mathcal{X} \rightsquigarrow \mathcal{I}$ are contravariant functors. It follows that $\varsigma_{\mathbf{A}} \colon \mathbf{A} \to \mathbb{I}\mathbb{X}(\mathbf{A})$ defined by $\varsigma_{\mathbf{A}}(a) = N_a$ is a natural dual transformation between the functor $\mathbb{I}\mathbb{X}$ and the functor $\operatorname{Id}_{\mathcal{I}}$, and that the mapping $\tau_{\mathbf{X}} \colon \mathbf{X} \to \mathbb{X}\mathbb{I}(\mathbf{X})$, defined by the formula $\tau_{\mathbf{X}}(x) = \{U \in \operatorname{Clop}(\mathbf{X}) : x \in U\}$, is a natural dual transformation between the functor $\mathbb{X}\mathbb{I}$ and the functor $\operatorname{Id}_{\mathcal{X}}$.

2   Three-valued Łukasiewicz Δ-implication algebras

An algebra $\mathbf{A} = \langle A, \to, \Delta, 1 \rangle$ of type $(2,1,0)$ is a *three-valued Łukasiewicz Δ-implication algebra* if

1. $x \rightarrow (y \rightarrow x) = 1$,

2. $(x \rightarrow y) \rightarrow ((y \rightarrow z) \rightarrow (x \rightarrow z)) = 1$,

3. $(x \rightarrow y) \rightarrow y = (y \rightarrow x) \rightarrow x$,

4. $((x \rightarrow y) \rightarrow (y \rightarrow x)) \rightarrow (y \rightarrow x) = 1$,

5. $((x \rightarrow (x \rightarrow y)) \rightarrow x) \rightarrow x = 1$,

6. $1 \rightarrow x = x$,

7. $\Delta x \rightarrow y = x \rightarrow (x \rightarrow y)$,

8. $\Delta(\Delta x \rightarrow y) = \Delta x \rightarrow \Delta y$.

The variety of all three-valued Łukasiewicz Δ-implication algebras will be denoted by $\mathcal{L}_3^{\{\rightarrow,\Delta\}}$. Note that $\langle A, \rightarrow, 1 \rangle$ is a three-valued Łukasiewicz implication algebra ([9]). It is known that the class of three-valued Łukasiewicz Δ-implication algebras is the class of all $\{\rightarrow, \Delta, 1\}$-subreducts of three-valued Łukasiewicz algebras. The $\{\rightarrow, \Delta, 1\}$-reduct of $\mathbf{L}_3$ will be denoted by $\mathbf{L}_3^{\rightarrow,\Delta}$. The subalgebra of $\mathbf{L}_3^{\rightarrow,\Delta}$ with subuniverse consisting of the 2-element chain will be denoted by $\mathbf{L}_2^{\rightarrow,\Delta}$. These algebras are the subdirectly irreducible algebras of $\mathcal{L}_3^{\{\rightarrow,\Delta\}}$, and they are simple. Besides, the congruences are determined by the implication filters.

Let $\mathbf{A} \in \mathcal{L}_3^{\{\rightarrow,\Delta\}}$, and let $\mathcal{M}$ the set of all maximal implication filters of $\mathbf{A}$. Then $\mathbf{A}$ is isomorphic to a subalgebra $\mathbf{A}'$ of $\mathbf{P} = \prod_{M \in \mathcal{M}} \mathbf{A}/M$. For each $M \in \mathcal{M}$, $\mathbf{A}/M$ is isomorphic to $\mathbf{L}_3^{\rightarrow,\Delta}$ or $\mathbf{L}_2^{\rightarrow,\Delta}$. Then, $\mathbf{P} \in \mathcal{L}_3$. From now on, we will identify $\mathbf{A}$ with $\mathbf{A}'$.

LEMMA 1. *Every three-valued Łukasiewicz Δ-implication algebra* $\mathbf{A}$ *is isomorphic to a union of implication filters of the three-valued Łukasiewicz algebra generated by A in* $\mathbf{P}$, *which we denote by* $\mathbf{L}_3(\mathbf{A})$.

Proof. Let us see first that if $b \in L_3(\mathbf{A})$ then

$$ b = \bigwedge_{k=1}^{r} \left[\left(\bigvee_{i \in I_k} g_i \right) \vee \left(\bigvee_{j \in J_k} \sim g_j \right) \right], $$

where $I_k \cap J_k = \emptyset$, for every k, $\bigcup_k^r (I_k \cup J_k) \neq \emptyset$ is a finite set and $g_i, g_j \in A$. For that, it is enough to prove that the set of elements b that are written as before is a subalgebra of $\mathbf{P} = \prod_{M \in \mathcal{M}} \mathbf{A}/M$. It is easy to see that the set is closed under $\vee$ and $\sim$. Since Δ is a lattice homomorphism, we have $\Delta b = \bigwedge_{k=1}^{r} \left[\left(\bigvee_{i \in I_k} \Delta g_i \right) \vee \left(\bigvee_{j \in J_k} \sim \nabla g_j \right) \right]$. Since $g_i \in A$, we have that $\Delta g_i \in A$. Since $\nabla g_j = (g_j \rightarrow \Delta g_j) \rightarrow g_j$, we have that $\nabla g_j \in A$.

Let $a \in A$ and $b \in L_3(\mathbf{A})$ such that $a \leq b$, and let us prove that $b \in A$. For that, it is enough to show that $b_k = \left(\bigvee_{i \in I_k} g_i \right) \vee \left(\bigvee_{j \in J_k} \sim g_j \right) \in A$, for any k.

If $J_k = \emptyset$ then $b_k = \bigvee_{i \in I_k} g_i \in A$, since $g_i \in A$. If $J_k \neq \emptyset$, then $b_k = \left(\bigvee_{i \in I_k} g_i \right) \vee \left(\bigvee_{j \in J_k} {\sim} g_j \right) = \left(\bigvee_{i \in I_k} g_i \right) \vee \left(\bigvee_{j \in J_k} {\sim} g_j \right) \vee \Delta a = \left(\bigvee_{i \in I_k} g_i \right) \vee \bigvee_{j \in J_k} ({\sim} g_j \vee \Delta a) = \left(\bigvee_{i \in I_k} g_i \right) \vee \bigvee_{j \in J_k} (g_j \to \Delta a) \in A$, since $a, g_i, g_j \in A$. So, A is an increasing subset of $\mathbf{L}_3(\mathbf{A})$ and then $\mathbf{A}$ is a union of implication filters of $\mathbf{L}_3(\mathbf{A})$. ∎

Let $F(\mathbf{A})$ be the implication filter generated by $\mathbf{A}$ in $\mathbf{L}_3(\mathbf{A})$. Observe that $F(\mathbf{A}) = F(\Delta\mathbf{A})$ and, since $\mathbf{A}$ is an increasing subset of $\mathbf{L}_3(\mathbf{A})$, $x \in F(\mathbf{A})$ if and only if $x = \bigwedge_{i=1}^n a_i$, $a_i \in A$.

LEMMA 2. *The three-valued Łukasiewicz algebra* $\mathbf{L}_3(\mathbf{A})$ *is equal to* $F(\mathbf{A}) \cup {\sim}F(\mathbf{A})$. *In particular, if* $F(\mathbf{A})$ *is proper then* $\mathbf{L}_3(\mathbf{A})/F(\mathbf{A}) \cong \mathbf{L}_2$.

Proof. It is routine to prove that $F(\mathbf{A}) \cup {\sim}F(\mathbf{A})$ is closed by ${\sim}$, $\wedge$, and Δ. Then, $F(\mathbf{A}) \cup {\sim}F(\mathbf{A})$ is the subalgebra generated by $F(\mathbf{A})$ in $\mathbf{P}$. So, $\mathbf{L}_3(\mathbf{A}) \subseteq F(\mathbf{A}) \cup {\sim}F(\mathbf{A})$. Then, $\mathbf{L}_3(\mathbf{A}) = F(\mathbf{A}) \cup {\sim}F(\mathbf{A})$. Finally, if $F(\mathbf{A})$ is proper then $F(\mathbf{A}) \cap {\sim}F(\mathbf{A}) = \emptyset$ and $\mathbf{L}_3(\mathbf{A})/F(\mathbf{A}) \cong \mathbf{L}_2$. ∎

For each $\mathbf{A} \in \mathcal{L}_3^{\{\to,\Delta\}}$, consider the following three-valued Łukasiewicz algebra, called the *Łukasiewicz closure* of $\mathbf{A}$,

$$\mathbf{CL}_3(\mathbf{A}) = \begin{cases} \mathbf{L}_3(\mathbf{A}) & \text{if } F(\mathbf{A}) \neq \mathbf{L}_3(\mathbf{A}) \\ \mathbf{L}_3(\mathbf{A}) \times \mathbf{L}_2 & \text{if } F(\mathbf{A}) = \mathbf{L}_3(\mathbf{A}) \end{cases}.$$

Observe that, from Lemma 2, if $F(\mathbf{A}) \neq \mathbf{L}_3(\mathbf{A})$, then $\mathbf{CL}_3(\mathbf{A})/F(\mathbf{A})$ is the 2-element chain. If $F(\mathbf{A}) = \mathbf{L}_3(\mathbf{A})$, identifying the elements of $F(\mathbf{A})$ with those of the form $(-, 1)$ in $\mathbf{L}_3(\mathbf{A}) \times \mathbf{L}_2$, we have that $F(\mathbf{A})$ is a maximal filter of $\mathbf{CL}_3(\mathbf{A})$ and again $\mathbf{CL}_3(\mathbf{A})/F(\mathbf{A}) \cong \mathbf{L}_2$. The next theorem asserts that $\mathbf{CL}_3(\mathbf{A})$ is the least three-valued Łukasiewicz algebra, up to isomorphism, in which $F(\mathbf{A})$ is maximal and $\mathbf{CL}_3(\mathbf{A})/F(\mathbf{A})$ is isomorphic to $\mathbf{L}_2$.

THEOREM 3. *Let* $\mathbf{A} \in \mathcal{L}_3^{\{\to,\Delta\}}$, *then:*

1. A *is increasing in* $\mathbf{CL}_3(\mathbf{A})$, *and* $\mathbf{CL}_3(\mathbf{A})/F(\mathbf{A}) \cong \mathbf{L}_2$,

2. *if* $\mathbf{L} \in \mathcal{L}_3$ *and* $h\colon \mathbf{A} \to \mathbf{L}$ *is an* $\{\to, \Delta\}$-*homomorphism, then there is a homomorphism* $\hat{h}\colon \mathbf{CL}_3(\mathbf{A}) \to \mathbf{L}$ *such that* $\hat{h}\restriction_A = h$, *i.e., the following diagram commutes:*

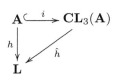

Proof. We have already proved (1). For (2), recall that if $b \in F(\mathbf{A})$ then there are $b_1, \cdots, b_n \in A$, such that $b = \bigwedge_{i=1}^{n} b_i$. We define $\hat{h}: \mathbf{CL}_3(\mathbf{A}) \to \mathbf{L}$ by $\hat{h}(b) = \bigwedge_{i=1}^{n} h(b_i)$ if $b \in F(\mathbf{A})$, and by $\sim\hat{h}(\sim b)$ if $b \notin F(\mathbf{A})$. It is straightforward to see that $\hat{h}$ is well defined, preserves $\to$, Δ and $\hat{h}(0) = 0$. (See the proof of [1, Theorem 2.2]). Then $\hat{h}$ is a homomorphism of Łukasiewicz algebras and by definition $\hat{h}\restriction_A = h$. ∎

COROLLARY 4. *Let* $\mathbf{A}_1, \mathbf{A}_2 \in \mathcal{L}_3^{\{\to,\Delta\}}$ *and a homomorphism* $h: \mathbf{A}_1 \to \mathbf{A}_2$. *Then there is a homomorphism* $\hat{h}: \mathbf{CL}_3(\mathbf{A}_1) \to \mathbf{CL}_3(\mathbf{A}_2)$ *such that* $\hat{h}\restriction_{A_1} = h$ *and* $\hat{h}^{-1}[F(\mathbf{A}_2)] = F(\mathbf{A}_1)$. *In particular, if* h *is an isomorphism then* $\hat{h}$ *is an isomorphism.*

Proof. The existence of a homomorphism $\hat{h}: \mathbf{CL}_3(\mathbf{A}_1) \to \mathbf{CL}_3(\mathbf{A}_2)$ such that $\hat{h}\restriction_{A_1} = h$ follows from Theorem 3 considering $\mathbf{A} = \mathbf{A}_1$ and $\mathbf{L} = \mathbf{CL}_3(\mathbf{A}_2)$. From definition of $\hat{h}$ we have that $F(\mathbf{A}_1) \subseteq \hat{h}^{-1}[F(\mathbf{A}_2)]$. Since $F(\mathbf{A}_1)$ is maximal and $\hat{h}^{-1}[F(\mathbf{A}_2)]$ is proper, we have the equality. If h is an isomorphism then $\hat{h}^{-1} = \widehat{h^{-1}}$. ∎

Two three-valued Łukasiewicz Δ-implication algebras may have the same Łukasiewicz closure, but they can be distinguished by means of the implication filters contained in them. If $\mathsf{M}(\mathbf{A})$ is the family of all maximal elements in the set of all implication filters of $\mathbf{CL}_3(\mathbf{A})$ contained in $\mathbf{A}$, then $\mathsf{M}(\mathbf{A})$ satisfies:

1. $A = \bigcup_{D \in \mathsf{M}(\mathbf{A})} D$, and $\mathsf{M}(\mathbf{A})$ is an antichain,

2. if M is an implication filter of $\mathbf{CL}_3(\mathbf{A})$ contained in A, then $M \subseteq D$ for some $D \in \mathsf{M}(\mathbf{A})$.

3 Three-valued implication spaces

Inspired by the Stone duality between Boolean algebras and totally disconnected compact Hausdorff spaces, M. Abad et al. introduced in [1] a topological representation for implication algebras. On the other hand, R. Cignoli and L. Monteiro extended the Stone duality in [5] in order to obtain a topological representation for MV_n-algebras. Combining these two representations and taking into account that the operator Δ recovers the properties of the Boolean elements in an MV-algebra, we introduce in this section the notion of three-valued implication space.

A three-valued Boolean space ([5]) is a pair $\langle \mathbf{X}, V \rangle$ such that $\mathbf{X}$ is a Boolean space and $V \subseteq X$ is closed. If the set L_3 is equipped with the discrete topology and $\langle \mathbf{X}, V \rangle$ is a three-valued Boolean space, then $\mathbf{C}_3(\mathbf{X}, V)$

denotes the three-valued Łukasiewicz algebra of all continuous functions $f\colon \mathbf{X} \to \mathbf{L_3}$ such that $f(V) \subseteq L_2$, with the algebraic operations defined pointwise. In [5], the authors prove that for each $\mathbf{A} \in \mathcal{L}_3$, there is a three-valued Boolean space $\langle X(\mathbf{A}), V_{\mathbf{A}} \rangle$ such that $\mathbf{A} \cong \mathbf{C_3}(X(\mathbf{A}), V_{\mathbf{A}})$. Moreover, $X(\mathbf{A})$ is homeomorphic to the Stone space of the Boolean algebra $C(\mathbf{A})$ of the complemented elements of A. Now we are going to explain briefly the construction of the space $\langle X(\mathbf{A}), V_{\mathbf{A}} \rangle$, and we will indicate the isomorphism between $\mathbf{A}$ and $\mathbf{C_3}(X(\mathbf{A}), V_{\mathbf{A}})$. For more details, see [5].

Let $X(\mathbf{A})$ be the set of all homomorphisms $\chi\colon \mathbf{A} \to \mathbf{L_3}$. Then, $X(\mathbf{A})$ becomes a Boolean space with the topology inherited from the product space $\mathbf{L}_3^A$, where L_3 is equipped with the discrete topology. The sets $\left\{ \chi \in X(A)\colon \chi(a) = \frac{j}{2} \right\}, a \in A, 0 \le j \le 2$, form a subbasis for this topology. Let $X(C(\mathbf{A}))$ be the set of all homomorphisms $\chi\colon C(\mathbf{A}) \to \mathbf{L_2}$, and consider the space $X(C(\mathbf{A}))$. We know that $X(C(\mathbf{A}))$ coincides with the Stone space of $C(\mathbf{A})$. Therefore, the map $\varphi_{\mathbf{A}}(\chi) = \chi^{-1}(\{1\}) \cap C(A)$ is a homeomorphism between $X(\mathbf{A})$ and $\mathrm{St}(C(\mathbf{A}))$. The closed subset $V_{\mathbf{A}}$ is defined by $V_{\mathbf{A}} = \{\chi \in X(\mathbf{A})\colon \chi(A) \subseteq L_2\}$. If we identify $X(\mathbf{A})$ with $\mathrm{St}(C(\mathbf{A}))$, we have that $V_{\mathbf{A}} = \{U \in \mathrm{St}(C(\mathbf{A})) : \mathbf{A}/F(U) \cong \mathbf{L_2}\}$. To each $a \in A$, let associate the function $\bar{a}\colon X(\mathbf{A}) \to \mathbf{L_3}$ defined by $\bar{a}(\chi) = \chi(a)$ for all $\chi \in X(\mathbf{A})$. It is known that $\bar{a}$ is continuous and the map

$$\alpha_{\mathbf{A}}\colon \mathbf{A} \to \mathbf{C_3}(X(\mathbf{A}), V_{\mathbf{A}})$$
$$a \mapsto \bar{a}$$

is an isomorphism.

Let $\langle \mathbf{X}, V \rangle$ be a three-valued Boolean space and consider $X(\mathbf{C_3}(\mathbf{X}, V)) \cong \mathrm{St}(C(\mathbf{C_3}(\mathbf{X}, V)))$. We will denote by γ_N the characteristic function of a given clopen $N \subseteq X$. Clearly, the correspondence $N \mapsto \gamma_N$ defines an isomorphism from $\mathrm{Clop}(\mathbf{X})$ onto $C(\mathbf{C_3}(\mathbf{X}, V))$. Then $X(\mathbf{C_3}(\mathbf{X}, V))$ is homeomorphic to $\mathrm{St}(\mathrm{Clop}(\mathbf{X}))$. Moreover,

$$t_3\colon \langle \mathbf{X}, V \rangle \to \langle \mathrm{St}(\mathrm{Clop}(\mathbf{X})), V_{\mathbf{C_3}(\mathbf{X},V)} \rangle$$
$$x \mapsto \{N \in \mathrm{Clop}(\mathbf{X}) : x \in N\}$$

is a homeomorphism such that $x \in V$ if and only if $t_3(x) \in V_{\mathbf{C_3}(\mathbf{X},V)}$.

Our next objective is to associate to any three-valued Łukasiewicz Δ-implication algebra a topological space endowed with certain distinguished objects.

Let $\mathbf{A} \in \mathcal{L}_3^{\{\to, \Delta\}}$, and consider $\mathbf{CL_3}(\mathbf{A})$, $F(\mathbf{A})$ and $\mathsf{M}(\mathbf{A})$ as indicated in §2. Recall that the lattice of implication filters of $\mathbf{CL_3}(\mathbf{A})$ is isomorphic to the lattice of filters of $C(\mathbf{CL_3(A)})$. Since there is a dual order isomorphism between filters of $C(\mathbf{CL_3(A)})$ and closed sets of $\mathrm{St}(C(\mathbf{CL_3(A)}))$, we have a correspondence between implication filters of $\mathbf{CL_3(A)}$ and closed sets of $\mathrm{St}(C(\mathbf{CL_3(A)}))$ given by

$$F \mapsto C_{\Delta F} = \{U \in \mathrm{St}(C(\mathbf{CL_3(A)})) : \Delta F \subseteq U\}.$$

We define

$$\mathbb{F}(\mathbf{A}) := \langle X(\mathbf{CL_3(A)}), V_{\mathbf{CL_3(A)}}, \Delta F(\mathbf{A}), \mathcal{C}(\mathbf{A})\rangle,$$

where $\langle X(\mathbf{CL_3(A)}), V_{\mathbf{CL_3(A)}}\rangle \cong \langle \mathrm{St}(C(\mathbf{CL_3(A)})), \{U \in \mathrm{St}(B(\mathbf{CL_3(A)})) : \mathbf{CL_3(A)}/F(U) \cong \mathbf{L_2}\}\rangle$ is the three-valued Boolean space indicated above, and $\mathcal{C}(\mathbf{A}) = \{C_{\Delta F} : F \in \mathsf{M}(\mathbf{A})\}$. From Theorem 3, and conditions of $\mathsf{M}(\mathbf{A})$, we have that $\mathbb{F}(\mathbf{A})$ has the following properties:

1. $\langle \mathrm{St}(C(\mathbf{CL_3(A)})), V_{\mathbf{CL_3(A)}}\rangle$ is a three-valued Boolean space, and $\Delta F(\mathbf{A}) \in V_{\mathbf{CL_3(A)}}$,

2. $\mathcal{C}(\mathbf{A})$ is an antichain, respect to inclusion, of closed sets of $\mathrm{St}(C(\mathbf{CL_3(A)}))$ such that $\bigcap \mathcal{C}(\mathbf{A}) = \{\Delta F(\mathbf{A})\}$,

3. if C is a closed subset of $\mathrm{St}(C(\mathbf{CL_3(A)}))$ such that for every $N \in \mathrm{Clop}(\mathrm{St}(C(\mathbf{CL_3(A)})))$, $C \subseteq N$ implies there is $D \in \mathcal{C}(\mathbf{A})$ such that $D \subseteq N$, then there is $D' \in \mathcal{C}(\mathbf{A})$ such that $D' \subseteq C$.

This motivates the following definition.

DEFINITION 5. A space $\langle \mathbf{X}, V, u, \mathcal{C}\rangle$ is a *three-valued implication space* if

1. $\langle \mathbf{X}, V\rangle$ is a three-valued Boolean space, and u is a fix element of X such that $u \in V$,

2. $\mathcal{C}$ is an antichain, respect to inclusion, of closed set of $\mathbf{X}$ such that $\bigcap \mathcal{C} = \{u\}$,

3. if C is a closed set of $\mathbf{X}$ such that for every $N \in \mathrm{Clop}(\mathbf{X})$, $C \subseteq N$ implies there is $D \in \mathcal{C}$ such that $D \subseteq N$, then there is $D' \in \mathcal{C}$ such that $D' \subseteq C$.

From the above, we have that $\mathbb{F}(\mathbf{A})$ is a three-valued implication space.

Now we will define the notion of morphism between two three-valued implication spaces.

Let $\mathbf{A_1}$, $\mathbf{A_2} \in \mathcal{L}_3^{\{\rightarrow, \Delta\}}$ and $h \colon \mathbf{A_1} \rightarrow \mathbf{A_2}$ be a homomorphism. Consider $\hat{h} \colon \mathbf{CL_3(A_1)} \rightarrow \mathbf{CL_3(A_2)}$ as in Corollary 4. We know that the restriction of $\hat{h}$ to $C(\mathbf{CL}_3(\mathbf{A}_1))$ is a Boolean homomorphism from $C(\mathbf{CL}_3(\mathbf{A}_1))$ to $C(\mathbf{CL}_3(\mathbf{A}_2))$. We define $\mathbb{F}(h) = \mathrm{St}(\hat{h})\colon \mathbb{F}(\mathbf{A}_2) \to \mathbb{F}(\mathbf{A}_1)$ by $\mathrm{St}(\hat{h})(U) = \hat{h}^{-1}(U)$, for every $U \in \mathrm{St}(C(\mathbf{CL}_3(\mathbf{A}_2)))$. It is easy to prove the following lemma.

LEMMA 6. *Let* $\mathbf{A}_1$, $\mathbf{A}_2 \in \mathcal{L}_3^{\{\to,\Delta\}}$ *and* $h\colon \mathbf{A}_1 \to \mathbf{A}_2$ *be a homomorphism. The function* $\mathbb{F}(h) = \mathrm{St}(\hat{h})\colon \mathbb{F}(\mathbf{A}_2) \to \mathbb{F}(\mathbf{A}_1)$ *defined by* $\mathrm{St}(\hat{h})(U) = \hat{h}^{-1}(U)$ *satisfies the following conditions:*

1. *it is i-continuous,*

2. $\mathrm{St}(\hat{h})(V_{\mathbf{CL}_3(\mathbf{A}_2)}) \subseteq V_{\mathbf{CL}_3(\mathbf{A}_1)}$.

DEFINITION 7. A function $f\colon \langle \mathbf{X}_1, V_1, u_1, \mathcal{C}_1 \rangle \to \langle \mathbf{X}_2, V_2, u_2, \mathcal{C}_2 \rangle$ is *i3-continuous* if f is i-continuous and $f(V_1) \subseteq V_2$. Moreover, if f is a homomorphism and its inverse is also *i3*-continuous then f is an *i3-homeomorphism*.

So we have that if $h\colon \mathbf{A}_1 \to \mathbf{A}_2$ is a homomorphism between two three-valued Łukasiewicz Δ-implication algebras, then the function $\mathbb{F}(h) = \mathrm{St}(\hat{h})\colon \mathbb{F}(\mathbf{A}_2) \to \mathbb{F}(\mathbf{A}_1)$ is *i3*-continuous.

3.1   The natural dual equivalence

Let $\mathcal{D}_3$ be the category of three-valued Łukasiewicz Δ-implication algebras with homomorphisms, and let $\mathcal{E}_3$ be the category of three-valued implication spaces with morphisms as *i3*-continuous functions. From the results indicated above and from [1, Sec. 4], the correspondence $\mathbb{F}\colon \mathcal{D}_3 \to \mathcal{E}_3$ defined by

$$
\begin{array}{ccc}
\mathbf{A}_1 & \overset{\mathbb{F}}{\rightsquigarrow} & \mathbb{F}(\mathbf{A}_1) \\
h \downarrow & & \uparrow \mathrm{St}(\hat{h}) \\
\mathbf{A}_2 & \underset{\mathbb{F}}{\rightsquigarrow} & \mathbb{F}(\mathbf{A}_2)
\end{array}
$$

is a contravariant functor. Now we are going to define a contravariant functor $\mathbb{G}\colon \mathcal{E}_3 \to \mathcal{D}_3$. Let $\langle \mathbf{X}, V, u, \mathcal{C} \rangle$ be a three-valued implication space. We define

$$
\mathbb{G}(\mathbf{X}) := \{ f \in \mathbf{C}_3(\mathbf{X}, V) : f \geq \gamma_N \text{ for some } N \in \mathbb{I}(\mathbf{X}) \},
$$

where $\mathbb{I}(\mathbf{X})$ was defined in §1. Since $\mathbb{G}(\mathbf{X})$ is increasing in $\mathbf{C}_3(\mathbf{X}, V)$ and $\Delta\gamma_N = \gamma_N$, for every $N \in \mathrm{Clop}(\mathbf{X})$, we have that $\mathbb{G}(\mathbf{X}) \in \mathcal{L}_3^{\{\to,\Delta\}}$.

LEMMA 8. *Let* $h\colon \langle \mathbf{X}_1, V_1, u_1, \mathcal{C}_1 \rangle \to \langle \mathbf{X}_2, V_2, u_2, \mathcal{C}_2 \rangle$ *be an i3-continuous function and, for each* $f \in \mathbb{G}(\mathbf{X}_2)$, *let* $\mathbb{G}(h)\colon \mathbb{G}(\mathbf{X}_2) \to \mathbb{G}(\mathbf{X}_1)$ *be defined by* $\mathbb{G}(h)(f) = f \circ h$. *Then* $\mathbb{G}(h)$ *is a homomorphism of* $\mathcal{L}_3^{\{\to,\Delta\}}$-*algebras.*

Proof. Let us see first that $\mathbb{G}(h)$ is well defined. Since f and h are continuous functions, $\mathbb{G}(h)$ is continuous. Since $h(V_1) \subseteq V_2$ and $f(V_2) \subseteq L_2$, we have that $\mathbb{G}(h)(f)(V_1) = f(h(V_1)) \subseteq L_2$. We are going to prove now that there is $U \in \mathbb{I}(\mathbf{X}_1)$ such that $\mathbb{G}(h)(f) \geq \gamma_U$. Since $f \in \mathbb{G}(\mathbf{X}_2)$, there is $S \in \mathbb{I}(\mathbf{X}_2)$ such that $f \geq \gamma_S$ and there is $C \in \mathcal{C}_2$ such that $C \subseteq S$. So, there exist $D \in \mathcal{C}_1$ such that $D \subseteq h^{-1}[C]$. Then, $h^{-1}[S] \supseteq h^{-1}[C] \supseteq D$, that is, $h^{-1}[S] \in \mathbb{I}(\mathbf{X}_1)$. Also, if $x \in h^{-1}[S]$ then $f(h(x)) = 1$, since $h(x) \in S$ and $f \geq \gamma_S$. Thus, $\mathbb{G}(h)(f) \geq \gamma_{h^{-1}[S]}$. Finally, as the operations in $\mathbb{G}(\mathbf{X}_1)$ and $\mathbb{G}(\mathbf{X}_2)$ are defined pointwise, we have that $\mathbb{G}(h)$ is a homomorphism. ∎

The correspondence $\mathbb{G}\colon \mathcal{E}_3 \to \mathcal{D}_3$ given by

$$
\begin{array}{ccc}
\mathbf{X}_1 & \overset{\mathbb{G}}{\rightsquigarrow} & \mathbb{G}(\mathbf{X}_1) \\
h \downarrow & & \uparrow \mathbb{G}(h) \\
\mathbf{X}_2 & \underset{\mathbb{G}}{\rightsquigarrow} & \mathbb{G}(\mathbf{X}_2)
\end{array}
$$

defines a contravariant functor.

Let us see now some properties of $\mathbb{G}(\mathbf{X})$. It is easy to see that for any $C \in \mathcal{C}$, the set

$$
F_C = \{ f \in \mathbf{C}_3(\mathbf{X}, V) : f \geq \gamma_N, N \in \mathrm{Clop}(\mathbf{X}) \text{ and } C \subseteq N \}
$$

is an implication filter of $\mathbf{C}_3(\mathbf{X}, V)$. Moreover, $\mathbb{G}(\mathbf{X}) = \bigcup_{C \in \mathcal{C}} F_C$.

LEMMA 9. *Let* $\langle \mathbf{X}, V, u, \mathcal{C} \rangle$ *be a three-valued implication space. The following properties hold in* $\mathbb{G}(\mathbf{X})$:

1. *the implication filter* $F[\mathbb{G}(\mathbf{X})]$ *generated by* $\mathbb{G}(\mathbf{X})$ *in* $\mathbf{C}_3(\mathbf{X}, V)$ *is maximal, and we have that* $\mathbf{C}_3(\mathbf{X}, V)/F[\mathbb{G}(\mathbf{X})] \cong \mathbf{L}_2$,

2. $\mathbf{CL}_3(\mathbb{G}(\mathbf{X})) = \mathbf{C}_3(\mathbf{X}, V)$,

3. $\mathsf{M}(\mathbb{G}(\mathbf{X})) = \{ F_C : C \in \mathcal{C} \}$.

Proof. Consider the implication filter of $\mathbf{C}_3(\mathbf{X}, V)$ given by

$$
F = \{ f \geq \gamma_N : N \in \mathrm{Clop}(\mathbf{X})) \text{ and } u \in N \}.
$$

Since $t_3(u) = \{N \in \mathrm{Clop}(\mathbf{X}) : u \in N\} \in \mathrm{St}(\mathrm{Clop}(\mathbf{X}))$, then $F \cap C(\mathbf{C}_3(\mathbf{X}, V))$ $= \{\gamma_N \in C(\mathbf{C}_3(\mathbf{X}, V)) : u \in N, N \in \mathrm{Clop}(\mathbf{X})\} \in X(C(\mathbf{C}_3(\mathbf{X}, V)))$. Also, since $u \in V$, we obtain F is maximal and $\mathbf{C}_3(\mathbf{X}, V)/F \cong \mathbf{L}_2$. If $f \in F_C$ then there is $N \in \mathrm{Clop}(\mathbf{X})$ such that $C \subseteq N$ and $\gamma_N \leq f$. Since $u \in C$, for any $C \in \mathcal{C}$, then $u \in N$ and $f \in F$. Thus, $F(\bigcup_{C \in \mathcal{C}} F_C) = F$. From this, we get $\mathbf{CL}_3(\mathbb{G}(\mathbf{X})) = \mathbf{C}_3(\mathbf{X}, V)$. Finally from condition (3) of the definition of three-valued implication space we have $\mathsf{M}(\mathbb{G}(\mathbf{X})) = \{F_C : C \in \mathcal{C}\}$. ∎

Now we can state the duality theorem.

THEOREM 10. *The functors* $\mathbb{F}$ *and* $\mathbb{G}$ *define a dual equivalence between the categories* $\mathcal{D}_3$ *and* $\mathcal{E}_3$. *That is,*

1. *if* $\mathbf{A}$ *is a three-valued Łukasiewicz* Δ*-implication algebra, then*

$$s_{\mathbf{A}} : \mathbf{A} \to \mathbb{G}(X(\mathbf{CL}_3(\mathbf{A})))$$

defined by $s_{\mathbf{A}}(a) = \bar{a}$, *is a natural transformation from the functor* $\mathbb{G}\mathbb{F}$ *to the identity functor,*

2. *for every three-valued implication space* $\langle \mathbf{X}, V, u, \mathcal{C} \rangle$, *the map*

$$\tau_{\mathbf{X}} : \langle \mathbf{X}, V, u, \mathcal{C} \rangle \to \langle \mathrm{St}(\mathrm{Clop}(\mathbf{X})), V_{\mathbf{C}_3(\mathbf{X}, V)}, F[\mathbb{G}(\mathbf{X})], \mathcal{C}(\mathbb{G}(\mathbf{X})) \rangle$$

defined by $\tau_{\mathbf{X}}(x) = t_3(x)$, *is a natural transformation from the functor* $\mathbb{F}\mathbb{G}$ *to the identity functor.*

Proof. (1)   Observe that by definition, $s_{\mathbf{A}} = \alpha_{\mathbf{CL}_3(\mathbf{A})} \upharpoonright_A$. Then to see that $s_{\mathbf{A}}$ is an isomorphism it is enough to prove that $\alpha_{\mathbf{CL}_3(\mathbf{A})}[A] = \mathbb{G}(X(\mathbf{CL}_3(\mathbf{A})))$. Indeed,

- if $a \in A$, there exists $M \in \mathsf{M}(\mathbf{A})$ such that $a \in M$. Thus, $C_{\Delta M} \subseteq N_{\Delta a}$. Since $\bar{a} \geq \overline{\Delta a} = \gamma_{N_{\Delta a}}$, it follows that $\bar{a} \in \mathbb{G}(X(\mathbf{CL}_3(\mathbf{A})))$.

- if $f \in \mathbb{G}(X(\mathbf{CL}_3(\mathbf{A})))$ then $f = \bar{a}$, for some $a \in \mathbf{CL}_3(\mathbf{A})$. We want to see that $a \in A$. But $f \geq \gamma_U$, for some $U \in \mathrm{Clop}(\mathrm{St}(C(\mathbf{CL}_3(\mathbf{A}))))$, and there exists $C \in \mathcal{C}_{\mathbf{CL}_3(\mathbf{A})}$ such that $C \subseteq U$. We know that there is $b \in C(\mathbf{CL}_3(\mathbf{A}))$ such that $\gamma_U = \bar{b} = \gamma_{N_b}$. Then $C \subseteq U = N_b$, and we have $b \in A$. Thus $b \leq a$, since $\alpha_{\mathbf{CL}_3(\mathbf{A})}$ is an isomorphism. So, $a \in A$.

(2)   Consider $\langle \mathbf{X}, V, u, \mathcal{C} \rangle$ and $\mathbb{F}(\mathbb{G}(\mathbf{X}))$. From Lemma 9 we know that $X(\mathbf{CL}_3(\mathbb{G}(\mathbf{X}))) = X(\mathbf{C}_3(\mathbf{X}, V)) \cong \mathrm{St}(\mathrm{Clop}(\mathbf{X}))$. Let $\tau_{\mathbf{X}} : \langle \mathbf{X}, V, u, \mathcal{C} \rangle \to \langle \mathrm{St}(\mathrm{Clop}(\mathbf{X})), V_{\mathbf{C}_3(\mathbf{X}, V)}, F[\mathbb{G}(\mathbf{X})], \mathcal{C}(\mathbb{G}(\mathbf{X})) \rangle$ defined by $\tau_{\mathbf{X}}(x) = t_3(x)$. Also, by the proof of (1) of Lemma 9, we know that $\tau_{\mathbf{X}}(u) = F[\mathbb{G}(\mathbf{X})]$, and by (3) we have that for a closed set C, $C \in \mathcal{C}$ if and only if $F_C \in \mathsf{M}(\mathbb{G}(\mathbf{X}))$ if and only if $\tau_{\mathbf{X}}[C] \in \mathcal{C}(\mathbb{G}(\mathbf{X}))$. Then $\tau_{\mathbf{X}}$ is an $i3$-homeomorphism. ∎

4   The i3-implication space of free algebras

We will denote by $\mathbf{F}_{\mathcal{K}}(X)$ the free algebra in a variety $\mathcal{K}$ with a set X of generators. Let $S^{\{\rightarrow,\Delta\}}[X^*]$ be the $\mathcal{L}_3^{\{\rightarrow,\Delta\}}$-subalgebra generated by X in $\mathbf{F}_{\mathcal{L}_3}(X)$. Since $\mathcal{L}_3^{\{\rightarrow,\Delta\}}$ is the $\{\rightarrow,\Delta\}$-subreduct of $\mathcal{L}_3$, the following lemma is clear.

LEMMA 11. *The algebra* $\mathbf{F}_{\mathcal{L}_3^{\{\rightarrow,\Delta\}}}(X)$ *is isomorphic to* $S^{\{\rightarrow,\Delta\}}[X^*]$.

It is easy to see that the set of minimals elements of $\mathbf{F}_{\mathcal{L}_3^{\{\rightarrow,\Delta\}}}(X)$ is $\{\Delta x : x \in X\}$ [6]. Note that if $x \in X$, then the implication filter of $\mathbf{F}_{\mathcal{L}_3}(X)$ generated by Δx is the increasing subset $[\Delta x) = \{a \in \mathbf{F}_{\mathcal{L}_3}(X) : a \geq \Delta x\}$. Thus, $\mathbf{F}_{\mathcal{L}_3^{\{\rightarrow,\Delta\}}}(X) = \bigcup_{x \in X}[\Delta x)$.

It is also known that $C(\mathbf{F}_{\mathcal{L}_3}(X))$ is $\mathbf{F}_\mathcal{B}(Y)$, the free Boolean algebra over the poset $Y = \{\Delta x, \nabla x : x \in X\}$ (see [4]). For each $x \in X$, $\{\Delta x, \nabla x\}$ is a chain called *principal chain*. Moreover, there is a bijection from the set of upwards closed subsets $S \subseteq Y$ onto the ultrafilters of $\mathbf{F}_\mathcal{B}(Y)$, given by $S \rightarrow F(S \cup \{\neg y : y \in Y\backslash S\}) = U_S$. Therefore, $\mathbf{F}_{\mathcal{L}_3}(X)$ is isomorphic to the algebra $\mathbf{C}_3(\mathrm{St}(\mathbf{F}_\mathcal{B}(Y)), V_{\mathbf{F}_{\mathcal{L}_3}(X)})$, where $U_S \in V_{\mathbf{F}_{\mathcal{L}_3}(X)}$ if and only if S is a union of principal chains ([4, Theorem 2.1]). From this, we have that $\mathbf{F}_{\mathcal{L}_3}(X)/F(X) \cong \mathbf{L}_2$, since $F(X) = F(U_Y)$. Then, $\mathbf{CL}_3(\mathbf{F}_{\mathcal{L}_3^{\{\rightarrow,\Delta\}}}(X)) \cong \mathbf{F}_{\mathcal{L}_3}(X)$.

LEMMA 12. $\mathrm{M}(\mathbf{F}_{\mathcal{L}_3^{\{\rightarrow,\Delta\}}}(X)) = \{F(\Delta x) : x \in X\}$.

Proof. Let us see that $F(\Delta x) = [\Delta x) \in \mathrm{M}(\mathbf{F}_{\mathcal{L}_3^{\{\rightarrow,\Delta\}}}(X))$. Let F be a proper implication filter of $\mathbf{F}_{\mathcal{L}_3}(X)$ such that $F \subseteq \mathbf{F}_{\mathcal{L}_3^{\{\rightarrow,\Delta\}}}(X)$, and let $x \in X$ such that $[\Delta x) \subseteq F$. Let $y \in F$. Then the element $y \wedge \Delta x$ exists and we have that $y \wedge \Delta x \in F$. But $y \wedge \Delta x \leq \Delta x$ and Δx is minimal. So $y \in [\Delta x)$. Consequently, $F = [\Delta x)$.

The argument that closes the proof is analogous to that in the proof of ([1, Lemma 5.1]).   ∎

Then $X(\mathbf{CL}_3(\mathbf{F}_{\mathcal{L}_3^{\{\rightarrow,\Delta\}}}(X))) \cong X(\mathbf{F}_{\mathcal{L}_3}(X)) \cong \mathrm{St}(C(\mathbf{F}_{\mathcal{L}_3}(X))) \cong \mathrm{St}(\mathbf{F}_\mathcal{B}(Y))$. For each $x \in X$, $C_{[\Delta x)} = \{U_S \in \mathrm{St}(\mathbf{F}_\mathcal{B}(Y) : \Delta x \in U_S\}$. Observe that $\bigcap C_{[\Delta x)} = \{U_Y\}$. Then,

$$\langle \mathrm{St}(\mathbf{F}_\mathcal{B}(Y)), V, \{C_{[\Delta x)} : x \in X\}, U_Y \rangle$$

is the i3-implication space of $\mathbf{F}_{\mathcal{L}_3^{\{\rightarrow,\Delta\}}}(X)$, where $U_S \in V$ if and only if S is a union of principal chains.

Acknowledgements: This paper is partially supported by Universidad Nacional del Sur and CONICET.

BIBLIOGRAPHY

[1] M. Abad, J. P. Díaz Varela and A. Torrens, *Topological Representation for Implication Algebras*, Algebra Universalis, 52(1), 39-48, 2004.

[2] J. C. Abbot, *Semi-boolean Algebras*, Mat. Vesnik, 4(19), 177-198, 1967.

[3] M. Baaz, *Infinite-valued Gödel logics with* 0-1-*projections and relativizations*, Gödel '96 (Brno, 1996), 23–33, Lecture Notes Logic, 6, Springer, Berlin, 1996.

[4] M. Busaniche and R.Cignoli, *Free Algebras in Varieties of BL-algebras Generated by a BL_n-chain* , J. Aust. Math. Soc., 80 (3), 419-439, 2006.

[5] R. Cignoli and L. Monteiro, *Maximal Subalgebras of MV_n-algebras. A Proof of a Conjecture of A. Monteiro*, Studia Logica, 84(3), 393-405, 2006.

[6] A. V. Figallo, *$I\Delta_3$-algebras*, Rep. Math. Logic, (24), 3-16, 1990.

[7] R. Grigolia, *Algebraic analysis of Lukasiewicz-Tarski's n-valued logical systems*. Selected papers on Lukasiewicz sentential calculi, pp. 81–92. Zaklad Narod. im. Ossolin., Wydawn. Polsk. Akad. Nauk, Wroclaw, 1977.

[8] P. Hájek, **Metamathematics of fuzzy logic**. Trends in Logic-Studia Logica Library, 4. Kluwer Academic Publishers, Dordrecht, 1998.

[9] L. Iturrioz and O. Rueda, *Algèbres Implicatives Trivalentes de Lukasiewicz Libres*, Discrete Math. 18(1), 35-44, 1977.

[10] A. Monteiro, *Algebras de Lukasiewicz trivalentes*, Lectures at the Universidad Nacional del Sur, 1963. Edited by L. Monteiro. Informe Interno. INMABB, Bahía Blanca, 2003.

[11] A. Monteiro, *Algebras implicativas trivalentes de Lukasiewicz*, Lectures at the Universidad Nacional del Sur, 1963.

[12] Gr. C. Moisil, *Recherches sur les Logiques non-chrysippiennes*, Ann. Sci. Univ. Jassy. Sect. I., 26, 431-466, 1940.

M. Abad
Departamento de Matemática
Universidad Nacional del Sur
Universidad Nacional del Comahue
Argentina
E-mail: imabad@criba.edu.ar

C. R. Cimadamore
Departamento de Matemática
Universidad Nacional del Sur
Argentina
E-mail: crcima@criba.edu.ar

J. P. Díaz Varela
Departamento de Matemática
Universidad Nacional del Sur
INMABB, CONICET.
Argentina
E-mail: usdiavar@criba.edu.ar

On Some Operations Using the min Operator

RODOLFO ERTOLA

ABSTRACT. In this paper the minimum operator is used to define some compatible and non-compatible operations on a Heyting algebra. We provide axiomatic systems, one of them simplifying the axiomatization of the connective G of Gabbay. We also provide sufficient conditions for operations to be equational and compatible.

1 Introduction

In what follows we assume that we are dealing with a Heyting algebra, though many facts already hold in a more general setting. The contents were motivated by [4], where operations in Heyting algebras give rise to the consideration of new connectives in intuitionistic logic. In particular, in the mentioned paper, the authors consider the notion of compatibility. In the present paper we study some unary operations $f(x)$ defined as $min\{y : Rxy\}$, some of them compatible and the others not compatible.

In [4] a function $f : H^n \to H$ (where H is a Heyting algebra) is called compatible iff it is compatible with all the congruence relations of H (such a function is compatible with a congruence relation θ of H iff $x_i \theta y_i$, for $i = 1, \ldots, n$ implies that $f(x_1, \ldots x_n)\theta f(y_1, \ldots y_n)$). It can easily be seen (see [4, Lemma 2.1]) that f is a compatible function of H iff $x \leftrightarrow y \leq f(x) \leftrightarrow f(y)$, for all x, y in H (in the case that f is unary).

2 On some non-compatible operations

In this section we study the following three examples of non-compatible unary operations, providing axiomatic systems for their corresponding connectives.

EXAMPLE 1. It is known that the dual pseudo complement $\rho(x)$ may be defined as the $min\{y : y \vee x = 1\}$. This can be seen equivalent to the following equations:

$$x \vee \rho(x) = 1,$$

$$x \vee \rho(x \vee y) = x \vee \rho(y),$$
$$\rho(x \to x) = 0.$$

These equations are the duals of the equations of the pseudo complement that appear e.g. in [10] and [1, VIII, 3] recalling that in this context $\rho(0) = 1$ is equivalent to $x \vee \rho(x) = 1$.

PROPOSITION 2. $\rho(x)$ *is not compatible.*

Proof. We do not have that $x \leftrightarrow y \leq \rho(x) \leftrightarrow \rho(y)$ (this is easily seen e.g. in the Heyting algebra H_3). ∎

Not being compatible, $\rho(x)$ cannot be given by a Heyting term. We also have the following fact, which is useful to prove that an extension is conservative:

PROPOSITION 3. $\rho(x)$ *is defined in every finite Heyting algebra.*

Proof. As $1 \vee x = 1$, $\{y : y \vee x = 1\} \neq \emptyset$. Also, if $y_1 \vee x = 1$ and $y_2 \vee x = 1$, then $(y_1 \wedge y_2) \vee x = (y_1 \vee x) \wedge (y_2 \vee x) = 1 \wedge 1 = 1$. ∎

Now, let us consider logical aspects of $\rho(x)$. Theorem 4.1 in [4] states that an axiomatic extension (meaning no other rules but *Modus Ponens* (MP)) of intuitionistic logic with a unary connective ∇ is strongly complete for the usual algebraic consequence relation iff $\vdash (\varphi \leftrightarrow \psi) \to (\nabla \varphi \leftrightarrow \nabla \psi)$. It follows that an axiomatic extension of intuitionistic logic with the dual pseudo complement is not strongly complete because we do not have $\vdash (\varphi \leftrightarrow \psi) \to (\nabla \rho \leftrightarrow \rho \psi)$, because ρ is not compatible.

But the mentioned Theorem 4.1 may be generalized to any extension, that is, to include extensions having also new rules and not just new axioms (this appears in [3, Theorem 1]). The generalization may be stated as follows: an extension of intuitionistic logic with the connective ∇ is strongly complete for the usual algebraic consequence relation iff $(\varphi \leftrightarrow \psi) \vdash (\nabla \varphi \leftrightarrow \nabla \psi)$ (the proof of this is almost the same as for Theorem 4.1 but now the Deduction Theorem is not needed).

So, in order to provide an axiomatic system for ρ, we will have to add some rule to MP. The dual pseudo complement may be axiomatized adding to any axiomatization of intuitionistic logic with MP as only rule either of the following systems:

System $\rho 1$:

$A\rho 1$. $\varphi \vee \rho\varphi$,
$A\rho 2$. $(\varphi \vee \rho(\varphi \vee \psi)) \leftrightarrow (\varphi \vee \rho\psi)$,

$A\rho3.$ $\rho(\varphi \to \varphi) \to \bot,$
$RF\rho.$ $\varphi \leftrightarrow \psi / \rho\varphi \leftrightarrow \rho\psi;$

System $\rho2$:

$A\rho1.$ $\varphi \vee \rho\varphi,$
$A\rho2.$ $(\varphi \vee \rho(\varphi \vee \psi)) \leftrightarrow (\varphi \vee \rho\psi),$
$RX\rho.$ $\varphi / \neg\rho\varphi;$

System $\rho3$:

$A\rho1.$ $\varphi \vee \rho\varphi,$
$RR\rho.$ $\varphi \vee \psi / \rho\varphi \to \psi.$

System $\rho1$ is just the incomplete system of axioms plus what we call the functionality rule. System $\rho2$ appears in [3, Ex. 4] and the rule $RX\rho$ appears in [9, p. 241] together with the pseudo-difference $\doteq$ (the dual of the relative pseudo-complement). The axiom $A\rho1$ and the rule $RR\rho$ in System $\rho3$ come naturally, respectively, from the facts that $\rho(x)$ belongs to $\{y : y \vee x = 1\}$ and that if $x \vee y = 1$, then $\rho(x) \leq y$ (in this system negation is not necessary). The only rule involving ρ in each system will be called the ρ-rule of the system.

Let us now prove that the three systems are equivalent. Note that in these extensions of intuitionistic logic the Deduction Theorem is no longer available, but in intuitionistic logic with the language including $\rho\varphi$, for all φ, the Deduction Theorem holds.

In the following derivations p stands for positive logic, i.e. the fragment $\{\wedge, \vee, \to\}$ of intuitionistic logic, m stands for Johansson's minimal logic and i stands for intuitionistic logic.

$\rho2 \Rightarrow \rho3$) It is enough to derive $RR\rho$:

1.	$\varphi \vee \psi$	Hyp
2.	$\neg\rho(\varphi \vee \psi)$	$RX\rho,$ 1
3.	$\rho\varphi \to (\psi \vee \rho(\varphi \vee \psi))$	p, $A\rho2$
4.	$(\psi \vee \rho(\varphi \vee \psi)) \to \psi$	i, 2
5.	$\rho\varphi \to \psi$	p, 3,4.

$\rho1 \Rightarrow \rho2$) It is enough to derive $RX\rho$:

1.	φ	Hyp
2.	$\varphi \to (\varphi \to \varphi)$	p
3.	$(\varphi \to \varphi) \to \varphi$	p,1
4.	$\varphi \leftrightarrow (\varphi \to \varphi)$	p, 2, 3
5.	$\rho\varphi \leftrightarrow \rho(\varphi \to \varphi)$	$RF\rho$, 4
6.	$\rho\varphi \to \rho(\varphi \to \varphi)$	p, 5
7.	$\rho(\varphi \to \varphi) \to \bot$	$A\rho3$
8.	$\rho\varphi \to \bot$	p, 6, 7
9.	$\neg\rho\varphi$	m, 8.

$\rho3 \Rightarrow \rho1$) In this case we have to derive $A\rho2$, $A\rho3$ and $RF\rho$:

1.	$\varphi \to (\varphi \vee \rho\psi)$	p
2.	$\psi \vee \rho\psi$	$A\rho1$
3.	$(\varphi \vee \psi) \vee (\varphi \vee \rho\psi)$	p, 2
4.	$\rho(\varphi \vee \psi) \to (\varphi \vee \rho\psi)$	$RR\rho$, 3
5.	$(\varphi \vee \rho(\varphi \vee \psi)) \to (\varphi \vee \rho\psi)$	p, 1, 4
6.	$\varphi \to (\varphi \vee \rho(\varphi \vee \psi))$	p
7.	$(\varphi \vee \psi) \vee \rho(\varphi \vee \psi)$	$A\rho1$
8.	$\psi \vee (\varphi \vee \rho(\varphi \vee \psi))$	p (asoc), 7
9.	$\rho\psi \to (\varphi \vee \rho(\varphi \vee \psi))$	$RR\rho$,8
10.	$(\varphi \vee \rho\psi) \to (\varphi \vee \rho(\varphi \vee \psi))$	p, 6, 9
11.	$(\varphi \vee \rho(\varphi \vee \psi)) \leftrightarrow (\varphi \vee \rho\psi)$	p, 5, 10.

1.	$\varphi \to \varphi$	p
2.	$(\varphi \to \varphi) \vee \bot$	m, 1
3.	$\rho(\varphi \to \varphi) \to \bot$	$RR\rho$, 2.

1.	$\varphi \leftrightarrow \varphi$	Hyp
2.	$\psi \to \varphi$	p, 1
3.	$\psi \vee \rho\psi$	$A\rho1$
4.	$\varphi \vee \rho\psi$	p, 2, 3
5.	$\rho\varphi \to \rho\psi$	$RR\rho$, 4.

(Analogously for $\rho\psi \to \rho\varphi$).

Intuitionistic connectives are usually expected to have certain properties. One of them is given in the next

DEFINITION 4. A unary connective ∇ is univocal iff $\vdash_{i+\nabla+\nabla'} \nabla\varphi \leftrightarrow \nabla'\varphi$, where ∇' is a new unary connective having analogous axioms and rules to ∇.

This concept already appears in a more general setting in [2]. Let us see that

PROPOSITION 5. *In any of the given systems ρ is univocal, i.e. $\vdash_{i+\rho}$ $\rho\varphi \leftrightarrow \rho'\varphi$, where ρ' has analogous axioms and rules to ρ.*

Proof. Consider the following derivation:

1. $\varphi \vee \rho'\varphi$ $A\rho'1$
2. $\rho\varphi \to \rho'\varphi$ $RR\rho$, 1.

(Analogously for the reciprocal.)   ∎

We will use the following simple fact:

PROPOSITION 6. $\vdash_{i+\rho} \neg\varphi \to \rho\varphi$.

Proof. Consider the following derivation:

1. $\varphi \vee \rho\varphi$ $A\rho1$ ∎
2. $\neg\varphi \to \rho\varphi$ i, 1.

PROPOSITION 7. *The axiomatic systems that come from the previous systems $\rho1 - \rho3$ replacing their ρ-rules by their corresponding conditionals, e.g. in system $\rho1$ we replace the rule $RF\rho$ by the axiom $(AF\rho)$: $(\varphi \leftrightarrow \psi) \to (\rho\varphi \leftrightarrow \rho\psi)$, are also equivalent to each other. Any of them will be called $i + \rho_\to$.*

Proof. This is easily seen because in this case we enjoy the Deduction Theorem as we are dealing with extensions of intuitionistic logic with MP as only rule.   ∎

PROPOSITION 8. *In the just mentioned axiomatic systems (i) ρ is a compatible connective. Also, in all of them (ii) $\vdash_{i+\rho_\to} \rho\varphi \leftrightarrow \neg\varphi$ and (iii) the logic becomes classical.*

Proof. It is easily seen that ρ is compatible in System $i + \rho_\to$, because this is what $AF\rho$ says. To see the second part, consider Proposition 6 and the following derivation (remember that now the Deduction Theorem is available):

1. $\rho\varphi$ Hyp
2. φ Sup
3. $\varphi \vee \bot$ p, 2
4. $\rho\varphi \to \bot$ $A\rho2$, 3, MP
5. $\bot$ MP, 4, 1
6. $\neg\varphi$ Deduction Theorem, 2-5
7. $\rho\varphi \to \neg\varphi$ Deduction Theorem, 1-6

To see that the logic becomes classical, consider the following derivation:

1. $\varphi \vee \rho\varphi$ Aρ1
2. $\rho\varphi \rightarrow \neg\varphi$ (from the first derivation of this proof)
3. $\varphi \vee \neg\varphi$ m, 1, 2.   ∎

EXAMPLE 9. In [7] a connective called "the strongest anticipator" was introduced. It may be defined algebraically as $\alpha(x) = min\{y : y \rightarrow x \leq x\}$. In algebraic contexts the strongest anticipator of x is called the minimum x-dense. It can be seen that α has similar properties to the dual pseudo complement. That is, α is equational (meaning that α can be characterized by equations), α is not compatible, α exists in every finite Heyting algebra and α may be axiomatized using at least one rule different from MP where it can be shown to be univocal and whose strengthening replacing the rules by the corresponding conditionals becomes classical logic, where $\alpha(x)$ collapses into $\neg x$. In what follows we see these items one by one.

PROPOSITION 10. $\alpha(x)$ can be given by the equations $\alpha(x) \rightarrow x \leq x$ and $\alpha(x \rightarrow y) \leq x$, i.e. $\alpha(x) = min\{y : y \rightarrow x \leq x\}$ iff $\alpha(x) \rightarrow x \leq x$ and $\alpha(x \rightarrow y) \leq x$ (Marta Sagastume's personal communication).

Proof. $\Rightarrow$) As $\alpha(x) \in \{y : y \rightarrow x \leq x\}$ we immediately get $\alpha(x) \rightarrow x \leq x$. We also have that if $y \rightarrow z \leq z$, then $\alpha(z) \leq y$. If we substitute $x \rightarrow y$ for z and x for y in the given conditional we get that if $x \rightarrow (x \rightarrow y) \leq x \rightarrow y$, then $\alpha(x \rightarrow y) \leq x$. But the antecedent holds in a Heyting algebra.

$\Leftarrow$) Let us first see that $\alpha(x) \in \{y : y \rightarrow x \leq x\}$. This follows immediately from $\alpha(x) \rightarrow x \leq x$. Now let us suppose that $y \rightarrow x \leq x$. Then $y \rightarrow x = x$. So $\alpha(y \rightarrow x) = \alpha(x) \leq y$.   ∎

PROPOSITION 11. $\alpha(x)$ is not compatible.

Proof. Consider the three element Heyting algebra H_3, with middle element b. It is enough to see that $b \leftrightarrow 1 \nleq \alpha(b) \leftrightarrow \alpha(1)$. We have that $b \leftrightarrow 1 = (b \rightarrow 1) \wedge (1 \rightarrow b) = 1 \wedge b = b$. But $\alpha(b) \leftrightarrow \alpha(1) = (\alpha(b) \rightarrow \alpha(1)) \wedge (\alpha(1) \rightarrow \alpha(b)) = (1 \rightarrow 0) \wedge (0 \rightarrow 1) = 0 \wedge 1 = 0$.   ∎

The following fact is useful to prove that an extension is conservative:

PROPOSITION 12. $\alpha(x)$ exists in every finite Heyting algebra.

Proof. $A(x) = \{y : y \rightarrow x \leq x\} \neq \emptyset$, because $1 \rightarrow x \leq x$. As $A(x)$ must be finite, then $\wedge A(x)$ exists. Also $\wedge A(x) \in A(x)$. To see this it is enough to apply induction and consider that $(((y_1 \wedge y_2) \rightarrow x) \wedge y_2) \wedge y_1 \leq x$, and so $((y_1 \wedge y_2) \rightarrow x) \wedge y_2 \leq y_1 \rightarrow x$. Then apply the hypothesis $y_1 \rightarrow x \leq x$ and transitivity to get $((y_1 \wedge y_2) \rightarrow x) \wedge y_2) \leq x$, and so $(y_1 \wedge y_2) \rightarrow x \leq$

$y_2 \to x$. Now apply the hypothesis $y_2 \to x \leq x$ and transitivity again to get $(y_1 \wedge y_2) \to x \leq x$. ∎

Now, let us consider logical aspects of $\alpha(x)$.

PROPOSITION 13. *The strongest anticipator may be axiomatized adding to any axiomatization of intuitionistic logic with MP as only rule either of the following systems:*

System $\alpha 1$:

$A\alpha 1$. $(\alpha\varphi \to \varphi) \to \varphi$,
$A\alpha 2$. $\alpha(\varphi \to \psi) \to \varphi$,
$RF\alpha$. $\varphi \leftrightarrow \psi / \alpha\varphi \leftrightarrow \alpha\psi$;

System $\alpha 2$:

$A\alpha 1$. $(\alpha\varphi \to \varphi) \to \varphi$,
$RX\alpha$. $(\psi \to \varphi) \to \varphi / \alpha\varphi \to \psi$.

System $\alpha 1$ is just the incomplete system of axioms plus the functionality rule. The axiom $A\alpha 1$ and the rule $RX\alpha$ in System $\alpha 2$ come naturally, respectively, from the facts that $\alpha(x)$ belongs to $\{y : y \to x \leq x\}$ and that if $y \to x \leq x$, then $\alpha(x) \leq y$. System $\alpha 2$ was suggested to the author by Xavier Caicedo. The only rule involving α in each system will be called the α-rule of the system.

Let us now prove that both systems are equivalent. Note that in these extensions of intuitionistic Logic the Deduction Theorem is no longer available, but in intuitionistic Logic with the language including $\alpha\varphi$, for all φ, the Deduction Theorem holds. Also neither $RF\alpha$ nor $RX\alpha$ make α compatible, they just make it what we call "functional".

$\alpha 1 \Rightarrow \alpha 2$) It is enough to derive $RX\alpha$:

1. $(\psi \to \varphi) \to \varphi$ Hyp
2. $\varphi \to (\psi \to \varphi)$ p
3. $\varphi \leftrightarrow (\psi \to \varphi)$ p, 1, 2
4. $\alpha\varphi \leftrightarrow \alpha(\psi \to \varphi)$ $RF\alpha$, 3
5. $\alpha\varphi \to \alpha(\psi \to \varphi)$ p, 4
6. $\alpha(\psi \to \varphi) \to \psi$ $A\alpha 2$
7. $\alpha\varphi \to \psi$ p, 5, 6

$\alpha 2 \Rightarrow \alpha 1$) It is enough to derive $A\alpha 2$ and $RF\alpha$:

1. $(\varphi \to (\varphi \to \psi)) \to (\varphi \to \psi)$ p
2. $\alpha(\varphi \to \psi) \to \varphi$ $RX\alpha, 1$

1. $\varphi \leftrightarrow \psi$ Hyp
2. $\varphi \to \psi$ p, 1
3. $\psi \to \varphi$ p, 1
4. $(\alpha\psi \to \varphi) \to (\alpha\psi \to \psi)$ p, 2
5. $(\alpha\psi \to \psi) \to \psi$ $A\alpha 1$
6. $(\alpha\psi \to \varphi) \to \psi$ p, 4, 5
7. $(\alpha\psi \to \varphi) \to \varphi$ p, 6, 3
8. $\alpha\varphi \to \alpha\psi$ $RX\alpha, 7$

(Analogously for the reciprocal.)

PROPOSITION 14. *In any of the given systems α is univocal, i.e.* $\vdash_{i+\alpha}$ *$a\varphi \leftrightarrow \alpha'\varphi$, where α' has analogous axioms and rules to α.*

Proof. Consider the following derivation:

1. $(\alpha'\varphi \to \varphi) \to \varphi$ $A\alpha'1$
2. $\alpha\varphi \to \alpha'\varphi$ $RX\alpha, 1$

(Analogously for the reciprocal.)   ∎

PROPOSITION 15. *The axiomatic systems that come from systems $\alpha 1$ and $\alpha 2$ replacing their α-rules by their corresponding conditionals, e.g. in system $\alpha 1$ we replace the rule $RF\alpha$ by the axiom $(AF\alpha)$: $(\varphi \leftrightarrow \psi) \to (\alpha\varphi \leftrightarrow \alpha\psi)$, are also equivalent to each other. Any of them will be called $i + \alpha_{\to}$.*

Proof. This is easily seen because in this case we enjoy the Deduction Theorem as we are dealing with extensions of intuitionistic logic with MP as only rule.   ∎

PROPOSITION 16. *(i) α is a compatible connective in $i + \alpha_{\to}$. Also, (ii)* $\vdash_{i+\alpha_{\to}} \alpha\varphi \leftrightarrow \neg\varphi$ *and (iii) $i + \alpha_{\to}$ is classical logic.*

Proof. It is easily seen that α is compatible in $i + \alpha_{\to}$, because this is what $AF\alpha$ says. To see the second part, consider the following derivations (remember that now the Deduction Theorem is available):

1. $\alpha\varphi$ Hyp
2. φ Sup
3. $(\bot \to \varphi) \to \varphi$ p, 2
4. $\alpha\varphi \to \bot$ $RX\alpha$, 3, MP
5. $\bot$ MP, 4, 1
6. $\neg\varphi$ Deduction Theorem, 2-5
7. $\alpha\varphi \to \neg\varphi$ Deduction Theorem, 1-6

To see that the logic becomes classical, consider the following derivation:

1. $\neg\varphi \to \varphi$ Sup
2. $\alpha\varphi$ Sup
3. $\neg\varphi$ (From the just given derivation)
4. φ MP, 1, 3
5. $\alpha\varphi \to \varphi$ Deduction Theorem 2-4
6. φ $A\alpha1$, 5, MP

Now we may get the other half of *(ii)*:

1. $\neg\varphi$ Hyp
2. $\alpha\varphi \to \varphi$ Sup
3. φ $A\alpha1$, 2, MP
4. $\bot$ m, 1, 3
5. $\neg(\alpha\varphi \to \varphi)$ Deduction Theorem, 2-4
6. $\alpha\varphi$ c, 5
7. $\neg\varphi \to \alpha\varphi$ Deduction Theorem, 1-6. ■

EXAMPLE 17. Let us define $\kappa x = min\{y : \neg y \le x\}$. It can be seen that this operation is somewhat similar to the operations in the two previous examples. It differs in not being defined in every finite Heyting algebra (e.g. it is not defined in the top of $\mathbf{2}^2$ in the Heyting algebra $\mathbf{2}^2 \oplus \mathbf{1}$).

PROPOSITION 18. κ *is equational.*

Proof. Consider the equations

$\kappa1.$ $\neg\kappa x \le x$,
$\kappa2.$ $\kappa(x \to y) \le x$ ($\kappa\neg x \le x$ is enough),
$\kappa3.$ $\kappa x \le \kappa(x \wedge y)$ ($\kappa x \le \kappa(x \wedge \neg y)$ is enough).

An alternative set of equations is given by $\neg\kappa x \le x$ and $\kappa(\neg y \vee x) \le y$. ■

PROPOSITION 19. κ *is not compatible.*

Proof. Consider the three element Heyting algebra H_3 with middle element a. Then, $a \leftrightarrow 1 = (a \to 1) \wedge (1 \to a) = a \nleq \kappa a \leftrightarrow \kappa 1 = a \leftrightarrow 0 = 0.$ ■

PROPOSITION 20. κ *may be univocally and strongly completely axiomatized in either of the following ways:*

System $\kappa 1$:

$A\kappa.\ \neg\kappa\varphi \to \varphi$,
$RR\kappa.\ \neg\psi \to \varphi / \kappa\varphi \to \psi$.

System $\kappa 2$:

$A\kappa.\ \neg\kappa\varphi \to \varphi$,
$A\kappa 1.\ \kappa(\neg\psi \vee \varphi) \to \psi$,
$RFk.\ \varphi \leftrightarrow \psi / \kappa\varphi \leftrightarrow \kappa\psi$.

System $\kappa 3$:

$A\kappa.\ \neg\kappa\varphi \to \varphi$,
$A\kappa 2.\ \kappa(\varphi \to \psi) \to \varphi$,
$A\kappa 3.\ \kappa\varphi \to \kappa(\varphi \wedge \psi)$,
$RF\kappa.\ \varphi \leftrightarrow \psi / \kappa\varphi \leftrightarrow \kappa\psi$.

Proof. The corresponding derivations are similar to the ones given for ρ and α.   ∎

Using either of the just given axiomatic systems we can syntactically prove that

PROPOSITION 21. κ *is univocal*

Proof. Consider the following derivation:

1. $\neg\kappa\varphi \to \varphi$ $A\kappa'1$
2. $\kappa\varphi \to \kappa'\varphi$ $RR\kappa, 1$

(Analogously for the reciprocal.)   ∎

Also, if we consider an axiomatic system with a functionality axiom instead of the given functionality rule $RF\kappa$ the system collapses into classical logic (As ρ and α , κ becomes classical negation).

3   On some compatible operations

In this section we study certain ways of obtaining equational and compatible operations using the minimum operator.

Let $\eta(x) = min\{y :$ there is a z such that $y = (x \to z) \vee (z \to x)\}$. It is immediate to see that η may be equivalently given by the following two conditions:

1) there is a z such that $\eta(x) = (x \to z) \vee (z \to x)$;
2) $\eta(x) \leq (x \to z) \vee (z \to x)$, for all z.

PROPOSITION 22. η *is compatible.*

Proof. It is enough to see that $x \leftrightarrow y \leq \eta(x) \to \eta(y)$, i.e. that $(x \leftrightarrow y) \wedge \eta(x) \leq \eta(y)$. But $\eta(x) \leq (x \to z) \vee (z \to x)$, where z is such that $\eta(y) = (y \to z) \vee (z \to y)$. Then $(x \leftrightarrow y) \wedge \eta(x) \leq (x \leftrightarrow y) \wedge ((x \to z) \vee (z \to x)) \leq (y \to z) \vee (z \to y) = \eta(y)$. ∎

As in the case of the constant $\delta = min\{y :$ there is a z such that $y = \neg z \vee \neg\neg z\}$ considered in [4, Ex. 3.3 in p. 1626], we also have the following

PROPOSITION 23. η *is not equational.*

Proof. Consider the same Heyting algebra as in Figure 1 in [4], but call j the top element and add two more elements (one called k and another as new top) in such a way that the resulting (finite) algebra is the one consisting of the first 10 elements of the free Heyting algebra of one generator ordered in the same way. Now in this algebra $\eta(c) = j$ but in the Heyting subalgebra $\{0, c, e, i, j, k, 1\}$ we have that $\eta(c) = 1$. ∎

As a matter of fact, Proposition 22 may be straightforwardly generalized substituting for $(x \to z) \vee (z \to x)$ in the definition of $\eta(x)$ any two variable formula $f(x, z)$ satisfying the condition which we introduce in the next

DEFINITION 24. We say that a binary function f is compatible in the first variable iff $x \leftrightarrow y \leq f(x, z) \leftrightarrow f(y, z)$ (note that e.g. all Heyting terms satisfy this condition).

Now, we have the following

PROPOSITION 25. *Let* $g(x) = min\{y :$ *there is a z such that* $y = f(x, z)\}$. *If* f *is compatible in the first variable, then* g *is compatible.*

Proof. First note that, as in the case of η, g can be equivalently defined by the conditions 1) there is a z such that $g(x) = f(x, z)$ and 2) $g(x) \leq f(x, z)$, for all z. To prove the proposition it is enough to see that $x \leftrightarrow y \leq g(x) \to g(y)$, i.e. that $(x \leftrightarrow y) \wedge g(x) \leq g(y)$. But $g(x) \leq f(x, z)$, where z is such that $g(y) = f(y, z)$. Then $(x \leftrightarrow y) \wedge g(x) \leq (x \leftrightarrow y) \wedge f(x, z) \leq$ (using compatibility of f) $f(y, z) = g(y)$.   ∎

COROLLARY 26. *No non-compatible function, e.g. the minimum x-dense* $= min\{y : y \rightarrow x \leq x\}$ *may be given as the* $min\{y :$ *there is a z such that* $y = f(x, z)\}$, *where* $f(x, z)$ *is a Heyting term.*

In the case of the operation known as the successor (see [8]) and defined by $S(x) = min\{y : y \rightarrow x \leq y\}$, the f in the previous proposition may either be $z \vee (z \rightarrow x)$ or $((z \rightarrow x) \rightarrow z) \rightarrow z$ or $((z \rightarrow x) \rightarrow x) \rightarrow (z \vee x)$ or $((z \rightarrow x) \rightarrow x) \rightarrow ((x \rightarrow z) \rightarrow z)$, providing a g that can be characterized by equations (to wit, the equations $Sx \rightarrow x \leq Sx$ and $Sx \leq y \vee (y \rightarrow x)$). So, the following question is natural: what conditions must f satisfy in order to have equations for g? Partially answering this question, let us consider the following

DEFINITION 27. We say that the values of a binary function f are fixed points at its second argument iff $f(x, f(x, y)) = f(x, y)$.

REMARK 28. (Joint work with Hernán San Martín) The three following conditions are equivalent:

(i) there is a z such that $y = f(x, z)$ iff $y = f(x, y)$ (from right to left this is trivially the case);

(ii) for all z, if $y = f(x, z)$, then $y = f(x, y)$;

(iii) $f(x, f(x, y)) = f(x, y)$.

Proof. To see the equivalence between (i) and (ii) it is enough to consider first order predicate logic.

(i) $\Rightarrow$ (iii): having trivially that there is a z such that $f(x, z) = f(x, z)$, we get (iii) using (i).

(iii) $\Rightarrow$ (i): Let us suppose that $y = f(x, z)$. But $f(x, z) =$ (using (iii)) $f(x, f(x, z)) =$ (by our supposition) $f(x, y)$.   ∎

Now we have the following

PROPOSITION 29. *Let* $g(x) = min\{y :$ *there is a z such that* $y = f(x, z)\}$. *If* $f(x, f(x, y)) = f(x, y)$, *then g is equational.*

Proof. Consider the equations $g(x) = f(x, g(x))$ and $g(x) \leq f(x, y)$.   ∎

What happens if, instead of using the scheme with the existential quantifier, we use the scheme $h(x) = min\{y : y = f(x, y)\}$? As in the case of the previous schema to get equationality we have the following

PROPOSITION 30. *Let* $h(x) = min\{y : y = f(x, y)\}$. *If* $f(x, f(x, y)) = f(x, y)$, *then h is equational.*

Proof. Consider the equations $h(x) = f(x, h(x))$ and $h(x) \leq f(x, y)$. ∎

But in the case of compatibility, as it does not seem enough to include the condition of compatibility in the first variable, we also include the condition that the values of f are fixed points at its second argument (the following proposition was inspired by Theorem 16 in [5]):

PROPOSITION 31. *Let* $h(x) = min\{y : y = f(x, y)\}$. *If* $f(x, f(x, y)) = f(x, y)$ *and* f *is compatible in the first variable, then* h *is compatible.*

Proof. We have the equation $h(x) \leq f(x, h(x))$. It follows that $(x \leftrightarrow y) \wedge h(x) \leq (x \leftrightarrow y) \wedge f(x, h(y)) \leq$ (using compatibility of f) $f(y, h(y)) = h(y)$. ∎

REMARK 32. The condition $f(x, f(x, y)) = f(x, y)$ in the previous two propositions is not necessary: consider $h(x) = min\{y : y = y \vee ((y \rightarrow x) \rightarrow (x \vee \neg x))\}$. It can be seen that $h(x) = \neg\neg x \rightarrow x$ (see Proposition 35). So, $h(x)$ is equational and compatible, but in the Heyting algebra $\mathbf{2}^2 \oplus \mathbf{1}$, i.e. the Heyting algebra built up from the four element Boolean algebra adding a fifth element as "new" top, we have that $f(a, 0) = c$ and $f(a, c) = 1$, where c is the "old" top and a is any of the two intermediate elements in the mentioned Boolean algebra.

What happens in the case of the schema $h(x) = min\{y : x = f(x, y)\}$? Analogously to the case of the schema $min\{y : y = f(x, y)\}$ we have the following

PROPOSITION 33. *Let* $h(x) = min\{y : x = f(x, y)\}$. *If* $f(f(x, y), y) = f(x, y)$, *then* h *is equational.*

Proof. Consider the equations $x = f(x, h(x))$ and $h(f(x, y)) \leq y$. ∎

Note that our previous operations α and κ in section 2 may be given, respectively, as $min\{y : x = y \rightarrow x\}$ and $min\{y : x = \neg y \vee x)\}$ and that taking $f(x, y)$ to be either $y \rightarrow x$ or $\neg y \vee x$ we have that $f(f(x, y), y) = f(x, y)$. So, the corresponding equations may be automatically obtained using the proposition we have just stated.

In [6] Gabbay introduced an intuitionistic connective with the following axiom schemas:

G1. $G\varphi \rightarrow (\psi \vee (\psi \rightarrow \varphi))$,
G2. $(\varphi \rightarrow \psi) \rightarrow (G\varphi \rightarrow G\psi)$,
G3. $\varphi \rightarrow G\varphi$,
G4. $G\varphi \rightarrow \neg\neg\varphi$,
G5. $(G\varphi \rightarrow \varphi) \rightarrow (\neg\neg\varphi \rightarrow \varphi)$.

Translating into the corresponding equations the natural question arises whether the corresponding operation Gx may be given as $min\{y : y = f(x,y)\}$. The answer (question and answer by Hernán San Martín simplified by the author) is yes, taking $f(x,y) = y \vee ((y \rightarrow x) \wedge \neg\neg x)$. Also, the values of f are fixed points at its second argument and, so, f may be given by the equations $(Gx \rightarrow x) \wedge \neg\neg x \leq x$ and $Gx \leq z \vee ((z \rightarrow x) \wedge \neg\neg x)$, as said in Proposition 30. Consequently, the axioms above (where G2 had already been proved redundant in [11]) may be reduced to just the following two:

G6. $((G\varphi \rightarrow \varphi) \wedge \neg\neg\varphi) \rightarrow G\varphi$,
G7. $G\varphi \rightarrow (\psi \vee ((\psi \rightarrow \varphi) \wedge \neg\neg\varphi))$.

As f is also compatible in the first variable, using Proposition 31 we have an alternative proof of the compatibility of G.

In [4] it is also proven that $G(x) = Sx \wedge \neg\neg x$. This suggests the following generalization:

PROPOSITION 34. *Let* $h(x) = min\{y : fx \circ (y \rightarrow x) \leq y\}$, *where* $\circ \in \{\wedge, \vee, \rightarrow\}$. *Then* h *is equational, compatible and* $h(x)$ *exists if* Sx *exists with* $h(x) = Sx \circ fx$, *if* $\circ \in \{\wedge, \vee\}$ *and* $h(x) = S(fx \rightarrow x)$, *if* $\circ = \rightarrow$.

Proof. 1) to see that h is equational consider the equations a) $fx \circ (h(x) \rightarrow x) \leq h(x)$ and b) $h(x) \leq z \vee (fx \circ (z \rightarrow x))$;

2) to see that h is compatible, using b) in 1) of this proof (taking $z = h(y)$) we get that $(x \leftrightarrow y) \wedge h(x) \leq (x \leftrightarrow y) \wedge (h(y) \vee (fx \circ (h(y) \rightarrow x))) \leq$ [using compatibility of $h(y) \vee (fx \circ (h(y) \rightarrow x))$] $h(y) \vee (fy \circ (h(y) \rightarrow y)) =$(using a) in 1) of this proof) $h(y)$;

3) Let us just consider the case $\circ \in \{\wedge, \vee\}$ as the case $\circ = \rightarrow$ is straightforward. Let us suppose that S exists. Firstly, we have that $Sx \circ fx \in \{y : fx \circ (y \rightarrow x) \leq y\}$ because using the equation $Sx \rightarrow x \leq Sx$ it follows that $fx \circ ((Sx \circ fx) \rightarrow x) \leq Sx \circ fx$. Secondly, let us take an y such that $fx \circ (y \rightarrow x) \leq y$. Then, using the equation $Sx \leq y \vee (y \rightarrow x)$ it follows that $Sx \circ fx \leq y$. ∎

In the case of $(y \rightarrow x) \rightarrow tx$ (note that this includes $\neg(y \rightarrow x)$ when $tx = \bot$), for tx a Heyting term, $h(x)$ always exists and if tx is respectively equal to $\bot$, x, $\neg x$, $\neg\neg x$, $\neg\neg x \rightarrow x$, $x \vee \neg x$ and $\neg\neg x \vee \neg x$, then $h(x)$ is respectively $\bot$, x, $\neg x$, $\neg\neg x$, $\neg\neg x \rightarrow x$, $\neg\neg x \rightarrow x$ (again) and 1. Only the last two cases are not easy (see the following two propositions). If (H) $\neg\neg x \vee \neg x \leq tx$, then it is easy to see that $h(x) = 1$: firstly, it is evident that $1 \in \{y : (y \rightarrow x) \rightarrow tx \leq y\}$; secondly, suppose (S) $(y \rightarrow x) \rightarrow tx \leq y$. Using (H), we have that $(y \rightarrow x) \rightarrow (\neg\neg x \vee \neg x) \leq (y \rightarrow x) \rightarrow tx$. Then, by (S), $(y \to x) \to (\neg\neg x \vee \neg x) \leq y$. Then $y \in M(x) = \{y : (y \to x) \to (\neg\neg x \vee \neg x) \leq y\}$. But $1 = minM(x)$ (see Proposition 36). Then $1 \leq y$.

In the case of $(y \to x) \to Sx$, $h(x)$ always exists and $h(x) = 1$. Proof: 1) $(1 \to x) \to S1 \leq 1$. 2) Suppose that $(y \to x) \to Sx \leq y$. Then $Sx \leq y$. Then $y \to x \leq Sx \to x$. But $Sx \to x \leq Sx$. Then $y \to x \leq Sx$. Then $(y \to x) \to Sx = 1$. Then $1 \leq y$.

PROPOSITION 35. $Min\{y : (y \to x) \to (x \vee \neg x) \leq y\} = \neg\neg x \to x$.

Proof. It is easily seen that $\neg\neg x \to x \in \{y : (y \to x) \to (x \vee \neg x) \leq y\}$. We also have to see that if (H) $(y \to x) \to (x \vee \neg x) \leq y$, then $\neg\neg x \to x \leq y$. If (H) it follows that $\neg x \leq y$. From this it follows that $y \to x \leq \neg\neg x$. Now it follows that $\neg\neg x \to x \leq (y \to x) \to (x \vee \neg x)$ and using (H) we get our goal. ∎

PROPOSITION 36. $Min\{y : (y \to x) \to (\neg\neg x \vee \neg x) \leq y\} = 1$.

Proof. It is easily seen that $1 \in \{y : (y \to x) \to (\neg\neg x \vee \neg x) \leq y\}$. We also have to see that if (H) $(y \to x) \to (\neg\neg x \vee \neg x) \leq y$, then $1 \leq y$. If (H) it follows that $\neg x \leq y$. From this it follows that $y \to x \leq \neg\neg x \vee \neg x$. Now it follows that $1 \leq (y \to x) \to (\neg\neg x \vee \neg x)$ and using (H) we get our goal. ∎

Acknowledgement

The author thanks Xavier Caicedo for several conversations.

BIBLIOGRAPHY

[1] Balbes, R. and Dwinger, P. *Distributive Lattices*. University of Missouri Press, Columbia, Missouri, 1974.

[2] Belnap, N. Tonk, Plonk and Plink. *Analysis*. vol. 22 (1962). pp. 130-134.

[3] Caicedo, X. Implicit Connectives of Algebraizable Logics. *Studia Logica*. vol. 78 (2003). pp. 155-170.

[4] Caicedo, X. and Cignoli, R. An Algebraic Approach to Intuitionistic Connectives, *The Journal of Symbolic Logic*. vol. 66 (2001). pp. 1620-1636.

[5] Castiglioni, J., Menni, M. and Sagastume, M. Compatible operations on commutative residuated lattices. *Journal of Applied Non-Classical Logics*. vol. 1 (2008). pp. 1-13.

[6] Gabbay, D. On some new intuitionistic propositional connectives, I. *Studia Logica*. vol. 33 (1977). pp. 127-139.

[7] Humberstone, L. The Pleasures of Anticipation: Enriching Intuitionistic Logic. *Journal of Philosophical Logic*. vol. 30 (2001). pp. 395-438.

[8] Kuznetsov, A. On the Propositional Calculus of Intuitionistic Provability, *Soviet Math. Dokl.*. vol. 32 (1985). pp. 18-21.

[9] Rauszer, C. Semi-Boolean algebras and their applications to intuitionistic logic wuth dual operations. *Fundamenta Mathematicae*. **83**(1974). pp. 219-249.

[10] Ribenboim, P. Characterization of the sup-complement in a distributive lattice with last element, *Summa Brasil. Math.* **2** (1949). pp. 43-49.

[11] Yashin, A. New solutions to Novikov's problem for intuitionistic connectives. *Journal of Logic and Computation*. vol. 8 (1998). pp. 637-664.

Rodolfo Ertola
Department of Philosophy
Universidad Nacional de La Plata
La Plata, Argentina.
E-mail: rudolf@fahce.unlp.edu.ar

Algebraic aspects of quantum indiscernibility

Décio Krause, Hércules de Araujo Feitosa

ABSTRACT. Quasi-set theory was proposed as a mathematical context to investigate collections of indistinguishable objects. After presenting an outline of this theory, we define an algebra that has most of the standard properties of an orthocomplete orthomodular lattice, which is the lattice of the closed subspaces of a Hilbert space. We call the mathematical structure so obtained Ɔ-lattice. After discussing, in a preliminary form, some aspects of such a structure, we indicate the next problem of axiomatizing the corresponding logic, that is, a logic which has Ɔ-lattices as its algebraic models. We suggest that the intuitions that the 'logic of quantum mechanics' would be not classical logic (with its Boolean algebra), is consonant with the idea of considering indistinguishability right from the start, that is, as a primitive concept. In other words, indiscernibility seems to lead 'directly' to Ɔ-lattices. In the first sections, we present the main motivations and a 'classical' situation which mirrors that one we focus on the last part of the paper. This paper is our first study of the algebraic structure of indiscernibility within quasi-set theory.

1 Introduction

Indiscernibility is a typical concept of quantum physics, and some facts implied by indiscernibility, as the properties of a Bose-Einstein condensate, have no parallel in classical physics. Without considering that quanta are indiscernible, no explanation of colors would be done, no vindication to the periodic table of elements would result, and, among other things, Planck would not arrive to his formula for the black body radiation. Some authors have sustained that quantum indiscernibility results from the raise of quantum "statistics" (really, ways of counting), while others think that they can *explain* quantum statistics without presupposing indiscernibility, but at the expenses of rejecting equiprobability.[1] That discussion is still alive,

---

[1] This is the particular van Fraassen's view; for instance, he supposes two particles 1 and 2 in two possible states A and B, and the possible cases are: (i) 1 and 2 in A; (ii)

and we have much to do in the philosophical, epistemological, logical, and on ontological aspects of quantum indiscernibility, mainly if we agree (with Arthur Fine) that philosophy of science should be engaged with on-going science (*apud* [14]). This leads us directly to the quantum field theories, and perhaps more, to string theories and to quantum gravitation. Acknowledging this naturalistic claim, we shall be here quite modest in discussing some algebraic aspects of a mathematical theory which was conceived to deal with indistinguishable objects, the *quasi-set theory*. Without revising all the details of such a theory (to which we refer to Chapter 7 of [15]), we shall keep the paper self-contained so that the reader can understand the basic ideas, although sometimes in the intuitive sense, and only the really necessary concepts and postulates are mentioned.

It should be recalled that indiscernibility enters in the standard quantum formalism by means of symmetry postulates. The relevant functions for systems of many quanta ought to be either symmetrical or anti-symmetrical, and this assumption makes the expectation values to assume the same values before and after a permutation of indiscernible elements. Thus, physicists (and philosophers accept that) say that "the individuality was lost," as if there would be something to lose. In this work, we enlarge our research program of providing a mathematical basis for quantum theory that takes indiscernibility "right from the start", as claimed by Heinz Post [25] (see [15]), with the algebraic discussion of indiscernibility. All the considerations are performed within quasi-set theory, which we revise in its main ideas below.

2   Quasi-sets

Quasi-set theory, denoted by $\mathfrak{Q}$, was conceived to handle collections of indistinguishable objects, and was motivated by some considerations taken from quantum physics, mainly in what respects Schrödinger's idea that the concept of identity cannot be applied to elementary particles [29, p. 17-18]. Of course, the theory can be developed independently of quantum mechanics, but here we shall have this motivation always in mind. Our way of dealing with indistinguishability is to assume that expressions like $x = y$ are not well formed in all situations that involve x and y. We express that by saying that the concept of identity does not apply to the entities denoted by x and y in these situations. Here, quantum objects do not mean

---

1 and 2 in B, both cases with probability $1/3$ each; (iii) 1 in A and 2 in B; (iv) 1 in B and 2 in A, both with probability $1/6$. According to this author, this way we can arrive at Bose-Einstein statistics [31]. But the problem is that situations (iii) and (iv) need to be distinguished from one another, and if the involved quanta are indiscernible, this can be done only either by the assumption of some kind of hidden variable or by some form of *substratum*, and we know that both possibilities conduce to well known problems.

necessarily *particles*, but ought to be thought as representing the basic objects of quantum theories, which although differ form one theory to another ([12, Chapter 6]), have some common characteristics, as those related to indiscernibility (with the exception of some hidden variable theories, like Bohm's, which will be not discussed here).[2] Due to the lack of sense in applying the concept of identity to certain elements, informally, a quasi-set (qset), that is, a collection of such objects, may be such that its elements cannot be identified by names, counted, ordered, although there is a sense in saying that these collections have a cardinal. This concept of cardinal is not defined by means of ordinals, as usual – see below. But we aim at to keep standard mathematics intact,[3] so the theory is developed in a way that ZFU (and hence ZF, perhaps with the axiom of choice, ZFC) is a subtheory of $\mathfrak{Q}$. In other words, there is a "copy" of ZFU into $\mathfrak{Q}$, that is, this theory is constructed so that it extends standard Zermelo-Fraenkel with *Urelemente* (ZFU). In this case, standard sets of ZFU can be viewed as particular qsets, that is, there are qsets that have all the properties of the sets of ZFU, and we call then $\mathfrak{Q}$-sets. The objects in $\mathfrak{Q}$ that correspond to the *Urelemente* of ZFU are termed M-atoms. But quasi-set theory encompasses another kind of *Urelemente*, the m-atoms, to which the standard theory of identity does not apply. More especially, expressions like $x = y$ are not well formed when m-atoms are involved.

When $\mathfrak{Q}$ is used in connection with quantum physics, these m-atoms are thought as representing quantum objects (henceforth, q-objects), and not necessarily they are 'particles', as mentioned above; waves or perhaps even strings (and whatever 'objects' sharing the property of indistinguishability of pointing elementary particles) can be also be values of the variables of $\mathfrak{Q}$. The lack of the concept of identity for the m-atoms make them *non-individuals* in a sense, and it is mainly (but not only) to deal with collections of m-atoms that the theory was conceived. So, $\mathfrak{Q}$ is a theory of generalized collections of objects, involving non-individuals. For details about $\mathfrak{Q}$ and about its historical motivations, see [5, p. 119], [9], [15, Chapter 7], [17], and [20].

In $\mathfrak{Q}$, the so called 'pure' qsets have only q-objects as elements (although these elements may be not always indistinguishable from one another), and

---

[2]Since such theories present difficulties due to results like Kochen-Specker theorem and Bell's inequalities, so as due to the fact that apparently they cannot be extended to quantum field theories, we shall leave them outside of our discussion.

[3]So respecting the quite strange rule what Birkhoff and von Neumann call "Henkel's principle of the 'perseverance of formal laws' ", explained by Rédei as "a methodological principle that is supposed to regulate mathematical generalizations by insisting on preserving certain laws in the generalization" [26]; of course we are 'preserving' all standard mathematics built in ZFC.

to them it is assumed that the usual notion of identity cannot be applied, that is, $x = y$, so as its negation, $x \neq y$, are not a well formed formulas if x and y stand for q-objects. Notwithstanding, there is a primitive relation $\equiv$ of indistinguishability having the properties of an equivalence relation, and a concept of *extensional identity*, not holding among m-atoms, is defined and has the properties of standard identity of classical set theories. Since the elements of a qset may have properties (and satisfy certain formulas), they can be regarded as *indistinguishable* without turning to be *identical* (being *the same* object), that is, $x \equiv y \not\Rightarrow x = y$.

Since the relation of equality, and the concept of identity, does not apply to m-atoms, they can also be thought as entities devoid of individuality. We further remark that if the 'property' $x = x$ (to be identical to itself, or *self-identity*, which can be defined for an object a as $I_a(x) =_{\mathsf{def}} x = a$) is included as one of the properties of the considered objects, then the so called Principle of the Identity of Indiscernibles (PII) in the form $\forall F(F(x) \leftrightarrow F(y)) \rightarrow x = y$ is a theorem of classical second-order logic, and hence there can not be indiscernible but not identical entities (in particular, non-individuals). Thus, if self-identity is linked to the concept of non-individual, and if quantum objects are to be considered as such, these entities fail to be self-identical, and a logical framework to accommodate them is in order (see [15] for further argumentation).

We have already discussed at length in the references given above (so as in other works) the motivations to build the quasi-set theory, and we shall not return to these points here,[4] but before to continue we would like to make some few remarks on a common misunderstanding about PII and quantum physics. People generally think that spatio-temporal location is a sufficient condition for individuality. Thus, an electron in the South Pole and another one in the North Pole *are* discernible, hence *distinct individuals*, so that we can call "Peter" one of them and "Paul" the another one. Leibniz himself prevented us about this claim (yet not directly about quantum objects of course), by saying that "it is not possible for two things to differ from one another in respect to place and time alone, but that is always necessary that there shall be some other internal difference" [21]. Leaving aside a possible interpretation for the word 'internal', we recall that even in quantum physics, where fermions obey the Pauli Exclusion Principle, which says that two fermions (yes, they 'count' as more than one) can not have all their quantum numbers (or 'properties') in common, two electrons (which are fermions), one in the South Pole and another one in the North Pole, *are not individuals in the standard sense*.[5] In fact, we can say that

---

[4]But see [6], [7], [8], [15], [18], [20].

[5]Without aiming at to extend the discussion on this topic here (see [15]), like an

the electron in the South Pole is described by the wave function ψ_S, while the another one is described by ψ_N (words like 'another' in the preceding phrase are just ways of speech). But the joint system is, in a simplified form, given by $\psi_{SN} = \psi_S - \psi_N$ (the function must be anti-symmetric in the case of fermions, that is, $\psi_{SN} = -\psi_{NS}$), a superposition of the two first wave functions, and this last function cannot be factorized. Furthermore, in the quantum formalism, the important thing is the square of the wave function, which gives the joint probability density; in the present case, we have $||\psi_{SN}||^2 = ||\psi_S||^2 + ||\psi_N||^2 - 2\mathrm{Re}\psi_S\psi_N$. This last term, called 'the interference term,' can not be dispensed with, and says that nothing, not even *in mente Dei*, can tell us which is the particular electron in the South Pole (and the same happens for the North Pole), that is, we never will know who is Peter and who is Paul, and in the limits of quantum mechanics, this is not a matter of epistemological ignorance, but it is rather an ontological question. As far as quantum physics is concerned with its main interpretations, they seem to be really and truly objects without identity.

In the next sections, we shall discuss from an algebraic point of view some issues of non-individuality. It should be interesting to recall that the 'qset'-operations of intersection ($\cap$), union ($\cup$), difference ($-$) work similarly in $\mathfrak{Q}$ as the standard ones in usual set theories.

3   Algebraic aspects: the lattice of indiscernibility

Quantum logic was born with Birkhoff and von Neumann's paper from 1936 [1]. Today it consists in a wide field of knowledge, having widespread to domains never thought by the two celebrated forerunners. For a look on the state of the art, see [10]. The main idea is that the typical algebraic structures arising from the mathematical formalism of quantum mechanics is not a Boolean algebra, but an orthocomplete (σ-orthocomplete in the general case [10, p. 39]) orthomodular lattice. We shall see below that in quasi-set theory, by considering indiscernibility right from the start, a similar structure 'naturally' arises. Let us provide the details before ending with some comments and conclusions.

Now we need of the concept of Tarski's system and topological space.

DEFINITION 1 (Tarski's Space). A Tarski's Space is a pair $(E,^{-})$ where E is a non empty set and $^{-}$ be a function $^{-} : \mathcal{P}(E) \to \mathcal{P}(E)$, called the Tarski's consequence operator, such that: (i) $A \subseteq \overline{A}$; (ii) $A \subseteq B \implies \overline{A} \subseteq \overline{B}$; (iii) $\overline{\overline{A}} \subseteq \overline{A}$.

THEOREM 2. *In any Tarski's Space* $(E,^{-})$ *we have that:*

---

individual we understand an object that obeys the classical theory of identity of classical logic (extensional set theory included).

(i) $\overline{\overline{A}} = \overline{A}$;

(ii) $\overline{A} \cup \overline{B} \subseteq \overline{A \cup B}$;

(iii) $\overline{A \cap B} \subseteq \overline{A} \cap \overline{B}$;

(iv) $\overline{\overline{A} \cup \overline{B}} = \overline{A \cup B}$;

(v) $\overline{A} \cap \overline{B} = \overline{\overline{A} \cap \overline{B}}$.

Proof. See [13].   ■

DEFINITION 3 (Closed and open sets). In a Tarski's Space $(E,^{-})$, a subset $A \subseteq E$ is closed when $\overline{A} = \overline{A}$ and A is open when its complement relative to E, denoted by A^C, is closed.

DEFINITION 4 (Closure and interior). Given $A \subseteq E$ in $(E,^{-})$, the set $\overline{A}$ is the closure of A and the set $\mathring{A} = (\overline{A^C})^C$ is the interior of A.

THEOREM 5. *In any Tarski's Space $(E,^{-})$ it follows that:* $\mathring{A} \subseteq A \subseteq \overline{A}$.

DEFINITION 6 (Topological Space). A Topological Space is a pair $(E,^{-})$ where E is a non empty set and $^{-}$ be a function $^{-} : \mathcal{P}(E) \to \mathcal{P}(E)$, such that: (i) - (iii) of Definition 3.1 hold plus (iv) $\overline{A \cup B} = \overline{A} \cup \overline{B}$; (v) $\overline{\emptyset} = \emptyset$.

  Hereafter, we shall be working in the theory $\mathfrak{Q}$, and use the equality symbol $=$ to stand for the extensional equality of $\mathfrak{Q}$. Intuitively speaking, $x = y$ holds when x and y are both qsets and have the same elements (in the sense that an object belongs to x iff it belongs to y) or they are both M-objects and belong to the same qsets. It can be proven that $=$ has all properties of standard identity of first-order ZFC. Qsets which may have m-atoms as elements are written (in the metalanguage) with square brackets "[" and "]", and $\mathfrak{Q}$-sets (qsets whose transitive closure have no m-atoms) with the usual curly braces "{" and "}".

  We start with the concept of cloud that will to point to the algebraic aspects whose we are involved.

DEFINITION 7 (Cloud). Let U be a non empty qset and A be a subqset of U. The *cloud* of A is the qset

$$\overline{A} =_{\mathsf{def}} [y \in U : \exists x(x \in A \wedge y \equiv x)].$$

Intuitively speaking, $\overline{A}$ is the qset of the elements of U (the universe) which are indistinguishable from the elements of A. If A is a $\mathfrak{Q}$-set, that is, a copy of a set of ZFU, then of course the only indistinguishable of a certain x is x itself, thus $\overline{A} = A$.

THEOREM 8. *The application[6] that associates to every subqset of U its cloud is a Tarski's operator and $(U,^-)$ is Tarski's Space.*

Proof. (i) $A \subseteq \overline{A}$: Let $t \in A$. Then, by the reflexivity of $\equiv$, we have $t \equiv t$, hence $t \in \overline{A}$. (ii) $A \subseteq B \Rightarrow \overline{A} \subseteq \overline{B}$: Let $A \subseteq B$, and let $t \in \overline{A}$. Then there exists $x \in A$ such that $t \equiv x$. Since $x \in B$, then $t \in \overline{B}$. (iii) $\overline{\overline{A}} \subseteq \overline{A}$: Let $t \in \overline{\overline{A}}$. Then there exists $x \in \overline{A}$ such that $t \equiv x$. But then there exists $y \in A$ such that $x \equiv y$. By the transitivity of $\equiv$, we have $t \equiv y$, hence $t \in \overline{A}$. ∎

From now on, we shall suppose that U is closed, that is, it contains all the indistinguishable objects of its elements. Some interpretations linked to physical situations are possible. For instance, $\overline{A}$ can be thought as the region where the wave function A of a certain physical system is different from zero. Another possible interpretation is to suppose that the clouds describe the systems plus the cloud of virtual particles that accompany those of the considered system. But in this paper we shall be not considering these motivations, but just to explore its algebraic aspects.

It is immediate to prove the following theorem:

THEOREM 9. $(U,^-)$ *is a topological space.*

Proof.

(i) $\overline{A} \cup \overline{B} = \overline{A \cup B}$: $A \subseteq A \cup B$, so $\overline{A} \subseteq \overline{A \cup B}$. In the same way, $B \subseteq A \cup B$ and $\overline{B} \subseteq \overline{A \cup B}$. Thus $\overline{A} \cup \overline{B} \subseteq \overline{A \cup B}$. Conversely, suppose $t \in \overline{A \cup B}$, then there is $x \in A \cup B$ such that $t \equiv x$. So there is $t \equiv x$ such that $x \in A$ or $x \in B$. In this way $t \in \overline{A}$ or $t \in \overline{B}$ and therefore $t \in \overline{A} \cup \overline{B}$.

(ii) It follows immediately from the definition of cloud that $\overline{\emptyset} = \emptyset$.

∎

---

[6]In $\mathfrak{Q}$, the concept of function must be generalized, for if there are m-atoms involved, a mapping in general does not distinguish between arguments and values. Thus we use the notion of q-function, which leads indistinguishable objects into indistinguishable objects, and which reduces to standard functions when there are no m-atoms involved. Thus, from the formal point of view, the defined mapping may associate to A whatever qset from a collection of indistinguishable qsets. But this does not matter. As in quantum physics, it is not the extension of the collections which are important; informally saying, *any* elementary particle of a certain kind serves for all purposes involving it. This is the principle of the invariance of permutations.

The next definition introduces the lattice operations on subqsets of a qset U, the universe.

DEFINITION 10 (ℑ-lattice operations). Let $A, B \subseteq U$. Then:

($\sqcap$) $A \sqcap B =_{\text{def}} \overline{A \cap B}$;

($\sqcup$) $A \sqcup B =_{\text{def}} \overline{A \cup B}$;

(0) $\mathbf{0} =_{\text{def}} \emptyset$;

(1) $\mathbf{1} =_{\text{def}} U$.

We note that even if $A \cap B = \emptyset$, may be that $\overline{A} \cap \overline{B} \neq \emptyset$.

THEOREM 11. *For any $A, B \in \mathcal{P}(U)$:*

(i) $A \sqcap B \subseteq \overline{(\overline{A} \cap \overline{B})}$;

(ii) $A \sqcap B \subseteq A \sqcup B$;

(iii) If A and B are closed, $A \cup B$ and $A \cap B$ are closed, and $A \sqcap B = \overline{A} \cap \overline{B}$.

Proof.

(i) Immediate, since $\overline{A \cap B} \subseteq \overline{A} \cap \overline{B}$ (Theorem 3.1 (iii));

(ii) $A \sqcap B = \overline{A \cap B} \subseteq$ (Theorem 3.1 (iii)) $\subseteq \overline{A} \cap \overline{B} \subseteq \overline{A} \cup \overline{B} = A \sqcup B$;

(iii) If $\overline{A} = A$ and $\overline{B} = B$, then $A \cup B = \overline{A} \cup \overline{B} = \overline{A \cup B}$ (Theorem 3.4 (i)). Furthermore, the same hypothesis entails that $\overline{A \cap B} = \overline{A} \cap \overline{B} =$ (Theorem 3.1 (v)) $\overline{A} \cap \overline{B} = A \cap B$. Finally, since $\overline{A} \cap \overline{B} = \overline{(\overline{A} \cap \overline{B})} = \overline{A \cap B} = A \sqcap B$ (Theorem 3.1 (v) and the hypothesis).

■

THEOREM 12. *Let $\mathcal{C}$ be the qset of all closed subqsets of U. Then the structure $\mathfrak{C} = \langle \mathcal{C}, \sqcap, \sqcup, \mathbf{0}, \mathbf{1} \rangle$ is a lattice with 0 and 1. But, if we consider also the sub-qsets of U that are not closed, then some of the properties of such a structure do not hold, as we emphasize in the proof below.*

Proof. In this case, for every $A \subseteq U$ holds $\overline{A} = A$. Firstly, it is immediate to see that if $U \neq \emptyset$, then $\mathcal{P}(U) \neq \emptyset$. Furthermore, we can prove that $A \cap (B \cap C) = (A \cap B) \cap C$ and $A \cup (B \cup C) = (A \cup B) \cup C$ for closed qsets.

(a) Idempotency (restricted to closed qsets): $A \sqcap A = \overline{A \cap A} = \overline{A} (= A$ when A is closed). Also, $A \sqcup A = \overline{A \cup A} = \overline{A} (= A$ when A is closed). If A is not closed, then $A \sqcap A = \overline{A} \neq A$ and $A \sqcup A = \overline{A} \neq A$;

(b) Commutativity (unrestricted): $A \sqcap B = \overline{\overline{A} \cap \overline{B}} = \overline{\overline{B} \cap \overline{A}} = B \sqcap A$. In the same way, $A \sqcup B = \overline{\overline{A} \cup \overline{B}} = \overline{\overline{B} \cup \overline{A}} = B \sqcup A$;

(c) Associativity (unrestricted): (we shall be using items (iii) and (iv) of Theorem 3.1 without mentioning):

(i) $A \sqcap (B \sqcap C) = A \sqcap (\overline{\overline{B} \cap \overline{C}}) = \overline{\overline{A} \cap \overline{\overline{B} \cap \overline{C}}} = \overline{\overline{A} \cap (\overline{B} \cap \overline{C})} = \overline{(\overline{A} \cap \overline{B}) \cap \overline{C}} = \overline{\overline{(\overline{A} \cap \overline{B})} \cap \overline{C}} = \overline{(\overline{A} \cap \overline{B})} \sqcap C = (A \sqcap B) \sqcap C$;

(ii) $A \sqcup (B \sqcup C) = A \sqcup (\overline{\overline{B} \cup \overline{C}}) = \overline{\overline{A} \cup \overline{\overline{B} \cup \overline{C}}} = \overline{\overline{A} \cup (\overline{B} \cup \overline{C})} =$ (Theorem 3.1 (i)) $\overline{\overline{A} \cup (\overline{B} \cup \overline{C})} = \overline{(\overline{A} \cup \overline{B}) \cup \overline{C}} =$ (Theorem 3.1 (i)) $\overline{\overline{(\overline{A} \cup \overline{B})} \cup \overline{C}} = \overline{(\overline{A} \cup \overline{B})} \cup \overline{C} = \overline{(\overline{A} \cup \overline{B})} \sqcup C = (A \sqcup B) \sqcup C$;

(d) Absorption (restricted):

(i) $A \sqcap (A \sqcup B) = A \sqcap (\overline{\overline{A} \cup \overline{B}}) = \overline{\overline{A} \cap (\overline{\overline{A} \cup \overline{B}})}$. But $A \subseteq \overline{A}$, so $A \subseteq \overline{A} \cup \overline{B}$, then $\overline{A} \cap (\overline{\overline{A} \cup \overline{B}}) = \overline{A}$ $(= A$ when A is a closed qset);

(ii) $A \sqcup (A \sqcap B) = A \sqcup (\overline{\overline{A} \cap \overline{B}}) = \overline{\overline{A} \cup \overline{\overline{A} \cap \overline{B}}} = \overline{A} \cup (\overline{\overline{A} \cap \overline{B}}) = \overline{A}$, for $A \sqcap B \subseteq A \subseteq \overline{A}$ $(= A$ when A is a closed qset);

(e) The properties of $\mathbf{0}$ and $\mathbf{1}$:

(i) $\mathbf{0} \sqcap A = \overline{\overline{\emptyset} \cap \overline{A}} = \overline{\emptyset} = \emptyset = \mathbf{0}$;

(ii) $\mathbf{0} \sqcup A = \overline{\overline{\emptyset} \cup \overline{A}} = \overline{\emptyset \cup \overline{A}} = \overline{A}$ $(= A$ when A is a closed qset);

(iii) $A \sqcap \mathbf{1} = \overline{\overline{A} \cap \overline{U}} = \overline{A} (= A$ when A is a closed qset);

(iv) $A \sqcup \mathbf{1} = \overline{\overline{A} \cup \overline{U}} = \overline{U} = U = \mathbf{1}$ (recall our initial hypothesis that U is closed).

∎

THEOREM 13. *The lattice $\mathfrak{C}$ of the closed qsets of U is distributive.*

Proof. We shall emphasize those passages which make use of the hypothesis that the qsets are closed.

(i) $A \sqcup (B \sqcap C) = A \sqcup (\overline{\overline{B} \cap \overline{C}}) = \overline{\overline{A} \cup \overline{\overline{B} \cap \overline{C}}} = $ [Th. 3.1 (ii)] $\overline{\overline{A} \cup \overline{(\overline{B} \cap \overline{C})}}$ $= \overline{\overline{A} \cup (\overline{B} \cap \overline{C})} = $ [Th. 3.1 (iv)] $\overline{\overline{A} \cup (\overline{B} \cap \overline{C})} = \overline{(\overline{A} \cup \overline{B}) \cap (\overline{A} \cup \overline{C})} = $ [Th. 3.1 (iii) and because $A \cup B$ and $A \cup C$ are both closed, for otherwise the equality does not hold] $\overline{(\overline{A} \cup \overline{B}) \cap (\overline{A} \cup \overline{C})} = $ [Th. 3.1 (ii)] $\overline{\overline{(\overline{A} \cup \overline{B})} \cap \overline{(\overline{A} \cup \overline{C})}}$ $= \overline{(\overline{A} \cup \overline{B})} \cap \overline{(\overline{A} \cup \overline{C})} = $ [Th. 3.1 (v)] $\overline{(\overline{A} \cup \overline{B})} \cap \overline{(\overline{A} \cup \overline{C})} = (\overline{A} \cup \overline{B}) \sqcap (\overline{A} \cup \overline{C})$ $= (A \sqcup B) \sqcap (A \sqcup C)$;

(ii) $(A \sqcap B) \sqcup (A \sqcap C) = \overline{(A \cap B)} \sqcup \overline{(A \cap C)} = \overline{\overline{(A \cap B)} \cup \overline{(A \cap C)}} =$
[Th. 3.1 (ii)] $\overline{\overline{(A \cap B)} \cup \overline{(A \cap C)}} = $ [for $A \cap B$ and $A \cap C$ are closed]
$\overline{(A \cap B) \cup (A \cap C)} = \overline{A \cap (B \cup C)}$ = (for closed qsets) $\overline{\overline{A} \cap \overline{(B \cup C)}} =$
[Th. 3.1 (i)] $\overline{\overline{A} \cap (\overline{B} \cup \overline{C})} = $ [since A is closed] $\overline{A \cap (\overline{B} \sqcup \overline{C})} = A \sqcap (B \sqcup C)$.

■

This result is not surprising, for we are dealing with set theoretical operations which, defined on the closed qsets of U, act as the usual set theoretical
properties on standard sets. But if we consider *all* qsets in U and not only
the closed ones, the distributive laws do not hold, as we can see from the
above proof, which makes essential use of the fact that the involved qsets
are closed (without such an hypothesis, the proof does not follow). Since the
corresponding structure $\mathfrak{I} = \langle \mathcal{P}(U), \sqcap, \sqcup, \mathbf{0}, \mathbf{1} \rangle$ has similarities with a lattice
with 0 and 1, we propose to call it *the lattice of indiscernibility*, or just
$\mathfrak{I}$-lattice for short. Other distinctive characteristics of this "quasi-lattice"
are obtained when we introduce other operations similar to those of order
and involution, or generalized complement [10, p. 11]. At Section 4, we sum
up the main properties of an $\mathfrak{I}$-lattice.

DEFINITION 14 ($\mathfrak{I}$-order). $A \leq B =_{\mathsf{def}} A \sqcup B = \overline{B}$.

THEOREM 15. *The order relation obeys the following properties:*

(i) $A \leq A$ and $A \leq \overline{A}$;

(ii) $A \leq B$ and $B \leq A \Rightarrow \overline{A} = \overline{B}$ (and $A = B$ if they are both closed);

(iii) $A \leq B$ and $B \leq C \Rightarrow A \leq C$

(iv) $A \sqcap B \leq A$, and $A \sqcap B \leq B$;

(v) $C \leq A$ and $C \leq B \Rightarrow C \leq A \sqcap B$

(vi) $A \leq A \sqcup B$, $B \leq A \sqcup B$;

(vii) $A \leq C$ and $B \leq C \Rightarrow A \sqcup B \leq C$;

(viii) $\mathbf{0} \leq A$, and $A \leq \mathbf{1}$ (recall that $\mathbf{1} = U$ is closed);

(ix) $A \leq B \Rightarrow A \sqcap B = \overline{A}$.

Proof.

(i) $A \sqcup A = \overline{A} \cup \overline{A} = \overline{A}$, so $A \leq A$; and $A \sqcup \overline{A} = \overline{A} \cup \overline{\overline{A}} = \overline{A} \cup \overline{A} = \overline{A}$, so
$A \leq \overline{A}$;

(ii) $A \leq B \Rightarrow \overline{A} \cup \overline{B} = \overline{B}$, while $B \leq A \Rightarrow \overline{B} \cup \overline{A} = \overline{A}$, since $\overline{A} \cup \overline{B} = \overline{B} \cup \overline{A}$, then $\overline{A} = \overline{B}$ ($A = B$ for closed qsets);

(iii) If $A \leq B$ and $B \leq C$, then $\overline{A} \cup \overline{B} = \overline{B}$ and $\overline{B} \cup \overline{C} = \overline{C}$, therefore $\overline{A} \cup \overline{C} = \overline{A} \cup (\overline{B} \cup \overline{C}) = (\overline{A} \cup \overline{B}) \cup \overline{C} = \overline{B} \cup \overline{C} = \overline{C}$, that is, $A \leq C$;

(iv) $A \sqcap B \leq A$ iff $\overline{(A \sqcap B)} \cup \overline{A} = \overline{A}$. But, by Theorem 3.4 (i), $\overline{(A \sqcap B)} \cup \overline{A} = \overline{(A \sqcap B) \cup A} = \overline{A}$. Equivalently, $A \sqcap B \leq B$ iff $\overline{(A \sqcap B)} \cup \overline{B} = \overline{B}$. But, by Theorem 3.4 (i), $\overline{(A \sqcap B)} \cup \overline{B} = \overline{(A \sqcap B)} \cup \overline{B} = \overline{B} = \overline{B}$;

(v) $C \leq A \sqcap B$ iff $\overline{C} \cup \overline{(A \sqcap B)} = \overline{(A \sqcap B)} = $ [Theorem 3.1 (v)] $\overline{A} \sqcap \overline{B}$. But the hypothesis tells us that $\overline{C} \cup \overline{A} = \overline{A}$ and $\overline{C} \cup \overline{B} = \overline{B}$, hence $\overline{C} \subseteq \overline{A}$ and $\overline{C} \subseteq \overline{B}$, that is, $\overline{C} \subseteq (\overline{A} \sqcap \overline{B})$. So, $\overline{C} \cup (\overline{A} \sqcap \overline{B}) = \overline{A} \sqcap \overline{B}$, that is, $C \leq A \sqcap B$;

(vi) $A \leq A \sqcup B$ iff $A \sqcup (A \sqcup B) = \overline{A \sqcup B} = \overline{(\overline{A} \cup \overline{B})} = \overline{A} \cup \overline{B}$ by Theorem 3.4 (i). But $A \sqcup (A \sqcup B) = A \sqcup (\overline{A} \cup \overline{B}) = \overline{A} \cup \overline{(\overline{A} \cup \overline{B})} = \overline{A} \cup \overline{(A \cup B)} = \overline{A} \cup (\overline{A} \cup \overline{B}) = \overline{A} \cup \overline{B}$, by the same theorem.;

(vii) The hypothesis says that $A \sqcup C = \overline{C}$ and $B \sqcup C = \overline{C}$, that is, $\overline{A} \cup \overline{C} = \overline{C}$, hence $\overline{A} \subseteq \overline{C}$. In the same vein, $\overline{B} \subseteq \overline{C}$. But these results entail that $\overline{A} \cup \overline{B} \subseteq \overline{C}$, hence $\overline{(\overline{A} \cup \overline{B})} \subseteq \overline{\overline{C}}$, then $\overline{(\overline{A} \cup \overline{B})} \cup \overline{C} = \overline{C}$;

(viii) $\mathbf{0} \sqcup A = \overline{\mathbf{0}} \cup \overline{A} = \overline{A}$, and $A \sqcup \mathbf{1} = \overline{A} \cup \overline{U} = U = \mathbf{1}$;

(ix) If $A \leq B$, then $A \sqcup B = \overline{B}$. But this entails that $\overline{A} \subseteq \overline{B}$. Thus $\overline{A \sqcap B} \subseteq \overline{A} \sqcap \overline{B} = \overline{A}$ [Theorem 3.1 (iii)], that is, $A \sqcap B = \overline{A}$.

■

Alternatively, we could define $A \leq_1 B$ iff $A \sqcap B = \overline{A}$. The theorem above follows, with the exception of item (ix), which should be substituted by $A \leq B \Rightarrow A \sqcup B = \overline{B}$. Really, assuming this definition, we have $A \sqcup B = \overline{A} \cup \overline{B} = \overline{B}$, for the hypothesis entails that $A \sqcap B = \overline{A}$, that is, $\overline{A} = \overline{A \sqcap B} \subseteq \overline{A} \sqcap \overline{B}$ by Theorem 3.1 (iii). So $\overline{A} \subseteq \overline{B}$, then $\overline{A} \cup \overline{B} = \overline{B}$, that is, $A \sqcup B = \overline{B}$. Item (ix) of the theorem and this result show that $A \leq B$ iff $A \leq_1 B$.

We have proved that $\leq$ is both reflexive and transitive ((i) and (iii) above), but only "partially" anti-symmetric, that is, $A \leq B$ and $B \leq A$ entail $\overline{A} = \overline{B}$. Thus, $\langle \mathcal{P}(U), \leq \rangle$ is a kind of "weak" poset. Since it contains $\mathbf{0}$ and $\mathbf{1}$ and since any two elements of U have a supremum (namely, $A \sqcup B$) and an infimum (namely, $A \sqcap B$). $\mathfrak{I}$ is a "weak lattice", but of course it is a lattice stricto sensu if we consider only closed qsets.

The complement of a qset A relative to the universe U is the sub-qset of U, termed $A^{\perp}$, which has no element indistinguishable from any element of A.

DEFINITION 16 (ℑ-involution, or Generalized ℑ-complement).

$$A^{\perp} =_{\mathsf{def}} U - \overline{A}.$$

THEOREM 17. *Let $A, B \in \mathcal{P}(U)$. Then:*

   (i) $\emptyset^{\perp} = U$;

   (ii) $U^{\perp} = \emptyset$;

   (iii) $U - A^{\perp} = \overline{A}$;

   (iv) $\overline{A^{\perp}} = A^{\perp} = \overline{A}^{\perp}$;

   (v) $A^{\perp\perp} = \overline{A}$ ($= A$ when A is a closed qset);

   (vi) $A \leq B \Rightarrow B^{\perp} \leq A^{\perp}$.

Proof.

   (i) $\emptyset^{\perp} = U - \overline{\emptyset} = U - \emptyset = U$;

   (ii) $U^{\perp} = U - \overline{U} = U - U = \emptyset$;

   (iii) $U - A^{\perp} = U - (U - \overline{A}) = \overline{A}$ because they are all closed;

   (iv) $\overline{A^{\perp}} = \overline{U - \overline{A}} = U - \overline{A} = A^{\perp} = \overline{A}^{\perp}$. Informally speaking, in $U - \overline{A}$ there are no elements indiscernible from the elements of $\overline{A}$ (according to definition of cloud). Thus, it is closed, and coincides with $\overline{A}^{\perp} = U - \overline{\overline{A}}$.

   (v) $A^{\perp\perp} = U - \overline{A^{\perp}} = U - A^{\perp} = \overline{A}$ ($= A$ for closed qsets);

   (vi) $A \leq B \Rightarrow A \sqcup B = \overline{B}$, hence $\overline{A} \cup \overline{B} = \overline{B}$ and $\overline{A} \subseteq \overline{B}$. But this implies that $U - \overline{B} \subseteq U - \overline{A}$, that is, $B^{\perp} \subseteq A^{\perp}$. So $B^{\perp} \cup A^{\perp} = A^{\perp}$, then, by Theorem 3.1 (i), $\overline{B^{\perp} \cup A^{\perp}} = \overline{A^{\perp}}$, hence $B^{\perp} \sqcup A^{\perp} = \overline{A^{\perp}}$ or $B^{\perp} \leq A^{\perp}$.

                                                        ■

Properties (v) and (vi) of the preceding theorem show that $\perp$ is an involution for closed qsets. For qsets in general, we shall call it ℑ-involution, in the spirit of the above discussion.

THEOREM 18. *If $A, B \in \mathcal{P}(U)$, then:*

(i) $A \sqcup A^\perp = \mathbf{1}$;

(ii) $A \sqcap A^\perp = \mathbf{0}$;

(iii) $A \sqcup (B \sqcap B^\perp) = \overline{A}$ ($= A$ for closed qsets);

(iv) $A \sqcap (B \sqcup B^\perp) = \overline{A}$ ($= A$ for closed qsets);

(v) (De Morgan) $(A \sqcup B)^\perp = A^\perp \sqcap B^\perp$;

(vi) ("Partial" De Morgan) $(A \sqcap B)^\perp \subseteq A^\perp \sqcup B^\perp$ (equality holds for closed qsets).

Proof.

(i) $A \sqcup A^\perp = \overline{A} \cup \overline{A^\perp} = \overline{A} \cup A^\perp = \overline{A} \cup (U - \overline{A}) = U = \mathbf{1}$;

(ii) $A \sqcap A^\perp = A \sqcap (U - \overline{A}) = \overline{A \cap (U - \overline{A})} = \overline{\emptyset} = \mathbf{0}$;

(iii) $A \sqcup (B \sqcap B^\perp) = A \sqcup \mathbf{0} = \overline{A} \cup \emptyset = \overline{A}$ ($= A$ for closed qsets);

(iv) $A \sqcap (B \sqcup B^\perp) = A \sqcap \mathbf{1} = \overline{A}$ ($= A$ for closed qsets);

(v) $(A \sqcup B)^\perp = (\overline{A} \cup \overline{B})^\perp = U - \overline{(\overline{A} \cup \overline{B})} = $ [Th. 3.1 (iv)] $U - \overline{(A \cup B)}$ $=$ [Th. 3.4 (i)] $U - (\overline{A} \cup \overline{B}) = (U - \overline{A}) \cap (U - \overline{B}) = $ [Th. 3.1 (ii) and the fact that the involved qsets are closed] $(U - \overline{A}) \cap (U - \overline{B}) = A^\perp \sqcap B^\perp$;

(vi) $(A \sqcap B)^\perp = U - \overline{(A \sqcap B)} = U - \overline{\overline{(A \cap B)}} = U - \overline{(A \cap B)} \subseteq $ [Th. 3.1 (iii); equality holds for closed qsets] $\subseteq U - (\overline{A} \cap \overline{B}) = (U - \overline{A}) \cup (U - \overline{B}) = A^\perp \cup B^\perp = $ [previous theorem (iv)] $\overline{A^\perp} \cup \overline{B^\perp} = A^\perp \sqcup B^\perp$. ∎

The lattice $\mathfrak{I}$ is $\mathfrak{I}$-orthomodular, that is, if $A \le B$, we have that $\overline{B} = A \sqcup (A \sqcup B^\perp)^\perp$.

THEOREM 19 ($\mathfrak{I}$-orthomodularity). *For all $A, B \in \mathcal{P}(U)$: $A \le B \Rightarrow A \sqcup (A \sqcup B^\perp)^\perp = \overline{B}$.*

Proof. $A \sqcup (A \sqcup B^\perp)^\perp = A \sqcup (B \sqcap A^\perp) = A \sqcup \overline{(B \cap A^\perp)} = \overline{A} \cup \overline{(B \cap A^\perp)} = $ [Th. 3.1 (iv)] $\overline{A} \cup \overline{(B \cap A^\perp)} = (A \cup B) \cap (A \cup A^\perp) = (A \cup B) \cap \mathbf{1} = \overline{A \cup B}$ [Th. 3.1 (i)] $\overline{A} \cup \overline{B} = A \sqcup B = $ (by the hypothesis) $\overline{B}$. ∎

DEFINITION 20 (Orthogonality). Let $A, B \subseteq U$. We say that A is orthogonal to B, and write $A \perp B$, when: $A \perp B =_{\mathsf{def}} A \leq B^{\perp}$. Furthermore, a collection S of elements of $\mathcal{P}(U)$ is called pairwise orthogonal iff for any $A, B \in S$ such that $A \neq B$, it results that $A \perp B$.

THEOREM 21. $A \perp B$ iff $A \cap \overline{B} = \emptyset$.

Proof. If $A \perp B$, then $A \leq B^{\perp}$, that is, $A \sqcup B^{\perp} = \overline{B^{\perp}}$. Thus, $\overline{A} \cup \overline{B^{\perp}} = \overline{B^{\perp}}$, so $\overline{A} \subseteq \overline{B^{\perp}}$, hence $A \subseteq \overline{B^{\perp}} = B^{\perp}$ [by Theorem 3.9 (iv)], so $A \cap \overline{B} = \emptyset$. Conversely, if $A \cap \overline{B} = \emptyset$, then $A \subseteq B^{\perp}$, hence $\overline{A} \cup \overline{B^{\perp}} = \overline{B^{\perp}} = B^{\perp}$ by Theorem 3.9 (iv), that is, $A \perp B$. ∎

Intuitively speaking, $A \cap \overline{B} = \emptyset$ (by the way, this could be an alternative definition) says that A has no element indistinguishable from elements of B.

In quantum logic, the operations $\leq$ and $\perp$ are usually understood as an *implication relation* and a *negation relation* respectively. Thus, we may introduce the concept of *logical incompatibility* just using the idea of orthogonality ([10, p. 12]): A is incompatible with B iff A implies the negation of B, that is iff they are orthogonal. The negation of the relation $\perp$ is called *accessibility* (*ibid.*), written $A \not\perp B$.

All of this show that our structure $\mathfrak{I}$ resembles a non-distributive orthocomplete orthonormal lattice, and it is a Boolean lattice if we consider only the closed qsets. Since every modular ortholattice is orthomodular [10, p. 15], it is an open question whether our lattice has some similarity with modular lattices, that is, $A \leq B$ entails $A \sqcup (C \sqcap B) = (A \sqcup C) \sqcap B$ (we still need to check this and other results).

4   Summing up

We resume here the properties of the quasi-lattice $\mathfrak{I} = \langle \mathcal{P}(U), \mathbf{0}, \mathbf{1}, \sqcap, \sqcup, ^{\perp}, \leq \rangle$:

($\mathfrak{I}$-idempotency) $A \sqcap A = \overline{A}$, $A \sqcup A = \overline{A}$

(Commutativity) $A \sqcap A = B \sqcap A$, $A \sqcup B = B \sqcup A$

(Associativity) $A \sqcap (B \sqcap C) = (A \sqcap B) \sqcap C$, $A \sqcup (B \sqcup C) = (A \sqcup B) \sqcup C$

($\mathfrak{I}$-absorption) $A \sqcap (A \sqcup B) = \overline{A}$, $A \sqcup (A \sqcap B) = \overline{A}$

($\mathfrak{I}$-minimum) $\mathbf{0} \sqcap A = \mathbf{0}$, $\mathbf{0} \sqcup A = \overline{A}$

($\mathfrak{I}$-maximum) $A \sqcap \mathbf{1} = \overline{A}$, $A \sqcup \mathbf{1} = \mathbf{1}$

($\mathfrak{I}$-involution - 1) $A^{\perp \perp} = \overline{A}$

($\Im$-involution - 2) $A \leq B \Rightarrow B^\perp \leq A^\perp$

(Complementation) $A \sqcap A^\perp = \mathbf{0}$, $A \sqcup A^\perp = \mathbf{1}$

($\Im$-absorption -1) $A \sqcup (B \sqcap B^\perp) = \overline{A}$

($\Im$-absorption-2) $A \sqcap (B \sqcup B^\perp) = \overline{A}$

($\Im$-De Morgan) $(A \sqcup B)^\perp = A^\perp \sqcap B^\perp$, $(A \sqcap B)^\perp \subseteq A^\perp \sqcup B^\perp$

($\Im$-orthomodularity) $A \sqcup (A \sqcup B^\perp)^\perp = \overline{B}$.

As we see, it is a rather unusual mathematical structure which resembles the non-distributive ortholattice of quantum mechanics. What the specific $\Im$-properties show is that sometimes we need to consider the closure of a certain qset for getting the desired result. If we interpret the qsets of elements of U as extensions of certain predicates, which might stand for physical properties, the necessity of considering the closure of the qsets show that some fuzzy characteristic of these properties are been shown. In fact, take for instance a qset A as the extension of a certain property P, that is, A should stand for the collection of objects having the property P^7. Then, for instance, if we transform A twice by the operation $^\perp$ ($\Im$-involution - 1), we do not obtain A anymore, but the qset of the indiscernible of its elements. It seems that something is changed when we operate with the collections of objects of the physical systems: we really *transform* them, as we really do with quantum systems. But we remark that the physical interpretation of such a structure and its consequences is still being investigated. For the moment, let us keep with its mathematical counterpart only.

5 The corresponding logic

In this section, we shall be dealing with the first ideas for an alternative axiomatization of a logic that has as its algebraic counterpart the $\Im$-lattice, based on the above assumptions and definitions. We remark once more that this is only a preliminary sketch, and maybe some modifications would need to be done, but let us continue even so. As before, we shall be working within the theory $\mathfrak{Q}$. The concepts introduced below, which mirror the standard ones, can be developed in the "standard part" of $\mathfrak{Q}$, so that we can use the usual mathematical terminology. Here, as before, the equality symbol "$=$" stands for the extensional equality of $\mathfrak{Q}$.

Let us take our algebra $\Im = \langle \mathcal{P}(U), \mathbf{0}, \mathbf{1}, \sqcap, \sqcup, ^\perp \rangle$. Now we shall introduce a generalized (or abstract) logic $\mathcal{L} = \langle F, \mathcal{T}, \curlywedge, \curlyvee, \sim, \rightarrow \rangle$ in the sense of [4],

---

[7]By the way, this is something that is lacking in the usual discussion on quantum theories, that is, a right "semantics", which would enable us to talk of the extension of the relevant predicates.

and we shall continue to use use $\rightarrow$ $\wedge$, $\vee$, $\neg$, $\forall$ and $\exists$ as metalinguistic symbols for implication, conjunction, disjunction, negation, the universal quantifier, and the existential quantifier, respectively. The elements of the $\mathfrak{Q}$-set F will be called *formulas*, and denoted by small Greek letters, while the elements of $\mathcal{T}$ ($\mathcal{T} \subseteq \mathcal{P}(F)$) are the *theories* of $\mathcal{L}$, and denoted by uppercase Greek letters (indices can be used in both cases).

To begin with, let us see how we link such a logic with the quasi-lattice $\mathfrak{I}$. Suppose that there is a valuation $v : F \mapsto \mathcal{P}(U)$ such that:

(i) For any $\alpha \in F$, $v(\alpha) \in \mathcal{P}(U)$;

(ii) $\curlywedge$ and $\curlyvee$ are binary operations on F, and we denote the corresponding images of the pair $\langle \alpha, \beta \rangle$ respectively by $\alpha \curlywedge \beta$ and $\alpha \curlyvee \beta$. These operations obey the following rules:

  (a) $v(\alpha \curlywedge \beta) = v(\alpha) \sqcap v(\beta)$

  (b) $v(\alpha \curlyvee \beta) = v(\alpha) \sqcup v(\beta)$;

(iii) $\sim$ is a mapping from F into F, and we define $v(\sim \alpha) = (v(\alpha))^{\perp}$, for any $\alpha \in F$. This means that if $v(\alpha) = A$, then $v(\sim \alpha) = U - \overline{A}$ according to the above definitions;

(iv) $F \in \mathcal{T}$, this is the trivial theory;

(v) If $\{\Gamma_i\}_{i \in I}$ is a collection of elements of $\mathcal{T}$, then $\bigcap \Gamma_i \in \mathcal{T}$.

It is clear that this definition is an algebraic characterization of our logic $\mathcal{L}$ by means of the lattice $\mathfrak{I}$. Some immediate consequences of this definition are: $v(\alpha \curlywedge \sim \alpha) = \mathbf{0}$, $v(\alpha \curlyvee \sim \alpha) = \mathbf{1}$, $v(\sim\sim \alpha) = v(\alpha)$, etc.

It is well known that in standard quantum logics there is an "implication-problem", to use Dalla Chiara *et al.*'s words [10, p. 164]. That is, all conditional connectives "that can be reasonably introduced" in quantum logics are "anomalous" (*ibid.*), and this was taken by some authors as a motive to criticize quantum logics as not being "real logics". As Dalla Chiara *et al.* say, there are some conditions that a conditional would satisfy to be classified as an implication[8]. These conditions are:

---

[8]Really, several "quantum implications" can be defined, as shown in [22], [23], [27], but we shall not continue with this discussion here. One of the first works (to our knowledge) that proposed an axiomatization of the lattice of quantum mechanics is [16], in which other conditionals are defined. We had no access to this paper, but know it from indirect sources, namely, [11] and [28].

Conditions for an Implication

(i) identity, that is, $\alpha \overset{*}{\to} \alpha$, being $\overset{*}{\to}$ the considered conditional;

(ii) *modus ponens*, that is, is α is true and $\alpha \overset{*}{\to} \beta$ is true, then β is true (*op. cit.*, p. 164);

(iii) In an algebraic semantics, a sufficient condition is: for any structure $\mathcal{A} = \langle A, v \rangle$, $\mathcal{A} \models \alpha \overset{*}{\to} \beta$ iff $v(\alpha) \leq v(\beta)$.

We say that a formula α is true in the structure $\mathfrak{I}$, and write $\mathfrak{I} \models \alpha$ iff $v(\alpha) = 1$, for any valuation v. In this case, $\mathfrak{I}$ is a model of α. We write $\Gamma \models \alpha$ to mean that every model of (the formulas of) Γ is model of α. Finally, α is valid iff it is true in every structure which is an $\mathfrak{I}$-lattice. In this case, we write $\models \alpha$. It is quite obvious that our aim is to prove a completeness theorem for our logic relative to the given semantic, but to do so we need to introduce the concept of deduction from a set of premises. To begin the issue we shall finish only in a forthcoming paper, let us define implication.

DEFINITION 22 ($\mathfrak{I}$-conditional). $\alpha \twoheadrightarrow \beta =_{\mathsf{def}} \beta \curlyvee (\sim \alpha \curlywedge \sim \beta)$

This conditional is quite similar to that one called "Dishkant implication" in [23]. Using the above definitions and Theorem 3.1 (ii), it is immediate to see that $v(\alpha \twoheadrightarrow \beta) = \overline{v(\beta) \cup (v(\alpha)^{\perp} \cap v(\beta)^{\perp})}$. Thus,

$$v(\alpha \twoheadrightarrow \alpha) = \overline{v(\alpha) \cup (v(\alpha)^{\perp} \cap v(\alpha)^{\perp})} = \overline{v(\alpha) \cup v(\alpha)^{\perp}} = \overline{1} = 1,$$

for $1 = U$ is closed. So, $\models \alpha \twoheadrightarrow \alpha$. Furthermore, if $v(\alpha) = 1$ and $v(\alpha \twoheadrightarrow \beta) = 1$, then $\overline{v(\beta) \cup (v(\alpha)^{\perp} \cap v(\beta)^{\perp})} = 1$ and, since $v(\alpha) = 1$, we get that $v(\beta) = 1$. Thus, our conditional obeys conditions (i) and (ii) of the Conditions for an Implication. In addition, we can see that condition (iii) is also fulfilled. In fact, by the hypothesis, we have $\models \alpha \twoheadrightarrow \beta$, so $v(\beta \curlyvee (\sim \alpha \curlywedge \sim \beta)) = 1$. Call $v(\alpha) = A$ and $v(\beta) = B$. Then $B \sqcup \overline{(A^{\perp} \cap B^{\perp})} = U$, that is, $\overline{B} \cup \overline{(A^{\perp} \cap B^{\perp})} = U$. For this equality to hold, we need that $\overline{(A^{\perp} \cap B^{\perp})} = \overline{B}^{\perp} = B^{\perp}$. Then $A^{\perp} \cap B^{\perp} = B^{\perp}$, so $B^{\perp} \subseteq A^{\perp}$, that is (for $\mathfrak{Q}$-sets), $A \subseteq B$. By Theorem 3.9 (vi), $A \leq B$, that is, $v(\alpha) \leq v(\beta)$.

Let us make a further remark on this definition. We say that $A \in \mathcal{P}(U)$ is *definable* by a formula $\alpha \in F$ if $v(\alpha) = A$. Let β be such that $v(\beta) = \overline{A}$. Is there such a β? The answer is in the affirmative. Since $A \subseteq \overline{A}$, then $v(\alpha) \leq v(\beta)$, hence by condition (iii) above, $\models \alpha \twoheadrightarrow \beta$. So, β is any formula implied by α. This affirmative makes sense, for $v(\alpha \twoheadrightarrow \beta) = 1$ and $v(\beta) = \overline{A}$ say that $\overline{\overline{A} \sqcup (A^{\perp} \sqcap \overline{A}^{\perp})} = U$, that is, $\overline{\overline{A} \cup \overline{(A^{\perp} \cap \overline{A}^{\perp})}} = A \cup (\overline{A^{\perp} \cap \overline{A}^{\perp}}) =$

$\overline{A} \cup \overline{A}^{\perp} = U$. This fact will be important for the definition of the connectives of the logic $\mathcal{L}$. Finally, let us say that $\alpha \longleftrightarrow \beta =_{\mathsf{def}} (\alpha \rightarrow \beta) \curlywedge (\beta \rightarrow \alpha)$.

Next we introduce the notion of syntactical consequence from a set of premises, written $\Gamma \vdash \alpha$, as follows, where $v(\Gamma) = \bigcup [v(\alpha) : \alpha \in \Gamma]$ (the terminology is from $\mathfrak{Q}$ – see again Section 2, if necessary).

DEFINITION 23 (Syntactical Consequence). $\Gamma \vdash \alpha$ iff any theory containing Γ (really, the formulas of Γ) contains α.

Let $\vdash \alpha$ abbreviates $\emptyset \vdash \alpha$, while $\alpha \vdash \beta$ abbreviates $\{\alpha\} \vdash \beta$ (recall that they are $\mathfrak{Q}$-sets, so the standard notation can be used), and $\Gamma \nvdash \alpha$ says that it is not the case that $\Gamma \vdash \alpha$. It is immediate to prove the following theorem:

THEOREM 24. *In $\mathcal{L}$, we have*

(i) $\alpha \in \Gamma \Rightarrow \Gamma \vdash \alpha$. In particular, $\alpha \vdash \alpha$;

(ii) $\Gamma \vdash \alpha \Rightarrow \Gamma \cup \Delta \vdash \alpha$;

(iii) If $\Gamma \vdash \alpha$ and for every $\beta \in \Gamma$, we have that $\Delta \vdash \beta$, then $\Delta \vdash \alpha$;

(iv) If $\{\Gamma_i\}_{i \in I}$ is a family of subqsets of F such that for every α, $\alpha \in \Gamma \leftrightarrow \Gamma_i \vdash \alpha$, then $\forall \alpha (\alpha \in \bigcap_{i \in I} \Gamma_i \leftrightarrow \bigcap_{i \in I} \Gamma_i \vdash \alpha)$.

Proof. Immediate, for the definition of consequence is standard (see [4], [24]). ∎

We shall not continue to develop the syntactical aspects of this logic (in algebraic terms, but see [4]), but just try to link it with the semantic aspects sketched above. The least theory containig α is denoted T_α, and it coincides with the intersection of all theories containing α (*op. cit.*). Thus, $\Gamma \vdash \alpha$ iff $v(\alpha) \subseteq v(T_\alpha)$. In particular, if Γ is a theory, that is, $\mathrm{Cn}(\Gamma) =_{\mathsf{def}} [\alpha : \Gamma \vdash \alpha] = \Gamma$, then $v(\alpha) \subseteq v(\Gamma)$, and in particular $v(\alpha) \subseteq \overline{v(\Gamma)}$. Finally, let us recall that since no deduction theorem holds in quantum logics [22], the same seems to happen here due to the nature of our implication (but this is still a open problem).

The last point of this paper, which conduces us to another work, is the question: how to characterize the logic $\mathcal{L}$ axiomatically? We shall follow the approach of generalized logics in the sense of [4], but not here.

Acknowledgements

This work has been sponsored by FAPESP. The article was written while Hércules Feitosa was at UFSC doing post-doctoral research.

BIBLIOGRAPHY

[1] Birkhoff, G. and von Neumann, J. The logic of quantum mechanics. *Annals of Mathematics*, v. 37, 1936, p. 823-843.

[2] Church, A. Review of 'The logic of quantum mechanics'. *Journal of Symbolic Logic*, v. 2, n. 1, 1937, p. 44-45.

[3] Castellani, E. (Ed.) *Interpreting Bodies: Classical and Quantum Objects in Modern Physics*. Princeton: Princeton Un. Press, 1998.

[4] da Costa, N. C. A. *Generalized Logics*, v. 2. Preliminary Version. Florianópolis: Federal University of Santa Catarina, 2006.

[5] da Costa, N. C. A. *Ensaio sobre os Fundamentos da Lógica*. São Paulo: Hucitec-EdUSP, 1980.

[6] da Costa, N. C. A. and Krause, D. Schrödinger logic. *Studia logica*, v. 53, n. 4, 1994, p. 533-50.

[7] da Costa, N. C. A. and Krause, D. An intensional Schrödinger logic. *Notre Dame Journal of Formal Logic*, v. 38, n. 2, 1997, p. 179-194.

[8] da Costa, N. C. A. and Krause, D. Logical and Philosophical Remarks on Quasi-Set Theory. *Logic Journal of the IGPL*, v. 15, 2007, p. 1-20.

[9] Dalla Chiara, M. L., Giuntini, R. and Krause, D. *Quasiset theories for microobjects: a comparision*. In Castellani 1998, p. 142-152.

[10] Dalla Chiara, M. L., Giuntini, R. and Greechie, R. *Reasoning in Quantum Theory: Sharp and Unsharp Quantm Logics*. Dordrech: Kluwer Ac. Pu., 2004.

[11] Drieschner, M. Review of J. Kotas 'Axioms for Birkhoff-von Neumann quantum logic'. *Journal of Symbolic Logic*, v. 40, n. 3, 1975, p. 463-464.

[12] Falkenburg, B. *Particle Metaphysics: A Critical Account of Subatomic Reality*. New York: Springer, 2007.

[13] Feitosa, H. A. Traduções conservativas (Conservative translations). Campinas: Tese (Doutorado em Lógica e Filosofia da Ciência) - Instituto de Filosofia e Ciências Humanas, Universidade Estadual de Campinas, 1997. 161 p.

[14] French, S. *On whitering away of physical objects*. In Castellani (Ed.), p. 93-113.

[15] French, S. and Krause, D. *Identity in Physics: A Historical, Philosophical, and Formal Analysis*. Oxford: Oxford Un. Press, 2006.

[16] Kotas, J. An axiom system for modular logic. *Studia Logica*, v. 21, n. 1, 1967, p. 17-37.

[17] Krause, D. On a quasi-set theory. *Notre Dame Journal of Formal Logic*, v. 33, 1992, p. 402-411.

[18] Krause, D. Axioms for collections of indistinguishable objects. *Logique et Analyse*, v. 153-154, 1996, p. 69-93.

[19] Krause, D. Why quasi-sets? *Boletim da Sociedade Paranaense de Matemática*, v. 20, n. 1/2, 2002, p. 73-92.

[20] Krause, D., Sant'Anna, A. S. and Sartorelli. A. On the concept of indentity in Zermelo-Fraenkel-like axioms and its relationships with quantum statistics. *Logique et Analyse*, v. 48, (189-192), 2005, p. 231-260.

[21] Leibniz, G. W. On the Principle of Indiscernibles. In Leibniz, G. W., *Philosophical Writings*. Vermont: Everyman, 1995, p. 133-135.

[22] Malinowski, J. The deduction theorem for quantum logic – some negative results. *Journal of Symbolic Logic*, v. 55, n. 2, 1990, p. 615-625.

[23] Megill, N. D. and Pavičić, M. Quantum implication algebras. *Int. J. Theor. Physics*, v. 42, n.12, 2003, p. 1-21.

[24] Mendelson, E. *Introduction to Mathematical Logic*. New York: Chapman & Hall, 4th. ed., 1997.

[25] Post, H. Individuality and physics. *The Listener*, v. 10, October, 1963, p. 534-537, reprinted in *Vedanta for East and West*, v. 132, 1973, p. 14-22.

[26] Rédei, M. The birth of quantum logic. *History and Philosophy of Logic*, v. 28, May 2007, p. 107-122.

[27] Román, L. A characterization of quantic quantifiers in orthomodular lattices. *Theory and Applications of Categories*, v. 16, n. 10, 2006, p. 206-217.

[28] Sánchez, C. H. La lógica de la mecánica cuántica. *Lecturas Mathemáticas*, v. 1, (6), n. 1, 2, 3, 1980, p. 17-42.

[29] Schrödinger, E. *Science and Humanism*. Cambridge: Cambridge Un. Press, Cambridge, 1952.

[30] Schrödinger, E. *What is an elementary particle?* Reprinted in Castellani 1998, p. 197-210.

[31] van Fraassen, B. *The problem of indistinguishable particles*. In Castellani 1998, p. 73-92.

Décio Krause
Department of Philosophy
UFSC
Brazil
E-mail: deciokrause@gmail.com

Hércules de Araujo Feitosa
Department of Mathematics
UNESP
Brazil
E-mail: haf@fc.unesp.br

SECTION 6

LOGIC AND THE QUESTION OF TRUTH-VALUES

Bivalent Semantics for De Morgan Logic (The Uselessness of Four-valuedness)

JEAN-YVES BÉZIAU

Dedicated to Newton da Costa for his 79th birthday

ABSTRACT.   In this paper we present a bivalent semantics for De Morgan logic in the spirit of da Costa's theory of valuation showing therefore the uselessness of four-valuedness - the four-valued Dunn-Belnap semantics being ordinarily used to characterize De Morgan logic. We also present De Morgan logic in the perspective of universal logic, showing how some general results connecting bivaluations to sequent rules and reducing many-valued matrices to non-truth functional bivalent semantics work.

1   De Morgan logic in the perspective of universal logic

In this paper we present a systematic study of a very simple and nice logical structure, De Morgan logic. This is a logic with a negation which is both paraconsistent and paracomplete, that is to say neither the principle of contradiction, nor the principle of excluded middle are valid for De Morgan negation, but all De Morgan laws hold, the reason for the name. This logic shows therefore the independence of the principle of contradiction and the principle of excluded middle relatively to De Morgan laws [1].

This logic is not new. It is connected with De Morgan lattices which can be traced back to Moisil [22] and which have been called quasi-boolean algebras by Rasiowa, [13], distributive i-lattices by Kalman [20], and have been especially studied by the school of Antonio Monteiro in Bahia-Blanca, Argentina [23].

---

[1]Negations which are both paraconsistent and paracomplete were called by da Costa *non-alethic*, we have proposed to used instead the adjective *paranormal* in order to keep the paraterminology. De Morgan negation is a good example of paranormal negation, it can reasonably be considered as a negation, due to the fact that it obeys De Morgan laws.

If we factor a De Morgan logic by the relation of logical equivalence, we get a De Morgan lattice, and an intuitive semantics for De Morgan logic is a matrix corresponding to a very De Morgan lattice, a four-valued lattice. This kind of semantics is connected with Dunn-Belnap four-valued semantics [3]. Michael Dunn especially made the connections between De Morgan lattices and logical systems [17]. This connection was later on studied by J.M.Font and V.Verdu [19] [18], O.Arielli and A.Avron [2] and A.P.Pynko [26].

We are facing here a phenomenon similar to the one happening with classical propositional logic: at the same time the factor structure of classical propositional logic is a Boolean algebra and its matrix semantics is the smallest Boolean algebra.

We show in this paper that it is also possible to construct a bivalent semantics for De Morgan logic, along the line of Newton da Costa's theory of valuation [16]: it is a non-truth functional bivalent semantics. And we also show how to establish the connection between this bivalent semantics and the four-valued De Morgan matrix.

In some previous papers we have shown how we can establish a close connection between non-truth functional bivalent semantics and sequent rules [12]. We also have proven some general results of reduction of semantics to bivalent semantics [7]. This is part of a general study of logical structures, we have called *universal logic* [5].

This paper is the opportunity to illustrate these general results of universal logic, as we already did for Łukasiewicz logic $L3$ [9] and also to point out some problems that could be usefully clarified by universal logic.

2   De Morgan logical structure

DEFINITION 1. A *De Morgan logic* is a structure $\mathcal{M} = \langle \mathcal{F}; \vdash \rangle$, where

- $\mathcal{F}$ is an is an absolutely free algebra $\langle \mathbb{F}; \wedge, \vee, \neg \rangle$ whose domain $\mathbb{F}$ is generated by the functions $\wedge, \vee, \neg$ from a set of atomic formulas $\mathbb{A} \subset \mathbb{F}$.

- $\vdash$ is a structural consequence relation obeying, besides the usual axioms for classical conjunction and disjunction, the following axioms:

$[\neg\wedge] \ \neg(a \wedge b) \dashv\vdash \neg a \vee \neg b$
$[\neg\vee] \ \neg(a \vee b) \dashv\vdash \neg a \wedge \neg b$
$[\neg\neg] \ a \dashv\vdash \neg\neg a$
where $x \dashv\vdash y$ means $x \vdash y$ and $y \vdash x$.

Defining a logic in this way is typical of the Polish approach [24], although in Poland people generally prefer to use the notion of consequence operator rather than the notion of consequence relation, but this is trivially equivalent. The notion of structural consequence operator is due to Łoś and Suszko [21] and it is the continuation of Tarski's theory of consequence operator which began at the end of the 1920s [27].

One has to be aware that such kind of definition is not proof-theoretical, it is not a system of deduction. The use of the word *axiom* and of the symbol ⊢ may lead to such kind of confusion. But as it is known the word *axiom* is also used in model theory. Such a definition has to be seen as a definition of for example the model-theoretical definition of structures of order. What we call a De Morgan logic is a model of the above group of axioms which defines the meaning of the relation ⊢ and of the functions ∧, ∨, ¬.

The kind of definition we are using here is of the same type for example as the definition of a De Morgan lattice. Moreover there is a strong connection between these two definitions, since the factorization of a De Morgan logic by the relation ⊣⊢, which is a congruence, is a De Morgan lattice.

Let us recall the definition of a De Morgan lattice.

DEFINITION 2. A *De Morgan lattice* is a distributive lattice $\langle \mathbb{E}; \cap, \cup, \sim \rangle$ where the unary operator obeys the two following axioms:

$\sim (a \cap b) = \sim a \cup \sim b$

$a = \sim \sim a$

3 Sequent System for De Morgan logic

DEFINITION 3. A *De Morgan sequent system* $\mathcal{LM}$ is a sequent system which has the same rules for conjunction and disjunction as the sequent system for classical propositional logic and which has the following rules for negation [2]:

$$\frac{\neg a \Rightarrow \qquad \neg b \Rightarrow}{\neg(a \wedge b) \Rightarrow} \; [\neg_l \wedge] \qquad\qquad \frac{\Rightarrow \neg a, \neg b}{\Rightarrow \neg(a \wedge b)} \; [\neg_r \wedge]$$

$$\frac{\neg a, \neg b \Rightarrow}{\neg(a \vee b) \Rightarrow} \; [\neg_l \vee] \qquad\qquad \frac{\Rightarrow \neg a \qquad \Rightarrow \neg b}{\Rightarrow \neg(a \vee b)} \; [\neg_r \vee]$$

$$\frac{a \Rightarrow}{\neg\neg a \Rightarrow} \; [\neg_l \neg] \qquad\qquad \frac{\Rightarrow a}{\Rightarrow \neg\neg a} \; [\neg_r \neg]$$

---

[2]We don't know who was the first to present such sequent rules, but they can be found in particular in [19] [1] [26].

To make things clearer we have written the rules without the contexts, but $\mathcal{LM}$ is contextually standard. It has also the same structural rules as the system for classical propositional logic, including the cut rule. Cut-elimination for $\mathcal{LM}$ can easily be proven, following a method similar to the one presented in [4].

The rules of $\mathcal{LM}$ don't have the subformula property, but they have something analogous: the subnegformula property. A subnegformula of a formula is a proper subformula or a negation of a proper subformula.

From cut-elimination and the subnegformula property it results the decidability of the logical structure generated in the usual way by $\mathcal{LM}$.

THEOREM 4. *A logical structure generated by a $\mathcal{LM}$ system of sequents is a De Morgan logic.*

Proof. We know that a logical structure generated by a structurally standard sequent system is a structural consequence relation [3]. We have to show furthermore that all axioms of Definiton 1 are valid in a logic generated by a $\mathcal{LM}$ system of sequents. We will study just the case of axiom $[\neg\wedge]$. It is enough to prove that the sequents $\neg(a \wedge b) \Rightarrow \neg a \vee \neg b$ and $\neg a \vee \neg b \Rightarrow \neg(a \wedge b)$ are derivable in a $\mathcal{LM}$ system of sequents.

$$\cfrac{\cfrac{\neg a \Rightarrow \neg a, \neg b}{\neg a \Rightarrow \neg a \vee \neg b} \qquad \cfrac{\neg b \Rightarrow \neg a, \neg b}{\neg b \Rightarrow \neg a \vee \neg b}}{\neg(a \wedge b) \Rightarrow \neg a \vee \neg b} \; [\neg_l \wedge]$$

$$\cfrac{\cfrac{\neg a \Rightarrow \neg a \qquad \neg b \Rightarrow \neg b}{\neg a \vee \neg b \Rightarrow \neg a, \neg b}}{\neg a \vee \neg b \Rightarrow \neg(a \wedge b)} \; [\neg_r \wedge]$$

∎

What is more difficult is to prove the converse of this theorem:

THEOREM 5. *A De Morgan logic can be generated by a $\mathcal{LM}$ system of sequents.*

Proof. To prove this theorem, there are two parts.

---

[3]*Structural* here is used in two different ways. A structural consequence relation is a relation invariant by substitutions. Using schema of rules we necessarily generate such a structural consequence relation. A structurally standard sequent system is a system having all the structural rules and standard contextual behaviour – see [12].

- The first part consists in a completeness theorem between the Definition 1 of De Morgan logic and a sequent systems close to it, $\mathcal{LM}_0$, which is a system of sequents with the standard structural rules, and with rules which are the most direct translations of the axioms for negation into sequent rules, for example the (model theoretical) axiom $[\neg\wedge]$ of Definition 1 is translated into the two following (proof-theoretical) axioms of the sequent system $\mathcal{LM}_0$: $\neg(a \wedge b) \Rightarrow \neg a \vee \neg b$ and $\neg a \vee \neg b \Rightarrow \neg(a \wedge b)$.

- The second part is just to prove that the rules of $\mathcal{LM}$ are derivable rules of $\mathcal{LM}_0$.

In this paper we will not deal with the first part of this proof. It is possible to prove a general completeness theorem connecting logical structures to systems of sequents. This is universal logic. Details of such connection will be given in another paper.

The second part of the proof is quite easy. We present the part showing how we can derive in $\mathcal{LM}$ the rules $[\neg_{l\wedge}]$ and $[\neg_{r\wedge}]$ from the axiom $[\neg\wedge]$.

$$\frac{\neg(a \wedge b) \Rightarrow \neg a \vee \neg b \; [\text{Axiom}] \qquad \dfrac{\neg a \Rightarrow \qquad \neg b \Rightarrow}{\neg a \vee \neg b \Rightarrow} \; [cut]}{\neg(a \wedge b) \Rightarrow}$$

$$\frac{\dfrac{\Rightarrow \neg a, \neg b}{\Rightarrow \neg a \vee \neg b} \qquad \neg a \vee \neg b \Rightarrow \neg(a \wedge b) \; [\text{Axiom}] \quad [cut]}{\Rightarrow \neg(a \wedge b)}$$

■

4 Bivalent semantics for De Morgan logic

4.1 A non-truth functional bivalent semantics

DEFINITION 6. We consider the set of functions $\mathbb{B}$ from $\mathbb{F}$ to the set $\{0, 1\}$ defined by the usual conditions for conjunction and disjunction and the following conditions for negation:

$[\![\neg\wedge]\!]$ $\beta(\neg(a \wedge b)) = 1$ iff $\beta(\neg a) = 1$ or $\beta(\neg b) = 1$
$[\![\neg\vee]\!]$ $\beta(\neg(a \vee b)) = 1$ iff $\beta(\neg a) = 1$ and $\beta(\neg b) = 1$
$[\![\neg\neg]\!]$ $\beta(\neg\neg a) = 1$ iff $\beta(a) = 1$

Note that this is typically a generalized bivalent semantics in the sense of da Costa [16] : this set of bivaluations is not generated by distributions of truth-values for atomic formulas, and it is not a set of homomorphisms between the set of formulas and some algebra of similar type of a logical matrix.

DEFINITION 7. The semantical consequence relation $\models_2$ is defined in the usual way:

$T \models_2 k$ iff for every $\beta \in \mathbb{B}$, if $\beta(j) = 1$ for every $j \in T$, then $\beta(k) = 1$.

4.2   Truth-Tables

Following the idea of da Costa and Alves [15], we can build some truth-tables based on this semantics.

DEFINITION 8. A truth-table is a table with a finite number of columns and lines such that that on the first line in each column we have a formula and on the other lines (proper lines) in each column we have 0 or 1 obeying the following conditions:

- each proper line of the table can be extended into a bivaluation β from $\mathbb{F}$ to $\{0, 1\}$

- for any bivaluation β, there is a proper line of the table such that $\beta(x) = y$, for any x, x being a formula given by the first line, and y being 0 or 1 according to the given line.

EXAMPLE 9. The following truth-table shows that for the atomic formulas p and q we have $\neg(p \wedge q) \models_2 \neg p \vee \neg q$ and $\neg p \vee \neg q \models_2 \neg(p \wedge q)$:

p	q	$\neg p$	$\neg q$	$p \wedge q$	$\neg(p \wedge q)$	$\neg p \vee \neg q$
0	0	0	0	0	0	0
0	0	0	1	0	1	1
0	0	1	0	0	1	1
0	0	1	1	0	1	1
0	1	0	0	0	0	0
0	1	0	1	0	1	1
0	1	1	0	0	1	1
0	1	1	1	0	1	1
1	0	0	0	0	0	0
1	0	0	1	0	1	1
1	0	1	0	0	1	1
1	0	1	1	0	1	1
1	1	0	0	1	0	0
1	1	0	1	1	1	1
1	1	1	0	1	1	1
1	1	1	1	1	1	1

This table has some additional features, corresponding to the following definition:

DEFINITION 10. A truth-table is said to be full if the set of formulas of the first line is closed by the subnegformula property, i.e. it contains all proper subformulas and negations of proper subformulas of this set of formulas.

THEOREM 11. *For any formula it is possible to construct a full truth-table having on the first line the formula and the set of its subnegformulas.*

Proof. We have first to give a method to build a table and then to show that this table is a full truth-table. This can been done along the same line as for the construction of truth-tables for the paraconsistent logic C1, see e.g. [4]. ∎

5 Adequacy of the bivalent semantics

THEOREM 12 (Soundness). *If $T \vdash k$ then $T \models_2 k$*

Proof. Straightforward, we leave it to the reader. ∎

THEOREM 13 (Completeness). *If $T \nvdash k$ then $T \not\models_2 k$*

Proof. If $T \nvdash k$ then there is, according to Lindenbaum-Asser theorem [25] [11] a relatively maximal extension of T in k, i.e. a set of formulas V such that

- $T \subseteq V$

- $V \not\vdash k$

- $W \vdash k$ for any strict extension W of V.

We just have to show that the characteristic function β_V of V is a bivaluation, then we will have $T \not\models_2 k$.
We show that the characteristic function β_V of V obeys the condition
  $[\![\neg\wedge]\!]\ \beta(\neg(a \wedge b)) = 1$ iff $\beta(\neg a) = 1$ or $\beta(\neg b) = 1$
leaving the other cases for the reader.
  If $\beta_V(\neg a) = 1$ or $\beta_V(\neg b) = 1$, then $V \vdash \neg a$ or $V \vdash \neg b$, then by the rule $[\neg_r\wedge]$, $V \vdash \neg(a \wedge b)$ then $\beta_V\neg(a \wedge b) = 1$
  If $\beta_V(\neg a) = 0$ and $\beta_V(\neg b) = 0$, then $V \not\vdash \neg a$ and $V \not\vdash \neg b$, then, since V is maximal in k, $V, \neg a \vdash k$ and $V, \neg b \vdash k$, then by the rule $[\neg_l\wedge]$, $V, \neg(a\wedge b) \vdash k$, therefore $V \not\vdash \neg(a \wedge b)$, therefore $\beta_V\neg(a \wedge b) = 0$.   ∎

6   Four-valued semantics for de Morgan logic

6.1   The four-valued De Morgan matrix

DEFINITION 14.   We consider a set of four values $\{0^-, 0^+, 1^-, 1^+\}$ partially ordered as following
  $0^- \prec 0^+ \prec 1^+$
  $0^- \prec 1^- \prec 1^+$
Using this partially order we define the following three functions from $\{0^-, 0^+, 1^-, 1^+\}$ to $\{0^-, 0^+, 1^-, 1^+\}$.

$\vec{\wedge}(x, y) = inf(x, y)$

$\vec{\vee}(x, y) = sup(x, y)$

$\vec{\neg}(x) = x$ if $x \in \{0^+, 1^-\}$
$\vec{\neg}(0^-) = 1^+$
$\vec{\neg}(1^+) = 0^-$

The structure $\langle \{0^-, 0^+, 1^-, 1^+\}; \vec{\wedge}, \vec{\vee}, \vec{\neg}\rangle$ is a finite De Morgan lattice, we will call it the four-valued De Morgan lattice.

The functions $\vec{\wedge}, \vec{\vee}, \vec{\neg}$ can be called truth-functions and be defined more visually by the following truth-tables.

$\bar{\wedge}$	0^-	0^+	1^-	1^+
0^-	0^-	0^-	0^-	0^-
0^+	0^-	0^+	0^-	0^+
1^-	0^-	0^-	1^-	1^-
1^+	0^-	0^+	1^-	1^+

TRUTH-TABLE FOR CONJUNCTION

$\bar{\vee}$	0^-	0^+	1^-	1^+
0^-	0^-	0^+	1^-	1^+
0^+	0^+	0^+	1^+	1^+
1^-	1^-	1^+	1^-	1^+
1^+	1^+	1^+	1^+	1^+

TRUTH-TABLE FOR DISJUNCTION

	$\bar{\neg}$
0^-	1^+
0^+	0^+
1^-	1^-
1^+	0^-

TRUTH-TABLE FOR NEGATION

DEFINITION 15. We consider the set of homomorphisms between the absolutely free algebra of formulas and the four-valued De Morgan lattice. This defines a set of quadrivaluations $\mathbb{T}$ from $\mathbb{F}$ to $\{0^-, 0^+, 1^-, 1^+\}$.

DEFINITION 16. We consider as a logical matrix the four-valued de Morgan lattice together with $\{1^-, 1^+\}$ as set of designated values.

DEFINITION 17. The semantical consequence relation $\models_4$ is defined in the usual way:

$T \models_4 k$ iff for every $\theta \in \mathbb{T}$, if $\theta(j)$ is designated for every $j \in T$, then $\theta(k)$ is designated.

This four-valued semantics is sound and complete for De Morgan logic. There are several ways to prove that. We will use here a translation between the four-valued semantics into the two-valued semantics.

What is interesting with this four-valued semantics is that it is based on an algebra which is similar to the algebra which is the factor structure of a De Morgan logical structure we get using the relation of logical equivalence. For the semantics however what is used is a finite De Morgan lattice, the four-valued one.

6.2 Translation of the four-valued semantics into the two-valued semantics

A way to reduce any semantics to a non-truth-functional bivalent semantics has been studied in particular in [7]. The example of reduction presented here is a straightforward applications of this method.

DEFINITION 18. Given a function β of the bivalent semantics, we define a quadrivaluation θ_β in the following way

$\theta_\beta(k) = 0^-$ iff $\beta(k) = 0$ and $\beta(\neg k) = 1$
$\theta_\beta(k) = 0^+$ iff $\beta(k) = 0$ and $\beta(\neg k) = 0$
$\theta_\beta(k) = 1^-$ iff $\beta(k) = 1$ and $\beta(\neg k) = 1$
$\theta_\beta(k) = 1^+$ iff $\beta(k) = 1$ and $\beta(\neg k) = 0$

THEOREM 19. *This is a bijection between* $\mathbb{B}$ *and* $\mathbb{T}$ *such that:*
$\theta_\beta(k)$ *is designated iff* $\beta(k) = 1$.

Proof. Straightforward on the complexity of formulas.   ∎

As a corollary we have the following result:

THEOREM 20. $T \models_2 k$ *iff* $T \models_4 k$.

The fact that the four-valued matrix semantics for De Morgan logic can be reduced to a bivalent semantics does not mean that the four-valued matrix semantics has no interest - for a discussion about this, see [14].

The subtitle of our paper is a kind of joke referring to the paper by O. Arielli and A. Avron, called "The value of four values" [2]. Nevertheless it is true that in some sense we don't need four-values to deal with De Morgan logic.

Acknowledgements: Work supported by a grant of CNPq-DCR-FUNCAP. We are grateful for comments of anonymous referees.

BIBLIOGRAPHY

[1] O. Arielli and A. Avron, 1994, "Logical billatices and inconsistent data", *Proceedings of the 9th IEEE Annual Symposium on Logic in Computer Science*, IEEE Press, pp.468–476..
[2] O. Arielli and A. Avron, 1998, "The value of four values", *Artificial Intelligence*, **102**, 97–141.

[3] N.D. Belnap, 1977, "A useful four-valued logic", in M.Dunn (ed), *Modern uses of multiple-valued logic*, Reidel, Boston, 1977, pp.8–37.

[4] J.-Y. Béziau, 1993, "Nouveaux résultats et nouveau regard sur la logique paraconsistante C_1", *Logique et Analyse*, **36**, 45–58.

[5] J.-Y. Béziau, 1994, "Universal Logic", in *Logica '94 - Proceedings of the 8th International Symposium*, T.Childers and O.Majers (eds), Czech Academy of Science, Prague, pp.73–93.

[6] J.-Y. Béziau, 1995, *Recherches sur la logique universelle.*, Ph.D. thesis, Department of Mathematics, University of Paris 7, Paris.

[7] J.-Y. Béziau, 1998, "Recherches sur la logique abstraite: les logiques normales", *Acta Universitatis Wratislaviensis*, **18**, 105–114.

[8] J.-Y. Béziau, 1998, "De Morgan lattices, paraconsistency and the excluded middle", *Boletim da Sociedade Paranaense de Matemática*, **18**, 169–172.

[9] J.-Y. Beziau, 1999, "A sequent calculus for Łukasiewicz's three valued logic based on Suszko's bivalent semantics", *Bulletin of the Section of Logic*, **28**, 89–97.

[10] J.-Y. Béziau, 1999, "What is paraconsistent logic ?", in *Frontiers of paraconsistent logic*, Research Studies Press, Baldock, 2000, pp.95-112.

[11] J.-Y. Béziau, 1999, "La véritable portée du théorème de Lindenbaum-Asser". *Logique et Analyse*, **167–168** (1999), 341–359.

[12] J.-Y. Béziau, 2001, "Sequents and bivaluations", *Logique et Analyse*, **176** (2001), 373–394.

[13] A. Bialynicki-Birula and H. Rasiowa, 1957, "On the representation of quasi-boolean algebra", *Bulletin de l'Académie Polonaise des Sciences, Classe III*, **5**, 259–261.

[14] C. Caleiro, W. Carnielli, M. Coniglio and J. Marcos, 2007, "The 'Humbug' of many logical values", in J.-Y. Beziau (ed), *Logica Universalis - Towards a general theory of logic*, Second Edition, Birkhäuser, Basel, 2007, pp.175–194

[15] N.C.A. da Costa and E.H. Alves, 1977, "A semantical analysis of the Calculi C_n", *Notre Dame Journal of Formal Logic*, **18**, 621–630.

[16] N.C.A. da Costa and J.-Y. Béziau, 1994, "Théorie de la valuation", *Logique et Analyse*, **37**, 95–117.

[17] J.M. Dunn, 1966 *The algebra of intensional logics*, Ph.D. thesis, University of Pittsburgh, Ann Arbor.

[18] J.-M. Font, 1997, "Belnap's four-valued logic and De Morgan lattices", *Logic Journal of the IGPL*, **5**, 413–440.

[19] J.-M. Font and V. Verdú, 1989, "Completeness theorems for a four-valued logic related to De Morgan lattices", *Mathematics Preprint Series*, **57**, Barcelona.

[20] J.A. Kalman, 1958, "Lattices with involution", *Transactions of the American Mathematical Society*, **87**, 485–491.

[21] J. Łoś and R. Suszko, 1958, "Remarks on sentential logics", *Indigationes Mathematicae* , **20**, 177–183.

[22] G.C. Moisil, 1935, "Recherche sur l'algèbre, de la logique", *Annales Scientifiques de l'Université de Jassy*, **22**, 1–117.

[23] A. Monteiro, 1960, "Matrices de Morgan caractéristiques pour le calcul propositionnel classique", *Anais da Academia Brasileira de Ciências*, **32**, 1–7.

[24] W.A. Pogorzelski and P. Wojtylak, 2001, "Cn-Definitions of Propositional Connectives", *Studia Logica*, **67**, 1–26.

[25] W.A. Pogorzelski and P. Wojtylak, 2007, *Completeness theory for propositional logics*, Birkhäuser, Basel.

[26] A.P. Pynko, 1995, "Characterizing Belnap's logic via De Morgan's laws", *Mathematical Logic Quarterly*, **44**, 442–454.

[27] A. Tarski, 1928, "Remarques sur les notions fondamentales de la méthodologie des mathématiques", *"Annales de la Société Polonaise de Mathématiques*, **6**, 270–271.

Jean-Yves Béziau
Department of Philosophy
University of Fortaleza
Fortaleza, Brazil
E-mail: jyb@ufc.br

A Four-Valued Logic for Reasoning about Finite and Infinite Computation Errors in Programs

BEATA KONIKOWSKA

ABSTRACT. The paper proposes a four-valued logic which can be employed in program specification and validation to support reasoning about two kinds of computation errors: finite, "machine" errors signalled by the computer and "infinite", undecidable errors resulting from infinite computations. The logic introduced here is underpinned by Kleene logic, used for handling machine errors and by McCarthy logic, employed for dealing with infinite errors. A sound and complete, decompositional proof system in Rasiowa-Sikorski style operating on sequences of signed formulas is developed for the logic, and then transformed into an equivalent Gentzen-style sequent calculus of ordinary formulas.

1 Introduction

To ensure that programs operate correctly and return the expected results, we need to be able to specify them correctly and to reason about their properties, like partial or total correctness, in a precise and reliable way. As in the practice of computing programs sometimes run into error states instead of producing useful results, any logic used for program specification and validation should also take into consideration the fact that a program sometimes fails to return the expected result, or any result at all, due to the occurrence of some computation error. Basically, this is done in two ways. The first is considering a partial logic, with formulas getting no value in case of a computation error — like in [BCJ84, Hog87, Owe85]. The second is to consider a three-valued logic with the third, "undefined" value representing a computation error — see e.g. [MC67, Bli91, KTB91, Ko93].

The second approach gives more flexibility in reasoning about computation errors, for it allows us to define the three-valued semantics of the considered program logic in the way best tailoured to the intended treatment of such errors. The drawback is that all possible computation errors are bundled together under the third logical value and are handled in the same way. However, in fact there are two distinct types of computation errors, of inherently different characters:

- so-called "machine" errors, signalled immediately by the computer itself (e.g. a function applied to an argument outside its domain, a syntax error, etc.); and

- infinite computations, which represent infinite loops generated by the program.

The difference between them is quite fundamental: machine errors can be seen as "finite", and by being signalled give us a chance of implementing some corrective action once we know they have occurred, while infinite computations are not recognizable by their very nature. Indeed, no matter how long the computation continues, in general we cannot tell for sure whether it will continue forever or whether it will finally terminate — maybe even in the next step. On a more abstract level, we know that the halting problem for programs is undecidable. In the present-day software terminology, the two kinds of errors have different severity levels, with infinite computations representing truly critical errors.

Clearly, the above implies that to ensure the optimum handling of computation errors, we should distinguish between machine errors and infinite errors and treat them in different ways. As the usual approach is to bundle them together and consider one type of error only, the aim of this paper is to remedy this deficiency by proposing a logic able to distinguish between both the above types of errors, and to handle them in distinct ways, suited to their diverse characters.

2 The logics for handling machine and infinite errors separately

Assume now that we are computing the values of logical expressions appearing in programs or program assertions in a sequential and lazy way — i.e., the computation proceeds from left to right, and stops if the value computed up to the current point determines the value of the whole expression. Then the practical consequences of the different natures of a machine error and an infinite error can be best explained on the example of computing $v(\alpha \vee \beta)$ for a valuation v.

If we encounter a machine error when computing $v(\alpha)$, then after getting an error signal we can drop α and go on to compute $v(\beta)$, and if the latter computation returns $\mathbf{t}$ (standing for *true*), we can put $v(\alpha \vee \beta) = \mathbf{t}$ too. However, if the computation of $v(\alpha)$ loops, then the evaluation of $\alpha \vee \beta$

loops too, for we will never get the error signal allowing us to move beyond α. As a result, even if $v(\beta) = \mathbf{t}$, we will never learn this fact, and so will not be able to put $v(\alpha \vee \beta) = \mathbf{t}$, like in case of a machine error. On the other hand, if $v(\alpha) = \mathbf{t}$, then we put $v(\alpha \vee \beta) = \mathbf{t}$ no matter what $v(\beta)$ is, even if its computation would in fact loop — for in a lazy computation we do not try to compute $v(\beta)$ once we have established that $v(\alpha)$ is $\mathbf{t}$. Analogous reasoning applies to conjunction, while negation simply preserves both the the errors.

Let us denote the machine error by $\mathbf{m}$, and the infinite error by $\mathbf{e}$. By the above remarks, for reasoning about programs involving the machine error $\mathbf{m}$ we can use the well-known three-valued *Kleene logic* [Kl52] with symmetric disjunction and conjunction, given by the following truth tables:

(1)

$\neg$	t	f	m
	f	t	m

$\vee$	t	f	m
t	t	t	t
f	t	f	m
m	t	m	m

$\wedge$	t	f	m
t	t	f	m
f	f	f	f
m	m	f	m

In turn, reasoning about programs involving the infinite error $\mathbf{e}$ can be best carried out using three-valued *McCarthy logic* [MC67] with asymmetric conjunction and disjunction, represented by the following truth tables:

(2)

$\neg$	t	f	e
	f	t	e

$\vee$	t	f	e
t	t	t	t
f	t	f	e
e	e	e	e

$\wedge$	t	f	e
t	t	f	e
f	f	f	f
e	e	e	e

Of the two, Kleene logic is by far the more popular one, being used also for describing undefinedness in general. In turn, McCarthy logic was originally developed for the purpose of describing the phenomenon of computation, including computation errors or infinite loops, and has found application in programming languages like Euclid, Ada and Algol-W. In [KTB91] and [Ko93], McCarthy connectives combined with Kleene quantifiers were used as a foundation for developing two versions of a three-valued logic for specification and validation of programs.

3   The logic for handling both machine and infinite errors

Our aim here is to combine the two approaches used for the two different error phenomena and develop a logic dealing both with machine errors and truly infinite errors, but treating them in two different ways, tailoured to their diverse criticality for computations. From the preceding considerations, it is clear that the most natural semantics underlying such a logic should employ four logical values: the classical ones: $\mathbf{t}$ and $\mathbf{f}$, as well as two non-classical values $\mathbf{m}$ and $\mathbf{e}$, representing the two different errors. What is more, the interpretations of the connectives restricted to $\mathbf{t}$, $\mathbf{f}$ and $\mathbf{m}$ should behave like Kleene connectives, and restricted to $\mathbf{t}$, $\mathbf{f}$ and $\mathbf{e}$ — like McCarthy ones.

3.1   Syntax and semantics of the language

We begin with defining the language of our logic, denoted by L_{2e}. Let PROP be a non-empty set of propositional variables. The set of well-formed formulas of L_{2e}, denoted by FORM, is the least set containing PROP and closed under $\neg, \vee, \wedge$.

The semantics of L_{2e} is based on the four-valued logical matrix $\mathcal{M}_{2e} = (\mathcal{T}, \mathcal{D}, \mathcal{I})$, where:

- $\mathcal{T} = \{\mathbf{f}, \mathbf{t}, \mathbf{m}, \mathbf{e}\}$ is the set of the logical values of $\mathcal{M}$,

- $\mathcal{D} = \{\mathbf{t}\}$ is the set of its designated values,

- $\mathcal{I}$ assigns to each k-ary connective $\diamond$ of L_{2e} its interpretation $\widetilde{\diamond} : \mathcal{T}^k \to \mathcal{T}$, where:

(3)

$\widetilde{\neg}$	t	f	m	e
	f	t	m	e

$\widetilde{\vee}$	t	f	m	e
t	t	t	t	t
f	t	f	m	e
m	t	m	m	e
e	e	e	e	e

$\widetilde{\wedge}$	t	f	m	e
t	t	f	m	e
f	f	f	f	f
m	m	f	m	e
e	e	e	e	e

It can be easily seen that after projecting the above truth tables to $\mathbf{f}, \mathbf{t}, \mathbf{m}$ we get the truth tables (1) for Kleene logic, while their projection to $\mathbf{f}, \mathbf{t}, \mathbf{e}$ yields the truth tables (2) for McCarthy logic. As a result, the matrix $\mathcal{M}_{2e}$ meets the requirements set out at the beginning of the paragraph, and provides an adequate semantic basis for a logic handling both machine and infinite errors.

The choice of $\mathbf{t}$ as the only designated value corresponds to considering a strong logic, which is usually the case in reasoning about programs. As to the interplay between the two kinds of errors, we have adopted here the assumption that $\mathbf{e}$ prevails over $\mathbf{m}$ whatever their order.

A *legal valuation* over $\mathcal{M}_{2e}$ is any function $v : \text{FORM} \to \mathcal{T} = \{\mathbf{t}, \mathbf{f}, \mathbf{m}, \mathbf{e}\}$ compliant with the truth tables of $\mathcal{M}_{2e}$ for $\neg, \vee, \wedge$, given in (3).

Let us denote the set of all legal valuations over $\mathcal{M}_{2e}$ by V_{2e}.

- A valuation $v \in V_{2e}$ is said to be:

  - a *model of* (*satisfy*) a formula $\varphi \in \text{FORM}$, written $v \models \alpha$, iff $v(\alpha) = \mathbf{t}$,

- a *model of* a set of formulas $F \subseteq \textsc{Form}$, written $v \models F$, if $v \models \varphi$ for every $\varphi \in F$.

• A formula $\varphi \in \textsc{Form}$ is said to be *valid* in L_{2e}, written $\models_{L_{2e}} \alpha$, iff $v \models \alpha$ for every $v \in V_{2e}$.

• The *consequence relation* of L_{2e} is a binary relation $\vdash_{L_{2e}} \subset \mathcal{P}(\textsc{Form}) \times \textsc{Form}$ such that, for any set of formulas $F \subseteq \textsc{Form}$ and any formula $\varphi \in \textsc{Form}$, $F \vdash_{L_{2e}} \varphi$ iff every model v of F is also a model of φ.

3.2   Finiteness operator

As we have already mentioned, a fundamental issue in handling the two diverse types of computation errors studied in this paper is the ability to distinguish between the finite machine error **m** and the infinite error **e**. The way of achieving this goal we propose in the paper is to equip L_{2e} with a kind of a dichotomous "finiteness" operator able to distinguish between the finite and infinite logical values operator", analogous to the well-known "definedness operator" Δ. This approach is quite novel — for the usual logics of programs distinguish only between the correct values and an error value, while the finiteness operator will allow us to treat the machine error in some respect like the regular logical values.

If we extend our language by adding a constant $\underline{t}$ representing the value **t**, then such an operator — denoted by $\circ$ — can be defined within the language L_{2e} itself as

(4) $\circ \alpha = \alpha \vee \underline{t}$

for any formula $\alpha \in \textsc{Form}$. Indeed: by the above definition, we have:

(5) $v(\circ \alpha) = \begin{cases} \mathbf{t} & \text{iff } v(\alpha) \in \{\mathbf{f}, \mathbf{t}, \mathbf{m}\} \\ \mathbf{e} & \text{iff } v(\alpha) = \mathbf{e}. \end{cases}$

for any formula $\alpha \in \textsc{Form}$ and any valuation $v \in V_{2e}$. Though the dichotomy of $\circ$ is of the "designated/non-designated" rather than the classical "true-false" type, like in case of Δ, this difference is, of course, immaterial from the viewpoint of reasoning, and $\circ$ will be quite adequate for the purposes of distinguishing and handling the two errors in the proof system of L_{2e} we shall develop here.

More precisely, as **t** is designated and **e** is not, (5) implies that $\circ$ allows us to distinguish between **m** and **e** on the formula satisfaction level. Namely, for any formula α and any valuation v, we have:

$$v(\alpha) = \mathbf{m} \Rightarrow v \models \circ \alpha, \qquad v(\alpha) = \mathbf{e} \Rightarrow v \not\models \circ \alpha$$

What is more, once we have ∘, we can express the logical value of each formula α in terms of satisfaction/non-satisfaction of $\alpha, \neg\alpha, \circ\alpha$ as follows:

$$
\begin{array}{llll}
(6) & v(\alpha) = \mathbf{t} & \text{iff} & v \models \alpha \\
 & v(\alpha) = \mathbf{f} & \text{iff} & v \models \neg\alpha \\
 & v(\alpha) = \mathbf{m} & \text{iff} & v \not\models \alpha, v \not\models \neg\alpha, v \models \circ\alpha \\
 & v(\alpha) = \mathbf{e} & \text{iff} & v \not\models \alpha, v \not\models \neg\alpha, v \not\models \circ\alpha
\end{array}
$$

This will allow us to develop a Rasiowa-Sikorski style proof system for our logic.

Accordingly, from now on we shall assume that L_{2e} contains the constant $\underline{\mathbf{t}}$, and that ∘ is the derived operator defined by (4).

In the proof system for our logic we will employ the following important properties of ∘:

FACT 1. For any formula $\alpha \in$ FORM and any valuation $v \in V_{2e}$:

$$
\begin{array}{llll}
(7) & (i) & v(\circ\alpha) \in \{\mathbf{t}, \mathbf{e}\} & \quad (ii) \quad v(\circ\alpha) = \mathbf{e} \text{ iff } v(\alpha) = \mathbf{e} \\
\\
 & (iii) & v(\circ\neg\alpha) = v(\circ\alpha) & \quad (iv) \quad v(\circ \circ \alpha) = v(\circ\alpha)
\end{array}
$$

Properties (i) and (ii) follow immediately from (5). In turn, (iii) follows from (i), (ii) and the fact that by (3) $v(\alpha) = \mathbf{e}$ iff $v(\neg\alpha) = \mathbf{e}$. Finally, (iv) follows directly from (4) and the idempotence of disjunction.

4   Rasiowa-Sikorski (R-S) deduction system for L_{2e}

To develop a proof system for L_{2e}, we will use the decompositional formalism of Rasiowa-Sikorski (R-S) deduction systems presented in [RS63], which is a well-proven, useful tool for formalizing various kinds of logics related to computer science applications. The big advantage of R-S systems is the simplicity of their development, and their particular suitability for proof search and automatic deduction. More information about the properties and applications of such systems can be found e.g. in [Or88, Or88, Or93, DOR94, KMO98, Ko99, Ko02, OOP03, GPO07].

4.1   Fundamentals of the R-S formalism

An R-S system operates on finite sequences of formulas, and consists of decomposition rules for such sequences and of axiomatic sequences[1] which are simple sequences of formulas guaranteed to be valid, e.g. $\alpha, \neg\alpha$ in classical logic. In other words, decomposition rules are the "inference rules" of the system, whereas axiomatic sequences represent its "axioms". Decomposition rules break down each complex formula into a sequence of simpler

---

[1] Referred to as "fundamental" in [RS63].

formulas (usually, subformulas of the original formula) whose validity is equivalent to validity of the original formula. Using decomposition rules, we construct a decomposition tree T for a sequence of formulas Ω (in particular, a single formula α), with vertices labelled by sequences of formulas. A branch of T terminates only if we encounter an axiomatic sequence or a sequence which cannot be decomposed any further. T is called a proof for Ω if it is finite and all its branches terminate with axiomatic sequences.

Sequences of formulas are interpreted disjunctively. Namely, a sequence Ω is said to be:

- *satisfied by a valuation* v, written $v \models \Omega$, iff v satisfies some formula in Ω;

- *valid* iff $v \models \Omega$ for any valuation v.

In other words,

(8) $v \models \varphi_1, \varphi_2, \ldots, \varphi_n$ iff $v \models \varphi_k$ for some $k, 1 \leq k \leq n$

and similarly for validity. A decomposition rule of an R-S system is an $(n+1)$-tuple $\Omega_0, \Omega_1, \ldots, \Omega_n$ of sequences of formulas, usually written as

$$\frac{\Omega_0}{\Omega_1 \mid \Omega_2 \mid \ldots \mid \Omega_n}$$

Ω_0 is called the *conclusion* of the rule, and $\Omega_1, \ldots, \Omega_n$ its *premises*. A rule is said to be *sound* provided its conclusion is valid iff all its premises are valid.

Thus, a decomposition rule is sound iff it leads from valid sequences to valid sequences both "downwards" and "upwards". Such a "two-way" notion of soundness, crucial for R-S systems, is stronger than the usual one, and amounts to the invertibility of rules. To emphasize this fact, we separate the premises from the conclusion of a rule by a double line instead of a single one.

Note also that the premises of a rule are interpreted conjunctively on the *validity level* (*not* on the level of satisfaction in a model). Indeed: in a sound rule, the conclusion is valid iff each of the premises is valid — but in general it need not be true that the conclusion is satisfied in a single model iff all the premises are satisfied in that model. Hence in the rule notation the comma between the individual formulas in sequences represents meta-disjunction at the level of satisfaction in a single model (here: a single valuation), while the vertical bar represents meta-conjunction on the level of validity, i.e. satisfaction in all models.

4.2   DRS$_{2e}$ - an R-S deduction systems for L_{2e}

In case of many-valued logics like L_{2e}, a standard way of developing an R-S system is to express, for every logical value a used in the semantics and any complex formula α, the condition for α to have the value a in terms of some simpler formulas having specific logical values. This involves first of all expressing (within the language of the logic) the fact that a formula has a particular logical value. In case of L_{2e}, the basic means for doing this are provided by Equation (6). However, to fully express (6) within the language of our logic, we must be able to express within that language satisfaction and non-satisfaction of a formula. For this purpose, we employ, similarly as in [Ko93], the well-known mechanism of dichotomous, two-valued "signed formulas", being a generalization of those introduced in [Be59] to the case of many-valued logics.

Thus we extend the language with two basic "signs": $\mathbf{T}$, meaning *is-true*, or *is-satisfied*, and $\mathbf{N}$, meaning *is-not-true*, or *is-not-satisfied*. The set of signed formulas SFORM is defined as

$$\text{SFORM} = \{\mathbf{T}\alpha : \alpha \in \text{FORM}\} \cup \{\mathbf{N}\alpha : \alpha \in \text{FORM}\}$$

We assume that any valuation $v \in V_{2e}$ is extended to SFORM by taking:

$$(9) \quad v(\mathbf{T}\alpha) = \begin{cases} \mathbf{t} & \text{iff } v(\alpha) = \mathbf{t} \\ \mathbf{f} & \text{iff } v(\alpha) \in \{\mathbf{f}, \mathbf{m}, \mathbf{e}\}, \end{cases} \qquad v(\mathbf{N}\alpha) = \begin{cases} \mathbf{t} & \text{iff } v(\alpha) \in \{\mathbf{f}, \mathbf{m}, \mathbf{e}\} \\ \mathbf{f} & \text{iff } v(\alpha) = \mathbf{t} \end{cases}$$

The notions of satisfaction and validity of signed formulas are defined identically as for the non-signed formulas in FORM.

The proof system for L_{2e} will operate on sequences of signed formulas, whose set will be denoted by SFORM*. Like we have already mentioned in Section 4.1, each such sequence will be interpreted disjunctively, in line with (8).

Now we can finally define the deduction system DRS$_{2e}$ for L_{2e}. As L_{2e} obeys DeMorgan laws, we will only present the decomposition rules for negation and disjunction, plus the derived definedness operator $\circ$. What is more, since the only reason for adding the constant $\underline{t}$ to L_{2e} was to define $\circ$, DRS$_{2e}$ will in fact be tailoured to the sublanguage of the extended L_{2e} limited to formulas where the only occurrences of $\underline{t}$ are those of the form $\alpha \vee \underline{t} = \circ\alpha$ for some $\alpha \in$ FORM.

The system DRS$_{2e}$ consists of the following:

- **Axiomatic sequences**

   All finite sequences Ω of signed formulas in SFORM which contain one of the pairs of formulas or singletons (i)–(v) below:

$$(10) \quad \begin{array}{lll} (i) \ \mathbf{T}\alpha, \mathbf{N}\alpha & (ii) \ \mathbf{N}\alpha, \mathbf{N}(\neg\alpha) & (iii) \ \mathbf{T}(\circ\alpha), \mathbf{N}\alpha \\ (iv) \ \mathbf{T}(\circ\alpha), \mathbf{N}(\neg\alpha) & (v) \ \mathbf{N}(\neg\circ\alpha) \end{array}$$

for any formula α in FORM

- **Decomposition rules**

$$(\mathbf{T}\neg\neg) \ \frac{\Omega', \mathbf{T}(\neg\neg\alpha), \Omega''}{\Omega', \mathbf{T}\alpha, \Omega''} \qquad\qquad (\mathbf{N}\neg\neg) \ \frac{\Omega', \mathbf{N}(\neg\neg\alpha), \Omega''}{\Omega', \mathbf{N}\alpha, \Omega''}$$

$$(\mathbf{T}\neg\circ) \ \frac{\Omega', \mathbf{T}(\neg \circ \alpha), \Omega''}{\Omega', \Omega''}$$

$$(\mathbf{T}\circ\circ) \ \frac{\Omega', \mathbf{T}(\circ \circ \alpha), \Omega''}{\Omega', \mathbf{T}(\circ\alpha), \Omega''} \qquad\qquad (\mathbf{N}\circ\circ) \ \frac{\Omega', \mathbf{N}(\circ \circ \alpha), \Omega''}{\Omega', \mathbf{N}(\circ\alpha), \Omega''}$$

$$(\mathbf{T}\circ\neg) \ \frac{\Omega', \mathbf{T}(\circ\neg\alpha), \Omega''}{\Omega', \mathbf{T}(\circ\alpha), \Omega''} \qquad\qquad (\mathbf{N}\circ\neg) \ \frac{\Omega', \mathbf{N}(\circ\neg\alpha), \Omega''}{\Omega', \mathbf{N}(\circ\alpha), \Omega''}$$

$$(\mathbf{T}(\alpha \vee \beta)) \ \frac{\Omega', \mathbf{T}(\alpha \vee \beta), \Omega''}{\Omega', \mathbf{T}(\circ\alpha), \Omega'' \mid \Omega', \mathbf{T}\alpha, \mathbf{T}\beta, \Omega''}$$

$$(\mathbf{N}(\alpha \vee \beta)) \ \frac{\Omega', \mathbf{N}(\alpha \vee \beta), \Omega''}{\Omega', \mathbf{N}\alpha, \Omega'' \mid \Omega', \mathbf{N}(\circ\alpha), \mathbf{N}\beta, \Omega''}$$

$$(\mathbf{T} \circ (\alpha \vee \beta)) \ \frac{\Omega', \mathbf{T}(\circ(\alpha \vee \beta)), \Omega''}{\Omega', \mathbf{T}(\circ\alpha), \Omega'' \mid \Omega', \mathbf{T}\alpha, \mathbf{T}(\circ\beta), \Omega''}$$

$$(\mathbf{N} \circ (\alpha \vee \beta)) \ \frac{\Omega', \mathbf{N}(\circ(\alpha \vee \beta)), \Omega''}{\Omega', \mathbf{N}\alpha, \Omega'' \mid \Omega', \mathbf{N}(\circ\alpha), \mathbf{N}(\circ\beta), \Omega''}$$

$$(\mathbf{T}\neg(\alpha \vee \beta)) \ \frac{\Omega', \mathbf{T}(\neg(\alpha \vee \beta)), \Omega''}{\Omega', \mathbf{T}(\neg\alpha), \Omega'' \mid \Omega', \mathbf{T}(\neg\beta), \Omega''}$$

$$(\mathbf{N}\neg(\alpha \vee \beta)) \ \frac{\Omega', \mathbf{N}(\neg(\alpha \vee \beta)), \Omega''}{\Omega', \mathbf{N}(\neg\alpha), \mathbf{N}(\neg\beta), \Omega''}$$

where α, β are any formulas in FORM and $\Omega', \Omega'' \in$ SFORM*.

As the R-S deduction formalism is based on the principle of decomposing each complex formula into some simpler formulas (usually, its subformulas), an R-S system operating on sequences of signed formulas must contain rules for decomposing any signed formula of the form $\mathbf{T}\alpha$ or $\mathbf{N}\alpha$ where α is obtained out of some other formulas using arbitrary operators of the language.

This surely holds true for L_{2e}. Namely, we have rules for decomposing $\mathbf{T}$ and $\mathbf{N}$ formulas underpinned by double negation and double $\circ$ (with the latter rule expressing the idempotence of $\circ$ and implied by (iv) in (7), and a rule for eliminating negation under $\circ$, justified by property (iii) in (7). The case of negation applied to $\circ$ is slightly different — as $\mathbf{N}(\neg \circ \alpha)$ is an axiom, we do not need any $(\mathbf{N}\neg\circ)$ rule, for axiomatic formulas are not subject to decomposition. In turn, as $\mathbf{T}(\neg \circ \alpha)$ is never satisfied, the $(\mathbf{T}\neg\circ)$ rule just removes this formula from the sequence under decomposition. Finally, we have six rules for disjunction: the $\mathbf{T}$ and $\mathbf{N}$ rules for disjunction alone, as well as for negated disjunction and disjunction preceded by $\circ$. This set of rules allows us to decompose any complex signed formula over L_{2e}, which is a prerequisite for the correct operation of the R-S proof mechanism.

Finally, DRS_{2e} satisfies the following two basic conditions required by the R-S formalism:

LEMMA 2.

1. *All axiomatic sequences of DRS_{2e} are valid;*

2. *All rules of DRS_{2e} are sound in the R-S sense, i.e. the conclusion of each rule is valid iff all its premises are valid.*

Proof. By (10), to prove Condition 1 above we have to show that, for any $\alpha \in$ FORM,

$$(i)\mathbf{T}\alpha, \mathbf{N}\alpha \quad (ii)\mathbf{N}\alpha, \mathbf{N}(\neg\alpha) \quad (iii)\mathbf{T}(\circ\alpha), \mathbf{N}\alpha \quad (iv)\mathbf{T}(\circ\alpha), \mathbf{N}(\neg\alpha) \quad (v)\mathbf{N}(\neg \circ \alpha)$$

are satisfied by any valuation $v \in V_{2e}$.

Let $v \in V_{2e}$. The satisfaction of (i) under v follows immediately from the semantics of $\mathbf{T}, \mathbf{N}$ — see (9). In turn, by (3) $v(\alpha)$ and $v(\neg\alpha)$ cannot be both equal to $\mathbf{t}$ — which by (9) implies that either $v \models \mathbf{N}(\alpha)$ or $v \models \mathbf{N}(\neg\alpha)$, whence v satisfies (ii). Further, by (5) and (9), if $v \not\models \mathbf{T}(\circ\alpha)$, then $v(\alpha) = \mathbf{e}$, whence also $v(\neg\alpha) = \mathbf{e}$, implying that $v \models \mathbf{N}(\circ\alpha)$ and $v \models \mathbf{N}(\neg\circ\alpha)$. Clearly, this proves that v satisfies both (iii) and (iv). Finally, as $v(\circ\alpha) \in \{\mathbf{t}, \mathbf{e}\}$ by (5), then $v(\neg \circ \alpha) \in \{\mathbf{f}, \mathbf{e}\}$, whence $v \models \mathbf{N}(\neg \circ \alpha)$, meaning that (v) is satisfied too. Hence Condition 1 holds.

To show Condition 2, we prove that for every rule of the form

$$(*) \quad \frac{\Omega_0}{\Omega_1 \mid \ldots \mid \Omega_k}$$

in DRS_{2e}, where $k \in \{1, 2\}$, and every valuation $v \in V_{2e}$, we have

$$v \models \Omega_0 \text{ iff } v \models \Omega_i \text{ for } i = 1, \ldots, k$$

which is in fact a stronger result. Like in case of Condition 1, the reasoning is based on the properties (9, 5) of the $\mathbf{T}, \mathbf{N}, \circ$ operators and on the truth tables (3) of the matrix $\mathcal{M}_{2e}$ underlying the semantics of our logics.

As an example, let us prove $(*)$ for the rule

$$(\mathbf{T}(\alpha \vee \beta)) \quad \frac{\Omega', \mathbf{T}(\alpha \vee \beta), \Omega''}{\Omega', \mathbf{T}(\circ\alpha), \Omega'' \mid \Omega', \mathbf{T}\alpha, \mathbf{T}\beta, \Omega''}$$

Using the notation employed in $(*)$, we have

$$\Omega_0 = \Omega', \mathbf{T}(\alpha \vee \beta), \Omega'' \qquad \Omega_1 = \Omega', \mathbf{T}(\circ\alpha), \Omega'' \qquad \Omega_2 = \Omega', \mathbf{T}\alpha, \mathbf{T}\beta, \Omega''$$

Assume first that $v \models \Omega_0$. Then by (8) $v \models \Omega_0'$ or $v \models \Omega''$ or $v \models \mathbf{T}(\alpha \vee \beta)$. As Ω', Ω'' are contained in Ω_i for $i = 1, 2$, then in the first two cases we obviously have $v \models \Omega_i$ for $i = 1, 2$. In turn, if $v \models \mathbf{T}(\alpha \vee \beta)$, then by the truth table for disjunction in (3) we have either (i) $v(\alpha) = \mathbf{t}$, or (ii) $v(\alpha) \neq \mathbf{e}$ and $v(\beta) = \mathbf{t}$. By (5, 9), (i) implies $v \models \mathbf{T}(\circ\alpha)$, whence $v \models \Omega_1$. Since $v \models \mathbf{T}(\alpha)$ as well, we also have $v \models \Omega_2$, whence the downward implication in $(*)$ holds.

To prove the upward implication, assume that $v \models \Omega_i$ for $i = 1, 2$. Obviously, if $v \models \Omega'$ or $v \models \Omega''$, then $v \models \Omega_0$. Accordingly, we are left with the case when (A) $v \models \mathbf{T}(\circ\alpha)$ and (B) $v \models \mathbf{T}\alpha, \mathbf{T}\beta$. In view of (5), (A) implies $v(\alpha) \neq \mathbf{e}$, while by (B) we have either $v(\alpha) = \mathbf{t}$ or $v(\beta) = \mathbf{t}$. Obviously, by the truth table for disjunction in (3), the latter combination of conditions yields $v \models \mathbf{T}(\alpha \vee \beta)$, whence $v \models \Omega_0$.

Thus $(*)$ indeed holds for rule $(\mathbf{T}(\alpha \vee \beta))$. The proofs for other rules are analogous. ∎

5   Soundness and completeness of the system DRS_{2e}

We begin with some formal preliminaries.

A signed formula in SFORM is said to be *indecomposable* in DRS_{2e} if it cannot be decomposed using any rule in DRS_{2e}.

From the form of the decompositions rules in DRS_{2e} it is easy to see that a signed formula is indecomposable iff it has of one of the following forms:

(11) $\mathbf{T}(p)$, $\mathbf{T}(\neg p)$, $\mathbf{N}(p)$, $\mathbf{N}(\neg p)$, $\mathbf{T}(\circ p)$, $\mathbf{N}(\circ p)$, $\mathbf{N}(\neg \circ p)$

where $p \in \text{PROP}$.[2]

A sequence of signed formulas $\Omega \in \text{SFORM}^*$ is said to be *indecomposable* iff each formula in Ω is indecomposable.

A *decomposition tree* for a finite sequence $\Omega \in \text{SFORM}$ in the system DRS_{2e} is any tree T labeled with sequences in SFORM^* such that:

- The root of T is labelled with Ω;

- Each vertex of T labelled with Ω_0 which is not a leaf of T has either:

  - a single child labelled with Ω_1, where $\dfrac{\Omega_0}{\Omega_1}$ is an instance of a rule in DRS_{2e}, or

  - two children labelled with Ω_1, Ω_2, where $\dfrac{\Omega_0}{\Omega_1 \mid \Omega_2}$ is an instance of a rule in DRS_{2e};

- Each leaf of T is labelled with either an axiomatic sequence of DRS_{2e} or an indecomposable sequence.

Clearly, a decomposition tree for Ω is obtained by decomposing Ω using the rules in DRS_{2e} and repeating this procedure for the resulting sequences until we obtain each time an either axiomatic or indecomposable sequence of formulas.

A *decomposition tree* for Ω is said to be *a proof* if it is finite and all its leaves are labelled with axiomatic sequences.

Since each of the decomposition rules in DRS_{2e} replaces the formula it actually decomposes with formula(e) of lower complexity, it can be easily seen that a finite number of applications of those rules transforms any sequence of formulas $\Omega \in \text{SFORM}^*$ into a set of either axiomatic or indecomposable sequences. Hence we have:

FACT 3. Each decomposition tree of a finite sequence $\Omega \in \text{SFORM}^*$ in DRS_{2e} is finite.

A sequence Ω of signed formulas is said to be *provable* in DRS_{2e}, in symbols $\vdash_{DRS_{2e}}$, if it has a decomposition tree in DRS_{2e} which is a proof.

THEOREM 4. *The system DRS_{2e} is sound and complete, i.e. a sequence Ω of signed formulas in* SFORM *is valid iff it is provable in DRS_{2e}.*

---

[2]Note that $\mathbf{T}(\neg \circ p)$ is trivially decomposable by elimination using Rule $(\mathbf{T}\neg\circ)$.

Proof. Soundness follows easily from the Lemma 2, which asserts (1) validity of axiomatic sequences and (2) strong, two-way soundness of the decomposition rules in DRS_{2e}. Indeed: Suppose Ω has a decomposition tree $DT(\Omega)$ in DRS_{2e} which is a proof. Then $DT(\Omega)$ is finite, and by the definition of the decomposition tree all the sequences of signed formulas labelling the leaves of $DT(\Omega)$ are obtained from Ω by a finite number of applications of the decomposition rules in DRS_{2e}. Hence by (2) (applied finitely many times) Ω is valid iff all those sequences are valid. However, since $DT(\Omega)$ is proof, then all such sequences are axiomatic, whence they are valid by (1) — and so is Ω.

Now it remains to prove completeness. We argue by contradiction. Suppose Ω is valid, but not provable, i.e. $\nvdash_{DRS_{2e}}$. By Fact3, Ω has a finite decomposition tree $DT(\Omega)$. Since $\nvdash_{DRS_{2e}}$, then one of the sequences labelling the leaves of $DT(\Omega)$ must be non-axiomatic. Let us denote it by Δ. By the definition of a decomposition tree, Δ can only contain indecomposable formulas of the form shown in (11), except for $\mathbf{N}(\neg \circ p)$, which is axiomatic.

Thus each formula in Δ must have one of the following forms

(12) $\mathbf{T}(p), \quad \mathbf{T}(\neg p), \quad \mathbf{N}(p), \quad \mathbf{N}(\neg p), \quad \mathbf{T}(\circ p), \quad \mathbf{N}(\circ p)$

where $p \in \text{PROP}$.

Let us define a valuation v as follows:

- For any $p \in \text{PROP}$,

$$(13) \quad v(p) = \begin{cases} \mathbf{t} & \text{if } \mathbf{N}(p) \in \Delta \\ \mathbf{f} & \text{if } \mathbf{N}(\neg p) \in \Delta \\ \mathbf{e} & \text{if } \mathbf{T}(\circ p) \in \Delta \\ \mathbf{m} & \text{otherwise} \end{cases}$$

- For any $\alpha \in \text{FORM}$, $v(\neg \alpha) = \widetilde{\neg} v(\alpha)$, $v(\alpha \vee \beta) = v(\alpha) \widetilde{\vee} v(\beta)$, where $\widetilde{\neg}, \widetilde{\vee}$ are given by (3).

Clearly, v is well-defined, since by (10) any two of the indecomposable formulas listed in (13) form an axiomatic sequence, so only one of them can appear in Δ. Since v is compliant with the truth tables of the matrix $\mathcal{M}_{2e}$, it is a legal valuation in V_{2e}. As before, we assume that v is extended in the usual way to signed formulas.

We will now show that v is a counter-model for Ω.

First we show that v is a counter-model for Δ. Let δ be any signed formula in Δ. Then by (12) there exists a $p \in \text{PROP}$ such that one of the following holds:

1. (i) $\delta = \mathbf{N}(p)$, (ii) $\delta = \mathbf{N}(\neg p)$ or (iii) $\delta = \mathbf{T}(\circ p)$. Then by (13) we have respectively (i) $v(p) = \mathbf{t}$, (ii) $v(p) = \mathbf{f}$ or (iii) $v(p) = \mathbf{e}$, whence $v \not\models \delta$.

2. $\delta \in \{\mathbf{T}(p), \mathbf{T}(\neg p), \mathbf{N}(\circ p)\}$. Then none of the formulas $\mathbf{N}(p), \mathbf{N}(\neg p)$, $\mathbf{T}(\circ p)$ is in Δ, for otherwise Δ would be axiomatic. Accordingly, by (13) we have $v(p) = \mathbf{m}$, whence $v \not\models \mathbf{T}(p), \mathbf{T}(\neg p), \mathbf{N}(\circ p)$. Thus $v \not\models \delta$.

Hence $v \not\models \Delta$, and so Δ is not valid. However, Δ is the label of a leaf in the finite decomposition tree of Ω — so by Lemma 2 and the reasoning used in the proof of soundness Ω cannot be valid either, which is a contradiction. ∎

LEMMA 5. DRS_{2e} provides a decision procedure for the logic L_{2e}, with the complexity $O(2^n)$.

Proof. It can be easily seen that, for any formula $\varphi \in \mathrm{SFORM}$, there is exactly one rule in DRS_{2e} which is applicable to φ. Hence, for every sequence of formulas $\Omega \in \mathrm{SFORM}^*$, we have a clear procedure for constructing its unique "left" decomposition tree $DT_l(\Omega)$. Namely, starting with Ω as the root of $DT_l(\Omega)$, in each construction step we decompose the leftmost decomposable formula in the current sentence using the single rule in DRS_{2e} which is applicable to that formula. By Fact 3, after a finite number of steps we obtain the desired (finite) $DT_l(\Omega)$, which is either a proof of Ω or contains an indecomposable, but non-axiomatic sequence at one of its leaves. In the first case, Ω is provable. We will show that in the second case it cannot be provable. Indeed, if Ω were provable, then by Theorem 4 it would also be valid. However, then the non-axiomatic, indecomposable sequence Σ_i labelling a leaf of $DT_l(\Omega)$ would be valid too as obtained out of Ω using the two-way sound rules of DRS_{2e}. Yet in that case Theorem 4 would imply that Σ_i has a proof in DRS_{2e} — which is impossible, for the only decomposition tree of Σ_i consists of a single vertex labelled by Σ_i, and that tree is not a proof, as Σ_i is not axiomatic.

Thus the construction of $DT_l(\Omega)$ is a decision procedure for provability of Ω; in particular, for $\Omega = \{\varphi\}$, where $\varphi \in \mathrm{SFORM}$, it is a decision procedure for provability of the formula φ. Due to the binary branching of the decomposition tree, it can be easily seen that the overall complexity of the decision procedure is exponential, and amounts to $O(2^n.)$ ∎

Let us note that for a non-provable sequence Ω, the left decomposition tree $DT_l(\Omega)$ constructed by the decision procedure described in the proof of Lemma 5 defines a refutation of Ω. Indeed, using the method given in the proof of Theorem 4, we can construct out of the non-axiomatic, indecomposable sequence labelling a leaf of $DT_l(\Omega)$ a valuation v such that $v \not\models \Omega$.

Obviously, Lemma 5 together with the Theorem 4 imply the decidability of satisfaction in L_{2e}.

The rough decision procedure we have presented above has a discouraging exponential complexity, but a reasonable conjecture would be that the L_{2e} provability/satisfaction problem can be handled in many practical instances using much more effective algorithms analogous to those employed with good results for SAT.

6   A Gentzen calculus system $\mathbf{SC}_{2e}$ for L_{2e}

Though the system DRS$_{2e}$ we have defined in Section 4.2 has great advantages from the viewpoint of automatic deduction, it is, formally speaking, a proof system for the language of signed formulas rather than for the original language of L_{2e} itself. What is more, we have proved its completeness for the validity of sequences of signed formulas, while our primary interest is in the consequence relation $\vdash_{L_{2e}}$ of L_{2e}, which represents our logic for reasoning about the two errors. In Section 3.1, we have defined $\vdash_{L_{2e}}$ as the binary relation on $\mathcal{P}(\text{FORM}) \times \text{FORM}$ such that:

(14) $F \vdash_{L_{2e}} \varphi$ iff, for any $v \in V_{2e}, v \models F$ implies $v \models \varphi$

Let us define a sequent over L_{2e} as $\Gamma \Rightarrow \Delta$, where Γ, Δ are any finite subsets of FORM. Then:

- A valuation $v \in V_{2e}$ is said to be *a model of* (*satisfy*) a sequent $\Gamma \Rightarrow \Delta$ iff either $v \not\models \Gamma$ or $v \models \Delta$; in other words

(15) $v \models (\varphi_1, \ldots, \varphi_k \Rightarrow \psi_1, \ldots, \psi_l)$ iff either $v \not\models \varphi_i$ for some i or $v \models \psi_j$ for some j

- A sequent S is said to be *valid* in L_{2e}, in symbols $\models_{L_{2e}} A$, iff S is satisfied by every valuation in V_{2e}.

Clearly, in view of (14) and (15), for a finite set of formulas F we have

$$F \vdash_{L_{2e}} \varphi \text{ iff } \models_{L_{2e}} (F \Rightarrow \varphi)$$

Hence the consequence relation of L_{2e} limited to finite sets of formulas is represented by validity of sequents over FORM — and the proof system best suited for reasoning directly about that relation is a calculus of such sequents.

However, the R-S formalism is proof-theoretically equivalent to the Gentzen sequent formalism, and there is a general method of transforming R-S systems into Gentzen sequent calculi, presented in detail in [Ko02]. Using that method, we will translate the sound and complete R-S system DRS_{2e} for sequences of signed formulas in SFORM to a sound and complete calculus SC_{2e} of sequents of ordinary, unsigned formulas in FORM. An additional advantage of the above will be presenting our system in the well-established sequent calculus formalism, which is used by the majority of logicians while R-S systems, despite their advantages for proof search, are much less popular. The translation is based on the following simple connection between validity of sequents and validity of sequences of signed formulas:

FACT 6. Let $\Gamma = \{\alpha_1, \ldots, \alpha_n\}, \Delta = \{\beta_1, \ldots, \beta_m\}$, where $\alpha_i, \beta_j \in$ FORM. Then the sequent $\Gamma \Rightarrow \Delta$ is valid whenever the sequence $\mathbf{N}(\Gamma), \mathbf{T}(\Delta)$ of signed formulas is valid, where

$$\mathbf{N}(\Gamma) = \mathbf{N}(\alpha_1), \mathbf{N}(\alpha_2), \ldots, \mathbf{N}(\alpha_n), \quad \mathbf{T}(\Delta) = \mathbf{T}(\beta_1), \mathbf{T}(\beta_2), \ldots, \mathbf{T}(\beta_n).$$

Hence by the completeness theorem for DRS_{2e} we have:

FACT 7. A sequent $\Gamma \Rightarrow \Delta$ over FORM is valid whenever the sequence $\mathbf{N}(\Gamma), \mathbf{T}(\Delta)$ of signed formulas has a proof in DRS_{2e}.

For any sequence Ω of signed formulas, let us denote

$$\Omega^+ = \{\alpha \in \text{FORM} : \mathbf{T}(\alpha) \in \Omega\}, \qquad \Omega^- = \{\alpha \in \text{FORM} : \mathbf{N}(\alpha) \in \Omega\}.$$

Then Fact 7 can be rephrased as follows:

FACT 8. For any finite sequence Ω of signed formulas, the sequent $\Omega^- \Rightarrow \Omega^+$ is valid iff Ω has a proof in DRS_{2e}.

The above fact suggests a simple method of transforming the system DRS_{2e} for sequences of signed formulas into a sequent calculus SC_{2e} (over FORM) for L_{2e}. Namely, it suffices to replace the axiomatic sequences and decomposition rules of DRS_{2e} by axioms and inference rules of the sequent calculus SC_{2e} according to the general algorithm given below (see [Ko02] for details):

- If Ω is an axiomatic sequence of DRS_{2e}, then $\Omega^- \Rightarrow \Omega^+$ is an axiom of SC_{2e}.

- If $\dfrac{\Omega}{\Omega_1}$ is a rule in DRS_{2e}, then $\dfrac{\Gamma, \Omega_1^- \Rightarrow \Delta, \Omega_1^+}{\Gamma, \Omega^- \Rightarrow \Delta, \Omega^+}$ is an inference rule in SC_{2e}.

- If $\dfrac{\Omega}{\Omega_1 \mid \Omega_2}$ is a rule in DRS_{2e}, then

$$\dfrac{\Gamma, \Omega_1^- \Rightarrow \Delta, \Omega_1^+ \qquad \Gamma, \Omega_2^- \Rightarrow \Delta, \Omega_2^+}{\Gamma, \Omega^- \Rightarrow \Delta, \Omega^+} \qquad \text{is an inference rule in SC}_{2e}.$$

As we can see, the above transformation turns the R-S rules upside down, i.e. the sequence(s) under/above the double line in an R-S rule give rise to the sequent(s) above/under the single line in the corresponding sequent calculus rule. However, considering the "upside-down" nomenclature for R-S rules, premises are still translated to premises, and conclusion — to conclusion, and the above vertical turnaround just reflects the converse direction of proof generation in the R-S and Gentzen deduction systems The transformation is best exemplified by the table below [Ko02], which gives sequent calculus rules corresponding to typical R-S decomposition rules:

R-S decomposition rule	Inference rule for sequents
$\dfrac{\Omega', \mathbf{T}(\alpha), \Omega''}{\Omega', \mathbf{T}(\alpha_1), \Omega''}$	$\dfrac{\Gamma \Rightarrow \Delta, \alpha_1}{\Gamma \Rightarrow \Delta, \alpha}$
$\dfrac{\Omega', \mathbf{N}(\alpha), \Omega''}{\Omega', \mathbf{N}(\alpha_1), \Omega''}$	$\dfrac{\Gamma, \alpha_1 \Rightarrow \Delta}{\Gamma, \alpha \Rightarrow \Delta}$
$\dfrac{\Omega', \mathbf{T}(\alpha), \Omega''}{\Omega', \mathbf{T}(\alpha_1), \mathbf{T}(\alpha_2), \Omega''}$	$\dfrac{\Gamma \Rightarrow \Delta, \alpha_1, \alpha_2}{\Gamma \Rightarrow \Delta, \alpha}$
$\dfrac{\Omega', \mathbf{N}(\alpha), \Omega''}{\Omega', \mathbf{N}(\alpha_1), \mathbf{N}(\alpha_2), \Omega''}$	$\dfrac{\Gamma, \alpha_1, \alpha_2 \Rightarrow \Delta}{\Gamma, \alpha \Rightarrow \Delta}$
$\dfrac{\Omega', \mathbf{T}(\alpha), \Omega''}{\Omega', \mathbf{T}(\alpha_1), \Omega'' \mid \Omega', \mathbf{T}(\alpha_2), \Omega''}$	$\dfrac{\Gamma \Rightarrow \Delta, \alpha_1 \qquad \Gamma \Rightarrow \Delta, \alpha_2}{\Gamma \Rightarrow \Delta, \alpha}$
$\dfrac{\Omega', \mathbf{N}(\alpha), \Omega''}{\Omega', \mathbf{N}(\alpha_1), \Omega'' \mid \Omega', \mathbf{N}(\alpha_2), \Omega''}$	$\dfrac{\Gamma, \alpha_1 \Rightarrow \Delta \qquad \Gamma, \alpha_2 \Rightarrow \Delta}{\Gamma, \alpha \Rightarrow \Delta}$

Hence the sequent calculus SC_{2e} obtained out of the R-S system DRS_{2e} for L_{2e} is as follows:

Axioms:

(i) $\quad \Gamma, \alpha \Rightarrow \Delta, \alpha$ $\qquad$ (ii) $\quad \Gamma, \alpha, \neg\alpha \Rightarrow \Delta$ $\quad$ (iii) $\quad \Gamma, \alpha \Rightarrow \Delta, \circ\alpha$

(iv) $\quad \Gamma, \neg\alpha \Rightarrow \Delta, \circ\alpha$ $\quad$ (v) $\quad \Gamma, \neg\circ\alpha \Rightarrow \Delta$

Inference rules:

$(\Rightarrow \neg\neg)$ $\qquad \dfrac{\Gamma \Rightarrow \Delta, \alpha}{\Gamma \Rightarrow \Delta, \neg\neg\alpha}$

$(\neg\neg \Rightarrow)$ $\qquad \dfrac{\Gamma, \alpha \Rightarrow \Delta}{\Gamma, \neg\neg\alpha \Rightarrow \Delta}$

$(\Rightarrow \neg\circ)$ $\qquad \dfrac{\Gamma \Rightarrow \Delta}{\Gamma \Rightarrow \Delta, \neg\circ\alpha}$

$(\Rightarrow \circ\circ)$ $\qquad \dfrac{\Gamma \Rightarrow \Delta, \circ\alpha}{\Gamma \Rightarrow \Delta, \circ\circ\alpha}$

$(\circ\circ \Rightarrow)$ $\qquad \dfrac{\Gamma, \circ\alpha \Rightarrow \Delta}{\Gamma, \circ\circ\alpha \Rightarrow \Delta}$

$(\Rightarrow \circ\neg)$ $\qquad \dfrac{\Gamma \Rightarrow \Delta, \circ\alpha}{\Gamma \Rightarrow \Delta, \circ\neg\alpha}$

$(\circ\neg \Rightarrow)$ $\qquad \dfrac{\Gamma, \circ\alpha \Rightarrow \Delta}{\Gamma, \circ\neg\alpha \Rightarrow \Delta}$

$(\Rightarrow \alpha \vee \beta)$ $\qquad \dfrac{\Gamma \Rightarrow \Delta, \circ\alpha \qquad \Gamma \Rightarrow \Delta, \alpha, \beta}{\Gamma \Rightarrow \Delta, \alpha \vee \beta}$

$(\alpha \vee \beta \Rightarrow)$ $\qquad \dfrac{\Gamma, \alpha \Rightarrow \Delta \qquad \Gamma, \circ\alpha, \beta \Rightarrow \Delta}{\Gamma, \alpha \vee \beta \Rightarrow \Delta}$

$(\Rightarrow \circ(\alpha \vee \beta))$ $\qquad \dfrac{\Gamma \Rightarrow \Delta, \circ\alpha \qquad \Gamma \Rightarrow \Delta, \alpha, \circ\beta}{\Gamma \Rightarrow \Delta, \circ(\alpha \vee \beta)}$

$(\circ(\alpha \vee \beta) \Rightarrow)$ $\qquad \dfrac{\Gamma, \alpha \Rightarrow \Delta \qquad \Gamma, \circ\alpha, \circ\beta \Rightarrow \Delta}{\Gamma, \circ(\alpha \vee \beta) \Rightarrow \Delta}$

$$(\Rightarrow \neg(\alpha \lor \beta)) \quad \frac{\Gamma \Rightarrow \Delta, \neg\alpha \quad \Gamma \Rightarrow \Delta, \neg\beta}{\Gamma \Rightarrow \Delta, \neg(\alpha \lor \beta)}$$

$$(\Rightarrow \neg(\alpha \lor \beta)) \quad \frac{\Gamma, \neg\alpha, \neg\beta \Rightarrow \Delta}{\Gamma, \neg(\alpha \lor \beta) \Rightarrow \Delta}$$

A sequent S over FORM is said to be *provable* in SC_{2e}, in symbols $\vdash_{SC_{2e}}$, if S can be deduced from the axioms of SC_{2e} using the inference rules of SC_{2e}.

THEOREM 9. *The system SC_{2e} is sound and complete for sequents over the language L_{2e}, i.e. for any sequent S over* FORM, $\models_{L_{2e}} S$ *iff* $\vdash_{SC_{2e}} S$.

The proof is a special case of the general proof for the completeness of a sequent calculus obtained out of a complete R-S system using the translation presented above, given in [Ko02]. It is based on the completeness theorem for DRS_{2e} and Fact 7, together with the fact that the translation of DRS_{2e} to SC_{2e} is an equivalent one. From the definition of the consequence relation of L_{2e}, we immediately conclude the following:

COROLLARY 10. *SC_{2e} is sound and complete for the consequence relation $\vdash_{L_{2e}}$ limited to finite sets of formulas, i.e. for any finite $F \subset$ FORM and any $\varphi \in$ FORM, $F \vdash_{L_{2e}} \varphi$ iff $F \Rightarrow \varphi$ is provable in SC_{2e}.*

As SC_{2e} has been obtained from DRS_{2e} using a translation of linear complexity, then from Lemma 5, Corollary 10 and Theorem 9 we immediately obtain:

COROLLARY 11. *Provability in SC_{2e} and the consequence relation $\vdash_{L_{2e}}$ are decidable, with complexity $O(2^n)$.*

7 Conclusions and future work

We have presented here a four-valued logic for reasoning about the two basic kinds of computation errors: a finite, "machine" error signalled by the computer itself and an infinite, inherently undecidable error representing an infinite computation. The logic has been obtained by combining two famous three-valued logics: Kleene logic, which can be shown to adequately handle the machine error, and McCarthy logic, which was originally designed for reasoning about machine computation, and whose asymmetric disjunction and conjunction are best suited for representing the infinite computation error.

Out of the two sound complete deduction systems we have developed for our logic, the Rasiowa-Sikorski style deduction system DRS_{2e} operates on sequences of signed formulas and is particularly suitable for proof search and automatic deduction, for it defines a decision procedure for the logic which constructs a proof for a provable formula, and a refutation for a non-provable one. In turn, the Gentzen-style sequent calculus SC_{2e} operates on ordinary formulas in the language of L_{2e}, gives a direct representation of the consequence relation of our logics, and is better understandable to the majority of logicians who are not cognisant with the R-S formalism.

The novelty of this logic is the dichotomous operator $\circ$, which partitions the four logical values into three finite ones (including machine error) and the single infinite value, representing infinite computation. Such a division reflects the fundamental difference in the criticality levels of machine errors and infinite computations: the former are practically unavoidable but can be dealt with using exception handling, while the latter are inherently unmanageable.

The logic considered in this paper is a simple propositional logic without any program constructs, so it would be difficult to either compare adequately it with any real program logic or give an in-depth example of its use in reasoning about programs here. However, it can be used as a building block in constructing a real logic for reasoning about programs or their specifications, e.g. a static logic following the lines of [KTB91] or [Ko93], a dynamic logic of programs in the Hoare style [Hoa95], or a variant of algorithmic logic similar to [Sa77]. It is easy to see that in such a logic the operator $\circ$ could be used for verifying exception handling — for example, by showing that a conditional instruction having α as a condition and equipped with an exception handling mechanism will terminate correctly if we can assert $\circ\alpha$ in the state before its execution.

More importantly, the rules for $\circ$ given here (extended to take into consideration functions, predicates, program constructs, etc.) could be used to prove that certain programs or procedures will not lead to infinite computations — which is of fundamental importance in e.g. mission-critical business applications. In that case, it would be advisable to consider a certain array of machine errors (possibly linked in a kind of hierarchy based on their character or severity) handled using a suitable exception handling mechanism.

BIBLIOGRAPHY

[BCJ84] Barringer H., Chang J.H., Jones C.B., *A logic covering undefinedness in program proofs*, Acta Informatica **21** (1984), pp. 251–269,

[Be59] Beth E.W. *The Foundations of Mathematics*, North-Holland, 1959,

[Bli91] Blikle A., *Three-valued predicates for software specification and validation*, Fundamenta Informaticae **14**(1991), pp. 387–410,

[DOR94] Demri S., Orlowska E., Rewitzky I., *Towards reasoning about Hoare relations*, Annals of Mathematics and Artificial Intelligence 12, 1994, pp. 265-289,

[GPO07] Golinska-Pilarek J., Orlowska E., *Tableaux and Dual Tableaux: Transformation of Proofs*, Studia Logica **85(3)** (2007), pp. 291-310,

[Hoa95] Hoare C.A.R., *An axiomatic basis for computer programming*, Comm. ACM. **12**(1969), pp. 576–583,

[Hog87] Hogevijs A., *Partial predicate logic in computer science*, Acta Informatica **24** (1987), pp. 381–393,

[Kl52] Kleene S.C., *Introduction to Metamathematics*, North-Holland, 1952,

[Ko93] Konikowska B., *Two over three: a two-valued logic for software specification and validation over a three-valued predicate calculus*, Journal for Applied Nonclassical Logic 3(1), 1993, pp. 39–71,

[KMO98] Konikowska B., Morgan C.G., Orlowska E., *A Relational Formalisation of Arbitrary Finite Valued Logics*, Logic Journal of the IGPL **6(5)** (1998), pp. 755-774,

[Ko99] Konikowska B., *Rasiowa-Sikorski deduction systems: a handy tool for Computer Science logic*, in: Proceedings WADT'98, Springer LNCS 1589, pp. 183-197, 1999,

[Ko02] Konikowska B., *Rasiowa-Sikorski deduction systems in computer science applications*, Theoretical Computer Science **286**, 2002, pp. 323–366,

[KTB91] Konikowska, B., Tarlecki, A., Blikle, J., *A three-valued logic for software specification and validation*, Fundamenta Informaticae 14(4), 1991, pp. 411–453,

[MC67] McCarthy J., *A basis for a mathematical theory of computation* Computer Programming and Formal Systems, North-Holland, 1967,

[OOP03] Omodeo E.G., Orlowska E., Policriti A., *Rasiowa-Sikorski Style Relational Elementary Set Theory*, RelMiCS 2003, pp. 215-226,

[Or88] Orlowska E., *Relational interpretation of modal logics*, in: Andreka, H., Monk, D. and Nemeti, I. (eds), Algebraic Logic, Colloquia Mathematica Societatis Janos Bolyai 54, North Holland, Amsterdam, 1988, pp. 443-471,

[Or88] Orlowska E., *Proof system for weakest prespecification and its applications*, Springer Lecture Notes in Computer Science 324, 1988, pp. 463-471,

[Or93] Orlowska E., *Dynamic logic with program specifications and its relational proof system*, Journal of Applied Non-Classical Logic 3, 1993, pp. 147-171,

[Owe85] Owe O., *An approach to program reasoning based on first order logic for partial functions*, Res. Rep. Institute of Informatics, University of Oslo, no. **89**, 1985,

[RS63] Rasiowa H., Sikorski R., *The mathematics of metamathematics*, Warsaw, PWN (Polish Scientific Publishers), 1963,

[Sa77] Salwicki A., *Algorithmic Logic: a tool for investigations of programs*, in: Butts, R.E., and Hintikka, K.J.J., eds., Logic, Foundations of Mathematics and Comoutability Theory (Reidel, Dordrecht 1977), pp. 281–295.

Beata Konikowska
Institute of Computer Science
Polish Academy of Sciences
Warsaw
E-mail: beatak@ipipan.waw.pl

Towards Fully Automated Axiom Extraction for Finite-Valued Logics

JOÃO MARCOS AND DALMO MENDONÇA

ABSTRACT. We implement an algorithm for extracting appropriate collections of classic-like sound and complete tableau rules for a large class of finite-valued logics. Its output consists of `Isabelle` theories.[1]

Keywords: many-valued logics, tableaux, automated theorem proving.

1 Introduction

This paper will report on the first developments towards the implementation of a fully automated program for the extraction of adequate proof-theoretical counterparts for sufficiently expressive logics characterized by way of a finite set of finite-valued truth-tables. The underlying algorithm was first described in [4]. Surveys on tableaux for many-valued logics can be found in [7, 1]. Our implementation has been performed in ML, and its application gives rise to an `Isabelle` theory (check [8]) formalizing a given finite-valued logic in terms of two-signed tableau rules.

The survey paper [7] points at a few very good theoretical motivations for studying tableaux for many-valued logics, among them:

- tableau systems are a particularly well-suited starting point for the development of computational insights into many-valued logics;

- a close interplay between model-theoretic and proof-theoretic tools is necessary and fruitful during the development of proof procedures for non-classical logics.

Section 2, right below, recalls the relevant definitions and some general results concerning many-valued logics as well as their homologous presentation in terms of bivalent semantics described by clauses of a certain format we call 'gentzenian'. An algorithm for endowing any sufficiently expressive finite-valued logic with an adequate bivalent semantics is exhibited and illustrated for the case of L_3, the well-known 3-valued logic of Łukasiewicz.

The main concepts concerning tableau systems in general and the particular results that allow one to transform any computable gentzenian semantics into a corresponding collection of tableau rules are illustrated in section 3, again for the case of L_3.

---

[1] A development snapshot of the code may be found at http://tinyurl.com/5cakro.

Section 4 discusses our current implementation, carefully explaining its expected inputs and outputs, and yet again illustrates its functioning for the case of L_3. Advantages and shortcomings of our program, in its present state of completion, as well as conclusions and some directions for future developments are mentioned in section 5.

2   Many-valued logics

Given a denumerable set At of *atoms* and a finite family $Cct = \{\textcircled{c}_j^i\}_{j \in J}$ of *connectives*, where $\mathsf{arity}(\textcircled{c}_j^i) = i$, let $\mathcal{S}$ denote the term algebra freely generated by Cct over At. Here, a *semantics* Sem for the algebra $\mathcal{S}$ will be given by any family of mappings $\{\S_k^{\mathcal{V}}\}_{k \in K}$ where $\mathsf{dom}(\S_k^{\mathcal{V}}) = \mathcal{S}$ and $\mathsf{codom}(\S_k^{\mathcal{V}}) = \mathcal{V}_k$, and where each collection of *truth-values* $\mathcal{V}_k$ is partitioned into sets of designated values, $\mathcal{D}_k$, and undesignated ones, $\mathcal{U}_k$. The mappings $\S_k^{\mathcal{V}}$ themselves may be called *(κ-valued) valuations*, where $\kappa = \mathsf{Card}(\mathcal{V}_k)$. A *bivalent semantics* is any semantics where $\mathcal{D}_k$ and $\mathcal{U}_k$ are singleton sets, for every $k \in K$. For bivalent semantics, valuations are often called *bivaluations*. The canonical notion of (single-conclusion) *entailment* $\models_{Sem} \subseteq \mathsf{Pow}(\mathcal{S}) \times \mathcal{S}$ induced by a semantics Sem is defined by setting $\Gamma \models_{Sem} \varphi$ iff $\S_k^{\mathcal{V}}(\varphi) \in \mathcal{D}_k$ whenever $\S_k^{\mathcal{V}}(\Gamma) \subseteq \mathcal{D}_k$, for every $\S_k^{\mathcal{V}} \in Sem$. The pair $\langle \mathcal{S}, \models_{Sem} \rangle$ may then be called a *generic κ-valued logic*, where $\kappa = \mathsf{Max}_{k \in K}(\mathsf{Card}(\mathcal{V}_k))$.

If one now fixes the sets of truth-values $\mathcal{V}$, $\mathcal{D}$ and $\mathcal{U}$, and fixes, for each connective $\textcircled{c}_j^i$ an interpretation $\widehat{\textcircled{c}}_j^i : \mathcal{V}^i \longrightarrow \mathcal{V}$, one may immediately build from that an associated algebra of truth-values $\mathcal{TV} = \langle \mathcal{V}, \mathcal{D}, \{\widehat{\textcircled{c}}_j^i\}_{j \in J} \rangle$ (in the present paper, whenever there is no risk of confusion, we shall not differentiate notationally between a connective symbol $\textcircled{c}$ and its operational interpretation $\widehat{\textcircled{c}}$). A *truth-functional* semantics is then defined by the collection of all homomorphisms of $\mathcal{S}$ into $\mathcal{TV}$. In the present paper, the shorter expression *κ-valued logic* (or, in general, *finite-valued logic*) will be used to qualify any generic κ-valued truth-functional logic, for some finite κ, where κ is the minimal value for which the mentioned logic can be given a truth-functional semantics characterizing the same associated notion of entailment.

It is interesting to observe that the canonical notion of entailment of any given semantics, and in particular of any given truth-functional semantics, may be emulated by a bivalent semantics. Indeed, let $\mathcal{V}^2 = \{F, T\}$ and $\mathcal{D}^2 = \{T\}$, and consider the 'binary print' of the algebraic truth-values produced by the total mapping $t : \mathcal{V} \longrightarrow \mathcal{V}^2$, defined by $t(v) = T$ iff $v \in \mathcal{D}$. For any κ-valued valuation $\S^{\mathcal{V}}$ of a given semantics Sem, consider now the characteristic total function $b_{\S} = t \circ \S^{\mathcal{V}}$. Now, collect all such bivaluations $b_{\S}$'s into a new semantics $Sem(2)$, and note that $\Gamma \models_{Sem(2)} \varphi$ iff $\Gamma \models_{Sem} \varphi$. As a matter of fact, the standard 2-valued notion of inference of Classical Logic is characterized indeed by a finite-valued semantics that is simultaneously bivalent and truth-functional. In general, nonetheless, if a logic is κ-valued, for $\kappa > 2$, a bivalent characterization of it will explore the trade-off between, on the one hand, the 'algebraic perspective' of many-valuedness, with its many 'algebraic truth-values' and its semantic characterization in terms of a set of homomorphisms, and, on the other hand, the classic-inclined 'logical perspective', with its emphasis on characterizations based on two 'logical values' (for more detailed discussions of this issue, check [4, 11]). Our interest in this paper is to probe some of the practical advantages of the bivalent classic-like perspective as applied to the wider domain of finite-valued truth-functional logics.

Our running example in the present paper will involve Łukasiewicz's well-known 3-valued logic Ł$_3$, characterized by the algebra of truth-values $\langle\{1, \frac{1}{2}, 0\}, \{1\}, \{\neg, \rightarrow, \vee, \wedge\}\rangle$, where the interpretation of the unary negation connective $\neg$ sets $\neg v_1 = 1 - v_1$ and the interpretation of the binary implication connective $\rightarrow$ sets $(v_1 \rightarrow v_2) = \mathsf{Min}(1, 1 - v_1 + v_2)$. The binary symbols $\vee$ and $\wedge$ can be introduced as primitive if we interpret them by setting $(v_1 \vee v_2) = \mathsf{Max}(v_1, v_2)$ and $(v_1 \wedge v_2) = \mathsf{Min}(v_1, v_2)$, but they can also, more simply, be introduced by definition just like classical disjunction and conjunction, setting $\alpha \vee \beta \overset{\text{def}}{=} (\alpha \rightarrow \beta) \rightarrow \beta$ and $\alpha \wedge \beta \overset{\text{def}}{=} \neg(\neg\alpha \vee \neg\beta)$. The binary prints of an arbitrary atom of Ł$_3$ and of its negation are illustrated in the table below.

v	$t(v)$	$\neg v$	$t(\neg v)$
0	F	1	T
$\frac{1}{2}$	F	$\frac{1}{2}$	F
1	T	0	F

(1)

Given some finite-valued logic $\mathcal{L}$ based on a set of truth-values $\mathcal{V}$, we say that $\mathcal{L}$ is *functionally complete* over $\mathcal{V}$ if any κ-valued n-ary operation, for $\kappa = \mathsf{Card}(\mathcal{V})$, is definable with the help of a suitable combination of its primitive operators $\{\textcircled{c}_j^i\}_{j\in J}$. When $\mathcal{L}$ is not functionally complete from the start, we may consider $\mathcal{L}^{\mathbf{fc}}$ as any functionally complete κ-valued conservative extension of $\mathcal{L}$. Given truth-values $v_1, v_2 \in \mathcal{V}$, we say that they are *separated*, and we write $v_1 \natural v_2$, in case v_1 and v_2 belong to different classes of truth-values, that is, in case either $v_1 \in \mathcal{D}$ and $v_2 \in \mathcal{U}$, or $v_1 \in \mathcal{U}$ and $v_2 \in \mathcal{D}$. Given a unary, primitive or defined, connective $\textcircled{s}$ of a given truth-functional logic, with interpretation $\widehat{\textcircled{s}}$, we say that $\textcircled{s}$ *separates* v_1 and v_2 in case $\widehat{\textcircled{s}}(v_1)\natural\widehat{\textcircled{s}}(v_2)$. Obviously, for any pair of truth-values in $\mathcal{V}$ it is possible to define in the term algebra of $\mathcal{L}^{\mathbf{fc}}$ an appropriate *separating connective* $\textcircled{s}$. When that separation can be done exclusively with the help of the original language of $\mathcal{L}$, we say that $\mathcal{V}$ is *effectively separable*, and the logic $\mathcal{L}$, in that case, will be considered to be *sufficiently expressive* for our purposes. It should be noticed that the vast majority of the most well-known finite-valued logics enjoy this expressivity property.

Notice in particular, from Table (1) above, how the negation connective of Ł$_3$ separates the two undesignated truth-values, $\frac{1}{2}$ and 0. Based on this table, one may in fact easily provide a unique identification to each of the 3 initial algebraic truth-values of L_3, by way of the 3 following statements:

$$
\begin{array}{llll}
v = 1 & \text{iff} & t(v) = T & \text{(I)} \\
v = \tfrac{1}{2} & \text{iff} & t(v) = F \text{ and } t(\neg v) = F & \\
v = 0 & \text{iff} & t(v) = F \text{ and } t(\neg v) = T &
\end{array}
$$

One can also use this separating connective $\circledS = \lambda u.\neg u$ in order to provide a bivalent description of each of the operators of the language. Consider for instance the cases of $\copyright_1 = \lambda vw.(v \to w)$ and $\copyright_2 = \lambda vw.\neg(v \to w)$ (that is, $\copyright_2$ is $\circledS\copyright_1$):

$\copyright_1$	0	$\tfrac{1}{2}$	1
0	1	1	1
$\tfrac{1}{2}$	$\tfrac{1}{2}$	1	1
1	0	$\tfrac{1}{2}$	1

$\copyright_2$	0	$\tfrac{1}{2}$	1
0	0	0	0
$\tfrac{1}{2}$	$\tfrac{1}{2}$	0	0
1	1	$\tfrac{1}{2}$	0

(2)

From table $\copyright_2$ it is clear for instance that:

$$
\circledS(\neg(\alpha \to \beta)) = 1 \qquad \text{iff} \qquad \circledS(\alpha) = 1 \text{ and } \circledS(\beta) = 0 \tag{II}
$$

Let's write $T{:}\varphi$ and $F{:}\varphi$, respectively, as abbreviations for $t(\circledS(\varphi)) = T$ and $t(\circledS(\varphi)) = F$. Then, the statement (II) may be described in bivalent form, with the help of (I), by writing:

$$
T{:}\neg(\alpha \to \beta) \qquad \text{iff} \qquad T{:}\alpha \text{ and } (F{:}\beta \text{ and } T{:}\neg\beta) \tag{III}
$$

In [4] an algorithm that constructively specifies a bivalent semantics for any sufficiently expressive finite-valued logic was proposed. The output of the algorithm is a computable class of clauses governing the behavior of all the bivaluations that together will canonically define an entailment that coincides with the original entailment defined with the help of the algebra of truth-values $\mathcal{TV}$ and the class of all the corresponding finite-valued homomorphisms of $\mathcal{S}$ into $\mathcal{TV}$. Moreover, all those clauses are in a specific format we call *gentzenian*, namely, they are conditional expressions (E) of the form $\Phi \Rightarrow \Psi$ where the symbol $\Rightarrow$ represents a meta-linguistic implication, and both Φ and Ψ are meta-formulas of the form $\top$ (top), $\bot$ (bottom) or a clause (G) of the form $\bigvee_{1 \leq i \leq m} \bigwedge_{1 \leq j \leq n_m} A(i, j, w)$, where each $A(i, j, w)$ has the form $b(\varphi_i^j) = w_i^j$, for some given $\varphi_i^j \in \mathcal{S}$ and $w_i^j \in \{F, T\}$. (Recall that we use $b : \mathcal{S} \to \mathcal{V}^2$ for bivaluations.) If we let $|$ represent disjunction and $\&$ represent conjunction in the meta-language, any clause (G) will have thus the extended format:

$$
b(\varphi_1^1) = w_1^1 \,\&\ldots\&\, b(\varphi_1^{n_1}) = w_1^{n_1} \mid \ldots \mid b(\varphi_m^1) = w_m^1 \,\&\ldots\&\, b(\varphi_m^{n_m}) = w_m^{n_m}.
$$

The meta-logic governing such meta-linguistic expressions is FOL, First-Order Classical Logic. Accordingly, an expression of the form $A_1|A_2 \Rightarrow A_3$ will be equivalent to $(A_1 \Rightarrow A_3)\&(A_2 \Rightarrow A_3)$, and an expression of the form $A_1\&A_2 \Rightarrow A_3$ will be equivalent to $A_1 \Rightarrow A_3|{\sim}A_2$, where the meta-linguistic negation $\sim$ is such that $\sim(b(\varphi) = T)$ denotes $b(\varphi) = F$, and $\sim(b(\varphi) = F)$ denotes $b(\varphi) = T$. We may also write $\Phi \Leftrightarrow \Psi$ as an abbreviation for $(\Phi \Rightarrow \Psi)\&(\Psi \Rightarrow \Phi)$.

With a slight notational change and using FOL, one can now see (III) as a description done in an abbreviated gentzenian format:

$$T{:}\neg(\alpha \to \beta) \quad \Leftrightarrow \quad T{:}\alpha \ \& \ F{:}\beta \ \& \ T{:}\neg\beta \qquad \text{(IV)}$$

Following the above line of reasoning, and considering now table $©_1$ instead of $©_2$, it is also correct to write, for instance, the clause:

$$
\begin{aligned}
F{:}(\alpha \to \beta) \quad \Leftrightarrow \quad & T{:}\alpha \ \& \ F{:}\beta \ \& \ F{:}\neg\beta \ | \qquad \text{(V)}\\
& T{:}\alpha \ \& \ F{:}\beta \ \& \ T{:}\neg\beta \ |\\
& F{:}\alpha \ \& \ F{:}\neg\alpha \ \& \ F{:}\beta \ \& \ T{:}\neg\beta
\end{aligned}
$$

According to the reductive algorithm described in [4], a sound and complete bivalent version of any sufficiently expressive finite-valued logic $\mathcal{L}$ is obtained if one performs the following two steps:

[step 1] Iterate the above illustrated procedure in order to obtain clauses describing exactly in which situations one may assert the sentences $T{:}©(\alpha_1, \dots , \alpha_n)$, $F{:}©(\alpha_1, \dots , \alpha_n)$, $T{:}\text{⑤}©(\alpha_1, \dots , \alpha_n)$ and $F{:}\text{⑤}©(\alpha_1, \dots , \alpha_n)$, for each primitive n-ary $© \in \mathsf{Cct}$ and each one of the separating connectives ⑤ of $\mathcal{L}$.

[step 2] Add to all those clauses the following extra axioms governing the behavior of the admissible collection of bivaluations:

(C1) $\top \ \Rightarrow \ T{:}\alpha \ | \ F{:}\alpha$

(C2) $T{:}\alpha \ \& \ F{:}\alpha \ \Rightarrow \ \bot$

(C3) $T{:}\alpha \ \Rightarrow \ \bigvee_{d\in\mathcal{D}} \bigwedge_{1\leq m<n\leq\mathsf{Card}(\mathcal{D})} w^d_{mn}{:}\text{⑤}_{mn}(\alpha)$

(C4) $F{:}\alpha \ \Rightarrow \ \bigvee_{u\in\mathcal{U}} \bigwedge_{1\leq m<n\leq\mathsf{Card}(\mathcal{U})} w^u_{mn}{:}\text{⑤}_{mn}(\alpha)$

for every $\alpha \in \mathcal{S}$, where ⑤_{mn} is the unary (primitive or defined) connective that we use to separate the truth-values m and n, and $w^v_{mn} = t(\text{⑤}_{mn}(v))$.

Notice that (C3) and (C4) have the role of recording the relevant information on binary prints, as it can be found, in the case of L_3, in Table 1.

3   Tableaux

Generic tableau systems for finite-valued logics are known at least since [5]. In the corresponding tableaux, however, formulas may receive as many labels as the number of truth-values in $\mathcal{V}$, and that somewhat obstructs the task of comparing for instance the associated notions of proof and of consequence relation to the corresponding classical notions. However, as it will be shown, with the help of the bivalent semantics illustrated in the previous section it is now straightforward to produce sound and complete collections of classic-like two-signed tableau rules (i.e., each formula appears with exactly one of *two* labels at the head of each rule).

The basic idea, explained in [4], is to dispose the gentzenian clauses governing the admissible bivaluations in an appropriate way. For that matter, clauses such as (IV) and (V) can be rendered, respectively, into the following tableau rules:

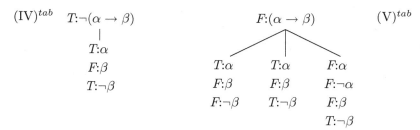

Each tableau branch can be seen as a non-empty sequence of signed formulas, $\Gamma, L{:}\varphi, \Delta$, where $L \in \{F, T\}$, $\varphi \in \mathcal{S}$ and Γ and Δ are sequences. To be sure, a rule such as $(\mathrm{V})^{tab}$ will be read thus as an instruction allowing one to transform a sequence $\Gamma, F{:}(\alpha \to \beta), \Delta$ into the three following new sequences:

[1] $\Gamma, T{:}\alpha, F{:}\beta, F{:}\neg\beta, \Delta$

[2] $\Gamma, T{:}\alpha, F{:}\beta, T{:}\neg\beta, \Delta$

[3] $\Gamma, T{:}\alpha, F{:}\neg\alpha, F{:}\beta, F{:}\neg\beta, \Delta$

If one recalls the first step of the algorithm described in the last section, there will be tableau rules with heads $L{:}\textcircled{c}\varphi$ and $L{:}\textcircled{s}\textcircled{c}\varphi$ for each sign $L \in \{F, T\}$, each primitive connective $\textcircled{c}$ and each separating connective $\textcircled{s}$. Rules $(\mathrm{IV})^{tab}$ and $(\mathrm{V})^{tab}$, above, are indeed examples of such tableau rules. To obtain soundness and completeness with respect to the original finite-valued semantics, in general, besides the rules produced in the first step one must also take into consideration the rules corresponding to axioms (C1)–(C4), mentioned in the second step. It is interesting to notice that in most practical cases, however, axioms (C3) and (C4) can often be directly proven from the remaining ones. Moreover, axiom (C2) expresses just the usual closure condition on tableau branches. On the other hand, axiom (C1) gives rise in general to the following *dual-cut* rule, for arbitrary α:

$$T : \alpha \qquad F : \alpha$$

All further definitions and concepts concerning the setup and construction of signed tableaux are standard (check [9]). In particular, a *branch* (sequence of signed formulas) is said to be *closed* if there is a formula that occurs in it both with the sign T and with the sign F, and a *closed tableau* for a given sequence of signed formulas is produced by a finite set of transformations allowed by the corresponding tableau rules to such an effect that at some point all branches that originate from the original branch are closed.

One could be worried, and with good reason, that the unrestrained use of the dual-cut rule might potentially make the corresponding tableaux non-analytic. We will discuss that issue in the conclusion. The tableau rules originated from the above procedure can naturally be used in order to prove theorems, check conjectures and suggest counter-models, but also, in the meta-theory, to formulate and prove derived rules that can be used to simplify the original presentation of the logic as originated by our algorithm. So, for instance, the above illustrated complex three-branching rule for $F : (\alpha \to \beta)$ can eventually be simplified into any one of the following equivalent two-branching rules:

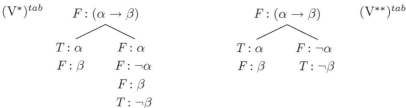

Tableau systems for the logic L_3, in particular, are known indeed at least since [10]. The latter have not been generated algorithmically, though, through an automated generic procedure such as the one we illustrate here.

4 Implementation, and applications

We used the functional programming language ML to automate the axiom extraction process. ML provides one, among other advantages, with an elegant and suggestive syntax, a compile-time type checking that is both modular and reliable, as well as a type inference mechanism that greatly helps both in preventing program errors and in the task of formal verification of correctness of the implemented algorithms.

The relevant inputs of our program include the detailed definition of a finite-valued logic, together with an appropriate set of separating connectives for that logic. Here's an example of an input for the logic L_3, presented above, where the functions CSym, CAri and CPre take a connective and return its symbol (syntactic sugar), arity and precedence/associativity rules, respectively. The primitive connectives are defined by their truth-tables, listed in CTabs. A truth-table of an n-ary connective ⓒ is represented as a list of all pairs ([$x_1, \ldots, x_n$], y) such that ⓒ($x_1, \ldots, x_n$) = y. Derived connectives should be defined by abbreviation in terms of the primitive ones, and these definitions are given by the function CDef.

```
(* PROGRAM INPUT, EXAMPLE OF L3 *)
   structure L3 : LOGIC =
struct
    val theoryName  = "TL3";
    val Values      = ["0", "1/2", "1"];
```

```
val Designated  = ["1"];
val Connectives = ["Neg", "Imp", "Disj", "Conj"];
val Primitives  = ["Neg", "Imp"];
val SeparatingD = [];
val SeparatingU = ["Neg"];

fun CSym "Neg"  = "~"
  | CSym "Imp"  = "-->"
  | CSym "Disj" = "|"
  | CSym "Conj" = "&";

fun CAri "Neg"  = 1
  | CAri "Imp"  = 2
  | CAri "Disj" = 2
  | CAri "Conj" = 2;

val CTabs = ref [
        ("Neg", [ (["0"],    "1"),
                  (["1/2"], "1/2"),
                  (["1"],    "0")] ) ,
        ("Imp", [ (["0", "0"],     "1"),
                  (["0", "1/2"],   "1"),
                  (["0", "1"],     "1"),
                  (["1/2", "0"],   "1/2"),
                  (["1/2", "1/2"], "1"),
                  (["1/2", "1"],   "1"),
                  (["1", "0"],     "0"),
                  (["1", "1/2"],   "1/2"),
                  (["1", "1"],     "1")] )
        ];

fun CDef "Disj" = ( "A0 | A1", "(A0 --> A1) --> A1" )
  | CDef "Conj" = ( "A0 & A1", "~(~A0 | ~A1)" );

fun CPre "Neg"  = "[40] 40"
  | CPre "Imp"  = "[34,35] 35"
  | CPre "Disj" = "[24,25] 25"
  | CPre "Conj" = "[29,30] 30";
end;
```

For every symbol from **Connectives** that may appear in both the list of **Primitives** and the list of rewrite rules **CDef**, our program calculates its corresponding truth-table in terms of those that can be found in **CTab**, and adds it to the input of the algorithm for extracting tableau rules. (A future version of the program shall handle truth-tables defined intensionally, by way of a lambda-calculus-like expression.)

To perform the extraction, our program first generates a list of heads for all the necessary rules, according to the algorithm explained and illustrated in sections 2 and 3.

```
val rulesList = [T "~(A0)",      T "A0 --> A1",
                 T "~(~(A0))",   T "~(A0 --> A1)",
                 F "~(A0)",      F "A0 --> A1",
                 F "~(~(A0))",   F "~(A0 --> A1)" ]
```

Obviously, in general, if there are c primitive connectives and s separating connectives, there will be exactly $2 \times c \times s$ rule heads in `rulesList`.

Next, each connective's truth-table is converted into a table where each value is exchanged by its binary print. The binary print of a value is calculated based on the separating connectives given as input (`SeparatingD` for designated values and `SeparatingU` for undesignated values). The program represents such tables as lists of pairs (`input, output`). For the case of `Neg`, the sole separating connective of our version of L_3, the table contains the following information:

```
             (* A0 *)                    (* ~(A0) *)
(* 0    *)   ( [F "A0", T "~(A0)"],  [T "~(A0)"] )
(* 1/2 *)    ( [F "A0", F "~(A0)"],  [F "~(A0)", F "~(~(A0))"] )
(* 1    *)   ( [T "A0"],             [F "~(A0)", T "~(~(A0))"] )
```

Now, for each expression in `rulesList`, a search through all tables generated in the last step is done, and all clauses in which the given formula appears on the right-hand side (as output) are returned. The left-hand side (input) of these clauses are the branches of the desired tableau rule. As an example, two of the clauses involving implication, and corresponding to the tableau rules $(\mathrm{IV})^{tab}$ and $(\mathrm{V})^{tab}$ from the last section, are calculated and recorded by the program as follows:

```
(* IV *) ([ [T "A0",F "A1",T "~(A1)"] ],      T "~(A0 --> A1)")
(* V  *) ([ [F "A0",F "~(A0)",F "A1",T "~(A1)"],
           [T"A0",F "A1",T "~(A1)"],
           [T"A0",F "A1",F "~(A1)"] ] ,      F "A0 --> A1")
```

The next, and final, steps include calculating axioms (C3) and (C4), and printing all definitions, syntactical details and rules into a theory file, ready to be used by `Isabelle`.

`Isabelle`, also written in `ML`, is a generic theorem-proving environment based on a higher-order meta-logic in which it is quite simple to create formal theories with rules and axioms for various kinds of deductive formalisms, and equally easy to define tacticals and to prove theorems about the underlying formal systems.

Here's an illustration of how we use `Isabelle`'s syntax for representing tableaux, extending the theory `Sequents.thy` that comes with `Isabelle`'s default library, and taking as example the file generated by our program as output for the logic L_3:

```
theory TL3

imports Sequents
begin
typedecl a
```

```
consts
  Trueprop :: "(seq'=>seq') => prop"
  TR       :: "a => o"            ("T:_" [20] 20)
  FR       :: "a => o"            ("F:_" [20] 20)
  Neg      :: "a => a"            ("~ _" [40] 40)
  Imp      :: "[a,a] => a"        ("_-->_" [24,25] 25)
syntax "@Trueprop"      ::  "(seq) => prop"  ("[_]" 5)

ML
{*
fun seqtab_tr  c [s] =  Const(c,dummyT) $ seq_tr s;
fun seqtab_tr' c [s] =  Const(c,dummyT) $ seq_tr' s;
*}

parse_translation {* [("@Trueprop", seqtab_tr "Trueprop")] *}
print_translation {* [("Trueprop", seqtab_tr' "@Trueprop")] *}

local
axioms

axC1:  "[| [ $H, T:A ] ; [ $H, F:A ] |] ==> [$H]"

axC21: "[ $H, T:A, $E, F:A, $G]"
axC22: "[ $H, F:A, $E, T:A, $G]"

axC3:  "[| [$H, T:A, $G ] |]
          ==> [ $H, T:A, $G ]"
axC4:  "[| [ $H, F:A, T:~(A), $G ] ;
           [ $H, F:A, F:~(A), $G ] |]
          ==> [ $H, F:A, $G ]"

ax0:   "[| [ $H, F:A0, T:~(A0), $G ] |]
          ==> [ $H, T:~(A0), $G ]"

ax1:   "[| [ $H, F:A0, T:~(A0), F:A1, T:~(A1), $G ] ;
           [ $H, T:A0, T:A1, $G ] ;
           [ $H, F:A0, F:~(A0), T:A1, $G ] ;
           [ $H, F:A0, F:~(A0), F:A1, F:~(A1), $G ] ;
           [ $H, F:A0, T:~(A0), T:A1, $G ] ;
           [ $H, F:A0, T:~(A0), F:A1, F:~(A1), $G ] |]
          ==> [ $H, T:A0 --> A1, $G ]"

ax2:   "[| [ $H, T:A0, $G ] |]
          ==> [ $H, T:~(~(A0)), $G ]"

ax3:   "[| [ $H, T:A0, F:A1, T:~(A1), $G ] |]
          ==> [ $H, T:~(A0 --> A1), $G ]"

ax4:   "[| [ $H, F:A0, F:~(A0), $G ] ;
           [ $H, T:A0, $G ] |]
          ==> [ $H, F:~(A0), $G ]"

ax5:   "[| [ $H, F:A0, F:~(A0), F:A1, T:~(A1), $G ] ;
           [ $H, T:A0, F:A1, F:~(A1), $G ] ;
           [ $H, T:A0, F:A1, T:~(A1), $G ] |]
          ==> [ $H, F:A0 --> A1, $G ]"
```

```
ax6:     "[| [ $H, F:A0, F:~(A0), $G ] |]
            ==> [ $H, F:~(~(A0)), $G ]"

ax7:     "[| [ $H, F:A0, F:~(A0), F:A1, T:~(A1), $G ] ;
            [ $H, T:A0, F:A1, F:~(A1), $G ] |]
            ==> [ $H, F:~(A0 --> A1), $G ]"

ML {* use_legacy_bindings (the_context ()) *}

(* Abbreviations *)
Conj_def:  "A0 & A1 == ~(~A0 | ~A1)"
Disj_def:  "A0 | A1 == (A0 --> A1) --> A1"

(* Structural rules *)
thin:     "[$H, $E, $G] ==> [$H, $A, $E, $B, $G]"
exch:     "[$H, $A, $E, $B, $G] ==> [$H, $B, $E, $A, $G]"

end
```

In this theory, `consts` lists the formula constructors. `TR :: "a => o"` means that the constructor `TR` takes a formula (typed `a`) and returns a labeled formula (typed `o`). The addition of the structural rules, unusual for tableau systems, is motivated by the fact that our formalism for tableaux was based on `Isabelle`'s `Sequents.thy`, but also because we want to be able to prove not just *theorems* but also *derived rules*. A detailed example of such proofs will be presented below.

In the generated axioms corresponding to the tableau rules, `T:X` and `F:X` are (labeled) formulas, `$X`, is any sequence of such formulas, or contexts, each sequence between square brackets represents a tableau branch, and a collection of branches is delimited by `[|` and `|]`. The symbol `==>` denotes `Isabelle`'s meta-implication, that may be used to write down object-language rules, and `==` denotes `Isabelle`'s meta-equality, that may be used for stating rewrite rules concerning the derived operators of our logics. In `Isabelle`, the application of a rule means that it's possible to achieve the goal (sequence on the right of the meta-implication) once it's possible to prove the hypotheses (sequences on the left of the meta-implication), which constitute the collection of new subgoals that take the place of the original goal after the rule is applied. The dual-cut rule corresponds to axiom `axC1`, and the closure rule for a branch of the tableau corresponds to the axioms `axC21` and `axC22`. We retain both the latter rules as primitive for reasons of efficiency, but clearly one is derivable from the other with the use of of `exch`. Notice in particular how, in effect, the tableau *rules* produced for the original finite-valued logic provided as input correspond to *axioms* in the higher-order language of `Isabelle`.

The axiom set generated by the generic algorithm that we implemented isn't necessarily the most smart or efficient one. Axiom `axC3`, for instance, is clearly ineffectual in the above theory, having the same sequences at both sides of the meta-implication. Moreover, it may also happen that the formula over which the rule is applied also appears in some of the resulting branches, yet one is likely to try to control that phenomenon when aiming at defining tacticals for automated theorem proving.

Note that ax5 corresponds to rule $(V)^{tab}$ from the last section. The simpler rule $(V^{**})^{tab}$, mentioned in the same section, can obviously be written in Isabelle as:

```
ax5SS: "[| [ $H, T:A0, F:A1, $E ] ;
           [ $H, F:~(A0), T:~(A1), $E ] |]
       ==> [$H, F:A0 --> A1, $E]"
```

The proof that ax5 and ax5SS are indeed equivalent tableau rules, in the sense that one can be derived from the other in the presence of the remaining rules of our theory, can now be done directly with the help of Isabelle's meta-logic. The rules mentioned below are the ones listed in the above theory. In what follows, verbatim text prefixed by > indicates user input entered at Isabelle's command line environment.

First we prove that ax5SS can be derived from the axioms of our theory.

```
> Goal " [| [ $H, T:A0, F:A1, $G ] ;
             [ $H, F:~(A0), T:~(A1), $G ] |]
        ==> [$H, F:A0 --> A1, $G]";
```

Let A0 and A1 be represented by the schemas α and β and the contexts $H and $G be represented by Γ and Δ. We start thus to construct our tableau from the sequence $\Gamma, F{:}\alpha \to \beta, \Delta$ and intend to extend it into the two branches $\Gamma, T{:}\alpha, F{:}\beta, \Delta$ and $\Gamma, F{:}\neg\alpha, T{:}\neg\beta, \Delta$.

Notice that we can apply rule ax5 over the initial sequence:

```
> by (resolve_tac [ax5] 1);
```

The following three branches originate then from $\Gamma, F{:}\alpha \to \beta, \Delta$, as new subgoals:

[1] $\Gamma, F{:}\alpha, F{:}\neg\alpha, F{:}\beta, T{:}\neg\beta, \Delta$

[2] $\Gamma, T{:}\alpha, F{:}\beta, T{:}\neg\beta, \Delta$

[3] $\Gamma, T{:}\alpha, F{:}\beta, F{:}\neg\beta, \Delta$

To obtain the two initially intended branches from those, we may now just apply the *thinning* structural rule:

```
> by (res_inst_tac [("A","<<F:A0>>"),("B","<<F:A1>>")] thin 1);
```

Notice how that forces the instantiation of the sequences $A and $B, respectively, with the singleton sequences constituted of the signed formulas F:A0 and F:A1. The new version of subgoal [1.1] that results from that transformation is:

[1.1] $\Gamma, F{:}\neg\alpha, T{:}\neg\beta, \Delta$

Similar transformations can be applied to subgoals [2] and [3]:

```
> by (res_inst_tac [("A","<<T:~A1>>")] thin 2);
> by (res_inst_tac [("A","<<F:~A1>>")] thin 3);
```

This originates, of course:

[2.1] $\Gamma, T{:}\alpha, F{:}\beta, \Delta$

[3.1] $\Gamma, T{:}\alpha, F{:}\beta, \Delta$

Notice how [2.1] and [3.1] coincide with the second branch of **ax5SS** and [1.1] corresponds to its first branch. The proof will be finished thus if one identifies the latter subgoals 'by assumption' with the intended branches that were entered as premises of the initial **Goal** command:

```
> by (REPEAT (assume_tac 1));
```

The message **No subgoals!**, issued by **Isabelle**, indicates that the proof is done. We can now record the result under the name **ax5SS**:

```
> qed "ax5SS";
```

Conversely, in the next step, we will prove **ax5** as a derived rule of our tableau system from the remaining axioms, together with **ax5SS**.

```
> Goal "[| [ $H, F:A0, F:~(A0), F:A1, T:~(A1), $G ] ;
           [ $H, T:A0, F:A1, T:~(A1), $G ] ;
           [ $H, T:A0, F:A1, F:~(A1), $G ] |]
         ==> [ $H, F:A0 --> A1, $G ]";
```

This time one can apply rule **ax5SS** over the initial sequence, $\Gamma, F{:}\alpha \to \beta, \Delta$:

```
> by (resolve_tac [ax5SS] 1);
```

The following two branches originate then as new subgoals:

[1] $\Gamma, T{:}\alpha, F{:}\beta, \Delta$

[2] $\Gamma, F{:}\neg\alpha, T{:}\neg\beta, \Delta$

We may apply **ax0** to subgoal [2]:

```
> by (resolve_tac [ax0] 2);
```

That will transform subgoal [2] into the new branch:

[2.1] $\Gamma, F{:}\neg\alpha, F{:}\beta, T{:}\neg\beta, \Delta$

Next, we may apply **ax4** to [2.1]:

```
> by (resolve_tac [ax4] 2);
```

This will transform [2.1] into the two new branches:

[2.1.1] $\Gamma, F{:}\alpha, F{:}\neg\alpha, F{:}\beta, T{:}\neg\beta, \Delta$

[2.1.2] $\Gamma, T{:}\alpha, F{:}\beta, T{:}\neg\beta, \Delta$

This time thinning alone won't do to finish the proof, though, as more branches and formulas will be needed to emulate **ax5**. It will be helpful thus to apply **axC4** to subgoal [1], or else **axC1** with instantiation over the cut-formula **A**.

```
> by (resolve_tac [axC4] 1);
```

From that we know that subgoal [1] will be transformed into the two new branches:

[1.1] $\Gamma, T{:}\alpha, F{:}\beta, T{:}\neg\beta, \Delta$

[1.2] $\Gamma, T{:}\alpha, F{:}\beta, F{:}\neg\beta, \Delta$

Observe that [1.1] coincides with [2.1.2], an intended branch of ax5, and the two other branches are provided by [1.2] and [2.1.1]. Thus, we can finish the proof using the premises:

```
> by (REPEAT (assume_tac 1));
```

With that we have proven that the two tableau systems for L_3, one with ax5 and the other one with ax5SS in its place, are *equivalent*. The above proofs could of course have been performed, more appropriately, directly inside a theory obtained by erasing axiom ax5 from the above theory TL3.

One last observation. For each symbol that does not appear in Connectives and Primitives a definition by abbreviation is expected in CDef. In that case one could use a strategy similar to the one above to propose and prove in Isabelle a list of derived rules involving that connective symbol.

5   Epilogue

The present paper has reported on the first concrete implementation of a certain constructive procedure for obtaining adequate two-signed tableau systems for a large number of finite-valued logics. Expressing a variety of logics in the *same* framework is quite useful for the development of comparisons between such logics, including their expressive and deductive powers. The general algorithm for tableaux for finite-valued logics proposed in [5], despite being more generally applicable in that it does not require the input logics to be 'sufficiently expressive', produces tableau rules having as many signs as the number of truth-values in the original semantical presentation of the given logic. Thus, for instance, even though a logic such as Łukasiewicz's L_5 is a deductive fragment of L_3 (in general, L_m is a deductive fragment of L_n iff $n-1$ is a divisor or $m-1$; check ch.8.5 of [6]), they will be hardly comparable inside such a multi-signed framework. In contrast, in our present framework, all two-signed rules of L_5 will be easily provable inside of L_3. We have indeed shown in the last section how it is already possible, with the help of the theories presently produced by our program, to use Isabelle's meta-logic to show that two different axiomatizations for the same logic are equivalent. To show equivalence between axiomatizations from without the local perspective of a given theory, Isabelle's *locales* (cf. [2]) will probably provide a better framework, and we hope to deal with that in the next version of our program.

There still remains some room for improvement and extension of both our algorithm (which should still, among other things, be upgraded in order to deal in general with first-order truth-functional logics) and its implementation. By way of an example, we have assumed from the start that the logics received as inputs to our program came together with a suitable collection of separating connectives. This second input, however, could be dispensed with, as the set of all definable unary connectives can in fact be automatically generated in finite time from any given initial set of operators of the input logic. That generation, however, may be costly for logics with a large number of truth-values and is not as yet performed by our program. Another direction that must be better explored, from the theoretical perspective, concerns the conditions for the admissibility or at least for the explicit control of the application of the dual-cut rule. On the one hand, the elimination of dual-cut has an obvious favorable effect on the definition of completely automated theorem-proving tacticals for our logics. If that result cannot be obtained in general but if we can at least guarantee, on the other hand, that this dual-cut rule will never be needed, in each case, for more than a finite number of known formulas —say, the ones related to the original goal as constituting its subformulas or being the result of applying the separating connectives to its subformulas— then again this will make it possible to devise tacticals for obtaining fully automated derivations using the above described tableaux for our finite-valued logics. An important recent advance in that direction has in fact been done in [3], where the axiom extraction algorithm has already been upgraded in order to originate entirely analytic tableau systems, without the dual-cut rule or any potentially 'complexifying' rules such as axC3 and axC4. A future version of our program had better be adapted to this new procedure, which will allow, moreover, for the algorithmic generation of fully automated proof tacticals, easily expressible in the framework of Isabelle.

Acknowledgments

The authors are indebted to the financial support provided by CNPq in the form of research grants, and by the FAPESP ConsRel Project.

BIBLIOGRAPHY

[1] Matthias Baaz, Christian G. Fermüller, and Gernot Salzer. Automated deduction for many-valued logics. In J. A. Robinson and A. Voronkov, editors, *Handbook of Automated Reasoning*, pages 1355–1402. Elsevier and MIT Press, 2001.

[2] Clemens Ballarin. Interpretation of locales in Isabelle: Theories and proof contexts. In J. M. Borwein and W. M. Farmer, editors, *Mathematical Knowledge Management*, (MKM 2006), LNAI 4108, pages 31–43. Springer, 2006.

[3] Carlos Caleiro and João Marcos. Classic-like analytic tableaux for finite-valued logics. 2009. Preprint available at:
http://wslc.math.ist.utl.pt/ftp/pub/CaleiroC/09-CM-ClATab4FVL.pdf.

[4] Carlos Caleiro, Walter Carnielli, Marcelo E. Coniglio, and João Marcos. Two's company: "The humbug of many logical values". In J.-Y. Béziau, editor, *Logica Universalis*, pages 169–189. Birkhäuser Verlag, Basel, Switzerland, 2005.
http://wslc.math.ist.utl.pt/ftp/pub/CaleiroC/05-CCCM-dyadic.pdf.

[5] Walter A. Carnielli. Systematization of the finite many-valued logics through the method of tableaux. *The Journal of Symbolic Logic*, 52(2):473–493, 1987.

[6] Roberto L. Cignoli, Itala M. L. D'Ottaviano and Daniele Mundici. *Algebraic Foundations of Many-Valued Reasoning*, volume 7 of *Trends in Logic*. Dordrecht: Kluwer, 1999.

[7] Reiner Hähnle. Tableaux for many-valued logics. In M. D'Agostino et al., editors, *Handbook of Tableau Methods*, pages 529–580. Springer, 1999.

[8] Tobias Nipkow, Lawrence C. Paulson, and Markus Wenzel. *Isabelle/HOL — A Proof Assistant for Higher-Order Logic*, volume 2283 of *LNCS*. Springer, 2002.

[9] Raymond M. Smullyan. *First-Order Logic*. Dover, 1995.

[10] Wojciech Suchoń. La méthode de Smullyan de construire le calcul n-valent de Łukasiewicz avec implication et négation. *Reports on Mathematical Logic*, 2:37–42, 1974.

[11] Heinrich Wansing and Yaroslav Shramko. Suszko's Thesis, inferential many-valuedness, and the notion of a logical system. *Studia Logica*, 88(3):405–429, 2008.

João Marcos and Dalmo Mendonça
DIMAp / CCET
Federal University of Rio Grande do Norte
Rio Grande do Norte, Brazil
E-mails: jmarcos@dimap.ufrn.br, dalmo3@gmail.com

SECTION 7

LOGIC, ABDUCTIVE AND DEFEASIBLE REASONING

Prioritized Dynamic Retraction Function on Non-monotonic Information Updates

Giuseppe Primiero

ABSTRACT. In this paper a model for updates on prioritized belief sets and retractions thereof is introduced, using the standard format of Adaptive Logics. The core of the update retraction procedure is represented by abnormal expressions derivable in the language: they express updates with information that contradict previously derived contents. The adaptive strategy aims at restricting the validity of these formulas by focusing at each decreasing degree on the update which is the most rational to retract in order to restore consistency as soon as possible.

1 Introduction

The formalization of the dynamics of logical reasoning is nowadays a very extensive area of research, exploring various aspects of knowledge processes. Ongoing research focuses on the formulation of successful methodologies to revise a knowledge base (or belief set) in the light of new inconsistent information received, and to define an appropriate contraction operation in view of previously performed updates.

The basic requirement of consistency demands a process of resolution which, on the one hand, does not represent agents as completely irrational, so that they be able to maintain non-contradictory beliefs; on the other hand, the rationality principle at the basis of the retraction process might be related to different criteria: in some cases consisting in preserving older information, in other cases in accepting more recent data, or being dictated by trust ascribed to information sources. A less explored and more complex task is the representation of the agent's ability to choose among possible different updates. In general, the representation of inconsistencies related to a temporal order is needed for a more realistic formulation of such processes.

The present paper defines a logic called **AIUR** (*Adaptive Information Update Retraction*) which simulates the choice of the most rational retraction to perform on possible updates, in order to restore consistency on a belief set. The adaptive framework represents an alternative approach to other systems known from the literature: in particular, it allows the representation of inconsistent information processing without trivilizing belief sets, for which the AGM-paradigm from [1] provides no realistic modelling.

The standard set of doxastic actions for belief change in AGM contains *expansion* $(K + A)$, *revision* $(K * A)$ and *contraction* $(K \div A)$ for a belief set K and a belief A. The standard contraction operator satisfies the well-known AGM-postulates:

K1 Closure: if K is logically closed, then so is $K \div A$ for all A ($K \div A = Cn(K \div A)$);

K2 Inclusion: the result of a contraction-operation should always be included in the belief set prior to that operation ($(K \div A) \subseteq K$);

K3 Vacuity: if some given content does not belong to the belief set, then contraction with respect to such content results in the belief set itself (If $A \notin K$ then $K \div A = K$);

K4 Success: no logical consequence of the set of beliefs can be contracted – or else if new information comes in, then it must be incorporated into our belief set (If $\nvDash A$ then $A \notin K \div A$);

K5 Recovery: the removal of a belief A followed by the reintroduction of the same belief should lead to the original belief set ($K \subseteq Cn((K \div A) \cup \{A\})$);

K6 Extensionality: logically equivalent sentences lead to equivalent contractions (If $\vDash A_1 \leftrightarrow A_2$ then $K \div A_1 = K \div A_2$).

The Inclusion Postulate is usually taken for granted in the AGM-paradigm. On the other hand, the Recovery Postulate is justified by an appeal to the following:

DEFINITION 1 (Principle of Informational Economy). Keep the loss of information to a minimum.

If this principle becomes overriding, the belief set resulting from contraction of K by A should be a maximal subset of K that does not imply A. This restriction is shown to lead to maxichoice contraction operations in [10], and usually it is considered too strong. In [21], the status of the Recovery postulate is analysed in view of the misapplication of maxichoice contraction operations to belief sets (theories): when contraction is applied to finite bases (rather than theories), recovery becomes vacuously true, and its body fails in general; on the other hand, when contraction is applied to sets closed under logical consequence, recovery becomes intuitive and morever useful.

Within the dominant AGM-paradigm, the difference between belief states assumed to be logically closed sets of sentences and non-closed sets (bases) is crucial in the treatment of inconsistent data. It has been argued at lenght in the literature that in real world situations knowing agents deal with inconsistent beliefs, without believing everything (explosion, see e.g. [16]). The case of inconsistencies in non-prioritized belief bases can be dealt with by enough applications of contraction on dispensable elements. One of the first formalization of this case is introduced in [15], where the operation of restoring consistency is called *consolidation* ($A!$), and it is performed by contraction on contradiction: $A! = A \div \perp$. The shortcoming is due to the underlying logic: consolidation cannot be used in a satisfactory way on inconsistent belief sets because all inconsistent sets are trivially equal in a classical setting, hence all distinctions are lost. A variation on the theme of inconsistent belief bases is given in [16], where a set of *local operators* for contraction, consolidation, revision and semi-revision is introduced. This is obtained by defining logical compartments in the belief base around a sentence, that is the subset of the base that is relevant to that sentence.[1] Consolidation is thus defined as an operation on the kernel for a belief base, determined by an inference operation which preserves consistency, inclusion and so-called core-retainment, corresponding to a form of compactness.[2]

A different approach to the localization of inconsistent data is introduced in [27]. The state of the beliefs on inconsistent information is considered a distinct element from the inconsistent data itself: even with respect to such data, beliefs are normally consistent. On the basis of this idea, the system introduced in [27] formulates operations to extract consistent beliefs form the represented information, thus differing from other treatments admitting inconsistencies by the use of a paraconsistent logic.[3] This holds – as in the logic AIUR here introduced – for beliefs closed under logical consequence, that is for belief sets, which eliminates the restriction from consolidation. To this aim, an inconsistency-tolerant 4-valued logic is used, the first degree entailment logic, which extends classical logic by admitting partial and inconsistent valuations keeping inconsistencies local. It defines methods to select only consistent three-valued valuations on finite information states. In particular, *extractors* are defined to obtain the consistent modulo of an inconsistent finite state: a consistency forcer function is defined to ignore contradictory valuation, another to take maximal consistent subvaluations of contradictory ones. The analogy with the adaptive logic introduced in this paper goes therefore at various levels: structurally, both use methods to select on the consequence relation to establish persistance of consistent data, and both admit these initial data to be inconsistent. The differences concern essentially the representation of such inconsistency in the language: whereas for the logic in [27] the valuations are the meta-theoretical means to represent inconsistent information, and minimization is the way to reduce it to relevant data; in the adaptive logic AIUR, inconsistencies will be given as derivable formulas in the language, and their form will implement the possible updates the agent is faced with. At the level of the selection method, the prioritized structure based on the temporal order of updates in AIUR allows to choose between the different consequence relations generated by an inconsistent premise set, whereas in the case of the extractors functions

---

[1] See also the related notion of *kernel* in [14] and the application of belief change strategies to model-based diagnosis in [31].

[2] See [16], pp. 60-62.

[3] See for example [23].

the resulting possibilities are treated on equal terms, and only some sort of meta-theoretical considerations can force to choose the "most consistent" valuations.[4]

It clearly appears from this analysis that the introduction of prioritized structure leads to a very different treatment of consistency-restore operations. In the AGM-paradigm, the formulation of priority relations is given by the introduction of the following principles:

DEFINITION 2 (Principle of Indifference). Contents held in equal regard should be treated equally.

DEFINITION 3 (Principle of Preference). Contents held in higher regard should be afforded a more favourable treatment.

Their combination with the Principle of Informational Economy requires the system to satisfy both the Inclusion and the Recovery Postulates. A contraction operator that satisfies the Inclusion Postulate but not the Recovery Postulate is usually called a *withdrawal* operator.[5] When retracting a belief A from a belief set K, there might be other beliefs in K that entail A, but one might not want to retrace all such beliefs back. By a result of Makinson which uses the Levi identity $K_A^* = (K_{\neg A}^-)_A^+$, the set of withdrawal functions generates a unique revision function satisfying all the AGM-postulates.

Along with AGM contraction and recovery, the literature on similar functions or variant models for prioritized bases/sets is very extensive: the possible-worlds interpretation of spheres in [13]; the epistemic entrenchment relation from [11]; its variant in preferential bases presented in [24] and the related axiomatization from [9]; the application of withdrawal on belief bases in [32].[6] One way to satisfy the Principle of Informational Economy is to preserve the use of prioritized sets of beliefs (semantically represented by a preorder on their interpretations), and to allow updates and retractions on the basis of a temporal order. This is the structure of the logic **AIUR**.

The AGM functions on the contraction operator can be notoriously represented by total preorders on a finite set of propositions (their interpretations), see e.g. [13]. An agent keeping track of the information received, and updating his basis first with $\neg A$ and then with A, can infer in a prioritized way and therefore reject $\neg A$ (assuming she has a preference for the most recent information). The retraction of the information A will natu-

---

[4]It is interesting to notice that in [27] the possible development of, and comparison with adaptive semantics is foreseen; nonetheless, it is also stressed how these semantics "differs widely". We provide here an adaptive logic for prioritized bases, which therefore implements further particular cases that were not included in the basic ideas from [27]. For some more relation of adaptive logics with standard belief revision theory, see [8]. For another model of retraction on inconsistent bases on the basis of a logic for default reasoning, see [6].

[5]See [21],[12] and [5].

[6]For a complete overview of the definitions of contraction and withdrawal functions, see [25].

rally lead to increase support in the belief that $\neg A$, which depicts therefore an internal dynamics with respect to the reasons held against other contents. This dynamics, called *liberation* in [5], is especially intuitive when retraction is performed on belief sets updated by contradictory information.[7] Other extensions of the AGM-paradigm explore the formulation of the standard postulates using modal logics ([30], [26]) or extending normal modal operators by non-normal information operators to perform iterated belief revision ([4]). In both cases, connections can be drawn to the present work, where updates are interpreted in the framework of a modal logic with a non-standard consequence relation.[8]

The aim of the present work is to combine the temporally-based retraction on inconsistent updates, with the Principle of Information Economy: the former allows to preserve the Principles of Indifference (for simultaneously received data) and of Preference (for older data); the latter is satisfied by retracting data back to the first stage at which consistency is restored. This procedure refers to the *retraction* operator $(K \ominus A)$ introduced in [18] which undoes the effect of a previous operation on a belief set. The operation defined in **AIUR** corresponds intuitively to a retraction on expansion, hence to a specific case of recovery of revision. A standard analysis of an update-retraction procedure is easily shown by the following diagram:

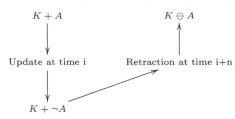

where the content of the retraction operation is based on a preference for the newer information. In [7] this form of update is implemented by the combination of two principles:

- *Primacy of new information*: the revised knowledge of the system should conform to the new information, which implies a complete relaince on its truth;

---

[7]The *retraction* operator from [5] does not satisfy Inclusion nor Recovery, but it satisfy Failure (if $\vDash A$, then $K \subseteq K \div A$).

[8]Another relevant paradigm for belief change is the Dynamic Epistemic Logic approach, [29]. In DEL an information update produces the elimination of irrelevant or redundant information; a Public Announcement $\phi!$ eliminates all the worlds which fail to satisfy ϕ; Knowledge after a Public Announcement $[\phi!K_jA]$ establishes that after the elimination in the current model of all the worlds that fail to satisfy ϕ, agent j knows that A; according to an Unsuccesful Update $\langle A \wedge \neg K_j A \rangle \neg(A \wedge \neg K_j A)$, the agent asserts that A and that agent j does not know that A, which becomes false because of the announcement. The extension to the temporal ordering is given by the framework of Epistemic Temporal Logic ([28]) and the introduction of past operators ([17]). Within this framework, neither a proper form of retraction or withdrawal on previously performed announcement, nor the mentioned combination of possible updates are considered.

- *Persistence of prior knowledge*: as much old knowledge as possible should be retained in the revised knowledge; the resulting state is obtained by some minimal change.

AIUR, which focuses on primacy of older data in view of new inconsistent information, models a specific case of the following further principle from [7]:

- *Fairness*: if there are many candidates for the revised knowledge that satisfy the above principles, then one of them should not be arbitrarily chosen.

It is the dynamics of selection on *possible updates*, whose combination is inconsistent in view of the starting belief set:

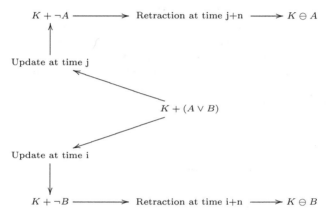

The temporally ordered update-retraction procedure induces a preorder on the models of the differently updated belief sets, revealing a relation of entrenchment: the older the update (that has not been yet retracted) the most entrenched becomes the related belief. A dynamic version of the principle of persistence of prior knowledge comes from recent literature in AI (see e.g. [33]):

DEFINITION 4 (Minimal Change Principle). During a state transition, the change between states should be as little as possible.

This principle becomes inappropriate when state change involves disjunctive descriptions of the resulting possible states. The present paper provides a resolution method for inconsistent outcomes involving the minimum amount of retracted information on disjunctive solutions.[9]

AIUR has the standard format of an Adaptive Logic (see [2, 3]):

DEFINITION 5. An Adaptive Logic (**AL**) in the standard format is defined as a triple $\mathbf{AL} = \langle LLL, \Omega, AS \rangle$, according to the following description:

---

[9] For an adaptive procedure that respect literally the principle from [7] on the primacy of newer information, see [22] and the comments in the conclusive section of this paper.

1. LLL is the lower limit logic, a monotonic, compact logic; for **AIUR** it is called $\mathbf{T}^+$ and consists of a set of superposed logics, each of them defined in terms of at least one temporally indexed abnormal formula; it corresponds to a multi-doxastic version of the modal logic $\mathbf{T}$;

2. Ω is a set of abnormal formulas; for **AIUR** the abnormalities express conflicts between some $A \in K$ and some update with information $\neg A$ obtained at later time;

3. AS is the Adaptive Strategy; in this case *Minimal Abnormality Strategy*, which selects valid formulas at each stage, hence establishing retractions of updates on a temporal basis.

I shall proceed as follows. After some preliminaries in Section 2, an intuitive characterization of the adaptive interpretation of Information Updates and their Retractions will be provided in Section 3. In Section 4, the core of the logic **AIUR** is introduced, with the semantic and proof-theoretical characterizations of the Lower Limit Logic $\mathbf{T}^+$ and the set of abnormalities Ω. In Section 5 and 6, respectively the proof theory and the semantics of **AIUR** as well as some examples are formulated. The final Section gives some remarks on the distinction with the standard retraction operator and possible extensions of this research.

2 Preliminaries

Let $\mathcal{L}$ be the standard language of classical propositional logic (henceforth **CL**) that is formed from a finite set of atoms in the usual way. A **CL**-model is a function from the set of atoms to $\{0, 1\}$. I shall use $M, M_1, \ldots$ as meta-variables for **CL**-models, $\mathcal{M}$ to denote the set of all **CL**-models. A model M is a model of a belief set K iff all the members of K are true in it; I shall use $\mathbf{M}$ as the subset of members of $\mathcal{M}$ in which all members of K are true. As usual, $M \vDash A$ will denote that M verifies A.

In the following, I define an operation of update in terms of an accessibility relation from the actual to some possible epistemic states, according to a temporal order. This relation preserves a defeasible interpretation of belief contents: the information held true by the agent in her actual epistemic state can still be rejected according to some other (later) state. It is in view of this property that we allow retractions on given updates. The frames of the possible worlds semantics of $\mathbf{T}^+$ are reflexive: according to this property each of the states is internally consistent, inconsistency arising by extensions to later states. The final belief set is obtained when temporal updates and related retractions have been completed.

The language $\mathcal{L}^I$ of $\mathbf{T}^+$ is the language $\mathcal{L}$ of **CL** extended with an appropriate set of indexed modalities for the operations of information update. The modal operator I behaves as a possibility operator. A belief set is intended as a finite set of sentences of $\mathcal{L}^I$, closed under logical consequence, i.e. such a set K corresponds to $Cn(K)$. K refers now to the (updated) belief set obtained by translating a finite set of sentences of $\mathcal{L}$ into the very same set of sentences in which each formula is prefixed by an occurence of the modal operator I_i. The index i is defined by membership to the set $\mathcal{T} = \{0, 1, \ldots, \}$ of temporal indices. An expression of the form $I_i A$ holding in the belief set K has the meaning:

$I_i A$: "the agent's belief set K is updated at time i with the information that A".

In view of the adaptive dynamic approach, at the end of the update-process the belief set will contain the result of finally accepted updates: to this aim the operator $\oplus I$ will be used, which behaves as a necessitation operator (validity in all states). By the formulation of indexed modalities, $\mathcal{L}^I$ is the language of a multi-modal version of **T**. In view of the meaning presently attached to such modalities, I will discard any expression formulating possible iterations of the form:

$I_i I_{i-n} A$: "the agent's belief set K is updated at time i with the information that she has received at time $i - n$ the information that A".

In other words, the modal language will be restricted to first degree modalities and only modal formulas in which no nested belief operators occur will henceforth be considered as well-formed. This simplifies the language and it is enough for the present aim. I moreover restrict myself to a one-agent based language, and leave the possible extension to a multi-agent formulation to another occasion.

The set of atoms in $\mathcal{L}^I$ shall be denoted by $\mathcal{W}^P$ and $\mathcal{W}^\pm$ will be used to refer to the set of literals (an atom or negation of an atom). Simply $\mathcal{W}$ is used to refer to all well-formed formulas of $\mathcal{L}^I$, composed by $\mathcal{W}^\pm$ in terms of the standard propositional connectives plus the modal operator: $\neg, \wedge, \vee, \supset$, I_i. The symbols $\vDash$ and $\vdash$ have the usual meanings of consequence and derivability relations (with abbreviations for the related logical languages attached to them to distinguish among **AL** and the **LLL**). The abbreviation $\bigvee \Delta$ will stand for the disjunction of the members of Δ, where Δ is a set of formulas.

3   Intuitive Characterization of Updates and Retractions

The notion of update appeared in the literature of theory change in relation to the operation of belief revision defined by the AGM paradigm. The distinction between revision and update was introduced in [20] and it was later formalized in [19]. By revision one formalizes changes due to new information in a static world; update refers instead to modifications that a knowledge base undergoes when the world of reference changes. An obvious extension of the notion of update is the addition of inconsistent information.

Consider the following example. An agent is informed that an important event she wants to attend to will be organised in one of the two theaters of her city, the Blue or the Red Theater. She receives later the information that the event will not be organised in the Blue Theater, so she simply infers that the event will take place at the Red Theater. At this stage a standard dynamic of information update and revision takes place: an update incoming from the external world forces the agent to derive some new content. Assume that, by a yet later announcement, she becomes informed (by a seemingly equally reliable source) that the event will not be organised in the Red Theater.[10] In this case, the doxastic dynamics is more complex in view of the aimed consistency: it does not concern the deterministic change in a belief state due to a certain AGM-style change; nor it involves some non-deterministic but immediate change of belief. It rather requires the description of the agent's internal dynamics: she deals with a set of incoming data and is required to perform a rational choice among possibly contradictory (but equally reliable) informations.

The formalisation of the example in $\mathcal{L}^I$ is obtained by the following premise set:

(1) $K = \{I_1(p \lor q), I_2 \neg p, I_3 \neg q\}.$

This set expresses a (initially empty) belief set updated at the initial stage 1 with the information that $(p \lor q)$: "the event will take place either at the Blue or at the Red Theater"; at the next stage 2, the set K is updated with the information that $\neg p$: "it is not the Blue Theater the place where the event will take place"; and at the final stage 3 the agent becomes informed that $\neg q$: "it is not the Red Theater the place where the event will take place". To formulate a consistent consequence set on the basis of these informations means to establish which update is the most rational to retract. Let start by giving some basic definitions.

DEFINITION 6 (Non-monotonic Update). An update at time j with information A $(I_j A)$ is non-monotonic with respect to K if $\neg A$ is $\mathbf{T}^+$-derivable from K by a previous update, i.e. if $K \vDash_{\mathbf{T}+} I_i \neg A(i < j \in T)$. The situation obtained after the non-monotonic update at time j is denoted by a formula of the form $I_j A \land \neg A$.

DEFINITION 7 (Combined Non-monotonic Update). A rational agent is faced with a combined non-monotonic update if for a belief set K it holds that $K \nvDash_{\mathbf{T}+} (I_i A \land \neg A), K \nvDash_{\mathbf{T}+} (I_j B \land \neg B)(i, j \in T)$, but it holds that $K \vDash_{\mathbf{T}+} (\bigvee \Delta)$ where formulas $(I_i A \land \neg A), (I_j B \land \neg B) \subseteq (\bigvee \Delta)$.

---

[10]The assumption on the equal reliability of the information sources is functional to performing the selection procedure uniquely on a temporal order. Different priority relations can be designed varying the relevance of sources.

In the logic $\mathbf{T}^+$, neither the formula $I_2\neg p \wedge p$ nor $I_3\neg q \wedge q$ are derivable from K, but their disjunction is. This means that the agent has to select either the validity of $\neg p$ provided that it is an update on p (and thus on $(p \vee q)$), *or* the validity of $\neg q$ provided it is an update on q (and thus on $(p \vee q)$). The retraction of one of these updates is necessary in those cases – as in our initial example – where the agent considers reliable some initial information (viz. that the event will take place, either in the Blue or in the Red Theater). On this basis, some later information is wrong or inaccurate.

Semantically, all three premises are verified in each of their models, and all models verify the disjunction $(I_2\neg p \wedge p) \vee (I_3\neg q \wedge q)$. In view of some crucial notions introduced in a later section, it will appear that **AIUR** selects the information that – if retracted – first makes it possible to come back to a consistent belief state. In the example this is obtained by selecting those models that verify $(I_3\neg q \wedge q)$ and falsify $(I_2\neg p \wedge p)$. In these models, also p and $\neg q$ are falsified, which means the updates $I_1 p$ and $I_3\neg q$ are retracted, whereas the formula $(q \wedge \neg p)$ is verified.[11]

The dynamic proof-theory for the retraction procedure is also based on the validity of updates depending on the order of time. This is translated by the typical procedure of derivation in the adaptive proof-theory, where contents are derivable on conditions. The disjunction $\bigvee(\Delta)$ of combined non-monotonic updates is $\mathbf{T}^+$-derivable. A derivation performed by the adaptive logic **AIUR** assumes that one disjunct $I_3\neg q \wedge q$ holds on condition that the older update $I_2\neg p \wedge p$ be false. This implies that any content assuming $I_3\neg q \wedge q$ being false is no longer derivable, so that the update with $\neg q$ is retracted.

4   A Logic for Non-monotonic Information Updates

The present section provides the semantic and syntactic characterizations of the Lower Limit Logic $\mathbf{T}^+$ and the definition of the derivable abnormal formulas that describe the non-monotonic updates.

As for any Adaptive Logic in standard format, $\mathbf{T}^+$ is a Tarski Logic (monotonic and compact). In this case, one is dealing with a prioritized multi-doxastic logic, whose consequence set for a premise set K is built from models verifying the various indexed abnormal formulas for each degree. The adaptive logic **AIUR** selects on the $\mathbf{T}^+$-consequence set according to the procedure defined by the Minimal Abnormality Strategy, in order to establish which disjunct(s) of combined non-monotonic updates hold(s): in turn this will tell which updates persist, and which retraction is needed.

---

[11]The result is thus an adaptive selection applied on a prioritized set of models. This recalls the models of *liberation* defined in [5]. For the model of σ-liberation, both the agent's set of beliefs and the way to remove beliefs are formed on the basis of the information received in "the course of its intellectual career" - the priority of the most recent information allows to extract consistent subsets out of the possibly inconsistent set of beliefs; for the linear liberation, different candidate belief sets are ordered, with the current belief set being the most preferred ones in the ordering.

$\mathbf{T^+}$ is characterized by a standard possible-world semantics. A $\mathbf{T^+}$-model is a quadruple $M = \langle W, w_0, \mathcal{R}, v \rangle$, where W is a set of possible worlds corresponding to epistemic states, in which formulas from the language $\mathcal{L}^I$ are valued; w_0 is the actual state of knowledge of the agent; $\mathcal{R}$ is a set of accessibility relations $R_i : w_0 \rightarrow W(i \in \mathcal{T})$ from the actual to the set of possible states; $v : \mathcal{W}^P \times W \rightarrow \{1, 0\}$ is the valuation function. Each possible state is time-indexed: the accessibility relation from one state to another simulates the update; reflexive relations allow for any such state only an atom or its negation to hold (viz. no inconsistent update is allowed in *one* world), whereas accessing various consecutive states the agent may face inconsistent beliefs.

The valuation of formulas in a model M is standardly characterized for logical connectives and with the update operator I_i defined as a possibility operator:

C1   where $A \in \mathcal{W}^P$, $v_M(A, w) = v(A, w)$
C2 $v_M(\neg A, w) = 1$ iff $v_M(A, w) = 0$
C3 $v_M(A \vee B, w) = 1$ iff $v_M(A, w) = 1$ or $v_M(B, w) = 1$
C4 $v_M(A \wedge B, w) = 1$ iff $v_M(A, w) = 1$ and $v_M(B, w) = 1$
C5 $v_M(A \supset B, w) = 1$ iff $v_M(A, w) = 0$ or $v_M(B, w) = 1$
C6 $v_M(I_i A, w) = 1$ iff $v_M(A, w') = 1$ for some w' such that $R_i w w'$

The standard semantic notions are defined as usual. A model M verifies A iff $v_M(A, w_0) = 1$; A is valid in $\mathbf{T^+}$ ($\vDash_{\mathbf{T^+}} A$) if it is verified by all its models; A is a consequence of a set of premises in $\mathbf{T^+}$ ($K \vDash_{\mathbf{T^+}} A$) if A is true in every model of K.

Proof-theoretically, the weakest characterization of $\mathbf{T^+}$ is given by the axioms of the modal logic $\mathbf{T}$. In the following, the description of the logic for non-monotonic updates will make use of the possibility fragment of $\mathbf{T}$, represented by the update operator I_i. In the syntactic formulation, the notions of Final Information Update and Final Information Update Retraction will be defined: these express the adaptive counterparts of update and retraction after any possible dynamics in the derivation has been performed. The abbreviation $\oplus I$ shall be used for the contents updated and no longer retractable after time n, where for all $I_i A \in K, i \in \mathcal{T} = \{1, \ldots, n\}$. In this way, the neccessitation fragment of $\mathbf{T}$ can be reconstructed so that $\mathbf{T^+}$ contains the axioms of $\mathbf{CL}$ plus:

Distribution $\oplus I(\varphi \rightarrow \psi)) \rightarrow (\oplus I\varphi \rightarrow \oplus I\psi)$;
Reflexivity $\oplus I\varphi \rightarrow \varphi$;
Modus Ponens $\varphi \rightarrow \psi; \varphi$, then ψ;
Necessitation $\varphi \rightarrow \oplus I\varphi$.

Non-monotonic updates derivable from a premise set K represent the second element in the definition of the adaptive logic $\mathbf{AIUR}$. For each index $i \in \mathcal{T}$, a formula of the form

DEFINITION 8 (Set of Abnormalities). $\Omega_i = \{I_i A \land \neg A \mid A \in \mathcal{W}^\pm\}$

is an abnormality. Disjunctions of abnormalities formally define expressions for combined non-monotonic updates:

DEFINITION 9 (Dab-Formula). $Dab(\Delta)$ stands for $\bigvee(\Delta)$ where $\Delta \subseteq \Omega$.

If Δ is a singleton, $Dab(\Delta)$ is simply an abnormality $(A \lor Dab(\emptyset))$, i.e. a member of Ω; if Δ is empty, $Dab(\Delta)$ is empty as well. In view of the indexed set of abnormalities, by the following definition one determines the position in time of combined non-monotonic updates:

DEFINITION 10 (Dab-formula at degree). A Dab-formula $Dab(\Delta)$ is said to be of degree i $(Dab^i(\Delta))$ iff $\Delta \subseteq \Omega_i$, and for any i' such that $Dab(\Delta')$ and $\Delta' \subseteq \Omega_{i'}$, then $i' < i \in \mathcal{T}$.

According to definitions 8 and 10, the temporal indices of the update operators are used to order derivable disjunctions of abnormalities by increasing degrees. The set of abnormalities at degree i is the union of sets $\Omega_1 \cup \Omega_2 \cup \ldots \cup \Omega_i$:

DEFINITION 11 (The Set of all indexed Abnormalities).

$$\Omega_n = \bigcup_{i=1}^{n} \Omega_i$$

In the case of our example $K = \{I_1(p \lor q), I_2\neg p$ and $I_3\neg q\}$, the set of abnormalities at degree 3 is given by the union of all disjuncts of abnormal formulas at previous degrees:

(2) $\Omega_3 = \{\{I_3\neg q \land q\} \cup \{I_2\neg p \land p\} \cup \{(I_1 p \land \neg p), (I_1 q \land \neg q)\}\}.$

In the following sections I shall first consider the proof-theoretical and next the semantic selection performed according to the third element of the adaptive logic **AIUR**, namely the Minimal Abnormality Strategy. The adaptive strategy defines a way to establish which disjuncts of $\mathbf{T}^+$-derivable Dab-formulas can be considered valid: this in turn means to accept some update and to allow some retraction.

5   Syntactic Selection of Retractions

In this section, the adaptive strategy of the logic **AIUR** is given in its proof-theoretical formulation. The Minimal Abnormality Strategy is the final element in the definition of the adaptive logic for update retractions.

The standard structure of a line in an adaptive derivation contains the following elements: (i) a line number; (ii) the derived formula; (iii) the line numbers of the formulas from which the element in (ii) is derived; (iv) the name of the rule(s) applied to derive the formula from previous lines; (v) the condition on which the second element is derived. In a **AIUR**-derivation the second element of a line is derived provided the elements of the condition are assumed to be false on the premise set. The adaptive rules for the derivation of a new line from previous ones are:

PREM   at any stage of a proof, for any $A \in K$, one may add to the proof a line consisting of:
  (i) an appropriate line number;
  (ii) A;
  (iii) a dash;
  (iv) PREM;
  (v) $\emptyset$;
by the *premise rule* premises are introduced in a line on the empty condition;

RU   at any stage of a proof, for any $B \in \mathcal{W}^P$, if $A_1, \ldots A_n \vdash_{\mathbf{T}+} B$, and $\Delta_1, \ldots, \Delta_n$ are the conditions for $A_1, \ldots A_n$, a line may be added consisting of:
  (i) an appropriate line number;
  (ii) B;
  (iii) the line numbers of the $A_1, \ldots A_n$;
  (iv) RU;
  (v) $\Delta_1 \cup \ldots \cup \Delta_n$;
by the *unconditional rule* a $\mathbf{T}^+$-derivable formula can be added to the proof, without any new condition but (if any) the conditions of the formulas to which the rule is applied;

RC   at any stage of a proof, for any $B \in \mathcal{W}^P$, if $A_1, \ldots A_n \vdash_{\mathbf{T}+} B \vee Dab(\Delta)$, and $\Delta_1, \ldots, \Delta_n$ are the conditions for $A_1, \ldots A_n$, a line may be added consisting of:
  (i) an appropriate line number;
  (ii) B;
  (iii) the line numbers of the $A_1, \ldots A_n$;
  (iv) RC;
  (v) $\Delta_1 \cup \ldots \cup \Delta_n \cup \Delta$;
by the *conditional rule* the derivability by $\mathbf{T}^+$ of a formula of the form $B \vee Dab(\Delta)$ allows the derivation of B on the assumption that all members of $Dab(\Delta)$ are false; in the new line they will be introduced as a new condition.

The derivation of a Dab-formula of a given degree is restricted by an extra condition. Provided that a disjunction of abnormalities expresses the possible non-monotonic updates the agent is faced with, the extra-condition establishes that one of the update is accepted assuming that updates indexed at earlier times are false. I shall call this the *Regularity Condition*: its intuitive meaning is that the analysis of a belief set which has become inconsistent requires each new update to make older ones false. The Regularity Condition for Dab-formulas is defined as follows:

DEFINITION 12 (Regular **AIUR** Dab-formulas). Given a **AIUR** proof, $Dab^i(\Delta)$ is a regular Dab-formula of degree i iff

(i) $Dab^i(\Delta)$ is derived on condition Θ;

(ii) $\Delta \subseteq \Omega_i$ and

(iii) $\Theta \subseteq \Omega_1, \dots, \Omega_{i-1}$.

If there is no other Δ for which $Dab^i(\Delta)$ is regular, then this Dab-formula is a called a *Minimal* **AIUR** *Dab-formula* of degree i:

DEFINITION 13 (Minimal **AIUR** Dab-formula). Given a **AIUR** proof, $Dab^i(\Delta)$ is a minimal Dab-formula of degree i iff $Dab^i(\Delta)$ is regular according to Definition 12, and there is no $\Delta' \subset \Delta$ for which $Dab^i(\Delta')$ is regular.

According to the Regularity Condition (which is a restriction on the Conditional Rule), the derivability of a minimal information update is accepted on condition of the falsity of previously performed updates. Informally, the minimal Dab-formulas for an **AIUR** premise-set explicitate all the updates that the agent has undergone at each consecutive temporal stage from her initially empty belief set. The adaptive selection can now be performed on these updates to show which one needs to be retracted.

The selection on minimal Dab-formulas has effect on the adaptive notion of derivability. Derivability in an adaptive logic is dynamic, which means that a formula derived at one stage of the proof can be later withdrawn. This implies first the definition of derivation at stage:

DEFINITION 14 (Derivability at Stage). A content A is derived from K at stage s of an **AIUR**-proof iff A is the second element of a line l which is not marked at stage s.

A Dab-formula of degree i can be derived at stage s of a proof iff it is regular. Accordingly, one defines the notion of minimal Dab-formula at stage :

DEFINITION 15 (Minimal Dab-formula at stage). Given a **AIUR** proof, $Dab^i_s(\Delta)$ is a minimal Dab-formula of degree i at stage s of the proof iff at that stage $Dab^i(\Delta)$ is regular and not marked, and there is no $\Delta' \subset \Delta$ for which $Dab^i(\Delta')$ is regular.

The marking procedure will now define the withdrawing of formulas previously derived at some stage of a derivation. A minimal Dab-formula derived at a stage s causes the marking of any previous line in which its conditions (at stage s assumed to be false) were derived. This is obtained by the following procedure. Let a choice set of $\Sigma = \{\Delta_1, \Delta_2, ...\}$ be a set that contains an element out of each member of Σ, and let a minimal choice set of Σ be a choice set of Σ of which no proper subset is a choice set of Σ. Consider now the minimal choice set $\Phi^i_s(K)$ of the minimal Dab-formulas at degree i of a premise set K at stage s of an **AIUR**-proof. Then a marked line is one whose content is retracted and considered (at that stage) no longer derived:

DEFINITION 16 (Marking for Minimal Abnormality). A line l of a **AIUR**-proof where a formula A is derived on condition Δ is marked at stage s iff

(i) there is no $\phi \in \Phi_s^i$ such that $\phi \cap \Delta = \emptyset$, or

(ii) for some $\phi \in \Phi_s^i$ there is no line at which A is derived on condition Θ and at that line $\phi \cap \Theta = \emptyset$.

The procedure goes on - if needed - stepwise at any next stage by looking at the minimal choice set of lower degree Φ_{s+i}^{i-1}, next at Φ_{s+j}^{i-2} and so on.

The dynamic notion of derivability at stage has its counterpart at the end of the marking procedure in the notion of final derivability: a line which is (and stays) *unmarked* is considered finally derived. This becomes for **AIUR** a specific definition of valid information update $\oplus I$:

DEFINITION 17 (Final Information Update $\oplus I$). An information update $I_i A$ for K is valid according to a **AIUR**-proof if A is finally derived from K in a **AIUR**-proof, i.e. iff

(i) A is the second element of a line l of the proof;

(ii) line l is not marked at that stage s and any extension of the proof in which line l is marked can be further extended such that l is unmarked, i.e. for any later stage $s+i$ at which A is derived on condition Θ, for any $\phi \in \Phi_{s+i}^{i-n}$, it holds that $\phi \cap \Theta = \emptyset$.

For lines that stays unmarked, there will be lines whose content shall remain *marked*. With respect to these contents, the dynamics of derivability of information updates allows the following definition of final update retraction $\ominus I$:

DEFINITION 18 (Final Information Update Retraction $\ominus I$). A non-monotonic information update $I_j B$ of K is retracted according to a **AIUR**-proof if B is marked at some stage of the derivation, i.e. iff

(i) B is the second element of a line l of the proof on the condition that an update $I_i A \wedge \neg A$ is false;

(ii) at some next stage of the proof, the update $I_i A \wedge \neg A$ is derived as a regular one;

(iii) at any next stage of the very same derivation $I_i A \wedge \neg A$ stays unmarked, which makes line l finally marked.

5.1 Examples

Let us consider a derivation from the mentioned premise set $K = \{I_1(p \vee q), I_2 \neg p, I_3 \neg q\}$:

1	$I_1(p \vee q)$	PREM	$\emptyset$	
2	$I_2 \neg p$	PREM	$\emptyset$	
3	$I_3 \neg q$	PREM	$\emptyset$	
4	$p \vee q$	1; RC	$\{I_1 p \wedge \neg p, I_1 q \wedge \neg q\}$	
5	$\neg p$	2; RC	$\{I_2 \neg p \wedge p\}$	
6	q	4, 5; RU	$\{I_1 p \wedge \neg p, I_1 q \wedge \neg q, I_2 \neg p \wedge p\}$	
7	$\neg q$	3; RC	$\{I_3 \neg q \wedge q\}$	$\checkmark^{10}$
8	p	4, 7; RU	$\{I_1 p \wedge \neg p, I_1 q \wedge \neg q, I_3 \neg q \wedge q\}$	$\checkmark^{10}$
9	$(I_2 \neg p \wedge p) \vee (I_3 \neg q \wedge q)$	1, 2, 3; RC	$\{I_1 p \wedge \neg p, I_1 q \wedge \neg q\}$	
10	$I_3 \neg q \wedge q$	9; RC	$\{I_1 p \wedge \neg p, I_1 q \wedge \neg q, I_2 \neg p \wedge p\}$	

Lines 7 and 8 are marked in view of line 10 so that their contents $\neg q$ and p are considered no longer derived, and in particular the update $I_3 \neg q$ is retracted. On the other hand, lines 5 and 6 stay unmarked, which means that in view of this **AIUR**-proof, content q holds and the update $I_2 \neg p$ is accepted. In the following line

11	$(I_1 p \wedge \neg p) \vee (I_1 q \wedge \neg q)$	1, 2, 3; RC	$\{I_2 \neg p \wedge p, I_3 \neg q \wedge q\}$

the Dab_{11}^1-formula is not minimal according to Definition 15 because its condition is of a higher degree and thus it is not regular. Hence, none of its disjuncts will be selected in a minimal choice set Φ_{11}^1 (in fact, there is no such minimal choice set of degree 1, because no Dab-formula of degree 1 can be minimal), and therefore no further marking is possible. This makes the derivability of the content at line 10 and the related retractions final.

Consider a different example, namely the premise set $K = \{I_1 \neg p, I_2(p \vee q), I_3 \neg q\}$; the following is a valid **AIUR**-derivation:

1	$I_1 \neg p$	PREM	$\emptyset$	
2	$I_2(p \vee q)$	PREM	$\emptyset$	
3	$I_3 \neg q$	PREM	$\emptyset$	
4	$\neg p$	1; RC	$\{I_1 \neg p \wedge p\}$	
5	$\neg q$	3; RC	$\{I_2 p \wedge \neg p, I_3 \neg q \wedge q\}$	$\checkmark^{10}$
6	$p \vee q$	2; RC	$\{I_2 p \wedge \neg p, I_3 \neg q \wedge q\}$	$\checkmark^{10}$
7	p	5; 6; RU	$\{I_2 p \wedge \neg p, I_3 \neg q \wedge q\}$	$\checkmark^{10}$
8	q	4, 6; RC	$\{I_1 \neg p \wedge p, I_2 p \wedge \neg p, I_2 q \wedge \neg q\}$	
9	$(I_2 p \wedge \neg p) \vee (I_3 \neg q \wedge q)$	1, 2, 3; RC	$\{I_1 \neg p \wedge p\}$	
10	$I_3 \neg q \wedge q$	1, 2, 3; 9 RC	$\{I_1 \neg p \wedge p, I_2 p \wedge \neg p\}$	

According to this derivation the contents at lines 5, 6 and 7 are marked: this means that the update $I_3 \neg q$ is retracted, because the content q stays unmarked at line 8. On the other hand, $\neg p$ and obviously $p \vee q$ are derivable, so that the updates with these contents are preserved. Hence, in this case, the quickest way to restore consistency is to retract the update at the last stage and to preserve the initial ones. As for the previous example, also in this derivation the following Dab-formula:

11	$(I_1 \neg p \wedge p) \vee (I_2 q \wedge \neg q)$	1, 2, 3; RC	$\{I_2 p \wedge \neg p, I_3 \neg q \wedge q\}$

is not regular and therefore does not allow for any further marking.

Consider now the following derivation from the premise set $K = \{I_1(p \vee q), I_2 \sim p, I_3 \sim q, I_4 p\}$:

1	$I_1(p \vee q)$	PREM	$\emptyset$	
2	$I_2 \neg p$	PREM	$\emptyset$	
3	$I_3 \neg q$	PREM	$\emptyset$	
4	$I_4 p$	PREM	$\emptyset$	
5	$p \vee q$	1; RC	$\{I_1 p \wedge \neg p, I_1 q \wedge \neg q\}$	
6	$\neg p$	2; RC	$\{I_2 \neg p \wedge p\}$	
7	q	5, 6; RU	$\{I_1 p \wedge \neg p, I_1 q \wedge \neg q, I_2 \neg p \wedge p\}$	
8	$\neg q$	3; RC	$\{I_3 \neg q \wedge q\}$	$\sqrt{}^{13}$
9	p	5, 8; RU	$\{I_1 p \wedge \neg p, I_1 q \wedge \neg q, I_3 \neg q \wedge q\}$	$\sqrt{}^{13}$
10	$(I_4 p \wedge \neg p) \vee (I_3 \neg q \wedge q)$	1, 2, 3, 4; RC	$\{I_2 \neg p \wedge p, I_1 \neg q \wedge q\}$	
11	$I_4 p \wedge \neg p$	10; RC	$\{I_3 \neg q \wedge q, I_2 \neg p \wedge p, I_1 \neg q \wedge q\}$	$\sqrt{}^{13}$
12	$(I_3 \neg q \wedge q) \vee (I_2 \neg p \wedge p)$	1, 2, 3; RC	$\{I_1 p \wedge \neg p, I_1 q \wedge \neg q\}$	
13	$I_3 \neg q \wedge q$	12; RC	$\{I_2 \neg p \wedge p, I_1 p \wedge \neg p, I_1 q \wedge \neg q\}$	

By this derivation the selection given by the minimal choice set Φ_{11}^4 does not have any marking effect, which means no retraction is performed and no consistency is reached. The next selection by Φ_{13}^3 leads to the marking of lines 8, 9 and 11, so that the updates $I_4 p$ and $I_3 \neg q$ are retracted, whereas the update $I_2 \neg p$ is valid and q stays derivable in view of update at time 1. This example shows that it is not always the latest update that must be retracted in order to restore consistency.

6 Semantic Selection of Retractions

In the present section the semantic selection of updates and retractions for the adaptive logic **AIUR** is introduced, consisting in a selection procedure on the indexed sets of models of the logic $\mathbf{T}^+$. According to the Minimal Abnormality Strategy, those $\mathbf{T}^+$-models are selected which are not more abnormal than what is required by the premises. The degree of abnormality is established again in terms of the temporal order of the abnormal formulas, directly imported in the models that verify them.

The consequence set of a premise set K in the combined adaptive logic **AIUR** is obtained by the super-position of increasing **LLL**-consequence sets (where i is the maximal number for which I_i occurs in the premise set):

$$Cn_{AIUR}(K) = Cn_{\mathbf{T}^+ i}(Cn_{\mathbf{T}^+ i-1}(\ldots(Cn_{\mathbf{T}^+ 1}))).$$

To each index i it corresponds a set of abnormalities. The semantics describes a selection on the related models, in which at least one abnormal formula is validated. The selection of valid abnormal formulas determines the valid retractions.

The selection is based on the indexing procedure for Dab-formulas. On this basis, a semantic definition for establishing the minimal abnormal degree of a model with respect to the premise set is formulated. Let us start by the definition of a Dab-consequence:

DEFINITION 19 (Dab-Consequence). $Dab^i(\Delta)$ is a Dab-consequence of a premise set K at degree i iff $K \vDash_{\mathbf{T}^+} Dab(\Delta)$, and $\Delta \subseteq \Omega_i$.

If $Dab(\Delta)$ is a Dab-consequence of a set K, then so is any $Dab(\Delta')$ such that $\Delta' \subset \Delta$. This is why a further definition is needed:

DEFINITION 20 (Minimal Dab-Consequence). $Dab^i(\Delta)$ is a *minimal* Dab-consequence of K at degree i iff $K \vDash_{\mathbf{T}^+} Dab(\Delta)$, $\Delta \subseteq \Omega_i$, and there is no $\Delta' \subset \Delta$ such that $K \vDash_{\mathbf{T}^+} Dab(\Delta')$ at that degree.

Where $Dab^1(\Delta), \ldots, Dab^i(\Delta)$, are the minimal Dab-consequences of a set of premises K at the various degrees, any $\mathbf{T}^+$-model of K will verify at least one member out of each disjunction of abnormalities. For any $\mathbf{T}^+$-model, its abnormal part is defined as follows:

DEFINITION 21 (Abnormal model). Provided M is a $\mathbf{T}^+$-model, $Ab^i(M) = \{A \in \Omega_i \mid M \vDash A\}$.

The set $\mathbf{M}$ of all $\mathbf{T}^+$-models of K at degree i contains abnormalities at the various degrees up to i: $\Delta \subseteq \Omega_1, \ldots, \Delta \subseteq \Omega_i$. At each degree a set of minimal Dab-consequences is verified. In order to perform a selection that resolves inconsistencies as soon as possible, abnormalities of degree i are in general preferred to those of degree $i - 1$. The selection starts therefore by taking into account all the abnormal models at degree i; then it considers those at degree $i - 1$, and so on.

DEFINITION 22 (Selected Models). For each degree i, the minimal Dab^i-formulas verified at that degree induce a selection on the set $\mathbf{M}$ of all models of K, performed stepwise as follows:

$$sel_0(\mathbf{M}) = \{M \mid M \vDash K\};$$
$$sel_{n+1}(\mathbf{M}) = \{M \in sel_n(\mathbf{M}) \mid \text{ for no } M' \in sel_n(\mathbf{M}) \text{ such that } Ab^{i-(n+1)}(M') \subset Ab^{i-(n+1)}(M)\}.$$

At step zero, one considers the entire set $\mathbf{M}$ of models of K and restricts this set stepwise going to each abnormal model of a lower degree. The selection stops when the selected models verify some abnormal formula but consistency is restored on the premise set. The selected models are called *minimally abnormal*:

DEFINITION 23 (Minimally Abnormal Model). A $\mathbf{T}^+$-model M of K is minimally abnormal iff it is included in all possible selection steps: $M \in sel_0(\mathbf{M}) \cap sel_1(\mathbf{M}) \cap \ldots$.

The minimally abnormal models establish the **AIUR**-consequence set of the premises:

DEFINITION 24 (**AIUR**-Consequence). $K \vDash_{\mathbf{AIUR}} A$ iff A is verified by all minimally abnormal models of K.

Consider, as an example, the premise set $K = \{I_1(p \vee q), I_2 \sim p, I_3 \sim q, I_4 p\}$. The consequence set $\mathbf{M}(K)$ contains four types of abnormal models:

$$Ab^4(M) = \{(I_4 p \wedge \neg p), (I_3 \neg q \wedge q), (I_2 \neg p \wedge p), (I_1 p \wedge \neg p), (I_1 q \wedge \neg q)\};$$

$$Ab^3(M) = \{(I_3 \neg q \wedge q), (I_2 \neg p \wedge p), (I_1 q \wedge \neg q)\};$$

$$Ab^2(M) = \{(I_2 \neg p \wedge p), (I_1 p \wedge \neg p)\};$$

$$Ab^1(M) = \{\emptyset\}.$$

The selection proceeds from the highest to the lower degree of abnormality. By the first step one obtains the entire collection of abnormal models $sel_0(\mathbf{M}) = \{\{Ab^4\}, \{Ab^3\}, \{Ab^2\}, \{Ab^1\}\}$, so that $K \models_{\mathbf{T}^+_4} (p, \neg q, \neg p, (p \vee q))$. The next selection step $sel_1(\mathbf{M}) = \{\{Ab^3\}, \{Ab^2\}, \{Ab^1\}\}$ will reduce the degree of abnormality by one, but the retraction $\ominus I_4 p$ is uneffective in view of the aimed consistency. Hence, it goes on with the next selection step $sel_2(\mathbf{M}) = \{\{Ab^2\}, \{Ab^1\}\}$ at degree 2 retracting $\ominus I_3 \neg q$. As a result $K \models_{\mathbf{T}^+_2} (\neg p, (p \vee q))$ and the selection stop with $K \models_{\mathbf{AIUR}} (\neg p, q)$, where consistency is restored.

7 Conclusion

In the logic **AIUR** a retraction operation for the adaptive logic of information updates has been defined. The given procedure restores the belief set at the time before the retracted update was performed. The adaptive retraction satisfies the Inclusion Postulate in a preferential structure; moreover, when one assumes logical closure of final updates and retractions, the Recovery Postulate is also satisfied. The related postulates and conditions are formulated with due modifications concerning the priority of the core (initial update), which is always the most safe from retractions. The most intuitive application for this selection procedure seems to be the dynamic form of reasoning proper of intelligent agents interacting with external information sources and able to retract previously accepted contents.

The very first advantage of using an adaptive frame for dealing with updates and retractions of inconsistent information is the possibility of generalizing to the predicative case. By the definition of a restricted form of abnormalities, one can moreover formulate retraction operations that are not temporally based, rather contentually restricted: this can lead to a definition of an *elimination* operator that erases from a propositional formula all the knowledge that involves a particular fact, i.e. all previous preconceptions on a set of propositional letters (see [18]).

In [22] an Adaptive Logic for simple Information Update, called **AIU**, is presented. It defines an update operation which simulates simple expansion with inconsistent information and preserves the most recent information when updates turn a belief state into inconsistency. In order to obtain this aim, it defines (both semantically and syntactically) operations of elimination of older information which has become unreliable in view of more recent updates. The analysis of the simple update procedure is apt to model mechanical procedures and it is presented for the specific application to database theory.

Acknowledgements

Research for this paper was supported by subventions from Ghent University. The author wishes to thank Joke Meheus for helpful discussions at early stages of this research, and a referee for comments that led to improvements of its presentation.

BIBLIOGRAPHY

[1] C.E Alchourrón, P. Gärdenfors, and D. Makinson. On the logic of theory change: partial meet contraction and revision functions. *Journal of Symbolic Logic,*, 50:510–530, 1985.

[2] D. Batens. A general characterization of adaptive logics. *Logique & Analyse*, 173–175:45–68, 2001.

[3] D. Batens. A universal logic approach to adaptive logics. *Logica universalis*, 1:221–242, 2007.

[4] G. Bonanno. Temporal interaction of information and belief. *Studia Logica*, 86:381–407, 2007.

[5] R. Booth, S. Chopra, A. Ghose, and T. Meyer. Belief liberation (and retraction). *Studia Logica*, Special issue on Reasoning about Action and Change(79):47–72, 2005.

[6] G. Brewka. Belief revision in a framework for default reasoning. In *The Logic of Theory Change Workshop, Konstanz*, volume 465 of *Lecture Notes In Computer Science*. Springer Verlag, 1991.

[7] M. Dalal. Investigations into theory of knowledge base revision. In *Proc. AAAI-88*, pages 475–479. St. Paul, MN, 1988.

[8] K. De Clerq. Maxichoice contraction and revision generalized to include the inconsistent case. *Logic Journal of the IGPL*, 2008. to appear.

[9] E. Ferme' and R.O. Rodriguez. A brief note about rott contraction. *Logic Journal of the IGPL*, 6(6):835–842, 1998.

[10] P. Gärdenfors. *Knowledge in Flux*. Cambridge University Press, 1988.

[11] P. Gärdenfors and D. Makinson. Revisions of knowledge systems using epistemic entrenchment. In M.Y. Vardi, editor, *Proceedings TARKII*, pages 83–95, 1988.

[12] S.M. Glaster. Recovery recovered. *Journal of Philosophical Logic*, 29:171–206, 2000.

[13] A. Grove. Two modelings for theory change. *Journal of Philosophical Logic*, 17:157–170, 1988.

[14] S.O. Hansson. Kernel contraction. *Journal of Symbolic Logic*, 59:845–859, 1994.

[15] S.O. Hansson. Semi-revision. *Journal of Applied Non-Classical Logics*, 7(2):151–175, 1997.

[16] S.O. Hansson and R. Wassermann. Local change. *Studia Logica*, 70:49–76, 2002.

[17] T. Hoshi and A. Yap. ETL, DEL and past operators. In J. van Benthem and E. Pacuit, editors, *Proceedings of the Workshop on Logic and Intelligent Interaction*, ESSLLI08, pages 132–142, 2008.

[18] H. Katsuno and A.O. Mendelzon. A unified view of propositional knowledge base updates. In *IJCAI 1989*, pages 1413–1419, 1989.

[19] H. Katsuno and A.O. Mendelzon. On the difference between updating a knowledge base and revising it. In Allen, Fikes, and Sandewall, editors, *Principles of Knowledge Representation and Reasoning*, pages 387–394. Morgan Kaufmann, 1991.

[20] A.M. Keller and M. Winslett Wilkins. On the use of an extended relational model to handle changing incomplete information. In *IEEE Transactions on Software Engineering*, volume SE-11, pages 620–633, 1985.

[21] D. Makinson. On the status of the postulate of recovery in the logic of theory change. *Journal of Philosohical Logic*, 16:383–394, 1987.

[22] G. Primiero. A model for processing updates with inconsistent information on propositional databases. Submitted for the Proceedings of the IV World Congress on Paraconsistency.

[23] G. Restall and J. Slaney. Realistic belief revision. Technical Report TR-ARP-2-95, Research School of Information Sciences and Engineering and Centre for Information Science Research, Australian National University, Canberra, Australia, 1995.

[24] H. Rott. Preferential belief change using generalized epistemic entrenchment. *Journal of Logic, Language and Information*, 1(1):45–78, 1992.

[25] H. Rott and M. Pagnucco. Severe withdrawal (and recovery). *Journal of Philosophical Logic*, 28:501–547, 1999. Corrected complete reprint in issue February 2000.

[26] K. Segerberg. Belief revision from the point of view of doxastic logic. *Bulletin of the Interest Group in Pure and Applied Logics*, pages 535–553, 1995.

[27] A.M. Tamminga. *Belief Dynamics - (Epistemo)logical Investigations*. PhD thesis, Universiteit van Amsterdam, 2001.

[28] J. van Benthem, J. Gerbrandy, and E. Pacuit. Merging frameworks for interaction: DEL and ETL. In *Theoretical Aspects Of Rationality And Knowledge*, Proceedings of the 11th conference on Theoretical aspects of rationality and knowledge, pages 72–81, 2007.

[29] H. van Ditmarsch, W. van der Hoek, and B. Kooi. *Dynamic Epistemic Logic*, volume 337 of *Synthese Library*. Springer, 2006.

[30] B. van Linder, W. van der Hoek, and J-J Ch. Meyer. Actions that make you change your mind. In *KI'95*, Proceedings of the 19th Annual German Conference on Artificial Intelligence, pages 185–196. Springer Verlag, 1995.

[31] R. Wassermann. Local diagnosis. In *Proceedings of the Eighth International Workshop on Nonmonotonic Reasoning*, 2000.

[32] M.A. Williams. On the logic of theory base change. In C. MacNish, D. Pearce, and L.M. Pereira, editors, *Logics in Artificial Intelligence*, volume 838 of *Lectures Notes in Artificial Intelligence*, pages 86–105. Springer Verlag, 1994.

[33] Y. Zhang and N.Y. Foo. On propositional knowledge base updates. *Australian Journal of Intelligent Information Processing Systems*, 2:20–29, 1995.

Giuseppe Primiero
Centre for Logic and Philosophy of Science
Philosophy and Moral Science Department
Ghent University, Belgium
E-mail: Giuseppe.Primiero@UGent.be

Conservative Contraction

Juliano S. A. Maranhão

ABSTRACT. Belief contraction is based on the principle of *minimal change* of the original set of beliefs. Based on Harman's principle of conservatism [7], we question if this is really the most economic way to deal with our belief stock, arguing that one should still be commited to at least some of the logical consequences of a rejected belief, if there is no special reason to reject them. An agent should be minimal not only with respect to her explicit beliefs, but also with respect to her original epistemic commitments. In the present paper we propose an operator called *conservative contraction*, satisfying Harman's principle of conservatism. The focus is on the logical consequences of the belief to be contracted, which may become explicit in the resulting contracted set. It turns out that the operator called *belief refinement*, explored in [9] is a particular case of conservative contraction, which also seems to have interesting relations with abductive reasoning.

1 Introduction: tenacious reasoners

When is one justified to hold a belief? And when is one justified to keep a belief? These are central questions to a theory of knowledge. The first is static: it relates to the structure of a given stock of beliefs and is concerned with how our actual beliefs are or should be related to each other. The second is dynamic: it relates to our disposition to change our beliefs and it is concerned with whether and how should one *give up* a belief when others have been abandoned.

Curiously enough the well known controversy in epistemology between foundationalism and coherentism has been for a long time confined to the static dimension.[1] Basically, foundationalist theories of knowledge defend that there are some basic beliefs, which do not depend on any other belief to be justified and which provide directly or indirectly the support for all other beliefs one may hold. Coherence theories advocate that no belief may have this foundational status and so every belief is supported directly or indirectly by each other. While foundationalist theories have been troubled with the problem of explaining the nature of such foundational beliefs, coherence theories face the challenge of explaining what exactly is this relation of mutual support, which is more than sheer consistency but less than the

---

[1] This remark was made by Andre Fuhrmann in a personal communication.

strong requirement of logical consequence (see [11] for a presentation and discussion of the subject).

It was only recently (considering the history of the debate) that the discussion benefited from a dynamic perspective. In the eighties, philosophical discussion was supported and inspired by the (at the time) fresh field of logical studies called "belief revision", whose landmark is the seminal paper by Alchourron, Gardenfors and Makinson [2].

The dynamic perspective has grown enormously and, as one would expect, it generated a lot of disputes. But there is a core intuition which seems to remain unshaken: reasoners are tenacious with respect to their original beliefs. Such tenacity is reflected by the principle of "minimal change", which has two claims: (i) one should give up beliefs only when "forced" to do so; (ii) in order to give up a belief one should abandon as few beliefs as possible ([6], p. 12).[2]

The first claim leaves open how *"forced"* should be interpreted. In a positive sense, it means that one should leave a belief whenever one believes one's reasons for keeping it are no good. In a negative sense, one should leave a belief whenever one does not find a good reason to keep it. According to Harman, within the dynamic perspective, this is the key notion to distinguish coherence from foundationalist theories of knowledge. In the latter case (foundationalism), it is assumed that in order to keep a belief one should keep track of its original justifications. In the former (coherentism) such demand is relaxed ([7], p. 39).

Harman interprets the coherentist view in light of a still stronger notion he calls the *"principle of conservatism"*: one's reasons for keeping a belief are no good only when there is a special reason against that belief. In his words, *"one is justified in continuing fully to accept something in the absence of a special reason not to"* ([7], p. 46). Following conservatism, even if one's original justification for holding a belief is undermined, one may still hold such belief, if it is still consistent or coherent with one's belief stock. Such belief perseverance is supported by empirical research and Harman attributes this phenomenon not to stubborness or bad habit, but to a rationalization process:

> ... the subjects do not see that the beliefs they have acquired have been discredited. They come up with all sorts of rationalizations (as we say) appealing to connections with other beliefs of a sort that the coherence theory, but not the foundations theory, might approve ([7], p. 38)

Harman takes a further step and states that not only this is how we *in fact* reason, but that is the policy we *should* follow with respect to the management of our beliefs. The claim is supported by the principle he calls *clutter avoidance*, according to which one should not clutter one's mind with

---

[2]Hansson picks up other terms such as minimal mutilation, conservativism and even Harman's term conservatism, which in his view express the same idea ([6], p. 12). As I argue below, Harman's conservatism embodies a stronger idea which calls for a different logical model.

trivialities. The idea is that we should spare room in our minds given that there is a limit to what one can remember. That is why one should not keep track of all intermediate steps to a conclusion and should store only key information for a later recall.

But clutter avoidance also implies that one should not believe explicitly in the logical consequences of one's belief. It introduces a distinction between explicit (basic) and implicit (logical consequences of the basic) beliefs. The distinction is crucial within the dynamic perspective, given that logically equivalent formulation of a set of beliefs may have different outcomes as a result of a revision (see the discussion by Hansson at [6], Chapter 1).

It is the claim of this paper that, due to the distincition between explicit and implicit beliefs, the principle of minimal change embodied in all belief revision models in the AGM tradition is not faithfull to Harman's principle of conservatism and, therefore, fails to do justice to tenacious reasoners. It indeed leads to unreasonable results of a belief set contraction. In order to represent conservatism, we propose a new contraction operator we call *conservative contraction*.

2   Minimal change *versus* Conservatism

The logical models of belief contraction in the AGM tradition are based on the principle of minimal change of the original set of beliefs. Consider a belief set K deriving a, a belief that should be abandoned. Minimality is satisfied in this contraction process by the construction of maximal subsets of K not deriving a. The so called *maxichoice contraction* is given by the choice of one of those subsets [1]. This model warrants that, in order to abandon a, only beliefs which are *relevant* for the inference of a are deleted (principle of relevance). Intuitively, the relevant beliefs for a are those within the set K which may contribute to the derivation of a. Let b be a belief such that $b \in K$. Then b is relevant for K if an only if there is $K' \subset K$ such that $a \notin Cn(K')$ but $a \in Cn(K' \cup \{b\})$. More than that, minimal change says that among those relevant beliefs one should abandon only what is enough to avoid a.

If one does not bother with cluter avoidance, then our set of beliefs should be closed under logical consequence (in this case the term belief set will be used). But clutter avoidance was already into play at [1], given that an alternative to maxichoice contraction, full meet contraction, is presented, which consists in the overlap of all maximal subsets of the original belief set. In [2], the innovation was the so called *partial meet contraction*, according to which the contraction should be the overlap of a selected subset of those maximal subsets. The example below shows the difference between these approaches:

EXAMPLE 1. Suppose the belief set is compose by things you would like to do and assume that you do not want to break the law. Suppose you would like to dive (d), use air tank (t) and to fish(f). Then you come to know that doing all these activities is forbidden, so you would contract the sentence $d \wedge t \wedge f$. Maxichoice contraction would make you abandon just one of the activities (for instance, fishing with air tank, dive and fish, dive with air tank). Full meet would make you abandon them all and do nothing. Partial meet would gives you more flexible options as abandoning two of these activities (for instance it may not make sense for you to use air tank if you gave up the idea of diving and decided to fish).

Maxichoice is the operator which takes minimality most seriously, but it may lead to unreasonable consequences. Suppose you believe that George Bush is the president of U.S.A (a). Your belief set is $Cn(\{a\})$. Then you learn that Obama is the president. Given that for every b (e.g. Homer Simpson is the president), $a \vee b \in Cn(\{a\})$, the result of a maxichoice contraction would abandon a but still keep the belief that either Bush or Homer is the president!

This is the problem of conciliating minimal change with clutter avoidance. One may argue that the problem was to work on belief sets in the first place. So in respect to clutter avoidance one should work on sets of beliefs which are not necessarily closed (which will be called belief bases). This was Makinson's argument in defense of maxichoice contraction [8].

But working on belief bases changes completely the balance between minimal change and clutter avoidance and it is my impression that this have not been taken fully into account.

Minimal change on belief sets pays due attention to our commitment to the logical consequences of our beliefs, but it is so cautious that it preserves too much clutter. Therefore clutter avoidance advises us to be more selective in the contraction process. But if we priviledge clutter avoidance by working on belief bases, then some relevant logical consequences of our beliefs run the risk of being disregarded. This is so because minimal change is then restricted to the base of explicit beliefs and is not applied to our original implicit commitments. This may lead to the suppression of beliefs that might be rationally worth retaining.

Note that if K is closed under logical consequence, then logical consequences of a belief may be kept. For instance, in the contraction of b from the set $Cn(\{a \wedge b\})$, a may be kept. But if we deal with a belief base, then it will be lost. Actually the result of that base contraction would be the empty set. This "disbelief propagation" effect would be justified, according to Hansson, by the intuition that *"merely derived beliefs are not worth retaining for their own sake. If one of them loses the support that it had in basic beliefs, then it will be automatically discarded"* ([6],p.18).

The intuition may work for the foundationalist account of knowledge. But is this really the most tenacious way to deal with our beliefs? Why should the agent abandon her previous epistemic comitments if she had no specific reasons against them? The principle of conservatism says she should not.

For the purposes of this paper conservatism is interpreted to the effect that logical consequences of our original explicit beliefs may be kept even if some of these beliefs supporting them are abandoned. The examples below indicate that the restriction of the principle of minimal change to bases seems inadequate and that conservatism should be the guiding principle of our base contraction policy.

EXAMPLE **2.** Suppose John believes his friend Paul is a reliable source of information. Paul says to John, very alarmed, that John's wife is kissing a blonde man at the park $(k \wedge y)$. John accepts that belief based on Paul's reliability, runs to the park and come back to Paul saying with sincere relief: "You lied to me! The man she was kissing has dark hair, I don't have to worry".

Why do we laugh about John's reaction? He abandoned his belief in $k \wedge y$ because he testified its negation $\neg(k \wedge y)$, therefore the justification to hold his previous belief is dropped: Paul became an unreliable source of information. The problem is that he neglected other possible justifications to keep part of the information, exactly the part which was relevant to the context of conversation. And that reaction is not rational, at least in a pragmatic sense. He should still hold k given that he had no specific evidence against it.

Legal interpretation brings several examples of this sort of reasoning. Take for instance the interpretation of the criminal law applied to abortion. Article 124 of Brazilian Criminal Law forbids abortion in any circumstance, while article 128 permits it when the mother's life is in danger. These rules are not taken as contradictory given an interpretive principle which says that *lex specialis derogat generalis*. The principle would have the effect of deleting art. 124 from the legal system *simpliciter*. But that is not what legal experts conclude. Instead there is another old interpretive principle originating from the *Justinianean Digest* according to which the most specific rule derogates the most general one *only in what it is specific*. This means that from the command present in art. 124 (abortion is forbiden in any circumstance) a logical consequence of it is rescued (abortion is forbiden if it is not the case that the mother's life is in danger).

It seems that legal interpretation is much closer to conservatism than to the idea of minimal change embodied in models of base revision. Indeed in constitutional interpretation by the Brazilian Supreme Court prevails a doctrine called *"interpretation of law in conformity with the constitution"* according to which if a law conflicts with the constitution, then only what is specifically against the constitution should be deleted, while the rest of the legal text may be preserved. For instance, the Supreme Court considered that a law creating a new tax with retroactive effects was held to offend the constitution. But it was just retroactivity, not the whole law that was considered invalid.

In the criminal law example the derived belief (original implicit commitment) preserved is a conditional. The resulting rule has a stronger condition of application. Formerly there was no condition and the interpretation resulted in a more refined or qualified rule. The rule may have furtuer qualifications as a process of revision (interpretation), such as the condition that the procedure should be done by a doctor. In many logical systems, but obviously not for all, considering *cond* to be the binary conditional conective, we have that for every formula b, $cond(a, b) \in Cn(b)$, for every a. For those systems it follows that, by keeping a conditional with the original deleted sentence as the consequent, the agent keeps an original implicit commitment. This is a trace of conservatism which may also be found in theorethical reasoning, as indicated in the example below.

EXAMPLE **3.** Suppose a Dutch believes that water turns into steam at 100^0C in any circumstance. She travels to La Paz, far above the sea level and boils water, which turns into steam at a different temperature. Instead of abandoning that belief, she wonders that she consider that some factor in that particular circumstance was responsible for the failure. There must be some refuting condition for such belief given that she has testified just that at La Paz, or above the sea level, water does not turn into steam at 100^0C. Hence, she gives up his original belief but keeps the belief that if one is at the sea level, then water turns into steam at 100^0C.

The relevant difference between conservatism and minimal change comes into light in the context of base contraction. Whenever we are dealing with belief sets, minimal change implies conservatism. But in the contraction of belief bases, minimal change warrants tenacity only with respect to explicit beliefs and not with respect to original original epistemic commitments. Within the static dimension a derived belief may be not worth retaining for its own sake, but in the dynamic dimension things are rather different, given that the derived belief may still be valid even if a stronger belief form which it was derived is discarded. Therefore, it seems interesting to develop a contraction operator for bases able to "rescue" consequences of the contracted sentence (implicit commitments) which were not explicit in the original set.

However, this is exactly what is avoided by the postulate of inclusion for bases, according to which $A - a \subseteq A$, where $A - a$ is the result of a contraction of a belief a from a set A. It is reasonable to expect that an operator satisfying conservatism will not satisfy this form of inclusion.

The formal questions to be faced then are: (i) how could the inclusion postulate be reasonably weakened to include in the resulting contracted set some of the consequences of the original belief base? (ii) what would be its corresponding construction? The present paper proposes a possible answer to these questions.

In the present paper the construction will be based on Kernel Contraction [5]instead of partial meet contraction. The intuition and the result of these operations are almost the same. Instead of choosing among maximal sets not deriving the sentence to be contracted (partial meet), kernel contraction first identifies those minimal subsets deriving the sentence to be contracted and then delete at least a sentence of each. Kernel contraction is better suited to our purposes given that it allows a better vision of the sentences to be deleted. Our interest, based on conservatism, is to change (to weaken) those sentences instead of deleting them.

First we make a very brief presentation of kernel contraction based on a slightly more general perspective, given that we do not define the revision operators on fixed sets. Then we introduce three versions of conservatism: maximal, partial and refinement.

3 Kernel contraction

We begin with some conventions. Let FOR denote the set of formulas of a propositional language, including at least negation ($\neg$) and implication ($\rightarrow$). Sets of sentences are denoted by capital letters and sentences by small letters. We are going to make extensive use of the tarskian consequence operator Cn. The notation of the consequence operator on unitary sets $Cn(\{a\})$ is here simplified by writing $Cn(a)$.

In standard presentations, the contraction operator is defined for a fixed set of formulas A and any formula a. Some clauses and restrictions are thus introduced on the construction and on the postulates of contraction in order to treat the limiting cases of a contraction of A by a where $a \notin Cn(A)$ or where $a \in Cn(\emptyset)$. Such restrictions allow the contracting set to remain unchanged in these cases.

We are going to exclude the above mentioned limiting cases *ab initio*, i.e. it is assumed that only *non-tautological* sentences which are *in* the original set are deleted, so we concentrate our efforts in the interesting ones, where some change must take place. We define all the relevant functions (that of contraction, kernel and incision) on pairs of a set A and a formula a. This allows the comparison of operations on distinct sets.

The interesting domain of the contraction function is defined as $INSET = \{(A, a) \in 2^{FOR} \times FOR : a \in Cn(A) \text{ and } a \notin Cn(\emptyset)\}$. Now we are in position to define the kernel function.

DEFINITION 1. The *kernel function* is a function $Ker : INSET \longrightarrow 2^{2^{FOR}}$ such that $X \in Ker(A, a)$ if and only if:

(i) $X \subset A$
(ii) $a \in Cn(X)$
(iii) if $X' \subset X$ then $a \notin Cn(X')$

Now each of these "chains" of derivation is broken by the incision function:

DEFINITION 2. An *incision* is a function $\sigma : INSET \longrightarrow 2^{FOR}$ such that:

(i) $\sigma(A, a) \subseteq \bigcup Ker(A, a)$

(ii) if $X \in Ker(A, a)$ then $X \cap \sigma(A, a) \neq \emptyset$

The kernel contraction is then an "incision" of the original set.

DEFINITION **3.** A *kernel contraction operator* s is a function $\ominus_\sigma : INSET$ $\longrightarrow 2^{FOR}$ such that $\ominus_\sigma(A, a) = A \ominus_\sigma a = A \setminus \sigma(A, a)$.

Kernel contraction may also be defined by postulates [5].

DEFINITION **4.** A *kernel contraction function* is a function $\ominus : INSET \longrightarrow$ 2^{FOR} such that $\ominus(A, a) = A \ominus a \subset FOR$ satisfying the following postulates:

success: $a \notin Cn(A \ominus a)$

inclusion: $A \ominus a \subset A$

core-retainment: if $b \in A \setminus A \ominus a$ then there is $A' \subset A$, such that $a \notin Cn(A')$ but $a \in Cn(A' \cup \{b\})$

The difference between partial meet contraction and kernel contraction is only noted on base contraction and it is concerns the last postulate of core-retainment. Partial meet satisfies a stronger property, called *relevance* which would read as follows: if $b \in A \setminus A \ominus a$ then there is A' such that $A \ominus a \subseteq A' \subset A$, such that $a \notin Cn(A')$ but $a \in Cn(A' \cup \{b\})$.

The following theorem confirms that these postulates characterize the construction based on kernel sets.

THEOREM **1.** (representation theorem [5]) Every kernel contraction operator based on an incision σ is a kernel contraction function. Let $\ominus$ be a kernel contraction function. Then there is an incision σ such that $\ominus = \ominus_\sigma$.

The operator thus defined is not uniform, in the sense that the same set of kernels (resulting, for instance, from the same set contracted by equivalent senteces) may generate different contractions, given the use of different and unconstrained incision fucntions. In order to obtain uniformity a postulate should be added in conjunction with a corresponding constraint in the incision function. We are not going to do it here, since our focus is only on what is essential to weaken the inclusion postulate.

4   Conservative Kernel Contraction

Let us now introduce what we call *conservative kernel contraction*. We interpret Harman's principle of conservatism by the directive that the agent is justified in keeping logical consequences of the rejected belief unless she has an specific reason not to. In the case of bases, this means that some of the consequences of the original belief base should be "rescued" when a sentence is deleted.

Of course, not all the consequences of the original sentence are the object of our concern and we should be restricted to what is relevant. But "what is relevant" is a question of its own, depending on some "conception" of relevance. We are going to explore three possible operations. Let us then introduce a choice function $*$ on the power set of formulas such that for every a, $Cn^*(a) \subseteq Cn(a)$. A broader conception of relevance would consider all the material (non-tautological) consequences of the deleted sentece, i.e. $Cn^*(a) = Cn(a) \backslash Cn(\emptyset)$. A less conservative option would be to recover just a part of them, i.e. to take $Cn^*(a) \subset Cn(a) \setminus Cn(\emptyset)$. The less conservative option would be to recover just one property as relevant, which means to take a weaker formula $w(a)$ and its complementary formula $w'(a)$, such that $\{w(a), w'(a)\} \subseteq Cn(a)$ and $a \in Cn(\{w(a), w'(a)\})$. Both are rescued as relevant and then it is decided which one should be kept. The conservative contraction will be based on the second option, given that it is the most general.

The construction here proposed count on the property of decomposibility of logical consequence, that is, for all a, there is $w'(a)$ and $w(a)$, such that $\{w(a), w'(a)\} \subset Cn(a)$ and $a \in Cn(\{w(a), w'(a)\})$, which holds, for instance, in classical logic. First, we use a choice function $*$ on the set of logical consequences satisfying the following condition: if $x \in Cn^*(a)$ then, $x' \in Cn^*(a)$ such that $a \in Cn(\{x, x'\})$. This property is trivial in the case $Cn^*(a) = Cn(a) \setminus Cn(\emptyset)$. Conservative kernel contraction is then built as follows.

DEFINITION 5. The *conservative kernel function* is a function $CKer : INSET \longrightarrow 2^{2^{FOR}}$ such that $X \in CKer(A, a)$ if and only if:

(i) $X \subset A \cup Cn^*(a)$
(ii) $a \in Cn(X)$
(iii) if $X' \subset X$ then $a \notin Cn(X')$

The incision function now operates on this extended version of a set of kernels, which we may also call *conservative kernels* or *c-kernels*.

DEFINITION 6. An *incision* is a function $\sigma : INSET \longrightarrow 2^{FOR}$ such that:

(i) $\sigma(A, a) \subseteq \bigcup CKer(A, a)$
(ii) if $X \in CKer(A, a)$ then $X \cap \sigma(A, a) \neq \emptyset$

Now, we introduce another selection function s on the set of conservative kernels, such that for every set A and sentence a, $s(CKer(A, a)) \subseteq CKer(A, a)$. We call such selection a *partial selection* if and only if $s(CKer(A, a)) = \{X : X \cap (Cn^*(a)) \neq \emptyset\}$. Note the following interesting property which is usefull:

OBSERVATION 1. Let $*$ be a choice and s a partial selection function. Then (i) $Cn^*(a) \subseteq \bigcup s(CKer(A, a))$ and (ii) $A \cup Cn^*(a) = A \cup (\bigcup s(CKer(A, a)))$.

PROOF. For the first part, by definition, if $x \in Cn(A)$, then $x' \in Cn(a)$ such that $a \in Cn(\{x, x'\})$. Hence $\{x, x'\}$ is a conservative kernel and, by definition of partial selection, $\{x, x'\} \in s(CKer(A, a))$ and thus $x \in \bigcup s(CKer(A, a)$. This also proves the left to right statement of the second part. To obtain the right to left, suppose $x \in A \cup \bigcup s(CKer, (A, a))$, then either $x \in A$ or $x \in X \in s(CKer(A, a))$. The first hypothesis leads to $x \in A \cup Cn^*(a)$. For the second hypothesis, we have by definition of partial selection that $X \cap Cn^*(a) \neq \emptyset$. Thus either $x \in X \cap Cn^*(a)$, in which case $x \in Cn^*(a)$, or $x \in A$ by the definition of c-kernel. $\qquad\square$

The conservative kernel contraction is taken as an incision of the original set which is extended by the recovery of some of its logical consequences.

DEFINITION 7. A *conservative kernel contraction operator* s is a function $\otimes_\sigma : INSET \longrightarrow 2^{FOR}$ such that $\otimes_\sigma(A, a) = A \otimes_\sigma a = A \cup (\bigcup s(CKer(A, a))) \setminus \sigma(A, a)$.

The function which characterizes maximal conservative kernel contraction is given below.

DEFINITION 8. A *conservative kernel contraction function* is a function $\otimes : INSET \longrightarrow 2^{FOR}$ such that $\otimes(A, a) = A \otimes a \subset FOR$ satisfying the following postulates:

success: $a \notin Cn(A \otimes a)$

inclusion: $A \otimes a \subset A \cup Cn^*(a)$

core-retainment: if $b \in A \setminus A \otimes a$ then there is $A' \subset A \cup Cn^*(a)$, such that $a \notin Cn(A')$ but $a \in Cn(A' \cup \{b\})$

THEOREM 2. (representation theorem) Every conservative kernel contraction operator based on an incision σ is a conservative kernel contraction function. Let $\otimes$ be a conservative kernel contraction function. Then there is an incision σ such that $\otimes = \otimes_\sigma$.

PROOF. For the first part, we have to show that the construction satisfies the proposed postulates. For inclusion, $A \otimes a = A \cup \bigcup s(CKer(A, a)) \setminus \sigma(A, a)$ and given observation 1 we have $A \otimes a = A \cup Cn^*(a) \setminus \sigma(A, a)$, what means that $A \otimes a \subseteq A \cup Cn^*(a)$. For success, suppose *ad absurdum* that $a \in Cn(A \cup \bigcup s(CKer(A, a)) \setminus \sigma(A, a))$. By compacity, there is $X \subseteq A \cup \bigcup s(CKer(A, a))$ such that $a \in Cn(X)$. By observation 1, $X \subseteq A \cup Cn^*(a)$. Thus there is $X' \subset X$, minimal, such that $X' \in CKer(A, a)$. But this contradicts the definition of incision function and the hypothesis, since $X' \cap \sigma(A, a) = \emptyset$. For core-retainment, suppose $b \in A \setminus A \otimes_\sigma a$. Hence $b \in \sigma(A, a)$. Since $\sigma(A, a) \subseteq \bigcup s(CKer(A, a))$, there is B, such that, $b \in B \in s(CKer(A, a))$. By observation 1, $B \subseteq A \cup Cn^*(a)$. Now, take $B' = B \setminus b$. Then $a \notin Cn(B')$ and $a \in Cn(B' \cup b)$, and we are done. For the second part, take $\sigma(A, a) = A \cup Cn^*(a) \setminus A \otimes a$. We have to prove that σ is an incision function and that $A \otimes a = A \otimes_\sigma a$. First, we prove that $\sigma(A, a) \subseteq \bigcup s(CKer(A, a))$. Suppose $b \in \sigma(A, a)$. Then $b \in A \cup Cn^*(A, a) \setminus A \otimes a$. By core-retainment, there is $B \subseteq A \cup Cn^*(A, a)$ such that $a \notin Cn(B)$, but $a \in Cn(B \cup b)$. By compactness there is a minimal set B', such that $a \notin Cn(B')$, but $a \in Cn(B' \cup b)$. Hence, by the definition of partial selection on c-kernel, $B' \cup b \in s(CKer(A, a))$ and thus $b \in \bigcup s(CKer(A, a))$. Now suppose $X \in s(CKer(A, a))$. We know by success that $a \notin A \otimes a$. Hence $X \nsubseteq A \otimes a$. Let x be such that $x \in X$ and $x \notin A \otimes a$. Since by observation 1, $s(CKer(A, a) \subseteq A \cup Cn^*(a)$, we have that $x \in A \cup Cn^*(a)$. Therefore, $x \in \sigma(A, a)$ and thus $X \cap \sigma(A, a) \neq \emptyset$. So σ is an incision function. Finally, it follows form inclusion, $A \otimes a \subseteq A \cup Cn^*(a)$ and the construction of $\sigma(A, a)$ that $A \otimes a = A \cup Cn^*(a) \setminus \sigma(A, a)$. Thus $\otimes = \otimes_\sigma$. $\qquad\square$

4.1 Maximal conservative kernel contraction

If we define the choice $*$ as $Cn^*(a) = Cn(a) \setminus Cn(\emptyset)$ we obtain an operator we may call *maximal conservative kernel contraction*. In this case, the construction actually does not need a partial selection function on the conservative kernels, given that they will correspond to $Cn^*(a)$, as the reader may verify rephrasing observation 1. It also satisfies an additional property: for any $b \notin Cn(\emptyset)$ we have that either $a \in Cn(A \otimes a \cup \{b\})$ or $a \in Cn(A \otimes a \cup \{\neg b\})$. It means that any further expansion of our beliefs will recover our original belief on a.

This operator meets our philosophical concern related to minimality providing formal expression to Harman's principle of conservatism. It supresses the sentence to be contracted but "rescues" *all* its material consequences which are not sufficient to recover the original sentence itself.

Note that conservatism is here pushed to its extreme and fails to pay due respect to clutter avoiance. Suppose we discover a belief to be mistaken, then we make an effort to supress only what from that sentece is to be blamed as mistaken, keeping all weaker information as at least hypothetically safe.

But such ideal situation is far from practical purposes. And it could even be questioned as a reasonable "ideal". Take, for instance, the experiment where water turns into steam at a temperature different from 100^0C, as originally believed. As a result of a conservative kernel contraction the agent will believe that water turns into steam at 100^0C, if the heating takes place at sea level, or if the sky is blue, or if Elvis is not dead, or anything else. This is so because for any sentence a and any set A we have that $\{a \vee b, a \vee \neg b\} \in CKer(A, a)$, for any b. This is a consequence of the decomposibility property we mentioned before.

Given this aparently counter-intuitive result, we may see maximal conservative contraction as unreasonable for practical purposes, even though it is the most faithfull to conservatism if the underlying logic satisfies decomposibility. It seems better to use partial selections. Another option located at the other extreme is to restrict our choice and abandon the original belief in order to recover only one weaker formula.

4.2 Kernel Refinement

A third operator satisfying the principle of conservatism would obtain by choosing just one kernel including a consequence of the rejected sentence.

Actually, this idea was explored by means of an operator called "refinement" [9]. Loosely speaking, refinement is an operation by which the agent qualifies her beliefs by a new condition. The belief to be qualified can be the original (internal refinement), the input (external refinement), or both may be qualified by muttually exclusive conditions (global refinement). If the underlying logic satisfies decomposibility, then external refinement is equivalent to selective revision [4]. Global refinement is built out of internal refinement by a simple expansion operator. Therefore, we are going to present only the operator of internal refinement, denoted by $\lhd$, which is also a particular case of contraction in the domain of belief sets.

In [9] refinement for bases was presented as follows. A conditional function h on the set of formulas was introduced so that for each sentence a to be contracted it selects the relevant condition $h(a)$ in which the belief in a does not hold (and therefore $\neg h(a)$ is the one which *confirms* the original belief). There are two ways to construct the internal refinement operator on bases:

(i) add to the base $\neg h(a) \to a$ and then simply contract it by a: $A \lhd_1 a = A \cup \{\neg h(a) \to a\} \div a$ ($\div$ is the partial meet contraction operator); or

(ii) contract the base by $h(a) \to a$ and then add $\{\neg h(a) \to a\}$: $A \lhd_2 a = A \div h(a) \to a \cup \{\neg h(a) \to a\}$.

Both operations are equivalent if the selection function used in the contraction and the conditional function satisfy some reasonable conditions [9].

The relevant postulates are the following:

root-cutting: $h(a) \to a \notin Cn(A \lhd a)$

success: $a \notin Cn(A \lhd a)$

inclusion: $A \lhd a \subset A \cup \{\neg h(a) \to a\}$

relevance: if $b \in A \setminus A \lhd a$, then there is A', $A \lhd a \subseteq A' \subseteq A \cup \{\neg h(a) \to a\}$, such that $a \notin Cn(A')$ and $a \in Cn(A' \cup \{b\})$

preservation: $\neg h(a) \to a \in A \lhd a$.

The first alternative for refinement, $\lhd_1$, is given by success, inclusion, relevance and preservation, while the second, $\lhd_2$, is given by root-cutting, relevance and inclusion, which are equivalent sets of postulates.

It is easy to see the correspondence between internal refinement and a particular case of conservative kernel contraction, where $Cn^* = s(CKer(A, a)) = \{h(a) \to a, \neg h(a) \to a\}$. Note first that the postulate of inclusion is weakened like in conservative contraction. Relevance is the counterpart of conservative core-retainment. Hence, by such restriction, we have that kernel refinement is characterized by success, inclusion and core-retainment. Kernel refinement would be given by the following postulates:

success: $a \notin Cn(A \lhd a)$

inclusion: $A \lhd a \subset A \cup \{h(a) \to a, \neg h(a) \to a\}$

core-retainment: if $b \in A \setminus A \lhd a$, then there is A', $A' \subseteq A \cup \{h(a) \to a, \neg h(a) \to a\}$, such that $a \notin Cn(A')$ and $a \in Cn(A' \cup \{b\})$

If one wants to specify which condition is the confirming or the refuting one for the original sentence, then one may add the postulate of preservation, which may correspond to a particular restriction on the incision function (for instance, an incision function is preservative if it selects $h(a) \rightarrow a$ to be discarded).

5   Some final notes on abductive reasoning and future work

The steps from maximal conservative kernel to refinement have a flavour of abductive reasoning.

In classical logic any weaker formula than a formula a may be expressed by a conditional formula $h \rightarrow a$, for some h (if h is weaker than a, then it is equivalent to $\neg h \rightarrow a$). Hence we may interpret the recovery of weaker formulas as the incorporation of new hypothesis qualifying the acceptance of the rejected sentence. Each conditional $h \rightarrow a$ may be seen as an explanation for a.

Harman considers abductive reasoning as an inference to the best explanation. Given an evidence e one infers an something to explain the evidence: *one starts believing e and comes to believe that e because h* ([7], p. 67). This would be an explanation in the context of expansion of the agent's belief. This idea was pursued by Pagnucco in a model of abductive reasoning based on belief revision [10]. In such model the agent would explain evidence e by expanding his belief set K until he reaches consistently that $e \in CnK \cup H$. Wasserman and Dias applied the model to belief bases [12].

According to Aliseda [3], abductive reasoning takes place not only when an agent wants to explain new data in a *surprising observation*, but also in presence of some *abnormality* refuting some previously accepted belief, i.e. when the agent has to explain why her original belief failed. She may abandon the contradicted belief, what is represented by contraction, but may also be tenacious and hold the original belief now qualified by a confirming condition h. That is, in front of a surprise forcing her to abandon her belief in e she may still hold that e under the condition that h. So abduction may occur in the context of a contraction of the belief base.

In this rationalization process, if one rejects e based on an observation where e did not obtain, she might look for relevant conditions in which e holds. Revision is still not the right strategy given that she might be not convinced that $\neg e$. If h is considered a relevant factor, this means that its presence or its absence makes a difference with respect to e. Either its presence or its absence will be a refuting condition for e. This process is captured in the conservative contraction model by the choice function on the set of logical consequences of the original belief base. In our model, the agent is justified in selecting the conservative kernel where $h \rightarrow e$ is present, i.e. $X = \{h \rightarrow e, \neg h \rightarrow e\}$. The incision function will then select which hypothesis is the refuting condition for e, thus remaining in the conservatively contracted set the confirming conditional.

Maximal conservative contraction may be interpreted as resulting in the set of all possible explanations to confirm e. Here we would have in the resulting conservatively contracted set either $h_i \rightarrow e$ or $\neg h_i \rightarrow e$ for every sentence h_i in the language, provided that $h_i \rightarrow e$ is not a tautology. Actually, this may represent an unreflected step where the agent simply does not know why e has failed. We may consider that at this level the agent just realized that e is not fully true, but she does not evaluate any of the possible explanations.

The next step, we may call partial conservative contraction, would represent a level of critical reasoning, where the agent would have rejected several unplausible explanations for the failure of e, or even where the agent has performed tests to exclude the relevance of some property h. This means that the resulting contracted set would now contain $\{h_1 \rightarrow e, h_2 \rightarrow e, ..., h_n \rightarrow e\}$. In this stage we have that $h_1 \vee h_2 \vee ... \vee h_n \rightarrow e \in Cn(A \otimes e)$.

The result of such tests would lead to a choice of which hypothesis may be considered the "best explanation" for the original rejected belief. That is where we come to the last version of conservative contraction, which we call refinement. This is the case when we choose just one kernel.

We may think of other reasonable assumptions for the selection of relevant hypothesis explaining the rejected sentence. For instance, all of our original beliefs should not be considered candidates to explain the rejected sentence, given that they were present when such sentence failed to obtain. This would provide two conditons on the selection function on conservative kernels. First, if $h \in A$, then $\{h \rightarrow e, \neg h \rightarrow e\} \notin s(CKer(A, a))$, given that h proved not to be relevant for e. But we have to consider the case of the c-kernel $\{h \rightarrow e, h\}$. If both are in the original set, then either one of them must be abandoned. If only h is in the original set, we may have a reason to keep it. This last case may be obtained by imposing a restriction on the incision function, for instance if $h \in A$ and $h \rightarrow e \notin A$, then $h \notin \sigma(A, e)$. Nevertheless, the exact connection between conservative contraction, refinement and abduction still have to be investigated.

Another interesting point would be to weaken not only the rejected sentence, but all those original beliefs which are responsible for deriving the rejected sentence. For instance, in the case $\{h \rightarrow e, h\} \subseteq A$ we may obtain a conservative contraction where a weaker formula than $h \rightarrow e$ or a weaker formula than h (or both) are in the resulting modified set.

A third idea which seems worth exploring would be a conservative contraction operator that "rescues" only those logical consequences which are subformulas of the rejected sentence. This is actually what happens in the example 1 and captures better the idea that one keeps just those parts of the sentence which are usefull, whithout introducing new propositions.

All these possibilities are left for future work.

BIBLIOGRAPHY

[1] Alchourrón, Carlos and Makinson, David. On the logic of theory change: contraction functions and their associated revision functions. *Theoria.* 48, 14-37, 1982.

[2] Alchourrón, Carlos, Gardenfors Peter, and Makinson, David. On the logic of theory change: partial meet contraction and revision functions. *Journal of Symbolic Logic.* 50, 510-530, 1985.

[3] Aliseda, Atocha. Abduction as epistemic change: a Peircean model in artificial inteligence. Flach, Peter A., and Antonis C. Kakas, Editors, *Abduction and Induction: Essays on Their Relation and Integration.* Dordrecht and Boston: Kluwer Academic Publishers, 2000.

[4] Fermé, Eduardo and Hansson, Sven Ove. Selective Revision. *Studia Logica,* 63, v.3, pp.331-342, 1999.

[5] Hansson, Sven Ove. Kernel Contraction. *Journal of Symbolic Logic,* 59, 845-859, 1994.

[6] Hansson, Sven Ove. *A textbook of belief dynamics:theory change and database updating.* Kluwer, 1999.

[7] Harman, Gilbert. *Change in view: principles of reasoning,* MIT Press, 1989.

[8] Makinson, D. On the status of the postulate of recovery in the logic of theory change. *Journal of Philosophical Logic,* 16, 383-394, 1987.

[9] Maranhão, Juliano S.A. Refining Beliefs, in J-Y. Béziau, A. Costa Leite (eds.) Perspectives on Universal Logic, 335-349, Polimetrica, 2007.

[10] Pagnucco, Maurice. *The Role of Abductive Reasoning Within the Process of Belief Revision,* Phd. Thesis, Department of Computer Science, University of Sydney, 1996.

[11] Sosa, E. The raft and the pyramid: coherence vs. foundations in the theory of knowledge. Midwest Studies in Philosophy, 5, 3-25, 1980.

[12] Wassermann, Renata and Dias, Wagner. Abductive Expansion of Belief Bases, *Proceedings of the IJCAI Workshop on Abductive Reasoning,* Seattle, August, 2001.

Juliano S. A. Maranhão
Faculty of Law
University of São Paulo
São Paulo, Brazil
E-mail: julianomaranhao@usp.br

The "Lottery Paradox" Paradox and other Self-Attacking Arguments from the Point of View of Defeasible Argumentation Frameworks

GUSTAVO A. BODANZA

ABSTRACT. The "lottery paradox" paradox –a problem derived from the well-known lottery paradox– arises as the result of chaining defeasible reasons that lead from the statement 'the lottery is fair' to the statement 'the lottery is not fair', conforming in consequence a self-attacking argument. This paper shows two simple semantics that capture the intuitions about what kinds of arguments should become justified in such scenarios. The semantics, called *sustainable* and *lax* respectively, are modeled upon argumentation frameworks and constructed on the common ground of the powerful concept of admissibility introduced by P.M. Dung. Both semantics result to be more "credulous" than preferred semantics, since they enable the acceptance of arguments that are attacked only by other arguments that in time are self-attacking. A special kind of self-attacks are those indirectly provoked by odd-lenght cycles of attack. While sustainable semantics overrule them all (in accordance with most semantics in the defeasble argumentation literature), we show some examples of practical reasoning where such cycles do not clearly imply self-attacks, so the involved arguments could be acceptable. Lax semantics is proposed here to capture also this alternative behavior.

1 Introduction

John Pollock ([8], [9], [11]) studied a particular instance of self-attacking arguments derived from the lottery paradox, an example that has been introduced by Kyburg ([7]) as a problem for reasoning with probabilities. Imagine a fair lottery with 1,000,000 tickets, so that each ticket has one in a million chances of winning. Given a particular ticket, we can tentatively conclude that it will not win, since its probability is very low. But the same can be concluded about every ticket, hence we can conclude that none of them will win. But given that the lottery is fair, one ticket must win. This is called "the lottery paradox". Upon this contradiction Pollock claimed that a further contradiction can be drawn, namely that since none of the tickets will win, the lottery cannot be fair. That is, the argument of the lottery paradox, based upon the premise that the lottery is fair, attacks itself since one of its consequences contradicts its premise. This is what Pollock called the ' "lottery paradox" paradox'.

This work is not about the lottery paradox, but about the "lottery paradox" paradox as a self-attacking argument. The study of the ways an argument can be deemed warranted with regard to its interaction with other arguments is known as *defeasible argumentation*. Roughly, an argument is warranted if it belongs to every *extension* of the argumentation framework, that is, to any set of arguments which members can all be successfully defended against the attacks by other arguments. For example, if an argument α attacks an argument β, and β in turn attacks an argument γ, we can say that γ is defended by α; moreover, if α has no attackers, then we can conclude that both α and γ can be successfully defended, so they both will belong to the (only) extension of the framework, and so they both will remain warranted.

Self-attacking arguments are special cases where the attack relation form a cycle of length one, and they are paradoxical from the point of view of warrant. If an argument α attacks itself, then α can be successfully defended only if α (which is the attacker of α) cannot be successfully defended. In the "lottery paradox" paradox, the argument concluding that the lottery is fair (β) is a subargument of the larger argument concluding that the lottery is not fair (α). Hence, argument α attacks its subargument β, and so it attacks itself. Provided that α is unacceptable as "successfully defended", it should be excluded from any extension. This lead us to expect that β remain warranted, since it does not receive any sensible attack. After all, the contradiction is clearly provoked by a chaining of reasons that goes from β's conclusion ('the lottery is fair') to the conclusion of the whole argument α ('the lottery is not fair'). The faulty character of α removes it from contention in any contest with conflicting arguments, in particular, with β. Then, it seems that the attack of α over β should be neglected. Surprisingly, there is no extension semantics -as far as we know- that yields this behavior, remaining sound in all other cases.

Although the "lottery paradox" paradox could seem to be a little bit artificial example, everyday life practical reasoning often involves us in argumentations that can lead to self-attacks. Consider, for example, a situation in which an individual has to decide whether to keep her savings or not in her account in Bank B during a financial crisis. She has a good argument (say β) to believe that Bank B is solvent: it always gave the money back to the savers in the past, even in times of financial crisis. This is a good reason for her to conclude that if she (or any other saver) comes to the bank now to get her money back, the bank will pay. This argument can be used to conclude that the bank will give the money back to her *and* to saver s_1 *and* to saver s_2 *and* ... etc.. But in this way Bank B will give back the money to a sufficiently big amount of savers as to run out of money, so that the bank will not be solvent. Stopping the reasoning here, the whole argument concluding that Bank B is not solvent (say α) attacks the argument concluding that it is solvent (β), and since the last is a subargument of the former a self-attack is obtained. So, the scheme of the attack relation on these arguments is the same as that of the "lottery paradox" paradox.

In this paper we argue that the problem of self-attacking arguments can be approached from an extension semantics view, upon the guiding idea of seeing an argumentation process as a strategic interaction between two agents (though we do not attach ontological weight to this assumption, since the agents may be the two sides that a single reasoner may take while pondering which arguments should be warranted). The idea behind the extensions we will propose is that an agent will select a maximum set of arguments that can be defended against any set of arguments that any "rational" opponent can select. In the "lottery paradox" paradox, no "rational" agent will choose the sets $\{\alpha\}$ or $\{\alpha, \beta\}$ (because they contain internal attacks), hence the set $\{\beta\}$ is the only maximum set of arguments that can be defended against any "rational" choice of a hypothetical opponent. Accordingly, an extension semantics should judge $\{\beta\}$ as the extension of that argumentation framework (after all, it seems sensible to conclude that the lottery is fair and that the bank is solvent in our motivating examples).

The solution takes up from a wide consensus among researchers in the argument systems' community, who seek to solve the problem of self-attacking arguments by rejecting any justification for arguments in any odd-length cycles of attack. We propose *sustainable semantics* (previously introduced in [3]) as the way of capturing this lead. The platform on which sustainable semantics is developed is the argumentation frameworks model of Dung ([5]). The main idea introduced there is the *admissibility* of sets of arguments. This notion defines the way a set of arguments can be defended from external attacks. We build upon this idea, stating that the defense of a set of arguments should not be defined in absolute terms but relative to other possible challenging sets of arguments. In this way we obtain a notion of *cogency* on which we base our sustainable semantics.

On the other hand, self-attacking arguments can arise by means of indirect attacks. The most simple case is given by a 3-length cycle of attacks:

α attacks β attacks γ attacks α. Since each one of these arguments attacks its defender, each one attacks itself in an indirect way. This is also a common hypothesis in the argument systems community. Nevertheless, one can imagine some scenarios for that framework in which the self-attacking character of the involved arguments is not clear. We will introduce here *lax extensions*, a semantics that enables the acceptance of such arguments, while avoids the undesirable interference of (direct) self-attacking arguments in a similar way as sustainable semantics.

The paper is organized as follows. The main features of Dung's system are presented in section 2. In section 3 we present the guiding intuitions of our proposal. Section 4 shows the formal concept of *cogency*, the common ground on which both sustainable and lax semantics are developed. Sections 5 and 6 discuss the treatment of self-attacking arguments by sustainable and lax semantics, respectively. Section 7 shows the relationships between the semantics. Finally, section 9 summarizes our conclusions.

2 Argumentation frameworks

The formal platform on which this work is based is Dung's argumentation frameworks (Dung ([5])). An argumentation framework is a pair $AF = \langle AR, \, attacks \rangle$, where AR is a set of abstract entities called 'arguments' and *attacks* is a binary relation $attacks \subseteq AR \times AR$ intended to denote attacks among arguments. The fundamental question about argument frameworks is which arguments remain "justified" under the attacks among them. A subset of "justified" arguments is called an *extension*, and alternative criteria yield different kinds of extensions. Two basic concepts allow to obtain all the extensions defined by Dung: 'acceptability' of arguments, and 'admissibility' of a set of arguments. The formal definitions of these notions are:

DEFINITION 1. *(Dung ([5])). In any argumentation framework AF an argument σ is said* acceptable *w.r.t. a subset S of arguments of AR, in case that for every argument τ such that τ attacks σ, there exists some argument $\rho \in S$ such that ρ attacks τ[1]. A set of arguments S is said* admissible *if each $\sigma \in S$ is acceptable w.r.t. S, and is conflict-free, i.e., the attack relation does not hold for any pair of arguments belonging to S. A* preferred extension *is any maximally admissible set of arguments of AF. A* complete extension *of AF is any conflict-free subset of arguments which is a fixed point of $F(\cdot)$, where $F(S) = \{\sigma : \sigma$ is acceptable w.r.t. $S\}$, while the* grounded extension *is the least (w.r.t. $\subseteq$) complete extension.*

---

[1] In order to simplify the notation we will say that σ *attacks* A (conversely, A *attacks* σ) iff there exists some $\tau \in A$ such that σ *attacks* τ (τ *attacks* σ, respectively).

Extensions are said to constitute the "semantics" of an argumentation framework, because they capture the notion (weaker than truthfulness) of *justifiability* of arguments. Different kinds of extensions give way to alternative semantics, that is, different intuitions about which arguments can be justified. The usual interpretation is that semantics justifying a single extension correspond to some kind of skeptical arguer. On the other hand, in a semantics that yields multiple extensions, to select their intersection also characterizes a kind of skeptical attitude, while the arbitrary choice of one of the extensions represents a credulous decision. For example, in Dung's system, preferred, complete and stable extensions usually yield several extensions, so they can be used to model alternative forms of credulity, while different conceptions of skepticism can be modeled by the grounded extension or the intersections (each for each notion of credulity) of the credulous extensions.

EXAMPLE 2. Assume $AF = \langle \{\alpha, \beta\}, \{(\alpha, \alpha), (\alpha, \beta)\} \rangle$. Then $\emptyset$ is the grounded extension and the only preferred and complete extension. There is no stable extension in this framework. Since AF represents here the structure of the "lottery paradox" paradox, all Dung's extensions represent the decision of refusing all the arguments and, hence, the abstention of concluding something about the fairness of the lottery, either from a skeptical or credulous point of view.

3 A strategical approach to argumentation

Dung's view of argument justification does not assign any role whatsoever to the utterer/s of the arguments. His framework can be interpreted as the decision-making setting of a single agent, who ponders the arguments in the framework. Another interpretation is to see the arguments as laid out by an agent to defend herself against the arguments presented by another agent. This distinction motivated some researchers to view defeasible argumentation from a dialogical point of view, modeling dialogues between two players (pro and con) who develop an argumentation strategy by advancing arguments one by one in alternation (e.g. Vreeswijk and Prakken ([12])).

While we take this view of argumentation as arising from the interaction of two agents, the novelty of our approach consists in that we focus on argumentation strategies instead of considering the properties of a sequential dialogue. An argumentation strategy can be viewed as the set of arguments that an intentional agent may lay out in response to the arguments of the other agent in a dialogue. As this notion involves all possible combinations of arguments, we consider that any subset of AR in an argumentation framework $AF = \langle AR, attacks \rangle$ can be seen as an argumentation strategy (this includes, of course, strategies that are not cleaver at all). In this context, our main working hypothesis is that the rationality of choosing an argumentation strategy is relative to the argumentation strategies that the rival may select. What matters here is not whether an argumentation strategy (as a subset of arguments) is "admissible" in the whole framework, but if it is so relative to the sub-frameworks formed by the strategy itself together with the possible strategies of the other agent.

Under this view we can rethink the main problems of argumentation. While usual questions about argumentation ask how arguments can be compared and justified, we want to know now how argumentation strategies can be compared and justified. Once that question is settled, we extend the justification to individual arguments. The main methodological steps in the characterization of justification are based on the following assumptions:

- argumentation strategies can be compared on the basis of the attack relation among the arguments that they include;

- argumentation strategies can be justified on the basis of their comparison; and

- the justification of an argument obtains from the justification of the argumentation strategies to which it belongs.

We will see that, in this way, the undesirable interferences of self-attacking arguments will disappear without any necessity of considerations ad hoc.

4   The notion of cogency

The first idea is to establish a comparison among argumentation strategies (just 'strategies' henceforth) using the notion of admissibility. We will consider a strategy A at least as cogent as a strategy B if and only if B is conflict-free and all its arguments can be defended against the arguments of B. This amounts to saying that A is an admissible set within the framework formed just by the arguments in $A \cup B$. In order to introduce this definition, we first need an auxiliary notion: a restriction of an argumentation framework AF by a subset of arguments $S \subseteq AR$, is the framework $AF|_S = \langle S, attacks|_S \rangle$, where $attacks|_S$ is the restriction of $attacks$ to S.

REMARK 3. From now on we will refer to an arbitrary but fixed argumentation framework AF, so all definitions and results will be relative to that framework.

DEFINITION 4. For any pair of strategies A, B, we say that A is *at least as cogent as* B, in symbols '$A \succeq_{Cog} B$' iff A is admissible in $AF|_{A \cup B}$.

It immediately follows that:

PROPOSITION 5. *A strategy A is admissible iff for every strategy B,*
$A \succeq_{Cog} B$.

Note that according to this result, it is easy to recover the preferred
semantics by requiring a strategy A to be a maximal strategy (for set-
theoretic inclusion) such that for every $B \subseteq AR$, $A \succeq_{Cog} B$. Admissibility
is sufficient for cogency, but cogency is weaker than admissibility: while
we understand that A is admissible if it can be defended against any other
strategy, A is just cogent if it can be defended against any other strategy
which can be defended against A. Formally:

DEFINITION 6. *A strategy A is* relatively cogent *iff for every strategy B,*
if $B \succeq_{Cog} A$ *then* $A \succeq_{Cog} B$.

This simple concept of relatively cogent strategies will enable us to define
two different kinds of extension semantics for dealing with the problem of
self-attacking arguments. The first one, which we will call 'sustainable',
will behave exactly the same as preferred semantics unless self-attacking
arguments are present, while the second one, which we will call 'lax', will
be more credulous than both preferred and sustainable semantics since it
will invoke a weaker criterion of defense. The following example will help
for understanding them.

EXAMPLE 7. *(Continuation of example 2)* In the argumentation frame-
work of example 2, the set of all possible strategies is $2^{\{\alpha,\beta\}} = \{\emptyset, \{\alpha\}, \{\beta\}, \{\alpha,\beta\}\}$. Then, $\{\beta\}$ and $\emptyset$ are the only relatively cogent strategies. It is
clear that $\{\alpha\}$ is not relatively cogent since $\emptyset \succeq_{Cog} \{\alpha\}$ but $\{\alpha\} \not\succeq_{Cog} \emptyset$.
Note also that $\{\alpha\} \not\succeq_{Cog} \{\beta\}$. In sum, interpretaing the framework in terms
of the "lottery paradox" paradox, the strategies that are cogent lead either
to conclude that the lottery is fair (strategy $\{\beta\}$) or to an abstention in
concluding something about its fairness.

5   Sustainable semantics

The first semantics just arises by asking for a strategy to be a maximal
relatively cogent strategy. We call it 'sustainable' suggesting a strategy
that holds against any articulated criticism.

DEFINITION 8. *A strategy A is a* sustainable extension *of AF iff A is a*
maximal (w.r.t. $\subseteq$) relatively cogent strategy in AF.

Sustainable extensions avoid the undesirable interference of self-attacking
arguments, as shown in the following example:

EXAMPLE 9. *(Continuation of example 2)* In the argumentation framework of example 2 $\{\beta\}$ is the only sustainable extension since, besides itself,
$\emptyset$ is the only strategy as cogent as $\{\beta\}$, but $\{\beta\}$ is as cogent as $\emptyset$ and is

larger. Hence, 'the lottery is fair' is the only conclusion supported by a sustainable strategy.

It is easy to see that sustainable extensions yield a more credulous semantics than preferred extensions. The framework of the above example is a case in which the only preferred extension, $\emptyset$, is strictly included in the only sustainable extension, so the latter is clearly more credulous than the preferred extension. We can generalize this fact:

THEOREM 10. *If A is a preferred extension then there exists some sustainable extension B such that $A \subseteq B$.*

Given a preferred extension, the sustainable extension to which it belongs may be the preferred extension itself.

EXAMPLE 11. *("The Nixon diamond")* In the famous benchmark problem known as "the Nixon diamond" there are two arguments: α: 'Nixon is pacifist, since he is quaker and quakers tend to be pacifist'; and β: 'Nixon is not pacifist, since he is republican and republicans tend not to be pacifist'. The argumentation framework representing the problem is $AF = \langle \{\alpha, \beta\}, \{(\alpha, \beta), (\beta, \alpha)\}\rangle$, where $\{\alpha\}$ and $\{\beta\}$ are both the preferred and sustainable extensions (the grounded extension is $\emptyset$).

EXAMPLE 12. Assume $AF = \langle \{\alpha, \beta, \gamma\}, \{(\alpha, \beta), (\beta, \gamma), (\gamma, \alpha)\}\rangle$. Then $\emptyset$ is the only extension at all, which is grounded, preferred and sustainable at the same time.

6   Lax semantics

We focus now on another kind of self-attacking arguments: those which attack themselves by means of indirect attacks. According to Dung ([5]), an argument α *indirectly attacks* an argument β iff there exists a sequence of arguments $\alpha_0, \ldots, \alpha_{2n+1}$ such that $\alpha = \alpha_0$, $\beta = \alpha_{2n+1}$ and α_{i+1} attacks α_i, for $0 \leq i \leq 2n$. It is easy to see that every self-attacking argument is also an argument that indirectly attacks itself. On the other hand, every argument which is involved in an odd-length cycle of attack poses an indirect attack on itself. Most extension semantics overrule indirect self-attacking arguments. Nevertheless, we will argue that arguments involved in cycles of attack of length greater than one are not self-attacking in essence and, at least in some cases, it seems to be reasonable to accept them in separate extensions. Accordingly, our next extension semantics, which we will call *lax*, will overrule direct self-attacking arguments enabling the acceptance of indirect self-attacking ones[2]. We will proceed as follows: first, we give our

---

[2]From now on, the expression 'self-attacking arguments' will refer to *direct* self-attacking arguments, unless the adjective 'indirect' be placed before.

guiding intuition for lax semantics, secondly, we introduce the semantics in formal terms and, finally, we discuss the problem of indirect self-attacking arguments with regards to the behavior of lax semantics.

The basic idea of lax semantics is to justify any strategy A such that each one of the strategies that "defeat" A are in time "defeated" by some other strategy. In other words, a strategy is deemed overruled just in case it is defeated by a non-defeated strategy. This idea, which does not work when applied to isolated arguments, do work well when applied to sets of arguments (provided some concessions, as we shall see). To see that it does not work for isolated arguments, let us consider an argumentation framework where we just have $attacks = \{(\alpha, \beta), (\beta, \gamma), (\gamma, \delta)\}$. If we expect to accept any argument which defeaters (i.e. attackers) are all in time defeated, then we would have to accept δ, since its only defeater, γ, is in time defeated by β; but this is counterintuitive since β is defeated by α, which is undefeated, and hence γ, the defeater of δ, is ultimately defended, so δ should remain unjustified. If we instead think in terms of strategies —that is, sets of arguments instead of arguments– then the idea works well. First, the notion of "defeat" among strategies can be understood in terms of cogency: a strategy A defeats a strategy B if A is at least as cogent as B but B is not at least as cogent as A (that is, A can be defended against B but B cannot be defended against A). So, if we expect to accept any strategy which defeaters are all defeated by some strategy in the above setting, we will see that the strategy $\{\alpha, \gamma\}$ is to be accepted, since there is no strategy that defeats it. On the other hand, the strategy $\{\delta\}$ is not to be accepted, since is defeated by $\{\alpha, \gamma\}$, which is not defeated by any strategy.

The following relation formalizes the "defeat" relation among strategies:

DEFINITION 13. A strategy A is *strictly more cogent* than a strategy B, in symbols $A \succ_{Cog} B$, iff $A \succeq_{Cog} B$ and $B \not\succeq_{Cog} A$.

Now, we take those strategies which defeaters are all defeated in time by some other strategies.

DEFINITION 14. A strategy A is *weakly cogent* iff for every strategy B, if $B \succ_{Cog} A$ then there exists some strategy C such that $C \succ_{Cog} B$.

Finally, we choose the maximal weakly cogent strategies as the winners.

DEFINITION 15. A is a *lax extension* iff A is maximally (w.r.t. $\subseteq$) weakly cogent.

EXAMPLE 16. *(Continuation of example 2)* In the argumentation framework corresponding to the "lottery paradox" paradox, the only weakly cogent strategies are $\{\beta\}$ and $\emptyset$ (no strategy is strictly more cogent than them). Then, $\{\beta\}$ is the only lax extension.

In this example, it is clear that no strategy A is such that $A \succ_{Cog} \{\beta\}$, hence the condition for $\{\beta\}$ to be weakly cogent is vacuously accomplished. In more intuitive terms, we can say that no strategy $-\{\alpha\}$ included$-$ is a "defeater" of $\{\beta\}$.

The very difference between lax and sustainable extensions arises in presence of attack cycles of length odd and greater than one.

EXAMPLE 17. Assume $AF = \langle \{\alpha, \beta, \gamma\}, \{(\alpha, \beta), (\beta, \gamma), (\gamma, \alpha)\} \rangle$. Then $\emptyset$ is the only sustainable extension while $\{\alpha\}$, $\{\beta\}$ and $\{\gamma\}$ are all lax strategies.

In order to recognize the rationale behind lax extensions in cases of this sort, let us consider an interpretation of the above example.

EXAMPLE 18. *(Interpretation of example 17)* Assume $AR = \{\alpha, \beta, \gamma\}$ where the arguments are:

α : "Symptoms x and y suggest the presence of disease d_1, so I should apply therapy t_1";

β : "Symptoms x and z suggest the presence of disease d_2, so I should apply therapy t_2";

γ : "Symptoms x and w suggest the presence of disease d_3, so I should apply therapy t_3".

Assume these are the main arguments considered by an agent $\mathcal{H}$ (say, a M.D.) having to make a decision on which therapy should be applied to some patient who has symptoms x, y, z and w. Then $\mathcal{H}$ establishes the attack relation as follows:

- α attacks β, since $\mathcal{H}$ thinks that therapy t_1 is less invasive than t_2;

- β attacks γ, since $\mathcal{H}$ thinks that symptom z is more clearly detected than symptom w;

- γ attacks α, since $\mathcal{H}$ thinks that the consequences of disease d_3 would be more serious than those of disease d_1.

This example shows that, whatever the decision of $\mathcal{H}$, it is clearly more prudent to select any of the arguments than to refuse all of them. Hence, the semantics given by lax extensions is more in accordance with this rationale than that of sustainable (and preferred) extensions.

Bodanza and Tohmé ([3]) introduced *tolerant* extensions, a semantics that is akin to lax semantics with respect to the treatment of odd-length cycles of attack.

DEFINITION 19. *(Bodanza and Tohmé ([3]))* A strategy $A \subseteq AR$ is *cyclically cogent* iff for every strategy C, if $C \succ_{Cog} A$ then there exists a sequence $D_1, \ldots, D_m$ of strategies such that $D_1 = C$, $D_m = A$ and $D_{k+1} \succ_{Cog} D_k$, for $1 \le k < m$. $A \subseteq AR$ is a *tolerant extension* iff A is a maximal (w.r.t. $\subseteq$) cyclically cogent strategy.

The behavior of both tolerant and lax semantics is at odds with Dung's conception of indirect attack. Regarding the arguments belonging to an odd-length cycle of attacks, all of them are indirectly self-attacking according to this definition. Nevertheless, our example 18 seems to challenge a general application of this view, inasmuch it is at least not clear at all that every argument of the framework posses an attack on itself, even though it were indirect.

On the other hand, the example shows the convenience of picking any one of the arguments in the cycle in such practical decision making scenarios. From this point of view the problem is not one of epistemic justification, since the argumentation tends not to justify what to believe but what to do. So both lax and tolerant semantics seem to fit well to this view.

As said above, the behavior of tolerant and lax semantics is akin, but it is not exactly the same. While the fact that tolerant extensions are always lax extensions is obvious, consider the following counterexample for the converse:

EXAMPLE 20. Assume $AF = \langle \{\alpha, \beta, \gamma\}, \{(\alpha, \beta), (\beta, \alpha), (\beta, \gamma), (\gamma, \alpha)\} \rangle$. Then $\{\alpha\}$ and $\{\beta\}$ are lax extensions but only $\{\beta\}$ is tolerant.

While tolerant semantics behaves as credulously as preferred semantics in this example, the behavior of lax semantics is clearly more credulous. The acceptance of $\{\alpha\}$ by lax semantics is due to the simple fact that its only "defeater", $\{\gamma\}$, is in time "defeated" by $\{\beta\}$ (this is the kind of concessions mentioned at the beginning of this section). This is, indeed, the reason why we call such extensions "lax": they are just strategies which potential "defeaters" are not outright warranted.

In sum, sustainable semantics on one hand and lax and tolerant semantics on the other offer two alternative sorts of tools for dealing with the problem of self-attacking arguments, the first one in accordance with Dung's point of view of indirect attacks, and the later in disagreement with that view but arguably more in accordance with practical decision making.

7   Relationships among different extension semantics

Sustainable and lax strategies are related among them and other semantics as follows:

THEOREM 21.

1. *Every preferred extension is contained in some sustainable extension.*

2. *Every sustainable extension is contained in some tolerant extension.*

3. *Every tolerant extension is a lax extension, but not vice versa.*

Proof.

1. Immediate from 5.

2. It is clear that every sustainable extension is cyclically cogent, hence it must be contained in some maximal cyclically cogent strategy, i.e., in a tolerant extension.

3. Obvious. Example 20 shows that lax extensions are not always tolerant.

∎

PROPOSITION 22. *In every argumentation framework which is free of self-attacking arguments the class of the preferred extensions and the class of the sustainable extensions coincide.*

Proof. Assume that A is a sustainable extension which is not preferred, and let us show that there exists at least one self-attacking argument. Note that A is not admissible, so there exists some argument α that attacks A but it is not attacked by A. Hence we have to cases: (i) $\{\alpha\} \succ_{Cog} A$, or (ii) α is self-attacking and so $\{\alpha\} \not\succ_{Cog} A$. But (i) contradicts the hypothesis that A is sustainable, hence (ii) is the case. ∎

On the other hand, the class of lax extensions does not necessarily coincide with those of preferred and sustainable semantics even in the absence of self-attacking arguments, as it is clear from example 17.

8   Discussion

Let us compare our proposal with some works in the same line. Besides the approach by Bodanza and Tohmé ([3]) discussed in section 6, a similar notion to lax and tolerant semantics have been introduced by Baroni and Giacomin in [1] and Baroni *et al.* in [2], which these authors call 'CF2 semantics'. The approach is inspired in the graph-representation of the relation of attack among arguments. In this view, an argumentation framework AF can be decomposed along its strongly connected components, $SCCS_{AF}$. Every strongly connected component $SCC \in SCCS_{AF}$ is a set of arguments, all of which are path-equivalent. This relation is reflexive, and two different arguments α and β are path-equivalent iff there is a path from α to β and there is a path from β to α. Since the SCC's can be ordered following the direction of the graph, the original idea is that the directionality can be used to yield recursively the semantics. They define a generic recursive function, for which the base case is ad-hoc for each semantics. The generic function for the semantics we are interested in is defined as follows: $E \in \mathcal{GF}(AF, C)$:

- in case $|SCCS_{AF}| = 1$, $E \in \mathcal{BF}(AF, C)$,

- otherwise $\forall S \in SCCS_{AF}(E \cap S) \in \mathcal{GF}(AF \downarrow_{UP(S,E)}, U(S,E) \cap C)$,

where $\mathcal{BF}(AF, C)$ is a base function such that given $|SCCS_{AF}| = 1$ and a set $C \subseteq AR$, yields a maximal conflict-free subset of AR. C is the set of nodes which have no attackers from outside S or whose outer attackers are all attacked by E. $U(S,E)$ is the subset of nodes of S that are not attacked by E from outside S and are defended by E (i.e., their attackers from outside S are all attacked by E). And $UP(S,E)$ consists of $U(S,E)$ united with the set of all the provisionally defeated nodes, that is, the nodes of S that are not attacked by E from outside S and are not defended by E (i.e., at least one of their attackers from outside S is not attacked by E). This semantics sanctions the same extensions as tolerant semantics in the frameworks of the "lottery paradox" paradox and those which simple odd-length cycles of attack. But in the following example the behavior is different:

EXAMPLE 23. Let $AF = \langle \{\alpha, \beta, \gamma, \delta, \epsilon, \eta, \phi\}, \{(\alpha,\beta), (\beta,\gamma), (\gamma,\alpha), (\alpha,\delta), (\delta,\epsilon), (\epsilon,\eta), (\eta,\phi), (\gamma,\phi), (\phi,\delta)\}\rangle$. Then $\{\alpha,\eta\}$ is a tolerant and lax extension but it is not an extension in the $CF2$ semantics.

To see that $\{\alpha,\eta\}$ is not sanctioned by the $CF2$ semantics, note that we have two SCC's: $S_1 = \{\alpha, \beta, \gamma\}$ and $S_2 = \{\delta, \epsilon, \eta, \phi\}$. Assume $E \cap S_1 = \{\alpha\}$, then for S_2 the parameters of the generic function are $AF \downarrow_{\{\epsilon,\eta,\phi\}}$ and $U(S_2, E) \cap C = \{\epsilon, \phi\}$. Hence, $SCCS_{AF\downarrow_{\{\epsilon,\eta,\phi\}}} = \{\{\epsilon\}, \{\eta\}, \{\phi\}\}$. Taking the strongly connected component $S' = \{\eta\}$, we have $(E \cap S') \in \mathcal{GF}(AF \downarrow_\emptyset, \emptyset) = \emptyset$, since $UP(S', E) = \emptyset$ and $C = \emptyset$. Therefore, $\{\alpha,\eta\}$ cannot be an extension in this semantics.

Despite the different behavior showed in this example —for which soundness we do not have a clear intuition– all lax, tolerant and $CF2$ semantics seem to behave similarly in the most usual examples. But in favor of the present proposal we can say that lax semantics is the simplest among them.

Sustainable semantics, on the other hand, is an alternative for dealing properly with the problem of self-attacking arguments, overruling arguments belonging to larger odd-length cycles of attack. This semantics is intuitively in accordance with Dung's concept of indirect attack while solves the behavior of his preferred semantics with regards to self-attacking arguments. We have not found any other approach yielding similar results. Sustainable semantics also agrees with Pollock's opinion on how to deal with the interpretations of example 17 ([9], [10]). On the other hand, the solution given by Pollock to the problem of self-attacking arguments in [8] and [9] consists in inhibiting *by definition* their interference in the warrant process. As far as we can see, that is the only way if one wants to preserve a skeptic behavior in all other settings. Our solution, instead, inhibits the interference of self-attacking arguments as an emergent ability of the model, not by fiat. Of course, it contributes the fact that we define a credulous semantics instead of a skeptic one. We also have to mention that in a later analysis made in [11], Pollock changes his view by accepting that self-defeat is really a necessary feature to get his non-monotonic entailment system working, provided that in some examples the set of conclusions derived from the warranted arguments may not result closed under logical consequences.

The treatment of these problems cannot be declared settled, however. The lack of more indisputably sound examples contributes to maintain the fuzziness of the concepts and the disagreement among authors. In [6], for instance, Jakobovits and Vermeir provide an extended version of the framework of the "lottery paradox" paradox (though not with this interpretation), where a third argument, say γ, is attacked by β. In their opinion, β must be disabled by the attack from α, while γ must be reinstated. This is clearly at odds with the generally accepted interpretation –to which we adhere– according to which α must be overruled, β reinstated and γ defeated by β.

Finally, another interesting issue to discuss is the occurrence of indirect attacks inside extensions. Sustainable and lax semantics cannot avoid such occurrences. Consider the following example taken from Coste-Marquis et al. ([4]): $AF = \langle \{\alpha, \beta, \gamma, \delta, \epsilon, \nu\}, \{(\gamma, \beta), (\beta, \alpha), (\gamma, \epsilon), (\epsilon, \delta), (\delta, \alpha), (\nu, \delta)\} \rangle$. Then $\{\gamma, \nu, \alpha\}$ is the only extension for any of Dung's semantics. It is also the only sustainable and lax extension. But, in terms of Dung, γ is *controversial* w.r.t. α (that is, α indirectly attacks and defends β); the sequence γ attacks ϵ attacks δ attacks α shows the indirect attack inside the extension. Coste-Marquis et al. define a set of arguments S as *p(rudent)-admissible* iff every argument that belongs to S is acceptable in S and there are no odd-length paths of attacks connecting two arguments that belong to S (i.e., S must be admissible and free of controversial arguments), and a *p-preferred extension* is a maximally p-admissible set of arguments. So the only prudent preferred extension in the example above is $\{\gamma, \nu\}$. In accordance with our analysis of indirect attacks, the occurrence of controversial arguments inside an extension should not be necessarily problematic. But anyway, "prudent" versions of sustainable and lax extensions could be obtained by replacing the term 'admissible' with 'p-admissible' in the definition of the cogency relation.

9 Conclusion

In this work we have analyzed the "lottery paradox" paradox from the point of view of extension semantics for argumentation frameworks, and showed two approaches to avoid the interference of a self-attacking argument in the acceptance of some of the subarguments that compound it. Sustainable semantics enables that outcome while its behavior is tantamount to that of preferred semantics in argumentation frameworks which are free of self-attacking arguments. On the other hand, we have defended some intuitions at odds with the conception of indirect attack, arguing that arguments involved in odd-length cycles of attacks are not essentially self-attacking, and so they could be accepted (specially when argumentation is about practical decisions). We have offered here lax extensions as a semantics capturing that intuition. The behavior is akin, but not equivalent, to other semantics in the literature like tolerant semantics ([3]) and CF2 semantics ([1]).

Acknowledgments

This work was supported by ANPCyT and Universidad Nacional del Sur, Argentina, under projects PICTO 731, PICT 693 and PGI 24/ZI29.

BIBLIOGRAPHY

[1] Baroni, P. and M. Giacomin: Solving Semantic Problems with Odd-Length Cycles in Argumentation *Proc. of 7th European Conference on Symbolic and Quantitative Approaches to Reasoning with Uncertainty (ECSQARU 2003)* LNAI 2711 Springer-Verlag, 440-451, Aalborg, Denmark, July 2003.

[2] Baroni, P., M. Giacomin and G. Guida: SCC-recursiveness: a general schema for argumentation semantics. *Artificial Intelligence* 168(1-2): 162–210, 2005.

[3] Bodanza, G. and F. Tohmé: Two approaches to the problems of self-attacking arguments and general odd-Length cycles of attack. *Journal of Applied Logic*, in press, 2008. Pre-published on line: doi:10.1016/j.jal.2007.06.012.

[4] Coste-Marquis, S., C. Devred and P. Marquis: Prudent Semantics for Argumentation Frameworks. *Proc. of the 17th IEEE International Conference on Tools with Artificial Intelligence (ICTAI'05)*, 568–572, Hong-Kong, 2005.

[5] Dung, P. M.: On the Acceptability of Arguments and its Fundamental Role in Non-Monotonic Reasoning, Logic Programming, and *n*-Person Games. *Artificial Intelligence*, 77: 321–357, 1995.

[6] Jakobovits, H. and D. Vermeir: Robust Semantics for Argumentation Frameworks. *J. Log. Comput.*, 9(2): 215–261, 1999.

[7] Kyburg, H.E.: *Probability and the Logic of Rational Belief.* Wesleyan University Press, Middletown, CT, 1961.

[8] Pollock, J.: *Nomic Probability and the Foundation of Induction.* Oxford University Press, New York, 1990.

[9] Pollock, J.: Self-defeating arguments. *Minds and Machines*, 1: 367–392, 1991.

[10] Pollock, J.: Justification and defeat. *Artificial Intelligence* 67: 377–407, 1994.
[11] Pollock, J.: *Cognitive Carpentry: A Blueprint for How to Build a Person.* The MIT Press, 1995.
[12] Vreeswijk, G., and Prakken, H.: Credulous and Sceptical Argument Games for Preferred Semantics. In *Proc. JELIA 2000*, 239–253, 2000. LNAI 1919.

Gustavo A. Bodanza
Logic and Philosophy of Science Research Center (CILF)
Artificial Intelligence Research and Development Laboratory (LIDIA)
Universidad Nacional del Sur, Bahía Blanca
CONICET
Argentina
E-mail: bodanza@gmail.com

Logic and Knowledge: Expectations via Induction and Abduction

ATOCHA ALISEDA

1 Introduction

This paper concerns the relationship between logic and knowledge. There are indeed very many ways to address this connection, which themselves depend on the epistemological issue at stake and the view of logic. To begin with, I am interested on the kind of knowledge regulated by logic, addressed through the following two questions:

1. What kind of Knowledge does Logic MANAGE?

2. What kind of Knowledge does Logic PRODUCE?

We may therefore reflect on the kind of knowledge dealt with by logic as well as that which results from it. Let us make a brief comparison with other disciplines. In Physics, the knowledge dealt with concerns theories which are exposed to the test of experiment, and the knowledge produced takes the form of universal laws of the physical world expressed mathematically. In the case of Chemistry, the knowledge managed takes the form of chemical formulae representing molecular life and its products are new substances, for the main goal in Chemistry is to synthesize new forms of matter.

In the case of logic, these questions turn out to lack a straightforward answer. As it turns out, it is not even clear what logic is about or whether it actually provides any kind of specific knowledge about the world. A radical view on this matter is Russell's:

> Logic and mathematics, useful as they are, are only intellectual training for the philosopher. They help him to know how to study the world, but they give him no actual information about it. They are the alphabet of the book of nature, not the book itself. [Russ74, p. 9].

But what kinds of things does logic deal with? We will suggest that cognitive operations, amongst other things, are regulated by logic. As we shall see, creating expectations as well as modifying them —when expectations fail— are two cognitive operations that can be modelled by induction and abduction, respectively. This type of question highlights the connection between logic and epistemology, but it also highlights another link to logic, namely from one of its closest and until recently neglected neighbours, psychology.

The recent study of cognitive operations as logical suggests a return of psychologism into logic, i.e. the view that logical rules are grounded on psychological facts.[1]   And the point of view of a well-known contemporary logician, Johan van Benthem, is as follows: *"Logic is not the static guardian of correctness, as we still find it defined in most textbooks, but rather the much more dynamic, and much more inspiring, immune system of the mind!"* [Lei08, p. 81]. Here we are at the centre of cognitive science, in which various disciplines and approaches converge to work on a question, this time concerning the underlying logical processes that model cognitive operations.

Regarding abduction and induction, these have a long history in philosophy. Their roots lie in the work of Aristotle under the names of *epagoge* and *apagoge*. They also find a place in contemporary philosophy of science, in which the focus is to view them as (more or less) logical inferences in their own right, apart from deductive logic, the most and best studied of all three. There is also recent work on the relation and integration of induction and abduction [FK00], in which a study of both forms of logical inference is carried out, all from a philosophical, logical and computational perspective. The integration of induction and abduction recognizes each as a separate computational process, but *"they synthesise an integrated form of reasoning that can produce more complex solutions, following a cyclic pattern with each form of reasoning feeding into the other"* [FK00, p. 27].

In recent times, however, despite the renewed interest in giving a logical account of cognitive human operations and the growing research on abduction and induction, there is no study —as far as we know— that presents a logical model for expectations of the kind we propose.[2]

---

[1]For a recent publication on this particular question, cf. [Lei08].

[2]The closest I know in this direction is the work by Dov Gabbay and John Woods [GW05], in which the authors characterize the notion of *practical agency*, in which abduction is the logical strategy to satisfy (rather than maximize) the response to the agent's cognitive target. Moreover, my own previous work [Ali06] is the foundation of the ideas in this paper, for there abduction is proposed as a process for epistemic change, in which cognitive theories are expanded or revised to account for new information, but there is no mention of the management of expectations as such: neither when presenting my logical models nor when the epistemic view on the logic of abduction is set forth.

As regards the philosophical literature, the pragmatist tradition is the most useful for building up our own proposal. Pragmatism, as initiated by Peirce and followed in certain respects by Russell, is our philosophical basis. For Peirce, all cognitive operations are logical, from those involved in perception to those immersed in thinking and discovery. Russell [Russ48], on the other hand, was interested in human knowledge, and found in nondeductive logic a model to give an account of it. He was particularly interested on investigating whether these forms of nondeductive inference do warrant knowledge. And in fact, he explored pragmatism for this purpose, something which also served to distance himself from the positivists.

More generally, our claim for this paper is that some objects of logic are cognitive operations. In particular we claim that the construction of expectations can be modelled in a logical way, via the logic of induction, and when these expectations fail, there is another logical way, this time the logic of abduction.

Under a broad view, logic deals with general forms of reasoning, be it human or automatic. That is, forms in which, from information we already have, the premises $(\alpha_1, \alpha_2, \ldots, \alpha_n)$, some more information, a conclusion (φ), is produced. Thus, knowledge is conceived as having an inferential format:

$$\alpha_1, \alpha_2, \ldots, \alpha_n \Rightarrow \varphi$$

Logic analyses and generates a series of methods to manipulate information given in an inferential format. The following question is then in order: is the product of an inference knowledge? And if so, this leads us back to our starting question: what kind of knowledge is it?

The way to tackle these questions is by first making a distinction between different kinds of inference, drawing on the classification made by Charles Peirce, the founder of Pragmatism: namely *deduction*, *induction* and *abduction*. The status of information that these three produce is, respectively, knowledge that is certain, probable or plausible.[3] But this knowledge status —whether probable or plausible— is only half the story. The other has to do with human cognitive operations. While induction models the workings of expectation generation —providing beliefs based on causal connections that make up the expectations— abduction represents the logic of dealing with surprising information, when expectations are absent or fail, and itself provides a way to construct explanations for the non-expected surprising phenomena. Both types of inferential reasoning, induction and abduction, serve to produce knowledge about the world, though —with a certain degree

---

[3]Notice that we are using knowledge and information interchangeably, since their putative differences are not relevant to my arguments.

of confidence— one of them yields probable expectations, while the other outputs only plausible explanations, when expectations are absent or fail. This is yet another connection between logic and epistemology, which I put forward in this article.

In the following section I shall argue for the threefold distinction between three kinds of reasoning, namely deduction, induction and abduction, and will point at several confusions in the use of the terms "induction" and "abduction". Section three is devoted to induction, presented as the logic of expectations. Induction produces a universal statement representing a law of nature, which fixes a belief generated by an expectation. Section four characterizes abduction as the logic of surprises. When expectations are absent or fail, surprises arise and the cognitive-logical way to deal which them is via abduction, a logical strategy for incorporating novel and anomalous information, in such a way that the body of knowledge remains coherent. Section five sums up our conclusions, and hopefully will get across our general message for this paper, namely that logic and epistemology are intertwined. Logical types of inference such as induction and abduction model cognitive operations which allow us to transit in a regular, but fallible world. Finally, this paper seeks to remind logicians that epistemological questions guide logical ones, and that insofar as logic is regarded as the science of reasoning, we should once in a while leave our theorems in favour of an epistemological reflection. This is but one of the many sides of logic addressed in this volume.

2   Logic and Reasoning

Before we analyse each kind of inference, I must first clarify that the existence of three kinds of logical inference it is not universally accepted; for some authors, there are only two kinds:

> *There are two sorts of logic: deductive and inductive. A deductive inference, if it is logically correct, gives as much certainty to the conclusion as the premises, while an inductive inference, even when it obeys all the rules of logic, only makes the conclusion probable even when the premises are deemed certain.*
> [Russ74, p. 38]

As we shall see, once we step beyond deductive logic, diverse terminologies come into use, and my preference is to mark a three-way distinction amongst deduction, induction and abduction. Of the three, deduction is the most studied. It is identified with mathematical reasoning and has evolved into what we conceive nowadays as formal modern logic. I shall not devote any further consideration to this type of inference, but only mention that the conclusions of deduction are certain knowledge. Once a theorem has been proved, there is no doubt of its validity and it remains certain even if additional results appear. However, deduction only claims the certainty of the conclusion on the condition of the truth of the premises:

> *Deduction tells you what follows from your premises, but does not tell you whether your premises are true.* [Russ74, p. 43]

Leaving behind deductive logic, the next challenge is to distinguish induction from abduction. There is indeed much confusion regarding these two, which we will deal with in the following subsection.

2.1   Abduction vs. Induction

Perhaps the most widely used term is inductive reasoning [Mill 58, Sal90, Tha88]. For C.S. Peirce, as we shall see, 'deduction', 'induction' and 'abduction' formed a natural triangle —but the literature in general shows many overlaps, and even confusions.

Since the time of John Stuart Mill (1806–1873), the technical name given to all kinds of non-deductive reasoning has been 'induction', though several *methods for discovery and demonstration of causal relationships* [Mill 58] were recognized. These included generalizing from a sample to a general property, and reasoning from data to a causal hypothesis (the latter further divided into methods of agreement, difference, residues, and concomitant variation). A more refined and modern terminology is 'enumerative induction' and 'explanatory induction'. Some instances of these are: 'inductive generalization', 'predictive induction', 'inductive projection', 'statistical syllogism' and 'concept formation'.

Such a broad connotation of the term 'induction' continues to the present day. For instance, in the so-called 'computational philosophy of science', induction is understood *"in the broad sense of any kind of inference that expands knowledge in the face of uncertainty"* [Tha88, page 54]. In artificial intelligence, 'induction' is used for the process of learning from examples— but also for creating a theory to explain the observed facts [Sha91]. Under this view, abduction is viewed as an instance of induction, when the observation is a single fact.

On the other hand, some authors regard abduction under the name of *inference to the best explanation*, as the basic form of non-deductive inference [Har65], and consider (enumerative) induction as a special case. But we must approach the term 'abduction' with care. Given a fact to be explained, there are often several possible explanations, but only one that counts as the *best one*. Thus, abduction is connected to both hypothesis generation and hypothesis selection. Some authors consider these processes as two separate steps, construction dealing with what counts as a possible explanation, and selection with applying some preference criterion over possible explanations to select the best one. Other authors regard abduction as a single process by which a single best explanation is constructed. While the latter view considers finding the best explanation as fundamental for abduction, abduction understood as the construction of explanations, regards the notion of explanation as more fundamental.

To clear up all these conflicts, which are terminological to a large extent, one might want to coin new terminology altogether. I have argued for a new term of *"explanatory reasoning"* in [Ali06], trying to describe its fundamental aspects without having to decide if they are instances of either abduction or induction. However, for the purposes of this paper, rather than introducing new terminology, I shall use the term 'abduction' for the basic type of explanatory reasoning.

My focus is on abduction as hypothesis construction. More precisely, I shall understand abduction as reasoning from *a single observation to its explanations*, and induction as *enumerative induction* from samples to general statements. Therefore, abduction (when properly generalized) encloses (some cases of) induction as one of its instances, when the observations are many and the outcome a universal statement.

3   Induction: The Logic of Expectations

The purpose here is to present the workings of inductive inference as modelling a cognitive operation performed by humans, when constructing expectations about the world. Many of our expectations about the world are based on beliefs we have constructed as a product of our repeated experiencing of events, such that allow us to draw inferences about the world. We see a door blowing in the wind and we expect to hear the noise of it slamming. We see a colleague walking across the corridor and expect her to say hello to us.

> *Induction starts, psychologically, from an animal propensity. An animal which has had experience of things happening in a certain way will behave as if it expected them to happen the same way next time. . . . Inductive logic is an attempt to justify this animal propensity, in so far as it can be justified. It cannot be justified completely, for after all, surprising things do sometimes happen.*
> [Russ74, pp. 46, 48]

If a dog receives its food regularly every day early in the morning, it will accordingly expect it at that time. The sign of us taking out its plate indicates food is on its way. The generated expectation minimizes the uncertainty about one of its vital needs: to eat. However, the same could very well happen to the dog as happened to Russell's inductivist chicken: after being fed for almost a year, it was killed to become the main dish for Christmas!

> *It would have been better for the chicken if its inductive inferences had been less crude. Inductive logic aims at telling you what kinds of inductive inferences are least likely to lead you to suffer the tragic disillusionment of the chicken.* [Russ74, p. 48]

Enumerative induction presupposes repeated events or a set of instances of the same kind, phenomena indicating regularity in nature, and the product of this inference is a generalization of those instances, which in itself constitutes an expectation for future events. This is a fallible inference which only asserts itself as probable. But it is the best we have regarding the construction of expectations about the world based on past experiences.

In the pragmatist philosophical tradition expectations are a kind of belief that produces habits, either of mental or action type. Expectations involve beliefs in causal laws; but in their most primitive form, they seem not to involve a belief, and this is what Russell calls "animal inference". There are three levels in the construction of expectations. In the case of the dog, which exemplifies the first level, the presence of its master taking out its plate, causes the expectation of eating soon afterwards, although the dog is not aware of the causal connection between the master taking its plate and the arrival of the food; it is therefore debatable whether the dog creates a belief from the expectation. At a second level, the belief *"A is present, therefore B will be as well"* is created, and only in the third level the following hypothetical generalization is created: *"If A is present, B will be as well"*. Therefore, the generation of expectations and beliefs in causal connections is a complex cognitive process, which induces a habit of action, even if the belief on which expectations are based has only a hypothetical status, though a probable one.

This suggests that habit acquisition —mental or of action— is a product of a logical capacity (innate or acquired?) put to work in our daily interaction with the world. In order to transit safely in a regular but fallible world, we need to perform inductive inferences and trust them.

Inductive inference has been a controversial topic in logic and philosophy. One main problem, the inductive problem, is precisely the absence of logical justification, in the sense in which deduction has that capacity, for it is an inference with fallible products. Moreover, there are even doubts of whether this method is indeed one of scientific inference ([Pop59]). On the logical side, it has been investigated as a method linked with probability, and attempts to construct an inductive logic have been performed ([Car55]), but again, without success as far as its logical justification is concerned. In fact, the only certain and well justified induction is mathematical induction. This is a method to prove properties over infinite sets. Its certainty is warranted because mathematical infinite sets over which the induction is made —like natural numbers— are regular; there is no room for surprises in their behaviour.

But the point here is epistemological, not in the sense concerned with truth or validity, but rather concerned with modelling a cognitive operation which we perform to construct beliefs: expectations about the world which allow us to behave safely, as far as the probable can permit.

4   Abduction: The Logic of Surprise

When it comes to abduction, as conceived by Peirce, the epistemological, the cognitive and the logical are intertwined. Peirce proposed abduction for dealing with failed expectations, and novelties, two kinds of surprises. But his project was much more ambitious.

The intellectual enterprise of Charles Sanders Peirce, in its broadest sense, was to develop a semiotic theory, in order to provide a framework to account for thought and language. With regard to our purposes, the fundamental question Peirce addressed was how synthetic reasoning was possible. Very much influenced by the philosophy of Immanuel Kant, Peirce's aim was to extend his categories and correct his logic:

> According to Kant, the central question of philosophy is 'How are synthetical judgments a priori possible?' But antecedently to this comes the question how synthetical judgments in general, and still more generally, how synthetical reasoning is possible at all. When the answer to the general problem has been obtained, the particular one will be comparatively simple. This is the lock upon the door of philosophy'. ([CP, 5.348], quoted in [Hoo92], page 18).

Peirce proposes abduction to be the logic for synthetic reasoning, that is, a method to acquire new ideas. He was indeed the first philosopher to give a logical form to abduction.

The development of a logic of inquiry occupied Peirce's thought since the beginning of his work. In the early years he thought of a logic composed of three modes of reasoning: deduction, induction and hypothesis, each of which corresponds to a syllogistic form. Later on, Peirce proposed these types of reasoning as the stages composing a method for logical inquiry, of which abduction is the beginning:

> *From its [abductive] suggestion deduction can draw a prediction which can be tested by induction.* [CP, 5.171].

Later on, Peirce refined his views on abduction by the more general conception of: *"the process of forming an explanatory hypothesis"* [CP, 5.171] and the syllogistic form is replaced by the following logical formulation of abduction ([CP, 5.189]):

The surprising fact, C, is observed.

But if A were true, C would be a matter of course.

Hence, there is reason to suspect that A is true.

In AI circles, Peirce's abductive formulation has been generally interpreted as the following logical argument–schema:

$$C$$
$$\underline{A \to C}$$
$$A$$

where the status of A is tentative (it does not follow as a logical consequence from the premises). In addition to this formulation which makes up the explanatory constituent of abductive inference, there are two other aspects to consider for an explanatory hypothesis, namely its being *testable* and *economic*. While the former sets up a requirement in order to give an empirical account of the facts, the latter is a response to the practical problem of having innumerable hypotheses to test and points to the need for a criterion to select the best explanation amongst the testable ones.

Abductive reasoning is essential in every human inquiry. Abduction plays a role in direct perceptual judgments, in which:

> *The abductive suggestion comes to us as a flash.* [CP, 5.181]

As well as in the general process of invention:

> *It [abduction] is the only logical operation which introduces any new ideas.* [CP, 5.171].

But, what is the cognitive purpose of abductive reasoning? Let us now turn to Peirce's epistemology for an answer. Thought is a dynamic process, essentially an interaction between two states of mind: doubt and belief. While the essence of the latter is the *"establishment of a habit which determines our actions"* [CP, 5.388], with the quality of being a calm and satisfactory state in which all humans would like to stay, the former *"stimulates us to inquiry until it is destroyed"* [CP, 5.373], and it is characterized by being a stormy and unpleasant state from which every human struggles to be freed:

> *The irritation of doubt causes a struggle to attain a state of belief.* [CP, 5.374].

Note that Peirce speaks of a state of belief and not of knowledge. Thus, the pair 'doubt-belief' is a cycle between two opposite states. While belief is a habit, doubt is its privation.

Peirce identifies doubt with surprise, and in fact, he seems to use these two terms interchangeably:

> *For belief, while it lasts, is a strong habit, and as such, forces the man to believe until some **surprise** breaks up the habit.* ([CP, 5.524], my emphasis).

Moreover, there are indeed two ways to break up a habit:

> *The breaking of a belief can only be due to some* novel experience *[CP, 5.524] or ...until we find ourselves confronted with some experience **contrary to those expectations**.* ([CP, 7.36], my emphasis).

Peirce's epistemic model proposes two varieties of surprise as the triggers for every inquiry, which we may call *novelty* and *anomaly*. Each of these induces an operation to incorporate a new belief and its explanation into the theory.

Under his view, the logical form of abduction is not that of an argument, but it is rather an epistemic process for belief change.[4] Thus, the abductive phenomenon has a general structure and is characterized by abductive novelty and anomaly, with their corresponding operations described as follows:

There are two triggers for abduction: *novelty* and *anomaly*:[5]

---

[4]This view is in fact an idea that naturally emerges from Peirce's work and has close connections with theories of belief change in artificial intelligence as orginally proposed in [Gär88]. See also my own proposal in detail[Ali06], which is only sketched below.

[5]As we have seen from the logical formulation of abduction, abductive reasoning is triggered by a surprising fact. However, the notion of surprise is relative, for a fact is surprising with respect to a background theory which provides expectations. What is surprising to me, may not be for you.

- **Abductive Novelty:** φ is novel. It cannot be explained: $\Theta \not\Rightarrow \varphi$, nor can its negation: $\Theta \not\Rightarrow \neg\varphi$

- **Abductive Anomaly:** φ is anomalous. It cannot be explained: $\Theta \not\Rightarrow \varphi$, and in fact the theory accounts for its negation: $\Theta \Rightarrow \neg\varphi$.

Abductive operations for epistemic change are those which induce each of the abductive triggers:

- **Abductive Expansion:** Given an abductive novelty φ, a consistent explanation α for φ is computed in such a way that $\Theta, \alpha \Rightarrow \varphi$. Thus, φ and α are both added to Θ by a simple expansion.

- **Abductive Revision:** Given an abductive anomaly φ, a consistent explanation α is computed as follows: the theory Θ is revised into Θ' so that it does not explain $\neg\varphi$. That is, $\Theta' \not\Rightarrow \neg\varphi$, where $\Theta' = \Theta - (\beta_1, \ldots, \beta_l)$.[6]

  Once Θ' is obtained, a consistent explanation α is calculated in such a way that $[\Theta', \alpha \Rightarrow \varphi$. Thus, the process of revision involves both contraction and expansion.

The overall cognitive process showing abductive inference as an epistemic process can be depicted as follows: a novel or an anomalous experience gives rise to a surprising phenomenon, generating a state of doubt which breaks up a belief habit and accordingly triggers abductive reasoning. In the case of a novel phenomenon, since it is novel but consistent with the theory its explanation is calculated and incorporated into the theory by an extension operation. In the second case, when the fact is anomalous, the revision operation is the one needed. The theory is revised in such a way that it is not in conflict with the fact to be explained, and then the explanation is calculated and incorporated into the revised theory by expansion.

The goal of this type of reasoning is precisely to explain the surprising fact and thus *soothe* the state of doubt. It is 'soothe' rather than 'destroy', for an abductive hypothesis has to be put to the test and must be economic, attending the further criteria Peirce proposed. The abductive explanation is simply a suggestion that has to be put to the test before converting itself into a belief.

---

[6]In many cases, several formulas and not just one must be removed from the theory. Sets of formulas which entail (explain) φ should be removed. E.g., given $\Theta = \{\alpha \to \beta, \alpha, \beta\}$ and $\varphi = \neg\beta$, in order to make $\Theta, \neg\beta$ consistent, one needs to remove either $\{\beta, \alpha\}$ or $\{\beta, \alpha \to \beta\}$. Note that the revision operation involves both contraction and expansion.

5   Logic and Knowledge: Final Reflections

I began this discussion on the connection between logic and knowledge by posing two questions which, as we have seen, expand as follows:

1. What kind of Knowledge does Logic MANAGE?

2. What kinds of Cognitive Operations does Logic MANAGE?

3. What kind of Knowledge does Logic PRODUCE?

4. What is the status of the Knowledge PRODUCED in Logic?

I proposed that the kind of knowledge logic manages (my first question) consists of cognitive operations: i.e., logical strategies to perform a cognitive operation like the creation of expectations. This is why the second question concerns the variety of cognitive operations that logic manages. I concentrated on the cognitive operations for dealing with expectations and then showed that induction and abduction are two logical strategies to deal with the construction of beliefs that make up expectations, and for dealing with their failure as well.

Turning to the third question, the products depend on the logical strategy, but they often have a logical form, or are themselves a logical process. Induction produces a universal statement representing a law of nature, an expectation. Abduction on the other hand, is a logical process for belief revision. When we encounter a novelty, finding an explanation will induce a process of theory expansion, whereas when facing a failure of expectations, we are compelled to revise our theory in order to construct a hypothesis of what has happened. And all this information can be structured in an inferential format.

As for the fourth question, given the problem of validating both induction and abduction, we know that they are both by nature deductively invalid. Strictly speaking, the products of neither induction nor abduction can be regarded as knowledge, for their fallibility is undeniable. However, they represent cognitive strategies that themselves cannot be avoided in daily human interaction with the world. We must, therefore, leave justification aside, and present the products of both induction and abduction as non-certain kinds of knowledge, or as probable and plausible knowledge, if you prefer.

Deduction corresponds to certain and necessary knowledge, a wonderful strategy when dealing with mathematical knowledge. Its central notion is that of truth, and inference is the means by which it warrants the transmission of truth: when the premises are true, the conclusion is necessarily so.

Induction corresponds to knowledge which is only probable. Its power relies on the frequent regularity of nature, presenting itself on repeated occasions. Its weakness, however is due to the fallibility of nature, which occasionally gives surprises and forces us to expand or modify our expectations. The central notion here is that of frequency, and it has been largely studied in the area of probability. There also work on this in the computational literature, as found in [FK00].

Abduction produces knowledge which is only plausible, mere beliefs based on a hypothetical explanation. Its job is to provide explanations for surprising facts: expectation absence which we characterized as a novelty, and expectation failure characterized as abductive anomaly. Abduction is the path for performing the changes in our body of beliefs necessary in order to account for a surprise. Plenty of work has been done towards its formalization and computer implementation, of which I only point to my own work and two other reference publications [Ali06, FK00, Tha88].

This view on the connection between logic and knowledge, with a particular notion of logic as managing cognitive operations, namely those used to deal with the construction of expectations and their failure (surprises), is by no means exhaustive. There may be some other cognitive operations to model, and nothing has been said of the logical models needed to operate all this. My message is much simpler: epistemology and logic are intertwined. One may relate logical types of inference with cognitive operations, and we have proposed to use induction and abduction for that purpose, modelling the construction of expectations and explanations, when expectations are absent or fail.

BIBLIOGRAPHY

[Ali06] A. Aliseda, *Abductive Reasoning. Logical Investigation into Discovery and Explanation*. Synthese Library/Volume 330. Springer. Dordrecht. 2006.

[Car55] R. Carnap. *Statistical and Inductive Probability*. Galois Institute of Mathematics and Art. Brooklyn, New York. 1955.

[Fan70] K.T. Fann. *Peirce's Theory of Abduction*. The Hague: Martinus Nijhoff. 1970.

[FK00] P. Flach and A. Kakas. *Abduction and Induction. Essays on their Relation and Integration*. Kluwer Academic Publishers. Applied Logic Series. Volume 18. Dordrecht, The Netherlands. 2000.

[Fra58] H. Frankfurt. 'Peirce's Notion of Abduction'. *The Journal of Philosophy*, 55 (p.594). 1958.

[GW05] D. Gabbay and J. Woods. *A Practical Logic of Cognitive Systems. Volume 2. The Reach of Abduction, Insight and Trial*. Elsevier. 2005.

[Gär88] P. Gärdenfors. *Knowledge in Flux: Modeling the Dynamics of Epistemic States*. MIT Press. 1988.

[Har65] G. Harman. 'The Inference to the Best Explanation'. *Philosophical Review*. 74 88-95 (1965).

[Hoo92] C. Hookway. *Peirce*. Routledge, London. 1992.

[Lei08] H. Leitgeb (ed.) *Studia Logica. An International Journal for Symbolic Logic. Special Issue: Psychologism in Logic?*. Volume 88, No. 1, 2008.

[Mill 58] J.S. Mill. *A System of Logic*. (New York, Harper & brothers, 1858). Reprinted in *The Collected Works of John Stuart Mill*, J.M. Robson (ed.), Routledge and Kegan Paul, London. 1958.

[CP] C.S. Peirce. *Collected Papers of Charles Sanders Peirce*. Volumes 1–6 edited by C. Hartshorne, P. Weiss. Cambridge, Harvard University Press. 1931–1935; and volumes 7–8 edited by A.W. Burks. Cambridge, Harvard University Press. 1958.

[Pop59] K. Popper. *The Logic of Scientific Discovery*. London: Hutchinson. (11th impression, 1979). 1959. (Originally published as *Logik der Forschung*, Springer. 1934).

[Russ48] B. Russell. *Human Knowledge: Its Scope and Limits*. London: Routledge. 1948.

[Russ74] B. Russell. *The Art of Philosophizing and Other Essays*. Littlefield, Adams and Co.

[Sal90] W. Salmon. *Four Decades of Scientific Explanation*. University of Minnesota Press, 1990.

[Sha91] E. Shapiro. 'Inductive Inference of Theories from Facts', in J.L. Lassez and G. Plotkin (eds.), *Computational Logic: Essays in Honor of Alan Robinson.*. Cambridge, Mass. MIT, 1991.

[Tha88] P.R. Thagard. *Computational Philosophy of Science*. Cambridge, MIT Press. Bradford Books. 1988.

Atocha Aliseda
Instituto de Investigaciones Filosóficas
U. N. A. M.
México
E-mail: atocha@filosoficas.unam.mx

Defining Inferential Contexts: Deduction and Abduction

ÁNGEL NEPOMUCENO-FERNÁNDEZ AND FERNANDO SOLER-TOSCANO

ABSTRACT. Representation of inferential contexts may need logical systems different from classical ones, particularly when the wish is to model how inferences are in concrete scientific practices, where reasoning may be sensitive to propositions that belong to the context. So many kinds of consequence relations should be considered, among which the closure ones constitute the classical style of inference. Other styles could be studied, as constituted by non monotonic consequence relations, but the classical style of reasoning keeps interesting properties to define a logical calculus. In this paper we shall discuss the advantage of using closure consequence relations to represent inferential contexts, particularly when the process of constructing the inferential context itself is studied from a logical point of view, likewise two problematical cases will be pointed out. From that conceptual tools, the use of deductive calculi to obtain solutions for abductive problems, when abduction is sensitive to contexts (classical style), is tackled, then a form of defining abductive calculi that are sound is justified.

1 Introduction

We consider that a logical system —to abbreviate, a logic— is given by a formal language L —it also represents its set of formulas, provided that ambiguity cannot arise— and a consequence relation, which is defined from $\mathcal{P}(L)$ to L, so it can have a syntactic or semantic nature. On the other hand, some advanced scientific theories can be axiomatized and formalized, then a formal correlate of such theories may be given by an initial set of postulates and the way of adding more and more sentences of the corresponding language. When an important number of phenomena can be explained by a theory, this can be taken as a part of a logical system, that is to say, as the set of its postulates and a set of its consequences or theorems, in accordance with the consequence relation that could be taken as semantical justification of such theorems. However, scientific practices are not monolithic and there is a richer variety of inferences that are difficult to codify as a conventional classical logical system, so contexts where scientific practices are developed may play an important role in a logical study of processes of discovery and justification of theories.

If abduction is the main form of reasoning to incorporate new propositions to an unfinished theory —it may be codified as sentences of a formal language—, deduction is the form of reasoning per excellence to justify the link of such propositions with the set of postulates and its consequences, so that an underlying (deductive) logic is a system of reasoning from which a context of justification is constructed as a global explanation of the context of discovery. In short, an inferential context can be represented by means of a consequence relation —or by means of a syntactical relation, which should keep certain relation with the former— in a way that abduction and deduction should be rightly related, then classical consequence relation is not the only one but one possibility among others. Our main concern is to study some inferential contexts and give a general justification of abductive calculi or, more specifically, how deductive calculi can be used to obtain solutions for abductive problems. However, we shall focus attention on the classical style of inference, which is characteristic of an important class of inferential contexts —those represented by means of consequence relations with determined structural rules, as it will be seen below—, by exploring exact conditions that such calculi should accomplish. This is in part a pragmatic decision, in fact we have taken into account the most important and known results in classical logic and that most of accepted standard treatments of abduction have often used classical logic after all —see [1]—, but this formal approach follows a simple schema, which is based on the duality between deduction and abduction: if from the pair (Θ, φ), where Θ is a set of sentences and φ a sentence of a given language, a new sentence α is abducted, then (and only then) φ can be deduced from Θ and α, in the specific sense of "deduction" given from a classical point of view, and this could be adapted to develop logics with different structural rules, logics whose concept of "deduction" is not exactly identical to the classical one.

2   Consequence relations

In modern logic, an important consequence relation is the classical one, which is commonly called *logical consequence relation* or *entailment relation* —to avoid mistakes when more relations are considered, we should add "in classical sense", though usually it is omitted to abbreviate—, whose structural properties are well known and it is an important reference to study other relations in which such properties may change.

Let L be a first order language with a semantics that is defined in terms of the model theory. $MOD(L)$ represents the class of all interpretative L-structures or L-models, to simplify, "models". If a formula $\varphi \in L$ is satisfied by $M \in MOD(L)$, this will be expressed as $M(\varphi) = 1$ or $M \models \varphi$. When $M \models \gamma$ for every $\gamma \in \Gamma$, then we shall say that M satisfies Γ and note that as $M \models \Gamma$. $MOD(\Gamma)$ denotes the set of models that satisfy Γ. If the cardinality of the universe of discourse of any $M \in MOD(L)$ is a natural number $n \geq 1$, then we shall note $\mid M \mid = n$. The classical consequence relation —our starting point— is defined as follows: for every $\Gamma \subset L$ and $\varphi \in L$, Γ entails φ, or φ is a logical consequence of Γ, in symbols $\Gamma \models \varphi$, if and only if — "iff", from now on—, for every $M \in MOD(L)$, if $M \models \Gamma$, then $M \models \varphi$. To simplify, we shall write $\Gamma, \Delta \models \varphi$ and $\Gamma, \alpha \models \varphi$ instead of $\Gamma \cup \Delta \models \varphi$ and $\Gamma \cup \{\alpha\} \models \varphi$, respectively.

The classical consequence relation, it should be noted, has a semantic character, and it is a *closure relation* because it accomplishes the following structural rules

- Reflexivity. For every $\Gamma \subset L$, $\Gamma \models \gamma$, for each $\gamma \in \Gamma$

- Monotonicity. For all $\Gamma, \Delta \subset L$ and $\varphi \in L$, if $\Gamma \models \varphi$, then $\Gamma, \Delta \models \varphi$

- Transitivity. For all $\Gamma \subset L$, $\varphi, \psi \in L$, if $\Gamma \models \varphi$ and $\varphi \models \psi$, then $\Gamma \models \psi$

When the relation is defined in terms of sequences of formulas, two rules more are settled: contraction and permutation, though as our definition is in terms of sets of formulas, both of them are trivially verified. On the other hand, other two properties are also verified, namely

- Compactness. For all $\Gamma \subset L$ and $\varphi \in L$, if $\Gamma \models \varphi$, then there exists $\Gamma' \subset \Gamma$ such that it is finite —so its cardinality is a natural number, in symbols $\mid \Gamma' \mid < \mid \mathbb{N} \mid$— and $\Gamma' \models \varphi$

- Uniform substitution. Taken a uniform substitution, defined as usual: $sub : L \longmapsto L$ with the corresponding restrictions, for $\Gamma \subset L$ and $\varphi \in L$, if $\Gamma \models \varphi$, then $sub(\Gamma) \models sub(\varphi)$, where $sub(\Gamma) = \{sub(\gamma) \mid \gamma \in \Gamma\}$

From that consequence relation, new ones can be defined. In accordance with the point of view expressed in [8, 9], which we adopt partially (in fact our goal is to investigate some inferential contexts and abduction instead of representing non monotonic consequence relations), three ways of getting more conclusions from given premises could be adopted: by using additional background assumptions, by restricting the set of models or by using additional rules, which cause new consequence relations whose properties may be (or not) the known ones. Let us see summarily the three kinds of consequence relations.

First, given $\Gamma, \Theta \subset L$, Θ countable and $\varphi \in L$, we say that φ is *logical consequence of* Γ *modulo the set* Θ iff all models of Θ that are models of Γ satisfy φ, in symbols $\Gamma \models_\Theta \varphi$. So, $\Gamma \models_\Theta \varphi$ iff $\Gamma, \Theta \models \varphi$. This is a closure relation, since for $\Gamma, \Theta, \Delta \subset L$ and $\gamma, \varphi, \psi \in L$ it is verified

- Reflexivity. $\Gamma \models_\Theta \gamma$ for every $\gamma \in \Gamma$, since $\Gamma, \Theta \models \gamma$ because of reflexivity of $\models$ and definition of $\models_\Theta$

- Monotonicity. Let $\models_\Theta$ be and $\Gamma \models_\Theta \varphi$. By definition, $\Gamma, \Theta \models \varphi$ and by monotonicity of $\models$ it is verified that $\Gamma, \Theta, \Delta \models \varphi$, because of which $\Gamma, \Delta \models_\Theta \varphi$

- Transitivity. Let $\models_\Theta$ be and $\Gamma \models_\Theta \varphi$ and $\varphi \models_\Theta \psi$. Then, according to definition, $\Gamma, \Theta \models \varphi$ and $\varphi, \Theta \models \psi$. Since $\models$ verifies transitivity, $\Gamma, \Theta \models \psi$, that is to say, $\Gamma \models_\Theta \psi$

With respect to the other two properties we can say that relations so defined verify compactness. In fact, for $\Gamma, \Theta \subset L$ and $\varphi \in L$, if $\Gamma \models_\Theta \varphi$, then $\Gamma, \Theta \models \varphi$ and, by compactness of $\models$, there exists $\Gamma' \subset (\Gamma \cup \Theta)$ such that $\mid \Gamma' \mid < \mid \mathbb{N} \mid$ and $\Gamma' \models \varphi$. Let $\Gamma' \cap \Gamma$ be, whose models that are models of Θ constitute a subclass of the class of models of Γ that are models of Θ, which satisfy φ, so that $\Gamma' \cap \Gamma \models_\Theta \varphi$ and $\mid \Gamma' \cap \Gamma \mid \leq \mid \Gamma' \mid < \mid \mathbb{N} \mid$. However, uniform substitution is not verified as it is shown in the following example. Consider the set $\Theta = \{Qa\}$, then $Pa \models_\Theta Pa \wedge Qa$ but $Pb \not\models_\Theta Pb \wedge Qb$ after substituting b for a in the premise and the conclusion. Such substitution, it should be noted, does not affect to the modulo, which is an essential part of the relation itself.

Two consistent sets of formulas Θ and Θ' are *deductively equivalent* in classical sense —two inconsistent sets of formulas are deductively equivalent *a fortiori*—, iff for every $\varphi \in L$, $\Theta \models \varphi$ iff $\Theta' \models \varphi$. Then $\models_\Theta$ and $\models_{\Theta'}$ are *equivalent*: for all $\Gamma \subset L$ and $\varphi \in L$, $\Gamma \models_\Theta \varphi$ iff $\Gamma \models_{\Theta'} \varphi$.

Another relation can be defined from classes of models. For $\mathcal{M} \subset MOD(L)$, $\Gamma \subset L$ and $\varphi \in L$, we say that φ is *logical consequence of* Γ *modulo* $\mathcal{M}$ iff for all $M \in \mathcal{M}$, if $M \models \Gamma$, then $M \models \varphi$, in symbols, $\Gamma \models_\mathcal{M} \varphi$. This is also a closure relation since it verifies reflexivity, monotonicity and cut, as one can see easily. However, it does not accomplish compactness in all cases as, for example, when $\mathcal{M}$ is the class of all finite models —models whose universe of discourse is finite—. It does not verify uniform substitution either. On the other hand, if $\mathcal{M}$ is characterizable by means of a set of formulas Θ, then $\models_\mathcal{M}$ and $\models_\Theta$ are equivalent, that is to say for all $\Gamma \subset L$ and $\varphi \in L$, $\Gamma \models_\mathcal{M} \varphi$ iff $\Gamma \models_\Theta \varphi$. The consequence relation modulo a given cardinality is a special case of the previous one. Let $\mathcal{M}_n \subset MOD(L)$ such that for every $M \in \mathcal{M}_n$, $\mid M \mid = n \geq 1$. To refer to $\models_{\mathcal{M}_n}$, we shall write $\models_n$ and it will be called *consequence modulo n*.

By means of a set of additional rules a new consequence relation can be defined. Let R be a set of pairs $\langle \Delta, \beta \rangle$, where $\Delta \subset L$ and $\beta \in L$ and are called *premises* and *conclusion*, respectively. For $\Gamma \subset L$,

$$R(\Gamma) = \{\beta \in L \mid \Delta \subseteq \Gamma \ \& \ \langle \Delta, \beta \rangle \in R\}.$$

Γ is closed under R iff for all $\Delta \subset L$ and $\beta \in L$, if $\Delta \subseteq \Gamma$ and $\langle \Delta, \beta \rangle \in R$, then $\beta \in \Gamma$, that is to say, iff $R(\Gamma) \subseteq \Gamma$. Given a set of rules R, $\Gamma \subset L$ and $\varphi \in L$, we say that φ is *logical consequence of* Γ *modulo* R —in symbols $\Gamma \models_R \varphi$— iff $\varphi \in \Gamma'$ for every Γ' such that $\Gamma \subseteq \Gamma'$ and $R(\Gamma') \subseteq \Gamma'$ and for every $\alpha \in L$, if $\Gamma' \models \alpha$ then $\alpha \in \Gamma'$. This consequence relation is also a closure one and verifies compactness but not uniform substitution.

Two sets of additional rules R and R' are *deductively equivalent* iff it is verified that for every $A \subseteq L$, A is closed under R iff it is closed under R'.

3   Inferential contexts and abductive processes

Intuitively and roughly speaking, an inferential context consists of those conceptual tools that allow inferences to be evaluated in scientific practices. Whatever the case may be, in scientific practices any inference must be obtained according to a system of reasoning, codified as a logical system, so a consequence relation represents formally an inferential context.

Every context may use a specific consequence relation though the same style of inference could be shared by several inferential contexts. A style of inference could be established, on the one hand, by means of the kind of new consequence relations, then there would be a style of inference corresponding to consequence relations modulo set of sentences, modulo set of models and modulo set of additional rules. On the other hand, as it is assumed in this work, a style of inference could be established by means of the structural rules of the consequence relations that represent contexts, then there is a style that works with closure relations, this is the classical style, but Reiter's systems, for example, like other substructural logics, are enclosed among the class of systems of reasoning that are called "non monotonic logics". The consequence relations in such logics, usually reperesented by $|\sim$, verifies certain rules —inclusion, idempotence, a form of cut, cautious monotony and cumulativity are accomplished—, because of which they are called *cumulative consequence relations* (see [3], chapter 13), but none is a closure relation. Inside the classical style of inference, several different contexts are well defined by means of the consequence relations studied above. $\models_X$, when the modulo X is a set of formulas, a set of models or a set of additional rules, can represent different inferential contexts but all of them are included in the classical style of inference, since it is a closure relation in all cases. Nevertheless, if two consequence relations are equivalent, then both of them become to represent the same inferential context.

Let us see some examples about different contexts in this style. Suppose two theories given by the following sets of axioms (at propositional level):

$$\Theta_1 = \{p \rightarrow q, q \vee r, \neg r \vee \neg p\}, \Theta_2 = \{r \rightarrow q, \neg r \vee \neg q\}.$$

Since Θ_1 and Θ_2 are not equivalent, $\models_{\Theta_1}$ and $\models_{\Theta_2}$ represent different contexts. In fact, it is verified that $p \models_{\Theta_1} q$ and $\neg r \models_{\Theta_1} q$ but $p \not\models_{\Theta_2} q$ and $\neg r \not\models_{\Theta_2} q$.

With respect to models, think the consequence relation modulo a natural number, then easily can be seen that $\{\forall x Px, \exists x Qx\} \models_1 \forall x(Px \wedge Qx)$, since in universes with only one element that satisfy $\forall x Px$ and $\exists x Qx$, the semantic values of P and Q coincide, namely the universe itself, so that

$\forall x(Px \wedge Qx)$ is also satisfied, however $\{\forall x Px, \exists x Qx\} \not\models_2 \forall x(Px \wedge Qx)$, because the universes with two elements can satisfy $\forall x Px$ and $\exists x Qx$ but the semantic value of Q may be different from the universe —the demanded condition is that such value must be different from $\emptyset$—.

The representation by means of sets of models could be reduced to the former if the modulo $\mathcal{M}$ can be characterized by means of a set of formulas Θ of L, in such cases $\models_{\mathcal{M}}$ and $\models_{\Theta}$ are equivalent in the sense that for all $\Gamma \subset L$ and $\varphi \in L$, $\Gamma \models_{\mathcal{M}} \varphi$ iff $\Gamma \models_{\Theta} \varphi$, so both of them would represent the same inferential context.

Finally, about additional rules, take $\Gamma = \{\forall x(Px \to Qx), \exists x Px\}$ and the sets of additional rules

$$R = \{\langle\{\exists x Px\}, \exists x Sx\rangle\}; \ R' = \{\langle\{\exists x Px\}, \forall x \neg Sx\rangle\}.$$

For $\Gamma' = \Gamma \cup \{\exists x Sx\}$, and $\Gamma'' = \Gamma \cup \{\forall x \neg Sx\}$, Γ' is closed under R but Γ'' is not closed under R, meanwhile Γ' is not closed under R' but Γ'' is closed under R'. So, $\Gamma \models_R \exists x Sx$, but $\Gamma \not\models_{R'} \exists x Sx$, and $\Gamma \models_{R'} \forall x \neg Sx$ but $\Gamma \not\models_R \forall x \neg Sx$. In this case, practically the same results could be achieved by working with Reiter's normal logics proposed in [12] —any rule $\langle \alpha, \beta \rangle$ can be considered an abbreviation of the original expression of a default rule $(\alpha : \mathbf{M}\beta/\beta)$ in a default normal logic (in [9] this simplification has been made)—. In short, given a default (normal) theory $\langle \Theta, \Delta \rangle$, where Θ is a (non empty) set of formulas and Δ a set of rules, after incorporating certain notions, specifically an inference operation Th_L and the operation Γ, E is called an extension of Θ under Δ iff it is a fixpoint under the operation Γ. With respect to our previous example, we could come to extensions E_1 and E_2 in a way that $\exists x Sx \in E_1$ and $\forall x \neg Sx \in E_2$. However, despite a new conceptual apparatus, to proceed with this kind of comparison is beyond our purpose in this paper, indeed to remain working in a classical style of inference we prefer to maintain the use of consequence relations (modulo a set of formulas, models or rules —the first and second ones without an inmediate relationship with Reiter's normal systems, except by forcing a reinterpretation of such modulo—), a uniform treatment, which is in keeping with our objectives of justifying the use of deductive calculi for abduction.

In general, for modules X and Z —set of formulas, models or additional rules—, if $\models_X$ and $\models_Z$ are equivalent, then both of them represent the same inferential context.

What have inferential contexts to do with abductive processes? Abduction is a kind of inference that is irreducible to induction or deduction —inferential thesis, in [7]—, but it is an inferential process after all, so they must be very related. To see that, we define the notions of *abductive problem* and *solution for an abductive problem*, the consistent and explicative abduction in [1]. First, given a finite $\Theta \subset$ L and $\varphi \in$ L, (Θ, φ) is an *abductive problem* —named as $AbdProb(\Theta, \varphi)$—iff $\Theta \not\models \varphi$ and $\Theta \not\models \neg\varphi$. Second, given $AbdProb(\Theta, \varphi)$, $\alpha \in L$ is an *abductive solution* for it iff:

1. $\Theta, \alpha \models \varphi$.

2. $\Theta \cup \{\alpha\}$ is satisfiable.

3. $\Theta \not\models \varphi$.

$Abd(\Theta, \varphi)$ denotes the set of abductive solutions for $AbdProb(\Theta, \varphi)$.

Of course this logical way of tackling abduction is based on the classical consequence relation, but that notions can be extended to the classical style, then defined with respect to other ones, so (Θ, φ) is an *abductive problem with respect to the consequence relation* $\models_X$ —now named as $AbdProb_{\models_X}(\Theta, \varphi)$— iff $\Theta \not\models_X \varphi$ and $\Theta \not\models_X \neg\varphi$. On the other hand, given $AbdProb_{\models_X}(\Theta, \varphi)$, $\alpha \in L$ is an *abductive solution* for it iff:

1. $\Theta, \alpha \models_X \varphi$.

2. $\Theta \cup \{\alpha\}$ is satisfiable.

3. $\Theta \not\models_X \varphi$.

When the modulo X is a set of formulas, the second condition could be specified as there is $M \in MOD(X)$ such that $M \models \Theta \cup \{\alpha\}$. If the modulo X is a set of models, then the condition should be that there is $M \in X$ such that $M \models \Theta \cup \{\alpha\}$. Finally, if X is a set of additional rules, then the second condition could be that no of such additional rules authorizes any contradiction, which can be expressed symbolically as $\Theta \cup \{\varphi\} \not\models_X \bot$. Now the set of abductive solutions of $AbdProb_{\models_X}(\Theta, \varphi)$ is represented by $Abd_{\models_X}(\Theta, \varphi)$

We could consider two ways of studying abduction in contexts. First defining abduction in any context, which may be representative of a specific style of inference. This will be tackled in the present work, though it is restricted to the classical style for convenience. Second taking into account a class of contexts, then a structure of possible contexts could be studied. In this case an interesting modal framework may be developed —put that forward in [13]—. On the other hand, inferential contexts could be seen as the result of constructing successive stages by means of abduction, as it is explained in the sequel.

Abduction with respect to classical consequence relation is a special case with respect to a consequence relation modulo X in which such modulo is $\emptyset$, that is to say $\models_\emptyset$ is $\models$ and we shall say that it is the *null inferential context* —with respect to sets of models, the classical consequence relation is equivalent to $\models_{MOD(L)}$—. Now an inferential context of the classical style of inference, particularly when represented by means of consequence modulo a set of formulas, can be analyzed not only as a class of contexts and an inclusion relation defined in that, but also as a process that begins from an initial context until a final one, the represented strictly speaking. Given $\Theta_0 \subset L$, an initial (consistent) context, and a set of formulas $\Phi = \{\varphi_1, ...,$

$\varphi_n\}$ that are consistent with Θ_0, $n <\mid \mathbb{N} \mid$, $\models_{\Theta_n}$ represents the context that results of obtaining the modulo Θ_n according to the following rule, for every $i < n$

1. If it is not the case that $AbdProb_{\models_{\Theta_i}}(\Theta_i, \varphi_{i+1})$, then $\Theta_{i+1} = \Theta_i$

2. In other case, $\Theta_{i+1} = \Theta_i \cup \{\varphi_{i+1}\} \cup \{f(\Theta, \varphi_{i+1})\}$, where

$$f : \Theta \times \{\varphi_{i+1}\} \longmapsto Abd(\Theta, \varphi_{i+1}), \text{ for } \Theta \in \mathcal{P}(L) \text{ and } \varphi_{i+1} \in \Phi$$

—given an abductive problem, f chooses the best explanation, according to a preferential relation defined in $Abd_{\models_{\Theta_i}}(\Theta_i, \varphi_{i+1})$—

So an interesting way of working abduction emerges, though this can not be developed in this paper. Only two characteristics of such process should be pointed out before the end of this section. First, downward nonmonotonicity of contexts: for $i < j < n$, it may be that $\Gamma \models_{\Theta_j} \varphi$ but $\Gamma \not\models_{\Theta_i} \varphi$. Consider $\Theta_i = \emptyset$ and $\Theta_j \neq \emptyset$, if $\Gamma \models_{\Theta_j} \varphi$, then $\Gamma \not\models \varphi$. Second, upward monotonicity of contexts: for $i < j < n$, let Θ_i and Θ_j be such that $\Theta_i \subset \Theta_j$, if $\Gamma \models_{\Theta_i} \varphi$, then $\Gamma \models_{\Theta_j} \varphi$. In fact, $\Gamma, \Theta_i \models \varphi$ implies (by monotonicity of $\models$) that $\Gamma, \Theta_j \models \varphi$ provided $\Theta_i \subseteq \Theta_j$.

4   Calculi conditions

In our previous comments consequence relations are implicitly semantic relations. What about syntactical ones? Another way of defining a logical system or logic is by means of a formal language and a set of rules of inference for obtaining (or "deducing") the truth of a statement (conclusion) from that of another ones (premises). Then the logical system is presented as a calculus. If $\varphi \in L$ is deduced from $\Gamma \subset L$ —nothing else is taking into account as premise apart from Γ— it is represented as $\Gamma \vdash \varphi$. When $\vdash$ represents a classical calculus, it is a closure relation, hence it verifies reflexivity, monotonicity and transitivity, and it accomplishes compactness and uniform substitution, and also it is verified that

$$\Gamma \models \varphi \text{ iff } \Gamma \vdash \varphi,$$

that is to say, $\vdash$ is a sound and complete syntactical relation that can represent the null context.

However, when we are in other context, inside the classical style of inference, things are not so easy to do and the calculus corresponding to the context (originally represented by $\models_X$) may be difficult from an abductive point of view. In fact, applications of deductive calculi to obtain solutions for abductive problems, even with respect to the classical consequence relation, have some limitations. Whatever the case may be, many inferential contexts should be representable by means of a syntactical relation as a suitable correlate of the corresponding semantic relation. To do that, each calculus must capture the essential characteristics of the pertinent consequence relation, so given $\models_X$, a calculus $\vdash_X$ has to be defined in a way such that it must be a closure relation, because of which it will verify reflexivity, monotonicity and transitivity.

In fact, for $\models_\Theta$, $\Theta \subset L$, we could take the calculus $\vdash_\Theta$ such that for $\Gamma \subset L$ and $\varphi \in L$, $\Gamma \vdash_\Theta \varphi$ iff $\Gamma, \Theta \vdash \varphi$. When a class of models is considered as the modulo, it would be restricted to classes that are representable by means of classes of formulas: for $\mathcal{M} \subset MOD(L)$ such that

$$\Theta = \{\beta \in L \mid M \in \mathcal{M} \text{ iff } M \models \beta\},$$

$\Gamma \vdash_{\mathcal{M}} \varphi$ iff $\Gamma, \Theta \vdash \varphi$. For a set of additional rules R and the context $\models_R$, $\vdash_R$ expresses that rules of $\vdash$ plus rules of R that must be taken into account for deducing. So, according to definition, a calculus $\vdash_X$, when X is a set of formulas or a set of additional rules —or a set of models with the mentioned condition—, verifies reflexivity, monotonicity and transitivity. That is to say, $\vdash_X$ is a closure relation and can represent an inferential context.

In spite of everything, some questions about applicability of calculi arise: What conditions constitute a minimal set of requirements to apply deductive calculi to abduction in contexts in the same way than in the null context? Has it to be sound? Must it be complete? Sound-completeness of $\vdash_X$ requires to accept the following proposition: for every $\Gamma \subset L$ and $\varphi \in L$,

$$\Gamma \models_X \varphi \text{ iff } \Gamma \vdash_X \varphi.$$

A possible criterion to use deductive calculi for abduction might be that they were sound and complete, then, since completeness implies compactness, if a consequence relation does not verify compactness, then the corresponding calculus —another correlated relation, subset of the corresponding Cartesian product, at last— can not be complete, so compactness is as essential as structural rules to obtain completeness. So that no standard calculus could be used to obtain abductive solutions when the context is represented by a consequence relation that does not verify compactness, that is to say, syntactic relations that do not verify compactness would be useless to solve abductive problems in some inferential contexts from a logical point of view. Therefore, with such criterion, we could only apply deductive calculi for abduction in contexts represented by consequence relations that are closure relations that verify compactness.

In [2] two formulations of sound-completeness theorem that are equivalent are presented, namely

1. there exists an axiomatic system $\vdash_\Sigma$ such that for every set of formulas $\Gamma \subset L$ and the formula $\varphi \in L$, $\Gamma \vdash_\Sigma \varphi$ iff $\Gamma \models \varphi$

2. there exists an axiomatic system $\vdash_\Sigma$ such that for every set of formulas $\Gamma \subset L$ and the formula $\varphi \in L$, $\Gamma \models \varphi$ iff

(a) $\vdash_\Sigma$ has *modus ponens*

(b) for $\alpha, \beta \in L$, $\vdash_\Sigma \alpha \rightarrow (\neg\alpha \rightarrow \beta)$
(contradiction implies triviality in $\vdash_\Sigma$)

(c) for all $\alpha \in L$, $\vdash_\Sigma (\neg\alpha \rightarrow \alpha) \rightarrow \alpha$
(it verifies *consequentia mirabilis*)

(d) the meta-theorem of deduction is verified in $\vdash_\Sigma$

(e) Γ is consistent in $\vdash_\Sigma$ —that is to say $\Gamma \nvdash_\Sigma \bot$— iff Γ is satisfiable

Though such forms are based on precise definitions of axiomatic system, calculus, etc., easily it can be adapted to classical systems that are sound and complete. For example, it is known that a deductive natural calculus (type Gentzen), expressed simply as $\vdash$, is sound and complete, so that the five indicated characteristics should be accomplished: in fact *modus ponens* and *deduction theorem* are basic rules, and satisfiability and consistency (in $\vdash$) are considered equivalent, in accordance with the semantics defined for formal languages from a classical point of view. Let us see the cases (b) and (c):

(b) Contradiction implies triviality:

1. α [hypothesis]

2. $\neg\alpha$ [hypothesis]

3. $\neg\beta$ [hypothesis]

4. $\alpha \wedge \neg\alpha$ [logical product 1, 2, hypothesis 3 is cancelled]

5. $\neg\neg\beta$ [introduction $\neg$, hypothesis 2 is cancelled]

6. β [double negation 5]

7. $\neg\alpha \rightarrow \beta$ [deduction theorem 2-5, hypothesis 1 is cancelled]

8. $\alpha \rightarrow (\neg\alpha \rightarrow \beta)$ [deduction theorem 1-6]

(c) Consequentia mirabilis

1. $\neg\alpha \rightarrow \alpha$ [hypothesis]

2. $\neg\alpha$ [hypothesis]

3. α [*modus ponens* 1, 2]

4. $\alpha \wedge \neg\alpha$ [logical product 2, 3, hypothesis 2 is cancelled]

5. $\neg\neg\alpha$ [introduction $\neg$ 2]

6. α [double negation 5, hypothesis 1 is cancelled]

7. $(\neg\alpha \to \alpha) \to \alpha$ [deduction theorem 1-6]

These two properties are characteristic of logical systems when the negation $\neg$ is assumed to be the classical one, which implies certain rules to regulate its (deductive) behaviour in the corresponding calculi (double negation, introduction of $\neg$ or *absurd*, etc.). Should us require that to use deductive calculi for abduction? In [11] minimal conditions have been proposed as a methodological way in searching a certain unity of science. For the moment the second formulation of soundness-completeness could be adopted as a criterion to choose a logic to be used in searching solutions for abductive problems, since those systems that can surpass this test are complete and have incorporated certain rules, as the indicated about negation, which facilitate the application of the calculus. However, we can immediately see that it is a very restrictive criterion if we want to work at any context inside the classical style of inference, which is shown in the next section.

5   Consequence relation without completeness in classical style

In this section we study shortly a couple of cases in contexts in which the use of calculi may be problematical with respect to the criterion provisionally accepted above, in the sense that such calculi are not complete. The first one is the *consequence relation modulo n*, for a natural number $n \geq 1$, in symbols $n < |\mathbb{N}|$. Of course, for the class $\mathcal{M}$ such that the cardinality of the universe of discourse of its models is n, $\mathcal{M} \subset MOD(L)$, so for any pair (Γ, φ), if $\Gamma \models \varphi$, then $\Gamma \models_n \varphi$ —all models that satisfy Γ also satisfy φ, enclosing the finite ones—, though when $\Gamma \models_n \varphi$, we are not forced to admit $\Gamma \models \varphi$, so that $\models \subset \models_n$, then easily it can be checked the character of closure relation of the new consequence relation, so that $\models_n$ verifies reflexivity, monotony and transitivity.

On the other hand, the existence of a sound deductive system comes in guaranteed by the fact that there is one that verifies the mentioned conditions when semantics is not restricted to a special class of models. So, for that semantics defined with respect to finite models, for all set of sentences Γ and a sentence φ, let $\vdash$ be that calculus, because of its soundness, it is accomplished that if $\Gamma \vdash \varphi$, then $\Gamma \models \varphi$, and by the mentioned inclusion, if $\Gamma \models \varphi$, then $\Gamma \models_n \varphi$, henceforth by hypothetic syllogism, if $\Gamma \vdash \varphi$, then $\Gamma \models_n \varphi$. However, the reciprocal can not be true. In fact a problem arises if we pay attention to the fifth clause of the above sound-completeness equivalence: suppose sets of sentences defined, for every natural number $n \geq 1$, as

$$\Gamma_n = \{\exists x_1, ..., x_n (\bigwedge_{i \neq j} x_i \neq x_j)\}, \text{ and } \Gamma = \bigcup \Gamma_n, \text{ for all } n < |\mathbb{N}|,$$

then $\Gamma_n \not\vdash \bot$ for every $n \geq 1$ henceforth $\Gamma \not\vdash \bot$, but there is no finite model M such that M satisfies all sentences of Γ and, as its consequence, it is not possible to state completeness, that is to say, for such set Γ and a sentence φ it could be the case that $\Gamma \models_n \varphi$ but $\Gamma \not\vdash \varphi$. This non-fulfilment is equivalent to the failure of the requisite of compactness, which was implicitly settled above as the additional condition for using deductive calculi in the search of solutions for abductive problems.

This was a predictable situation, since in fact some results are well know in finite model theory, namely

1. failure of completeness,

2. the set of finitely satisfiable sentences is not recursive,

3. the set of finitely valid sentences is not recursively enumerable,

4. the upward Löwenheim-Skolem theorem cannot be formulated for finite models, since this establishes that if any set of sentences has an infinite model, then it has arbitrary large models. At most a "finite" upward theorem could be stated, namely if $\varphi \in L$ is n-satisfiable — there is a model M such that $|M| = n$ and $M \models \varphi$—, $n \geq 1$, then φ is m-satisfiable for all $m \geq n$ (see [11], we omit the proof to abbreviate).

A similar case turns up when second order logic is considered, provided standard semantics is adopted. Despite any discussion about the character of second order systems, if they must be considered pure classical logic systems or a kind of extensions, we can distinguish two ways of putting forward second order logic. To simplify, let L^2 be a second order language, extension of L, where quantification is not restricted to individual variables, then it contains predicate variables and if X is one of them —of certain arity—, then $\exists X \beta$, $\forall X \beta \in L^2$, for any $\beta \in L$. L^2-models are the same L-models, although the range of second order quantification must be explicit. In first order semantics the range of quantification is implicitly the universe of discourse of corresponding model, that is to say, an L-model M satisfies the sentence $Qx\varphi$ iff M satisfies $\varphi(x/a)$ for an/all (according to be Q existential or universal) $\mathbf{a}$ of universe of M with $a^M = \mathbf{a}$. An L^2-model M is an L-model, the universe is $\mathbb{M}$ but for arity 1 it is considered a set of subsets of the universe $\mathbb{M}$ and for every arity $n \geq 2$, subsets of $\mathbb{M}^n$. So $\mathbb{M}_1$, $\mathbb{M}_2$,..., sets of subsets of $\mathbb{M}$, $\mathbb{M}^2$, ..., respectively, are ranges of second order quantification. To abbreviate, let $\exists X \beta$ be a sentence and a M an L^2-model and consider the range of quantification $\mathbb{M}_{ari(X)}$, where $ari(X)$ expresses the natural number that is arity of the predicate variable X, then $M \models_H \exists X \beta$ iff there exists $\mathbb{R} \in \mathbb{M}_{ari(X)}$, such that for the constant R (of that arity), it is verified that $R^M = \mathbb{R}$ and $M \models \beta(X/R)$. In a similar way for universal quantifier —the index H should remember us that non standard semantics for second order logic (higher order logic, in general) was first introduced by

Henkin in [6]—. When $\mathbb{M}_i$, for all $i \geq 1$, contains *all* predicates (or relations) defined in $\mathbb{M}$, then M is a *full model* and the consequence relation is the standard one: if M is a full model, $R^M = \mathbb{R} \in \mathbb{M}_{ari(X)}$ and $M \models \beta(X/R)$, then it is represented $M \models \exists X \beta$, without index.

In order to represent a context, to distinguish these two semantics for second order languages, it should be noted, may be crucial. When it is taken the consequence relation $\models_H$ the situation is the same than in first order, *mutatis mutandi*, so by means of sets of sentences, a variety of contexts could be represented. Let Θ be a set of sentences of L^2, for any set $\Gamma \subset L^2$ and a sentence $\varphi \in L^2$, φ is logical consequence of Γ modulo the set Θ (in the sense of Henkin's semantics) iff all L^2-Henkin's-models of Θ that are L^2-Henkin's-models of Γ satisfy φ, in symbols $\Gamma \models_{H/\Theta} \varphi$. When the semantics is with full models, in a similar way new consequence relations modulo sets of sentences can be defined, so φ is logical consequence in a standard sense of Γ modulo Θ iff all full L^2-models that are full models of Θ satisfy φ, in symbols $\Gamma \models_\Theta \varphi$. In both cases one deals with a closure relation, that is to say $\models_H$ and $\models$ verify reflexivity, monotonicity and transitivity, then $\models_\Theta$ and $\models_{H/\Theta}$ also accomplish that three properties, though the latter ones do not verify uniform substitution, and they represent two contexts in the classical style of inference.

In accordance with our previous remarks, we can represent approximately different inferential contexts by means of these second order consequence relations, although in the same style of inference: the classical one (or a simple extension one), then abduction could be treated in some of such contexts. However, the use of calculi has limitations once again. In fact the two semantics turn out very different because of several properties, as it is shown in the following resume of features of second order logic —following the given in [14]—:

1. Completeness, for $\Gamma \subset L^2$, $\varphi \in L^2$ and a calculus $\vdash$. Two different results are obtained according to the adopted semantics, namely

   (a) $\Gamma \vdash \varphi$ iff $\Gamma \models_H \varphi$

   (b) no given deductive calculus is complete for full models (it can be sound, but not complete)

2. Compactness. That depends on the semantics again

   (a) any set of sentences, every finite subset of which has a Henkin model, has itself a Henkin model, but

   (b) failure of compactness for full models is well known

3. The Löwenheim-Skolem theorem. As in previous cases,

   (a) it is verified when Henkin's semantics is considered, though

(b) it fails with full models

In this case a similar question arises, namely, what should be the criterion to take that semantics according to which the use of calculi for abduction were possible? An option is to delimit the consequence relation in a way that completeness would be guaranteed. For example, by choosing certain classes of sentences, though the semantics might be the full one. On this matter Σ_1^1 is an interesting fragment of second order logic: $\alpha \in L^2$ is a Σ_1^1-sentence iff α is $\exists X_1, ..., X_n \beta$, $n \geq 1$, where X_i is a predicate variable of certain arity for every $i \leq n$ and $\beta \in L$ —it is quantifier free, since β is a first oder one—.

6   Framework for defining abductive calculi

In order to define abductive calculi that take advantage of properties of deductive calculi in the classical style of inference, we should definitely revise the sound-completeness criterion. In order to do that, a way is to weaken this requisite as an application of the following concept. In classical style, therefore in contexts with a consequence relation that is a closure one, as it accomplishes compactness as it does not, we shall say that a (deductive) calculus $\vdash_X$ is *suitable* for a consequence relation $\models_X$ iff it is verified that

1. $\vdash_X$ is sound

2. $\vdash_X$ has the following rules

   (a) *modus ponens*
   (b) contradiction implies triviality
   (c) *consequentia mirabilis*
   (d) (meta) theorem of deduction

By means of this notion, a part of the specification of sound-completeness in [2] is adopted instead of the exact sound-completeness criterion. In fact, adopting this suitability criterion we have only left the 2.(e) clause out, which was equivalent to compactness property, then any sound (deductive) calculus will be able to be used for abduction provided certain behaviour of connectives is observed ("negation" —in classical sense— and "material implication"), that is to say, such calculus is sufficient to capture what may be called the most relevant of a consequence relation, hence the corresponding inferential context could also be represented by this syntactic relation. All of that, of course, when the context is one of those that make up the classical style of inference. So that, to obtain solutions for abductive problems with respect to a consequence relation $\models_X$, an abductive calculus $\vdash_{ABD(X)}$ is definable if there exists a deductive calculus $\vdash_X$ that is suitable for $\models_X$.

Given a consequence relation $\models_X$ and a suitable calculus $\vdash_X$ for that, $\Theta \subset L$, $\Theta = \{\theta_1, ..., \theta_n\}$, for the natural number $n \geq 1$, and $\alpha, \varphi \in L$, $REL_{\vdash_X}(\Theta, \alpha, \varphi)$ expresses that it is verified

1. $\Theta, \alpha \not\vdash_X \bot$

2. $\alpha \not\vdash_X \varphi$

3. at least one of the following conditions

  (a) $\alpha \vdash_X \theta_1 \wedge \ldots \wedge \theta_n \to \varphi$
  (b) $\Theta, \neg\varphi \vdash_X \neg\alpha$

Let Θ and φ be called "first" and "second" element (or premise) of the abductive problem, alternatively "theory" and "fact" to be explained, and α the "abducted" or "explanation". So, $REL_{\vdash_X}(\Theta, \alpha, \varphi)$ is relative to a set Θ that is finite, $\Theta \cup \{\alpha\}$ is consistent, φ is not deducible in this calculus from α only and 3.(a) and 3.(b) are related to the deduction theorem and contraposition, respectively —see proof of the proposition below—. Now we can settle a general framework for defining abductive calculi in the classical style of inference. The only framework rule says that if an abductive problem is given (with respect to the consequence relation $\models_X$) and its theory, explanation and fact are in the relation $REL_{\vdash_X}$, then the explanation can be concluded, in symbols

$$\frac{AbdProb_{\models_X}(\Theta, \varphi), \; REL_{\vdash_X}(\Theta, \alpha, \varphi)}{\alpha}$$

As we have pointed before, $\vdash_{ABD(X)}$ represents an abductive calculus defined from a deductive one according to such rule and $\Theta, \varphi \vdash_{ABD(X)} \alpha$ expresses that α is obtained as an explanation from the premises Θ and φ —theory and fact, respectively— of a given abductive problem with respect to a certain consequence relation. The following proposition says that abductive calculi so defined are sound (in the specified sense)

  For $\Theta \subset L$, $\varphi, \alpha \in L$, if $\vdash_{ABD(X)}$ is defined in accordance with the abductive framework rule, then it is sound, that is to say, all consequents of such rule are solutions for the corresponding abductive problems.

This proposition can be simplified as follows

$$\text{if } \Theta, \varphi \vdash_{ABD(X)} \alpha, \text{ then } \alpha \in Abd_{\models_X}(\Theta, \varphi).$$

To prove that proposition, suppose $\Theta \subset L$, $\Theta = \{\theta_1, \ldots, \theta_n\}$, for a natural number $n \geq 1$, and $\varphi, \alpha \in L$ such that it is verified that $\Theta, \varphi \vdash_{ABD(X)} \alpha$, which is equivalent to state

$$AbdProb_{\models_X}(\Theta, \varphi) \text{ and } REL_{\vdash_X}(\Theta, \alpha, \varphi),$$

hence $\Theta \not\models_X \varphi$ and $\Theta \not\models_X \neg\varphi$. On the other hand, as a consequence from hypothesis, it is accomplished at least one of the following requisites

1. $\alpha \vdash_X \theta_1 \wedge ... \wedge \theta_n \rightarrow \varphi$

2. $\Theta, \neg\varphi \vdash_X \neg\alpha,$

then we can obtain, if it is verified the first,

1. $\alpha \vdash_X \theta_1 \wedge ... \wedge \theta_n \rightarrow \varphi$ [Premise]

2. $\alpha \vdash_X \theta_1 \rightarrow (... \rightarrow (\theta_n \rightarrow \varphi)...)$ [equivalent]

3. $\alpha, \theta_1 \vdash_X \theta_2 \rightarrow (... \rightarrow (\theta_n \rightarrow \varphi)...)$ [deduction theorem]

$\vdots$

n. $\alpha, \theta_1, ..., \theta_n \vdash_X \varphi$ [deduction theorem]

alternatively

1. $\alpha \vdash_X \theta_1 \wedge ... \wedge \theta_n \rightarrow \varphi$ [Premise]

2. $\alpha, \theta_1 \wedge ... \wedge \theta_n \vdash_X \varphi$ [deduction theorem]

3. $\Theta, \alpha \vdash_X \varphi$ [equivalent]

or, if it is verified the second,

1. $\Theta, \neg\varphi \vdash_X \neg\alpha$ [Premise]

2. $\Theta, \alpha, \neg\varphi \vdash_X \neg\alpha$ [monotonicity]

3. $\Theta, \alpha, \neg\varphi \vdash_X \alpha$ [reflexivity]

4. $\Theta, \alpha, \neg\varphi \vdash_X \alpha \rightarrow (\neg\alpha \rightarrow \varphi)$ [contradiction implies triviality]

5. $\Theta, \alpha, \neg\varphi \vdash_X \neg\alpha \rightarrow \varphi$ [*modus ponens* 3, 4]

6. $\Theta, \alpha, \neg\varphi \vdash_X \varphi$ [*modus ponens* 2, 5]

7. $\Theta, \alpha \vdash_X \neg\varphi \rightarrow \varphi$ [deduction theorem]

8. $\Theta, \alpha \vdash_X (\neg\varphi \rightarrow \varphi) \rightarrow \varphi$ [*consequentia mirabilis*]

9. $\Theta, \alpha \vdash_X \varphi$ [*modus ponens* 7, 8]

Whatever the case may be, $\Theta, \alpha \vdash_X \varphi$, but $\vdash_X$ is sound by hypothesis, therefore $\Theta, \alpha \models_X \varphi$ and, by definition, $\alpha \in Abd_{\models_X}(\Theta, \varphi)$.

7 Concluding remarks

Some representations of inferential contexts show certain utility of using new consequence relations to study abduction in contexts, though applicability of deductive calculi to obtain solutions for abductive problems only to very specific representations of contexts, namely those in which the corresponding consequence relation is a closure one and verifies compactness, is too restrictive. The classical style of inference goes beyond a consequence relation defined with respect to a first order language and several kinds of consequence relations modulo (a finite cardinality of universe of discourse, for example) are closure relations but do not verify compactness, the same happens when the language is a second order one (with standard semantics), since for second order languages it is not possible compactness though we could have a closure relation. But the case of second order logic itself suggests that calculi could be restrictively used to tackle abduction in contexts or study precesses of constructing them as abductive constructions. In fact, the point of view according to which a logic is defined in semantic terms is not the only one, a fundamental conception looks at a syntactic consequence relation as the best reference to characterize a logic and the known (first order) calculi represent closure relations that verify compactness, which is irrelevant when the environment is the class of finite models and it fails in second order (with standard semantics). Is it not enough soundness and certain additional requisites to use a calculus, as suggested in [11]? Is completeness indispensable for a (logical) life after all? About classical style of inference, the suitability criterion is weaker than the mentioned one and it proves to be effective to reach abductive calculi, so that a framework abductive rule has been proposed, which somehow defines such abductive calculi in all contexts of classical style of inference. On the other hand, all abductive calculi defined according to such rule are sound, hence it justifies the most known proposals of using classical calculi for abduction, among which the given in [1, 10, 11, 13] should be mentioned —in short, such proposals, which could be seen as instances of the so called *AKM model of abduction* in [5], point out the relevance of the deduction theorem and the character of the negation, in the last resort they make use of contraposition and deduction theorem to obtain abductible formulas and both of them are embraced by our notion of suitability—.

As it has been suggested, other styles of inference should be explored, since from an intuitive point of view it is not the case that all inferential contexts are governed (and representable) by a logic whose consequence relation verifies the studied structural rules. In fact, in some scientific practices the underlying logic is non monotonic or verifies a restricted form of monotonicity (there are important non monotonic logics, cumulative and not cumulative, as summareized in [4] for example), so it is interesting to explore how abduction works in these cases. To tell the truth, *inferential context* has been taken as a notion whose features are not determined in all details and, in order to enrich the formal apparatus, it may be necessary to incorporate more conceptual elements —an example may be the mentioned case of default normal systems—, then the logical treatment may require other logical systems. Besides our mention to a modal approach by considering a set of possible contexts, to take into account agents of knowledge, for example, may be very convenient ([5], from chapter 3, is interesting in this sense), then to develop representations by means of some epistemic logics. In short, the variety of existent deductive logics constitutes a challenger to work in refining methods to study abduction.

Aknowledgements

Research for this paper was supported by subventions from *Ministerio de Educación y Ciencia (Spanish Govern)*, project HUM2007/65053, and *Consejería de Innovación, Ciencia y Empresa de la Junta de Andalucía*, project P06-HUM-01538.

We want to say a few words of thanks to an anonymous referee. Her/his criticism and comments on our proposal helped us to improve this paper and provide an incentive to start new lines of investigations about abductive logics.

BIBLIOGRAPHY

[1] Aliseda, A.: Abductive Reasoning: Logical Investigations into Discovery and Explanation. Volume **330** of Synthese Library. Springer (2006)

[2] Amor, J. A.: Compacidad en la lógica de primer orden y su relación con el Toerema de Completud. Facultad de Ciencias de la U. N. A. M. (2006)

[3] Antoniou, G.: Nonmonotonic Reasoning. The MIT Press (1997)

[4] Carnota, R. J.: "Lógica e Inteligencia Artificial". Lógica. Enciclopedia Iberoamericana de Filosofía, vol. **7**. Trotta (1995) 143–183

[5] Gabbay, D., Woods, J.: A Practical Logic of Cognitive Systems, Volume **2**: The Reach of Abduction: Insight and Trial. Elsevier (2006)

[6] Henkin, L.: "Completeness in the theory of types". The Journal of Symbolic Logic, **15** (1950) 81–91

[7] Hintikka, J.: "What is abduction? The fundamental problem of contemporary epistemology". Transactions of the Charles S. Peirce Society, **34** (3) (1998) 503–533.

[8] Makinson, D.: Bridges between Classical Logic and Nonmonotonic Logic Logic Journal of the IGPL **11**, 1 (2003)

[9] Makinson, D.: Bridges between Classical Logic and Nonmonotonic Logic King's College London, Texts in Computing **5**, (2005)

[10] Mayer, M. C.; Pirri, F.: "First order abduction via tableau and sequent calculi" *Bulletin of the IGPL*, vol. I (1993) 99–117

[11] Nepomuceno, A., Soler, F., Aliseda, A.: "Searching the Unity of Science: from Classical Logic to Abductive Logical Systems" The Unity of Science: Non Traditional Approaches. Kluwer, Serie Logic, Epistemology, and the Unity of Science (forthcoming)

[12] Reiter, R.: "A logic for default reasoning", *Artificial Intelligence* **13** (1980) 81–132.

[13] Soler, F. "A modal interpretation of explicative reasoning", Paper presented in *II Jornadas Ibéricas de Lógica y Filosofía de la Ciencia*, Lisbon (2007)

[14] Väänänen, J.: "Second order logic and foundation of mathematics" The Bulletin of Symbolic Logic **7**, 4 (2001) 504-520

Ángel Nepomuceno-Fernández and Fernando Soler-Toscano
Dpto. de Filosofía, Lógica y Filosofía de la Ciencia
Universidad de Sevilla
Sevilla, Spain
E-mails: {nepomuce, fsoler}@us.es

SECTION 8

LOGIC AND COMPUTING

The Complexity of $AUTOSAT(\Sigma^i_m)$

FLAVIO A. FERRAROTTI AND JOSÉ M. TURULL TORRES

Abstract: The language $AUTOSAT(F)$, for a given logic F, is the class of the F sentences which in a certain encoding as relational structures satisfy themselves. This problem was introduced by J. A. Makowsky and Y. B. Pnueli in 1996 where, among other subjects, the complexity of the problem for some fragments of First and Second Order Logic was studied. In this paper, we show that, for every $i, m \geq 1$, $AUTOSAT(\Sigma^i_m)$ is complete for Σ^{i+1}_m under polynomial-time reductions. We prove that $AUTOSAT(\Sigma^i_m)$ is hard for Σ^{i+1}_m by giving a polynomial-time reduction to $AUTOSAT(\Sigma^i_m)$ from a higher-order version of QSAT which was proved to be complete for the corresponding prenex fragments of higher-order logics by L. Hella and J. M. Turull Torres in 2006.

1 Introduction

We studied in previous papers [FT05, FT06, FT07] the expressive power of different fragments of higher-order logics over finite relational structures (or equivalently, relational databases). Among others results, we proved in those articles the following: Let $HAA^i(r, m)$ be the class of $(i + 1)$-th order logic formulae in $\Sigma^i_m \cup \Pi^i_m$ such that the maximal-arity (a generalization of the concept of arity, not just the maximal of the arities of the quantified variables) of the higher-order variables is bounded by r. As usual, Σ^i_m and Π^i_m denote the well known existential and universal, respectively, prenex fragments of higher-order logic formulae of order $i+1$ in which the quantifiers of order $i + 1$ are arranged into at most m alternating blocks. It follows that for every order $i \geq 3$, the resulting HAA^{i-1} hierarchy of formulae of i-th order logic is proper. More precisely, it follows that for every $i \geq 2$ and every $r, m \geq 1$, there are classes of finite structures not definable in $HAA^i(r, m)$ but definable in $HAA^i(r + c(r), m + 2)$ where $c(r) = 1$ for $r > 1$ and $c(r) = 2$ for $r = 1$.

In particular, the problem that allowed us to separate the levels of the HAA^i hierarchies of arity and alternation, is called $AUTOSAT$ and was first introduced by Makowsky and Pnueli in [MP96]. They used $AUTOSAT$ to prove the properness of an arity-alternation hierarchy in second-order logic which they called the AA hierarchy. Interestingly, it is still open whether the corresponding version for second-order logic of the arity-alternation hierarchies defined above, i.e., the HAA^1 hierarchy, is proper. Note that the HAA^1 hierarchy is the hierarchy denoted as SAA in [MP96].

Roughly speaking, $AUTOSAT(F)$ is the set of formulae of a given logic F which, encoded as finite structures, satisfy themselves. $AUTOSAT$ was also studied in [MP96] from the point of view of its complexity. Let Σ_m^0 denote the set of first-order formulae in prenex normal form which start with an existential block of quantifiers and have up to m alternating blocks. They proved the following.

THEOREM 1 ([MP96]). *Under polynomial-time reductions, it holds that:*

i. $AUTOSAT$(FO) is complete for PSPACE.

ii. For every $r, m \geq 1$, $AUTOSAT(HAA^1(r, m))$ is complete for PSPACE.

iii. For every $m \geq 1$, $AUTOSAT(\Sigma_m^0)$ is complete for the class Σ_m^p of the polynomial-time hierarchy.

The fact that for first-order logic $AUTOSAT$ is hard for PSPACE, was proved by reduction from the well known problem of *quantified satisfiability* (QSAT) to $AUTOSAT$(FO). Note that QSAT, also known as quantified boolean formulae (QBF), is complete for PSPACE (see [Pap94, BDG95] among other sources). Since for every $r, m \geq 1$, it holds that FO $\subset HAA^1(r, m)$, the same reduction from QSAT to $AUTOSAT$(FO), also applies from QSAT to $AUTOSAT(HAA^1(r, m))$. Regarding $AUTOSAT(\Sigma_m^0)$, they gave a polynomial-time reduction from $QSAT_m$ (*quantified satisfiability with m alternations of quantifiers*). Since $QSAT_m$ is complete for the corresponding level Σ_m^p of the polynomial-time hierarchy, this proves that $AUTOSAT(\Sigma_m^0)$ is hard for Σ_m^p.

In this paper, we extend this result establishing the complexity of $AUTOSAT$ for each prenex fragment Σ_m^i of higher-order logic formulae of order $i + 1$. Notice that the argument used in [MP96] to prove that $AUTOSAT(HAA^1(r, m))$ is in PSPACE, does not apply to $AUTOSAT(\Sigma_m^1)$. The main obstacle is that the arity of the variables which appear in the formulae in Σ_m^1 is not bounded by a fixed r as is the case for $HAA^1(r, m)$.

We prove instead that for every $i, m \geq 1$, $AUTOSAT(\Sigma_m^i)$ is in the complexity class $\text{NTIME}(\exp_i(\mathcal{O}(n^c)))^{\Sigma_{m-1}^p}$, where Σ_{m-1}^p is the $(m-1)$-th level of the polynomial-time hierarchy, and $\exp_i(f(n))$ is the following i-fold exponential function: $\exp_0(f(n)) = f(n)$, and for $i \geq 1$, $\exp_i(f(n)) = 2^{\exp_{i-1}(f(n))}$.

Since by the characterization in [HT03, HT06b], the prenex fragment Σ^i_m of higher-order logic formulae of order $i+1$ captures $\mathrm{NTIME}(\exp_{i-1}(\mathcal{O}(n^c)))^{\Sigma^p_{m-1}}$, and for every $m \geq 1$, we also give in this paper a polynomial-time reduction from a complete problem in Σ^{i+1}_m to $AUTOSAT(\Sigma^i_m)$, we get the following result.

THEOREM 2. *For every $i, m \geq 1$, it holds that $AUTOSAT(\Sigma^i_m)$ is complete for Σ^{i+1}_m under polynomial-time reductions.*

The paper is organized as follows. In Section 2 we first stablish the notation and give the syntax and semantics of higher-order logics. Then in Subsection 2.2 we formally define the $AUTOSAT$ problem. In Subsection 2.3 we define the Σ^i_m theory of the Boolean model which is known to be a complete problem for higher-order logics [HT05, HT06a], and which is essentially a generalization of QSAT. In Sections 3 and 4, we show that for every $i, m \geq 1$, $AUTOSAT(\Sigma^i_m)$ is hard for the prenex fragment Σ^{i+1}_m of the higher-order logic of order $i+1$. We do this by a polynomial-time reduction from the Σ^i_m theory of the Boolean model to $AUTOSAT(\Sigma^i_m)$. For the sake of clarity, we present this reduction in two steps. First, we gave in Section 3 a polynomial-time reduction from the Σ^1_m theory of the Boolean model to $AUTOSAT(\Sigma^1_m)$. Then in Section 4, we extend this reduction to each order $i \geq 1$. Finally, we complete our proof of Theorem 2, showing in Section 5 that for every $i, m \geq 1$, $AUTOSAT(\Sigma^i_m[\rho])$ is in Σ^{i+1}_m. We conclude this paper in Section 6 where we present a few suggestions for future work.

2   Preliminaries

We use the notion of a *logic* in a general sense. A formal definition would only complicate the presentation and is unnecessary for our work. As usual in finite model theory, we regard a logic as a language, that is, as a set of formulas (see [EF99]). We do require that the syntax and the notion of satisfaction for any logic $\mathcal{L}$ be *decidable*. We only consider vocabularies which are purely *relational*, and for simplicity we do not allow constant symbols. If σ is a relational vocabulary, we denote by $\mathcal{L}[\sigma]$ the set of $\mathcal{L}$-formulae over σ. We consider *finite* structures only. Consequently, the notion of *satisfaction*, denoted as $\models$, is related to only finite structures. If $\mathbf{I}$ is a finite structure of some vocabulary σ, we denote its domain as I and sometimes also as $dom(\mathbf{I})$. If R is a relation symbol in σ of arity r, for some $r \geq 1$, we denote as $R^{\mathbf{I}}$ the (second-order) relation of arity r which interprets the relation symbol R in $\mathbf{I}$, with the usual notion of interpretation.

We denote as $\mathcal{B}_\sigma$ the class of *finite* structures of vocabulary σ.

By $\varphi(x_1, \ldots, x_r)$ we denote a formula of some logic whose free variables are *exactly* $\{x_1, \ldots, x_r\}$. If $\varphi(x_1, \ldots, x_r) \in \mathcal{L}[\sigma]$, $\mathbf{I} \in \mathcal{B}_\sigma$, $\bar{a}_r = (a_1, \ldots, a_r)$ is a r-tuple over $\mathbf{I}$, let $\mathbf{I} \models \varphi(x_1, \ldots, x_r)[a_1, \ldots, a_r]$ denote that φ is TRUE, when interpreted by $\mathbf{I}$, under a valuation v where for $1 \leq i \leq r$ $v(x_i) = a_i$. Then we consider the set of all such valuations as follows:

$$\varphi^{\mathbf{I}} = \{(a_1, \ldots, a_r) : a_1, \ldots, a_r \in I \wedge \mathbf{I} \models \varphi(x_1, \ldots, x_r)[a_1, \ldots, a_r]\}$$

That is, $\varphi^{\mathbf{I}}$ is the relation defined by φ in the structure $\mathbf{I}$, and its arity is given by the number of free variables in φ. Let φ be a sentence in $\mathcal{L}[\sigma]$, we denote by $Mod(\varphi)$ the class of finite σ-structures $\mathbf{I}$ such that $\mathbf{I} \models \varphi$. A class of finite σ-structures $\mathcal{C}$ is *definable* by a $\mathcal{L}$-sentence if $\mathcal{C} = Mod(\varphi)$ for some $\varphi \in \mathcal{L}[\sigma]$.

2.1   Syntax and Semantics of Finite-Order Logic

Finite-order logic is an extension of first-order logic which allows us to quantify over higher-order relations. We define here its syntax and semantics following the account in [Lei94]. We emphasize the fact that the set of formulae of finite-order logic can be viewed as a set of strings over a *finite* alphabet, i.e., as a formal language. This plays an important role in the encoding of formulae as finite structures which we define in Subsection 2.2.

DEFINITION 3. We define the set of *types*, as the set **typ** of strings over the alphabet $\{\iota; (;); ,\}$ inductively generated by: $\iota \in \mathbf{typ}$ and if $\tau_1, \ldots, \tau_r \in \mathbf{typ}$ $(r \geq 1)$, then $(\tau_1, \ldots, \tau_r) \in \mathbf{typ}$. If $\tau_1 = \cdots = \tau_r = \iota$, then $(\tau_1, \ldots, \tau_r)$ is denoted by ι^r.

The set of types can be naturally stratified into *orders* which are inductively defined, as follows: $order(\iota) = 1$ and $order((\tau_1, \ldots, \tau_r)) = 1 + max(\{order(\tau_1), \ldots, order(\tau_r)\})$.

For $\tau = (\tau_1, \ldots, \tau_r)$, r is the *arity* of the type τ. We associate a non-negative integer, the *maximal-arity* (ma), with each type, as follows: $ma(\iota) = 0$ and $ma((\tau_1, \ldots, \tau_r)) = max(\{r, ma(\tau_1), \ldots, ma(\tau_r)\})$. Clearly, if $order(\tau) = 2$, then the maximal-arity of τ coincides with its arity.

We denote as $typ(i, r)$ the subset of types of order $\leq i$ and maximal-arity $\leq r$. Note that each subset $typ(i, r)$ is finite.

The intended interpretation is that objects of type ι are individuals, i.e., elements of the universe of a given model, whereas objects of type $(\tau_1, \ldots, \tau_r)$ are r-ary relations, i.e., sets of r-tuples of objects of types $\tau_1, \ldots, \tau_r$, respectively.

DEFINITION 4. Given a set U, the set U_τ of *objects* of type τ over U is defined by

$$U_\iota = U \qquad \text{and} \qquad U_{(\tau_1,\ldots,\tau_r)} = \mathcal{P}(\prod_{i=1}^{r} U_{\tau_i}) = \mathcal{P}(U_{\tau_1} \times \cdots \times U_{\tau_r})$$

Over a relational vocabulary σ, each formula of finite-order logic is a string of symbols taken from the *alphabet* $A = \{\neg; \vee; \wedge; \exists; \forall; (;); = ; x; X; |; \iota;, \} \cup \sigma$. The words that belong to the language $\{x|^n : n > 0\}$ are called *individual variables*, while the words that belong to the language $\{X\tau|^n : \tau \in \mathbf{typ} \setminus \{\iota\}$ and $n > 0\}$ are called *higher-order variables*. We call the higher-order variables of the form $X\tau|^n$, for $i = order(\tau)$ and $r = ma(\tau)$, *i-th order variables* of *maximal-arity* r. To simplify the notation we denote strings of the form $|^n$, $n > 0$, as subscripts, e.g., we write x_3 for $x|||$. In addition, we write the types of the higher-order variables as superscripts, e.g., we write $X_2^{(\iota)}$ for $X(\iota)||$. Sometimes, we omit the superscript when we denote second-order variables (i.e., variables of type ι^r, for some $r \geq 1$) if their arity is clear from the context. We use V^τ to denote any variable of type τ. So, if $\tau = \iota$ then V^τ stands for an individual variable, otherwise V^τ stands for a higher-order variable of type τ.

DEFINITION 5. We define the set of *well-formed formulae* (wff) of finite-order logic over a relational vocabulary σ (here we do not allow constant symbols), as the extension of the set of wff of first-order logic plus equality with the following formation rules:

i. If V^τ is a higher-order variable of type $\tau = (\tau_1, \ldots, \tau_r)$ and $V_1^{\tau_1}, \ldots, V_r^{\tau_r}$ are variables of types $\tau_1, \ldots, \tau_r$, respectively, then $V^\tau(V_1^{\tau_1}, \ldots, V_r^{\tau_r})$ is a wff.

ii. If φ is a wff and V^τ is a higher-order variable, then $\exists V^\tau(\varphi)$ and $\forall V^\tau(\varphi)$ are wff's.

The free occurrence of a variable (either an individual variable or a higher-order variable) in a formula of finite-order logic is defined in the obvious way. Thus, the set $free(\varphi)$ of *free variables* of a formula φ is the set of both individual and higher-order variables which do not occur in φ under the scope of a quantifier which binds them.

The *semantics* of formulae of finite-order logic is similar to the semantics of formulae of second-order logic, except that a valuation over a structure with universe U maps higher-order variables of type τ to objects in U_τ.

DEFINITION 6. Let σ be a relational vocabulary. A *valuation val* on a σ-structure $\mathbf{I}$ with domain I, is a function which assigns to each individual variable an element in I and to each higher-order variable V^τ, for some type $\tau \neq \iota$, an object in I_τ. Let val_0, val_1 be two valuations on a σ-structure $\mathbf{I}$, we say that val_0 and val_1 are V^τ-*equivalent* if they coincide in every variable of whichever type, with the possible exception of variable V^τ. We also use the notion of equivalence w.r.t. sets of variables.

Let $\mathbf{I}$ be a σ-structure, and let val be a valuation on $\mathbf{I}$. The notion of *satisfaction* in finite-order logic extends the notion of satisfaction in first-order logic with the following rules:

i. $\mathbf{I}, val \models V^\tau(V_1^{\tau_1}, \ldots, V_r^{\tau_r})$ where $\tau = (\tau_1, \ldots, \tau_r)$ iff
   $(val(V_1^{\tau_1}), \ldots, val(V_r^{\tau_r})) \in val(V^\tau)$.

ii. $\mathbf{I}, val \models \exists V^\tau(\varphi)$ where V^τ is a higher-order variable and φ is a well-formed formula, iff there is a valuation val', which is V^τ-equivalent to val, such that $\mathbf{I}, val' \models \varphi$.

iii. $\mathbf{I}, val \models \forall V^\tau(\varphi)$ where V^τ is a higher-order variable and φ is a well-formed formula, iff for every valuation val', which is V^τ-equivalent to val, it holds that $\mathbf{I}, val' \models \varphi$.

The restriction of finite-order logic to formulae whose variables are all of order $\leq i$, for some $i \geq 1$, is called *i-th order logic* and is denoted by HO^i. Note that, for $i = 1$ this is first-order logic (FO), and for $i = 2$ this is second-order logic (SO). The logics of order $i \geq 2$ are usually known as *higher-order logics* (HO).

As in many other extensions of first-order logic, in second-order logic we can naturally associate a non-negative integer, the *arity*, with each formula. Usually, the arity of a formula of second-order logic is defined as the biggest arity of a second-order variable occurring in that formula. Taking a similar approach, we define the *maximal-arity* of a HO^i formula, $i \geq 2$, as the biggest maximal-arity of any higher-order variable occurring in that formula.

An easy induction using renaming of variables and equivalences such as $\neg \exists V^\tau(\varphi) \equiv \forall V^\tau(\neg \varphi)$ and $(\phi \vee \forall V^\tau(\psi)) \equiv \forall V^\tau(\phi \vee \psi)$ if V^τ is not free in ϕ, shows that each HO^i formula is logically equivalent to an HO^i formula in *prenex normal form*, i.e., to a formula of the form $Q_1 V_1 \ldots Q_n V_n(\varphi)$, where $Q_1, \ldots, Q_n \in \{\forall, \exists\}$, and where $V_1, \ldots, V_n$ are variables of order $\leq i$ and φ is a quantifier-free HO^i formula. Moreover, for every $i \geq 2$, each HO^i formula is logically equivalent to one in prenex normal form in which the quantifiers of order i precede all the remaining quantifiers in the prefix (among others, see [HT03, HT06b] for a detailed proof of this fact). Such normal form is known as *generalized Skolem normal form*, or GSNF. The formulae of finite-order logic which are in GSNF comprise well known hierarchies whose levels are denoted Σ_m^i and Π_m^i. The class Σ_m^i consists of those HO^{i+1} formulae in GSNF in which the quantifiers of order $i+1$ are arranged into at most m alternating blocks, starting with an existential block. Π_m^i is defined dually. Clearly, every HO^{i+1} formula is equivalent to a Σ_m^i formula for some m, and also to a Π_m^i formula.

We define next for each order $i \geq 2$ a hierarchy in HO^i defined in terms of both, alternation of quantification and maximal-arity of the higher-order variables.

DEFINITION 7. (HAA^i **hierarchies**) Let $i, r, m \geq 1$, $HAA^i(r, m)$ is the class of $\Sigma_m^i \cup \Pi_m^i$ formulae in which the higher-order variables of all orders up to $i + 1$ have maximal-arity at most r.

THEOREM 8 ([FT07]). *For every $i \geq 2$ and every $r, m \geq 1$, there are classes of finite structures not definable in $HAA^i(r, m)$ but definable in $HAA^i(r + c(r), m + 2)$ where $c(r) = 1$ for $r > 1$ and $c(r) = 2$ for $r = 1$.*

The problem that allowed us to prove the previous theorem is called *AUTOSAT* and was first introduced by Makowsky and Pnueli in [MP96]. Next we formally define this problem and latter on we study its complexity for the prenex fragments of higher-order logics.

2.2 The *AUTOSAT* Problem

Using *word models* (see [EF99]), it is easy to see that every formula of finite-order logic over a given relational vocabulary σ can be viewed as a finite relational structure of the following vocabulary.

$$\pi(\sigma) = \{<, R_\neg, R_\vee, R_\wedge, R_\exists, R_\forall, R_(, R_), R_=, R_x, R_X, R_|, R_\iota, R_.\} \cup \{R_a : a \in \sigma\}$$

EXAMPLE 9. If $\sigma = \{E\}$ is the vocabulary of graphs and φ is the sentence

$$\exists X_1^{(\iota)}(\exists x_2(X_1^{(\iota)}(x_2) \vee E(x_2, x_2))),$$

which using our notation for the variables corresponds to

$$\exists X(\iota)|(\exists x||(X(\iota)|(x||) \vee E(x||, x||))),$$

then the following $\pi(\sigma)$-structure $\mathbf{I}$ encodes φ.

$$\mathbf{I} = \langle I, <^{\mathbf{I}}, R_\neg^{\mathbf{I}}, R_\vee^{\mathbf{I}}, R_\wedge^{\mathbf{I}}, R_\exists^{\mathbf{I}}, R_\forall^{\mathbf{I}}, R_(^{\mathbf{I}}, R_)^{\mathbf{I}}, R_=^{\mathbf{I}}, R_x^{\mathbf{I}}, R_X^{\mathbf{I}}, R_|^{\mathbf{I}}, R_\iota^{\mathbf{I}}, R_.^{\mathbf{I}}, R_E^{\mathbf{I}} \rangle$$

where $<^{\mathbf{I}}$ is a linear order on I, $|I| = length(\varphi)$, and for each $R_a \in \pi(\sigma)$, $R_a^{\mathbf{I}}$ contains the positions in φ carrying an a,

$$R_a^{\mathbf{I}} = \{b \in I : \text{for some } j \ (1 \leq j \leq |I|), a \text{ is the } j\text{-th symbol in } \varphi, \text{ and}$$

b is the j-th element in the order $<^{\mathbf{I}}$}

Moreover, instead of having a different vocabulary $\pi(\sigma)$ depending on the vocabulary σ of the formulae which we want to encode as relational structures, we can have a vocabulary ρ rich enough to describe formulae of finite-order logic for any arbitrary vocabulary σ. That is, we can fix a vocabulary ρ such that every formula of finite-order logic of whichever vocabulary σ can be viewed as a finite ρ-structure. This can be done as follows. Let ρ be the vocabulary

$$\{<, R_\neg, R_\vee, R_\wedge, R_\exists, R_\forall, R_(, R_), R_x, R_X, R_|, R_\iota, R_,, R_P\}$$

We first identify every formula φ of finite-order logic over an arbitrary vocabulary σ, with a formula φ' over the vocabulary $\sigma' = \{P|^i : 1 \leq i \leq |\sigma| + 1\}$, where, for a predefined bijective function f from $\sigma \cup \{=\}$ to σ', φ' is the formula obtained by replacing in φ each occurrence of a relation symbol $R \in \sigma \cup \{=\}$ by the word $f(R) \in \sigma'$. We then identify every formula φ with the ρ-structure $\mathbf{I}_{\varphi'}$ corresponding to the word model for φ'.

Note that, following the previous schema, even the formulae of finite-order logic over ρ can be viewed as finite ρ-structures. This motivates the following important definition.

DEFINITION 10. Let F be a set of formulae of finite-order logic of vocabulary ρ, let $\rho' = \{P|^i : 1 \leq i \leq 15\}$ and let f be the following bijective function from $\rho \cup \{=\}$ to ρ'.

$$\{< \mapsto P_1, = \mapsto P_2, R_\neg \mapsto P_3, R_\vee \mapsto P_4, \ldots, R_\iota \mapsto P_{13}, R_, \mapsto P_{14}, R_P \mapsto P_{15}\}$$

where, for $1 \leq i \leq 15$, P_i denotes a string of the form $P|^i$. For every formula φ of finite-order logic over ρ, let φ' be the formula obtained by replacing in φ each occurrence of a relation symbol $R \in \rho \cup \{=\}$ by the word $f(R) \in \rho'$. We identify every formula φ with the ρ-structure $\mathbf{I}_{\varphi'}$, where the cardinality of I is the length of φ', $<^{\mathbf{I}_{\varphi'}}$ is a linear order on I, and for each $R_a \in \rho$, $R_a^{\mathbf{I}_{\varphi'}}$ corresponds to the positions in φ' carrying an a, i.e., we identify φ with the word model for φ'.

- We denote by $AUTOSAT(F)$ the set of finite ρ-structures $\mathbf{I}_{\varphi'}$ such that φ is in F and there is a valuation val for which $\mathbf{I}_{\varphi'}, val \models \varphi$.

2.3 A Complete Problem for Higher-Order Logics

Next, we define a problem which is known from [HT05, HT06a] to be complete for the prenex fragments of higher-order logics. We use this problem to show that, for every $i, m \geq 1$, $AUTOSAT(\Sigma_m^i)$ is hard for the fragment

Σ_m^{i+1} of higher-order logic formulae of order $i+2$. The problem is a generalization for higher-order logics of the well known QSAT_m problem which is known to be complete for the corresponding level Σ_m^p of the polynomial-time hierarchy. Note that, by the correspondence between the prenex fragments of second-order logic and the polynomial-time hierarchy (see [Sto76]), for every $m \geq 1$, QSAT_m is also complete for the Σ_m^1 fragment of second-order logic.

DEFINITION 11. Let $\mathbf{B}$ be the *Boolean model* $\langle \{0,1\}, 0^{\mathbf{B}}, 1^{\mathbf{B}} \rangle$ of the *Boolean vocabulary* $\{0,1\}$, i.e., a vocabulary with no relation (or function) symbols, and which has two constant symbols which are interpreted by the two elements, respectively. For $i, m \geq 1$, we denote $\Sigma_m^i\text{-Th}(\mathbf{B})$, the Σ_m^i theory of the Boolean model $\mathbf{B}$, i.e., $\Sigma_m^i\text{-Th}(\mathbf{B})$ is the set of Σ_m^i-formulae of the Boolean vocabulary which are satisfied by $\mathbf{B}$.

THEOREM 12 ([HT05, HT06a]). *For $i, m \geq 1$, it holds that $\Sigma_m^i\text{-}Th(\mathbf{B})$ is complete for Σ_m^{i+1} under polynomial-time reductions.*

Although we do not use it here, we should mention that as shown also in [HT05, HT06a], there are complete problems for all Σ_m^i fragments even under *quantifier-free first-order reductions*, and without assuming the existence of an order in the input structures in such reductions. By using those problems as the interpretation of (relativized) Lindström quantifiers, it is shown there that every fragment Σ_m^i of higher-order logics can be captured with a first-order logic Lindström quantifier. Moreover, a normal form is given showing that each Σ_m^i sentence is equivalent to a single occurrence of the quantifier plus a tuple of quantifier-free first-order formulae.

3 $AUTOSAT(\Sigma_m^1)$ is hard for Σ_m^2

By a polynomial-time reduction from $\Sigma_m^1\text{-Th}(\mathbf{B})$ to $AUTOSAT(\Sigma_m^1[\rho])$, we show in this section that $AUTOSAT(\Sigma_m^1)$ is hard for Σ_m^2. Here ρ is the vocabulary used in the definition of $AUTOSAT$ (see Definition 10).

First, we fix a translation from second-order formulae of Boolean vocabulary to second-order formulae of vocabulary ρ.

DEFINITION 13. Let φ be a any second-order formula of the Boolean vocabulary, we denote $\hat{\varphi}$ the following second-order formula of vocabulary ρ:

$$\exists w_0 w_1 (\psi_{min}(w_0) \wedge \psi_{max}(w_1) \wedge w_0 \neq w_1 \wedge \varphi'(w_0, w_1))$$

where $\psi_{min}(x) \equiv \forall y(\neg y < x)$, $\psi_{max}(x) \equiv \forall y(\neg x < y)$, w_0 and w_1 are first order variables which do not appear in φ, and $\varphi'(w_0, w_1)$ is defined by induction on φ, as follows:

- Let φ be an atomic formula.

    — If φ has the form $s = t$, then $\varphi'(w_0, w_1)$ is

$$f(s) = f(t) \wedge (f(s) = w_0 \vee f(s) = w_1) \wedge (f(t) = w_0 \vee f(t) = w_1)$$

where $f(t)$ ($f(s)$) denotes w_0, w_1 or t (s), depending on whether t (s) is the constant symbol 0, 1 or a first-order variable, respectively.

    — If φ has the form $V(x_1, \ldots, x_r)$, then $\varphi'(w_0, w_1)$ is

$$V(x_1, \ldots, x_r) \wedge (x_1 = w_0 \vee x_1 = w_1) \wedge \ldots \wedge (x_r = w_0 \vee x_r = w_1)$$

- Let φ be a compound formula.

    — If φ has the form $\neg\psi$, then $\varphi'(w_0, w_1)$ is $\neg\psi'(w_0, w_1)$.

    — If φ has the form $\psi \vee \gamma$, then $\varphi'(w_0, w_1)$ is $\psi'(w_0, w_1) \vee \gamma'(w_0, w_1)$.

    — If φ has the form $\psi \wedge \gamma$, then $\varphi'(w_0, w_1)$ is $\psi'(w_0, w_1) \wedge \gamma'(w_0, w_1)$.

    — If φ has the form $\exists x(\psi)$, then $\varphi'(w_0, w_1)$ is

$$\exists x((x = w_0 \vee x = w_1) \wedge \psi'(w_0, w_1))$$

    — If φ has the form $\forall x(\psi)$, then $\varphi'(w_0, w_1)$ is

$$\forall x((x = w_0 \vee x = w_1) \rightarrow \psi'(w_0, w_1))$$

    — If φ has the form $\exists X(\psi)$, then $\varphi'(w_0, w_1)$ is

$$\exists X(\forall x_1 \ldots x_r (X(x_1, \ldots, x_r) \rightarrow (x_1 = w_0 \vee x_1 = w_1) \wedge \ldots \wedge$$
$$(x_r = w_0 \vee x_r = w_1)) \wedge \psi'(w_0, w_1))$$

    — If φ has the form $\forall X(\psi)$, then $\varphi'(w_0, w_1)$ is

$$\forall X(\forall x_1 \ldots x_r (X(x_1, \ldots, x_r) \rightarrow (x_1 = w_0 \vee x_1 = w_1) \wedge \ldots \wedge$$
$$(x_r = w_0 \vee x_r = w_1)) \rightarrow \psi'(w_0, w_1))$$

Now, for each valuation v on the Boolean model $\mathbf{B}$ and each ρ-structure $\mathbf{A}$ with at least two elements, we define a corresponding valuation v_A on $\mathbf{A}$. We define v_A in such a way that for every second-order formula φ, it holds that $\mathbf{B}, v \models \varphi$ iff $\mathbf{A}, v_A \models \hat{\varphi}$.

DEFINITION 14. We define $\mathcal{B}_\rho|_{\geq 2} = \{\mathbf{A}_i \in \mathcal{B}_\rho : |dom(\mathbf{A}_i)| \geq 2\}$, i.e., $\mathcal{B}_\rho|_{\geq 2}$ is the class of ρ-structures with at least two elements. We denote $min(\mathbf{A})$ and $max(\mathbf{A})$ the least and last element, respectively, in the linear order $<^{\mathbf{A}}$ of the ρ-structure $\mathbf{A}$. Let v be a valuation on $\mathbf{B}$, and let $\mathbf{A} \in \mathcal{B}_\rho|_{\geq 2}$, we define the valuation v_A on $\mathbf{A}$, as follows:

- If x is a first-order variable, then

$$v_A(x) = \begin{cases} min(\mathbf{A}) & \text{if } x \text{ is the variable } w_0 \\ max(\mathbf{A}) & \text{if } x \text{ is the variable } w_1 \\ min(\mathbf{A}) & \text{if } x \text{ is a variable other than } w_0 \text{ and } w_1, \text{ and} \\ v(x) = 0^{\mathbf{B}} \\ max(\mathbf{A}) & \text{if } x \text{ is a variable other than } w_0 \text{ and } w_1, \text{ and} \\ v(x) = 1^{\mathbf{B}} \end{cases}$$

- If X is a second-order variable of arity r, then v_A assigns to X an r-ary relation $R \subseteq \{min(\mathbf{A}), max(\mathbf{A})\}^r$ such that $(a_1, \ldots, a_r) \in R$ iff $(g(a_1), \ldots, g(a_r)) \in v(X)$ for $g(a_i) = 0^{\mathbf{B}}$ if $a_i = min(\mathbf{A})$ and $g(a_i) = 1^{\mathbf{B}}$ if $a_i = max(\mathbf{A})$.

It is not difficult to see that a valuation v on the Boolean model satisfies a second-order formula φ iff in every ρ-structure $\mathbf{A}$ with at least two elements the corresponding valuation v_A satisfies $\hat{\varphi}$. We can prove this fact with the following two lemmas.

LEMMA 15. *Let φ be any second-order formula of the Boolean vocabulary, let v be a valuation on $\mathbf{B}$, let $\hat{\varphi}$ be as in Definition 13, and let $\mathcal{B}_\rho|_{\geq 2}$ and v_A be as in Definition 14. If there is an $\mathbf{A} \in \mathcal{B}_\rho|_{\geq 2}$ such that $\mathbf{A}, v_A \models \hat{\varphi}$, then for every $\mathbf{A}_i \in \mathcal{B}_\rho|_{\geq 2}$, it holds that $\mathbf{A}_i, v_A \models \hat{\varphi}$.*

Note that the converse of Lemma 15 is trivially true since $\mathcal{B}_\rho|_{\geq 2}$ is *not* empty. Thus, if for every $\mathbf{A}_i \in \mathcal{B}_\rho|_{\geq 2}$, $\mathbf{A}_i, v_{A_i} \models \hat{\varphi}$, then there is an $\mathbf{A} \in \mathcal{B}_\rho|_{\geq 2}$ such that $\mathbf{A}, v_A \models \hat{\varphi}$.

LEMMA 16. *Let φ be any second-order formula of the Boolean vocabulary, let v be a valuation on $\mathbf{B}$, let $\hat{\varphi}$ be as in Definition 13, and let $\mathcal{B}_\rho|_{\geq 2}$ and v_A be as in Definition 14. Then, it holds that $\mathbf{B}, v \models \varphi$ iff there is an $\mathbf{A} \in \mathcal{B}_\rho|_{\geq 2}$ such that $\mathbf{A}, v_A \models \hat{\varphi}$.*

Because of space restrictions, we omit here the proofs of these lemmas. Nevertheless, both lemmas can be easily proved by induction on φ as shown in [Fer08].

Finally, we show that $AUTOSAT(\Sigma_m^1)$ is hard for Σ_m^2 by a reduction from Σ_m^1-Th($\mathbf{B}$) to $AUTOSAT(\Sigma_m^1[\rho])$. The strategy is to show that for each Σ_m^1-sentence φ of the Boolean vocabulary, we can build in polynomial time $\hat{\varphi}$, a Σ_m^1-sentence of vocabulary ρ, which is equivalent to $\hat{\varphi}$. Since $\hat{\varphi}$ and $\hat{\varphi}$ are equivalent, we get from Lemma 15, Lemma 16, and the fact that $\mathcal{B}_\rho|_{\leq 2}$ is not empty, the following: Let $\varphi \in \Sigma_m^1$ be a sentence of the Boolean vocabulary,

- $\mathbf{B} \models \varphi$ iff, for every ρ-structure $\mathbf{A} \in \mathcal{B}_\rho|_{\leq 2}$, it holds that $\mathbf{A} \models \hat{\varphi}$, and

- $\mathbf{B} \not\models \varphi$ iff, for every ρ-structure $\mathbf{A} \in \mathcal{B}_\rho|_{\leq 2}$, it holds that $\mathbf{A} \not\models \hat{\hat{\varphi}}$.

Given that in particular the ρ-structure $\mathbf{I}_{\hat{\hat{\varphi}}'}$ which encodes $\hat{\hat{\varphi}}$ has more than two elements, $\mathbf{B} \models \varphi$ iff $\mathbf{I}_{\hat{\hat{\varphi}}'} \models \hat{\hat{\varphi}}$.

PROPOSITION 17. *For every $m \geq 1$, $AUTOSAT(\Sigma^1_m[\rho])$ is hard for Σ^2_m under polynomial time reductions.*

Proof. (By a reduction from Σ^1_m-Th$(\mathbf{B})$ to $AUTOSAT(\Sigma^1_m[\rho])$). Given φ, a Σ^1_m-sentence of the Boolean vocabulary, we produce in polynomial time $\hat{\hat{\varphi}}$, a sentence in $\Sigma^1_m[\rho]$, such that $\varphi \in \Sigma^1_m$-Th$(\mathbf{B})$ iff $\mathbf{I}_{\hat{\hat{\varphi}}'} \in AUTOSAT(\Sigma^1_m[\rho])$, or equivalently, $\mathbf{B} \models \varphi$ iff $\mathbf{I}_{\hat{\hat{\varphi}}'} \models \hat{\hat{\varphi}}$.

Let $\hat{\varphi}$ be the sentence obtained from φ as indicated in Definition 13. We build $\hat{\hat{\varphi}}$ from $\hat{\varphi}$ and φ. First, we eliminate all second-order quantifiers in $\varphi'(w_0, w_1)$, obtaining $\varphi''(X_{11}, \ldots, X_{1s_1}, X_{21}, \ldots, X_{2s_2}, \ldots, X_{m1}, \ldots, X_{ms_m}, w_0, w_1)$, where for $1 \leq j \leq m$, $s_j \geq 1$, and $X_{11}, \ldots, X_{ms_m}$ is the set of second-order variables which appear free in φ''. Note that, the set of second-order variables which appear free in φ'' (i.e., all variables in φ''), coincides with the set of all second-order variables in $\varphi'(w_0, w_1)$, which in turn coincides with the set of second-order variables in the Σ^1_m-sentence φ.

We then write $\hat{\hat{\varphi}}$ as

$$\exists X_{11} \ldots \exists X_{1s_1} \forall X_{21} \ldots \forall X_{2s_2} \ldots Q X_{m1} \ldots Q X_{ms_m} \left(\exists w_0 w_1 (\psi_{min}(w_0) \wedge \psi_{max}(w_1) \right.$$

$$\wedge w_0 \neq w_1 \wedge \varphi''(X_{11}, \ldots, X_{1s_1}, X_{21}, \ldots, X_{2s_2}, \ldots, X_{m1}, \ldots, X_{ms_m}, w_0, w_1)),$$

where $\exists X_{11} \ldots \exists X_{1s_1} \forall X_{21} \ldots \forall X_{2s_2} \ldots Q X_{m1} \ldots Q X_{ms_m}$ is the prefix of second-order quantifiers in φ.

Clearly $\hat{\hat{\varphi}}$ can be built in polynomial time, and given φ followed by $\hat{\varphi}$ as input, $\hat{\hat{\varphi}}$ can be built in linear time.

Also clearly $\hat{\hat{\varphi}}$ is a $\Sigma^1_m[\rho]$-sentence. Furthermore, $\hat{\hat{\varphi}}$ is equivalent to $\hat{\varphi}$. This can be seen by noting that, if α is a formula in $\Sigma^1_m[\rho] \cup \Pi^1_m[\rho]$, then for every ρ-structure $\mathbf{A}$ and valuation v on $\mathbf{A}$,

$$\mathbf{A}, v \models \alpha'(w_0, w_1) \text{ iff } \mathbf{A}, v \models Q_1 X_1 \ldots Q_k X_k(\alpha''(X_1, \ldots, X_k, w_0, w_1)), \quad (1)$$

where $\alpha''(X_1, \ldots, X_k, w_0, w_1)$ is the formula obtained by eliminating all second-order quantifiers in $\alpha'(w_0, w_1)$, and $Q_1 X_1 \ldots Q_k X_k$ is the prefix of second-order quantifiers of α.

To prove (1), we use induction on α. If α is an atomic formula, or α is of the form $\neg\psi$, $\psi \vee \gamma$, $\psi \wedge \gamma$, $\exists x(\psi)$ or $\forall x(\psi)$, then $Q_1 X_1 \ldots Q_k X_k(\alpha''(X_1, \ldots, X_k, w_0, w_1))$ and $\alpha'(w_0, w_1)$ are exactly the same formula, and our proposition is trivially true.

If α is of the form $\exists X(\psi)$, then $\alpha'(w_0, w_1)$ is $\exists X(\forall x_1 \ldots x_r(X(x_1, \ldots, x_r) \rightarrow (x_1 = w_0 \vee x_1 = w_1) \wedge \ldots \wedge (x_r = w_0 \vee x_r = w_1)) \wedge \psi'(w_0, w_1))$. We want to show that this is equivalent to $\exists X Q_1 X_1 \ldots Q_k X_k(\alpha''(X, X_1, \ldots, X_k, w_0, w_1))$, where $\alpha''(X, X_1, \ldots, X_k, w_0, w_1)$ is the formula obtained by eliminating all second-order quantifiers in $\alpha'(w_0, w_1)$, and $\exists X Q_1 X_1 \ldots Q_k X_k$ is the prefix of second-order quantifiers of α. If $\mathbf{A}, v \models \alpha'(w_0, w_1)$ then there is a valuation v', which is $\{X\}$-equivalent to v, such that $\mathbf{A}, v' \models \forall x_1 \ldots x_r(X(x_1, \ldots, x_r) \rightarrow (x_1 = w_0 \vee x_1 = w_1) \wedge \ldots \wedge (x_r = w_0 \vee x_r = w_1)) \wedge \psi'(w_0, w_1)$. Since $\mathbf{A}, v' \models \psi'(w_0, w_1)$, it follows by induction hypothesis that, $\mathbf{A}, v' \models Q_1 X_1 \ldots Q_k X_k(\psi''(X_1, \ldots, X_k, w_0, w_1))$, where $\psi''(X_1, \ldots, X_k, w_0, w_1)$ is the formula obtained by eliminating all second-order quantifiers in $\psi'(w_0, w_1)$, and $Q_1 X_1 \ldots Q_k X_k$ is the prefix of second-order quantifiers of ψ. Given that, none of the second order variables $X_1, \ldots, X_k$ appear free in the formula $\forall x_1 \ldots x_r(X(x_1, \ldots, x_r) \rightarrow (x_1 = w_0 \vee x_1 = w_1) \wedge \ldots \wedge (x_r = w_0 \vee x_r = w_1))$, we have that $\mathbf{A}, v' \models Q_1 X_1 \ldots Q_k X_k(\forall x_1 \ldots x_r (X(x_1, \ldots, x_r) \rightarrow (x_1 = w_0 \vee x_1 = w_1) \wedge \ldots \wedge (x_r = w_0 \vee x_r = w_1)) \wedge \psi''(X_1, \ldots, X_k, w_0, w_1))$. Therefore, $\mathbf{A}, v \models \exists X Q_1 X_1 \ldots Q_k X_k(\alpha''(X, X_1, \ldots, X_k, w_0, w_1))$.

On the other hand, if $\mathbf{A}, v \models \exists X Q_1 X_1 \ldots Q_k X_k(\alpha''(X, X_1, \ldots, X_k, w_0, w_1))$, then there is a valuation v', which is $\{X\}$-equivalent to v, such that $\mathbf{A}, v' \models Q_1 X_1 \ldots Q_k X_k(\forall x_1 \ldots x_r(X(x_1, \ldots, x_r) \rightarrow (x_1 = w_0 \vee x_1 = w_1) \wedge \ldots \wedge (x_r = w_0 \vee x_r = w_1)) \wedge \psi''(X_1, \ldots, X_k, w_0, w_1))$, where $\psi''(X_1, \ldots, X_k, w_0, w_1)$ is the formula obtained by eliminating all second-order quantifiers in $\psi'(w_0, w_1)$, and $Q_1 X_1 \ldots Q_k X_k$ is the prefix of second-order quantifiers of ψ. It follows that $\mathbf{A}, v' \models Q_1 X_1 \ldots Q_k X_k(\psi''(X_1, \ldots, X_k, w_0, w_1))$, and by induction hypothesis that $\mathbf{A}, v' \models \psi'(w_0, w_1)$. Since the valuation v' on $\mathbf{A}$ also satisfies, $\forall x_1 \ldots x_r(X(x_1, \ldots, x_r) \rightarrow (x_1 = w_0 \vee x_1 = w_1) \wedge \ldots \wedge (x_r = w_0 \vee x_r = w_1))$, we have that $\mathbf{A}, v \models \alpha'(w_0, w_1)$.

If α is a formula of the form $\forall X(\psi)$, the proof is completely analogous to the previous case.

Summarising, for each Σ_m^1-sentence φ of the Boolean vocabulary, we can build in polynomial time $\hat{\varphi}$, a Σ_m^1-sentence of vocabulary ρ, which is equivalent to $\hat{\varphi}$. Since $\hat{\varphi}$ and $\hat{\hat{\varphi}}$ are equivalent, we get by Lemma 16 that $\mathbf{B} \models \varphi$ iff there is an $\mathbf{A} \in \mathcal{B}_\rho|_{\geq 2}$, such that $\mathbf{A} \models \hat{\hat{\varphi}}$. And by Lemma 15 and the fact that $\mathcal{B}_\rho|_{\leq 2}$ is not empty, we also get that there is an $\mathbf{A} \in \mathcal{B}_\rho|_{\geq 2}$, such that $\mathbf{A} \models \hat{\hat{\varphi}}$ iff for every $\mathbf{A}_i \in \mathcal{B}_\rho|_{\geq 2}$, $\mathbf{A}_i \models \hat{\hat{\varphi}}$. Given that for every $\varphi \in \Sigma_m^1[\rho]$, we have that $\mathbf{I}_{\hat{\varphi}'} \in \mathcal{B}_\rho|_{\geq 2}$, it clearly follows that $\mathbf{B} \models \varphi$ iff $\mathbf{I}_{\hat{\varphi}'} \models \hat{\hat{\varphi}}$. ∎

4 $AUTOSAT(\Sigma_m^i)$ is hard for Σ_m^{i+1}

In this section, we show how to extend to each order $i \geq 1$, the result proved in the previous section. That is, for $i \geq 1$, we sketch a polynomial-time reduction from Σ_m^i-Th($\mathbf{B}$) to $AUTOSAT(\Sigma_m^i[\rho])$.

Let $\tau_{j,k} = \underbrace{(\tau_{i,r}, \ldots, \tau_{i,r})}_{r}$ where, for $j, k \geq 1$,

- $\tau_{j,k} = \iota$ if $j = 1$ and

- $\tau_{j,k} = \underbrace{(\tau_{j-1,k}, \ldots, \tau_{j-1,k})}_{k}$ if $j > 1$.

Note that by the definition of higher-order logic used in [HT05, HT06a] (where Theorem 12 was proved), we do not need to consider higher-order variables of a higher-order type other than the types in $\{\tau_{i,r} : i, r \geq 1\}$, since these are the only kind of higher-order variables allowed in the formulae in Σ_m^i in those papers.

The corresponding translation from higher-order formulae of Boolean vocabulary to higher-order formulae of vocabulary ρ is as follows.

DEFINITION 18. Let φ be any higher-order formula of the Boolean vocabulary such that for every higher-order variable V^τ in φ, we have that $\tau = \tau_{j,k}$ for some $j, k \geq 1$. We denote $\hat{\varphi}$ the following higher-order formula of vocabulary ρ:

$$\exists w_0 w_1(\psi_{min}(w_0) \wedge \psi_{max}(w_1) \wedge w_0 \neq w_1 \wedge \varphi'(w_0, w_1))$$

where $\psi_{min}(x) \equiv \forall y(\neg y < x)$, $\psi_{max}(x) \equiv \forall y(\neg x < y)$, w_0 and w_1 are first order variables which do not appear in φ, and $\varphi'(w_0, w_1)$ is defined by induction on φ by the rules in Definition 13 for second-order formulae extended with the following rules:

- If φ has the form $\mathcal{X}^{\tau_{i,r}}(X_1^{\tau_{i-1,r}}, \ldots, X_r^{\tau_{i-1,r}})$, for some $i \geq 3$ and $r \geq 1$, then $\varphi'(w_0, w_1)$ is $\mathcal{X}^{\tau_{i,r}}(X_1^{\tau_{i-1,r}}, \ldots, X_r^{\tau_{i-1,r}}) \wedge \psi_{\tau_{i-1,r}}(X_1^{\tau_{i-1,r}}) \wedge \cdots \wedge \psi_{\tau_{i-1,r}}(X_r^{\tau_{i-1,r}})$, where for $i = 3$,

  $\psi_{\tau_{i-1,r}}(X^{\tau_{i-1,r}}) \equiv \forall x_1 \ldots x_r(X^{\tau_{i-1,r}}(x_1, \ldots, x_r) \rightarrow$
  $((x_1 = w_0 \vee x_1 = w_1) \wedge \ldots \wedge (x_r = w_0 \vee x_r = w_1)))$,

  and for $i > 3$,

  $\psi_{\tau_{i-1,r}}(X^{\tau_{i-1,r}}) \equiv \forall X_1^{\tau_{i-2,r}} \ldots X_r^{\tau_{i-2,r}}(X^{\tau_{i-1,r}}(X_1^{\tau_{i-2,r}} \ldots X_r^{\tau_{i-2,r}}) \rightarrow$
  $(\psi_{\tau_{i-2,r}}(X_1^{\tau_{i-2,r}}) \wedge$
  $\cdots \wedge \psi_{\tau_{i-2,r}}(X_r^{\tau_{i-2,r}})))$.

- If φ has the form $\exists \mathcal{X}^{\tau_{i,r}}(\alpha)$, for some $i \geq 3$ and $r \geq 1$, then $\varphi'(w_0, w_1)$ is

$$\exists \mathcal{X}^{\tau_{i,r}} \left(\forall X_1^{\tau_{i-1,r}} \ldots X_r^{\tau_{i-1,r}} \left(\mathcal{X}^{\tau_{i,r}}(X_1^{\tau_{i-1,r}}, \ldots, X_r^{\tau_{i-1,r}}) \rightarrow \right. \right.$$
$$(\psi_{\tau_{i-1,r}}(X_1^{\tau_{i-1,r}}) \wedge$$
$$\left. \left. \cdots \wedge \psi_{\tau_{i-1,r}}(X_r^{\tau_{i-1,r}}))) \wedge \right. \right.$$
$$\alpha'(w_0, w_1)),$$

were $\psi_{\tau_{i-1,r}}(X^{\tau_{i-1,r}})$ is defined exactly as in the previous rule.

- If φ has the form $\forall \mathcal{X}^{\tau_{i,r}}(\alpha)$, for some $i \geq 3$ and $r \geq 1$, then $\varphi'(w_0, w_1)$ is

$$\forall \mathcal{X}^{\tau_{i,r}} \left(\forall X_1^{\tau_{i-1,r}} \ldots X_r^{\tau_{i-1,r}} \left(\mathcal{X}^{\tau_{i,r}}(X_1^{\tau_{i-1,r}}, \ldots, X_r^{\tau_{i-1,r}}) \rightarrow \right. \right.$$
$$(\psi_{\tau_{i-1,r}}(X_1^{\tau_{i-1,r}}) \wedge$$
$$\left. \left. \cdots \wedge \psi_{\tau_{i-1,r}}(X_r^{\tau_{i-1,r}}))) \rightarrow \right. \right.$$
$$\alpha'(w_0, w_1)),$$

were again $\psi_{\tau_{i-1,r}}(X^{\tau_{i-1,r}})$ is defined as in the previous rules.

Now, we extend Definition 14 of v_A in such a way that for every higher-order formula φ of the Boolean vocabulary in which the higher-order variables have the form $V^{\tau_{j,k}}$ for some $j, k \geq 1$, it holds that $\mathbf{B}, v \models \varphi$ iff $\mathbf{A}, v_A \models \hat{\varphi}$ in every ρ-structure $\mathbf{A}$ with at least two elements.

DEFINITION 19. Let $\mathcal{B}_\rho|_{\geq 2}$, $min(\mathbf{A})$ and $max(\mathbf{A})$ be as in Definition 14. Let v be a valuation on $\mathbf{B}$, let $\mathbf{A} \in \mathcal{B}_\rho|_{\geq 2}$ and let $U = \{min(\mathbf{A}, max(\mathbf{A}))\}$. For $i, r \geq 1$ and $a \in U_{\tau_{i,r}}$ an object of type $\tau_{i,r}$, let $g_{\tau_{1,r}}(a)$ denote $0^{\mathbf{B}}$ if $a = min(\mathbf{A})$ and $1^{\mathbf{B}}$ if $a = max(\mathbf{A})$, and let $g_{\tau_{i,r}}(a)$ denote $\{(b_1, \ldots, b_r) \in \{0^{\mathbf{B}}, 1^{\mathbf{B}}\}_{\tau_{i,r}} : \text{there is a } (c_1, \ldots, c_r) \in a \text{ such that, for } 1 \leq j \leq r, g_{\tau_{i-1,r}}(c_j) = b_j\}$ for $i > 1$. We define the valuation v_A on $\mathbf{A}$, by adding the following case to Definition 14:

- If $\mathcal{X}^{\tau_{i,r}}$ is a higher-order variable of order $i \geq 2$ and type $\tau_{i,r}$, then v_A assigns to $\mathcal{X}^{\tau_{i,r}}$ an r-ary higher-order relation $\mathcal{R} \in U_{\tau_{i,r}}$ such that

$$(a_1, \ldots, a_r) \in \mathcal{R} \quad \text{iff} \quad (g_{\tau_{i-1,r}}(a_1), \ldots, g_{\tau_{i-1,r}}(a_r)) \in v(\mathcal{X}^{\tau_{i,r}}).$$

Under these definitions, it is a routine task to prove the following analogous results to Lemma 15 and 16.

LEMMA 20. *Let φ be any higher-order formula of the Boolean vocabulary such that for every higher-order variable V^τ in φ, we have that $\tau = \tau_{j,k}$ for some $j, k \geq 1$. Let v be a valuation on $\mathbf{B}$, let $\hat{\varphi}$ be as in Definition 18, and let $\mathcal{B}_\rho|_{\geq 2}$ and v_A be as in Definition 19. If there is an $\mathbf{A} \in \mathcal{B}_\rho|_{\geq 2}$ such that $\mathbf{A}, v_A \models \hat{\varphi}$, then for every $\mathbf{A}_i \in \mathcal{B}_\rho|_{\geq 2}$, it holds that $\mathbf{A}_i, v_{A_i} \models \hat{\varphi}$.*

As with Lemma 15, the converse of this lemma is trivially true since $\mathcal{B}_\rho|_{\geq 2}$ is *not* empty. Thus, if for every $\mathbf{A}_i \in \mathcal{B}_\rho|_{\geq 2}$, $\mathbf{A}_i, v_{A_i} \models \hat{\varphi}$, then there is an $\mathbf{A} \in \mathcal{B}_\rho|_{\geq 2}$ such that $\mathbf{A}, v_A \models \hat{\varphi}$.

LEMMA 21. *Let φ be any higher-order formula of the Boolean vocabulary such that for every higher-order variable V^τ in φ, we have that $\tau = \tau_{j,k}$ for some $j, k \geq 1$. Let v be a valuation on $\mathbf{B}$, let $\hat{\varphi}$ be as in Definition 18, and let $\mathcal{B}_\rho|_{\geq 2}$ and v_A be as in Definition 19. Then, it holds that $\mathbf{B}, v \models \varphi$ iff there is an $\mathbf{A} \in \mathcal{B}_\rho|_{\geq 2}$ such that $\mathbf{A}, v_A \models \hat{\varphi}$.*

Thus, if φ is a higher-order formula of the Boolean vocabulary such that for every higher-order variable V^τ in φ it holds that $\tau = \tau_{j,k}$ for some $j, k \geq 1$ (recall that by the results in [HT05, HT06a] these are the only higher-order formulae that we need to consider here), then valuation v on the Boolean model satisfies φ iff in every ρ-structure $\mathbf{A}$ with at least two elements the corresponding valuation v_A satisfies $\hat{\varphi}$.

The strategy to show that $AUTOSAT(\Sigma_m^i)$ is hard for Σ_m^{i+1} is also completely analogous to the strategy for second-order logic that we gave in the previous section.

PROPOSITION 22. *For every $m, i \geq 1$, $AUTOSAT(\Sigma_m^i[\rho])$ is hard for Σ_m^{i+1} under polynomial time reductions.*

Proof. (Sketch). We proceed by polynomial time reduction from Σ_m^i-Th($\mathbf{B}$) to $AUTOSAT(\Sigma_m^i[\rho])$.

The strategy is to show that for each Σ_m^i-sentence φ of the Boolean vocabulary such that for every higher-order variable V^τ in φ it holds that $\tau = \tau_{j,k}$ for some $1 < j \leq i + 1$ and $k \geq 1$ (again recall that by the results in [HT05, HT06a] these are the only higher-order formulae that we need to consider here), we can build in polynomial time $\hat{\varphi}$, a Σ_m^i-sentence of vocabulary ρ, which is equivalent to $\hat{\varphi}$. Since $\hat{\varphi}$ and $\hat{\varphi}$ are equivalent, we get from Lemma 20, Lemma 21, and the fact that $\mathcal{B}_\rho|_{\leq 2}$ is not empty, the following: Let $\varphi \in \Sigma_m^i$ be a sentence of the Boolean vocabulary such that for every higher-order variable V^τ in φ it holds that $\tau = \tau_{j,k}$ for some $1 < j \leq i + 1$ and $k \geq 1$.

- $\mathbf{B} \models \varphi$ iff, for every ρ-structure $\mathbf{A} \in \mathcal{B}_\rho|_{\leq 2}$, it holds that $\mathbf{A} \models \hat{\varphi}$, and

- $\mathbf{B} \not\models \varphi$ iff, for every ρ-structure $\mathbf{A} \in \mathcal{B}_\rho|_{\leq 2}$, it holds that $\mathbf{A} \not\models \hat{\varphi}$.

Given that in particular the ρ-structure $\mathbf{I}_{\hat{\varphi}'}$ which encodes $\hat{\varphi}$ has more than two elements, $\mathbf{B} \models \varphi$ iff $\mathbf{I}_{\hat{\varphi}'} \models \hat{\varphi}$.

So, we only need to show that given φ, a Σ_m^i-sentence of the Boolean vocabulary, we can produce $\hat{\varphi}$ in polynomial time. Let $\hat{\varphi}$ be the sentence obtained from φ as indicated in Definition 18. We build $\hat{\hat{\varphi}}$ from $\hat{\varphi}$ and φ. First, we eliminate all quantifiers of order $i+1$ in $\varphi'(w_0, w_1)$, obtaining $\varphi''(\mathcal{X}_{11}, \ldots, \mathcal{X}_{1s_1}, \mathcal{X}_{21}, \ldots, \mathcal{X}_{2s_2}, \ldots, \mathcal{X}_{m1}, \ldots, \mathcal{X}_{ms_m}, w_0, w_1)$, where for $1 \leq j \leq m$, $s_j \geq 1$, and $\mathcal{X}_{11}, \ldots, \mathcal{X}_{ms_m}$ is the set of variables of order $i+1$ which appear free in φ''. Note that, the set of all variables of order $i+1$ which appear free in φ'', coincides with the set of all variables of order $i+1$ in $\varphi'(w_0, w_1)$, which in turn coincides with the set of variables of order $i+1$ in the Σ_m^i-sentence φ.

We then write $\hat{\hat{\varphi}}$ as

$$\exists \mathcal{X}_{11} \ldots \exists \mathcal{X}_{1s_1} \forall \mathcal{X}_{21} \ldots \forall \mathcal{X}_{2s_2} \ldots Q\mathcal{X}_{m1} \ldots Q\mathcal{X}_{ms_m} \left(\exists w_0 w_1 (\psi_{min}(w_0) \wedge \psi_{max}(w_1) \right.$$

$$\left. \wedge w_0 \neq w_1 \wedge \varphi''(\mathcal{X}_{11}, \ldots, \mathcal{X}_{1s_1}, \mathcal{X}_{21}, \ldots, \mathcal{X}_{2s_2}, \ldots, \mathcal{X}_{m1}, \ldots, \mathcal{X}_{ms_m}, w_0, w_1) \right),$$

where $\exists \mathcal{X}_{11} \ldots \exists \mathcal{X}_{1s_1} \forall \mathcal{X}_{21} \ldots \forall \mathcal{X}_{2s_2} \ldots Q\mathcal{X}_{m1} \ldots Q\mathcal{X}_{ms_m}$ is the prefix of quantifiers of order $i+1$ in φ.

Clearly $\hat{\varphi}$ can be built in polynomial time, and given φ followed by $\hat{\varphi}$ as input, $\hat{\hat{\varphi}}$ can be built in linear time.

Also clearly $\hat{\hat{\varphi}}$ is a $\Sigma_m^i[\rho]$-sentence. Furthermore, $\hat{\hat{\varphi}}$ is equivalent to $\hat{\varphi}$. This can be seen by noting that, if α is a formula in $\Sigma_m^i[\rho] \cup \Pi_m^i[\rho]$, then for every ρ-structure $\mathbf{A}$ and valuation v on $\mathbf{A}$,

$$\mathbf{A}, v \models \alpha'(w_0, w_1) \text{ iff } \mathbf{A}, v \models Q_1 \mathcal{X}_1 \ldots Q_k \mathcal{X}_k (\alpha''(cal X_1, \ldots, \mathcal{X}_k, w_0, w_1)), \quad (1)$$

where $\alpha''(\mathcal{X}_1, \ldots, \mathcal{X}_k, w_0, w_1)$ is the formula obtained by eliminating all quantifiers of order $i+1$ in $\alpha'(w_0, w_1)$, and $Q_1 \mathcal{X}_1 \ldots Q_k \mathcal{X}_k$ is the prefix of quantifiers of order $i+1$ of α.

Note that (1) can be proved using induction on α and following the same strategy as in the proof of Proposition 17. ∎

5 $AUTOSAT(\Sigma_m^i)$ is in Σ_m^{i+1}

To prove that for every $i, m \geq 1$, $AUTOSAT(\Sigma_m^i[\rho])$ is in Σ_m^{i+1}, we build a *nondeterministic* Turing machine with an oracle in Σ_{m-1}^p which decides in time $\mathcal{O}(\exp_i(n^c))$ whether an arbitrary input ρ-structure $\mathbf{I}_{\varphi'}$ is in $AUTOSAT(\Sigma_m^i[\rho])$.

We need the following fact for our proof.

FACT 23. Fix $r \geq 1$ and a finite set U with $|U| = n \geq 1$. For any $i \geq 1$ and any type τ such that $2 \leq order(\tau) \leq i+1$ and $ma(\tau) \leq r$, the number of objects of type τ over U (i.e., $|U_\tau|$), is bounded above by $\exp_i(c \cdot n^r)$ for some constant c.

Proof. By induction on i. Let $i = 1$. For every type $\tau \in typ(i+1, r)$, it holds that $|U_\tau| \leq |U_{\iota^r}|$. Since $U_{\iota^r} = \mathcal{P}(\prod_{j=1}^r U_\iota) = \mathcal{P}(\prod_{j=1}^r U)$, we get that $|U_{\iota^r}| = 2^{n^r} = \exp_i(n^r)$.

For the inductive step $(i > 1)$, let $\tau' = (\underbrace{\tau_{i,r}, \ldots, \tau_{i,r}}_{r})$ where, for $j, k \geq 1$,

- $\tau_{j,k} = \iota$ if $j = 1$ and

- $\tau_{j,k} = (\underbrace{\tau_{j-1,k}, \ldots, \tau_{j-1,k}}_{k})$ if $j > 1$.

Clearly, for every type $\tau \in typ(i+1, r)$, it holds that $|U_\tau| \leq |U_{\tau'}|$. Since $U_{\tau'} = \mathcal{P}(\prod_{j=1}^{r} U_{\tau_{i,r}})$ and by inductive hypothesis, $|U_{\tau_{i,r}}| \leq \exp_{i-1}(c \cdot n^r)$ for some constant c, we get that $|U_{\tau'}| \leq 2^{r \cdot \exp_{i-1}(c \cdot n^r)} \leq \exp_i(c' \cdot n^r)$ for some constant c'. ∎

Note that by the results in [HT03, HT06b], Σ_m^i captures $\bigcup_{c \in \mathbb{N}} \text{NTIME}(\exp_{i-1}(n^c))^{\Sigma_{m-1}^p}$, where n is the size of the input and Σ_m^p the m-th level of the polynomial-time hierarchy. Not surprisingly then the proof of the following result is similar to the proof of Theorem 3.3 in [HT06b].

PROPOSITION 24. *For every $i, m \geq 1$, $AUTOSAT(\Sigma_m^i[\rho])$ is in Σ_m^{i+1}.*

Proof. We define a Turing machine $\mathcal{M}$ which decides whether an arbitrary input ρ-structure $\mathbf{I}_{\varphi'}$ encodes a Σ_m^i-sentence φ of vocabulary ρ such that $\mathbf{I}_{\varphi'} \models \varphi$, and which works in $\text{NTIME}(\exp_i(n^c))^{\Sigma_{m-1}^p}$.

Note that the number of variables in φ as well as the arity of the quantified higher-order variables, is $\mathcal{O}(n)$, where n is the size of (the domain of) the input structure $\mathbf{I}_{\varphi'}$.

Let φ be a Σ_m^i-formula, we can think of φ as a formula where all higher-order quantifiers are existential, and are grouped together at the beginning with $m - 1$ interleaving negation symbols. This is clearly possible by the well known relationship between existential and universal quantifiers.

Let the ρ-structure $\mathbf{I}_{\varphi'}$ be the input to $\mathcal{M}$ and n the size of $\mathbf{I}_{\varphi'}$. The machine works as follows:

1. $\mathcal{M}$ writes in its work tape the formula φ encoded by $I_{\varphi'}$. This requires space $\mathcal{O}(n)$ and can be done working deterministically in polynomial time.

2. $\mathcal{M}$ deterministically checks that φ is a well formed sentence in $\Sigma_m^i[\rho]$. Again this takes polynomial time.

3. Let φ be $\exists X_1^{\tau_1} \ldots \exists X_{s_1}^{\tau_{s_1}}(\psi)$, where $s_1, r \geq 1$, for $1 \leq j \leq s_1$ $X_j^{\tau_j}$ is a higher-order variable of type τ_j with $order(\tau_j) = i+1$ and $ma(\tau_j) \leq r$, and ψ is either an i-th order formula or the *negation* of a formula

which starts with a block of existential quantifiers of order $i + 1$. $\mathcal{M}$ guesses a sequence of (higher-order) relations as possible values for the (higher-order) variables $X_1^{\tau_1}, \ldots, X_{s_1}^{\tau_{s_1}}$, respectively.

By Fact 23, if we use the following inductively defined representation of objects of type τ as binary strings, then the space needed for the guessed relations is $\mathcal{O}(s_1 \cdot \exp_{i-1}(c \cdot n^r))$ for some constant c.

For the induction base, we use the order $<^{\mathbf{I}_{\varphi'}}$ to encode the elements of $dom(\mathbf{I}_{\varphi'})$ using the usual encoding as binary strings of length $\log n$; and we encode a second-order relation $R^{\mathbf{I}_{\varphi'}}$ in $dom(\mathbf{I}_{\varphi'})^k$ ($k \geq 1$) as a bit string of length n^k, where "1" in a given position indicates that the corresponding tuple in the lexicographical order of tuples induced by $<^{\mathbf{I}_{\varphi'}}$ in $dom(\mathbf{I}_{\varphi'})^k$ is in $R^{\mathbf{I}_{\varphi'}}$. For $\tau = (\tau_1, \ldots, \tau_k)$, assume that we have already defined a representation of objects of type τ_m ($m = 1, \ldots, k$) as binary strings. This induces a lexicographical ordering on the set of k-tuples $(R_1, \ldots, R_k) \in dom(\mathbf{I}_{\varphi'})_{\tau_1} \times \cdots \times dom(\mathbf{I}_{\varphi'})_{\tau_k}$ (i.e., set of k-tuples $(R_1, \ldots, R_k)$ such that for $1 \leq m \leq k$, R_m is an object of type τ_m). An object $\mathcal{R}$ of type τ, can therefore be encoded by a bit string $s_{\mathcal{R}}$ where again "1" in a given position indicates that the corresponding tuple in the lexicographical order of tuples in $dom(\mathbf{I}_{\varphi'})_{\tau_1} \times \cdots \times dom(\mathbf{I}_{\varphi'})_{\tau_k}$ is in $\mathcal{R}$. By induction, it is possible to prove that the lengths of the strings encoding higher-order relations are as claimed.

Since the length of φ is bounded by n, we know that r and s_1 are also bounded by n. Therefore, the sequence of guessed (higher-order) relations needs space bounded by $\mathcal{O}(n \cdot \exp_{i-1}(c \cdot n^n))$, i.e., $\mathcal{M}$ needs space $\mathcal{O}(\exp_i(n^{c'}))$ for some constant c'. The nondeterministic time needed for this step is also $\mathcal{O}(\exp_i(n^{c'}))$.

4. Now $\mathcal{M}$ evaluates the formula ψ substituting the guessed relations for the higher-order variables $X_1^{\tau_1}, \ldots, X_{s_1}^{\tau_{s_1}}$, respectively. According to what we said above regarding the formula ψ, for its evaluation we have to consider two different cases:

(a) ψ is a i-th order formula:
Then $\mathcal{M}$ evaluates ψ deterministically, which takes time polynomial in the size of the guessed relations. Thus, this step takes time $(\mathcal{O}(\exp_i(n^c)))^{c_2}$ for some constants c and c_2, which is still $\mathcal{O}(\exp_i(n^{c'}))$ for some constant c'.
Note that, to evaluate ψ we do not need to use oracles (as we do in the second case, see below), since as the quantified variables in ψ are variables of order i, we can afford the time needed

to evaluate the negation of an existential quantifier *deterministically* as a universal quantifier (note that by Fact 23 there are $\mathcal{O}(\exp_{i-1}(c \cdot n^r))$ higher-order relations of order i and arity r). So, the evaluation of the formula ψ is done completely in a deterministic way, and still in time $\mathcal{O}(\exp_i(n^c))$ for some constant c.

(b) ψ is the *negation* of a formula which starts with a block of existential quantifiers of order $i + 1$:

Say, $\psi \equiv \neg(\exists Y_1^{\tau_1'} \ldots \exists Y_{s_2}^{\tau_{s_2}'}(\psi'))$, where $s_2, r' \geq 1$, for $1 \leq j \leq s_2$ $X_j^{\tau_j'}$ is a higher-order variable of type τ_j' with $order(\tau_j) = i + 1$ and $ma(\tau_j) \leq r'$.

Then, $\mathcal{M}$ calls an oracle Turing machine $\mathcal{M}_\psi$ which will evaluate the sub-formula $\exists Y_1^{\tau_1'} \ldots \exists Y_{s_2}^{\tau_{s_2}'}(\psi')$ over the structure $\mathbf{I}_{\varphi'}$ extended with the relations of order $i + 1$ guessed by $\mathcal{M}$. When $\mathcal{M}_\psi$ ends, the machine $\mathcal{M}$ proceeds by inverting the result of the computation of $\mathcal{M}_\psi$. That is, $\mathcal{M}$ accepts the input structure $\mathbf{I}_{\varphi'}$ iff $\mathcal{M}_\psi$ rejects its input.

The way in which the oracle machine $\mathcal{M}_\psi$ works is exactly the same as the way in which the original machine $\mathcal{M}$ works. First, $\mathcal{M}_\psi$ guesses a sequence of (higher-order) relations as possible values for the (higher-order) variables $Y_1^{\tau_1'}, \ldots, Y_{s_2}^{\tau_{s_2}'}$, respectively. Then, $\mathcal{M}_\psi$ evaluates the formula ψ' substituting the guessed (higher-order) relations for the variables $Y_1^{\tau_1'}, \ldots, Y_{s_2}^{\tau_{s_2}'}$, respectively. If in the quantifier prefix of the sub-formula ψ' there is a negation symbol before an existential quantifier of order $i + 1$, then $\mathcal{M}_\psi$ calls in turn another oracle Turing machine which will work like $\mathcal{M}$ and $\mathcal{M}_\psi$.

This process is followed until there are no more negations before existential quantifiers of order $i+1$ in the remaining sub-formulae. Note that the oracle machine $\mathcal{M}_\psi$ needs the same space and time as the original machine $\mathcal{M}$, i.e., $\mathcal{O}(\exp_i(n^c))$ for some constant c, where n is the size of $\mathbf{I}_{\varphi'}$, which is the input structure to the machine $\mathcal{M}$. However, it is not necessary for $\mathcal{M}_\psi$ to work in $\mathrm{NTIME}(\exp_i(n^c))$. We define $\mathcal{M}_\psi$ in such a way that it works in $\mathrm{NTIME}(n^c)$ (i.e., in NP).

To evaluate ψ, $\mathcal{M}_\psi$ needs to know the input structure $\mathbf{I}_{\varphi'}$, as well as the values for the relation variables $X_1^{\tau_1}, \ldots, X_{s_1}^{\tau_{s_1}}$ of order $i + 1$ which were guessed by the machine $\mathcal{M}$. Therefore, the input to the oracle machine $\mathcal{M}_\psi$ is as follows: a) the formula ψ, which requires space $\mathcal{O}(n)$, b) the input structure $\mathbf{I}_{\varphi'}$, which also requires space $\mathcal{O}(n)$, and c) the guessed values for the variables $X_1^{\tau_1}, \ldots, X_{s_1}^{\tau_{s_1}}$, which requires space bounded by $\exp_i(n^c)$ as seen above.

So, the necessary space for the query to the oracle in the oracle tape is $\mathcal{O}(\exp_i(n^c))$. In fact, we use padding to make sure that the input to the oracle is of size $\exp_i(n^c)$. That is, the query to the oracle consists of the formula ψ, the input structure $\mathbf{I}_{\varphi'}$ and the guessed values for $X_1^{\tau_1}, \ldots, X_{s_1}^{\tau_{s_1}}$ padded with enough "quasiblanks" to make the total length of the input string to the oracle to be $\mathcal{O}(\exp_i(n^c))$.

Hence, if the oracle machine $\mathcal{M}_\psi$ works in polynomial time, it will work actually in $\mathrm{NTIME}((\exp_i(n^c))^{c'})$, where n is the size of $\mathbf{I}_{\varphi'}$. Then an oracle in NP is enough.

As to the other oracle machines, which could eventually be called in turn by $\mathcal{M}_\psi$ or any other oracle machine in the chain, the input in the oracle tape should be built in the same way, except for the input structure, which should be extended with the guessed values for all the higher-order variables of order $i + 1$ which are quantified in the prefix of the original formula φ, before the sub-formula which is to be evaluated by the given oracle. The space required in the oracle tape, though, is still $\mathcal{O}(\exp_i(n^c))$.

So, we have got a non deterministic Turing machine $\mathcal{M}$ which decides $AUTOSAT(\Sigma_m^i[\rho])$ in time $\mathcal{O}(\exp_i(n^c))$, for some constant c, and which calls a chain of oracle machines, each belonging to the class NP. Clearly, the depth of nesting of the chain of successive calls to oracles is given by the number of negations which appear in the prefix of the formula φ, minus 1.

Therefore, $\mathcal{M}$ is in the class $\bigcup_{c \in \mathbb{N}} \mathrm{NTIME}(\exp_i(n^c))^{\Sigma_{m-1}^p}$. ∎

6 Future work

We know from [MP96] that if we bound the arity of the second-order variables, then $AUTO\text{-}SAT(HAA^1(r,m))$ is complete for PSPACE. Our conjecture to this regard is that this result can also be generalized to higher orders.

CONJECTURE 25. For every $i \geq 2$ and every $r, m \geq 1$, $AUTOSAT(HAA^i(r,m))$ is complete for $\mathrm{DSPACE}(\exp_{i-1}(n^c))$.

To the best of our knowledge, there is no known complete problem for $\mathrm{DSPACE}(\exp_i(n^c))$, except for the first few values of i. However, we think it is quite possible that in the same way as QSAT (quantified satisfiability) is complete for PSPACE, the union for every j of the Σ_j^i theory of the Boolean model ($i \geq 1$), i.e., $\bigcup_{j \geq 1} \Sigma_j^i - \text{Th}(\mathbf{B})$, could be complete for DSPACE($\exp_i(n^c)$). If that is the case, then it should be relatively straightforward to reduce $\bigcup_{j \geq 1} \Sigma_j^i - \text{Th}(\mathbf{B})$ to $AUTOSAT(HAA^{i+1}(r, m))$ for every $r, m \geq 1$. Also, space bounded by $\mathcal{O}(\exp_i(n^r))$ should be enough to decide $AUTOSAT(HAA^{i+1}(r, m))$ since the maximal-arity of the variables in $HAA^{i+1}(r, m)$ is restricted by r.

BIBLIOGRAPHY

[BDG95] José Luis Balcázar, Joseph Díaz, and Joaquim Gabarró. *Structural Complexity I.* Texts in Theoretical Computer Science, EATCS. Springer, Berlin Heidelberg New York, 2 edition, 1995.

[EF99] Heinz-Dieter Ebbinghaus and Jörg Flum. *Finite Model Theory*. Perspectives in Mathematical Logic. Springer, Berlin Heidelberg New York, 2nd edition, 1999.

[Fer08] Flavio Antonio Ferrarotti. *Expressibility of Higher-Order Logics on Relational Databases: Proper Hierarchies.* PhD thesis, Department of Information Systems, Massey University, Wellington, New Zealand, June 2008.

[FT05] Flavio Antonio Ferrarotti and Jose Maria Turull Torres. Arity and alternation of quantifiers in higher order logics. Technical Report 10/2005, Department of Information Systems, Massey University, New Zealand, September 2005.

[FT06] Flavio Antonio Ferrarotti and José María Turull Torres. Arity and alternation: A proper hierarchy in higher order logics. In Jürgen Dix and Stephen J. Hegner, editors, *Proceedings of the 4th International Symposium on Foundations of Information and Knowledge Systems*, volume 3861 of *Lec. Notes Comput. Sci.*, pages 92–115. Springer, February 2006.

[FT07] Flavio Antonio Ferrarotti and José María Turull Torres. Arity and alternation: A proper hierarchy in higher order logics. *Ann. Math. Artif. Intell.*, 50(1–2):111–141, 2007.

[HT03] Lauri Hella and José María Turull Torres. Expressibility of higher order logics. *Electr. Notes Theor. Comput. Sci.*, 84, 2003.

[HT05] Lauri Hella and Jose Maria Turull Torres. Complete problems for higher order logics. Technical Report 12/2005, Department of Information Systems, Massey University, New Zealand, September 2005.

[HT06a] Lauri Hella and José María Turull Torres. Complete problems for higher order logics. In *Proceedings of the 15th Annual Conference of the EACSL (and 20th International Workshop) on Computer Science Logic*, volume 4207 of *Lec. Notes Comput. Sci.*, pages 380–394, Szeged, Hungary, September 2006. Springer.

[HT06b] Lauri Hella and José María Turull Torres. Computing queries with higher-order logics. *Theor. Comput. Sci.*, 355(2):197–214, 2006.

[Lei94] Daniel Leivant. Higher order logic. In Dov M. Gabbay, Christopher J. Hogger, J. A. Robinson, and Jörg H. Siekmann, editors, *Handbook of Logic in Artificial Intelligence and Logic Programming*, volume 2, pages 229–322. Oxford University Press, 1994.

[MP96] Johann A. Makowsky and Yachin B. Pnueli. Arity and alternation in second-order logic. *Ann. Pure Appl. Logic*, 78(1-3):189–202, 1996.

[Pap94] Christos H. Papadimitriou. *Computational Complexity*. Addison-Wesley, Reading, MA, 1994.

[Sto76] Larry J. Stockmeyer. The polynomial-time hierarchy. *Theor. Comput. Sci.*, 3(1):1–22, 1976.

Flavio A. Ferrarotti
Departamento de Ingeniería Informática
Facultad de Ingeniería, Universidad de Santiago de Chile
E-mail: flavio.ferrarotti@gmail.com

José M. Turull Torres
School of Engineering and Advanced Technology
College of Sciences, Massey University
Wellington, New Zealand
E-mail: J.M.Turull@massey.ac.nz

Intersection Type System with de Bruijn Indices

DANIEL LIMA-VENTURA, MAURICIO AYALA-RINCÓN, FAIROUZ KAMAREDDINE

ABSTRACT. The λ-calculus in de Bruijn notation avoids α-conversion using indices instead of variable names. Intersection types provide finitary type polymorphism and characterise normalisable λ-terms, that is a term is normalisable if and only if it is typable. To be closer to computations and to simplify the formalisation of the atomic operations involved in β-contractions several calculi of explicit substitution were developed and most of them are written in de Bruijn notation. Versions of explicit substitutions calculi without types and with simple type systems are well investigated in contrast to versions with more elaborated type systems such as intersection types. Besides the application in real implementations, the study of a system in de Bruijn's notation is of interest in proof theory, since the type-contexts, usually treated as sets, are changed to sequences. As a first step, a λ-calculus in de Bruijn notation with an intersection type system is introduced in this paper and it is proved that this system satisfies the subject reduction property and its typed λ-terms preserve theirs types under β-reduction. The proof of subject reduction is done in a standard way, through a generation and substitution lemmas. For doing this, the proper definition of a *free index* is given and properties corresponding to those in the λ-calculus with names related to free variables are proved.

1 Introduction

The λ-calculus à la de Bruijn [dB72] was introduced by the Dutch mathematician N.G. de Bruijn in the context of the project Automath [NGdV94], one of the leading projects on automated deduction which still influences modern proof assistants [Kam03]. Variables are represented by indices instead of names, assembling each α-class of terms in the λ-calculus with names in a unique term in de Bruijn's notation. Although there is a common sense that de Bruijn's notation is unreadable, it is machine-friendly and has been adopted for several calculi of explicit substitutions (e.g. [dB78],

[ACCL91], [KR95]) in which operations related to β-reductions are atomized in order to create calculi closer to actual implementations of the λ-calculus. Type free and simply typed versions of the λ-calculus with/without explicit substitutions have been investigated, but to the best of our knowledge there is no work on more elaborated type systems for these calculi in de Bruijn notation.

In this paper a version of the λ-calculus in de Bruijn notation with an intersection types system is introduced. Intersection types were introduced to provide a characterization of strongly normalizing λ-terms [CDC78, CDC80, Pot80]. In programming, the intersection type discipline is of interest because λ-terms not typable in the standard Curry type assignment system ([CF58]) or in extensions allowing some sort of polymorphism, as the one present in programming languages such as ML ([Mil78]), are typable with intersection types. For instance, $\lambda x.(x\ x)$ is typable, assigning two different types to x ($x : \sigma \to \varphi\ \cap\ \sigma$). The intersection type system presented in [BCDC83] is closed under β-equality, a property that does not hold for simply typed systems. However, the typability problem (Given a λ-term t, is there a context Γ and a type σ such that $\Gamma \vdash t\ :\ \sigma$?), decidable in the Curry type assignment system, is undecidable in [BCDC83]. This is a consequence of the fact that all terms having normal form can be characterized by their assignable types. In [CW04] Carlier and Wells presented the exact correspondence between the inference mechanism for their intersection type system and β-reduction. They introduce *expansion variables* to perform *Expansion*, an operation used during type inference (see [CW04.2]).

The type system in this paper is based on the one given in [KN07]. The version in de Bruijn's notation is proved to preserve subject reduction, that is it preserves types under β-reduction: whenever $\Gamma \vdash t\ :\ \sigma$ and t β-reduces into s, then $\Gamma \vdash s\ :\ \sigma$.

Section 2 presents the λ-calculus in de Bruijn's notation and introduces the formal definition of a *free index*, giving some lemmas about syntactic properties regarding the update of free indices (free variables), substitution and β-reduction. In section 3 the intersection type system is introduced and properties about the shape of types and contexts (ordered environments) are presented, in a similar manner to those given in [KN07]. Section 4 proves the property of subject reduction, following the standard sketch establishing a generation and substitution lemmas. Finally, we conclude with a discussion on future work.

2 The type free calculi

2.1 λ-calculus in de Bruijn notation

DEFINITION 1 (Set Λ_{dB}). The syntax of the λ-calculus in de Bruijn notation, **the λdB-calculus,** is defined inductively by:

Terms $\quad M ::= \underline{n} \mid (M\ M) \mid \lambda.M$ where $n \in \mathbb{N}^* = \mathbb{N} \smallsetminus \{0\}$

DEFINITION 2.

1. We define $FI(M)$, the set of free indices of $M \in \Lambda_{dB}$, by:

$$
\begin{array}{rcl}
FI(\underline{n}) &=& \{\underline{n}\} \\
FI(\lambda.M) &=& \{\underline{n-1}, \forall \underline{n} \in FI(M), n > 1\} \\
FI(M_1\ M_2) &=& FI(M_1) \cup FI(M_2)
\end{array}
$$

2. A term M is called closed if $FI(M) \equiv \emptyset$.

3. The greatest value of a free index in M, denoted by $sup(M)$, is defined by:
$$
sup(M) = \left\{ \begin{array}{ll} 0 & \text{if } FI(M) \equiv \emptyset \\ n \text{ where } \underline{n} \in FI(M) \text{ and } n \geq i, \forall \underline{i} \in FI(M) & \text{otherwise} \end{array} \right.
$$

LEMMA 3.

1. $sup(M_1\ M_2) = max(sup(M_1), sup(M_2))$.

2. If $sup(M)=0$, then $sup(\lambda.M)=0$. Otherwise, $sup(\lambda.M)=sup(M)-1$.

Proof.

1. If $sup(M_1\ M_2) = 0$, nothing to prove. Otherwise, $sup(M_1\ M_2) = n$, where $n \geq i$, $\forall \underline{i} \in FI(M_1\ M_2) = FI(M_1) \cup FI(M_2)$ and $\underline{n} \in FI(M_1)$ or $\underline{n} \in FI(M_2)$. Suppose, w.l.o.g., that $\underline{n} \in FI(M_1)$. Hence, $n \geq sup(M_1)$ and $sup(M_1) \geq n$, thus, $n = sup(M_1)$ and $n \geq sup(M_2)$.

2. If $sup(M)=0$, then $FI(\lambda.M)=FI(M)=\emptyset$, hence, $sup(\lambda.M)=0$. Let $sup(M)=m>0$. Hence, $m \geq i$, $\forall \underline{i} \in FI(M)$ and $\underline{m} \in FI(M)$. If $m=1$, then $FI(M)=\{\underline{1}\}$, thus, $FI(\lambda.M)=\emptyset$ and $sup(\lambda.M) = 0$. Otherwise, $FI(\lambda.M)=\{\underline{n-1}, \forall \underline{n} \in FI(M), n > 1\}$. Thus, $\underline{m-1} \in FI(\lambda.M)$ and $m - 1 \geq i - 1$, $\forall \underline{i-1} \in FI(\lambda.M)$.

∎

Terms like $((\dots((M_1 \ M_2) \ M_3)\dots) \ M_n)$ are written as $(M_1 \ M_2 \ \dots \ M_n)$, as usual. The β-contraction definition in this notation needs a mechanism which detects and updates free indices of terms. It follows an operator similar to the one presented in [ARK01].

DEFINITION 4. Let $M \in \Lambda_{dB}$ and $i \in \mathbb{N}$. The **i-lift** of M, denoted as M^{+i}, is defined inductively by:

1. $(M_1 \ M_2)^{+i} = (M_1^{+i} \ M_2^{+i})$ 3. $\underline{n}^{+i} = \begin{cases} \underline{n+1}, & \text{if } n > i \\ \underline{n}, & \text{if } n \leq i. \end{cases}$

2. $(\lambda.M_1)^{+i} = \lambda.M_1^{+(i+1)}$

The **lift** of a term M is its 0-lift, denoted by M^+. Intuitively, the lift of M corresponds to an increment by 1 of all free indices occurring in M. The next lemma states general relations between the i-lift and the free indices of M.

LEMMA 5.

1. If $i \geq sup(M)$, then $M^{+i} \equiv M$.

2. $FI(M^{+i}) = \{\underline{n} \mid \underline{n} \in FI(M), n \leq i\} \cup \{\underline{n+1} \mid \underline{n} \in FI(M), n > i\}$.

3. If $sup(M) > i$, then $sup(M^{+i}) = sup(M)+1$.

4. If $sup(M) \leq i$, then $sup(M^{+i}) = sup(M)$.

Proof. 1 and 2: By induction on the structure of M.
3: If $sup(M) = m$, then $m \geq n$, $\forall \underline{n} \in FI(M)$ and $\underline{m} \in FI(M)$. Since $m > i$, by Lemma 5.2, $\underline{m+1} \in FI(M^{+i})$ and $\forall \underline{j} \in FI(M^{+i})$, either $j = n$ or $j = n+1$, where $\underline{n} \in FI(M)$. One has $m+1 \geq n+1 > n, \forall \underline{n} \in FI(M)$, thus, $m+1 \geq j, \forall \underline{j} \in FI(M^{+i})$.
4: From Lemma 5.1, $M^{+i} \equiv M$, thus, $sup(M^{+i}) = sup(M)$. ∎

Using the i-lift, we are able to present the definition of the substitution used by β-contractions, similarly to the one presented in [ARK01].

DEFINITION 6. Let $m, n \in \mathbb{N}^*$. The **β-substitution** for free occurrences of $\underline{n}$ in $M \in \Lambda_{dB}$ by term N, denoted as $\{\underline{n}/N\}M$, is defined inductively by

1. $\{\underline{n}/N\}(M_1 \ M_2) = (\{\underline{n}/N\}M_1 \ \{\underline{n}/N\}M_2)$ 3. $\{\underline{n}/N\}\underline{m} = \begin{cases} \underline{m-1}, & \text{if } m > n \\ N, & \text{if } m = n \\ \underline{m}, & \text{if } m < n \end{cases}$

2. $\{\underline{n}/N\}\lambda.M_1 = \lambda.\{\underline{n+1}/N^+\}M_1$

Observe that in item 2 of Definition 6, the lift operator is used to avoid captures of free indices in N. We present the β-contraction as defined in [ARK01].

DEFINITION 7. β-**contraction** in λdB is defined by $(\lambda.M\,N) \to_\beta \{\underline{1}/N\}M$.

Notice that item 3 in Definition 6, for $n=1$, is the mechanism which does the substitution and updates the free indices in M as consequence of the lead abstractor elimination.

LEMMA 8.

1. If $\underline{i} \notin FI(M)$, then
   $FI(\{\underline{i}/N\}M) = \{\underline{n} \mid \underline{n} \in FI(M), n < i\} \cup \{\underline{n-1} \mid \underline{n} \in FI(M), n > i\}$.

2. Otherwise,
   $FI(\{\underline{i}/N\}M) = FI(N) \cup \{\underline{n} \mid \underline{n} \in FI(M), n < i\} \cup \{\underline{n-1} \mid \underline{n} \in FI(M), n > i\}$.

3. If $i > sup(M)$, then $\{\underline{i}/N\}M \equiv M$.

Proof. By induction on the structure of M. ∎

In particular, if $FI(M) = \{\underline{i}\}$, then $\{\underline{n} \mid \underline{n} \in FI(M), n < i\} \equiv \emptyset$ and $\{\underline{n-1} \mid \underline{n} \in FI(M), n > i\} \equiv \emptyset$, thus, $FI(\{\underline{i}/N\}M) = FI(N)$.

COROLLARY 9. If $\underline{1} \in FI(M)$, then $FI(\{\underline{1}/N\}M) = FI(\lambda.M\ N)$. Otherwise, $FI(\{\underline{1}/N\}M) = FI(\lambda.M)$.

LEMMA 10. Let M be a term such that $sup(M) = m$:

1. If $i < m$ and $\underline{i} \notin FI(M)$, then $sup(\{\underline{i}/N\}M) = m-1$.

2. If $i > m$, then $sup(\{\underline{i}/N\}M) = m$.

3. Suppose $\underline{i} \in FI(M)$. If $FI(M) = \{\underline{i}\}$, then $sup(\{\underline{i}/N\}M) = sup(N)$. Otherwise, $sup(\{\underline{i}/N\}M) = max(sup(N), m-1)$.

Proof.

1. One has that $m \geq n$, $\forall \underline{n} \in FI(M)$ and $\underline{m} \in FI(M)$. Since $m > i$, by Lemma 8.1, $\underline{m-1} \in FI(\{\underline{i}/N\}M)$ and $\forall \underline{j} \in FI(\{\underline{i}/N\}M)$, either $j = n < i$ or $j = n-1$, where $\underline{n} \in FI(M)$. Thus, $m-1 \geq n-1 \geq i, \forall \underline{n} \in FI(M)$ such that $n > i$, hence, $m-1 \geq j, \forall \underline{j} \in FI(\{\underline{i}/N\}M)$.

2. If $i > m$, then, by Lemma 8.3, $\{\underline{i}/N\}M \equiv M$, thus, $sup(\{\underline{i}/N\}M) = sup(M)$.

3. By Lemma 8.2 one has $FI(\{\underline{i}/N\}M) = FI(N) \cup A$, where $A \equiv \{\underline{n} \mid \underline{n} \in FI(M), n < i\} \cup \{\underline{n-1} \mid \underline{n} \in FI(M), n > i\}$. If $FI(M) = \{\underline{i}\}$, then $A \equiv \emptyset$, thus $FI(\{\underline{i}/N\}M) = FI(N)$. Otherwise, A is not empty and, similarly to case 1, one has that $m - 1 \geq j, \forall \underline{j} \in A$.

∎

LEMMA 11. $sup(\{\underline{1}/N\}M) \leq sup(\lambda.M \ N)$.

Proof. If $\underline{1} \in FI(M)$, then $sup(\{\underline{1}/N\}M) = sup(\lambda.M \ N)$. Otherwise, one has two possibilities. If $sup(M) = 0$, then, by Lemma 10.2, $sup(\{\underline{1}/N\}M) = 0 \leq max(0, sup(N)) = sup(\lambda.M \ N)$. If $sup(M) > 1$, then, by Lemma 10.1, $sup(\{\underline{1}/N\}M) = sup(M) - 1 = sup(\lambda.M) \leq max(sup(\lambda.M), sup(N))$. ∎

DEFINITION 12. β-**reduction** in λdB is defined by:

$$\frac{(\lambda.M \ N) \to_\beta \{\underline{1}/N\}M}{(\lambda.M \ N) \longrightarrow_\beta \{\underline{1}/N\}M} \qquad \frac{M \longrightarrow_\beta N}{\lambda.M \longrightarrow_\beta \lambda.N}$$

$$\frac{M_1 \longrightarrow_\beta N_1}{(M_1 \ M_2) \longrightarrow_\beta (N_1 \ M_2)} \qquad \frac{M_2 \longrightarrow_\beta N_2}{(M_1 \ M_2) \longrightarrow_\beta (M_1 \ N_2)}$$

THEOREM 13. *If* $M \longrightarrow_\beta N$ *then* $FI(N) \subseteq FI(M)$ *and* $sup(N) \leq sup(M)$.

Proof. By induction on the derivation $M \longrightarrow_\beta N$.

- If $M \equiv (\lambda.M_1 \ M_2)$ and $N \equiv \{\underline{1}/M_2\}M_1$ then, by corollary 9, $FI(\{\underline{1}/N\}M_1) \subseteq FI(\lambda.M_1 \ M_2)$.

- Let $M \equiv (M_1 \ M_2)$ and $N \equiv (M_1 \ N_2)$, where $M_2 \longrightarrow_\beta N_2$, then, by IH, $FI(N_2) \subseteq FI(M_2)$. Thus, $FI(N) = FI(M_1) \cup FI(N_2) \subseteq FI(M_1) \cup FI(M_2) = FI(M)$.

- Case $M \equiv (M_1 \ M_2)$ and $N \equiv (N_1 \ M_2)$, where $M_1 \longrightarrow_\beta N_1$, is similar.

- If $M \equiv \lambda.M'$, then $N \equiv \lambda.N'$, where $M' \longrightarrow_\beta N'$. By IH, $FI(N') \subseteq FI(M')$, hence, $\forall \underline{n} \in FI(N')$, $\underline{n} \in FI(M')$. Thus, $\forall \underline{n-1} \in FI(\lambda.N')$, $\underline{n-1} \in FI(\lambda.M')$.

∎

3 The Type System

DEFINITION 14.

1. Let $\mathcal{A}$ be a denumerably infinite set of atomic types. The **intersection types** are defined by:

$$\mathbb{T} ::= \mathcal{A} \mid \mathbb{U} \to \mathbb{T} \qquad\qquad \mathbb{U} ::= \omega \mid \mathbb{U} \sqcap \mathbb{U} \mid \mathbb{T}$$

   The types are quotiented by taking $\sqcap$ to be commutative, associative, idempotent and to have ω as neutral.

2. Contexts are ordered lists of types $U \in \mathbb{U}$, defined by: $\Gamma ::= nil \mid U.\Gamma$

   Let Γ be some context and $n \in \mathbb{N}$. Then $\Gamma_{<n}$ denotes the first $n-1$ types of Γ. Similarly we define $\Gamma_{>n}$, $\Gamma_{\leq n}$ and $\Gamma_{\geq n}$. Note that, for $\Gamma_{>n}$ and $\Gamma_{\geq n}$ the final nil element is included. For $n=0$, $\Gamma_{\leq 0}.\Gamma = \Gamma_{<0}.\Gamma = \Gamma$. The i-th element of Γ is denoted by Γ_i. The length of Γ is defined as $|nil|=0$ and, if Γ is not nil, $|\Gamma|=1+|\Gamma_{>1}|$. For any $i > m = |\Gamma|$, let $\Gamma_{\geq i} = \Gamma_{>i} = \Gamma_{>m}$ and $\Gamma_{\leq i} = \Gamma_{<i} = \Gamma_{\leq m}$.

   For a term M, we denote env_ω^M the context Γ such that $|\Gamma| = sup(M)$ and $\Gamma = \omega.\omega.\cdots.\omega.nil$.

   The extension of $\sqcap$ for contexts is done by $nil \sqcap \Gamma = \Gamma \sqcap nil = \Gamma$ and $(U_1.\Gamma) \sqcap (U_2.\Delta) = (U_1 \sqcap U_2).(\Gamma \sqcap \Delta)$. Hence, $\sqcap$ is commutative, associative and idempotent on contexts.

Some properties over contexts follow from the above definitions.

LEMMA 15. *Let Γ and Δ be contexts, where neither Γ nor Δ are nil:*

1. *If $|\Gamma| \geq sup(M)$, then $\Gamma \sqcap env_\omega^M = \Gamma$*

2. *$\Gamma \sqcap \Delta = (\Gamma_1 \sqcap \Delta_1).(\Gamma_{>1} \sqcap \Delta_{>1})$*

3. *If $i \leq |\Gamma|, |\Delta|$, then $(\Gamma \sqcap \Delta)_i = \Gamma_i \sqcap \Delta_i$.*

4. *$(\Gamma \sqcap \Delta)_{<i} = \Gamma_{<i} \sqcap \Delta_{<i}$ and $(\Gamma \sqcap \Delta)_{>i} = \Gamma_{>i} \sqcap \Delta_{>i}$. The same for $(\Gamma \sqcap \Delta)_{\leq i}$ and $(\Gamma \sqcap \Delta)_{\geq i}$.*

5. *$|\Gamma \sqcap \Delta| = max(|\Gamma|, |\Delta|)$.*

DEFINITION 16. The typing rules are given as follows:

$$\frac{}{\underline{1}:\langle T.nil \vdash T\rangle}\ var \qquad\qquad \frac{M:\langle nil \vdash T\rangle}{\lambda.M:\langle nil \vdash \omega \rightarrow T\rangle}\ \rightarrow'_i$$

$$\frac{\underline{n}:\langle \Gamma \vdash U\rangle}{\underline{n+1}:\langle \omega.\Gamma \vdash U\rangle}\ varn \qquad\qquad \frac{M_1:\langle \Gamma \vdash U\rightarrow T\rangle \quad M_2:\langle \Gamma' \vdash U\rangle}{M_1\ M_2:\langle \Gamma \sqcap \Gamma' \vdash T\rangle}\ \rightarrow_e$$

$$\frac{}{M:\langle env_\omega^M \vdash \omega\rangle}\ \omega \qquad\qquad \frac{M:\langle \Gamma \vdash U_1\rangle \quad M:\langle \Gamma \vdash U_2\rangle}{M:\langle \Gamma \vdash U_1 \sqcap U_2\rangle}\ \sqcap_i$$

$$\frac{M:\langle U.\Gamma \vdash T\rangle}{\lambda.M:\langle \Gamma \vdash U\rightarrow T\rangle}\ \rightarrow_i \qquad\qquad \frac{M:\langle \Gamma \vdash U\rangle \quad \langle \Gamma \vdash U\rangle \sqsubseteq \langle \Gamma' \vdash U'\rangle}{M:\langle \Gamma' \vdash U'\rangle}\ \sqsubseteq$$

where the binary relation $\sqsubseteq$ is defined by the following rules:

$$\frac{}{\Phi \sqsubseteq \Phi}\ ref \qquad\qquad \frac{\Phi_1 \sqsubseteq \Phi_2 \quad \Phi_2 \sqsubseteq \Phi_3}{\Phi_1 \sqsubseteq \Phi_3}\ tr$$

$$\frac{}{U_1 \sqcap U_2 \sqsubseteq U_1}\ \sqcap_e \qquad\qquad \frac{U_1 \sqsubseteq V_1 \quad U_2 \sqsubseteq V_2}{U_1 \sqcap U_2 \sqsubseteq V_1 \sqcap V_2}\ \sqcap$$

$$\frac{U_2 \sqsubseteq U_1 \quad T_1 \sqsubseteq T_2}{U_1 \rightarrow T_1 \sqsubseteq U_2 \rightarrow T_2}\ \rightarrow \qquad\qquad \frac{U_1 \sqsubseteq U_2}{\Gamma_{\leq i}.U_1.\Gamma_{>i} \sqsubseteq \Gamma_{\leq i}.U_2.\Gamma_{>i}}\ \sqsubseteq_c$$

$$\frac{U_1 \sqsubseteq U_2 \quad \Gamma' \sqsubseteq \Gamma}{\langle \Gamma \vdash U_1\rangle \sqsubseteq \langle \Gamma' \vdash U_2\rangle}\ \sqsubseteq_{\langle\rangle}$$

Φ, Φ', $\Phi_1,\ldots$ are used to denote $U \in \mathbb{U}$, contexts Γ or typings $\langle \Gamma \vdash U\rangle$. Note that in $\Phi \sqsubseteq \Phi'$, Φ and Φ' belong to the same sort.

Type judgements will be of the form $M : \langle \Gamma \vdash U\rangle$, meaning that term M has type U provided Γ for $FI(M)$. Briefly, M has type U in Γ.

The next lemmas states some properties about the shape of types and contexts, and their link with the subtyping relation defined by $\sqsubseteq$.

LEMMA 17.

1. *If* $U \in \mathbb{U}$, *then* $U = \omega$ *or* $U = \sqcap_{i=1}^n T_i$ *where* $n \geq 1$ *and* $\forall\, 1 \leq i \leq n$, $T_i \in \mathbb{T}$.

2. $U \sqsubseteq \omega$.

3. *If* $\omega \sqsubseteq U$, *then* $U = \omega$.

Proof. See [KN07] ∎

Observe that, from $\underline{2} : \langle \omega.T.nil \vdash T \rangle$ and the $\sqsubseteq$ relation we have that $\underline{2} : \langle U.T.nil \vdash T \rangle$, for any U. This allows some sort of weakening in the type system, which is not allowed in the type system given in [KN07]. This happens because ω's are needed in the context first positions to give the proper type for some free index $\underline{i}$. Although, in Lemma 20 we prove this weakening is limited by the term itself.

LEMMA 18. *Let $V \neq \omega$.*

1. *If $U \sqsubseteq V$, then $U = \sqcap_{j=1}^{k} T_j$, $V = \sqcap_{i=1}^{p} T_i'$ where $p, k \geq 1$, $\forall 1 \leq j \leq k$, $1 \leq i \leq p$, $T_j, T_i' \in \mathbb{T}$, and $\forall 1 \leq i \leq p$, $\exists 1 \leq j \leq k$ such that $T_j \sqsubseteq T_i'$.*

2. *If $U \sqsubseteq V' \sqcap a$, then $U = U' \sqcap a$ and $U' \sqsubseteq V'$.*

3. *Let $p, k \geq 1$. If $\sqcap_{j=1}^{k}(U_j \rightarrow T_j) \sqsubseteq \sqcap_{i=1}^{p}(U_i' \rightarrow T_i')$, then $\forall 1 \leq i \leq p$, $\exists 1 \leq j \leq k$ such that $U_i' \sqsubseteq U_j$ and $T_j \sqsubseteq T_i'$.*

4. *If $U \rightarrow T \sqsubseteq V$, then $V = \sqcap_{i=1}^{p}(U_i \rightarrow T_i)$ where $p \geq 1$ and $\forall 1 \leq i \leq p$, $U_i \sqsubseteq U$ and $T \sqsubseteq T_i$.*

5. *If $\sqcap_{j=1}^{k}(U_j \rightarrow T_j) \sqsubseteq V$ where $k \geq 1$, then $V = \sqcap_{i=1}^{p}(U_i' \rightarrow T_i')$ where $p \geq 1$ and $\forall 1 \leq i \leq p$, $\exists 1 \leq j \leq k$ such that $U_i' \sqsubseteq U_j$ and $T_j \sqsubseteq T_i'$.*

Proof. See [KN07] ∎

LEMMA 19.

1. *If $\Gamma \sqsubseteq \Gamma'$ and $U \sqsubseteq U'$, then $U.\Gamma \sqsubseteq U'.\Gamma'$.*

2. *$\Gamma \sqsubseteq \Gamma'$ iff $|\Gamma| = |\Gamma'| = m$ and, if $m > 0$ then $\forall 1 \leq i \leq m$, $\Gamma_i \sqsubseteq \Gamma_i'$.*

3. *If $|\Gamma| = sup(M)$, then $\Gamma \sqsubseteq env_\omega^M$.*

4. *If $env_\omega^M \sqsubseteq \Gamma$, then $\Gamma = env_\omega^M$.*

5. *$\langle \Gamma \vdash U \rangle \sqsubseteq \langle \Gamma' \vdash U' \rangle$ iff $\Gamma' \sqsubseteq \Gamma$ and $U \sqsubseteq U'$.*

6. *If $\Gamma \sqsubseteq \Gamma'$ and $\Delta \sqsubseteq \Delta'$, then $\Gamma \sqcap \Delta \sqsubseteq \Gamma' \sqcap \Delta'$.*

Proof.

1. By induction on the derivation $\Gamma \sqsubseteq \Gamma'$ we have that if $\Gamma \sqsubseteq \Gamma'$, then $V.\Gamma \sqsubseteq V.\Gamma'$. Using tr we have the result.

2. Only if) By induction on the derivation $\Gamma \sqsubseteq \Gamma'$. If) By induction on m using 1.

3. By Lemma 17.2 and 2.

4. By 2, $|\Gamma| = sup(M) = m$. If $m = 0$, them $env_\omega^M = \Gamma = nil$. Otherwise, for every $1 \leq i \leq m$, $\omega \sqsubseteq \Gamma_i$. Hence, by Lemma 17.3, $\forall 1 \leq i \leq m$, $\Gamma_i = \omega$.

5. Only if) By induction on the derivation $\langle \Gamma \vdash U \rangle \sqsubseteq \langle \Gamma' \vdash U' \rangle$. If) By $\sqsubseteq_{\langle \rangle}$.

6. This is a corollary of 2.

■

The following lemma shows the strict relation in a type judgement between the length of a context Γ and the free indices of term M, where $M : \langle \Gamma \vdash U \rangle$ for some type U.

LEMMA 20.

1. *If $M : \langle \Gamma \vdash U \rangle$, then $|\Gamma| = sup(M)$.*

2. *For every Γ and M such that $|\Gamma| = sup(M)$, we have $M : \langle \Gamma \vdash \omega \rangle$.*

Proof.

1. By induction on the derivation $M : \langle \Gamma \vdash U \rangle$.

2. By ω, $M : \langle env_\omega^M \vdash \omega \rangle$. By Lemma 19.3, $\Gamma \sqsubseteq env_\omega^M$. Hence, by $\sqsubseteq_{\langle \rangle}$ and $\sqsubseteq$, $M : \langle \Gamma \vdash \omega \rangle$.

■

Consequently, the weakening allowed in the system is limited by the maximum value of a free index occurring in a term.

The following lemma shows that another version of the var and $\sqcap_i$ rules, axiom and intersection introduction respectively, are derivable from the typing rules and subtyping relation, presented in Definition 16.

LEMMA 21.

1. *The rule* $\dfrac{M : \langle \Gamma \vdash U_1 \rangle \quad M : \langle \Delta \vdash U_2 \rangle}{M : \langle \Gamma \sqcap \Delta \vdash U_1 \sqcap U_2 \rangle}$ $\sqcap_i'$ *is derivable.*

2. *The rule* $\dfrac{}{1 : \langle U.nil \vdash U \rangle}$ *var$'$ is derivable.*

Proof.

1. Let $M : \langle \Gamma \vdash U_1 \rangle$ and $M : \langle \Delta \vdash U_2 \rangle$. By Lemma 20.1, $|\Gamma| = |\Delta| = m$. Thus, $|\Gamma \sqcap \Delta| = m$ and $(\Gamma \sqcap \Delta)_i = \Gamma_i \sqcap \Delta_i$, $\forall 1 \leq i \leq m$. By rule $\sqcap_e$ and Lemma 19.2, $\Gamma \sqcap \Delta \sqsubseteq \Gamma$ and $\Gamma \sqcap \Delta \sqsubseteq \Delta$. Hence, by rules $\sqsubseteq_{\langle \rangle}$ and $\sqsubseteq$, $M : \langle \Gamma \sqcap \Delta \vdash U_1 \rangle$ and $M : \langle \Gamma \sqcap \Delta \vdash U_2 \rangle$. Thus, by rule $\sqcap_i$, $M : \langle \Gamma \sqcap \Delta \vdash U_1 \sqcap U_2 \rangle$.

2. By Lemma 17.1:

   - Either $U = \omega$, then by rule ω the result holds.
   - Or $U = \sqcap_{i=1}^{k} T_i$ where $\forall 1 \leq i \leq k$, $T_i \in \mathbb{T}$, then, by rule var, $\underline{1}$: $\langle T_i.nil \vdash T_i \rangle$ and, by $k-1$ applications of rule $\sqcap_i'$, $\underline{1} : \langle U.nil \vdash U \rangle$.

■

4 The subject reduction property

4.1 Subject reduction for β

The subject reduction property is proved in the standard way, with a generation and substitutions lemmas (Lemmas 22 and 24, respectively) being the properties to be established first.

LEMMA 22 (Generation).

1. *If $\underline{n} : \langle \Gamma \vdash U \rangle$, then $\Gamma_n = V$ where $V \sqsubseteq U$.*

2. *If $\lambda.M : \langle \Gamma \vdash U \rangle$ and $sup(M) > 0$, then $U = \omega$ or $U = \sqcap_{i=1}^{k}(V_i \to T_i)$ where $k \geq 1$ and $\forall 1 \leq i \leq k$, $M : \langle V_i.\Gamma \vdash T_i \rangle$.*

3. *If $\lambda.M : \langle \Gamma \vdash U \rangle$ and $sup(M) = 0$, then $\Gamma = nil$, $U = \omega$ or $U = \sqcap_{i=1}^{k}(V_i \to T_i)$ where $k \geq 1$ and $\forall 1 \leq i \leq k$, $M : \langle nil \vdash T_i \rangle$.*

Proof.

1. By induction on the derivation $\underline{n} : \langle \Gamma \vdash U \rangle$. By Lemma 20.1, $|\Gamma| = n$.

   - If $\dfrac{}{\underline{1} : \langle T.nil \vdash T \rangle}$, nothing to prove.
   - If $\dfrac{}{\underline{n} : \langle env_{\omega}^{n} \vdash \omega \rangle}$, nothing to prove.
   - Let $\dfrac{\underline{n} : \langle \Gamma \vdash U \rangle}{\underline{n+1} : \langle \omega.\Gamma \vdash U \rangle}$. One has that $(\omega.\Gamma)_{n+1} = \Gamma_n$ and, by IH, $\Gamma_n = V$ where $V \sqsubseteq U$.
   - Let $\dfrac{\underline{n} : \langle \Gamma \vdash U_1 \rangle \quad \underline{n} : \langle \Gamma \vdash U_2 \rangle}{\underline{n} : \langle \Gamma \vdash U_1 \sqcap U_2 \rangle}$. By IH, $\Gamma_n = V$ where $V \sqsubseteq U_1$ and $V \sqsubseteq U_2$. Then, by rule $\sqcap$, $V \sqsubseteq U_1 \sqcap U_2$.

- Let $\dfrac{\underline{n}:\langle\Gamma\vdash U\rangle \quad \langle\Gamma\vdash U\rangle\sqsubseteq\langle\Gamma'\vdash U'\rangle}{\underline{n}:\langle\Gamma'\vdash U'\rangle}$. By IH, $\Gamma_n=V$ where $V\sqsubseteq U$. By Lemma 19.5, $\Gamma'\sqsubseteq\Gamma$ and $U\sqsubseteq U'$. Thus, by Lemma 19.2, $\Gamma'_n=V'\sqsubseteq V$. By rule tr, $V'\sqsubseteq U'$.

2. By induction on the derivation $\lambda.M:\langle\Gamma\vdash U\rangle$.

- If $\dfrac{}{\lambda.M:\langle env^{\lambda.M}_\omega\vdash\omega\rangle}$, nothing to prove.

- If $\dfrac{M:\langle U.\Gamma\vdash T\rangle}{\lambda.M:\langle\Gamma\vdash U\to T\rangle}$, nothing to prove.

- Let $\dfrac{\lambda.M:\langle\Gamma\vdash U_1\rangle \quad \lambda.M:\langle\Gamma\vdash U_2\rangle}{\lambda.M:\langle\Gamma\vdash U_1\sqcap U_2\rangle}$. By IH, one has the following cases:

  - If $U_1=U_2=\omega$, then $U_1\sqcap U_2=\omega$.
  - If $U_1=\omega$, $U_2=\sqcap_{i=1}^k(V_i\to T_i)$ where $k\ge 1$ and $\forall 1\le i\le k$, $M:\langle V_i.\Gamma\vdash T_i\rangle$, then, $U_1\sqcap U_2=U_2$
  - If $U_2=\omega$, $U_1=\sqcap_{i=1}^k(V_i'\to T_i')$ where $k\ge 1$ and $\forall 1\le i\le k$, $M:\langle V_i'.\Gamma\vdash T_i'\rangle$, then, $U_1\sqcap U_2=U_1$
  - If $U_1=\sqcap_{i=1}^k(V_i\to T_i),U_2=\sqcap_{i=k+1}^{k+l}(V_i\to T_i)$, where $k,l\ge 1$ and $\forall 1\le i\le k+l$, $M:\langle V_i.\Gamma\vdash T_i\rangle$, then $U_1\sqcap U_2=\sqcap_{i=1}^{k+l}(V_i\to T_i)$.

- Let $\dfrac{\lambda.M:\langle\Gamma\vdash U\rangle \quad \langle\Gamma\vdash U\rangle\sqsubseteq\langle\Gamma'\vdash U'\rangle}{\lambda.M:\langle\Gamma'\vdash U'\rangle}$. By Lemma 19.5, $\Gamma'\sqsubseteq\Gamma$ and $U\sqsubseteq U'$. By IH, one has the following:

  - If $U=\omega$, then, by Lemma 17.3, $U'=\omega$.
  - Otherwise, $U=\sqcap_{i=1}^k(V_i\to T_i)$ where $k\ge 1$ and $\forall 1\le i\le k$, $M:\langle V_i.\Gamma\vdash T_i\rangle$. By Lemma 17.1, either $U'=\omega$, and then nothing to prove, or, by Lemma 18.5, $U'=\sqcap_{i=1}^p(V_i'\to T_i')$ where $p\ge 1$ and $\forall 1\le i\le p$, $\exists 1\le j_i\le k$ such that $V_i'\sqsubseteq V_{j_i}$ and $T_{j_i}\sqsubseteq T_i'$. By Lemmas 19.1 and 19.5, $\langle V_{j_i}.\Gamma\vdash T_{j_i}\rangle\sqsubseteq\langle V_i'.\Gamma'\vdash T_i'\rangle$, for each $1\le i\le p$, then, $M:\langle V_i'.\Gamma'\vdash T_i'\rangle$.

3. By Lemma 3.2, $sup(\lambda.M)=0$ and, by Lemma 20.1, $|\Gamma|=nil$, thus, $\lambda.M:\langle nil\vdash U\rangle$. The proof is the same as for 2, where $\to_i'$ is used on induction step, instead of $\to_i$.

■

It follows an auxiliary lemma for the Substitution Lemma (Lemma 24), stating a property relating to type judgements and the index update mechanism.

LEMMA 23. *If* $M : \langle \Gamma \vdash U \rangle$ *and* $0 \leq i < sup(M)$, *then* $M^{+i} : \langle \Gamma_{\leq i}.\omega.\Gamma_{>i} \vdash U \rangle$.

Proof. By induction on the derivation $M : \langle \Gamma \vdash U \rangle$.

- Let $\dfrac{}{\underline{1} : \langle T.nil \vdash T \rangle}$. For $i = 0$, $\underline{1}^+ = \underline{2}$ and, by rule varn, $\underline{2} : \langle \omega.T.nil \vdash T \rangle$.

- If $\dfrac{}{M : \langle env_{\omega}^M \vdash \omega \rangle}$, nothing to prove.

- Let $\dfrac{\underline{n} : \langle \Gamma \vdash U \rangle}{\underline{n+1} : \langle \omega.\Gamma \vdash U \rangle}$. If $i = 0$, then by rule varn $\underline{n+2} : \langle \omega.\omega.\Gamma \vdash U \rangle$. Otherwise, note that $\underline{n}^{+i} + 1 = \underline{n+1}^{+(i+1)} = \underline{n+2}$. By IH one has $\underline{n}^{+i} : \langle \Gamma_{\leq i}.\omega.\Gamma_{>i} \vdash U \rangle$. By rule varn, $\underline{n+2} : \langle \omega.\Gamma_{\leq i}.\omega.\Gamma_{>i} \vdash U \rangle$.

- Let $\dfrac{M : \langle U.\Gamma \vdash T \rangle}{\lambda.M : \langle \Gamma \vdash U \rightarrow T \rangle}$. By Lemma 3.2 one has $sup(M) > i+1$, hence, by IH, $M^{+(i+1)} : \langle U.\Gamma_{\leq i}.\omega.\Gamma_{>i} \vdash T \rangle$. Hence, by rule $\rightarrow_i$ and i-lift definition, $(\lambda.M)^{+i} : \langle \Gamma_{\leq i}.\omega.\Gamma_{>i} \vdash U \rightarrow T \rangle$.

- Let $\dfrac{M_1 : \langle \Gamma \vdash U \rightarrow T \rangle \quad M_2 : \langle \Delta \vdash U \rangle}{M_1 \ M_2 : \langle \Gamma \sqcap \Delta \vdash T \rangle}$. By Lemma 3.1 one has $sup(M_1) > i$ or $sup(M_2) > i$. Suppose w.l.o.g. that $i < sup(M_1), sup(M_2)$. By IH, $M_1^{+i} : \langle \Gamma_{\leq i}.\omega.\Gamma_{>i} \vdash U \rightarrow T \rangle$ and $M_2^{+i} : \langle \Delta_{\leq i}.\omega.\Delta_{>i} \vdash U \rangle$. Thus, by $\rightarrow_e$ and observing that $(\Gamma_{\leq i}.\omega.\Gamma_{>i}) \sqcap (\Delta_{\leq i}.\omega.\Delta_{>i}) = (\Gamma \sqcap \Delta)_{\leq i}.\omega.(\Gamma \sqcap \Delta)_{>i}$, $(M_1 \ M_2)^{+i} : \langle (\Gamma \sqcap \Delta)_{\leq i}.\omega.(\Gamma \sqcap \Delta)_{>i} \vdash T \rangle$.

- Let $\dfrac{M : \langle \Gamma \vdash U_1 \rangle \quad M : \langle \Gamma \vdash U_2 \rangle}{M : \langle \Gamma \vdash U_1 \sqcap U_2 \rangle}$. By IH, $M^{+i} : \langle \Gamma_{\leq i}.\omega.\Gamma_{>i} \vdash U_1 \rangle$ and $M^{+i} : \langle \Gamma_{\leq i}.\omega.\Gamma_{>i} \vdash U_2 \rangle$. Thus, by rule $\sqcap_i$, $M^{+i} : \langle \Gamma_{\leq i}.\omega.\Gamma_{>i} \vdash U_1 \sqcap U_2 \rangle$.

- Let $\dfrac{M : \langle \Gamma \vdash U \rangle \quad \langle \Gamma \vdash U \rangle \sqsubseteq \langle \Gamma' \vdash U' \rangle}{M : \langle \Gamma' \vdash U' \rangle}$. By IH, $M^{+i} : \langle \Gamma_{\leq i}.\omega.\Gamma_{>i} \vdash U \rangle$ and, by Lemma 19.5, $\Gamma' \sqsubseteq \Gamma$ and $U \sqsubseteq U'$. Hence, by Lemma 19.2, $\Gamma'_{\leq i}.\omega.\Gamma'_{>i} \sqsubseteq \Gamma_{\leq i}.\omega.\Gamma_{>i}$. Thus, by rules $\sqsubseteq_{\langle \rangle}$ and $\sqsubseteq$, $M^{+i} : \langle \Gamma'_{\leq i}.\omega.\Gamma'_{>i} \vdash U' \rangle$.

∎

LEMMA 24 (Substitution). *Let* $M : \langle \Gamma \vdash U \rangle$ *and* $i \in \mathbb{N}^*$, *for* $i \leq sup(M)$, *and* $N : \langle \Delta \vdash \Gamma_i \rangle$:

1. If $\underline{i} \notin FI(M)$, *then* $\{\underline{i}/N\}M : \langle \Gamma_{<i}.\Gamma_{>i} \vdash U \rangle$.

2. *Otherwise, if* $sup(N) \geq i-1$, *then* $\{\underline{i}/N\}M : \langle (\Gamma_{<i}.\Gamma_{>i}) \sqcap \Delta \vdash U \rangle$.

Proof. By induction on the derivation $M : \langle \Gamma \vdash U \rangle$.

1. Observe that $i < |\Gamma| = sup(M)$:

   - If $\dfrac{}{\underline{1} : \langle T.nil \vdash T \rangle}$, nothing to prove.

   - Let $\dfrac{}{M : \langle env_\omega^M \vdash \omega \rangle}$. By Lemma 10.1, $sup(\{\underline{i}/N\}M) = sup(M) - 1$. Thus, $env_\omega^{\{\underline{i}/N\}M} = (env_\omega^M)_{<i}.(env_\omega^M)_{>i}$ and the result holds trivially by rule ω.

   - Let $\dfrac{\underline{n} : \langle \Gamma \vdash U \rangle}{\underline{n+1} : \langle \omega.\Gamma \vdash U \rangle}$. By Lemma 20.1, $|\omega.\Gamma| = n+1$, hence, $i < (n+1)$ and $\{\underline{i}/N\}\underline{n+1} = \underline{n}$. Note that $(\omega.\Gamma)_i = \Gamma_{(i-1)}$, thus, by IH one has $\{\underline{i-1}/N\}\underline{n} : \langle \Gamma_{<(i-1)}.\Gamma_{>(i-1)} \vdash U \rangle$. Since $(i-1) < n$, $\{\underline{i-1}/N\}\underline{n} = \underline{n-1}$, hence, by rule varn, $\underline{n} : \langle \omega.\Gamma_{<(i-1)}.\Gamma_{>(i-1)} \vdash U \rangle$.

   - Let $\dfrac{M : \langle U.\Gamma \vdash T \rangle}{\lambda.M : \langle \Gamma \vdash U \to T \rangle}$. If $sup(N) = 0$, then, by Lemma 5.1, $N^+ \equiv N$, otherwise, by Lemma 23, $N^+ : \langle \omega.\Delta \vdash \Gamma_i \rangle$. By IH, $\{\underline{i+1}/N^+\}M : \langle U.\Gamma_{<i}.\Gamma_{>i} \vdash T \rangle$, thus, by $\to_i$, $\lambda.\{\underline{i+1}/N^+\}M : \langle \Gamma_{<i}.\Gamma_{>i} \vdash U \to T \rangle$.

   - Let $\dfrac{M_1 : \langle \Gamma \vdash U \to T \rangle \quad M_2 : \langle \Gamma' \vdash U \rangle}{M_1\,M_2 : \langle \Gamma \sqcap \Gamma' \vdash T \rangle}$. Suppose, w.l.o.g., $i < sup(M_1)$ and $i < sup(M_2)$, thus, $(\Gamma \sqcap \Gamma')_i = \Gamma_i \sqcap \Gamma'_i$. By rules $\sqcap_e$, $\sqsubseteq_{\langle\rangle}$ and $\sqsubseteq$ one has $N : \langle \Delta \vdash \Gamma_i \rangle$ and $N : \langle \Delta \vdash \Gamma'_i \rangle$. Hence, by IH, $\{\underline{i}/N\}M_1 : \langle \Gamma_{<i}.\Gamma_{>i} \vdash U \to T \rangle$ and $\{\underline{i}/N\}M_2 : \langle \Gamma'_{<i}.\Gamma'_{>i} \vdash U \rangle$. Thus, by rule $\to_e$, $(\{\underline{i}/N\}M_1\,\{\underline{i}/N\}M_2) : \langle (\Gamma_{<i} \sqcap \Gamma'_{<i}).(\Gamma_{>i} \sqcap \Gamma'_{>i}) \vdash T \rangle$.

   - Let $\dfrac{M : \langle \Gamma \vdash U_1 \rangle \quad M : \langle \Gamma \vdash U_2 \rangle}{M : \langle \Gamma \vdash U_1 \sqcap U_2 \rangle}$. By IH, $\{\underline{i}/N\}M : \langle \Gamma_{<i}.\Gamma_{>i} \vdash U_1 \rangle$ and $\{\underline{i}/N\}M : \langle \Gamma_{<i}.\Gamma_{>i} \vdash U_2 \rangle$. Thus, by rule $\sqcap_i$, one has that $\{\underline{i}/N\}M : \langle \Gamma_{<i}.\Gamma_{>i} \vdash U_1 \sqcap U_2 \rangle$.

   - Let $\dfrac{M : \langle \Gamma \vdash U \rangle \quad \langle \Gamma \vdash U \rangle \sqsubseteq \langle \Gamma' \vdash U' \rangle}{M : \langle \Gamma' \vdash U' \rangle}$. By Lemma 19.5, $\Gamma' \sqsubseteq \Gamma$ and $U \sqsubseteq U'$, hence, by Lemma 19.2, $\Gamma'_i \sqsubseteq \Gamma_i$ and $\Gamma'_{<i}.\Gamma'_{>i} \sqsubseteq \Gamma_{<i}.\Gamma_{>i}$. Thus, by rules $\sqsubseteq_{\langle\rangle}$ and $\sqsubseteq$, $N : \langle \Delta \vdash \Gamma_i \rangle$, and, by IH, $\{\underline{i}/N\}M : \langle \Gamma_{<i}.\Gamma_{>i} \vdash U \rangle$. By rules $\sqsubseteq_{\langle\rangle}$ and $\sqsubseteq$, $\{\underline{i}/N\}M : \langle \Gamma'_{<i}.\Gamma'_{>i} \vdash U' \rangle$.

2. - If $\dfrac{}{\underline{1} : \langle T.nil \vdash T \rangle}$, nothing to prove.

- Let $\dfrac{}{M:\langle env_\omega^M \vdash \omega \rangle}$. One has the following cases:

  - If $FI(M) = \{\underline{i}\}$, then $|env_\omega^M| = i$, thus, $(env_\omega^M)_{<i}.(env_\omega^M)_{>i} = env_\omega^{M'}$, where M' is any term such that $sup(M') = i - 1$. Hence, $env_\omega^{M'} \sqcap \Delta = \Delta$. By Lemmas 10.3 and 20.1, $sup(\{\underline{i}/N\}M) = sup(N) = |\Delta|$, hence, by Lemma 20.2, $\{\underline{i}/N\}M:\langle \Delta \vdash \omega \rangle$.

  - Otherwise, by Lemma 10.3 and 20.1, $sup(\{\underline{i}/N\}M)$ is given by $max(sup(N), sup(M)-1) = max(|\Delta|, |env_\omega^M| - 1)$, which is equivalent to $|\Delta \sqcap ((env_\omega^M)_{<i}.(env_\omega^M)_{>i})|$. Thus, by Lemma 20.2, $\{\underline{i}/N\}M:\langle \Delta \sqcap ((env_\omega^M)_{<i}.(env_\omega^M)_{>i}) \vdash \omega \rangle$.

- Let $\dfrac{\underline{n}:\langle \Gamma \vdash U \rangle}{\underline{n+1}:\langle \omega.\Gamma \vdash U \rangle}$. For $i = n+1$, $\{\underline{n+1}/N\}\underline{n+1} = N$ and, by Lemma 20.1, $|\Gamma| = n$. By Lemma 22, $\Gamma_n = V$, where $V \sqsubseteq U$. Thus, by rule $\sqcap_e$ and Lemma 19.2, $(\omega.\Gamma_{<n}.nil) \sqcap \Delta \sqsubseteq \Delta$ and, by rules $\sqsubseteq_{\langle \rangle}$ and $\sqsubseteq$, $N:\langle (\omega.\Gamma_{<n}.nil) \sqcap \Delta \vdash U \rangle$.

- Let $\dfrac{M:\langle U.\Gamma \vdash T \rangle}{\lambda.M:\langle \Gamma \vdash U \to T \rangle}$. Observe that $(U.\Gamma)_{(i+1)} = \Gamma_i$. If $sup(N) = 0$, then, by Lemma 5.1, $N^+ \equiv N$, otherwise, by Lemma 23, one has $N^+:\langle \omega.\Delta \vdash \Gamma_i \rangle$. By IH, $\{\underline{i+1}/N^+\}M:\langle (U.\Gamma_{<i}.\Gamma_{>i}) \sqcap \Delta' \vdash T \rangle$, where Δ' is either nil or $\omega.\Delta$. If $\Delta' \equiv \omega.\Delta$, then $(U.\Gamma_{<i}.\Gamma_{>i}) \sqcap \Delta' = U.((\Gamma_{<i}.\Gamma_{>i}) \sqcap \Delta)$. Thus, by rule $\to_i$, one has that $\lambda.\{\underline{i+1}/N^+\}M : \langle (\Gamma_{<i}.\Gamma_{>i}) \sqcap \Delta \vdash U \to T \rangle$. The case where $\Delta' \equiv nil$ is trivial.

- Let $\dfrac{M_1:\langle \Gamma \vdash U \to T \rangle \quad M_2:\langle \Gamma' \vdash U \rangle}{M_1\, M_2:\langle \Gamma \sqcap \Gamma' \vdash T \rangle}$. If $\underline{i} \in FI(M_1)$ and $\underline{i} \in FI(M_2)$, then, $(\Gamma \sqcap \Gamma')_i = \Gamma_i \sqcap \Gamma'_i$, and, by rules $\sqcap_e$, $\sqsubseteq_{\langle \rangle}$ and $\sqsubseteq$, $N:\langle \Delta \vdash \Gamma_i \rangle$ and $N:\langle \Delta \vdash \Gamma'_i \rangle$. By IH, $\{\underline{i}/N\}M_1:\langle (\Gamma_{<i}.\Gamma_{>i}) \sqcap \Delta \vdash U \to T \rangle$ and $\{\underline{i}/N\}M_2:\langle (\Gamma'_{<i}.\Gamma'_{>i}) \sqcap \Delta \vdash U \rangle$. Note that $(\Gamma_{<i}.\Gamma_{>i}) \sqcap \Delta \sqcap (\Gamma'_{<i}.\Gamma'_{>i}) \sqcap \Delta = ((\Gamma \sqcap \Gamma')_{<i}.(\Gamma \sqcap \Gamma')_{>i}) \sqcap \Delta$. Thus, by rule $\to_e$, $\{\underline{i}/N\}(M_1\, M_2):\langle ((\Gamma \sqcap \Gamma')_{<i}.(\Gamma \sqcap \Gamma')_{>i}) \sqcap \Delta \vdash T \rangle$. The cases $\underline{i} \notin FI(M_1)$ and $\underline{i} \notin FI(M_2)$ are similar, using 1 on the induction step whenever necessary.

- Let $\dfrac{M:\langle \Gamma \vdash U_1 \rangle \quad M:\langle \Gamma \vdash U_2 \rangle}{M:\langle \Gamma \vdash U_1 \sqcap U_2 \rangle}$. By IH one has that $\{\underline{i}/N\}M:\langle (\Gamma_{<i}.\Gamma_{>i}) \sqcap \Delta \vdash U_1 \rangle$ and $\{\underline{i}/N\}M : \langle (\Gamma_{<i}.\Gamma_{>i}) \sqcap \Delta \vdash U_2 \rangle$. Thus, by rule $\sqcap_i$, $\{\underline{i}/N\}M:\langle (\Gamma_{<i}.\Gamma_{>i}) \sqcap \Delta \vdash U_1 \sqcap U_2 \rangle$.

- Let $\dfrac{M:\langle \Gamma \vdash U \rangle \quad \langle \Gamma \vdash U \rangle \sqsubseteq \langle \Gamma' \vdash U' \rangle}{M:\langle \Gamma' \vdash U' \rangle}$. By Lemma 19.5, $\Gamma' \sqsubseteq \Gamma$ and $U \sqsubseteq U'$, hence, by Lemma 19.2, $\Gamma'_i \sqsubseteq \Gamma_i$ and $\Gamma'_{<i}.\Gamma'_{>i} \sqsubseteq$

$\Gamma_{<i}.\Gamma_{>i}$. Thus, by rules $\sqsubseteq_{\langle\rangle}$ and $\sqsubseteq$, $N:\langle\Delta \vdash \Gamma_i\rangle$ and, by IH, one has $\{\underline{i}/N\}M:\langle(\Gamma_{<i}.\Gamma_{>i}) \sqcap \Delta \vdash U\rangle$. By Lemma 19.6, $(\Gamma'_{<i}.\Gamma'_{>i}) \sqcap \Delta \sqsubseteq (\Gamma_{<i}.\Gamma_{>i}) \sqcap \Delta$, thus, by rules $\sqsubseteq_{\langle\rangle}$ and $\sqsubseteq$, $\{\underline{i}/N\}M:\langle(\Gamma'_{<i}.\Gamma'_{>i}) \sqcap \Delta \vdash U'\rangle$.

■

As a consequence of Lemma 20 and the possibility of some free indices being eliminated during a β-reduction, we need the following definition.

DEFINITION 25. Let M be a term and $sup(M)=m$. For a context Γ, let $\Gamma\!\restriction_M$ be the restriction of Γ to $FI(M)$, given by $\Gamma_{\leq m}.nil$.

The definition above will allow us to type the resulting term from a β-reduction in a shorter context, related to the original one. First, we prove some properties about the restriction on contexts.

LEMMA 26.

1. *If $sup(N) \leq sup(M)$, then $env_\omega^M\!\restriction_N = env_\omega^N$.*

2. *If $|\Gamma| \leq sup(M)$, then $(\Gamma \sqcap \Delta)\!\restriction_M = \Gamma \sqcap \Delta\!\restriction_M$.*

3. *If $sup(N) > 0$, then $(U.\Gamma)\!\restriction_N = U.\Gamma\!\restriction_{(\lambda.N)}$.*

Proof.

1. Straightforward from Definition 25 and the definition of env_ω^M.

2. Let $sup(M) = m$. Thus, $(\Gamma \sqcap \Delta)\!\restriction_M = (\Gamma \sqcap \Delta)_{\leq m}.nil = (\Gamma_{\leq m} \sqcap \Delta_{\leq m}).nil = (\Gamma_{\leq m}.nil) \sqcap (\Delta_{\leq m}.nil) = \Gamma \sqcap (\Delta_{\leq m}.nil) = \Gamma \sqcap \Delta\!\restriction_M$.

3. If $sup(N) > 0$, by Lemma 3.2, $sup(\lambda.N) = sup(N) - 1$. Thus, $(U.\Gamma)\!\restriction_N = (U.\Gamma)_{\leq sup(N)}.nil = U.\Gamma_{\leq (sup(N)-1)}.nil = U.\Gamma\!\restriction_{(\lambda.N)}$.

■

Finally, we have Theorem 27 stating the proof for β-contraction and then Theorem 28 for any β-reduction.

THEOREM 27 (SR for β-contraction). *If $(\lambda.M\ N):\langle\Gamma \vdash U\rangle$ then $\{\underline{1}/N\}M:\langle\Gamma\!\restriction_{\{\underline{1}/N\}M} \vdash U\rangle$*

Proof. By induction on the derivation $(\lambda.M\ N):\langle\Gamma \vdash U\rangle$.

- Let $\dfrac{}{(\lambda.M\ N):\langle env_\omega^{(\lambda.M\ N)} \vdash \omega\rangle}$. By Lemma 11, one has $sup(\{\underline{1}/N\}M) \leq sup(\lambda.M\ N)$, hence, by Lemma 26.1, $env_\omega^{\lambda.M\ N}\!\restriction_{\{\underline{1}/N\}M} = env_\omega^{\{\underline{1}/N\}M}$. By rule ω the result is obtained trivially.

- Let $\dfrac{\lambda.M:\langle\Delta\vdash U\rightarrow T\rangle \quad N:\langle\Delta'\vdash U\rangle}{(\lambda.M\ N):\langle\Delta\sqcap\Delta'\vdash T\rangle}$. One has the following cases:

  - Case $sup(M)=0$. By Lemmas 22.3 resp. 8.3, $\Delta=nil$, $M:\langle nil\vdash T\rangle$ resp. $\{\underline{1}/N\}M\equiv M$. Thus, $\Delta\sqcap\Delta'=\Delta'$ and $\Delta'\lfloor_{\{\underline{1}/N\}M}=\Delta'\lfloor_M=nil$.

  - Case $sup(M)>0$. By Lemma 22.2, $M:\langle U.\Delta\vdash T\rangle$:
    - If $\underline{1}\notin FI(M)$, then, by Lemma 24.1, $\{\underline{1}/N\}M:\langle\Delta\vdash T\rangle$. By Lemma 26.2, $(\Delta\sqcap\Delta')\lfloor_{\{\underline{1}/N\}M}=\Delta\sqcap(\Delta'\lfloor_{\{\underline{1}/N\}M})$, hence, by rule $\sqcap_e$ and Lemma 19.2, $(\Delta\sqcap\Delta')\lfloor_{\{\underline{1}/N\}M}\sqsubseteq\Delta$. Thus, by rules $\sqsubseteq_{\langle\rangle}$ and $\sqsubseteq$, $\{\underline{1}/N\}M:\langle(\Delta\sqcap\Delta')\lfloor_{\{\underline{1}/N\}M}\vdash T\rangle$.
    - Otherwise, by Lemma 24.2, $\{\underline{1}/N\}M:\langle\Delta\sqcap\Delta'\vdash T\rangle$. By Lemma 20.1, $|\Delta\sqcap\Delta'|=sup(\{\underline{1}/N\}M)$, thus, $(\Delta\sqcap\Delta')\lfloor_{\{\underline{1}/N\}M}=\Delta\sqcap\Delta'$.

- Let $\dfrac{(\lambda.M\ N):\langle\Gamma\vdash U_1\rangle\ (\lambda.M\ N):\langle\Gamma\vdash U_2\rangle}{(\lambda.M\ N):\langle\Gamma\vdash U_1\sqcap U_2\rangle}$. By IH one has $\{\underline{1}/N\}M:\langle\Gamma\lfloor_{\{\underline{1}/N\}M}\vdash U_1\rangle$ and $\{\underline{1}/N\}M:\langle\Gamma\lfloor_{\{\underline{1}/N\}M}\vdash U_2\rangle$. Thus, by rule $\sqcap_i$, $\{\underline{1}/N\}M:\langle\Gamma\lfloor_{\{\underline{1}/N\}M}\vdash U_1\sqcap U_2\rangle$.

- Let $\dfrac{(\lambda.M\ N):\langle\Gamma\vdash U\rangle\ \langle\Gamma\vdash U\rangle\sqsubseteq\langle\Gamma'\vdash U'\rangle}{(\lambda.M\ N):\langle\Gamma'\vdash U'\rangle}$. By IH, one has $\{\underline{1}/N\}M:\langle\Gamma\lfloor_{\{\underline{1}/N\}M}\vdash U\rangle$. By Lemma 19.5, $\Gamma'\sqsubseteq\Gamma$ and $U\sqsubseteq U'$, hence, by Lemma 19.2, $\Gamma'\lfloor_{\{\underline{1}/N\}M}\sqsubseteq\Gamma\lfloor_{\{\underline{1}/N\}M}$. Thus, by rules $\sqsubseteq_{\langle\rangle}$ and $\sqsubseteq$, $\{\underline{i}/N\}M:\langle\Gamma'\lfloor_{\{\underline{1}/N\}M}\vdash U'\rangle$. ∎

THEOREM 28 (SR for β-reduction). *If* $M:\langle\Gamma\vdash U\rangle$ *and* $M\longrightarrow_\beta N$, *then* $N:\langle\Gamma\lfloor_N\vdash U\rangle$.

Proof. By induction on the derivation $M:\langle\Gamma\vdash U\rangle$

- Let $\dfrac{}{M:\langle env_\omega^M\vdash\omega\rangle}$. One has $FI(N)\subseteq FI(M)$, hence, $sup(N)\leq sup(M)$. By Lemma 26.1, $env_\omega^M\lfloor_N=env_\omega^N$, thus, by rule ω, $N:\langle env_\omega^N\vdash\omega\rangle$.

- Let $\dfrac{M':\langle V.\Gamma\vdash T\rangle}{\lambda.M':\langle\Gamma\vdash V\rightarrow T\rangle}$. By IH, $N':\langle(V.\Gamma)\lfloor_{N'}\vdash T\rangle$, where $M'\longrightarrow_\beta N'$.

  If $sup(N')=0$, then $N':\langle nil\vdash T\rangle$. By $\rightarrow'_i$, $\lambda.N':\langle nil\vdash\omega\rightarrow T\rangle$, hence, by rules $\rightarrow$, $\sqsubseteq_{\langle\rangle}$ and $\sqsubseteq$, $\lambda.N':\langle nil\vdash V\rightarrow T\rangle$.

  If $sup(N')>0$, then, by Lemma 26.3, $(V.\Gamma)\lfloor_{N'}=V.\Gamma\lfloor_{\lambda.N'}$. Thus, by rule $\rightarrow_i$, $\lambda.N':\langle\Gamma\lfloor_{\lambda.N'}\vdash V\rightarrow T\rangle$.

- Let $\dfrac{M':\langle nil \vdash T\rangle}{\lambda.M':\langle nil \vdash \omega \to T\rangle}$. Thus, $M' \longrightarrow_\beta N'$ and, by Theorem 13, $sup(N') \leq sup(M') = 0$. By IH, $N':\langle nil \vdash T\rangle$, hence, by rule $\to'_i$, $\lambda.N':\langle nil \vdash \omega \to T\rangle$.

- Let $\dfrac{M_1:\langle \Delta \vdash U \to T\rangle \quad M_2:\langle \Delta' \vdash U\rangle}{M_1\ M_2:\langle \Delta \sqcap \Delta' \vdash T\rangle}$. Suppose that $N \equiv (N_1\ M_2)$, where $M_1 \longrightarrow_\beta N_1$, hence, by IH, $N_1:\langle \Delta\!\downarrow_{N_1} \vdash U \to T\rangle$. By rule $\to_e$, $(N_1\ M_2):\langle \Delta\!\downarrow_{N_1} \sqcap \Delta' \vdash T\rangle$.

  - If $sup(N_1) \geq sup(M_2)$, then $sup(N) = sup(N_1)$ and, by Lemma 26.2, $(\Delta \sqcap \Delta')\!\downarrow_{N_1} = \Delta\!\downarrow_{N_1} \sqcap \Delta'$.

  - If $sup(M_2) > sup(N_1)$, then $sup(N) = sup(M_2)$ and, by Lemma 26.2, $(\Delta \sqcap \Delta')\!\downarrow_{M_2} = \Delta\!\downarrow_{M_2} \sqcap \Delta'$. By rule $\sqcap_e$ and Lemma 19.2, one has that $(\Delta\!\downarrow_{M_2})_{>sup(N_1)} \sqcap \Delta'_{>sup(N_1)} \sqsubseteq \Delta'_{>sup(N_1)}$, thus, by Lemma 19.2, $(\Delta \sqcap \Delta')\!\downarrow_{N_1} \cdot ((\Delta\!\downarrow_{M_2})_{>sup(N_1)} \sqcap \Delta'_{>sup(N_1)}) \sqsubseteq (\Delta \sqcap \Delta')\!\downarrow_{N_1} \cdot \Delta'_{>sup(N_1)}$. Observe, by Lemma 15.4 and Definition 25, that $(\Delta \sqcap \Delta')\!\downarrow_{N_1} \cdot \Delta'_{>sup(N_1)} = \Delta\!\downarrow_{N_1} \sqcap \Delta'$ and that $(\Delta \sqcap \Delta')\!\downarrow_{N_1} \cdot ((\Delta\!\downarrow_{M_2})_{>sup(N_1)} \sqcap \Delta'_{>sup(N_1)}) = \Delta\!\downarrow_{M_2} \sqcap \Delta'$. Thus, by rules $\sqsubseteq_{\langle\rangle}$ and $\sqsubseteq$, $N:\langle \Delta\!\downarrow_{M_2} \sqcap \Delta' \vdash T\rangle$.

- Let $\dfrac{M:\langle \Gamma \vdash U_1\rangle \quad M:\langle \Gamma \vdash U_2\rangle}{M:\langle \Gamma \vdash U_1 \sqcap U_2\rangle}$. By IH, one has $N:\langle \Gamma\!\downarrow_N \vdash U_1\rangle$ and $N:\langle \Gamma\!\downarrow_N \vdash U_2\rangle$, thus, by rule $\sqcap_i$, $N:\langle \Gamma\!\downarrow_N \vdash U_1 \sqcap U_2\rangle$.

- Let $\dfrac{M:\langle \Gamma' \vdash U'\rangle \quad \langle \Gamma' \vdash U'\rangle \sqsubseteq \langle \Gamma \vdash U\rangle}{M:\langle \Gamma \vdash U\rangle}$. By IH, $N:\langle \Gamma'\!\downarrow_N \vdash U'\rangle$ and, by Lemma 19.5, $\Gamma \sqsubseteq \Gamma'$ and $U' \sqsubseteq U$. Thus, by Lemma 19.2, $\Gamma\!\downarrow_N \sqsubseteq \Gamma'\!\downarrow_N$ and, by rules $\sqsubseteq_{\langle\rangle}$ and $\sqsubseteq$, $N:\langle \Gamma\!\downarrow_N \vdash U\rangle$.

■

5 Conclusions and Future Work

We introduced an intersection type system in de Bruijn's notation and proved it to preserve subject reduction. One particular difference between the type system presented in Definition 16 and the one in [KN07] is that the former allows some kind of weakening, while the latter does not. This characteristic may be relevant while investigating the principal typing property [Wel02]. A type inference algorithm for the type system in Definition 16 might need Expansions to be performed [CW04.2]. Apparently, the way to achieve this is by adding expansion variables to the type system [CW04, CW04.2].

The investigation of type inference, principal types, principal typings and other relevant properties in this system of intersection types as well as its adaptation for explicit substitution calculi in de Bruijn's notation is an interesting work to be done.

Acknowledgments

Daniel Lima Ventura is currently supported by a PhD sandwich scholarship of the CNPq at the Heriot-Watt University. Mauricio Ayala-Rincón was partially supported by the CNPq.

BIBLIOGRAPHY

[ACCL91] M. Abadi, L. Cardelli, P.-L. Curien, and J.-J. Lévy. Explicit Substitutions. *Journal of Functional Programming*, 1(4):375–416, 1991.

[ARK01] M. Ayala-Rincón and F. Kamareddine. Unification via the λs_e-Style of Explicit Substitution. *The Logical Journal of the Interest Group in Pure and Applied Logics*, 9(4):489–523, 2001.

[BCDC83] H. Barendregt, M. Coppo, and M. Dezani-Ciancaglini. A filter lambda model and the completeness of type assignment. *Journal of Symbolic Logic*, 48:931–940, 1983.

[CDC78] M. Coppo and M. Dezani-Ciancaglini. A new type assignment for lambda-terms. *Archiv für Mathematische Logik und Grundlagenforschung*, 19:139–156, 1978.

[CDC80] M. Coppo and M. Dezani-Ciancaglini. An Extension of the Basic Functionality Theory for the λ-Calculus. *Notre Dame Journal of Formal Logic*, 21(4):685–693, 1980.

[CF58] H. B. Curry and R. Feys. *Combinatory Logic*, volume 1. North Holland, 1958.

[CW04] S. Carlier and J. B. Wells. Type Inference with Expansion Variables and Intersection Types in System E and an Exact Correspondence with β-reduction. In *Proceedings of the 6th ACM SIGPLAN International Conference on Principles and Practice of Declarative Programming - PPDP 2004*, pages 132–143. ACM, 2004.

[CW04.2] S. Carlier and J. B. Wells. Expansion: the Crucial Mechanism for Type Inference with Intersection Types: a Survey and Explanation. In *Proceedings of the 3rd International Workshop on Intersection Types and Related Systems - ITRS 2004*, volume 136 of *ENTCS*, pages 173-202. Elsevier, 2005.

[dB72] N.G. de Bruijn. Lambda-Calculus Notation with Nameless Dummies, a Tool for Automatic Formula Manipulation, with Application to the Church-Rosser Theorem. *Indagationes Mathematicae*, 34(5):381–392, 1972.

[dB78] N.G. de Bruijn. A namefree lambda calculus with facilities for internal definition of expressions and segments. T.H.-Report 78-WSK-03, Technische Hogeschool Eindhoven, Nederland, 1978.

[Kam03] F. Kamareddine, editor. *Thirty Five Years of Automating Mathematics*. Kluwer, 2003.

[KN07] F. Kamareddine and K. Nour. A completeness result for a realisability semantics for an intersection type system. *Annals of Pure and Applied Logic*, 146:180–198, 2007.

[KR95] F. Kamareddine and A. Ríos. A λ-calculus à la de Bruijn with Explicit Substitutions. In *Proceedings of the 7th International Symposium on Programming Languages: Implementations, Logics and Programs - PLILP 1995*, volume 982 of *LNCS*, pages 45–62. Springer, 1995.

[Mil78] R. Milner. A theory of type polymorphism in programming. *Journal of computer and System Science*, 17(3):348–375, 1978.

[NGdV94] R. P. Nederpelt, J. H. Geuvers, and R. C. de Vrijer. *Selected papers on Automath*. North-Holland, 1994.

[Pot80] G. Pottinger. A type assignment for the strongly normalizable λ-terms. In J.P. Seldin and J. R. Hindley, editors, *To H. B. Curry: Essays on Combinatory Logic, Lambda Calculus and Formalism*, pages 561–578. Academic Press, 1980.

[Wel02] J.B. Wells. The essence of principal typings. In *Proceedings of 29th International Colloquium on Automata, Languages and Programming - ICALP 2002*, volume 2380 of *LNCS*, pages 913–925. Springer, 2002.

Daniel Lima Ventura and Mauricio Ayala-Rincón
Grupo de Teoria da Computação, Dep. de Matemática
Universidade de Brasília
Brasília D.F., Brasil
E-mail: {ventura,ayala}@mat.unb.br

Fairouz Kamareddine
School of Mathematical and Computer Sciences
Heriot-Watt University
Edinburgh, Scotland
E-mail: fairouz@macs.hw.ac.uk

www.ingramcontent.com/pod-product-compliance
Lightning Source LLC
LaVergne TN
LVHW012325060326
832902LV00011B/1731